ROUTLEDGE HANDBOOK OF WINE TOURISM

Wine tourism, enotourism, oenotourism, winery tourism or vinitourism is a special interest tourism that empowers local culture and spawns business opportunities for the local community. The comprehensive *Routledge Handbook of Wine Tourism* offers a thorough inquiry into both regular and emerging issues of wine tourism.

Modern wine tourism extends beyond the mere cultivation of grapes and the production and selling of wine. *Routledge Handbook of Wine Tourism* examines the complex interplay of market profiling, sustainable regional development, and innovative experiential marketing constructs which, when successful, contribute to the growth and sustainable evolution of global wine tourism. This handbook examines how the success of various enotourism events such as vineyard visits, winery tours, wine festivals and wine trails can stimulate the development of wine-producing regions and territories.

Incorporating the latest philosophies and research themes, this handbook will be an essential reference for students, researchers, academics and industry practitioners of hospitality and tourism, gastronomy, management, marketing, cultural studies, development studies, international business, and for encouraging dialogue across disciplinary boundaries.

Saurabh Kumar Dixit is Professor and founding Head of the Department of Tourism and Hotel Management, North-Eastern Hill University, Shillong (Meghalaya), India. His research interests include consumer behavior, gastronomic tourism, and experiential management and marketing in hospitality and tourism contexts. He has 15 books to his credit, including *The Routledge Handbook of Consumer Behavior in Hospitality and Tourism* (2017), *The Routledge Handbook of Gastronomic Tourism* (2019), *The Routledge Handbook of Tourism Experience Management and Marketing* (2020), *Tourism in Asian Cities* (2020), *Tourism in India: Marketing Perspectives* (2021), *Consumer Behavior in Hospitality & Tourism* (2021), and *The Emerald Handbook of Luxury Management for Tourism and Hospitality* (2021). Presently he is editing *Routledge Handbook of Collaborative Consumption in Tourism* and *Teaching Cases in Tourism, Hospitality and Events* (CABI), to be published in 2022. He can be contacted at saurabh5sk@yahoo.com.

'There is a periodic need for review and update of academic fields of study, a task performed admirably by this new *Routledge Handbook of Wine Tourism*. With seven sections and 57 chapters it is both comprehensive and timely, including a chapter on post-Covid implications. I am particularly pleased to see a wider coverage geographically, for example, India and Japan, and expansion of the field theoretically and methodologically. In addition to traditional themes of wine tourism such as marketing and regional development, readers will gain new insights on experiences - including virtual - sustainability, cultural constructs, social media and other technological developments. The editor summarizes progress and sets a new research agenda to move the field forward.'

Prof. Donald Getz, PhD, Emeritus Professor, The University of Calgary, Canada

'This handbook provides a wide-ranging, multidisciplinary examination of wine tourism. Chapters draw from both established and emerging scholars in the field and span traditional wine tourism destinations and relative newcomers around the world. Topics engage with diverse perspectives and exciting new directions in wine tourism research, from studies of underrepresented winery visitors to the wine tourism developments in the post-COVID-19 environment. This state-of-the-art volume provides essential reading for scholars and students interested in this dynamic area of tourism.'

Prof. Velvet Nelson, PhD, Department of Environmental and Geosciences, Sam Houston State University. Huntsville, Texas, USA

'This handbook is an extensive and multidisciplinary effort to frame the complexity of wine tourism, with an interesting focus on its role to boost regional development, local economy, and community empowerment. 57 up-to-date chapters covering theoretical and practical contents through the contribution of renowned scholars, emerging talented researchers and practitioners. Prof. Saurabh Kumar Dixit, in his 4th experience as editor of a Routledge Handbook, has been able to set up an outstanding review of the main challenges that wine tourism stakeholders have to face, as well as appropriate methods and tools to address them.'

Prof. Alessio Cavicchi, PhD, Department of Agriculture, Food and Environment, University of Pisa, Pisa, Italy

ROUTLEDGE HANDBOOK OF WINE TOURISM

Edited by

Saurabh Kumar Dixit

Routledge
Taylor & Francis Group
LONDON AND NEW YORK

Cover image: © Getty Images

First published 2023
by Routledge
4 Park Square, Milton Park, Abingdon, Oxon OX14 4RN

and by Routledge
605 Third Avenue, New York, NY 10158

Routledge is an imprint of the Taylor & Francis Group, an informa business

British Library Cataloguing-in-Publication Data
A catalogue record for this book is available from the British Library

Library of Congress Cataloging-in-Publication Data
Names: Dixit, Saurabh Kumar, 1979- editor.
Title: Routledge handbook of wine and tourism / edited by Saurabh Kumar Dixit.
Other titles: Handbook of wine and tourism
Description: First edition. | Abingdon, Oxon; New York, NY: Routledge, 2022. | Includes bibliographical references and index. | Contents: The growth and evolution of global wine tourism / Manoj Kumar Yadav and Saurabh Kumar Dixit – Nature and significance of wine tourism: an Australian perspective – Genevieve d'Ament, Anthony J. Saliba and Tahmid Nayeem – Conceptualization of a winescape framework / Vander Valduga, Sarah Marroni Minasi, Gui Lohman. | Identifiers: LCCN 2022010385 (print) | LCCN 2022010386 (ebook) | ISBN 9780367698607 (hbk) | ISBN 9780367698645 (pbk) | ISBN 9781003143628 (ebk)
Subjects: LCSH: Wine tourism. Classification: LCC TP548.5.T68 R68 2022 (print) | LCC TP548.5.T68 (ebook) | DDC 663/.2—dc23/eng/20220718
LC record available at https://lccn.loc.gov/2022010385
LC ebook record available at https://lccn.loc.gov/2022010386

ISBN: 978-0-367-69860-7 (hbk)
ISBN: 978-0-367-69864-5 (pbk)
ISBN: 978-1-003-14362-8 (ebk)

DOI: 10.4324/9781003143628

Typeset in Bembo
by SPi Technologies India Pvt Ltd (Straive)

Dedicated to my wonderful family

&

In the profound memory of late Dr. Byron Marlowe, whose contribution to the Handbook and to the domain of wine tourism research is truly irreplaceable.

CONTENTS

FIGURES

TABLES

ACKNOWLEDGEMENTS

The *Routledge Handbook of Wine Tourism* is going to take the final shape now after about one and a half years of hard work and dedication. I can finalize the contents of this title due to the enduring assistance, support, and enthusiastic collaboration of many people. Although it is impossible to mention everyone here, I particularly want to express my appreciation to the following.

Firstly, I would like to thank all colleagues and the researchers working in tourism, hospitality, and other disciplines who have kindly volunteered their time and efforts to put this incredible Handbook together. They incorporated the suggestions and amendments with full enthusiasm within the stipulated timeframe. I, therefore, extend my sincere gratitude to all the contributors of the Handbook. Secondly, I wish to extend my heartfelt thanks to Faye Leerink, Commissioning Editor, and Esha Gupta, Editorial Assistant at Routledge, for their patience and sincere support from the Handbook proposal to the full manuscript submission. I would also like to express my word of appreciation for all other Routledge editorial, production, and marketing teams for successfully executing this Handbook project. I would also like to acknowledge the constant encouragement and motivation offered by my heavenly father, mother, and other family members. Last but not least, I wish to express special thanks to my lovely wife, Seema, and my wonderful kids, Rakshit and Rayaan, for their constant support and patience during the entire Handbook project.

Saurabh Kumar Dixit
Shillong (India)
2022

CONTRIBUTORS

Maria Alebaki is a researcher at the Agricultural Economics Research Institute (AGRERI) of the Hellenic Agricultural Organization DIMITRA, Greece. She holds a PhD in agricultural economics from the Aristotle University of Thessaloniki. Her academic interests lie in rural and wine tourism areas, examining networks, education, and sustainability/resilience issues. Her work appears in peer-reviewed journals, book chapters, and conference proceedings, while her academic experience also includes teaching in higher education institutes, as well as participation in research and development projects. Currently, she serves as an adjunct faculty member at the Hellenic Open University, the International Hellenic University, and the Agricultural University of Athens.

Marcello Atzeni, PhD, is Assistant Professor in Marketing at the Department of History, Human Sciences and Education, University of Sassari (Italy), where he teaches cultural marketing and management. His research is related to tourist behavior, tourism authenticity, cultural tourism, sustainable tourism, museum marketing, ICT, and cultural heritage sites.

Adela Balderas-Cejudo is Research Fellow at the Oxford Institute of Population Ageing, University of Oxford, and Professor at Deusto Business School and at the bachelor's degree in Gastronomy and Culinary Arts, Basque Culinary Center-University of Mondragon, Spain. Dr. Balderas-Cejudo has a master's in marketing, an executive MBA, a master's in career coaching, and a PhD in business administration. She is also specialized in professional coaching with studies at New York University, Stanford University, and Harvard University. Her research interests include senior tourism, food and wine marketing, gastronomy, business innovation, and talent management.

Carla Barbieri, PhD, is Professor in Sustainable Tourism at North Carolina State University (USA), where she leads the Agritourism and Societal Well-Being Lab. She primarily investigates the intersection between tourism and agricultural systems, seeking to enhance the well-being of farmers, tourists, and local residents sustainably. Barbieri conducts research from regional to international levels, undertaking focal and systems approaches.

Wided Batat is Professor at EM Normandie Business School, Metis Lab and University of Lyon 2, Paris, France. She is an expert in experiential marketing and the ultimate leading expert in youth marketing, customer experience, and digital transformation applied to tourism, luxury, and food. As Entrepreneur and International keynote speaker, Dr. Batat directs and carries out significant research in experiential marketing in France, Europe, the United States, Asia, and the Middle East. With over 15 years of experience in the field and combining scientific research and field experience, Dr. Batat transforms organizations by providing innovative market research and consumer insights. Her expertise can help businesses have an eternal perspective of innovation, profitability, progress, and performance. She has also published many books in the field, and her 2019 best-selling book *Experiential Marketing* is invaluable for consumer behavior, marketing experts, and executive teams.

Janez Bogataj, PhD, born in 1947, doctor of ethnological sciences and B.A. in art history, from 2011 retired honorary professor of the Department of Ethnology and Cultural Anthropology at the Faculty of Arts at the University of Ljubljana, Slovenia. He researches the questions of cuisine and gastronomy, tourism arts and crafts, cultural heritage, and its relations to modern society. He published more than 80 books and 2,000 papers. He is a member of Slovene Ethnological Society, IUEAS, IGCAT, a Conseiller Gastronomique National de Slovénie of Chaîne des Rôtisseurs, the ambassador of EDEN (European Destinations of Excellency) in Slovenia, as well as a founding member of the Honorable Wine Convent of St. Urbanus in Ljubljana.

Ana Brochado is Professor of Management at ISCTE-IUL and a researcher at DINAMIA'CET, Portugal. She has a BSc in economics, an MSc in quantitative methods, and a PhD in management. Her main research interests are hospitality and tourism management. She has authored or co-authored over 70 publications in top international journals, which have received over 1,700 citations.

Aylin Caliskan received her bachelor's in international logistics and transportation from the Yeditepe University, Istanbul (2011). She obtained her master's and PhD in maritime business administration from the Dokuz Eylül University, Izmir in 2013 and 2017, respectively. She worked as a full-time research assistant at the Department of International Logistics Management in the Faculty of Business between 2012 and 2018. She is now a full-time lecturer in vocational school at the Department of Transportation Services, Yasar University, Izmir, Turkey. Her areas of interest include marketing and maritime logistics.

Magali Canovi received her PhD from the University of Chester (UK) and holds a master's degree in international hospitality management from the University of Derby (UK). Her research interests include corporate entrepreneurship and strategic behavior in family firms, social identity, and identity formation, as well as wine tourism development and diversification. She has worked and published on a number of projects in Italy, focusing on how family firms engage in strategic decision-making, notably within the context of wine tourism.

Jodie Conduit is Professor of Marketing in the Business School at the University of Adelaide, Australia. Her research interests lie in understanding how to engage consumers with organizations, and each other, to enable them to work together to achieve meaningful and relevant outcomes. This underpins her research agenda in the areas of customer engagement, value co-creation, and service research. Her research has been published in the *Journal of Marketing*, *Journal of Service Research*, *Industrial Marketing Management*, and *Journal of Business Ethics*, among others.

Ricardo Jorge da Costa Guerra, PhD in tourism, leisure, and culture from the University of Coimbra, is Assistant Professor and Vice Director at the School of Tourism and Hospitality of the Polytechnic Institute of Guarda (Portugal). He is also a researcher at CETRAD, CITUR, and UDI/IPG. His main interest areas of investigation are tourism, hospitality, sustainability, health and wellness, wine tourism, and marketing.

Nicholas Cradock-Henry is Senior Scientist at Manaaki Whenua Landcare Research, New Zealand, where he leads work on climate change adaptation, risk, and resilience.

Namércio Cunha, PhD, is Researcher at Lusófona University (Portugal). His primary interest areas of investigation are marketing, advertisement, and wine tourism.

Genevieve d'Ament, BPsych (Hons), is a PhD candidate at Charles Sturt University, Australia, researching a holistic understanding of the co-created cellar door experience from staff and consumer perspectives. She is supported through a National Wine and Grape Industry Centre scholarship, and her research involves cellar doors from several of Australia's wine regions. She lives in the Shoalhaven Coast Wine region on the south coast of New South Wales. When not researching, she works in the cellar door of a local winery, who were finalists in the 2018 Australian Tourism Awards. Wine tourism is a growing industry in this region, predominantly visited for its white sand beaches.

Giacomo Del Chiappa, PhD, is Associate Professor of Marketing at the Department of Economics and Business, University of Sassari (Italy) and Senior Research Fellow at the School of Tourism & Hospitality, University of Johannesburg (South Africa). He teaches tourism marketing and destination management, and his research is related to destination branding, consumer/tourist behavior, food and wine tourism, and digital marketing.

Shenrui (Demi) Deng, MTour Mgt, is a doctorate student in the School of Hospitality Business Management at Washington State University, Pullman, Washington, USA. Her research interests include F&B management, wine, and wine tourism.

Saurabh Kumar Dixit is Professor and founding Head of the Department of Tourism and Hotel Management, North-Eastern Hill University, Shillong (Meghalaya), India. His research interests include consumer behavior, gastronomic tourism, and experiential management and marketing in hospitality and tourism contexts. He has 15 books to his credit, including *The Routledge Handbook of Consumer Behavior in Hospitality and Tourism* (2017), *The Routledge Handbook of Gastronomic Tourism* (2019), *The Routledge Handbook of Tourism Experience Management and Marketing* (2020), *Tourism in Asian Cities* (2020), *Tourism in India: Marketing Perspectives* (2021), *Consumer Behavior in Hospitality & Tourism* (2021), and *The Emerald Handbook of Luxury Management for Tourism and Hospitality* (2021). Presently he is editing *Routledge Handbook of Collaborative Consumption in Tourism*, and *Teaching Cases in Tourism, Hospitality and Events* (CABI) to be published in 2022. His detailed profile can be seen at https://www.routledge.com/authors/i15903-saurabh-dixit.

Jane Eastham, from Harper Adams University, Newport, UK, has a background in tourism initiatives and venues and consultancy in logistics and supply chain operations. Her particular research interests lie in small business management, cooperative activity, and supply chain management in the Oeno Gastronomy tourism sectors. Particular interests relate to the implications of market conditions and structures on the effective delivery of economic and social value to businesses operating in the varied collective models. She has published key texts and papers on food supply chain management and collective activities in food. Her PhD thesis looked at the success of marketing cooperatives in the UK agricultural sector.

Lázaro Echegaray Eizaguirre PhD from the University of the Basque Country, is Professor in the Cámarabilbao University Business School and Coordinator of the Research Department in the same institution. As an academic researcher, Lazaro Echegaray specializes in communication, social analysis, and media analysis with a high interest in everything related to audience analysis and reception processes. Within the private field, he has more than 15 years of experience as a market researcher aimed at consumption matters. He is a specialist in qualitative social research.

Alet C. Erasmus, PhD, is appointed as a research associate at the Gordon Institute of Business Science, University of Pretoria, South Africa, since retirement as Head of the Department of Consumer Science at the same institution. Her primary interest is consumers' decision-making, particularly dealing with uncertainty, the evaluation of complex commodities, as well as consumers' behaviour in different retail contexts. She has supervised 10 PhD studies and more than 50 master's studies to date, producing more than 60 international publications including textbooks for tertiary institutions.

Michele F. Fontefrancesco is Assistant Professor of Cultural Anthropology at the University of Gastronomic Sciences in Italy. His research is focused on local development in Italy and Eastern Africa, with an interest in food tourism. The analysis of the local impact of food festivals is further investigated in his recent monograph, *Food Festivals and Local Development in Italy* (2020).

Joanna Fountain is Senior Lecturer in Tourism Management at Lincoln University, New Zealand. She has diverse research interests, much of which are related to agri-food and wine value chains, including tourism and analysis of change and resilience in rural regions.

Francesc Fusté-Forné is a Lecturer and researcher at the University of Girona, Spain, and a collaborating professor and tutor at the Open University of Catalonia, Spain. His research is focused on rural and food tourism and food media. Particularly, he has studied the food tourism phenomenon by bringing connections between authenticity, food heritages, landscapes and landscapers, regional development, rural activities, and tourist experiences.

Wagner Otávio Gabardo is a PhD student in geography at the Federal University of Parana, Brazil. He has a master's in tourism and another in viticulture and oenology. He is also a graduated sommelier. His research interests include sommelllerie, wine tourism experience, and cultural approaches to geography.

Roberta Garibaldi is Professor of Tourism Management at the University of Bergamo, Italy. Her current research covers the field of gastronomy tourism. She is on the Board of Directors of the World Food Travel Association and serves on the Advisory Board of World Gastronomy Institute and the Italian Society of Tourism Science. She is currently President of the Italian Association of Gastronomy Tourism and author of the Report on Gastronomy Tourism in Italy.

Philippa Golding completed her BSc in tourism at the University of Applied Sciences of the Grisons in Chur, Switzerland. She subsequently earned her MSc BA with a major in tourism at the same university. She currently works in marketing and business development for the EHL Swiss School of Tourism and Hospitality, Switzerland.

Anne Gombault is Professor of Management at Kedge Business School, France, where she leads the MSc ARTS and Creative Industries Management in Paris and the Sino-French Institute of Arts & Design Management in Shanghai. She is a member of the Creative Industries & Culture research center that she founded. Her research work on the identity and strategy of cultural and creative organizations has been published in France and internationally.

Eduardo Cândido Cordeiro Gonçalves, PhD in tourism and leisure from the University of Rovira I Virgili (Spain), is Associate Professor at the University Institute of Maia (Portugal) and Head at CEDTUR–Tourism Development Studies Center, and Researcher at CETRAD (Portugal). His primary interest areas of investigation are tourism, hospitality, sustainability, governance, and wine tourism.

Karen Goodlad is Associate Professor, New York City College of Technology, City University of New York, USA. She is also a certified specialist of wine with the Society of Wine Educators and is an active member of Les Dames d'Escoffier International. She was named by *Total Food Service* as a Top Woman in Metro New York Foodservice & Hospitality in 2019 and 2020. Her publications include "Place-Based Learning across the Disciplines: A Living Laboratory Approach to Pedagogy" in *InSight: A Journal of Scholarly Teaching*, and "Comparing Faculty and Student Perception of Academic Performance, Classroom Behavior, and Social Interactions in Learning Communities" in *College Teaching*.

Ulrike Gretzel is Senior Fellow at the Center for Public Relations, University of Southern California, USA, and Director of Research at Netnografica. Her research focuses on persuasion in human–technology interactions, smart technologies and intelligent systems, tourism experiences, online and social media marketing, and adoption and use of technologies, as well as non-adoption and digital detox.

Alexandre Guedes is Assistant Professor at the University of Trás-os-Montes and Alto Douro (UTAD), Portugal. His area of research interest includes tourism planning, spatial analysis, wine tourism, and river tourism.

Coralie Haller is Associate Professor of management of information system and entrepreneurship at EM Strasbourg Business School, Strasbourg University, France. She is in charge of the Master of International Wine Management and Tourism and the Master in Tourism Management programs, and she founded the Chair in 'Wine and Tourism.' Her research has been published in the *International Journal of Entrepreneurship and Small Business*, *International Business Review*, *International Journal of Wine Business Research*, and *Journal of Gastronomy and Tourism*, among others.

Robert J. Harrington, PhD, is Professor and Director in the School of Hospitality Business Management at Washington State University, USA. He is the Ivar Haglund Chair in Hospitality with research interests in strategic management, innovation management, and culinary tourism. He is the author of *Food and Wine Pairing: A Sensory Experience* and has a secondary research interest in the sensory aspects of food and wine.

Sylvie Hertrich is Associate Professor of Marketing at EM Strasbourg Business School, Strasbourg University, France. She earned a PhD in management from the University of Paris Panthéon-Assas. She has been a visiting professor on the five continents. Prior to her current appointment, Dr. Hertrich served for seven years as a CEO with M.K.I GmbH. She created and developed the Master 2 in Marketing and Event Management at EM Strasbourg Business School. She is a co-author of books, book chapters, and over 50 business cases.

Rory Hill is a Lecturer in wine business and innovation at the Eastern Institute of Technology in New Zealand. He has research interests in wine, terroir, sense of place, gastronomy, and rural economy.

Clare Hindley is Professor in the Language and Communication Department at IUBH International University of Applied Sciences in Bad Honnef–Bonn, Germany. She lectures primarily in research methods and communication modules and is a university research coordinator. Her international career in business and education has been diverse in location and focus, but language, culture, and communication have remained the guiding lights with interest in cross-disciplinary studies. Her recent research publications focus on areas of hospitality, tourism, management education, culture, and sociology.

Zainub Ibrahim is Professor at Algonquin College's School of Hospitality and Tourism, Canada. She holds a PhD in geography from the University of Waterloo, Canada, and specializes in issues surrounding tourism planning. Her research interests can broadly be stated as how tourism is negotiated in the spaces it occupies. Some of her research studies include the topics of retirement migration, evaluation of tourism plans, and coastal tourism sustainability.

Annamma Joy is Professor of Marketing, Faculty of Management, at the University of British Columbia, Okanagan campus, Canada. Her research interests are primarily in the area of consumer behavior and branding with a special focus on luxury brands, fashion brand experiences, consumer behavior in the People's Republic of China, sustainability in the wine and fashion industries, and aesthetic consumption.

Amnaj Khaokhrueamuang is Associate Professor of Tourism at the School of Management and Information, University of Shizuoka, Japan. His research interests are rural, tourism-related issues associated with community development, agricultural extension, culture and heritage, and sustainability.

Anupama S. Kotur (Kaddi), PhD, is Assistant Professor at the Symbiosis Centre for Management Studies, Symbiosis International (Deemed University), Pune (India). She is an expert in E and Mobile learning for travel and tourism professionals with experience from working with several global leisure and tourism brands. She has also developed and delivered a number of training programs for luxury tourism and cruise line brands. Her research interests include wine tourism, luxury tourism, and destination imagery studies. Her upcoming edited publication is *The Emerald Handbook of Luxury Management for Hospitality and Tourism* (2022).

Kuan-Huei Lee is Associate Professor at Singapore Institute of Technology (SIT) in the Design and Specialised Businesses Cluster. Dr. Lee graduated from the University of Queensland with a PhD in tourism management and holds an MBA from the University of Illinois at Urbana-Champaign. She accumulated many years of industry experience in the restaurant, air transport, and hospitality businesses. Her research interest includes food tourism, food and beverage management, and cross-cultural studies of tourists.

Willy Legrand is Professor of Hospitality Management at the IUBH International University of Applied Sciences located in Bad Honnef–Bonn, Germany. He is the lead author of *Sustainability in the Hospitality Industry: Principles of Sustainable Operations* as well as *Social Entrepreneurship in Hospitality: Principles and Strategies for Change*, and he is co-editor of the *Routledge Handbook of Sustainable Food and Gastronomy*. Professor Legrand is Guest Chief Editor of the *Hotel Yearbook Sustainable Hospitality 2018* and *2020 Special Editions*. He chairs the *HospitalityNet World Panel on Sustainability in Hospitality*, an online panel supported by more than 85 experts across many fields, including food, energy, water, waste, facilities, communication, social, and strategy.

Andrea Levis completed his bachelor's program in hospitality management at the IUBH International University of Applied Sciences in Bad Honnef, Germany. He had previously worked in a luxury hotel, Belmond Castello di Casole, that thrived on wine tourism and is currently working for Timbers Resorts, a luxury brand hotel, as a reservations agent.

Gui Lohmann, PhD in tourism management, is Professor in Aviation and Tourism Management at Griffith University (Australia). He is Founder and previous Executive Director of ABRATUR, the International Academy for the Development of Tourism Research in Brazil. His research interests include tourism destination development, tourism transport, transport geography, tourism geography, and tourism theory.

Maria del Pilar Leal Londoño is Professor and researcher in tourism at CETT Barcelona School of Tourism, Hospitality and Gastronomy, University of Barcelona, Spain. She combines teaching with the position of Director and Coordinator of master's and postgraduate programs. Her particular research interests range from sustainable tourism studies to food and wine tourism. She has published books and papers particularly in food tourism and new tourism phenomena.

Sandra Maria Correia Loureiro, PhD, is Associate Professor at Instituto Universitário de Lisboa in Portugal (ISCTE-IUL) and Researcher at Business Research Unit (BRU-IUL). Her research interests include tourism marketing and the connection between marketing and VR, AR, and AI issues. Her papers have been published in a variety of peer-reviewed journals, including *Journal of Marketing*, *Journal of Retailing*, *Tourism Management*, *Journal of Business Research*, and *International Journal of Hospitality Management*.

Andy Lyon is Senior Lecturer in International Tourism at Liverpool John Moores University, UK. He has been teaching and researching tourism for over 20 years. His research focuses on destination management and sustainable tourism development, and he has published papers in a number of top tourism journals. He is also a board member of Marketing Cheshire, the regional destination management organization, and has also carried out consultancy projects in Tenerife and the Gambia.

Keith H. Mandabach is Professor in the School of Hotel Restaurant and Tourism Management at New Mexico State University, with an Agricultural Experiment Station appointment. He completed his doctorate and master's degrees at the University of Houston and his bachelor's degree at St. Edward's University, USA. His research interests include gastronomy/culinary arts, food and beverage management, rural tourist motivation, hotel management, technology in culinary arts, and customer engagement.

Byron Marlowe, PhD, is Director of Washington State University's Wine Beverage Business Management program and is the Don Smith Distinguished Professor of Hospitality Business Management. He is a co-author of *Wine Sales and Distribution: The Secrets to Building a Consultative Selling Approach*, winner of the 2020 Gourmand Book Award for best wine book for professionals in the United States. Byron's research interests include food and beverage, wine, tourism, and hospitality.

Jorge García Martín, PhD in information sciences, is Professor of Marketing at the Cámarabilbao University Business School and Professor in the Degree of Communication at the International University of La Rioja (UNIR). His current lines of research are wine tourism as a source of competitive advantage, and skills required in the market for marketing professionals.

Fiona McCarthy-Kennedy is a hospitality and tourism management student at Algonquin College of Applied Arts and Technology, Canada. She holds a diploma in culinary management from George Brown College, where she received the Unilever Food Solutions College Entrance Scholarship from the Canadian Hospitality Foundation in 2012. Her current interests lie in oenotourism and culinary tourism.

Dan McCole is Associate Professor in the Department of Community Sustainability at Michigan State University, USA, where he conducts research and engagement involving wineries, wine consumers, and communities that attract wine tourists. Dan has written extensively about the wine industry and is a frequent speaker at wine industry conferences in the United States and abroad. He holds a PhD from the University of Minnesota and an MBA from Michigan State University.

Thorsten Merkle is Professor for Hospitality Management and serves as Head of the Expert Group Hospitality Management at the Zurich University of Applied Sciences (ZHAW), Switzerland. Before joining ZHAW, he served as Director of Studies in Service Innovation and Design and as Director of Studies in Tourism at the University of Applied Sciences of the Grisons in Chur, Switzerland. His chapter in this *Handbook* emerged from his employment at the UAS Grisons. He holds a PhD from the University of Gloucestershire. His research interests lie in service innovation and consumer behavior studies as well as in hospitality-related operations and experiences. Prior to joining the University of Applied Sciences of the Grisons, Professor Merkle has been involved professionally in food and beverage-related projects on a global scale.

Sarah Marroni Minasi has a PhD in tourism and hospitality management and a master's in regional development. She is a member of the research group TerroirTur – tourism, wine tourism, gastronomy, wine heritage, and development – and the International Academy for the Development of Tourism Research in Brazil. Her research interests include regional development, management, territorial planning, big data, and geotechnologies in tourism and hospitality.

Caroline Morrow is a tourism development professional specializing in food tourism, agritourism, and rural tourism. Her wide-ranging experience includes strategy planning, product development, and visitor experience design for destinations around the world. She regularly speaks on these topics and takes great pleasure connecting with the farmers, vintners, chefs, and small business owners that make each destination unique. Caroline has a master's in urban and regional planning and a bachelor's in international development studies. She is currently pursuing a master's in wine tourism innovation to explore connections between agritourism and rural community development.

Khairul Anuar Muhamad is a Lecturer at UOW Malaysia KDU University College, Malaysia. He holds a BSc in culinary arts and food service management (minor in hospitality and oenology studies) from Johnson & Wales University, Providence, Rhode Island, USA. He pioneered the implementation of culinary tourism into private higher education in Malaysia. His research areas include culinary arts, gastronomy tourism, and wine tourism.

Ester Napolitano, PhD, is working as an expert junior in territorial and tourism marketing for the MedArtSal project with the CUEIM (University Consortium for Industrial and Managerial Economics) Foundation (Italy). Her research interests include tourism marketing, consumer behavior, food, and wine tourism.

Tahmid Nayeem is Senior Lecturer at Charles Sturt University, Australia. His major research interest areas and ongoing studies are consumer awareness and attitudes, branding, community engagement, and cultural influences in consumer decision-making in the profit and nonprofit sectors. Dr. Nayeem has published in several highly ranked marketing and communication journals, including *Journal of Retailing and Consumer Services*, *Marketing Intelligence and Planning*, *Australasian Marketing Journal*, and *Journal of Environmental Management*, among others.

Christy Yen Nee Ng is Assistant Professor at the Institute for Tourism Studies (IFT), Macau. She holds a PhD in hospitality administration from Oklahoma State University, USA. Dr. Ng has research interests in relationship marketing, customer satisfaction, job satisfaction, and service quality.

Fernando Oliveira has a BSc in psychology and two PhDs, in social sciences and in tourism management. His primary research interests are food tourism and wine tourism. He is currently a continuous improvement manager at ISCTE-IUL, Portugal.

Humberto Rivas Ortega, MSc (environment studies) and PhD (in economics and law, is Professor at the School of Engineering in Expedition Management and Ecotourism, Universidad San Sebastián, Chile. Professor Rivas worked previously at the National Tourism Service. He has been involved in consulting projects at a national and international level for organizations such as the UNWTO, OAS, and the IDB. His publications include *Guidelines for the Design of Tourism Development Plans* (SERNATUR, 2008, 2015), *Quality Systems in Tourism and Its Relationship with Sustainability Tourism Laws* (UNWTO, 2010), and *Sustainable Tourist Destinations. Proposal for a System of Distinction* (CORFO, 2012).

Georgia Papadopoulou is a master's student at the Department of Tourism Business Administration of the Hellenic Open University (School of Social Sciences), Greece. She holds a bachelor's in mathematics from the Aristotle University of Thessaloniki. Her academic interests lie in the areas of wine tourism and social media, networking, and strategic marketing. Currently, she serves as an energy consultant at Volton Hellenic Energy.

Camilo Peña-Moreno, MSc (sustainable development), is a PhD candidate at the University of British Columbia, Okanagan campus, Canada. His research interests include consumer behavior and sustainability, with case studies in the coffee and wine industries.

Susan Phillip is Associate Professor, New York City College of Technology, City University of New York, USA. She teaches courses in sustainable, cultural, and special-interest tourism. She recently published 'Using Monuments to Teach About Racism, Colonialism, and Sexism,' which appears in the book *Interdisciplinary Team Teaching*. She has also written about tourism in Africa, Europe, and Asia. She is currently conducting research on tourism in Grenada. Her background includes museum-sponsored educational travel. She earned a Master of Science in tourism management from New York University and a Bachelor of Arts from Howard University.

Ige Pirnar is Chair of the Department of Business Administration, Yasar University, Turkey. She earned her MBA from Bilkent University (1989) and her PhD from Ankara University (1998). She has many articles, conference papers in English and Turkish and has eight books in Turkish (three edited and two with co-authors) and one in English. Her areas of expertise are international business, marketing management, F&B management, hospitality marketing, and international tourism management.

Andrea Pozzi is a Lecturer in Marketing and fellow researcher at the University of Bergamo, Italy. His main field of research is food tourism. He is attending a Ph.D. course at the University of Balearic Islands (Spain). Previously he worked as a project manager in the EU co-funded project in regional bodies.

Bruce Prideaux is Program Director of the Master of Sustainable Tourism Management program at Central Queensland University, Australia. His research interests including sustainability issues, climate change, protected area tourism, rural tourism, remote area tourism, and river and canal tourism. Other active areas of research include mobilities research and implications of the transition to net-zero. He has authored over 220 journal articles and book chapters and authored or co-authored 13 books, the most recent of which is a co-authored book on tourism theories, concepts, and models. His next co-edited book will look at tourism in agricultural areas.

Paulo Ramos, PhD, is Assistant Professor at Fernando Pessoa University and Lusíada University, Portugal, and a former marketing course coordinator at both universities. He is a member of the editorial board of the *Journal of Organizational Studies and Innovation*, Management and Business Academy, UK. He is also an EU project expert and evaluator by appointment of the Northern Regional Coordination Commission (CCRN).

Olga Rauhut Kompaniets, PhD (marketing), is Senior Lecturer and researcher at the School of Business, Innovation and Sustainability, Halmstad University, Halmstad, Sweden. Her research interests cover place and destination marketing, and wine tourism. She has performed several studies on the emerging wine tourism destination in southern Sweden. Her publications contain several contributions to theory and methodology with the relevance to her research interests.

Elena S. Rotarou, MSc (urban and regional studies) and Doctor in Economics, is Associate Professor and researcher at the Department of Public Health, Universidad San Sebastián, Chile. Professor Rotarou is an interdisciplinary researcher who studies the factors influencing economic development, such as sustainability, local development policies, and economic activities.

Concerning tourism, she has published extensively on sustainable tourism practices, environmental impacts of tourism development, and niche tourism, such as enotourism. She has participated in research projects related to local perception of tourism development and the environmental sustainability of tourism destinations.

Namita Roy is a researcher at the School of Management, Operations and Marketing, University of Wollongong, Australia. She is an interdisciplinary researcher and studies tourism experiences using post-humanist ontologies. Her interests include studying the role of emotions, space, and culture on consumption.

Ayşe Şahin is a doctoral researcher at Akdeniz University, Tourism Faculty in Turkey. She holds a PhD from the Gastronomy and Culinary Arts Programme. Her research interests include food experiential values, food experience, and emotions and taste preferences. She has published several refereed articles and international conference papers. Her research has been published in *Journal of Tourism and Gastronomy Studies*, the *Journal of International Social Research*, and the *Journal of Foodservice Business Research*.

Anthony J. Saliba is Professor of Perceptual Psychology at Charles Sturt University, Australia. His research focuses on how humans combine sensory information to form perceptions. Professor Saliba has published over 100 peer-reviewed journal articles and government reports and managed over $8 million in research funding. Prof Saliba has broad experience with a range of foods, beverages, and technologies, across the university, government, and private sectors.

Vasco Santos, PhD, is Coordinator Professor of Tourism and Marketing at ISLA Santarém, Portugal. He is Coordinator of the Degree in Tourism Management and a researcher in CiTUR – Center for Tourism Research, Development, and Innovation. He is on the editorial advisory board of peer-reviewed scientific journals and the scientific committee of several conferences. He has authored scientific papers and book chapters. His research areas are wine marketing, wine tourism, wine tourist behavior, employer branding.

Eduardo Moraes Sarmento, PhD, is a full Professor at Lusófona University (Portugal) in tourism and a researcher at CEsA-CSG (ISEG–Lisbon University). His primary interest areas of investigation are tourism, hospitality, sustainability, development, and wine tourism.

Pasquale Sasso has a PhD in entrepreneurship and innovation and collaborates with the chair in Management of Cultural Organization at the Economics Department of the University of Campania Luigi Vanvitelli, Italy. His research interests concern the management and marketing of wine and food companies and the cultural and tourism sector.

Maite Sena is a marketing manager. She has a BSc in marketing management and an MSc in international management from ISCTE-IUL, Portugal. Her main research interest is wine marketing.

Ludovico Solima is Full Professor of Business Management and holder of the chair in Management of Cultural Organization at the Economics Department of the University of Campania Luigi Vanvitelli, Italy. He is also holder of the Chair in Management for Museums at the Suor Orsola Benincasa University in Naples. He has published more than 120 works nationally and internationally, and his research interests concern the management of the cultural sector.

Bruno Sousa, PhD, is Coordinator Professor at the Polytechnic Institute of Cavado and Ave (IPCA, Portugal) and Head of the Master Programme in Tourism Management. He is a researcher at the Centre for Tourism Research, Development and Innovation (CiTUR) and the Applied Management Research Unit (UNIAG). He is the author or co-author of several papers, and his research interests include tourism management, marketing, and strategy.

Albert F. Stöckl is Program Director of the Bachelor studies 'International Wine Business' at IMC Krems, Austria. He started working in wine taverns, inns, and restaurants as a teenager and concluded an apprenticeship as a waiter and wine butler after his A-levels. In 1998 he graduated from Heilbronn University in wine management. Subsequently, he concluded a double-degree master's program in tourism management in Sweden, France, and the United Kingdom. His PhD (Kiel University) deals with consumer emotional attachments to wine brands. His research focus is on the wine business, culinary tourism, and consumer behavior with regard to wine.

Matthew J. Stone is Associate Professor of Recreation, Hospitality, and Parks Management at California State University, Chico, USA, and a 2019 Fulbright Scholar at Haaga-Helia University of Applied Sciences in Helsinki, Finland. His primary area of research is consumer behavior in tourism, with a focus on food and drink tourism. He is a senior research advisor to the World Food Travel Association.

Paul Strickland is a Lecturer in Tourism and Hospitality at La Trobe University in Melbourne, Australia. He has taught in undergraduate courses in Australia, Singapore, and Bhutan in mainly hotel management–related subjects. His research interests include ethnic restaurants, Bhutanese studies, space tourism, wine tourism, wine marketing, and business education.

Jon Sundbo is Professor of Business Administration and Innovation at Roskilde University, Denmark. He is a leader of Roskilde University's Centre for Experience and Tourism Research. He has published over 200 articles, books, and book chapters, primarily on innovation, service business, experience, and tourism. In addition, he has been an advisor to the European Union, the Danish government, and several firms and trade organizations within service and tourism.

Ilinka Terziyska, PhD, has been Associate Professor at the Tourism Department of the South-West University 'Neofit Rilski,' Bulgaria, since 2007. She is currently teaching several courses, including Introduction to Tourism, Cultural Tour Design, Cultural Tourism, Ethno Tourism, and Tourist Behavior. She is a founding member of the Euro-Asia Tourism Studies Association (EATSA) and part of the organizing committees of two international and six national scientific conferences. Her research interests include cultural tourism, wine tourism, event tourism, and tourism behavior. She is the author of more than 40 publications, most of which are in cultural tourism, food and wine tourism, and festivals.

Michelle Thompson is Lecturer–Tourism at Central Queensland University, Cairns, Australia. She has a PhD from James Cook University (2015), which modeled the drivers and barriers to tourism development in agricultural regions. Her research interests are in the areas of food, wine, and agritourism, as well as aspects of regional and remote area tourism development and tourism sustainability. She is currently working on a project examining drive tourism, and her next project is a co-edited book on tourism in agricultural areas.

Fabrice Thuriot is 'Dr. Hab' in public law, Lecturer and Researcher in cultural policies and local authorities, Coordinator of the Centre of Research on Law and Territories, Director of Development in SSH at the University of Reims Champagne-Ardenne, France, and a member of the Institute of Vine and Wine in Champagne, France. He has published almost 120 articles and books, nationally and internationally.

Stephanie Tischler is a researcher and Lecturer at the IMC University of Applied Sciences Krems (Austria). She holds a master's degree in business administration focusing on tourism and leisure management from the Vienna University of Economics and Business and is a PhD student at Christian-Albrecht University of Kiel (Germany), Department of Agricultural Economics. Her dissertation deals with memorable culinary tourism experiences and the role of destination restorative capacities. Her research interests include culinary and wine tourism, health and spa tourism, tourism and destination marketing, as well as experiential marketing.

Marcos Alberto Torres is Professor of Geography at the Federal University of Parana, Brazil. His research interests include landscapes, spatial representations, and cultural approaches to geography.

Rémy Tremblay is Professor at TÉLUQ, University of Québec, Canada. He holds a PhD in geography from the University of Ottawa, Canada, and specializes in issues surrounding tourism and social geography. His research interests include retirement migration, Canada–US migrations, attraction, and retention of highly-skilled labor.

Vander Valduga, PhD in geography, is Professor in Tourism at the Department of Tourism and in the Postgraduate Program in Tourism, Federal University of Paraná, Curitiba, Brazil. He coordinates the research group TerroirTur – tourism, wine tourism, gastronomy, wine heritage, and development and a member of CEPAVIN–Center of Wine Heritage and Culture. His research interests include the geography of food and beverages, wine tourism, and spatial relations in hospitality and tourism.

Renée van Wyk is the co-founder of a consumer behavior start-up company, Telligent (Pty Ltd). Before starting this company, she worked as a culinary assistant on two seasons of My Kitchen Rules, South Africa, and as a junior lecturer and assistant at the Department of Consumer Science, University of Pretoria, South Africa. Renée's master's degree combined her passion for food, wine, and consumer behavior, and some of the results form part of her chapter.

Roland Vogels completed his bachelor's program in hotel management in the Netherlands and continued with a master's program in hospitality management at the IUBH International University of Applied Sciences in Bad Honnef, Germany. He obtained his Wine & Spirit Education Trust (WSET) Level 1 to 3 Diploma and is currently completing his sommelier studies at the German Hotel Academy in Cologne. Mr. Vogels also works in wine sales at a young and innovative local winery. He is also involved in wine-related events for numerous wineries in the Middle Rhine area.

Chuanfei Wang earned her PhD in global studies from Sophia University, Tokyo, Japan. She is a postdoctoral research fellow at the same university and an adjunct faculty member teaching sociology and global studies at several universities in Tokyo. She is the author of numerous articles and book chapters on the Japanese wine industry, the globalization of Japanese culinary culture, and Japanese wine tourism.

Kim M. Williams is a Lecturer in the Faculty of Higher Education at William Angliss Institute, Melbourne, Australia. Kim lectures in event management and human resources. Her research background is diverse but tends to focus on human resources issues, with a prime emphasis on professional development and training. She is also interested in event management, gastronomy, fashion, and wine tourism.

Erik Wolf is the founder of the food travel trade industry and Founder and Executive Director of the World Food Travel Association. He's also the publisher of *Have Fork Will Travel* and author of *Culinary Tourism: The Hidden Harvest.* He is a highly sought speaker on food and beverage tourism.

Brian Kee Mun Wong is a Professor who holds a PhD in Tourism Management from the University of Malaya, Kuala Lumpur. As an active academic, a researcher, and a practitioner, he specialises in marketing, management and entrepreneurship, particularly in the travel and tourism industry. He is also a reviewer for top-tier journals in these areas.

Shuangyu Xu, PhD, is Assistant Professor in the School of Natural Resources at the University of Missouri, USA. Her research is primarily focused on developing and promoting sustainable tourism (with an emphasis in wine tourism and nature-based tourism) for communities using a combined stakeholder theory and GIS approach.

Manoj Kumar Yadav, PhD (travel and tourism management), completed his bachelor's and master's degrees in hotel management from Bundelkhand University, Jhansi (India). He is currently working as a Lecturer at the Institute of Hotel Management, Catering Technology & Applied Nutrition, Gwalior (India), since 2010. He has 15 years of experience in teaching, and he currently teaches front office management and has also taught food and beverage service. Previously, Dr. Yadav also worked with Gujarat Institute of Hotel Management, Vadodara (under the aegis of AH&LA), then moved to JP-Institute of Hotel Management, Meerut. His area of specialization is food and beverage service. Dr. Yadav has already written three textbooks and several research papers. Additionally, he has contributed four chapters in *Front Office Management* and *Food & Beverage Management* textbooks published by Uttarakhand Open University (India).

Gökhan Yılmaz is a doctoral researcher at Akdeniz University, Tourism Faculty in Turkey. His research interests include gastronomy, food tourism, tourist behavior, restaurant management, experiential management, and marketing in food, hospitality, and tourism contexts. He has published several peer-reviewed articles and books. His research has been published in the *International Journal of Contemporary Hospitality Management*, *International Journal of Tourism Cities*, *Journal of Culinary Science & Technology*, *Journal of Gastronomy and Tourism*, *Anatolia: A Journal of Tourism Research*, and *Journal of Foodservice Business Research.* He is an active reviewer of many journals.

Izaskun Zurbitu-Aldama is Senior Lecturer in Gastronomy Marketing and Entrepreneurship at Basque Culinary Center, University of Mondragon, Spain. Her research interests include gastronomy, place branding, food and wine marketing, and business innovation. She is presently pursuing her doctoral program at Deusto Business School, University of Deusto, Spain.

INTRODUCTION

Saurabh Kumar Dixit

Wine tourism, interchangeably denoted by the terms 'enotourism,' 'oenotourism,' 'vinitourism,' and 'winery tourism,' is considered as a specialized niche of tourism focusing on the visitors' activities at wineries and wine-growing regions. Since the 1990s, this specialized tourism form has developed in different parts of the world, extending opportunities at both an individual/business and a community/destination level (Carlsen 2004). The conceptualization and understanding of 'wine tourism' is still evolving, and in most instances, it relates to the tourist's motivation and experiences.

Mitchell and Hall (2006) classified wine tourism research into the following areas: wine tourism product; wine tourism and regional development; quantification on demand; segmentation of wine tourists; visitors' behavior; nature of the visit to wineries; and food safety and wine tourism. Getz (2000) stressed three significant perspectives of wine tourism: wine producers, tourism agencies (representing the destinations), and consumers. The challenge for wine tourism researchers is integrating wine tourism products and consumer knowledge to develop a more comprehensive view of wine tourism's strategic and pragmatic issues. Wine tourism, therefore, involves engagement with a specific group of tourists, popularly termed as wine tourists. Wine tourists are betrothed primarily in wine-related travel to experience a holiday with wine and wine tourism activities.

Byrd et al. (2016) presented the idea of the core product and other ancillary services anticipated by wine tourists or offered by the wine producer to augment observed benefits and product value from wine tourism. Byrd et al. (2016), as presented in Figure 0.1, further elaborated on the *core, augmented, and ancillary services* model applicable for contemporary wine tourism. The *core product* (inner circle) in wine tourism is wine consumption and procurement, which may involve the quality, tasting, or purchasing of wine. *Augmented services* (the middle ring) may include a vineyard, winemaking activity, tasting rooms, customer service, wine education, and socializing/wine club offerings. *Ancillary services* (the outer ring) may consist of relaxation and recreation opportunities, entertainment and events, and other typical tourism and hospitality services contributing to wine tourism in the region. Wines contribute immensely to spread the local traditions and culture of the places besides serving as a major motivating factor for visitations by tourists. Wine, food, tourism resources, and cultural activities form the core elements of the wine tourism product and offer the opportunity to wine tourists to experience the wine, terroirs, and community closely.

DOI: 10.4324/9781003143628-1

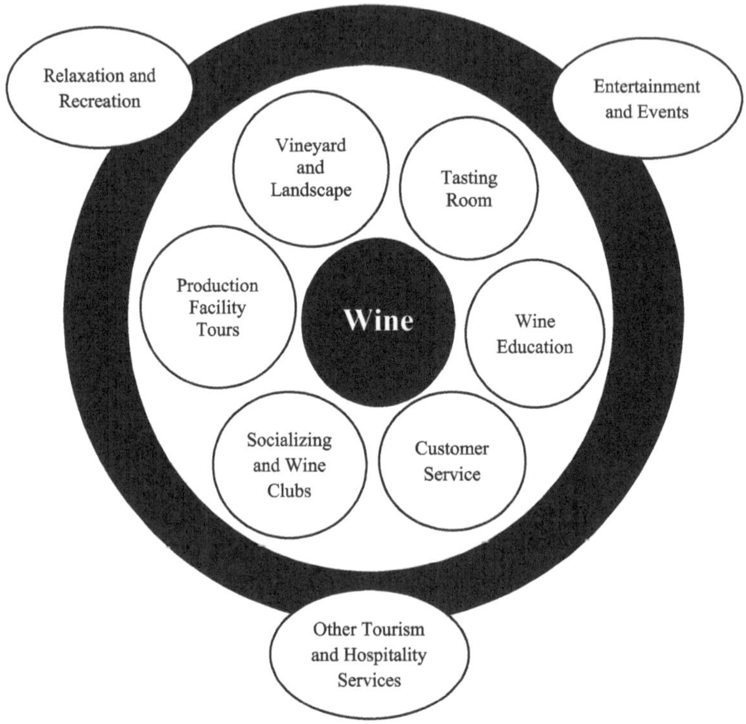

Figure 0.1 Core, augmented, and ancillary services in wine tourism.

Source: Byrd et al. (2016)

Wine tourism researchers, wine industry managers, and destination planners need to understand how the integration of the primary (agriculture), secondary (wine industry), and tertiary (tourism) sectors diverge and converge across the economic, cultural, geographic, and vocational facets. Winemaking is a predominantly primary and partially secondary industry characterized as a supply-and-demand-driven industry (Carlsen 2004). Mitchell, Hall, and McIntosh (2000) presented the demand and supply framework of the wine tourism experience. The demand perspective studies wine tourists' motivations, perceptions, previous experiences, preferences, information, and expectations for travel. To meet the demand for a quality wine tourism experience, destinations should be able to supply a number of amenities, services, and resources. Hence, the supply perspectives of wine tourism involve exploring distinctive wine tourism resources and attribute at a destination, including wine tasting and visitation to wineries, wine trails, festivals, and 'winescapes.'

Wine tourism experiences are one of the major catalysts in fabricating a wine-growing region as a wine tourism destination. Wine tourism has been identified as the outcome of a combination of factors, and wine tourists as individuals seeking experiences that combine landscapes, wines, gastronomy, culture, history, and human relations (Charters 2006; Demossier 2010; Dixit 2019). These facets are pertinently reflected in the winescape tourism concept (Quadri-Felliti and Fiore 2012), which relates wine tourism attractions with wine tourism activities, to assimilate wine, art, music, and cultural heritage (Williams 2001). According to Bruwer and Gross (2017), a winescape's main dimensions are the infrastructure, social environment, natural and cultural resources, atmosphere, social setting, layout, signage, and people.

Terziyska and Damyanova (2020) have identified six major attributes of the winescape as perceived by organized wine tourists as tour guide, tour planning and logistics, core wine product, complementary activities, food and dining, and nature and scenery. Wine tourism research has classified the wine tourism experience or winescape via both a micro and a macro approach (Bruwer and Gross 2017). The micro-level wine tourism experience looks at the winescape 'as the environment at a specific winery' (Quintal, Thomas, and Phau 2015, p. 597). On the contrary, the macro-level wine tourism experience looks at the winescape by observing the attributes of the wine region to predict wine tourist consumption behavior (Bruwer and Gross 2017). Therefore, over time, the conceptualization of the winescape has evolved from the core vineyard and/or winery facility to the greater wine region as a tourist destination (Byrd et al. 2016). (Further details are in Chapter 3.)

Owing to its hedonic and sensory nature, wine tourism is rationally one of the most appropriate areas that offer diverse experiential products. Wine marketers and wine tourism destinations should also make consumers aware of the aspects of wine 'flavor' and 'taste' to understand wine attributes closely. Unique attributes of a winery, themes and activities, landscape, hosts' attitude, environment, education, and novelty contribute to a memorable wine tourism experience. If a winery can offer its visitor an opportunity to participate or co-create activities such as picking grapes, bottling the wine, or making one's own wine, those activities clearly contribute to a memorable experience (Saayman and van der Merwe 2015; Dixit 2020). The delivery of meaningful, personalized, co-created, and transformational wine tourism experiences have become the new benchmark for competitiveness and differentiation in wine tourism destinations (Sigala 2012).

As per Getz (2000), the wine tourism research priorities should be focused on wine consumers, wine tourists, marketing effectiveness, and success factors for wineries or destinations. Wine tourism is assumed to be embracing three production levels (see details in Chapter 2): the primary level, including vineyard management for grape cultivation; the secondary level, involving winery and cellar door operations; and the tertiary level, comprising of accommodations and restaurants and other allied services at wine tourism destinations (Brochado et al. 2020). The success of wine tourism at a destination depends on the effective implementation of the wine tourism system's vital components, comprising the value chain of wine tourism, basic tourism resources, viticulture/wineries operations, and wine exports (Getz and Brown 2006). Wine tourism plays a notable role in regional development and tourism plans by sustaining regions' economic and social attributes.

As wine has an agricultural base, it can offer horizontal and vertical linkages within the rural tourism environment. In contrast to other agricultural commodities, local wines are promoted by the geographical location of production, i.e., its terroir. Winemaking can help rural areas redesign wine production frameworks and offering enduring hospitality to tourists. Rural territories in wine-growing regions may endorse local products, culinary delicacies, scenery, and other outdoor activities, emphasizing rural lifestyles and landscapes. Many wine regions and tourism destinations have realized that the benefits of wine tourism may be extended well beyond the vineyards and cellar doors, directly or indirectly, to the local economy and to community empowerment. Due to the diverse and extensive range of players involved, wine tourism is considered a complex ecosystem, and Salvado and Kastenholz (2017) have proposed a model of the coopetitive wine tourism ecosystems sustained on three interdependent pillars or subsystems: (1) wine culture, (2) territory/landscape, and (3) tourism. The three pillars conglomerate a range of stakeholders, involving divergent business strategies (related to main activities, product design, prices and profit formation, value creation, and distribution channels) to achieve territorial competitiveness and sustainability (Salvado and Kastenholz 2017).

The contemporary wine tourism sector is moving to the vibrant technological tools and social media for its marketing and promotion endeavors. Web 2.0 – constituted by blogs and video; Facebook, Twitter, and Pinterest; social media tools; and apps such as WineryLovers, Vivino, and UberVINO – is becoming increasingly crucial to wine tourism marketing. Social media platforms and user-generated content contribute to arousing or increasing wine consumers' interest and engagement for a wine-producing destination, a winery, a wine variety, or a wine brand (Sigala, Evangelos, and Gretzel 2012). Social media significantly changes how wine consumers communicate and engage with wine retailers, wineries, and other wine consumers. Post-COVID-19 wine tourism is beginning to adopt Web 3.0, which integrates with natural language processing to make computers understand human language and provide faster and better results, in order to be constantly in touch, virtually, with all its clients. The content and functionality elements of Web 2.0 and Web 3.0 enhance the destination image and demand for real-life experiences and improve the destination branding process of wine tourism (see details in Chapters 43 and 44). Martins et al. (2017) offer a conceptualization of a multisensory virtual wine experience that includes a conceptual perspective and a technological solution proposal.

The present *Handbook* has brought together a contemporary selection of researches to construct a sound underpinning for understanding the truly global perspectives of the wine tourism experience via impartial and rational coverage of up-to-date philosophies, practices, and global success stories. Therefore, the volume aims to bridge the gap between wine tourism researchers' existing and future knowledge base. The present *Handbook* is an authoritative resource encompassing theoretical and applied themes of wine tourism. The *Handbook* offers a synthesis of chapters authored by educators, industry practitioners, and eminent experts involved in wine tourism research, industrial practices, and operations from different parts of the world. The *Handbook* will help researchers and practitioners to refine their research and thinking about wine tourism.

The structure of the handbook

The 57 chapters of the *Handbook* are divided into seven parts to offer a significant contribution to wine tourism research, management, and marketing. The volume epitomizes an exceptional scholarly attempt to cherish the synergetic interface of wine tourism with sustainable regional development, destination planning, contemporary trends, marketing, segmentation, behavior, and experiences. The overall content of the volume should be of great interest to researchers and stakeholders working in the wine and tourism sectors.

- Part I, 'Setting the scene for wine tourism,' endeavors to illuminate the nature, significance, winescape, and contributions of wine routes and trails in the wine tourism experience. This part expounds on the ethical and regulatory concerns, the co-creation process, and challenges that global wine tourism faces.
- Part II, 'Profiling and segmentation of the wine tourism market,' amplifies the understanding of the global wine tourism market, market segmentation, wine tourists' behavior, and wine tourists' motivations/preference. It also offers chapters on wine consumers' knowledge and on first-time winery and senior travelers' wine tourism experience.
- Part III, 'Wine tourism as an instrument for regional development,' discusses strategies for developing wine tourism destinations, as well as both the host community's and winery owners' perspectives in managing wine tourism destinations. This part looks further at rural wine tourism, wine gentrification, and the contribution of the great wine capitals network and local wine events in regional wine tourism promotion.

- Part IV, 'New approaches and practices in wine tourism marketing,' focuses on marketing destinations via wine tourism, the role of wine festivals, the emergence of fruit wine tourism and creative star wine districts. The part also appraises the readers on folk wine festivals, matching business travel with wine tourism.
- Part V, 'The wine tourism experience: management and success stories,' highlights the wine tourism experience framework, staging of the wine tourism experiences, and the antecedents, elements, and outcomes of wine tourism experiences. Part V also makes the reader aware of multisensory wine experiences, wine tasting experiences, authenticity in wine tourism, and experiential aspects in winery visitors' TripAdvisor reviews.
- Part VI, 'Innovations and technological advancements in wine tourism,' embodies chapters on the role of technological innovations in wine tourism and how the technical media promotes wine tourism. It further highlights the impacts of the COVID-19 outbreak on wine tourism.
- Part VII, 'Terroir sustainability and cultural constructs in wine tourism,' reflects on sustainable operations and management of the wine tourism business. It further enlightens readers about the integration of wine tourism, territory, culture, and rural tourism. The multidisciplinary nature of the *Handbook* will enable its readers to appreciate diverse perspectives of global wine tourism.

Part I: Setting the scene for wine tourism

Part I includes eight chapters to build the theoretical base for readers to construct the underpinning of wine tourism. Chapter 1, authored by Manoj Kumar Yadav and Saurabh Kumar Dixit, throws light on the historical evolution of global wine tourism. The chapter also discusses the cyclic patterns for the development and growth of any destination to become a wine tourism hotspot. The authors conclude the chapter by discussing the emergent trends of global wine tourism. Chapter 2, by Genevieve d'Ament, Anthony J. Saliba, and Tahmid Nayeem, examines the Australian wine tourism industry's multidimensional nature and evolution. The chapter embodies two cases; the first compares wine tourism in non-wine-region wineries and wine region wineries from a senior management perspective. The second case study relates to the experiences of emerging and established wineries from a business owner's perspective. Vander Valduga, Sarah Marroni Minasi, and Gui Lohmann, authors of Chapter 3, propose a theoretical and analytical panorama of winescape by considering the geographical and historical perspectives as a starting point to establish a winescape framework. To appreciate the winescape framework diligently, the chapter also offers the case study of Vale dos Vinhedos, Brazil. Chapter 4, authored by Namita Roy and Ulrike Gretzel, focuses on discussing the role of wine trails/routes in creating the tourist experience using the Actor-Network Theory (ANT). ANT highlights three translations or conditions of engagement (symbolic, material, affective) within the assemblage that give rise to particular wine tourism experiences. A case study of the Santa Barbara Urban Wine Trail website is also included to explain how these conceptual dimensions and wine trail websites create and shape the online and offline wine trail experience.

Chapter 5, authored by Anupama S. Kotur (Kaddi), offers a discourse on existing policies and regulations applicable to the wine industry in Nashik (India) and explores if the ethical predicaments have any influence on policymaking concerning the wine industry in Nashik. Chapter 6, by Eduardo Moraes Sarmento, Sandra Maria Correia Loureiro, and Namércio Cunha, explore how wine distributors and producers describe personal and organizational social capabilities as facilitating facets of relationship quality. The study revealed that organizational social capabilities comprise of organizational capability, social events promoted by the partners, and the history of the partners. Zainub Ibrahim, Rémy Tremblay, and Fiona McCarthy-Kennedy, in Chapter 7, examine the history of wine-making and prominent wine-growing regions in Canada and present its comparison with the rest of the world. The chapter's case study of Nk'Mip Cellars showcases how an indigenous community used wine tourism to make themselves economically self-sufficient and to protect the culture. Vasco Santos, Paulo Ramos, and Bruno Sousa, in Chapter 8, address the context and relevance of niche tourism and its linkages with wine tourism. The chapter further highlights the challenges, trends, and opportunities for worldwide wine tourism with the support of the Porto Wine Cellars case study.

Part II: Profiling and segmentation of the wine tourism market

Part II includes eight chapters related to the global wine market, wine consumer segmentation, and appreciation of wine tourist behavior. Brian Kee Mun Wong and Khairul Anuar Muhamad, in Chapter 9, throw light on the 'Old World,' 'New World,' and 'Emerging World' wine-growing markets and the unique characteristics of the wine grown over there. The chapter also offers insights into the distinguishing features of wine culture and the winescape of the three wine markets. Chapter 10, authored by Roberta Garibaldi, Matthew J. Stone, Erik Wolf, and Andrea Pozzi, investigates wine tourists' consumption behavior and explores further the factors that affect wine purchase decision-making. The chapter extends the wine tourism theory and considers it within the context of other food products available at a destination. Kuan-Huei Lee, in Chapter 11, presented the wine consumer behavior and preferences using the Consumer Involvement Profile (CIP), Maslow's Hierarchy of Needs, and personality traits. The author highlights the theme by unfolding two major wine events in the world: Slow Wine and ProWein or Pro Wine. Alexandre Guedes, in Chapter 12, presents wine tourism´s institutional framework and governing system in Portugal and six international reference markets as benchmarks. The discussed theories will help to explain the broader context of the wine tourism governance system on an international institutional benchmark.

Brian Kee Mun Wong and Christy Yen Nee Ng, in Chapter 13, explore wine tourists' motivations to enable an in-depth understanding of their expectations, aims, and requirements relating to wine tourism. The authors propose the operational definition of motivation as 'a set of psychological and physiological needs that are triggered by curiosity and the urge to gain new experiences and knowledge in a less familiar destination.' Chapter 14, authored by Alet C. Erasmus and Renée van Wyk, ponders upon consumers' wine self-confidence (WSC). The acquaintance of WSC has become very pertinent as the range of wines to choose from and the characteristics of wine are continuously changing and expanding globally. The chapter converses on South African consumers' wine knowledge elements by a quantitative study. Chapter 15, authored by Karen Goodlad and Susan Phillip, evaluates young, multicultural, urban undergraduates having their first winery experience using an exploratory qualitative approach. The study underlines the potential market of younger multicultural persons in wine tourism. Izaskun Zurbitu-Aldama and Adela Balderas-Cejudo, in Chapter 16, offer insights into the importance of experiences, the Experience Economy model to wine tourism relating to seniors. The family-owned winery Vivanco Winery in Rioja (Spain) is presented as a case study in the chapter.

Part III: Wine tourism as an instrument for the regional development

Part III comprises eight chapters focusing on the emerging tools for sustainable regional development, predominantly for rural areas. Chapter 17, authored by Caroline Morrow, highlights how the wine tourism destination's narrative acts as an essential tool for transmitting the local value of wine and building appeal for the destination. The chapter further explores the contribution of foodways in wine tourism development, with a case study from Ontario, Canada. Eduardo Cândido Cordeiro Gonçalves and Ricardo Jorge da Costa Guerra, in Chapter 18, aims to design a tourism destination management strategy to be applied to the Douro Demarcated Region (DDR) in Portugal. To propose the appropriate strategy, the authors adopted qualitative and/or mixed methodologies and followed an exploratory perspective of the case study. The authors look for a composite wine tourism product strategy in DDR with local traditions and culture. The resident community participation is vital for achieving community support for tourism development and essential for the success and sustainability of any tourism endeavor. Chapter 19, by Shuangyu Xu and Carla Barbieri, critically evaluates residents' attitudes toward wine tourism development and communities' wine tourism-related social capital in North Carolina. Magali Canovi and Andy Lyon, in Chapter 20, examine winery owners' perceptions of and motivations towards wine tourism as a business using qualitative evidence from the North Italian wine region of *Langhe*. The study concludes that the motivations for engaging in the wine tourism business are considerably more complex and are inextricably linked to their place attachment/place identity and the family's situation.

Chapter 21, authored by Chuanfei Wang, inspects wine tourism as an instrument, integrating urban and rural social resources to stimulate rural economic development in Japan. Looking into the case of wine tourism in Japan's Yamanashi Prefecture, it presents how indigenous wine tourism may offer an impetus to the local community's economy. Chapter 22, authored by Shenrui (Demi) Deng, Byron Marlowe, and Robert J. Harrington, investigates gentrification in rural areas of the United States due to the development of wine tourism in a region. For this analysis, the chapter focuses on analyzing tourism gentrification in three US wine-specific regions, i.e., the Napa Valley AVA (California), the Willamette Valley AVA (Oregon), and the Walla Walla Valley AVA (Washington). Adela Balderas-Cejudo, Jorge García Martín, and Lázaro Echegaray Eizaguirre, in Chapter 23, offer insights on how to position wine tourism destinations using the Great Wine Capitals (GWC), an international network of global wine regions. The chapter also delves into the understanding of the contribution of the GWC network by a case study of Bilbao-La Rioja, one of the wine regions of Spain. Chapter 24, authored by Francesc Fusté-Forné, provide a discourse about food and wine events as a tool to promote the pairing of 'food' and 'place' and analyzes how wine events contribute to the preservation of local wine heritage. To illustrate it, the author presents the case study of the event 'Reus viu el vi' (Reus lives wine), which is held in Reus, southern Catalonia, in northeastern Spain.

Part IV: New approaches and practices in wine tourism marketing

Part IV's cluster of seven chapters examine the emerging trends in global wine tourism marketing. Chapter 25, authored by Stephanie Tischler and Albert F. Stöckl, elucidates the strategic approaches of tourism destination marketing and ascertains relevant destination attributes and experiences that can promote wine tourism. The chapter also presents three case studies of the Austrian Wine Marketing Board, Wein Burgenland (an official wine marketing organization), and wine experience center SANDGRUBE 13 wein.sinn, all contributing to make Austria a global wine tourism destination. Olga Rauhut Kompaniets, in Chapter 26, reviews the role of wine festivals in Swedish wine tourism marketing. The conceptual framework for the study is grounded on the Tourism Destination Life Cycle and Experience Economy model for the Taste Swedish Wines festival. Chapter 27, authored by Jon Sundbo, discusses the entrepreneurial initiatives for Danish fruit wine using the market-creation process. This chapter also highlights the entrepreneurial process and the challenges encountered by entrepreneurs in establishing this emerging wine tourism activity. The chapter also presents a Cold Hand Winery case study to showcase wine production and experiential elements of winemaking. Pasquale Sasso, Anne Gombault, Fabrice Thuriot, and Ludovico Solima authored Chapter 28, which describes the main features of the creative star wine districts and shows how the wine industry can cooperate with other creative industries and local institutions to develop this ambitious project. The chapter also presents three case studies of three creative star wine districts: Saint-Emilion, Langhe-Roero Monferrato and Chianti.

Chapter 29, authored by Michele F. Fontefrancesco, highlights the characteristics and history of folk wine festivals, often called *sagra* (in Italian), and their contributions to wine tourism and the local communities. To emphasize its content, the chapter offers the case studies of The Grape Harvest Festival of Lu and the Erbaluce Festival in Caluso. Chapter 30, authored by Philippa Golding and Thorsten Merkle, explores the key themes of wine tourism and business tourism and inspects the case of wine tourism in the *Bündner Herrschaft* in Switzerland. Giacomo Del Chiappa, Ester Napolitano, and Marcello Atzeni, in Chapter 31, examine whether perceived memorability can be appropriately employed to profile different types of wine visitors. Moreover, it seeks to observe whether subgroups differ in their sociodemographics, visit-related aspects, overall satisfaction, WOM intention, and purchasing behavior.

Part V: The wine tourism experience: management and success stories

Part V encompasses 11 chapters addressing the evolving themes and success stories relating to wine tourism experiential management and marketing. Wided Batat, in Chapter 32, explores wine tourism experiences' drivers, forms, and outcomes to propose an integrative framework that may help in understanding the mechanism of designing a comprehensive and engaging wine tourism experience. Chapter 33, authored by Paul Strickland and Kim M. Williams, explores the staging of wine tourism experiences, including relaxed wine tasting rooms, interactive wine notes and wine rating apps on smartphones, degustation and wine matching classes, and winery tours. The chapter further stresses the revival of repeat patronage with the help of the case study of the *La Dolce Vita* wine and food festival from Australia. The wine tourism experience is sensitive, intangible, and endures sensitivities observed in an individual and subjective way by one who experiences it. Wagner Otávio Gabardo, Vander Valduga, and Marcos Alberto Torres, in Chapter 34, appreciate the wine tourism experience from its aromas, flavors, and tasting that evokes sensitivities. Chapter 35, authored by Gökhan Yılmaz and Ayşe Şahin, discuss the wine experience of tourists at a destination. More specifically, the chapter strives to identify the antecedents, elements, and outcomes of the wine experience at that destination.

Chapter 36, authored by Anthony J. Saliba and Genevieve d'Ament, refers to the multisensory experience of wine tourism focusing on the cellar door. With the help of two case studies, the authors compare the link between sensory experience, emotion, and wine. Understanding the impact of the multisensory cellar door experience offers insight into buying behavior, post-visit recollection, future revisit intentions, and pre-visit expectations of wine tourism enthusiasts. Chapter 37, authored by Joanna Fountain, explores the concept of involvement and its role in understanding engagement with the wine tourism experience. The study's findings conclude that wine festival attendees differ in their level of engagement with wine and the centrality of wine in their lives and lifestyles. Ana Brochado, Maite Sena, and Fernando Oliveira, in Chapter 38, study Azeitão village (Portugal) to identify main winery experience dimensions in narratives shared online by the winery and/or wine cellar visitors and how wineries differentiate their wine tourism offers. The chapter offers managers an insight to analyze user-generated content to expand understanding of the key dimensions of visitors' wine tourism experience.

Chapter 39, authored by Sylvie Hertrich, Coralie Haller, and Jodie Conduit, envisages an extraordinary wine tourism experience with luxury boarding and lodging facilities with the case of the Louise, a luxury vineyard accommodation located in the Barossa Valley, a wine region in South Australia. Chapter 40 by Ilinka Terziyska proposes a conceptual model of authenticity in wine tourism focusing on its constituents. The proposed conceptual model is verified through a case study of the small Bulgarian town of Melnik, known for its wine production and the local wine. Georgia Papadopoulou and Maria Alebaki, in Chapter 41, intend to improve understanding of the structural components that constitute the wine tourism experience. The study used a cross-case study approach and applied the netnographic method to explore winery visitors' comments shared on TripAdvisor to propose the framework for experiential wine tourism. Chapter 42, authored by Janez Bogataj, deliberates on the development and structure of wine tourism in Slovenia along with the typology of wine (gastronomic) tourists. The chapter further assesses the linkage between gastronomic and wine tourism in Slovenia.

Part VI: Innovations and technological advancements in wine tourism

Part VI shares a collection of seven chapters examining the technological interface and novelties happening in wine tourism mainly due to the COVID-19 pandemic. Roberta Garibaldi in Chapter 43 offers a preview of selected international case studies related to technological innovations in wine tourism. The chapter further advocates the three main scopes of technology in creating wine tourism experience – i.e., supporting/assisting, enabling, and partly replacing. Chapter 44, authored by Sandra Maria Correia Loureiro, offers an overview of virtual technologies such as virtual reality (VR) and augmented reality (AR) in the wine tourism experience. The chapter also presents the case study on port wine virtual simulation from the Douro valley region. Chapter 45, authored by Ige Pirnar and Aylin Caliskan, reviews social media platforms and tourism social media marketing applications that are available for wine tourism marketing. The chapter also sheds light on industry-specific social media strategies, success criteria, and terminologies used in wine tourism marketing.

E-storytelling adds to interpreting the value of social interaction and the value co-creation of wine products and the destination wine region. Chapter 46, authored by Amnaj Khaokhrueamuang, inspects the contents of e-storytelling about wine as an innovative integrated marketing communication approach. To illustrate the chapter contents, the case study of Wine Prefecture, Yamanashi, is also discussed. Chapter 47, authored by Dan McCole, throws light on the development of hybrid wine grapes and the subsequent emergence of a new category of the wine experience regions. The chapter also presents the case study of North America's Upper Mississippi River Valley AVA, which relies on newly developed, cold-hardy hybrid wine grapes. Jane Eastham, in Chapter 48, explores the impacts of COVID-19 on wine tourism and wine tourism destinations. The chapter also touches upon the management of risk in supply chains due to COVID-19. The chapter further presents the case studies to offer the Argentinian and French experiences in this context. Chapter 49, authored by Joanna Fountain, Rory Hill, and Nicholas Cradock-Henry, summarizes the impacts and repercussions of the COVID-19 pandemic on the global wine industry and presents a situational analysis of the New Zealand experience of, and response to, COVID-19, focusing on wine producers' and wine tourism operators' responses.

Part VII: Terroir sustainability and cultural constructs in wine tourism

Part VII incorporates seven chapters detailing the sustainability and cultural perspectives of wine terroir and tourism. Chapter 50, authored by Elena S. Rotarou and Humberto Rivas Ortega, accentuates the interrelationship between enotourism, local community, governance aspects, and sustainability issues as a key component for the successful development of a wine tourism destination. To illustrate emphatically, the chapter presents the case of Chile as an example of the sustainability challenges involved in wine tourism development. Chapter 51, authored by Willy Legrand, Clare Hindley, Roland Vogels, and Andrea Levis, familiarizes readers with the concept of terroir fluidity due to a cyclical chain reaction based on macro and micro feedback loops. The chapter uses case studies from the wine regions of Tuscany, Italy, and the Rhine valley, Germany, to epitomize its discussion. Camilo Peña-Moreno and Annamma Joy, in Chapter 52, probe the sustainability, nature, and terroir concepts underlying natural wine tourism. As natural wines are low-intervention products made from organic or biodynamically grown grapes and free of synthetic additives, winemakers can and should promote these wines as honest representations of specific wine regions. Chapter 53, authored by Michelle Thompson and Bruce Prideaux, examines how the interactions between the rural landscape, early settler traditions (internal culture), and growing demand for wine-related tourism (external culture) have influenced the emergence of Barossa and the Margaret River as two culturally different Australian wine tourism destinations.

Chapter 54, authored by Rory Hill and Joanna Fountain, emphasizes that wine lists are key sites of encounter for wine tourists, the local hospitality offering, and the wine region within which it takes place. The chapter uses the case study of the tourism destination of Queenstown, a wine region in New Zealand, to inspect it further. Maria del Pilar Leal Londoño, in Chapter 55, explores the concepts of innovation, knowledge, creativity, and learning as crucial elements for nurturing territorial development based on wine tourism and offers theoretical and practical insights for territorial dynamization. Chapter 56, authored by Keith Mandabach, elaborates how wine tourism is integrated into the rural agricultural tourism experience. The chapter further highlights the phenomenon of wine, heritage, cultural, rural, scenic, landscape, and agritourism tourism and their interrelationships in marketing approaches to promote wine tourism.

Saurabh Kumar Dixit concludes the *Handbook* with a chapter entitled 'Conclusion: Uncorking the Emerging Realms for Wine Tourism Research,' in which the outcomes and constructs of the handbook chapters are reviewed and synthesized, and future research directions for wine tourism research is revealed.

Therefore, this *Handbook* is organized to offer an indispensable understanding of segmentation and profiling of wine tourism markets, appreciating wine tourism as an instrument of regional development, and highlighting novel approaches in wine tourism marketing and other technological innovations contributing to the terroir sustainability and the wine tourism industry. It will be an essential resource for wine tourism business operators, customers, and destinations to enable them to create, manage, market, and avail enriching wine tourism experiences. This *Handbook* will also offer theoretical and practical insights to overcome the challenges and grab the prospects in the arena of wine tourism. Therefore, the *Handbook* aims to provide the updated comprehensive volume to give conceptual, theoretical, and applied advancements concerning to global wine tourism business. The *Handbook* will not merely be a collection of chapters or case studies. Each chapter contributes to the in-depth understanding of one or more aspects of the subject, supported by a range of suitable examples from global wine tourism contexts.

The *Handbook* is intended for undergraduates, master's, and research students, as well as practitioners and industry consultants. The interdisciplinary organization and diverse content of the title also make it interesting for even non-tourism researchers such as marketing, consumer behavior, management, geography, psychology, anthropology, international business, sociology, food studies, and cultural studies.

I am indebted to all the chapter contributors for their valuable support and inputs for this *Handbook*. It has been a pleasure and an excellent learning experience working with you all!

As editor of the volume, I expect this *Handbook* will serve as the foundation resource for a deeper inquiry into the several unexplored facets of wine tourism in the future. I hope that the reader will pair their favorite wine while reading these aromatic chapters.

References

Brochado, A., Troilo, M., Rodrigues, H., and Oliveira-Brochado, F. (2020). Dimensions of wine hotel experiences shared online. *International Journal of Wine Business Research*, 32(1): 59–77. https://doi.org/10.1108/IJWBR-12-2018-0072

Bruwer, J., and Gross, M. J. (2017). A multi-layered macro approach to conceptualizing the winescape construct for wine tourism. *Tourism Analysis*, 22(4): 497–509. https://doi.org/10.3727/108354217X15023805452059

Byrd, E. T., Canziani, B., Hsieh, Y.-C., Debbage, K., and Sonmez, S. (2016). Wine tourism: Motivating visitors through core and supplementary services. *Tourism Management*, 52: 19–29. https://doi.org/10.1016/j.tourman.2015.06.009

Carlsen, J. (2004). A review of global wine tourism research. *Journal of Wine Research*, 15(1): 5–13. https://doi.org/10.1080/0957126042000300281

Charters, S. (2006). *Wine and Society: The Social and Cultural Context of a Drink*. London: Routledge.

Demossier, M. (2010). *Wine Drinking Culture in France: A National Myth or a Modern Passion?* French and Francophone Studies. Cardiff: University of Wales Press.

Dixit, S. K. (ed.). (2019). *The Routledge Handbook of Gastronomic Tourism*. Oxon: Routledge.

Dixit, S. K. (ed.). (2020). *The Routledge Handbook of Tourism Experience Management and Marketing*. London: Routledge.

Getz, D. (2000). *Explore Wine Tourism: Management, Development and Destinations*. New York: Cognizant Communication Corporation.

Getz, D., and Brown, G. (2006). Critical success factors for wine tourism regions: A demand analysis. *Tourism Management*, 27(1): 146–158. https://doi.org/10.1016/j.tourman.2004.08.002

Martins, J., Gonçalves, R., Branco, F., Barbosa, L., Melo, M., and Bessa, M. (2017). A multisensory virtual experience model for thematic tourism: A Port wine tourism application proposal. *Journal of Destination Marketing & Management*, 6(2): 103–109. https://doi.org/10.1016/j.jdmm.2017.02.002

Mitchell, R., Hall, C. M., and McIntosh, A. (2000). Wine tourism and consumer behavior. In Hall, C. M., Sharples, L., Cambourne, B., and Macionis, N. (eds.), *Wine Tourism Around the World: Development, Management and Markets* (pp. 115–135), Oxford, UK: Butterworth-Heinemann.

Mitchell, R., and Hall, M. (2006). Wine tourism research: The state of play. *Tourism Review International*, 9(4): 307–332. https://doi.org/10.3727/154427206776330535

Quadri-Felliti, D., and Fiore, A. M. (2012). Experience economy constructs as a framework for understanding wine tourism. *Journal of Vacation Marketing*, 18(1): 3–15. https://doi.org/10.1177/1356766711432222

Quintal, V., Thomas, B., and Phau, I. (2015). Incorporating the winescape into the theory of planned behaviour: Examining 'new world' wineries. *Tourism Management*, 46: 596–609. https://doi.org/10.1016/j.tourman.2014.08.013

Saayman, M., and van der Merwe, A. (2015). Factors determining visitors' memorable wine-testing experiences at wineries. *Anatolia: An International Journal of Tourism and Hospitality Research*, 26(3): 372–383. https://doi.org/10.1080/13032917.2014.968793

Salvado, J., and Kastenholz, E. (2017). Sustainable wine tourism eco-systems via coopetition. *Revista Turismo & Desenvolvimento*, 27–28(1): 1917–1931. https://doi.org/10.34624/rtd.v1i27/28.10335

Sigala, M. (2012). Managing and marketing wine destinations with and through art: A framework for designing wine experiences. In Sigala, M., and Robinson, R. N. S. (eds.), *Wine Tourism Destination Management and Marketing Theory and Cases* (pp. 221–243). Cham, Switzerland: Palgrave Macmillan.

Sigala, M., Evangelos, C., and Gretzel, U. (2012). *Social Media in Travel, Tourism and Hospitality*. England: Farnham.

Terziyska, I., and Damyanova, R. (2020). Winescape through the lens of organized travel – a netnography study. *International Journal of Wine Business Research*, 32(4): 477–492. https://doi.org/10.1108/IJWBR-09-2019-0050

Williams, P. (2001). Positioning wine tourism destinations: An image analysis. *International Journal of Wine Marketing*, 13(3): 42–58. https://doi.org/10.1108/eb008726

PART I

Setting the scene for wine tourism

1

THE GROWTH AND EVOLUTION OF GLOBAL WINE TOURISM

Manoj Kumar Yadav and Saurabh Kumar Dixit

An introduction to wine tourism

Wine tourism can be defined as visitation to vineyards, wineries, wine festivals, and wine shows where grape wine tasting and/or experiencing the attributes of a grape wine region are the prime motivating factors for visitors (Hall and Macionis 1998). Tourism is important for many wineries to sell wine either directly to visitors or offer a marketplace where such visitors can get on a direct mail order list. People often consider wine tourism just as the journey to wineries and vineyards. But a winery visit is just one component that formulates the overall wine tourism experience.

Wine tourism is an emerging phenomenon gaining popularity among domestic and international tourists. New, old, and emerging wine-producing countries are transforming themselves into wine tourism destinations in order to uncork their wine tourism potential. In the current scenario, tourists have more disposable income than ever before, with more leisure time to travel in different parts of the world. Tourism is the only available component to tourists that permits them to escape from the ordinary course to immerse themselves in the booming world of freedom and novelty. Now more and more tourists are embarking on the quest for memorable travel experiences; in this endeavor, the gastronomic experience is a highly sought-after motivation among tourists, enabling them to enjoy destinations and their food and wine experiences. Figure 1.1 reveals major components of wine tourism that play an imperative role in mobilizing wine tourists.

Wine tourism is a close companion of agritourism, sustainable tourism, ecotourism, and culinary tourism. But wine tourism particularly encourages sustainable tourism, as it allows the local wine producers to earn some revenue from them, and it has its own pull factors. Wine tourism's major pull factors include wine tasting sessions, wine knowledge tours, wine setting experiences, pleasant rural settings, food and wine pairing knowledge workshops, and fun and recreation around vineyards. Wine tourists also visit wine tourism destinations to explore wine culture, health aspects of wine, to experience the green tourism around vineyards, to explore the architectural beauty of wine tourism destinations and so forth.

The concept of wine tourism, therefore, revolves around wine tasting, purchase of wine, wine tours to wineries and vineyards, attending festive wine occasions, and other wine-related events. Identical to any other tourism, wine tourism is also gaining popularity and making

DOI: 10.4324/9781003143628-3

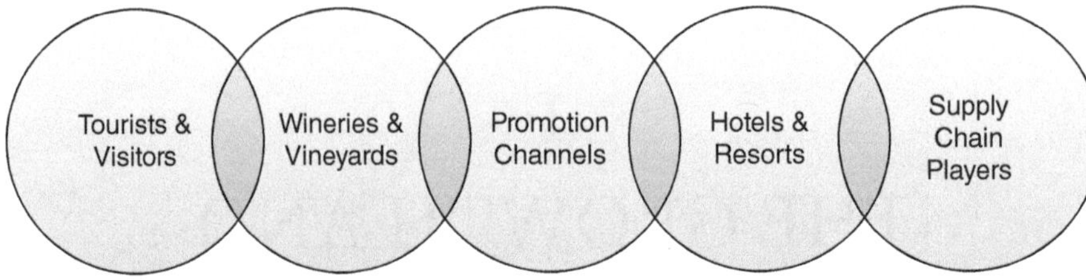

Figure 1.1 Components of wine tourism.

Source: Authors

its value-place in the travel market. These days, wine tourism has achieved popularity around the globe and has opened its wings to soar high. Wine tourism has a significant impact on vine-growing fields (or vineyards), winery locations, methods of harvesting and cultivation, amount of time for wine maturation, the process of corking wine bottles, techniques involved in wine fermentation, and so forth. These elements often become the pull factors for wine tourists, corporate players, wine researchers, consultants, and other professionals.

Historical evolution of wine tourism

The footprints of wine can be seen in many cultures and traditions. The earliest known traces of wine are available in the history from the literature of Georgia (6000 BC), Iran (5000 BC) and Sicily (4000 BC) (Decanter 2011; Johnson 2019). Indian Vedic scriptures too, have the footmarks of wine culture, known as *Somras*. But one common fact detected throughout wine's history is that it has been flashed and featured for its intoxicating effects. During the last several centuries, the major credit for wine's popularity has been given to European countries such as France, Italy, Spain, Portugal, and Germany. Wine cultivation then moved across America, Australia, Argentina, Chile, Switzerland, and New Zealand. The journey of wine cultivation did not end there; it further prospered, and new countries such as Brazil, Russia, India, China, and South Africa, collectively known as BRICS, are spreading their wings. Wine practice and tradition have given birth to wine tourism – an alternate but widespread tourism phenomenon.

The history of wine tourism varies from region to region. The beginning of the modern wine tourism era was attributed to the German wine routes, laid during the 1920s (Cambourne et al. 2000). But the door of real wine tourism growth was opened in the 1960s, when California vineyards invited tourists around the globe to explore wine cultures and practices of their terroir (Karagiannis and Metaxas 2020). Upon seeing the benefits, South Africa and Australia, the countries of the New Wine world, followed the concept and allowed their tourists to experience wine destinations. Meanwhile, the Old Wine–producing countries did not permit their tourists to visit the wine destinations at that time, since they wanted to keep their viticulture and vinification practices under the kimonos.

However, wine tourism started flourishing after the successful winning of Californian vintages over French wines in 1976 at the Judgement of Paris (Edwards 2016). This unexpected victory popularized California wine destinations with world-honored status and blew the trumpet of the success of wine tourism not only across California but throughout the world. It opened the gates for this alternative tourism, wherein travelers and tourists could relish their tastebuds with wine in its natural habitat. Looking after wine tourism's immeasurable potential, several old wine countries of the European Union started transforming their vineyards into wine destinations (Karagiannis and Metaxas 2020). In earlier centuries, John Locke and Thomas Jefferson visited the French vineyards, particularly Bordeaux (Gabler 1995; Unwin 1991), and explored their immemorial experience with the rest of the world. When such visits started attracting business, the venture transformed into *business with pleasure* for wine tourism operators. Further promotion happened when countries like Germany, Austria, Slovenia, Spain, France, Greece, Hungary, Italy, and Portugal started celebrating the second Sunday of November each year as an *Enotourism Day*.

With the growth of viticulture globally, emerging and new wine-producing countries have too embarked on this journey. Chile has rapidly grown its wine tourism in the last two decades and opened several tourist routes throughout the country. The Indian subcontinent has also unclipped its wings to honor wine tourism. As a result, more than ten thousand tourists have started visiting India's famous 'wine carnival festival' every year (Chandra 2015). It is difficult to quantify the number of consumers in any one market; however, in the context of wine tourism, it has been suggested that drinkers comprise about 33% of the total market (Lockshin and Spawton 2001). Earlier, the wine concept was limited to grape production and considered a primary industry for grape growing. Later it became a secondary industry for winemaking, and now it has become a territorial industry for wine tourism operators (Carlsen 2004). In addition, it opened the doors of education discipline for researchers and knowledge aspirants. This multi-identity feature of wine tourism makes it a prime fiddle for wine tourism operators and stakeholders.

Developing wine tourism destinations

Countries involved in the production and trade of wine have divided themselves on many paradigms to corner the market competition and to earn the highest possible revenue. Among all criteria, the world-honored wine classification criteria are *terroir*-based wine taxonomy. In the terroir-based wine taxonomy, wine production is divided into three broad groups: the *new wine world*, the *old wine world*, and the *emerging wine world* (Hall et al. 2009). This fact reveals that only countries involved in wine production and trade can be promoted as the destination spot for wine tourism development. The term *wine world* is concerned with various wine-producing countries of the world, which are located at different altitudes, longitudes, and latitude of the equator. The wine grows mostly between the latitudes of 30–50° north and 30–50° south of the equator. Within these equator bands, the annual mean temperature prevails between 10° and 20°C. Latitude has a profound effect on a region's overall climatic conditions before *microclimate* could impact the vineyards (Julyan 1991).

Since the most suitable latitude for vine growing lies between 30° north and 50° north in the Northern Hemisphere and 30° south and 50° south in the Southern Hemisphere (Julyan 1991), most of the old wine world countries are located in the European continent. These European countries benefit from the cool climatic conditions because of their high altitudes and latitudes, such as Germany, the Loire Valley, and Champagne, followed by some new wine world countries such as Switzerland and Australia (Tasmania). However, most of the new wine world regions are located in warmer climates because of their low altitude and latitude points, such as Argentina, India, and South Africa. Equator band differences differentiate the wine tourism seasons. In the Northern Hemisphere (such as in North America and Europe), the grape growing season starts in March and ends in October, whereas in the Southern Hemisphere (such as in Argentina and New Zealand), the grape growing season starts in the month of September and ends in April. These seasonal periods are the best suitable tenures for promoting wine tourism. The seasonal dimension has a great role and impact on wine tourism since it differentiates the flow of wine visitors during peak and weak seasons.

The core differences between the wines from the Old World and the New World lie in their *winemaking practices* and the impact of climate and land on grapes (or *terroir*). The term 'new and emerging wine world' designates to all those wine-producing countries that are outside the traditional or old wine-growing countries of Europe, the Middle East, and North Africa. The term mainly refers to the United States, Central and South America (primarily Chile, Brazil, Mexico, and Argentina), South Africa, Algeria, India, Australia, England, Switzerland, and New Zealand. Like the old wine world, new/emerging wine-producing countries have their own wine designation system, wine laws, wine regions/subregions, principal grape varieties, system for wine branding soil and climatic conditions, viticulture, and vinification practices. It creates learning opportunities for wine tourists to know how and why wines are differentiated. Wine tourists can experience the *wine development cycle* and *wine production protocol* while visiting these wine worlds.

Old World wines are traditionally more *terroir* and *structure driven*, whereas New World wines are typically more fruity and *varietal driven*. Because of *fruit-driven* characteristics, wine labels often are based on the varietal concept (or *varietal wines*). The term *varietal* denotes specific wine features, such as the particular grape variety from which the wine is made; *varietal labeling* is concerned with the practice of naming the wine based on the grape variety. Most varietal wines are required to contain at least 85% of the grape variety stated on the bottle label. In wine tourism, wine tourists have more chances to learn such technical aspects of wine alongside wine laws and wine labels. Wine tourists could apply these technical aspects during dining and drinking in haute cuisine restaurants.

Several wineries have transformed their vineyards into beautiful *wine tourism destinations*. Some wineries have also dedicated their space for wine learning by founding wine institutes in the existing premises themselves. To make wine destinations more appealing, wineries have started organizing wine events and wine festivals. Emerging winemaking countries, of which India is one, have also started navigating to wine tourism after identifying its everlasting potential. After the analyses and interpretation of wine tourism, wine economists determined that wine production and wine tourism are standing at two opposite ends of the industrial framework because economic conditions applied in wine and tourism industries are falling apart. Many emerging wine-producing countries have been facing the same trouble, and it impedes the growth rate at which wine tourism could flourish in wine destinations. These confronting reasons need to be addressed profoundly to develop this new emerging wine tourism concept on a global map.

Figure 1.2 on the *wine tourism outline model* defines the range of functional activities in a cyclic pattern for the development and growth of any wine tourism destination. *Quadrilateral A* shows the series of activities involved in the development of vineyards as wine tourism destinations; the journey starts from the initial stage of viticulture and ends with vinification. *Quadrilateral B* depicts the flow of activities from the guest's perspective in creating a wine destination to meet and/or exceed the expectation level of prospective identified market segments. *Quadrilateral C* allows one to see wine tourism activities from the perspective of service providers in an attempt to make the destination package a *sweetheart deal*. And *Quadrilateral D* demonstrates the stages in the life cycle of wine tourism development. When all quadrilateral squares are handled with great care and attention, only imminent goals and objectives of wine tourism can be transformed into a profitable deal.

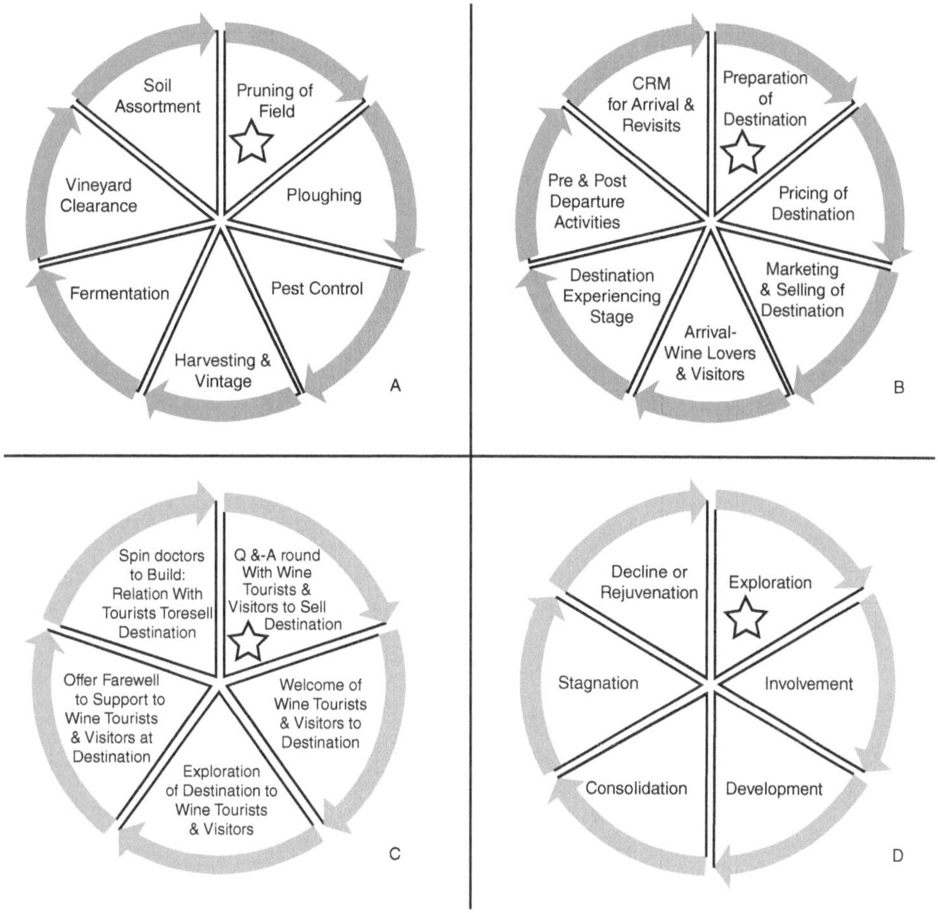

Figure 1.2 Wine tourism outline model (★ shows initialization stage of activities). A: Vineyard activity cycle; B: Wine tourism cycle (from the guest perspective); C: Wine tourism activities (from the service provider's perspective); D: Wine tourism life cycle.

Source: Authors

Figure 1.3 shows wine tourism stakeholders who extend their support directly and indirectly in the successful organization of wine tourism activities. Wine stakeholders such as hospitality professionals and tourism service providers can enrich and ornament their knowledge reservoir on wine and its technical production protocol by learning these *wheels within the wheels technicalities*. These wine differences are vital to know for stakeholders since they aid in every sphere of knowledge transformation to aspirant scholars and in the satisfactory handling of wine tourists. Today, wine tourists are interested in obtaining wine knowledge on tips, since now people are more health-conscious than ever before. Such a deep theoretical wine knowledge framework and profound pragmatic knowledge adds value to the wine tourism concept.

Many wine regions from new, old, and emerging wine worlds have perfected the art and science of viticulture and vinification practices and protocols. Wine tourism destinations are now extensively marketing their range of products and services among global tourists. They predominantly focus on earning monetary gains and international recognition. So, insightful

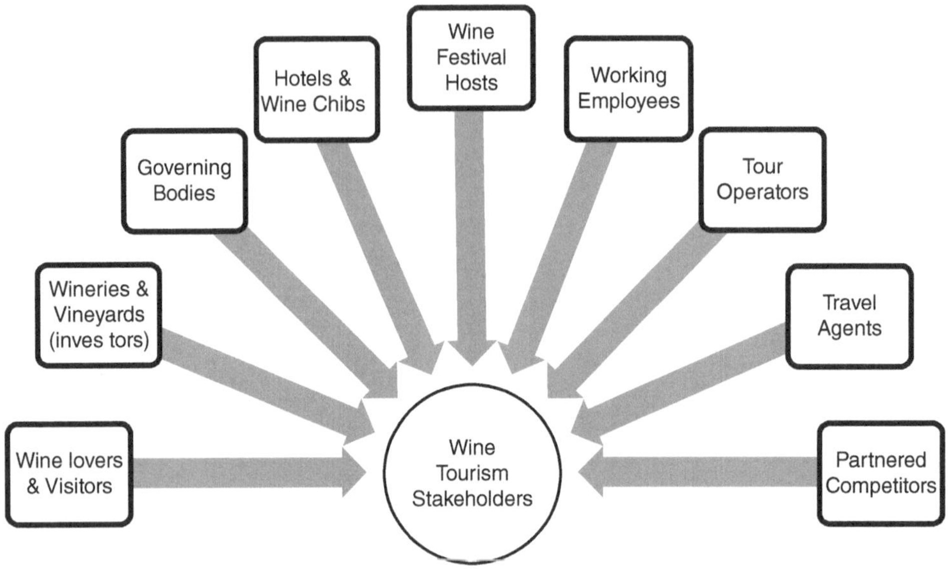

Figure 1.3 Wine tourism stakeholders.

Source: Authors

wine knowledge, supported through wine pamphlets, will incorporate a great advantage to a wine publicity campaign. In addition, almost every wine tourism stakeholder involves itself in some or another way amongst the plethora of publicity campaigns to attract wine tourists from the global marketplace. Efforts of one stakeholder create opportunities for others due to the commonality of market segment(s).

Emerging trends in wine tourism

Due to technological advancement and exponential growth rate, new concepts and possibilities and forms of tourism are emerging every day. Wine tourism is one such emerging concept, and it is also evolving to keep up with changes. Now planners and stakeholders are incorporating new trends to make wine tourism more gainful and advantageous. The following suggested emerging trends in the global wine tourism marketplace are proven as highly gainful and money-making deals for wine destinations.

Experiential learning programs

It is an experiential education-based wine tourism trend in which educators and knowledge aspirants are involved. The aim and purpose of such programs are to attract wine researchers and scholars to explore information about wine tourism and trade, so they can develop new theories, search for challenges and opportunities, find hidden potential, and suggest strategies and marketing campaigns to promote wine tourism. Their published articles play a significant role in transforming an ordinary tourist into a wine tourist. On the contrary, wineries support wine education programs by organizing wine events and festivals to attract wine tourists. Programs may include wine blending seminars, winemakers' booth camp, personal membership programs, and ownership opportunities. Here, wine tourists receive practical exposure to viticulture and vinification practices at vineyards and wineries. Wine educators and scholars apply this firsthand knowledge for a better understanding of the wine discipline. Likewise, wine tourists utilize their experiential learning during meal occasions wherein wine tasting and wine pairing become an integral part of the dining experience (Dixit 2019; Gunasekar et al. 2021). It fosters the interest of wine seekers and offers an opportunity for a deep reflection of wine tourism. Experiential learning programs offer wine lovers a unique experience in a pollution-free atmosphere.

Wine villages and museums

Wine villages (or *chateaux* or *domaines* in French) are remote places where vine growers undertake wine cultivation and harvesting activities. Wine tourists visit these remote places to experience the exotic beauty and systematic work pattern of a vineyard. For example, France's Bordeaux and Burgundy wine province is the most popular wine chateau and domaine. At these destinations, wine museums testify to the richness and diversity of winemaking craft through exposing visitors to tools and objects utilized in activities from vine growing to wine bottling and labeling. Desmond Castle (Kinsale), Port Wine Museum (Portugal), Universal Exposition of Wines and Spirits (France), and Vinseum-Catalan Wine Cultures Museum (Spain) are some globally recognized wine villages and museums wherein wine tourists and travelers can enrich their experience with explored winemaking ethics and protocols.

Wine festivals and events

These days, wine festivals are common events in most of the wine-producing countries around the globe. A wine festival celebrates successful viticulture and is usually organized after the harvest of the grapes, usually at the end of September in old wine-producing countries. Wine lovers can enjoy these festivals in terms of wine drinking, grape pressing, tasting regional specialties, enlightening them with cultural traditions, folklore, different kinds of music, and similar activities. Amongst all old wine-producing countries, France is a primary nation that celebrates wine festivals every year. People from varying countries visit during these festivals to perceive the country's rich opulence and wine heritage. In new wine-producing countries like India, wine lovers can enjoy the same experience at cost-effective rates.

Unique tour options

The concept of unique tour options revolves around the phenomenon of *off the beaten track* – discovering the undiscovered and unexplored destinations to promote regional and cultural heritage specialties. Wine tourists can undertake these unique journeys to explore the wine culture and drinking tradition. Generally, allocentric and mid-centric tourists are the primary market segment of unique tour options. *Allocentric tourists* are those holiday makers who actively seek out new destinations to travel, whereas *mid-centric tourists* prefer to visit lesser-known but discovered destinations. In the middle, *psycho-centric* tourists like these holiday-makers seek out familiar destinations to travel, but they rarely like this most modern wine trend. These unique tours can be organized through many channels and modes, most prominently via jeep, cycle, tractor, horseback, and hot air balloon. In unique tour options, service providers offer rides through biodynamically farmed vineyards, wine walks of wineries, and sometimes virtual tours from the comfort of the room. Vine parks are also trending these days wherein vineyards are transformed into vine parks with the arrangement of benches, tents, and a few sports activities.

Wine cruise liners

Similar to other world-honored cruise liners such as Star Cruises and Royal Caribbean, wine cruise liners these days are part of a newly emerging trend for the popularity of wine tourism. SeaDream Yacht Club, Silversea, Cunard, and Oceania are the most popular wine cruise liners around the globe. Wine tourists can perceive the experience of wine tourism destinations during the voyage over the surface of sea waters. These cruise liners organize wine tasting sessions while on board and then visit ashore to taste the great wines at wine production units. Additionally, these tourists can have the benefit of gourmet lunches and dinners with superb quality wines in nearby restaurants. The glimpses of wine cruise liners can be seen in the Mediterranean Sea, Galapagos Islands, and South Africa. Generally, family clienteles and group travelers are the primary market segment of these wine cruise liners. To make this trend more authentic many cruise liners are now entirely dedicated to travel for *wine cities*. But not all wine-producing countries are fortunate enough to receive this benefit since they are not situated on the coastal belt, like around the ocean, seas, and bay.

Innovative collaboration

Under the innovative collaboration trend, two or more independent service providers come together to cater to wine tourists through more than one venture. For instance, most of the luxury hotels and *haute cuisine* restaurants have a partnership with wineries and vineyards to work together. As a result, they offer discount coupons to visitors to visit wineries and receive free wine tasting sessions or to avail hefty discounts on the purchase of wine. This trend makes the property more competitive for other establishments. Likewise, partner wineries offer their visitors coupons for local restaurants and hotels. Such innovative collaborations create an atmosphere of warmth that appeals to visitors and also allows exploration of all avenues for the betterment and growth of involved partners and stakeholders. Nowadays, many ventures are running on the equation of innovating collaborations, such as Chateau Tumbleweed and the Mine Café of France. After seeking the benefits, many modern-day tourism service providers have adopted the concept of innovative collaboration.

Wine clubs

Wine clubs are perfect examples of the innovating concept of wine selling. Wine clubs provide a wide range of wines with attractive offers to wine tourists. Many wine clubs flash their regular patrons on media pages, while others make their tourists brand ambassadors. The purpose of these efforts is to propagate their venture to others through *word-of-mouth* advertisement. Wine clubs want to strikingly market their outlet and its range of products; therefore, they appoint either professionals or regular clienteles for this purpose. These days it's a very popular merchandising tool for new wine-producing countries to attract more and more tourists and generate revenue. Revenue from wine club visitors is always higher than the average contribution of any other field of visitors. At the same time, the cost of investment is also quite low in wine clubs' foundations. The owner may need to invest in POS/e-commerce solutions only; everything else is already available to customers, like vineyards and wineries. Presently, around 37% of wineries offer wine club options to their patrons. Cellar 503, Orange Glou, the California Wine Club, Dry Wine Club, Winc, and Wine Awake are a few popular wine clubs around the globe that cater to wine tourists.

Wine tour operators (or intermediaries)

Wine tour operators create additional space for tourist visits on their *tour itinerary*. Many such tour operators can be identified in the United Kingdom, such as Grape Escapes and Arblaster & Clark Wine Tours. This is a perfect way to create a new market segment and to generate high business flow for revenue betterment. Wine tourists are an elementary target market segment for these wine tour operators since identified market segment has a sound *average spending capacity* (ASC). The higher ASC improves the outlet's revenue performance through higher average sales per cover. Besides tour operators, service providers can also involve other intermediaries for selling wine tourism destinations. Figure 1.4 better defines the intermediaries for promoting wine tourism as they form the bridge between tourism service providers and tourists. Some notable and remarkable names in the realm of intermediary channels include 101 Wine Tours, 290 Wine Shuttle, A Great Oregon Wine Tour, and A Life Well Drunk.

Wine trails are designed to offer a rich textured wine experience to wine tourists; wineries feature their vineyard/domaine/chateau conditions on publicity materials. All wine-growing countries have different climatic conditions that modify grape features like sugar content, shape, color, catalytic agent, and texture of grapes. For instance, viticulture protocols of old wine-producing countries are more terroir and structure driven, whereas the New World's wines are typically more fruity and varietal driven. Under wine trails, many wineries have conveniently clustered the barrels of a variety of wines for tourists to sample them on their single trip. Wineries organize these trips through vineyard walks, winery walks, or bike riding of wine spots. Nowadays, the trend of bike riding or walking is more in fashion, since it is a health-conscious exercise and a safe path to vineyards or wineries themselves. The wine walk is usually organized in the wee hours of the morning and dawn hours of the evening. In the United Kingdom, the WineTrails company offers wine walks and bike rides in the vineyards and wineries of France, Spain, Italy, Portugal, Bulgaria, Hungary, Cyprus, Madeira, Switzerland, and other countries of the new wine world as well.

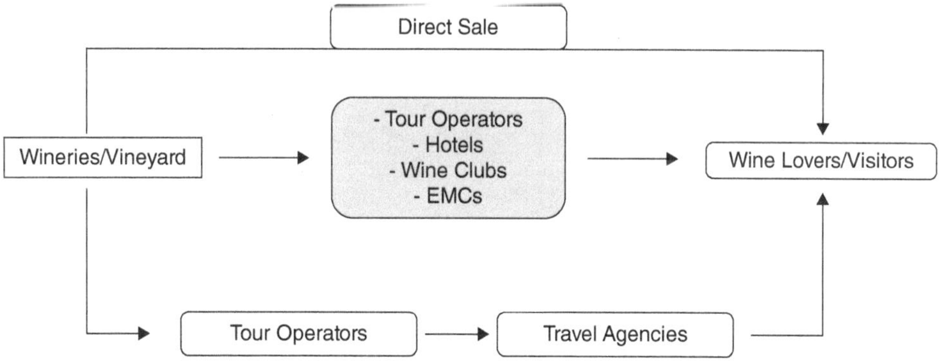

Figure 1.4 Wine tourism sales and distribution channel.

Source: Authors

Direct to consumers (DTC)

Many wineries have also been involved in directly supplying wines to consumers at their home addresses through door-to-door sale services. To make this endeavor successful, it is vital to have a cellar door service provision, if feasible, that also involves the wine producer/owner in this initiative. *A cellar door* is a place or outlet that is used to sell wine by retail outlet and is mostly situated in the vicinity of vineyards and wineries. It is a part of the winery from where wine may be sampled or purchased. Nowadays, these cellar doors have provisions for door-to-door wine sales. Currently, 85% of wineries ship wine directly to consumers. Almost all wineries use the provision of cellar doors to sell wines to visited tourists. This trend is oriented towards greater revenue generation, and it shows a wide range of market segments about the wine tourism destination.

Wine hotels

A new hotel concept that is blooming and blossoming in the international marketplace, wine hotels are usually large wineries and vineyards which have been transformed into hotels and resorts. These properties are primarily targeted towards wine tourists, regardless of the transient or non-transient segment. This newly emerged wine hotel concept offers a scenic view of vineyards with learning facilitation about the 3S's – Swirl, Sniff, and Sip. It has opened the gateway of wine regions for global tourists. The conglomeration of wine with travel and accommodation elevates the expectation level and augments the tour experience of wine tourists and travelers. Such wine resorts and wine hotels also offer many indoor and outdoor recreational activities like Vino Spa and wine baths for value addition to the wine destination. Rosewood Castiglion Del Bosco (Montalcino, Italy), Villa La Coste Provence (France), and Dew Drop Boutique Retreat are a few notable wine resorts and hotels around the world.

Other floating trends in wine tourism

Wineries and vineyards may additionally make it easy for wine tourists to find information about them on blogging internet sites, in books and magazines, and on social media platforms. In addition to such efforts, wineries may also involve themselves in activities like organizing wine tasting sessions with personalized touches for tourists to differentiate their range of wines from competitors. They may, additionally, offer exclusive perks for people who revisit in order to buy wine. When executed and implemented properly, wine tourism activities can become an additional source of income and a brand enhancement tool for most of the wineries. To make the venture more adventurous and gainful, wineries can dedicate a certain amount of space for wine bars, wine restaurants, and wine-themed boutique hotel development.

Middle-order supply chains have a significant role in overall wine tourism development. Their sincere efforts may bridge the gap between the wineries and wine tourists for wine tourism promotion. A certain amount of gain can be shared with grape agrarians in the form of remunerative prices to enthuse them. To make wine tourism a gainful venture and to structure the *moment of truth* of tourists as a remarkable and unforgettable experience, services of knowledgeable wine connoisseurs and free promotional wine campaigns should also be incorporated in the effort list. These efforts may divert the interest of regular tourists of the mass tourism segment into a wine tourist candidature; otherwise, they would continuously become part of *traditional tourism forms*.

Case Study 1.1 Growth and evolution of fruit wine tourism in Meghalaya (India)

Meghalaya, popularly known as the abode of clouds, is one of the most beautiful states of North-Eastern India. It offers tourists a variety of scenic views, rich handicrafts, vibrant indigenous festivals, and culinary delicacies (in the form of local foods, wine, tea, and coffee plantations). Meghalaya is blessed with diverse agro-climatic conditions favoring the cultivation of a range of horticultural crops, such as fruits, vegetables, and flowers. The horticultural produce of Meghalaya, like grapes, gooseberry mandarin orange, peach, plum, pear, pineapple, banana, and jackfruit, are used by the indigenous winemakers and fruit growers to create a better market for their harvest (see Table 1.1).

Some of the popular homemade fruit wine brands of Meghalaya include the following:

1 Asame (manufactured since 2003)
2 Khasi Wines (started in 2010) (Figure 1.5)
3 Mawlot and Sons Wine (Figure 1.6)
4 Na Lyngwiar Wine (established in May 2015)
5 Nes-Ki Wine (started in 2015) (Figure 1.7)
6 Old Red Wine (available from 1995)
7 Daly's Wine

After a long, 18-year struggle, the State Government of Meghalaya legalized the production of homemade fruit wines through enacting the *Manufacture and Sale of Home Made Fruit Wines Rules, 2020*, on September 24, 2020. The Shillong wine festival (organized annually from 2002) and other indigenous fruit wine festivals contribute to the preservation, promotion, and awareness about the art of winemaking and to unleashing its potential for wine tourism. The commercialization of Meghalaya's homemade fruit wines is expected

Table 1.1 Fruit wine varieties made in Meghalaya

S.N.	*Name of the fruit wine*	*Fruit used (Botanical names)*
1	Cashewnut apple wine	Cashew Nut Apple (Anacardium occidentale)
2	Cherry wine	Cherry (Prunus cerasus L.)
3	Orange wine	Orange (Citrus sinensis)
4	Ginger wine	Ginger (Zingiber Officinale)
5	Mulberry wine	Mulberry (Morus)
6	Strawberry wine	Strawberry (Fragaria)
7	Passion fruit wine	Passion fruit (Passiflora)
8	Blackberry wine	Blackberry (Rubus)
9	Banana wine	Banana (Musa)
10	Jackfruit wine	Jackfruit (Artocarpus Heterophyllus)
11	Peach wine	Peach (Prunus Persica)
12	Pear wine	Pear (Pyrus)
13	Pineapple wine	Pineapple (Ananas Comosus)
14	Kiwi wine	Kiwi fruit (Actinidia Deliciosa)

Source: Kumar and Kulshreshtha (2020)

Figure 1.5 Khasi Wine, Meghalaya.

Source: Saurabh Kumar Dixit

Figure 1.6 Mawlot and Sons Wine, Meghalaya.

Source: Saurabh Kumar Dixit

Figure 1.7 Nes-Ki Wine, Meghalaya.

Source: Saurabh Kumar Dixit

to provide a much-desired boost to the state's economy besides escalating the prospects of promoting wine tourism in the tiny mountain state of Meghalaya. However, commercial fruit winemaking and wine tourism in Meghalaya is still in the infancy stage and needs a proper tourism development plan.

Conclusion and implications

Wine tourism is a strikingly flourishing trend around the globe that attracts tourists and travelers from every walk of life. To compete in the market and attract more and more tourists, properties have classified themselves as old/new/emerging wine world countries. Wine tourism stakeholders and the wine tourism component model reveal their imperative role in the growth and development of wine tourism. Many old wine-producing countries of the European continent claim to have a significant role in wine tourism upliftment. In contradiction, new winemaking countries justify that their fruit-driven varietal wine concept put forward the concept of wine tourism. The wine tourism outline model describes a vineyard's activities from the perspective of tourists (in the wine tourism cycle) and the service provider (in the wine tourism activity cycle) through the wine tourism life cycle. Trends like experiential learnings, wine trails, innovating collaborations, unique tour options, and the emergence of wine hotels and resorts, wine museums, and wine clubs add fuel to the growth of global wine tourism development. The Indian subcontinent is also emerging as a stiff competitor in wine tourism amongst the existing wine tourism destinations of New World and Old World countries. As with other international wine-growing destinations, India has started presenting itself as a wine destination. In the Indian subcontinent, the roots of wine culture could be traced in Vedic Scriptures as well, where wine is designated as a divine drink by the name *somras*. The efforts of public and private sector partnerships will transform the undergrowth stage of wine tourism in India into a developed wine tourism destination.

The popularity and growth of wine tourism at the global level would bring this tourism form as an austere competitor before every other form of tourism. Since wine tourism is directly concerned with the eating and drinking habits of people and also plays an imperative role in enhancing the meal experience of diners. Therefore, more and more people are now becoming part and parcel of this newly merged wine tourism concept. By engaging in wine tourism, tourists perceive the beauty of nature, explore new knowledge, experience the richness and opulence of wine tradition, and simultaneously relish their palate with sovereign recipes and international delicacies. Wine tourism makes us aware of different food and drink cultures, gastronomic cuisines, dining etiquettes, and eating habits. In addition, it aids in exploring the music, folklore, attire, art, craft, and heritage of the destination point. To accentuate it, this chapter offered a case study of the indigenous fruit wines of Meghalaya (India). It also helps in maintaining the healthy Forex, generating employment, upgrading the standard of living, mixing with foreign culture, and so forth. If addressed accurately, wine tourism could become the barometer of any economy in tourism exploration and multiply income at breakneck speed.

References

Cambourne, B., Hall, C. M., Johnson, G., Macionis, N., Mitchell, R., and Sharples, L. (2000), The maturing wine tourism product: an international overview. In Hall, C. M., Sharples, L., Cambourne, B., and Macionis, N. (eds.), *Wine Tourism around the World: Development, Management and Markets* (pp. 24–66). Oxford, UK: Butterworth-Heinemann.

Carlsen, J. (2004). A review of global wine tourism research. *Journal of Wine Research*, 15(1): 5–13.

Chandra, A. (2015). Celebrating wines. *Business Standard*, March 7. https://www.business-standard.com/article/beyond-business/celebrating-wines-115030600617_1.html

Decanter. (2011). *Decanter* magazine, 36(5–8). February.

Dixit, S. K. (ed.). (2019). *The Routledge Handbook of Gastronomic Tourism*. Oxon: Routledge.

Edwards, O. (2016).That revolutionary May day in 1976 when California wines bested France's finest. *Smithsonian Magazine*, May 3. https://www.smithsonianmag.com/smithsonian-institution/that-revolutionary-May-day-1976-when-california-wines-bested-france-finest-180958971/

Gabler, J. M. (1995). *How to Be a Wine Expert*. Emeryville, CA: Bacchus Press.

Gunasekar, S., Das, P., Dixit, S. K., Mandal, S., and Mehta, S. R. (2021). Wine-experienscape and tourist satisfaction: through the lens of online reviews. *Journal of Foodservice Business Research*. https://doi.org/10.1080/15378020.2021.2006039

Hall, C. M., and Macionis, N. (1998). Wine tourism in Australia and New Zealand. In Butler, R. W., Hall, C. M., and Jenkins, J. M. (eds.), *Tourism and Recreation in Rural Areas* (pp. 267–298). Sydney: John Wiley & Sons.

Hall, C. M., Sharples, L., Cambourne, B., and Macionis, N. (2009). *Wine Tourism Around the World* (3rd ed.). New York: Routledge.

Johnson, H. (2019). *Hugh Johnson's Pocket Wine Book*. London: Octopus Publishing Group.

Julyan, K. B. (1991). *Wines of the World, Beverage Sales and Services* (2nd ed., pp. 39–89). Oxford, UK: Butterworth Heinemann.

Karagiannis, D., and Metaxas, T. (2020). *Sustainable Wine Tourism Development: Case Studies from the Greek Region of Peloponnese*. Greece: Department of Economics, University of Thessaly.

Kumar, A., and Kulshreshtha, S. K. (2020). Fruit wines of Meghalaya: A sustainable approach for rural livelihoods. *Journal of Applied Management and Investments*, 9(4): 169–182.

Lockshin, L., and Spawton, T. (2001). Using involvement and brand equity to develop a wine tourism strategy. *International Journal of Wine Marketing*, 13(1): 72–81. https://doi.org/10.1108/eb043371

Unwin, T. (1991). *Wine and Vine: An Historical Geography of Viticulture and the Wine Trade*. London: Routledge.

2

NATURE AND SIGNIFICANCE OF WINE TOURISM

An Australian perspective

Genevieve d'Ament, Anthony J. Saliba, and Tahmid Nayeem

During discussions with the owner of an established winery in a premium wine region of Australia on the nature and significance of wine tourism, she stated, 'It's everything, to be honest; I think if we didn't have wine tourism, we wouldn't be here.' Plentiful research validates this statement of the significance of the wine tourism (WT) sector for continued growth and the profitability of the wine industry (Carlsen and Boksberger 2015). In 2019, the wine producer organization Australian Grape and Wine defined WT as 'visitation to wineries and wine regions to experience the unique qualities of contemporary Australian lifestyle associated with the enjoyment of wine at its source – including wine and food, landscape and cultural activities' (Battaglene 2019). WT is predominantly self-drive couples or groups of friends in Australia, with little planning required beyond selecting a region to explore. Large organized tours of international tourists can be actively discouraged; some wineries have gates too narrow for buses to enter. Before COVID-19, there was rarely a need to book tastings for groups of fewer than eight, and cellar door staff would provide an enthusiastic welcome, always ready to share the tastes and stories of their winery and region.

Throughout this chapter, the term *winery* refers mainly to the wine business rather than the winemaking, which can be missing in some wine tourist destinations (i.e., cellar door amongst vineyards). The chapter will explore the nature and significance of WT with a brief review of research perspectives, followed by the emergent themes of two comparative case studies. The first evaluates WT in non-wine and wine tourism regions from a senior management perspective. The second study compares the experiences of emerging and established wineries from a business owner's perspective. We conclude by discussing emergent themes from both studies.

WT is a multisensory, multilevel contribution to the Experience Economy. WT experiences are a dyadic relationship incorporating at least two actors (i.e., grower, winemaker, cellar door staff, and visitors). This co-created relationship showcases wine in an array of bucolic, architectural, and interior settings, governed by individual winery business models, all of which are unique and often created by non-business personnel (i.e., winemakers). Being co-created, each actor brings their interpretations and expectations from previous experience.

WT incorporates the three production levels of a wine business, including primary, secondary, and tertiary. The *primary level* is the viticultural aspect of wine tourism and crosses over with agritourism. WT offerings in this level include tours of vineyards and information sessions talking about soil, aspect, and climate with viticulturists, who may also be the winemaker.

DOI: 10.4324/9781003143628-4

Winemakers are found in the second *production level* during vintage, pressing grapes, fermenting the juices, and tending them as they macerate on skins in the winery. Secondary-level WT provides winery and barrel room tours with the winemaker, providing opportunities to share knowledge of their winemaking processes. Most WT begins at the *tertiary level*, in the cellar door, incorporating wholesale and retail sales and marketing. Winemakers can be found here, too, sharing their wine and telling the whole story of the vintage, from pruning to bottling.

The significance of the cellar door to WT is well supported in the literature. Bruwer (2002) recognized the emergence of WT as a significant contribution to special-interest tourism. More so, the role of the cellar door in the development and success of WT. The cellar door experience provides a sense of place or belonging, where a relationship builds brand awareness and customer loyalty. Ball and Stolle (2019) further support the cellar door becoming the iconic element in Australian WT due to its ability to convey a winery's unique contributions and history, and it is the space where business and personal connect, co-creating experiences. They found customers treasured the experience and connections they co-created with cellar door staff.

Cellar door staff are often tertiary-educated but in various fields not immediately associated with the wine industry. Due to fluctuating visitor numbers, cellar doors offer predominantly casual positions, attracting university students or graduates working to pay off student loans before finding permanent jobs in their chosen careers. Also, there is no consistent title and little understanding of a job description for these staff members. Interestingly, a case study analysis of winery staff in recognized wine regions of Australia found a higher number of participants than expected intended to remain in the wine industry (Williams 2013). However, it was not possible to determine if this decision was made before commencing employment in the industry or the result of a subsequently developed passion that staff share during co-created WT experiences.

WT research is multidisciplinary, with marketing, geography, economics, consumer psychology, and tourism disciplines. WT experiences are viewed as hedonic, pleasurable activities, where an educational component increases the enjoyment, brand loyalty, and connection with the tourist (Crespi-Vallbona and Mascarilla-Miró 2020). Research has shown positive cellar door and overall winery experiences improve the perception of wine quality and value, strengthening the tourist's brand attitude (Bruwer et al. 2013). WT offerings such as the co-created cellar door experience is the temporal space in the lives of the actors involved, where wine knowledge is shared, where the opportunity to offer the tourist a place in the winery's story, history, and heritage is presented, and a connection to the winery is created (Frost et al. 2020). WT provides opportunities for viticulturalists, winemakers, oenologists, and cellar door staff to make this connection and share their passion for the wine industry. However, engaging and building a relationship with the wine tourist or co-creating a wine tourism experience is largely unexplored in the literature. Research defining the wine tourist by more than demographics, providing an understanding of their interaction with staff, and how expectations and experiences impact future involvement with wine and wine tourism is scarce.

Over the past few decades, perceptions of wine tourists have evolved from being feared as mobile drunks interested only in free wine to being acknowledged as educated and wine-involved consumers, to be embraced as enduring customers by wineries around the globe (Carlsen 2004; Pratt 2014; Gunasekar et al. 2021). Research has shown wine tourists need to 'connect' with wine in the location or region where it is produced (Bruwer, Prayag, and Disegna 2018). However, wine tourists are often researched from a marketing perspective, resulting in a high, medium, or low wine involvement classification. Highly involved wine tourists might belong to several wine clubs, interested in the calamities of the current vintage, or eager to share their vast knowledge, whereas low involvement is associated with tourists who do not consider themselves connoisseurs, consuming wine only occasionally. However, tourists in the low-involvement segment might be amenable to 'connecting' with a winery, developing the passion associated with high involvement, if they found wine experiences more approachable. Whilst practical, segmentation gives little information about the attitudes and behaviors of wine tourists, an understanding of which would allow staff to develop WT experiences that engage with medium- to low-involvement tourists. Galloway et al. (2008) studied the personality trait 'sensation seeking' in wine tourists finding the trait a good predictor of higher consumption, spend, and wine tourism. However, whilst the number of studies is increasing, a consumer neuropsychology focus on the wine tourist remains limited in the research.

WT enriches and accelerates the development of a region, providing economic, cultural, and psychosocial benefits to the surrounding areas (Sigala 2020). Crick, Crick, and Chaudhry (2020a) conducted an in-depth study of wineries in a region of New Zealand. They found it essential to frame comments about competitors and services in the region positively. Each winery contributes to the region's status; diminishing a competitor diminishes the region in the mind of the wine tourist. A further study found increased 'coopertition' (cooperative competition) between wineries, in the form of sharing equipment, knowledge, and staff, similarly benefited the region as a whole (Crick, Crick, and Tebbett 2020b). WT provides opportunities for regional economic and cultural growth, with different dynamics between local and out-of-state visitors (Bruwer et al. 2017).

New World wineries have embraced the wine tourist, with a majority offering accessible WT experiences, recognizing WT as a valuable direct-to-consumer income stream. Old World wineries seem to adopt a more formal approach to WT, with predicated wine routes for wine vacations, and Italy has two national associations dedicated to WT (Brandano, Osti, and Pulina 2019). New World wine regions developing alongside wine emerging as a vital tourism commodity could explain this difference. Therefore, each winery naturally develops individual WT offerings as part of its business model, with no need to rely on tourists on a wine vacation. Alternatively, Old World wineries have well-developed sales channels with little need to establish WT experiences as a sales avenue when all the wine is sold through established trade partnerships (Mattiacci et al. 2019). Perhaps New World wineries embrace WT as a significant revenue stream.

In contrast, Old World wineries view wine tourism as an avenue to promote the centuries of wine culture in their region. Crespi-Vallbona and Mascarilla-Miró (2020) studied a walking tour of historic wine vats to determine the benefits of cultural additions to the hedonistic attributes of WT offerings in the region. Their findings support the importance of including the local cultural context in WT experiences, promoting regional tourism.

Gaps exist in the literature, and Bruwer, Prayag, and Disegna (2018) maintain 'there is no single theory that defines tourist experience' (p. 356). The literature has been unable to define even the temporal space of the WT experience. Bruwer et al. (2013) suggest experiences begin upon arrival at a winery; however, current research considers the pre-visit decision, often based upon an internet search, extending experiences to post-visit engagement through online review and purchase platforms (Brandano, Osti, and Pulina 2019). This brief review has provided an idea of the nature and significance of WT. However, future research could travel deeper into WT experiences and provide valuable knowledge to aid in developing hedonistic, educational, profitable, and thoroughly enjoyable WT experiences.

The following case studies are unapologetically exploratory, striving to understand the nature and significance of WT from the perspective of owners and employees with many years of experience in and passion for the wine industry. One interviewee learnt winemaking from their mother; another's passion was kindled during a childhood spent watching their wine merchant father taste wines; another family has been integral in developing the infrastructure to support WT in their region. Each interview explored unique experiences which gave a depth of understanding and support to the themes that emerged.

Case Study 2.1 Evaluation of WT from the perspective of senior management in a non-wine tourism region and a recognized wine tourism region

The regions selected for this case study are located in two Australian states. The non-grape-growing and non-wine-region wineries (NWRW) compete with better-known beach and surf attractions, attracting families for summer holidays or weekend getaways by the beach. The region is not a premium wine-producing area, and it is a two-hour drive from metropolitan centers. The wine region wineries (WRW), a mere 30-minute drive from metropolitan centers, thrive in an internationally recognized premium wine region. All wineries offered food, but only two had fine dining options. A total of seven interviews were conducted with staff holding senior management roles in the cellar door, business management, and human resources. A thematic analysis exposed the following perceptions of the nature and significance of WT.

Not every visitor to NWRW is considered a wine tourist, with a clear distinction that visitors who stop to look at the view, or use the facilities, and maybe try a wine are not wine tourists and are not looking for a WT experience. Although with more generous support and market exposure from tourism bodies rather than WT, their offerings are developing away from WT, becoming more general tourism with a wine option. As a region known for beaches and surfing, NWRW fight a legitimacy issue resulting in surprise and delight at the high quality of the wines produced.

WRW consider every visitor a wine tourist, with expectations of high-quality wine. WRW understand the value and need for WT and the benefit of keeping experiences wine-focused. WT offerings need to reflect the recognition of the region as a producer of premium wine.

Visitors to NWRW display a general assumption that all cellar doors sit amongst the vineyards producing the fruit made into the wine on offer. Most NWRW grow a small number of varietals on their properties but offer up to 14 different varietals from their winemakers, making wines from fruit grown in other regions. Tourist assumptions and questions about where all the vines are suggested a desire to taste the wine made from fruit grown in the region, allowing them to distinguish unique characteristics. It also makes it challenging for NWRW to attract and engage with wine tourists. It is up to the winery, rather than the region, to attract the wine tourist. The price point becomes critical here, as NWRW are located a few hours' drive from major metropolitan centers. The experiences offered have to be worth the distance traveled, and a $5 wine tasting doesn't meet that expectation. Food offerings provide another option to attract tourists and complement wine tourism. Food culture in the region has risen to prominence, attracting celebrity chefs. Immersive offerings, such as accommodation on the property, lead to enduring customers and revisitation. However, the quality of alternative attractions needs to complement, not compete with or diminish, the wines. NWRW business models have evolved to make wine accessible, appealing to a broader demographic. 'because we offer a diverse range of products ... there's a lot of ticking of boxes for what different people want.'

Figure 2.1 Biosecurity measures to keep vines healthy (a good wine starts with good fruit).

Source: Genevieve d'Ament

WRW cellar doors are sited amongst substantial vineyards, with a majority of wineries producing estate-grown wines. However, while the vines make for a pleasant winescape, the vines themselves cannot be used in WT experiences due to biosecurity (see Figure 2.1). Visitors are actively discouraged from being in proximity of the vines. While WT sits comfortably within agritourism, developing experiences from a tourism perspective is challenging without a vineyard tour. With no winery to showcase production, experience focus centers on the cellar door and maybe a barrel room. WRW benefit from tourists planning to visit the region; however, high expectations put pressure on management to deliver exceptional experiences to match the wine (see Figure 2.2); 'people want to feel as though the experiences we offer are more bespoke and tailored.'

The cellar door provides a direct-to-consumer sales avenue for NWRW; being unrecognized as a wine region, consumers are less likely to take a risk buying unknown wine from an unknown region in a bottle shop. Both WRW and NWRW find brand loyalty develops during the cellar door experience more than at any other point of the WT experience. The winescape anchors the experience in the customer's wine tourism schema, but the human connection is made during the cellar door experience. Visitors want to hear the history of the winery and the region while tasting the wines of the region, becoming part of the winery's history and feeling invited into the family. Wine club memberships begin in the cellar door, and these visitors, who became club members during their cellar door experience, ordered extra wine through COVID-19. They did so not just because they were drinking more, but to support the wineries and ensure their survival for wine tourism to continue after COVID-19 lockdowns.

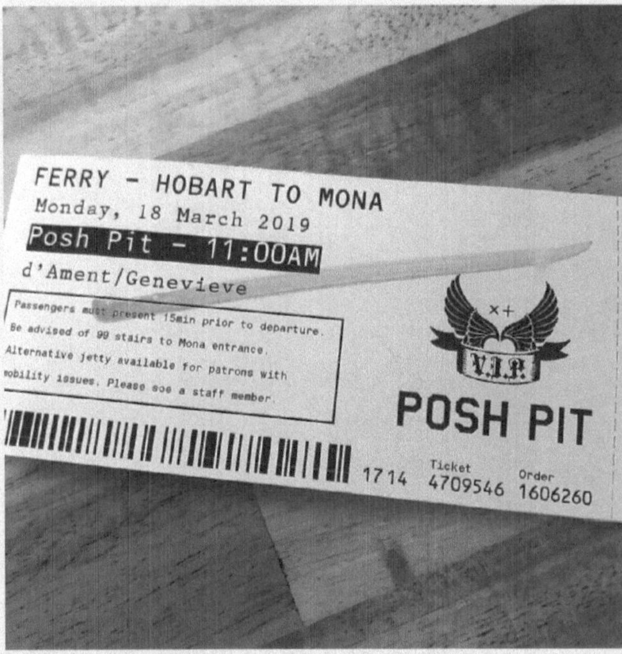

Figure 2.2 Ticket to ride. The ferry to Hobart's Museum of Old and New Art showcases the wines produced by the winery in the Museum's grounds, so while you can't walk through their vines, you can enjoy bespoke wine tourism activities.

Source: Genevieve d'Ament

COVID-19 has limited the options for all tourists, and NWRW find they now operate at capacity all year. Fluctuations of summer and weekend peaks have disappeared. However, these seasonal fluctuations were already flattening, possibly because accessible and inclusive WT experiences broaden their appeal. They were able to adapt to new consumer demands, quickly flourishing under COVID-19 restrictions.

WRW still find winter a low season. However, COVID-19 has impacted management's focus on WT for WRW. With borders closed, domestic and international tourists were conspicuously absent, and the need to understand, develop, and nurture WT was brought into sharp relief.

NWRW have invested significant time and money in staff education and training, as gaining good wine knowledge prior to employment in the industry is rare. A Viticulture 101 diploma was suggested to reduce training costs, introducing an industry to many school leavers, especially in NWRW areas, who may not have the ability to move to metropolitan centers.

WRW felt an opportunity lost where 'this industry has this magnetism where you find people who end up making this a career.' Appropriate and accessible courses would promote the tertiary level of the business as a first choice.

The NWRW find support geared to more prominent regions, with information available but not always accessible. There's a feeling of being left to connect the dots or being too small to be considered 'the national and state wine associations we're not really on their radar.' Seminars tend toward being export-ready or have an inbound tourism focus; 'the focus is on export because … the biggest opportunity for growth.'

Bookable experiences had to be developed for NWRW to be included in tourism marketing maps. Before COVID-19, these were rare outside of the largest wineries, with consumers still unhappy to pay if there was a charge for wine tastings in small wineries. Their regional association is governed somewhat by its membership, but with state and federal funding and support being rare, they're approaching WT with bold new ideas. Moving away from organizing one annual festival event allows them to dedicate their inadequate resources to marketing accessible experiences available all year.

The WRW found state-based tourism agencies better understood WT. However, there was a clear need for a dedicated WT association, given the complex nature of winery experiences and WT.

Wine tourists are hungry for knowledge. The older demographics in NWRW want to know where the grapes originate, as do the younger demographics in WRW. Younger visitors to WRW are also interested in sustainability, the impact of climate change, and the processes from vineyard to glass. However, a delicate balance is required when engaging with wine tourists. As an information-dense industry, it is easy to bombard visitors with technical information or show insufficient appreciation of the tourist's immense knowledge.

NWRW have a broad range of visitors, but not all are wine tourists. The bigger spenders and enduring consumers come from Boomers and Gen X. Boomers tend to flash cash and treat friends to lunch. Gen X is more about a treat or reward. Gen X is amenable to spending on quality and develop a loyal following, becoming valuable ambassadors who utilize social media. Word-of-mouth recommendation is the best form of marketing from any demographic. However, a more diverse range of visitors is drawn in by social media posts from younger groups.

WRW find similarly that younger consumers are willing to spend and develop a loyal following. Authenticity is essential to all wine tourists; however, the younger demographic requires sustainability to be demonstrated, more aware of life cycle analysis, appreciating and rewarding cradle-to-cradle production.

> [Y]ounger people are developing a palette and understanding a little bit more about their wines … interested in drinking wine, also how the product was made, is it organic, are you recycling your bottles … I get asked those questions many times.

Case Study 2.2 Comparing wine tourism experiences in emerging and established wine businesses from the perspective of the business owners

The WT region selected for this case study is internationally recognized as a premium wine-producing region located 30 minutes from metropolitan centers. The established winery (ESW), 20 years or more, is a family-owned-and-run vineyard and cellar door, emphasizing wine tasting experiences. Their winemaking is conducted offsite. The emerging winery (EMW), established for less than five years, is a husband-and-wife winemaking team with vineyards, winery, and cellar door onsite and under their management. Their emphasis is on immersive sensory wine experiences. A total of three interviews were conducted with business owners. The thematic analysis exposed the following perceptions of the nature and significance of WT.

The ESW is adamant that WT centers on visitation to the winery. Once on the property, customer service and providing a quality experience is the focus. They let the wines speak for themselves; every winery in their region is award-winning; it's a premium region, so that doesn't set them apart. Their business model is proven and is about the profitability of tourism possibilities fitting that model. There's a sense that experimental days are behind them; however, they are willing to make bold moves forward with a certainty coming from many years working with wine tourists and embracing WT.

The EMW is aware of their need to understand and engage their market. Their understanding of the wine tourist is more profound due to a need to understand their market and establish their winery. Even though they too have decided on a business model, there's a willingness to explore different wine-related experiences to enhance experiences beyond wine tourist desires. This different approach may be due to being winemakers, consciously aware of the multisensory aspect of wine, wine tasting, and wine experiences.

There was no doubt about the significance of WT to the success of their business and the region. ESW and EMW agree; WT needs to be an everyday focus, not just one annual festival event or harvest lunch.

The wineries are very aware of the value of authenticity and of being family businesses. Wine tourists enjoy speaking with people who understand wine and the winemaking process. Not all cellar doors have working wineries attached, and not all wineries have cellar doors; some vineyards have cellar doors and no winery. An enviable promotional drawcard is being able to offer the complete vineyard, winery, and cellar door wine experience. Being all three doesn't guarantee better-quality wine, but it does speak to the authenticity of the experience.

Tourists are eager to understand the wine region and the effect soil and climate have on the wine. The opportunity to discern the unique qualities, or the taste of the region, is provided with the winemaker on hand to provide a trusted confirmation of their discovery; 'it's so refreshing to be in this area because you're talking to someone who understands the wine.'

Both the ESW and the EMW agreed that without a cellar door, there is no business. Wine tourists come to taste wines and learn about the wine and the region, which happens during the cellar door experience. They become part of the winery story, and the winery becomes part of theirs. This involvement creates wine club members and enduring customers. You don't get that exporting wine or with inbound tourism. Some business is derived from supplying wines to restaurants and bottle shop retail outlets. But without the cellar door to showcase the full range of wines, these customers may only be aware of whichever bottle is on the wine list or shop shelf.

The ESW commented on their domestic WT market:

> [T]hey experience what we have on offer, and then they fall in love, they come in, sit down, do a tasting, you're with them for up to an hour with their platter and

> chatting, and oh I love this, and then they join, and that's how we get our memberships, so if we don't get our WT, we don't get our wine club signups, and we don't get the visitation and the wine sales.

Explaining why an international tourism focus from Wine Australia is flawed and doesn't generate enduring customers nor contribute to profitability; 'they don't join the wine club, they don't buy wine, they can't take wine home.'

Due to the pandemic being well managed in Australia, COVID has been a boom time for Australian wineries, with a cellar door. WT has allowed wine businesses to grow their direct-to-consumer base through mailing lists and wine club memberships. These contacts were invaluable to wineries through lockdown, enabling record sales for this period. Domestic tourists were forced to look to their backyard for holiday entertainment with state and international borders closed. Visitation rates increased 20%, and locals discovered quality wine on their doorstep. Wine tourists discovered more regions in their state and the different wine regions in Australia. COVID has also provided the opportunity to educate Australians, for example, on cool-climate varietals (i.e., not all Shiraz is grown in the Barossa).

Every experience must now be booked at the ESW, and cellar doors that have introduced the bookable experience have no intention to return to the very casual Australian WT style where bookable experiences were rare. Bookable experiences streamline staffing and have reduced the number of nonpurchasing visitors from the non-wine-involved segment. Bookable experiences necessitate having an online presence, and all wineries understand the importance of their website, Instagram, and Facebook. However, nothing beats the authenticity of a word-of-mouth recommendation, usually generated through a cellar door experience, for the EMW:

> It's everything, it's all-important ... word of mouth is very important because if a customer refers a friend, that's so much more powerful than when we post on Instagram saying come to our winery when the consumer tells them there's a lot more legitimacy ... and it's not just one person, it's their 300 followers.

The ESW and the EMW invested heavily in education and training for cellar door staff. Training costs are high, and it's an ongoing expense with high turnover due to an attitude toward jobs being stopgap or paying for studies before embarking on a real career. It is not considered a career in Australia, and there are no apprenticeships, traineeships, or courses at the consumer/WT level. Bachelor's degrees are available for wine science and winemaking, and sommelier courses can extend personal wine knowledge. However, sommeliers are not suitable for the customer service component of WT, being more suited to the restaurant trade. Without structured courses, WT is not seen as a career option, although visitors comment on how lucky we are to work in a winery; 'they think we skip through the vines and drink wine all day.'

The EMW covers the work in all three levels of the business with only a couple of family members. The high workload leaves little time for further business or tourism education, not to mention the expense of privately run courses. Any time outside of running the winery is devoted to promoting WT in the region to strengthen and build a customer base. Any money left over is reinvested in the business, either in the cellar door, winery equipment, vines, or staffing.

The ESW found their staff learnt a lot from their customer base, perhaps due to their business model being more attuned to an older, more wine-involved demographic. They also invested in external staff training, which could engender an openness to learning opportunities in their staff. The ESW benefits from their commercial gloss but emphasizes the importance of retaining the authenticity attached to family-owned-and-run wineries. The EMW relishes the opportunity to share wine knowledge and showcase their region, an advantage of being an EMW where the winemaker runs the cellar door.

The ESW and EMW recognize the importance of federal-, state-, and regional-level agencies and associations. The ESW seems to have a less frustrating relationship with the various agencies, resulting from learning how to engage over the years. Unfortunately, many of the grants and incentives are export-focused or exclude wineries, with an annual turnover of less than $1 million. A portion of this could be explained through contributions levied on total crush; therefore, larger wineries pay higher contributions to the associations. However, EMW felt a benefit of the federal- and state-level agencies providing targeted assistance based on annual crush rather than wine region.

> It would be good if there was a body that looked after instead of a region an industry size grouping. Because we're just a little one on the side that's saying what about us and they say oh we haven't got time for you … would be a huge benefit because these little struggling businesses … there isn't a lot of money in a winery, and yet they provide such a great service to the community and inspire a lot of people.

Neither the ESW nor the EMW found the wine and tourism industry worked well together and felt neither had an understanding of the other sector. The wine industry research focuses on vines and production, whereas tourism was all about a unique bookable experience. Neither was able to provide ready access to demographic or psychographic information on the wine tourist to aid the development of marketing and promotional activities.

The ESW found their most profitable tourist to be the older demographic. The EMW found the younger demographic more engaged and involved with learning about the winery and the region. The younger consumer is not interested in commercial gloss, preferring to support wineries at which they discuss the wines with the winemaker and then follow that winemaker on Instagram. A younger demographic may be intimidated by the cathedral-like cellar doors of established wineries. In contrast, the older wine tourist may find the commercial gloss reassuring, as 20 to 30 years ago in Australia you would drink only commercially produced wines. The ESW was happy to rely on information provided by industry agencies and trust they understand the wine tourist, even though their research is predominantly first- and second-level focused. The EMW, like their clientele, question everything 'for a lot of people and wine tourists, it's a discovery tour they're trying to find something new and exciting and different.'

Discussion

The need for a dedicated WT association and an expansion of education options, including TAFE-level training courses and consumer information, were the most passionately discussed elements in every interview. WT has become a significant revenue stream, especially for smaller wineries in Australia, and a dedicated association would better benefit the growth and profitability of the industry. There was also agreement that basing all information relevant to wineries on total crush rather than region would similarly benefit the growth and profitability of the industry. For example, seminars, online courses, and assistance could be grouped into total crush categories of up to 250 tonnes, up to 500 tonnes, over 1,000 tonnes, and over 10,000 tonnes. Education and innovation have been key drivers to the success of the Australian wine industry. However, this focus has been and remains on the first two levels of the business. Investment in the tertiary level would provide opportunities to lift current WT experiences to the excellent quality of the wines showcased.

Wineries are multilevel businesses operating in an information-dense industry, and whilst wine and food tourism are often paired, they are two very different products. The food and hospitality industries are well catered for concerning education and training. Apprenticeships are offered for chefs and restaurant management. Similar approaches could be considered for the tertiary level of the wine industry, the introduction of TAFE courses, and dedicated electives in relevant bachelor degrees (i.e., WT).

Consumer education is a crucial element highlighted in experience tourism research. Wine tourists want to taste the characteristics of the region, but where this isn't possible, sharing the where, why, and how of winemaking processes will also satisfy the consumer. However, the balance of information communicated to the consumer relies on authenticity and pitching the information to the consumer's level of understanding. Tailoring the delivery of information during a WT experience requires education and training. Educational opportunities are lacking, possibly due to insufficient recognition of tertiary level staff as essential to business success, with no recognized job description or title beyond staff. Training and education are high-cost investments, and where monetary support in the form of apprenticeships or government-funded courses would be most welcomed. Such an approach to education opens up new career paths not currently offered to school leavers in Australia.

Defining the wine tourist was not a focus of these case studies; however, as found in the literature, they are a diverse group with common needs. Authenticity was mentioned whenever WT experiences were discussed. Wine tourists enjoy the educational aspects of their experiences. However, their interests travel beyond maceration and new French barriques, wine tourists want to know the provenance of the winery, know the family story, meet the maker, taste the soil. Wine businesses are multilevel, and wine is multisensory; the literature has found our tourists to be educated consumers. Therefore, while further research is required for a holistic understanding of the co-created experience, involved and involving experiences acknowledging the wine tourists' eagerness to learn are necessary to engage with wine tourists and to gain trust and loyalty.

Necessity is the forerunner of invention, perfectly illustrated by the success of NWRW wine tourism offerings. NWRW have fought for recognition from wine tourists and the various industry agencies. Being too small for recognition by current wine industry agencies, without marketable products for tourism agencies, forced NWRW into developing WT offerings with no certainty of their being successful. This approach to WT can embolden WRW and EMW to branch out with WT offerings and be flexible with offerings, finding their unique place to stand out in the market.

These studies were conducted at the end of 2020, and we must acknowledge the impact of COVID-19. The coronavirus pandemic dominated lives, caused upheaval, and rewrote our expectations. However, Australia was spared the worst, and with international borders closed, the resulting enforced domestic tourism allowed WT to flourish, securing its worthy place in the Experience Economy. Wineries could rely on their loyal customer base, initially created through WT, to maintain a sales avenue during lockdowns. As restrictions eased, new tourists ventured into surrounding regions, enjoying the experiences on offer. During this time, delivering enjoyable and educational experiences enabled wineries to boost their wine club memberships and recover from the initial downturn due to the economic uncertainty accompanying the first wave of global lockdowns.

Festivals and large-scale events have become a staple of wine tourism. However, comments from all interviewees indicate a move away from big annual festivals to more intimate wine experiences offered at all times. In small regions, organizing wine festivals can monopolize funds and time, reducing the focus on growing the region and communicating this progress to the everyday wine tourist. Regional associations are rarely well funded, often requiring extra funding for large-scale events, and escalating political tensions. The notion of coopertition is embraced and thriving in WRW. However, the NWRW are yet to fully realize the benefits of regional cooperation beyond one big annual WT event, possibly due to constant competition with other tourism and large retailers for market share. Further research in this area is needed to fully evaluate the benefits to winery profitability, regional growth, and expansion of WT offerings.

The benefits of the cellar door to profitability and consumer loyalty has been well established in the literature, and this was emphasized in both case studies. However, tourism agencies criticize cellar door experiences, urging wineries to develop unique tourism offerings to set them apart. An approach that risks losing the benefits of cellar door experiences highlights the need for a dedicated Wine Tourism Association in Australia. Another significant benefit of cellar door experiences, and by extension WT, is word-of-mouth marketing, a highly valuable marketing tool consistently rated as authentic and low cost. The cellar door experience is a co-created experience providing the opportunity to engage the tourist, providing an enjoyable experience *and* a script for another brand ambassador. Recommendations from social contacts carry greater authenticity and more personally relevant information than the most well-planned website or inclusion of winery offerings on a regional wine trail map. Gaps in the research exist here, determining the benefits of various marketing and communication techniques. Such research would benefit winery management, providing informed marketing choices rather than trusting a sometimes costly suck-it-and-see style of research.

The rise of social media has added a dimension to word-of-mouth marketing, extending its reach to an audience of possibly thousands on platforms such as Instagram. Individual winemakers are developing Instagram followings, taking their followers on their winemaking journeys. An opportunity exists here to educate the consumer in an informal and approachable manner, enhancing their future WT experiences and improving the accessibility of younger generations to WT experiences. Here is a communication platform that could dispel a few assumptions; educating the consumer, so they understand it *is* possible to grow and ripen Shiraz fruit in cool climate areas, the reasons why not all wine is estate-grown, and the need for developed relationships with growers providing access to good fruit when vintage catastrophes occur such as smoke taint from bush fires or extreme weather events destroying fruit.

Interestingly, wine quality is rarely mentioned. As almost every wine and winery are award-winning, promotion of wine quality is almost considered irrelevant, letting the wines speak for themselves. Also, to the less wine-involved, high-quality wine might be considered too acidic, or tannins too prominent, perhaps less fruity than expected, or they don't have the vocabulary to label their sensory perceptions. Again, there is a balance to be achieved when matching WT offerings to the wine while being mindful of the expectations of the wine tourist. Bruwer et al. (2013) found positive cellar door and WT experiences improve the perception of wine quality. Future research could focus on what constitutes this positive experience, understanding the expectations of wine tourists beyond customer service, exploring the relationship between the style of wine and experiences offered. Such research should also endeavor to broaden the scope beyond single-region case studies.

Conclusion

WT is a growth industry, highlighted by the emergence of wineries in non-grape-growing tourism regions. COVID has forced domestic tourists to explore their backyard, leading to the discovery of established and emerging wineries and wine regions. Visitation has increased and continues to grow, leaving winery management in no doubt of the importance of WT to the continued growth of their business. However, there is a perceived gap in structured education available for the tertiary level of the business (i.e., vocational courses, apprenticeships). It would appear that Australian WT has outgrown the support industry associations can provide, and a dedicated WT association could alleviate some of the pressures, especially on smaller wineries. These conclusions were limited to the experiences of four regions in Australia.

To summarize the nature and significance of WT, we felt it best left to the words of a WT event creator of a winery in an internationally recognized premium Australian wine region:

> [B]rand loyalty doesn't come from bottle shops, doesn't come from restaurants, it comes from physically being at a time and a place of where this fruit has come from, seeing the production, meeting the makers, I think wine tourism has accelerated where the wine industry is compared to where it would be if we didn't have it. The industry would not be where it is without wine tourism. WT has accelerated decisions, it's accelerated growth, and it's allowed for wineries to look at themselves, not just as a product, but as an experience and as a cultural icon within the areas that they operate.

Implications

The complex nature of winery experiences gave rise to assertions from within the industry that Australian WT could benefit from a dedicated association. An industry-based team combining vast experience from the tertiary level would possess high-level skills to develop and implement accessible WT education modules for staff and consumers. A grassroots understanding of the business allows for considered restructuring of access to assistance, based on total crush rather than regions. Also, continued support for focused research ensures that WT enjoys continued growth and development in a competitive market.

References

Ball, J., and Stolle, M. (2019). Cellar doors: the impact of cellar doors; design, experiences, and maximising customer relationships. *Wine & Viticulture Journal*, 34(4): 60.

Battaglene, T. (2019). Australian grape and wine pre-budget submission 2019–20. https://treasury.gov.au/sites/default/files/2019-03/360985-Australian-Grape-and-Wine.pdf

Brandano, M. G., Osti, L., and Pulina, M. (2019). How motivations and satisfaction influence wine tourists' loyalty? An analysis of the Italian case. *International Journal of Culture, Tourism and Hospitality Research*, 13(1): 55–69. https://doi.org/10.1108/IJCTHR-04-2018-0054

Bruwer, J. (2002). The importance and role of the winery cellar door in the Australian wine industry: some perspectives. *Australian and New Zealand Grapegrower and Winemaker*, (463): 96.

Bruwer, J., Coode, M., Saliba, A., and Herbst, F. (2013). Wine tourism experiences effects of the tasting room on consumer brand loyalty. *Tourism Analysis*, 18(4): 399–414. https://doi.org/10.3727/108354213X13736372325957

Bruwer, J., Pratt, M. A., Saliba, A., and Hirche, M. (2017). Regional destination image perception of tourists within a winescape context. *Current Issues in Tourism*, 20(2): 157–177. https://doi.org/10.1080/13683500.2014.904846

Bruwer, J., Prayag, G., and Disegna, M. (2018). Why wine tourists visit cellar doors: segmenting motivation and destination image. *International Journal of Tourism Research*, 20(3): 355–366. https://doi.org/10.1002/jtr.2187

Carlsen, J. (2004). A review of global wine tourism research. *Journal of Wine Research*, 15(1): 5–13.

Carlsen, J., and Boksberger, P. (2015). Enhancing consumer value in wine tourism. *Journal of Hospitality & Tourism Research*, 39(1): 132–144. https://doi.org/10.1177/1096348012471379

Crespi-Vallbona, M., and Mascarilla-Miró, O. (2020). Wine lovers: Their interests in tourist experiences. *International Journal of Culture, Tourism and Hospitality Research*, 14(2): 239–258. https://doi.org/10.1108/IJCTHR-05-2019-0095

Crick, J. M., Crick, D., and Chaudhry, S. (2020a). The dark-side of coopetition: it's not what you say, but the way that you do it. *Journal of Strategic Marketing* (3). https://doi.org/10.1080/0965254X.2019.1642936

Crick, J. M., Crick, D., and Tebbett, N. (2020b). Competitor orientation and value co-creation in sustaining rural New Zealand wine producers. *Journal of Rural Studies*, 73: 122–134. https://doi.org/10.1016/j.jrurstud.2019.10.019

Frost, W., Frost, J., Strickland, P., and Maguire, J. S. (2020). Seeking a competitive advantage in wine tourism: heritage and storytelling at the cellar-door. *International Journal of Hospitality Management*, 87: 102460. https://doi.org/10.1016/j.ijhm.2020.102460

Galloway, G., Mitchell, R., Getz, D., Crouch, G., and Ong, B. (2008). Sensation seeking and the prediction of attitudes and behaviours of wine tourists. *Tourism Management*, 29(5): 950–966. https://doi.org/10.1016/j.tourman.2007.11.006

Gunasekar, S., Das, P., Dixit, S. K., Mandal, S., and Mehta, S. R. (2021). Wine-experienscape and Tourist satisfaction: Through the lens of online reviews. *Journal of Foodservice Business Research*. https://doi.org/10.1080/15378020.2021.2006039

Mattiacci, A., Bruni, A., Magno, F., and Cassia, F. (2019). Sales capabilities in the wine industry: an analysis of the current scenario and emerging trends. *British Food Journal*, 121(12): 3380–3395. https://doi.org/10.1108/BFJ-04-2019-0231

Pratt, M. (2014). Four wine tourist profiles. Academy of Wine Business Research, *8th International Conference*, Geisenheim (Alemania).

Sigala, M. (2020). The transformational power of wine tourism experiences: the socio-cultural profile of wine tourism in South Australia. In Forbes, S. L., DaSilva, T.-A., and Gilinsky, A. (eds.), *Social Sustainability in the Global Wine Industry: Concepts and Cases* (pp. 57–73). London: Palgrave Macmillan.

Williams, K. M. (2013). A profile of cellar door personnel: an exploratory case study analysis of two Australian wine regions. *Journal of Human Resources in Hospitality & Tourism*, 12(1): 91–108. https://doi.org/10.1080/15332845.2013.723268

3
CONCEPTUALIZATION OF THE WINESCAPE FRAMEWORK

Vander Valduga, Sarah Marroni Minasi, and Gui Lohmann

Introduction

The term *winescape* unfolds from the concept of landscape in the geography discipline. It was then incorporated into the heritage debate when UNESCO gave protection to landscapes. This debate first occurred during the General Conference of the United Nations Educational, Scientific and Cultural Organization (UNESCO), held in Paris in 1972. This event is a milestone in the history of the protection of cultural goods, when a list of World Heritage sites was established for the inscription of goods of exceptional value, including cultural and natural criteria (UNESCO 1972).

From the perspective of cultural geography, the debate on landscape dates to 1924 with the publication of the paradigmatic work 'The Morphology of Landscape' by Carl Sauer. According to him, culture is the agent that, over time, would transform the natural framework (cradle) of the landscape. His views opposed the contemporary American geographers' deterministic approaches while providing a more structural characteristic for landscape analysis. The cultural approach made it possible to understand the landscape according to its simultaneities, dynamic and evolutionary characteristics, subjectivities, and intentionalities (Besse 2014). Additionally, prospects for the future made cultural approaches the object of studies in different areas. This approach has occurred particularly since the 1970s, with contributions from humanities and environmental studies, including new cultural geography. Since then, significant advances have been made to conceptualize the understanding of winescapes, with contributions from several other disciplines: geography, history, economics, and marketing. In a fluid way, the tourism and hospitality fields were supported in these debates, with the heritage perspective of winescapes as a guiding axis.

In this chapter, the landscape is examined as a cultural system related to multiplicities and coexistences (e.g., economics, culture, society, public policy) designed by the process of historical formation itself. For the viticultural landscape, the interpretation from the materiality of the viticultural territory highlights the importance of the cultural system. In this particular case, because of the tourist demand, it fosters its cultural and economic relevance.

DOI: 10.4324/9781003143628-5

This kind of cultural system brings together a productive, effective, touristic, and economic bond, constructing meanings and real and symbolic structures that make up the wine territories (Saquet 2007).

This chapter proposes a conceptual framework of the winescape. It would be problematic and unwise to ignore the changes in society during the current times while these landscapes are configured. Even in rural settings, the contemporary period is characterized by a fast pace and fluid daily activities (e.g., work-life balance challenges, home office, networks, and connections made available because of technology and social media). Likewise, daily work, study, and leisure activities are the product of the interaction between space and time. The social relations that have been established reveal the power that exists over them. Those forces that dominate the territory also control the politics of the place (Dixit 2020; Harvey 2005). New work, practical, and cultural relationships affect individuals and their relationships with territories and their social groups, with the technological context as a mediator; a prominent example is social media use. We live in hypermodern times (Lipovetsky 2007; Lipovetsky and Charles 2004) that are fluid or, in slightly more radical terms, the time of dilution of certainties (Bauman 2001). The question that arises, therefore, is the following: How is scientific knowledge of winescapes structured?

The relevance of studies and academic research can contribute to forming a conceptual framework where the winescape concept is presented. The literature produced on winescape opens the possibility of dealing with the complexity that surrounds the idea. The academic literature tackling the cultural landscape's interrelationship, viticulture, and wine tourism is still nascent. Therefore, this chapter aims to propose the construction of a winescape framework. Methodologically, this chapter uses a literature review from two international databases: Web of Science and Scopus. The theoretical corpus arising from these databases comprises 26 journal articles addressing the concept of winescapes. The data analysis was conducted from a theoretical and critical perspective. IRaMuTeQ software was used to support the analysis, enabling different types of textual data analysis. In this case, scientific papers collected in the databases were processed as textual corpora. Classification analysis (Reinert method) was deployed using the hierarchical descending classification (CHD) Simple on Text Segment (ST). This tool generates a word graph that offers a similarity graph representing the links between the chosen word and the others in the same class. Similitude analysis was also deployed, which resulted in the maximum similitude tree graph. This graph illustrates the relationships among the main words that make up the processed texts.

Winescape literature

The winescape as a neologism first appeared in the literature in the late 1990s in Peters's (1997) study. The concept refers to the rural scenario where grape culture predominates, with other crops present on a smaller scale (Peters 1997). Joliet (2005) considers the viticultural landscape as the result of complex natural and human processes, comprising the terrain, technical and cultural dimensions, and constituting memories. From Elías's (2014) perspective, the heritage aspect must be considered since it involves generations of work endowed with knowledge and techniques linked to the territories.

Landscapes are restricted not to what is seen, but to what is represented. The landscape does not reside in only the subject or in only the object; rather, it results from the complex interaction between the two. It is at this complex intersection that landscape study is fixed (Berque 2004).

Summarizing the cultural dimensions of the landscape, Claval (1997) states that places not only have shape, color, and economic functionality but also are loaded with meaning for those who inhabit or frequent them. It is a cultural system (Bonnemaison 2005) related to a spatial interpretation based on multiplicities and coexistences with their trajectories and stories (Massey 2008).

The literature on winescapes appeared in the early 2010s. However, its origins permeated the production of knowledge of the cultural landscape since the beginning of the 20th century. Currently, a growing number of publications consider the concept, featuring a recent research topic.

A winescape comes from agrarian landscapes' local and regional configuration, specifically vitiviniculture (Visentin and Vallerani 2018). It is permeated by changes in the landscape and land use generated by planting vineyards. It refers to long-term historical factors that led to the emergence of the specific wine landscape (Basso 2018).

Viticulture is one of the activities that has become an essential element of cultural and landscape development (Greinert, Kostecki, and Vystavna 2019). Although grape cultivation areas occupy small areas globally, they create striking agricultural landscapes (Arnáiz and Hernando 2019). The rural landscape's quality is recognized as an essential element of winescapes (Sottini et al. 2019). There is currently a tendency to showcase unique characteristics that portray local history and traditions that have often been long forgotten (Greinert, Kostecki, and Vystavna 2019).

Cultural landscape analysis provides an understanding of the processes of economic activities, particularly considering the distinct geophysical forms of a particular area (Minasi, Lohmann, and Valduga 2020). At the same time, it is necessary to consider the multidimensionality of the continuous experience pursued at the emotional level (Sottini et al. 2019; Sundbo and Dixit 2020). These aspects are engendered in heritage development as critical for promoting rural tourism in the context of winescapes (Visentin and Vallerani 2018). As a result, private initiatives that consider the landscape as an economic opportunity have emerged to reinforce wines' identity (Arnáiz and Hernando 2019).

The modernization of grape culture, the growing demand for quality wines, and the emergence of wine tourism have created a new agrarian landscape enhanced by institutional and social recognition (Arnáiz and Hernando 2019). In addition, the wine industry offers a sensorial experience that includes various cultural and economic opportunities for tourists worldwide (McFarlane et al. 2017). Thus, the aim is to develop a new strategy to improve and preserve wine tourism in the future. The development of wine tourism needs to be analyzed from the node, network, and landscape perspectives (Ferreira and Hunter 2017).

From a market viewpoint, wine tourism is one of the many activities offered by wineries. It represents a diversification opportunity for a typical winery, including increased wine production and income (Remeňová, Skorková, and Jankelová 2019). Therefore, activities linked to wine tourism that promote a sustainable revenue base for wineries are a significant concern when establishing wine tourism activities (Byrd et al. 2016).

The literature presents how winescapes affect wine tourists' behavior (Quintal, Thomas, and Phau 2015). Visitation with measuring scales points to the winescape among the main dimensions of the wine tourism experience: (1) wine storytelling, (2) wine-tasting excitement,

(3) wine involvement, and (4) winescape (Thomas, Quintal, and Phau 2018; Santos et al. 2020). It is also amenable to use as a diagnostic tool and has been indicated for increasing profitability (McFarlane et al. 2017; Thomas, Quintal, and Phau 2018).

The measurement of a wine region image by identifying several elements of the winescape has been widely discussed (Bruwer and Gross 2017). Research conceptualizes the winescape framework using the image of a wine region as perceived by wine tourists. The construction of the winescape is identified from eight dimensions: nature-related; wineries and vineyards; heritage-related towns; wine and other products; signage and layout; ambient factors; service staff and locals; and fun activities. The winescape's most critical dimension is the natural beauty/geographical setting (Bruwer, Gross, and Lee 2016; Bruwer and Joy 2017; Bruwer and Lesschaeve 2012; Bruwer et al. 2017; Bruwer and Rueger-Muck 2019).

The winescape also emerges as an attribute of the hedonic pricing model, including increased accommodation prices for lodges with winescape aesthetic features (Guedes and Rebelo 2020). This economic effect appears beyond the tourist experience in the intention of buying wines. When consumers have options, they are more likely to pay for some innovative wine attributes, including more sustainable production, which is directly related to the quality of winescapes (Nassivera et al. 2020), as settled by UNESCO protection certification.

In this chapter, a case study of Vale dos Vinhedos in Brazil shows the hedonic enhancement of the landscape as a contributor to increased accommodation prices. The increase in land value in a rural setting occurred after the geographical indication and designation of origin was granted because of the heritage and increased tourism appeal.

Towards a proposed winescape framework

Winescape is an original concept of geographical interpretation. However, as the literature attests, it is an open, interdisciplinary concept. The landscape expresses the techniques used in rural production, in this case, grapes and wine. These techniques incorporate the knowledge of those who inhabit, design, and transform the territory, notably winegrowers. These techniques add value to the landscape, and they express the historical past, as evident from historical photographs. However, the landscape, as it is not static and is a living object, undergoes a transformation into a heritage (anchoring) process that protects the future (objectification) (Gravari-Barbas 1996; Lazzarotti 2003). In the patrimonialization process, communication and marketing mechanisms are designed for others' eyes (i.e., the tourist and anyone from outside the region) to value the internal practices of those who build the landscape. As a result, tourist flows are integrated with the winescape, the local economy, and other historical and hospitality facilities.

The discussion dynamics proposed do not necessarily obey any particular order or hierarchy (Figure 3.1). Traditional wine regions have undergone heritage and touristification processes. Emerging wine regions tend to be structured on aesthetic logics and tourist rationalities (e.g., the case of Serra Catarinense in Brazil, as presented in Minasi, Lohmann, and Valduga 2020) but later anchor their heritage processes in the territory and its cultural systems, as analyzed by Gabardo and Valduga (2019).

Grape cultivation and wine production bring notable changes in the local and regional configuration of rural landscapes. Traditional viticulture has quickly transformed into an intensive monoculture with significant investments and the fostering of new winery developments (Visentin and Vallerani 2018). World-renowned wine-producing regions characterize this

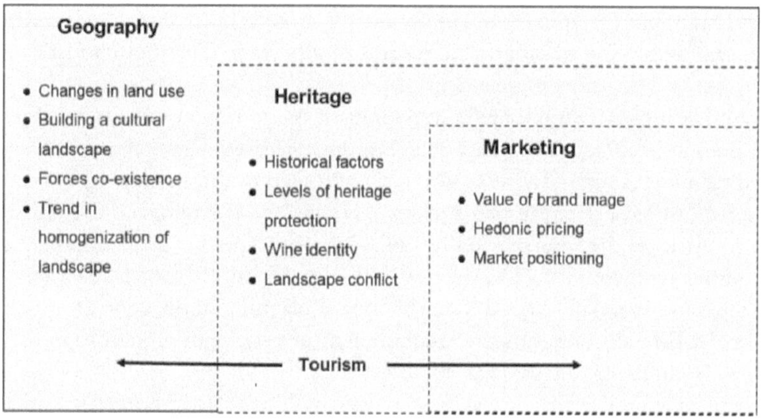

Figure 3.1 Winescape framework proposal.

Source: Authors

process, such as the prosecco region in Italy (Basso 2018; Visentin and Vallerani 2018) and the Douro, Porto, and Madeira regions in Portugal (Arnáiz and Hernando 2019; Guedes and Rebelo 2020; Santos et al. 2020).

Several forces need to coexist in the territoriality of winescapes, including the productive increase due to growing local and global demands, a rural representation based on local habits, and the landscape itself (Visentin and Vallerani 2018). However, it is necessary to effectively harmonize wine growers' production needs with the requirements of other land uses to protect historical landscape patterns and ecological biodiversity (Basso 2018).

Crowley (1993) points to the wine market as the main force behind changes in the landscape. Areas without previous cultivation or with a different crop, pastures, or woodlands converted into vineyards (Basso 2018) tend to experience increased urbanization levels and homogenization of the landscape (Visentin and Vallerani 2018). Therefore, this alters the physical attributes of winescapes. However, changes in the landscape and land use cannot be attributed only to planting new vineyards. Combined with wine production, wine tourism coexists in the rural landscape. The popularization and sophistication of wine tourism have transformed many rural regions worldwide (Minasi, Lohmann, and Valduga 2020).

Wine tourism is a consolidated practice of tourism in wine regions on all continents. It follows some protocols with few variations in its fundamental structures. Ricolfe, Barrio, and Vera (2010) identify, through a bibliographic review, which wine tourism activities are the most cited in the literature and constitute a motivating interest for wine tourism: buying wine, tasting wine at the winery, and receiving information about the wine. Additionally, wine tourism highlights the wine elaboration process: visit a winery, visit its vineyard and landscapes, participate in wine fairs or parties, eat at the winery, visit a wine museum, visit natural sites in the region, make friends, and visit monuments, among other activities.

Gálvez, Fernández, and López-Guzmán (2015), when analyzing the motivation and satisfaction of visitors, identified that the degree of satisfaction of the visit increases among those with a punctual interest in enjoying the flavors of regional enogastronomy, followed by those who attend the festivals in search of entertainment and pleasure. Lavandoski and Lanzer (2008), from the analysis of photographs, found that vineyards have a strong appeal in the wine tourism experience, with numerous photographic records of vineyards. The vineyard landscape would therefore constitute a legitimizing element of the wine tourism experience.

When considering that wine tourism development is also responsible for transforming rural landscapes, consolidated and emerging wine regions need high-level protection mechanisms towards natural and cultural resources. Winescape protection can be offered through an agricultural reserve or the declaration of a national heritage site (Ferreira and Hunter 2017). The starting point discusses the interrelationship of wine landscapes, land use and land cover, and their relationship with tourism. This matter finds support in a database (e.g., land use, land cover, and tourist services) developed to enable spatial planning in the context of wine tourism regions. The database also supports the establishment of geographical indications of origin for wines (Minasi, Lohmann, and Valduga 2020).

It is critical to consider that the winescape is not limited to the external aspects of the vineyards, the winery, or the topography. The interior of a wine building, its underground cellars, maturation cellars, and wooden barrels where the wines ripen are also components of this landscape. The winescape interpretation contemplates other nonmaterial elements, such as aroma, flavors, and sounds. Wines are not just representations of a place but an integral part of it, and the subjective aspects of the landscape that strictly escape the dimension of the gaze, or the field of vision, are constituents of the wine landscape (Elías 2014; Lugindühl 2005; Peters 1997).

The safeguarding of winescapes as patrimonial assets implies protecting their natural and cultural features, with repercussions in the relationship with the tourist/consumer, mainly because of growing concerns about environmental issues. Nassivera et al. (2020) highlighted an increasing willingness of consumers to pay for the unique attributes of the products originating from particular regions, including wine landscape, certification, carbon footprint declaration, and price.

In the marketing literature, winescapes are widely considered one of the most significant attributes influencing wine tourism (Quintal, Thomas, and Phau 2015). As Byrd et al. (2016) stated, winescapes are among the most influential factors in visiting a winery. However, winescapes are often reduced to scenery, and excluding other factors might imply that winescapes have only economic (monetary) importance. Winescapes are offered as an amenity based on an agricultural product that meets the tourist's desire for leisure, increasing the profitability of producers and intermediaries. From a marketing perspective, winescapes gain objectivity and are incorporated by the idea of adding value to a product, be it wine or tourism. With the advancement of wine tourism, a strong aesthetic appeal of the landscape emerged that was quickly incorporated as an opportunity to reinforce the growing value of the brand (Arnáiz and Hernando 2019). Brochado et al. (2019) describe the wine tourism experience as being wine, accommodation facilities, food service (e.g., restaurant and meals), scenery (including views and vineyards), staff, transportation, and recommendation. The motivations that guide wine tourists' behavior are mainly hedonic (Bruwer et al. 2017); the winescape is often determined only by its role in promoting wine tourism, disregarding that it transcends its origin in the vineyard.

In the context of tourism development, it is necessary to articulate the environmental and cultural characteristics of a territory's landscape with the concept of winescapes (Sottini et al. 2019). Tourism is desirable for wineries and local agents, as it helps in the rural socioeconomic context and enhances the landscape as a heritage and cultural value of the territory. As a synthesis, Figure 3.2 shows the relationships of words and their proximity from the literature analysis, with the winescape as its central axis.

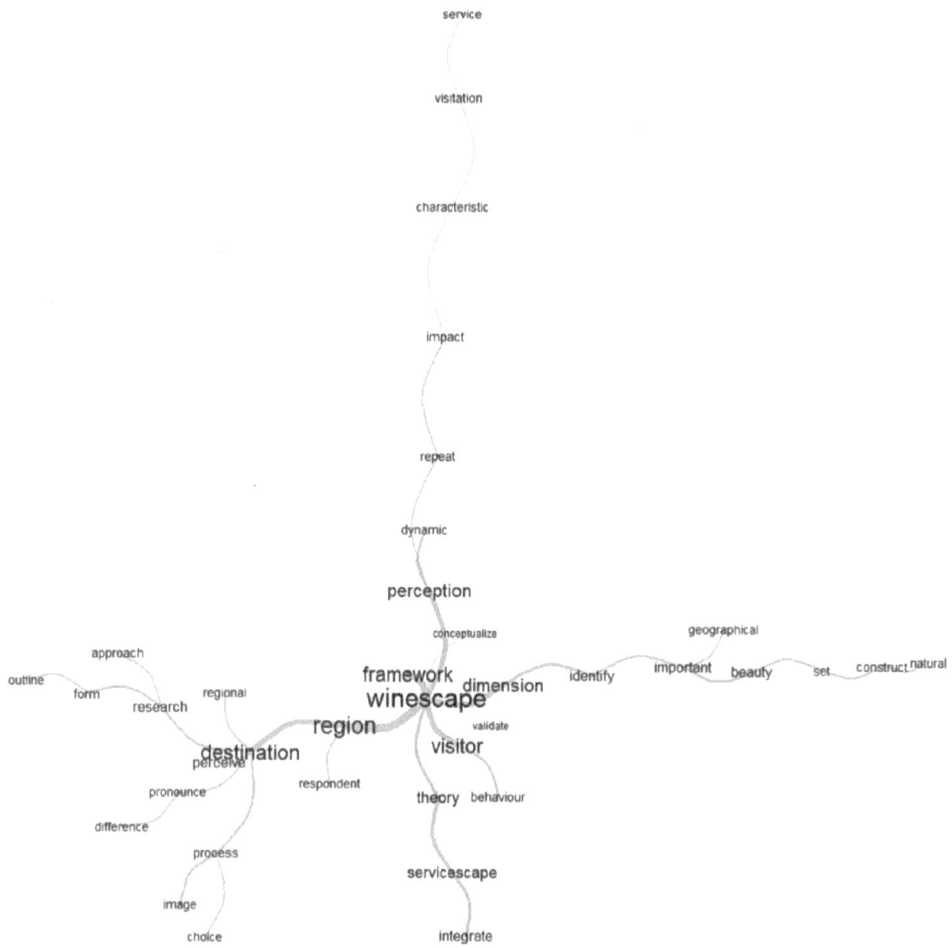

Figure 3.2 Winescape literature analysis.

Source: Authors

Case Study 3.1 Vale dos Vinhedos, Brazil – Denomination of Origin

The Vale dos Vinhedos region, in the southern part of Brazil, is located in the Serra Gaúcha region in the northeastern state of Rio Grande do Sul. Vale dos Vinhedos comprises the municipalities of Bento Gonçalves, Garibaldi, and Monte Belo do Sul. With a humid and hot subtropical climate, the dense native forests of the Atlantic Forest, and altitudes ranging from 450 to 650 meters of rugged topography, the region represents a challenge for cultivating grapes and configuring a unique landscape context.

The region received a massive flow of Italian-origin immigrants, mainly from Veneto, Lombardia, Trento, and Friuli (Falcade 2011). The Brazilian government-financed these immigrants' arrival starting in 1875 and for more than 30 years, with more than 600,000 Italians migrating to Brazil in this period. Upon arrival in Brazil, they received farm equipment, lots to be transformed into self-sustaining land, and an allowance for acquiring essential goods for their subsistence in the first year (Dal Pizzol and Souza 2014). This government policy allowed a territorial configuration of small properties, different from the reality of other Brazilian rural regions in the colonial period, marked by the large areas of monoculture, with a hierarchical, patriarchal structure and slave labor (Holanda 1995).

European immigrants that settled in the south of the country were part of a historical process of territorial occupation characterized by the distribution of small portions of land acquired at affordable prices with loans offered for up to 15 years after their settlement (Dal Pizzol and Souza 2014). In addition, Schneider (2006) observed that colonists engaged in other nonagricultural activities, such as logging and infrastructure work, created during the colonization process to settle these debts.

Until the 1950s, the regional economy was based on multiple rural activities, emphasizing viticulture, which was practically exclusive to this region until the end of the 20th century. However, economic rationalization focusing on viticulture began to predominate in the late 20th century, and Vale dos Vinhedos is a relevant example of this historical process.

Wine tourism in the region began in the 1930s, with regional festivals related to grapes and wine. In addition, there was an intense movement of immigrants to strengthen the country's regionalism and nationalism, as occurred in countries such as Italy and Germany.

The intensification of tourism became possible in the 1970s, when large cooperatives and family wineries that produced quality wines opened their doors to visitors. This change established a process of the touristification and heritage of regional wine activities that would consolidate from the early 2000s.

Vale dos Vinhedos had a series of private investments during its development, resulting in more than 80 projects in a geographical area of just more than 84 km2. The regional organization took place after the constitution of the Association of Wine Producers of Vale dos Vinhedos (Aprovale), which is also the main gateway for tourism.

Beginning in 2000, Vale dos Vinhedos received a series of private investments in hotels, inns, restaurants, small regional businesses, and new wineries that added to the traditional ones from the 1970s. This process resulted in more than 70 tourist developments in an area of just more than 84 km2. These private ventures worked together to create Aprovale, an organization that locally manages the geographical indication and designation of origin for wine production and operates in the organization of local tourism in the promotion and marketing of this tourism destination.

The Vale dos Vinhedos is a wine region that constituted the first designation of origin of wines in Brazil. Certification was granted in 2012, after the title of geographical indication was received in 2002 (Falcade 2011). With the certification of geographical indication, which was the first in Brazil for agri-food origin products, there was national recognition of the quality of production and a significant prominence in the media, generating a constant increase in tourist demand. Figure 3.3 shows the increase in the demand flow since 2001, when the measurements started.

The accelerated investment and touristification process, driven by certification, generated a series of spatial, landscape, and real estate conflicts between private investors and

Figure 3.3 Flow of tourist demand on Vale dos Vinhedos, Brazil (2001–2019).

Source: Authors

regional council codes. With the ultimate development of luxury rural condos built in certified areas, hotel chains established their presence, with the rural regions' valorization ranging from 200% to 500% between 2000 and 2010. Landscape protection mechanisms were later created, such as the rural master plans in the municipalities that make up the Vale dos Vinhedos. However, with different rules among cities, these efforts were only slightly effective in slowing the urbanization process, along with the lack of inspection and the corruption of private and public agents (for a better understanding of the challenges of tourism development and corruption in Brazil, see Lohmann et al. 2021).

In the Vale dos Vinhedos, traditional landscape cultural systems coexist (Figure 3.4), reminiscent of the origins of territorial occupation, an expression of conventional viticulture in northern Italy in the 19th century, with new landscapes of wine aligned with productive rationality. Added to this context are tourist developments, and sometimes vertical construction, that impose themselves before the established cultural landscape.

Figure 3.5 shows an international luxury hotel with spa services whose accommodation prices are associated with the viticultural landscape. The daily rates, in general, cost more than double those charged in other lodging facilities in the region. Its vertical construction, imposing itself on the landscape, with an architecture reinterpretation that clashes with the regional context, was the object of several criticisms from architects, historians, and researchers. Other similar conflicts exist in the region, which has become an object of desire for tourists and wine lovers.

In this sense, Vale dos Vinhedos is an emblematic case in Brazil. It anticipated a series of issues in the country's new wine regions and required attention in structuring the viticultural landscape. Wine tourism is anchored in wine-growing landscapes, and Vale dos Vinhedos experiences intense social, economic, and cultural transformation processes. In this sense, it is necessary to recognize the value of the wine landscape and its cultural structures so that new generations can enjoy their history, knowledge, and culture.

Figure 3.4 Traditional winescape in the Vale dos Vinhedos.

Source: Authors

Figure 3.5 New landscape in the Vale dos Vinhedos.

Source: Authors

Conclusion

This chapter started by posing the following question: How is the scientific knowledge of winescapes structured? Figure 3.1 shows the structure of the winescape as addressed by the literature, in this case, composed of a corpus of 26 papers. Different approaches are based on the authors' interests and disciplinary backgrounds, including geography, economics, marketing, management, and tourism.

We identified that different perspectives lead to varying interpretations, either from the point of view, for example, of the wine market as the main force behind landscape changes, or of cultural aspects of cultivation established in previous times preserved under some legal requirements. Common to these approaches, the urbanization process still considers the tendency to homogenize wine territories' landscapes. This process is significantly accelerated by touristification or economic rationality in wine-producing regions, leaving other competing crops out.

Wine tourism, along with the wine market, has been shown to be a driving force in the economy. It also links to landscape valuation because of the increase in hotel rates, the movement of associated production chains, and the direct export of resources generated by the expenditures of wine tourists. Wine tourism also expands to other contexts the perspective of interpreting winescapes involving tourism activity, such as visits to wine cellars, wooden barrel rooms, tasting, and shopping spaces. Thus, winescapes constitute, in addition to the topographic sphere, a sensitive landscape to be interpreted by wine tourism experts and those who appreciate aspects related to the world of wine.

The case of Vale dos Vinhedos illustrates the complexity of winescapes. Multiple actors are essential in wine territories, including real estate agents, tourism and hospitality agents, winemakers (whether large or small producers), and the public sector. Furthermore, as evidenced by the literature, touristified wine regions where the landscape is the main attraction remain challenged to expand the economic activities and the tourist experiences while preserving the wine landscape. Therefore, Vale dos Vinhedos should be analyzed among Brazilian wine regions.

These aspects are fundamental to consider winescapes as heritage, legacy, and central elements of tourism and hospitality. The implications of geographical, heritage, and marketing approaches allow a reflection on land use, territorial planning, rural transformations, urbanization, consumption, wine tourism destinations, and the mechanisms for protecting landscapes, either via geographical indications and denomination of origin or through heritage nationals.

Implications

Analyzing winescapes from different aspects is an academic imperative for wine tourism. Despite the predominance of monocultures of grapes, winescapes are a complex reality. As a productive process, it includes a series of structured activities, especially tourism and hospitality. The winescape framework evidenced in Figure 3.2 illustrates the aspects relevant to wine tourism regions. The proposed framework can help tourism agents, regional governments, the public, and mixed bodies consider the geographic, patrimonial, and marketing aspects of these regions' composition of development plans. In addition, it supports managers and planners in incorporating new elements into the territorial analysis, opening a broad scope to minimize the negative consequences. The aim is to overcome a sectarian and imposed spatial analysis from special or particular interests. Finally, it highlights several implications for land use, rural transformations, and protection and promotion mechanisms, such as geographical indications and denominations of origin.

References

Arnáiz, M. M., and Hernando, F. M. (2019). El valor patrimonial del paisaje como imagen de calidad en la estrategia comercial de los territorios vitivinícolas del Duero. *Cuadernos geográficos de la Universidad de Granada*, 58(3): 169–194. https://doi.org/10.30827/cuadgeo.v58i3.8837

Basso, M. (2018). From daily land-use practice to global phenomenon: on the origin and recent evolution of prosecco's wine landscape (Italy). *Miscellanea Geographica*, 22(2): 109–115. https://doi.org/10.2478/mgrsd-2018-0013

Bauman, Z. (2001). *Modernidade Líquida*. Rio de Janeiro: Zahar.

Berque, A. (2004). Milieu et identité humaine/Milieu and human identity. *Annales de Géographie*, 113(638): 385–399. https://doi.org/10.3406/geo.2004.21630

Besse, J. M. (2014). *O Gosto do Mundo: Exercícios de paisagem*. Rio de Janeiro: Editora UERJ.

Bonnemaison, J. (2005). *Culture and Space: Conceiving a New Cultural Geography*. New York: Bloomsbury Publishing.

Brochado, A., Troilo, M., Rodrigues, H., and Oliveira-Brochado, F. (2019). Dimensions of wine hotel experiences shared online. *International Journal of Wine Business Research*, 32(1): 59–77. https://doi.org/10.1108/IJWBR-12-2018-0072

Bruwer, J., and Gross, M. J. (2017). A multilayered macro approach to conceptualizing the winescape construct for wine tourism. *Tourism Analysis*, 22(4): 497–509. https://doi.org/10.3727/108354217X15023805452059

Bruwer, J., Gross, M. J., and Lee, H.C. (2016). Tourism destination image (TDI) perception within a regional winescape context. *Tourism Analysis*, 21(2–3): 173–187. https://www.ingentaconnect.com/content/cog/ta/2016/00000021/f0020002/art00004

Bruwer, J., and Joy, A. (2017). Tourism destination image (TDI) perception of a Canadian regional winescape: A free-text macro approach. *Tourism Recreation Research*, 42(3): 367–379. https://doi.org/10.1080/02508281.2017.1318482

Bruwer, J., and Lesschaeve, I. (2012). Wine tourists' destination region brand image perception and antecedents: conceptualization of a winescape framework. *Journal of Travel & Tourism Marketing*, 29(7): 611–628. https://doi.org/10.1080/10548408.2012.719819

Bruwer, J., Pratt, M. A., Saliba, A., and Hirche, M. (2017). Regional destination image perception of tourists within a winescape context. *Current Issues in Tourism*, 20(2): 157–177. https://doi.org/10.1080/13683500.2014.904846

Bruwer, J., and Rueger-Muck, E. (2019). Wine tourism and hedonic experience: A motivation-based experiential view. *Tourism and Hospitality Research*, 19(4): 488–502. https://doi.org/10.1177/1467358418781444

Byrd, E. T., Canziani, B., Hsieh, Y.-C., Debbage, K., and Sonmez, S. (2016). Wine tourism: Motivating visitors through core and supplementary services. *Tourism Management*, 52: 19–29. https://doi.org/10.1016/j.tourman.2015.06.009

Claval, P. (1997). As abordagens da geografia cultural. In I. E. Castro, P. C. Gomes, and R. L. Corrêa (eds.), *Explorações Geográficas - percursos no fim do século* (pp. 89–117). Rio de Janeiro: Bertrand Brasil,

Crowley, W. K. (1993). Changes in the French winescape. *Geographical Review*, 83(3): 252–268. https://doi.org/10.2307/215728

Dal Pizzol, R. D., and Souza, S. I. (2014). *Memórias do vinho gaúcho*. Porto Alegre: AGE.

Dixit, S. K. (ed.). (2020). *Tourism in Asian Cities*. London: Routledge.

Elías, L. V. (2014). El paisaje del viñedo: su papel en el enoturismo. *Revista Iberoamericana de Viticultura, Agroindustria y Ruralidad*, 1(3): 12–32.

Falcade, I. (2011). *A paisagem como representação espacial: a paisagem vitícola como símbolo das indicações geográficas de vinhos do Vale dos Vinhedos, Pinto Bandeira e Monte Belo (Brasil)* [Unpublished doctoral dissertation]. Universidade Federal do Rio Grande do Sul.

Ferreira, S. L., and Hunter, C. A. (2017). Wine tourism development in South Africa: A geographical analysis. *Tourism Geographies*, 19(5): 676–698. https://doi.org/10.1080/14616688.2017.1298152

Gabardo, W. O., and Valduga, V. (2019). Los sistemas culturales y el paisaje del viñedo brasileño: recursos para el enoturismo. *Estudios y Perspectivas en Turismo*, 28 (3): 759–779.

Gálvez, J. C. P., Fernández, G. A. M., and Guzmán, T. L. G. (2015). Motivación y satisfacción turística en los festivales del vino: XXXI ed. cata del vino Montilla-Moriles, España. *Tourism & Management Studies*, 11(2): 7–13. https://doi.org/10.18089/tms.2015.11201

Gravari-Barbas, M. (1996). Le 'sang' et le 'sol': le patrimoine, facteur d'appartenance à un territoire urbain. *Géographie et Cultures*, 20: 55–67.

Greinert, A., Kostecki, J., and Vystavna, Y. (2019). The history of viticultural land use as a determinant of contemporary regional development in Western Poland. *Land Use Policy*, 85: 249–258. https://doi.org/10.1016/j.landusepol.2019.04.010

Guedes, A., and Rebelo, J. (2020). Winescape's aesthetic impact on lodging room prices: a spatial analysis of the douro region. *Tourism Planning & Development*, 17(2): 187–206. https://doi.org/10.1080/21568316.2019.1597761

Harvey, D. (2005). *Condição Pós-Moderna*. São Paulo: Edições Loyola.

Holanda, S. B. (1995). *Raízes do Brasil*. São Paulo: Companhia das Letras.

Lavandoski, J., and Lanzer, R. 2008. Percepção da Paisagem: Uma análise a partir de fotografias do Vale dos Vinhedos, RS: Seminário de pesquisa em turismo do Mercosul. *SeminTUR*, 5: 1–15.

Lazzarotti, O. (2003). Tourisme et patrimoine: ad augusta per angustia/Tourism and heritage: ad augusta per angustia. *Annales de Géographie*, 112(629): 91–110. https://doi.org/10.3406/geo.2003.893

Lipovetsky, G. (2007). *A felicidade paradoxal: Ensaio sobre a sociedade de hiperconsumo*. São Paulo: Companhia das Letras.

Lipovetsky, G., and Charles, S. (2004). *Os Tempos Hipermodernos*. São Paulo: Barcarolla.

Lohmann, G., Lobo, H. A. S., Trigo, L. G. G., Valduga, V., Castro, R., Coelho, M. D. F., Cyrillo, M. W., Dalonso, Y., Gimenes-Minasse, M. H., Gosling, M. D. S., Lanzarini, R., Leal, S. R., Marques, O., Mayer, V. F., Moreira, J. C., de Moraes, L. A., Panosso Netto, A., Perinotto, A. R. C., Queiroz Neto, A., Trentin, F., and Raimundo, S. (2021). Tourism in Brazil: from politics, social inequality, corruption and violence towards the 2030 Brazilian tourism agenda. *Tourism Review*. Vol. ahead-of-print No. ahead-of-print. https://doi.org/10.1108/TR-07-2020-0323

Lugindühl, Y. 2005. *Les paysages culturel viticoles. Étude thématique*. Icomos, Paris, France: Convention du patrimoine mondiale UNESCO.

Massey, D. B. (2008). *Pelo Espaço: Por uma nova política da espacialidade*. Rio de Janeiro: Bertrand.

McFarlane, J., Grant, B., Blackwell, B., and Mounter, S. (2017). Combining amenity with experience: Exploring the hidden capital of a winescape experience. *Tourism Economics*, 23(5): 1076–1095. https://doi.org/10.1177/1354816616665754

Minasi, S. M., Lohmann, G., and Valduga, V. (2020). Geographic information systems are critical tools to manage wine tourism regions. *Tourism Geographies*, 22(1): 1–22. https://doi.org/10.1080/14616688.2020.1861081

Nassivera, F., Gallenti, G., Troiano, S., Marangon, F., Cosmina, M., Bogoni, P., Campisi, B., and Carzedda, M. (2020). Italian millennials' preferences for wine: An exploratory study. *British Food Journal*, 122(8): 2403–2423. https://doi.org/10.1108/BFJ-05-2019-0306

Peters, G. L. (1997). *American Winescapes: The Cultural Landscapes of America's Wine Country*. New York: Routledge.

Quintal, V. A., Thomas, B., and Phau, I. (2015). Incorporating the winescape into the theory of planned behaviour: Examining 'new world' wineries. *Tourism Management*, 46: 596–609. https://doi.org/10.1016/j.tourman.2014.08.013

Remeňová, K., Skorková, Z., and Jankelová, N. (2019). Wine tourism as an increasingly valuable revenue stream of a winery's business model. *Економика пољопривреде*, 66(1): 23–34. https://doi.org/10.5937/ekoPolj1901023R

Ricolfe, J. S. C., Barrio, J. E. R., and Vera, J. M. B. (2010). Análisis del mercado potencial enoturístico en Valencia. *Papers de Turisme*, 6(47–48): 93–108.

Santos, V., Ramos, P., Almeida, N., and Santos-Pavón, E. (2020). Developing a wine experience scale: A new strategy to measure holistic behaviour of wine tourists. *Sustainability*, 12(19): 8055. https://doi.org/10.3390/su12198055

Saquet, M. A. (2007). Abordagens e concepções de território e territorialidade. *Revista Geográfica de América Central*, 2(47): 1–16.

Schneider, S. (2006). Agricultura familiar e desenvolvimento rural endógeno: Elementos teóricos e um estudo de caso. In Froehlich, J. M., and Vivien, D. (eds.), *Desenvolvimento Rural: Tendências e debate contemporâneo*. (pp. 42–61). Ijuí: Unijuí.

Sottini, V. A., Barbierato, E., Bernetti, I., Capecchi, I., Fabbrizzi, S., and Menghini, S. (2019). Winescape perception and big data analysis: An assessment through social media photographs in the Chianti Classico region. *Wine Economics and Policy*, 8(2): 127–140. https://doi.org/10.1016/j.wep.2019.07.001

Sundbo, J., and Dixit, S. K. (2020). Conceptualisations in tourism experience. In Dixit, S. K. (ed.), *The Routledge Handbook of Tourism Experience Management and Marketing* (pp. 15–26). London: Routledge.

Thomas, B., Quintal, V. A., and Phau, I. (2018). Wine tourist engagement with the winescape: scale development and validation. *Journal of Hospitality & Tourism Research*, 42(5): 793–828. https://doi.org/10.1177/1096348016640583

Visentin, F., and Vallerani, F. (2018). A countryside to sip: Venice Inland and the Prosecco's uneasy relationship with wine tourism and rural exploitation. *Sustainability*, 10(7): 2195. https://doi.org/10.3390/su10072195

4

CONTRIBUTIONS OF WINE ROUTES AND TRAILS TO WINE TOURISM EXPERIENCES

Namita Roy and Ulrike Gretzel

Introduction

Wine trails or routes are marketing tools that connect different businesses and services in a wine region following the trajectory of a road or trail network (Hall, Cambourne, Macionis, and Johnson 1997). The idea behind this geographic alignment is to direct tourists on a travel itinerary, leading them to visit multiple wineries in the region, translating into greater economic activity. Wine trails or routes are popular in wine-producing European countries where they are particularly planned and developed through the use of on-ground signage, markers, incorporating physical attractions and services (Bruwer 2003). Several other countries have followed suit and have developed wine routes/trails to direct tourist visitation to and through their wine regions. Examples are the Foxen Canyon Wine Trail in California, the network of Chilean Wine Routes, the Cape 62 wine route in South Africa, and the Riesling Trail in South Australia.

Literature on wine trails/routes suggests multifarious advantages of wine routes/trails. Wine trails bring together businesses, in turn capitalizing on the combined knowledge and the social capital of the region (Xu, Barbieri, and Seekamp 2020). This confluence facilitates a regional sociopolitical integration whereby businesses come together for the common social good of the regional community. As wine businesses contribute towards the regional economy, the development of a wine trail brings about the consciousness of a regional community, which pushes businesses to work together and strive for common infrastructure development (roads, signage, cycle routes) and co-marketing strategies.

Another key advantage of a wine route/trail is business network development (Boyne, Williams, and Hall 2002; Hall and Gössling 2016; Hojman and Hunter-Jones 2012; Naipaul, Wang, and Okumus 2009). Much previous research has focused on the benefits of network formation through trails and routes, which facilitates a competitive advantage by creating a brand identity for the region that, ultimately, encourages tourist visitation. Network development through routes further helps small wine businesses in the region to gain visitation, and it provides access to technical knowledge and resource pools, all of which allow them to flourish (Lewis, Byrom, and Grimmer 2015).

Though the research on wine trails/routes mostly indicates purposeful development following social or economic objectives achieved through network development, little research

DOI: 10.4324/9781003143628-6

has focused on how these trails/routes facilitate the tourist experience. Research suggests that tourists indulge in various activities in the wine region, such as fine dining, visiting heritage attractions, and sports (Corigliano 2016; López-Guzmán, Vieira-Rodríguez, and Rodríguez-García 2014; Mason and O'Mahony 2007). This indicates that a larger role of the wine route is in not only assimilating wine networks but also creating the tourist experience.

This chapter presents a conceptual framework that facilitates understanding the role of wine trails/routes in creating the tourist experience. We discuss how following the experiential perspective forwards the selection of the relational theoretical framework of the Actor-Network Theory (ANT), which enables a detailed understanding of the contributions of the wine trail or route towards tourist experiences.

Research on the tourist experience

Experience research in the tourism domain is informed by multiple disciplines. Following the progress of understanding the 'experience' in the marketing discipline, the tourist experience has been conceptualized as subjective (Cutler, Carmichael, and Morgan 2010; Mannell and Iso-Ahola 1987), lived (Cohen 1979), embodied/sensory (Cutler, Carmichael, and Doherty 2014; Edensor 2000; Husemann and Eckhardt 2018; Rakić and Chambers 2012), and emotional and felt (Buda, d'Hauteserre, and Johnston 2014; Knobloch, Robertson, and Aitken 2017). Recent research on experiences articulates its emergence 'in the moment,' which is cognitively evaluated as well as affectively sensed and embodied (Holmqvist, Diaz Ruiz, and Peñaloza 2020; Schmitt 2020; Vo Thanh and Kirova 2018). This notion of 'momentariness' of the experience has been acknowledged by several tourism experience design researchers who have argued for materialities and affectivities in the creation of the experience (Conway 2013; Higgins and Hamilton 2020; Jepson and Sharpley 2015; Mattila and Gao 2017; Moscardo 2017; Roy and Gretzel 2021). This poststructural conceptual understanding of the tourist experience is helpful in identifying the material and affective contributions of the wine trail, which have been overlooked in previous research.

Another key characteristic of the experience is its assembled nature (Canniford and Bajde 2015; Hoffman and Novak 2018). Several researchers have argued for an important role of interactions between the human–nonhuman environment in the creation of a tourist experience (Bærenholdt, Haldrup, and Urry 2017; Van der Duim, Ren, and Thór Jóhannesson 2013). Acknowledging the importance of interactions with the contextual environment gives agency to things, roads, buildings, vines, and sceneries, highlighting the 'coming together' or assembled nature of the experience. This particular characteristic helps thinking about the wine route/trail from an agentic perspective, which enables the assemblage of the tourist experience.

Conceptual framework

Following recent conceptualizations of the tourist experience points towards a poststructural relational conceptualization of the wine trail/route. Acknowledging the material, cognitive, affective, and assembled characteristics of the experience and drawing on Latour's (2005) Actor-Network Theory (ANT), we forward a conceptual framework that helps explain the contributions of the wine trail.

Though ANT has been extensively used in consumer culture studies, its use in understanding the tourism experience has been limited. Tourism researchers have used ANT to highlight the agency of assembled relations that construct the tourism marketplaces (Franklin 2014;

Ren 2009; Van der Duim 2007; Van der Duim, Ren, and Thór Jóhannesson 2013), achieve sustainable business orchestrations (Dedeke 2017), and develop relations that constitute mega-events (Dawson and Jöns 2018). Very few researchers have used ANT to study the tourism experience and found it useful to highlight hidden actors and relations and interactions that impact tourist emotions and experiences (Benali and Ren 2019; Stinson and Grimwood 2019).

Actor-network theory

Actor-network theory is a relational ontological framework that prioritizes the interactions and relations between humans, nonhumans, and their environment that give rise to a social experience (Latour 2005). ANT recognizes that the relations among things, discourses, people, and technologies are all crucial for an experience to emerge (Bettany 2007; Bettany and Daly 2008; Bettany, Kerrane, and Hogg 2014; Borgerson 2005). The key objective of ANT is to move the conceptual focus away from the 'human' and the 'psychological' by highlighting the materiality and situatedness of the 'social.' This is achieved by following a flat ontological stance whereby agency is shared by human and nonhuman entities and is analyzed based on the effect a particular relation or interaction has on the overall experience.

The two key concepts of ANT which help understand the role of wine trails in the tourism experience are actor-networks and translations. For most constructivist approaches, the human has been the quintessential actor who exerts agency to make things happen. However, ANT uses the term *actant* to acknowledge the agency of things, discourses, bodies, affects, and other elements that construct experiences (Van der Duim, Ren, and Jóhannesson 2017). Actor-networks or actant-networks highlight the heterogeneous hybrid of human and nonhuman elements that interact or are assembled during a particular experience. For example, Van der Duim (2007) discusses how considering destinations as actants helps highlight their networks to consist of tourism operators, attractions, services, and infrastructure. Considering the wine trail as an actant helps highlight the different actants within its network extending beyond the wine businesses, such as the road, signage, map, vines, hotels, restaurants, and marketing agencies and collateral.

The second ANT concept that is helpful in understanding the contributions of the wine trail in tourism experiences is translation. Translation is understood as an event or delineation of a scenario which defines and situates the roles and relations of the actant-networks (Callon 1984; Giesler 2012; Latour 2005). Translation is often understood as a transformation process or a particular alignment of relations or conditions of interactions amongst actors (Martin and Schouten 2014). For example, nine translations (processes or phases) are identified by Dedeke (2017) to establish a sustainable tourism business. On the other hand, Arsel (2016) highlights the assemblage of different heterogeneous actors that are involved in the valuation of a lipstick in the particular translation of 'exchange'. Translations are helpful in identifying actant networks which facilitate the understanding of the scope of the phenomenon. For instance, in the case of the wine trail, translations help in identifying which actant networks are established and situated in tourist experiences, and in what ways.

In the following paragraphs, we conceptualize the wine trail as an actant connected to other network actants under three different translations derived from our understanding of tourism experiences: symbolic, material, and affective. These translations are stages or conditions of engagement that help trace the different facets of the experience. Figure 4.1 highlights the different translations, the role played by the wine trail in each translation, as well as the network actants involved in each translation.

Figure 4.1 Translations and actants within the wine trail assemblage.

Source: Authors

Symbolic translation – theme generator

Symbolic translation refers to the tourist experience whereby tourists symbolically relate to the theme of wine through the wine trail. Much research has established that tourists are attracted to the wine trail following its symbolic reference to the theme of wine (Demossier 2011; Getz and Brown 2006; Joy, Belk, Charters, Wang, and Peña 2018; Mason and O'Mahony 2007). Most visual representations of wine regions are in the form of wine trails which make the wine trail synonymous with wine regions in the minds of the tourists (Sigala and Robinson 2019). In the symbolic translation of the wine trail, different actants such as the website, trail map, signage, and marketing collateral, as well as physical elements such as vines and buildings, play an important role in creating the theme of wine for the tourist. The trail is printed in the marketing collateral such as on websites, brochures, and magazines, as well as the map whereby the marking of wineries resonates the theme of wine production and consumption. In addition, the trail description is put together with descriptions and images of wines and vineyards in the online and offline marketing collateral, thus triggering discourses and competencies of wine with which tourists are already familiar. The marketing collateral, including the website, map, and signage, thus extends the symbolism of the theme of wine to the wine trail. Further, the activities encountered in the wine region, such as horse riding, fine dining, and heritage accommodations, also get intermingled with the symbolism of wine when they are reflected together in the wine trail map.

In this symbolic translation, the wine trail acts as a theme generator. A theme is understood as a connecting order which is used extensively in tourism marketing to attract tourists because of its symbolism of discourses (Åstrøm 2017; Jolliffe and Piboonrungroj 2020). Themes use motifs, cues, and other symbols that help tourists cognitively relate to the tourism servicescape (Beardsworth and Bryman 1999; Dixit 2017; Lukas 2016). Timothy and Boyd (2014) emphasize the thematic interpretive function of the trail, which directs the tourism experience. The wine trail's contribution as a theme generator is emphasized through images of wines, names of wineries on the website, and other marketing collateral, along with advertisements of fine dining restaurants. Together, these elements create a thematic story of wine exploration in the region (Mora and Livat 2013; Moscardo 2020).

Material translation – movement guide

Material translation allows an understanding of the tourist experience that is shaped owing to specific material and spatial arrangements of the trail and its network actants. The wine trail road or trail network plays a crucial role in shaping the tourist space and visitation by physically directing tourists towards and through a region (MacLeod 2016). Being linear spaces themselves, the trail or road networks organize other elements around them which guide tourist movement and other tourist practices. The network actants that have an important role to play in the material translation include the road alignment, entryways, signages, parking spaces, vehicle, and building structures. Tourists are known to visit adjacent wineries following the alignment of the road (Corigliano 2016). Similarly, the trail/route map on the website and in the brochure assisted by the road signage help shape the tourist experience as they enable planning the trip and decision-making while being influenced along the way by materialities of parking and winery entryways (Miles 2016; Wagner and Minca 2016). The relation between the vehicle, driver, and the geography of the road is also known to influence tourist mobility, whereby tourists may decide to visit a particular winery based on the road condition and the competency of their vehicle (Merriman 2004; Roy and Hannam 2013). In addition, the trail also strings other places of interest together, such as nearby accommodations and restaurants, thereby triggering performances of other social practices while following the trail.

The contribution of the wine trail as a movement guide in the tourism experience is realized in the material translation. Previously, wine trails and routes have been studied as linear geographic spaces that bundle destinations/services (Timothy and Boyd 2014). However, from a poststructural lens, movement is understood as the result of the intersection of material affordances of the environment and tourist practices (Edensor 2000, 2010). Driving on the wine trail encompasses embodied practices which are shaped by the material assemblage of the vehicle, road, traffic, and driver (Merriman 2004; Thrift 2004). This is especially important in the case of wine trails or routes where the social practices are influenced as well as shaped by the materialities of other actants. For example, while on the route, tourists may stop for petrol at a service station and end up visiting the adjacent winery. Thus, the material affordances of the wine trail interact with other actants such as the vehicle, road infrastructure, signage, and map, all acting together in directing tourist movement and, in turn, experience. Thus, the contribution of the wine trail in shaping tourist experience is also reflected in its role as a movement director.

Affective translation – comforter

The affective translation is witnessed when emotions and feelings are triggered and channeled by the wine trail. The affective translation records the sense of comfort and familiarity that is communicated by the wine trail through the map and the signage. The wine trail provides a tool for tourists to explore a region while reducing decision-making and increasing a sense of familiarity and comfort, especially for tourists who may be visiting for the first time (Bagnoli 2009; Becken and Wilson 2007; Rossetto 2012; Wang 2006). This sense of comfort emerges as the tourists interact with the website, the map, and the signage while planning the trip and is reinstated when they are able to locate the buildings, names, or wine businesses as markers while traveling on the trail. In this case, the trail in the form of the map, the scenery that is encountered, the signage, and several other actants connect together for the affect of comfort to emerge. In this case, the wine trail acts as a comforter in the wine trail experience.

The role of the wine trail as a comforter is reflected in the emergence of the affect of comfort. Affect is understood as an in-betweenness between emotion and cognition, which is more sensed than cognitively deciphered (Deleuze and Guattari 1994). Affect is encountered in everyday preconscious interactions and interplays with consciousness in human decision-making (Thrift 2008). For example, the affect of comfort is sensed as tourists rely on the wine trail map to visualize and perceive the wine region (Roy and Gretzel 2021). The trail marked through colors, lines, names of the wineries, and other businesses creates a mental image of the region, which produces the affect of comfort for the traveler (Del Casino and Hanna 2000). Further, simply following the map, or the manifestation of the trail, allows tourists to get into a rhythmic flow, which further enhances their experience (Roy 2019).

Case Study 4.1 Santa Barbara Urban Wine Trail

A website exhibits most elements of the wine trail. The website of the Santa Barbara Urban Wine Trail (urbanwinetrailsb.com) serves as an example of how nonhuman actant themes move and comfort – and, thus, significantly shape – the online and offline experience of the wine trail. Specifically, the website immediately and neatly shapes the spatial perception of the trail by assigning wineries to neighborhoods. It does so on its home page by prominently displaying the neighborhood to which the winery belongs in the picture reel that dominates the page. When scrolling down, the following description appears: 'To make the Trail easier to navigate, we've identified five neighborhoods of Santa Barbara, and Goleta, Montecito, and Summerland. We've color-coded each of the wineries on the map to help identify their neighborhood.' This immediately creates a special schema of the trail in terms of neighborhoods. At the bottom of the page, five color-coded and hyperlinked descriptions of the neighborhoods invite the website user to explore each neighborhood separately. It is difficult to escape this spatial framing of the trail. Even when one clicks on the list of all wineries along the trail, the wineries are displayed using color blocks that correspond to the respective neighborhoods. The trail map itself also highlights the neighborhoods through color blocks, further encouraging a neighborhood-based navigation through the available information. It is easy to imagine that this translates into neighborhood-based consumption of the trail, given that the website stresses which wineries belong spatially together rather than highlighting how to combine wineries across neighborhoods. The latter would encourage much more spatially diverse movements. The only element that encourages movement across neighborhoods is the 'Tasting Card,' a membership-based promotion that offers individuals free wine tastings at every winery, and promotes discounts and priority access for special events.

The website also projects themes, not only by describing the neighborhoods in specific ways, but by its design and logo. The logo highlights the word 'urban' and displays the outlines of a wine bottle and a glass in funky shapes. The neighborhoods are represented through vivid colors. The page that displays the individual wineries uses so-called tiles to structure the text entries, which are combined with square pictures. Together, this creates a look that resembles that of Instagram and communicates a very young and vibrant experience.

Last, the website comforts in two specific ways. First, it offers an FAQ section that seeks to answer frequent questions in relation to opening hours, transportation needs, and other travel planning requirements. The website, therefore, anticipates potential points of friction and offers assistance. Second, it provides users with a tab called 'Lingo' that explains important wine-related terms. By doing so, it equips potential wine trail visitors with knowledge that will facilitate their experience of the wine trail.

It is easy to recognize how the website not only promotes but actively shapes the kind of experience a potential visitor might have on the trail. As a likely entry point into the experience, the website frames expectations and affect. Accordingly, it needs to be seen as an important actant in the wider wine trail network rather than as just a simple marketing tool.

Conclusion

While wine trails have been revered as socio-economic tools for creating collaborative networks and regional development, this research studied their importance in shaping the tourist experience. The experience was conceptualized as affective, symbolic, and material at the same time, and a poststructuralist relational ontological lens of ANT was applied to study the contributions of the wine trail to this experience. The proposed conceptual framework articulated the wine trail as an actant within an assembled and situated network of other actants that come together under different translations making the trail agentic as a theme generator, movement director, and comforter. This framework helps identify which actants support, as well as which actants undermine the overall agency of the wine trail, in turn facilitating or impeding its specific symbolic, material, and affective roles. As such, it supports recent research on wine trails that emphasize their development not as homogeneous but as heterogeneous assemblages by presenting an experiential perspective (Bregoli et al. 2016). This conceptualization is significant, as it extends the role of the wine trail beyond its use as a collaboration tool and points towards its contributions to the larger gastronomic service ecosystem that operates in the wine region (Roy et al. 2019).

The presented conceptual framework facilitates future research on the role of wine trails in creating and shaping the tourism experience by highlighting the importance of symbolic, material, and affective dimensions and their related actants. Effectiveness of marketing materials, for example, then needs to be established in terms of their ability to connect, move, and comfort, rather than being focused on their persuasiveness. When investigating the experiences on wine trails, it becomes important to realize that innovative approaches are needed to capture experiential dimensions beyond the obvious, cognitively accessible factors. Photo-elicitation studies, the use of drawings, and mobile ethnographies, for example, can be used to identify the role of various actants and the intricate translations that take place.

From a practical point of view, conceptualizing wine trails as complex assemblages with human and nonhuman actants and as multiple translations through which experiences emerge requires new and more systematic approaches toward wine trail design (Fesenmaier and Xiang 2017). In general, the ANT perspective enables tourism practitioners to articulate what is often only intuitively known. This allows for more careful management of specific relations among actants and more explicit facilitation of desirable translations.

References

Arsel, Z. (2016). Assembling markets and value. In Canniford, R., and Bajde, D. (eds.), *Assembling Consumption: Researching Actors, Networks and Markets* (pp. 32–41). Oxon: Routledge.

Åstrøm, J. K. (2017). Theme factors that drive the tourist customer experience. *International Journal of Culture, Tourism and Hospitality Research*, 11(2): 125–141. https://doi.org/10.1108/IJCTHR-07-2015-0070

Bærenholdt, J. O., Haldrup, M., and Urry, J. (2017). *Performing Tourist Places*. London: Taylor & Francis Group.

Bagnoli, A. (2009). Beyond the standard interview: the use of graphic elicitation and arts-based methods. *Qualitative Research*, 9(5): 547–570. https://doi.org/10.1177/1468794109343625

Beardsworth, A., and Bryman, A. (1999). Late modernity and the dynamics of quasification: the case of the themed restaurant. *The Sociological Review*, 47(2): 228–257.

Becken, S., and Wilson, J. (2007). Trip planning and decision making of self-drive tourists. *Journal of Travel and Tourism Marketing*, 20(3–4): 47–62. https://doi.org/10.1300/J073v20n03_04

Benali, A., and Ren, C. (2019). Lice work: non-human trajectories in volunteer tourism. *Tourist Studies*, 19(2): 238–257. https://doi.org/10.1177/1468797619832311

Bettany, S. (2007). The material semiotics of consumption or where (and what) are the objects in consumer culture theory? In Belk, R. W., and Sherry, J. F. (eds.), *Consumer Culture Theory (Research in Consumer Behavior, Vol. 11)* (pp. 41–56), Bingley: Emerald Group Publishing Limited. https://doi.org/10.1016/S0885-2111(06)11003-0

Bettany, S., and Daly, R. (2008). Figuring companion-species consumption: A multi-site ethnography of the post-canine Afghan hound. *Journal of Business Research*, 61(5): 408–418.

Bettany, S. M., Kerrane, B., and Hogg, M. K. (2014). The material-semiotics of fatherhood: the co-emergence of technology and contemporary fatherhood. *Journal of Business Research*, 67(7): 1544–1551. https://doi.org/10.1016/j.jbusres.2014.01.012

Borgerson, J. L. (2005). *Materiality, Agency, and the Constitution of Consuming Subjects: Insights for Consumer Research*. Paper presented at the *Advances in Consumer Research*, Minnesota.

Boyne, S., Williams, F., and Hall, D. (2002). On the trail of regional success: tourism, food production and the Isle of Arran Taste Trail. In Hjalager, A.-M., and Richards, G. (eds.), *Tourism and Gastronomy* (pp. 91–114). London: Routledge.

Bregoli, I., Hingley, M., Del Chiappa, G., and Sodano, V. (2016). Challenges in Italian wine routes: managing stakeholder networks. *Qualitative Market Research*, 19(2): 204–224. https://doi.org/10.1108/QMR-02-2016-0008

Bruwer, J. (2003). South African wine routes: some perspectives on the wine tourism industry's structural dimensions and wine tourism product. *Tourism Management*, 24(4): 423–435. https://doi.org/10.1016/S0261-5177(02)00105-X

Buda, D. M., d'Hauteserre, A.-M., and Johnston, L. (2014). Feeling and tourism studies. *Annals of Tourism Research*, 46: 102–114. https://doi.org/10.1016/j.annals.2014.03.005

Callon, M. (1984). Some elements of a sociology of translation: domestication of the scallops and the fishermen of St Brieuc Bay. *The Sociological Review*, 32(1_suppl): 196–233. https://doi.org/10.1111/j.1467-954X.1984.tb00113.x

Canniford, R., and Bajde, D. (2015). *Assembling Consumption: Researching Actors, Networks and Markets*. Oxon: Routledge.

Cohen, E. (1979). A phenomenology of tourist experiences. *Sociology*, 13(2): 179–201. https://doi.org/10.1177/003803857901300203

Conway, D. (2013). Slow tourism: experiences and mobilities. *Annals of Tourism Research*, 40(0): 444–446. https://dx.doi.org/10.1016/j.annals.2012.11.002

Corigliano, M. A. (2016). Wine routes and territorial events as enhancers of tourism experiences. In Marta, P.-O., Rama, D. R., de la Cruz, M., and Carlos, R.-A. (eds.), *Wine and Tourism* (pp. 41–56). Cham: Springer.

Cutler, S. Q., Carmichael, B., and Doherty, S. (2014). The Inca Trail experience: does the journey matter? *Annals of Tourism Research*, 45: 152–166. https://doi.org/10.1016/j.annals.2013.12.016

Cutler, S. Q., Carmichael, B. A., and Morgan, M. (2010). The dimensions of the tourist experience. In Morgan, M., Lugosi, P., and Ritchie, J. R. B. (Eds.), *The Tourism and Leisure Experience: Consumer and Managerial Perspectives* (pp. 3–26). Bristol: Channel View Publications

Dawson, J., and Jöns, H. (2018). Unravelling legacy: a triadic actor-network theory approach to understanding the outcomes of mega events. *Journal of Sport and Tourism*, 22(1): 43–65. https://doi.org/10.1080/14775085.2018.1432409

Dedeke, A. N. (2017). Creating sustainable tourism ventures in protected areas: an actor-network theory analysis. *Tourism Management*, 61: 161–172.

Del Casino, V. J., and Hanna, S. P. (2000). Representations and identities in tourism map spaces. *Progress in Human Geography*, 24(1): 23–46. https://doi.org/10.1191/030913200673388638

Deleuze, G., and Guattari, F. (1994). *What Is Philosophy?* New York: Columbia University Press.

Demossier, M. (2011). Beyond *terroir*: territorial construction, hegemonic discourses, and French wine culture. *Journal of the Royal Anthropological Institute*, 17(4): 685–705. https://doi.org/10.1111/j.1467-9655.2011.01714.x

Dixit, S. K. (ed.). (2017). *The Routledge Handbook of Consumer Behaviour in Hospitality and Tourism*. London: Routledge.

Edensor, T. (2000). Walking in the British countryside: reflexivity, embodied practices and ways to escape. *Body and Society*, 6(3–4): 81–106. https://doi.org/10.1177/1357034x00006003005

Edensor, T. (2010). Walking in rhythms: place, regulation, style and the flow of experience. *Visual Studies*, 25(1): 69–79. https://doi.org/10.1080/14725861003606902

Fesenmaier, D. R., and Xiang, Z. (2017). *Design Science in Tourism*. Cham: Springer.

Franklin, A. (2014). On why we dig the beach: tracing the subjects and objects of the bucket and spade for a relational materialist theory of the beach. *Tourist Studies*, 14(3): 261–285. https://doi.org/10.1177/1468797614536331

Getz, D., and Brown, G. (2006). Critical success factors for wine tourism regions: a demand analysis. *Tourism Management*, 27(1): 146–158.

Giesler, M. (2012). How doppelgänger brand images influence the market creation process: longitudinal insights from the rise of Botox cosmetic. *Journal of Marketing*, 76(6): 55–68.

Hall, C. M., Cambourne, B., Macionis, N., and Johnson, G. (1997). Wine tourism and network development in Australia and New Zealand: review, establishment and prospects. *International Journal of Wine Marketing*, 9(2): 5–31.

Hall, C. M., and Gössling, S. (2016). From food tourism and regional development to food, tourism and regional development: themes and issues in contemporary foodscapes. In Hall, C. M., and Gössling, S. (eds.), *Food Tourism and Regional Development: Networks, Products and Trajectories* (pp. 3–59). London: Routledge.

Higgins, L., and Hamilton, K. (2020). Pilgrimage, material objects and spontaneous communitas. *Annals of Tourism Research*, 81: 102855. https://doi.org/10.1016/j.annals.2019.102855

Hoffman, D. L., and Novak, T. P. (2018). Consumer and object experience in the internet of things: an assemblage theory approach. *Journal of Consumer Research*, 44(6): 1178–1204. https://doi.org/10.1093/jcr/ucx105

Hojman, D. E., and Hunter-Jones, P. (2012). Wine tourism: Chilean wine regions and routes. *Journal of Business Research*, 65(1): 13–21. https://doi.org/10.1016/j.jbusres.2011.07.009

Holmqvist, J., Diaz Ruiz, C., and Peñaloza, L. (2020). Moments of luxury: hedonic escapism as a luxury experience. *Journal of Business Research*, 116: 503–513. https://doi.org/10.1016/j.jbusres.2019.10.015

Husemann, K. C., and Eckhardt, G. M. (2018). Consumer deceleration. *Journal of Consumer Research*, 45(6): 1142–1163. https://doi.org/10.1093/jcr/ucy047

Jepson, D., and Sharpley, R. (2015). More than sense of place? Exploring the emotional dimension of rural tourism experiences. *Journal of Sustainable Tourism*, 23(8–9): 1157–1178. https://doi.org/10.1080/09669582.2014.953543

Jolliffe, L., and Piboonrungroj, P. (2020). The role of themes and stories in tourism experiences. In Dixit, S. K. (ed.), *The Routledge Handbook of Tourism Experience Management and Marketing*, (pp. 218–228). Oxon: Routledge.

Joy, A., Belk, R. W., Charters, S., Wang, J. J. F., and Peña, C. (2018). Performance theory and consumer engagement: Wine-tourism experiences in South Africa and India. In Cross, S. N. N., Ruvalcaba, C., Venkatesh, A., and Belk, R. W. (eds.), *Consumer Culture Theory* (pp. 163–187). Bingley: Emerald Publishing Limited.

Knobloch, U., Robertson, K., and Aitken, R. (2017). Experience, emotion, and eudaimonia: a consideration of tourist experiences and well-being. *Journal of Travel Research*, 56(5): 651–662. https://doi.org/10.1177/0047287516650937

Latour, B. (2005). *Reassembling the Social: An Introduction to Actor-Network-Theory*. Oxford: Oxford University Press.

Lewis, G. K., Byrom, J., and Grimmer, M. (2015). Collaborative marketing in a premium wine region: the role of horizontal networks. *International Journal of Wine and Business Research*, 27(3): 203–219. https://doi.org/10.1108/IJWBR-06-2014-0028

López-Guzmán, T., Vieira-Rodríguez, A., and Rodríguez-García, J. (2014). Profile and motivations of European tourists on the Sherry wine route of Spain. *Tourism Management Perspectives*, 11(Supplement C): 63–68. https://doi.org/10.1016/j.tmp.2014.04.003

Lukas, S. A. (2016). *A Reader in Themed and Immersive Spaces*. Morrisville, NC: Lulu Press.

MacLeod, N. (2016). The role of trails in the creation of tourist space. *Journal of Heritage Tourism*, 12(5): 1–8. https://doi.org/10.1080/1743873X.2016.1242590

Mannell, R. C., and Iso-Ahola, S. E. (1987). Psychological nature of leisure and tourism experience. *Annals of Tourism Research*, 14(3): 314–331.

Martin, D. M., and Schouten, J. W. (2014). Consumption-driven market emergence. *Journal of Consumer Research*, 40(5): 855–870. https://doi.org/10.1086/673196

Mason, R., and O'Mahony, B. (2007). On the trail of food and wine: the tourist search for meaningful experience. *Annals of Leisure Research*, 10(3–4): 498–517. https://doi.org/10.1080/11745398.2007.9686778

Mattila, A. S., and Gao, L. Y. (2017). Atmospherics and the touristic experience. In Fesenmaier, D. R., and Xiang, Z. (eds.), *Design Science in Tourism* (pp. 151–160). Cham: Springer.

Merriman, P. (2004). Driving places: Marc Augé, non-places, and the geographies of England's M1 motorway. *Theory, Culture and Society*, 21(4–5): 145–167. https://doi.org/10.1177/0263276404046065

Miles, S. (2016). Remembrance trails of the great war on the Western Front: routes of heritage and memory. *Journal of Heritage Tourism*, 12(5): 441–451. https://doi.org/10.1080/1743873X.2016.1242589

Mora, P., and Livat, F. (2013). Does storytelling add value to fine Bordeaux wines? *Wine Economics and Policy*, 2(1): 3–10. https://doi.org/10.1016/j.wep.2013.01.001

Moscardo, G. (2017). Stories as a tourist experience design tool. In Fesenmaier, D. R., and Xiang, Z. (eds.), *Design Science in Tourism* (pp. 97–124). Cham: Springer.

Moscardo, G. (2020). Stories and design in tourism. *Annals of Tourism Research*, 83: 102950. https://doi.org/10.1016/j.annals.2020.102950

Naipaul, S., Wang, Y., and Okumus, F. (2009). Regional destination marketing: a collaborative approach. *Journal of Travel and Tourism Marketing*, 26(5–6): 462–481. https://doi.org/10.1080/10548400903162998

Rakić, T., and Chambers, D. (2012). Rethinking the consumption of places. *Annals of Tourism Research*, 39(3): 1612–1633. https://doi.org/10.1016/j.annals.2011.12.003

Ren, C. (2009). Constructing the tourist destination: a socio-material description. *European Journal of Tourism Research*, 2(1): 134.

Rossetto, T. (2012). Embodying the map: tourism practices in Berlin. *Tourist Studies*, 12(1): 28–51. https://doi.org/10.1177/1468797612444192

Roy, N. (2019). *Following the Theme: Conceptualizing Value as a More-than-Human Achievement* [Doctor of Philosophy]. University of Wollongong, Wollongong.

Roy, N., and Gretzel, U. (2021). Feeling opulent: adding an affective dimension to symbolic consumption of themes. *Tourism Geographies*, 1–23. https://doi.org/10.1080/14616688.2020.1867885

Roy, N., Gretzel, U., Waitt, G., and Yanamandram, V. (2019). Gastronomic trails as service ecosystem. In Dixit, S. K. (ed.), *The Routledge Handbook of Gastronomic Tourism* (pp. 179–188). Oxon: Routledge.

Roy, S., and Hannam, K. (2013). Embodying the mobilities of the Darjeeling Himalayan railway. *Mobilities*, 8(4): 580–594.

Schmitt, B. (2020). Sensuality and experience. In Morhart, F., Wilcox, K., and Czellar, S. (eds.), *Research Handbook on Luxury Branding* (pp. 208–223). Cheltenham, UK: Edward Elgar Publishing.

Sigala, M., and Robinson, R. (2019). *Wine Tourism Destination Management and Marketing*. Cham: Springer.

Stinson, M. J., and Grimwood, B. S. R. (2019). On actor-network theory and anxiety in tourism research. *Annals of Tourism Research*, 77: 141–143. https://doi.org/10.1016/j.annals.2018.12.003

Thrift, N. (2004). Driving in the city. *Theory, Culture and Society*, 21(4–5): 41–59.

Thrift, N. (2008). *Non-Representational Theory: Space, Politics, Affect*. London: Routledge.

Timothy, D. J., and Boyd, S. W. (2014). *Tourism and Trails: Cultural, Ecological and Management Issues* (Vol. 64): Bristol: Channel View Publications.

Van der Duim, R. (2007). Tourismscapes: an actor-network perspective. *Annals of Tourism Research*, 34(4): 961–976. https://doi.org/10.1016/j.annals.2007.05.008

Van der Duim, R., Ren, C., and Jóhannesson, G. T. (2017). ANT: A decade of interfering with tourism. *Annals of Tourism Research*, 64: 139–149. https://doi.org/10.1016/j.annals.2017.03.006

Van der Duim, R., Ren, C., and Thór Jóhannesson, G. (2013). Ordering, materiality, and multiplicity: enacting Actor–Network Theory in tourism. *Tourist Studies*, 13(1): 3–20. https://doi.org/10.1177/1468797613476397

Vo Thanh, T., and Kirova, V. (2018). Wine tourism experience: a netnography study. *Journal of Business Research*, 83: 30–37. https://doi.org/10.1016/j.jbusres.2017.10.008

Wagner, L., and Minca, C. (2016). Topographies of the Kasbah Route: Hardening of a heritage trail. *Tourist Studies*, 16(3): 296–322. https://doi.org/10.1177/1468797615618307

Wang, N. (2006). Itineraries and the tourist experience. In Minca, C., and Oakes, T. (eds.), *Travels in Paradox: Remapping Tourism* (pp. 65–76). Oxford: Rowman and Littlefield Publishers.

Xu, S., Barbieri, C., and Seekamp, E. (2020). Social capital along wine trails: spilling the wine to residents? *Sustainability*, 12(4): 1592.

5

REGULATORY AND ETHICAL INFLUENCES AND PREDICAMENTS OF WINE TOURISM DEVELOPMENT

A case from Nashik, India

Anupama S. Kotur (Kaddi)

Introduction

A mature wine industry is imperative to the growth and development of wine tourism. The development of the wine industry and the subsequent growth of wine tourism commenced in India over three decades ago, and Nashik in Maharashtra emerged as the premier winescape in India. Nashik, known as the 'wine bowl' of India, accounts for close to 90% of wine production in the country. Nashik is home to a number of wineries that have put Indian wines on the global map. While India's wine story, slowly and steadily, continues to gain momentum worldwide, the quantum of growth leaves much to be desired. The Indian wine and wine tourism market is a vast yet largely untapped one (Gore 2008) due to the rather complex regulatory environment in which the industry operates. Nashik is also home to a number of sites of religious importance and therefore occupies a prominent place among the country's religious tourist destinations.

India shares a rather complex relationship with alcohol consumption and religion. This has raised some questions pertaining to ethical predicaments in the promotion of wine tourism in Nashik, Maharashtra, given its religious identity. The author, through this qualitative investigation, attempts to (a) review the existing policies imposed on the wine industry in Nashik and its impact on wine tourism development; and (b) explore if the ethical predicaments have any influence on policymaking with respect to the wine industry in Nashik. The tourism industry thrives within a complex environment encompassing sociocultural, economic, ecological, and legal factors. The unique socioreligious and sociocultural fabric of destinations may present complexities in the development of the tourism industry. Through this explorative study, the author sets out to investigate the interplay between regulatory and ethical factors on wine tourism policymaking.

 DOI: 10.4324/9781003143628-7

Review of the literature

The need for a regulatory system in wine tourism

The regulatory environment of the wine region plays a pivotal role in the growth and development of wine tourism. The success of any wine region depends largely on the effective implementation of 'regional policies that sustain landscapes conducive to producing top-quality wines and attracting appropriate numbers of visitors' (Williams, Graham, and Mathias 2006). Wine, being categorized as an alcoholic beverage, has been subject to a number of government regulations worldwide. However, the nature of the regulations tends to vary significantly within and between countries (Meloni et al. 2019). These regulations are usually in the form of policies on quality standards, licenses, permits, taxes, tariffs, and restrictions on the production of wines (Anderson 2020; Anderson, Meloni, and Swinnen 2018). Regulations may also be applicable for the sale and distribution of wines in some cases.

Wine tourism regions often use distinct imagery as part of their product narrative and marketing efforts in order to convey the attractiveness of the wine destination. Wine tourism regions gain their unique appeal due to distinct and complex socio-ecological systems (Alebaki and Ioannides 2018). The ecological distinctiveness of wine regions is reflected in the unique sensory attributes (Van Leeuwen and Seguin 2006) of the wines produced there. Appellations of wine regions legally define and regulate the type of grapes grown in the region and production methods adopted in winemaking. Appellations, to a large extent, have protected the specificity and reputation of the wines (European Federation of Origin Wines 2021). The rules governing appellations vary from one country to another. Good-quality wines are central to any wine tourism development and in ensuring a positive tourist experience. Well-defined wine quality control regulations can be instrumental in averting any adverse impacts of poor-quality wines on the overall tourist experience (Skinner 2000; Williams, Graham, and Mathias 2006).

Wine tourism is often seen as complementary to the wine business by wineries. Wine regions look to generate greater economic benefits through wine tourism activities. The economic benefits are in the form of employment creation (Tafel and Szolnoki 2020), rural and community development (Montella 2017; Poitras and Donald 2006; Trišić et al. 2020), tax revenue for the regional government (Gál-Czékus and Nagy-Kovács 2017; Hall et al. 2009) and overall sustainable development of the region (Poitras and Donald 2006). However, the extent of development and the benefits such development is able to impart to the region depends largely on the policy guiding wine tourism development (Williams, Graham, and Mathias 2006).

Further, wine tourism development may also have an adverse impact on the destination manifesting in the form of an influx of a large number of tourists, thereby increasing the pressure on local infrastructure, conflict between tourists and local communities, unregulated increase in the land price, and marginalization of local communities (Skinner 2000), among others. Often, these are caused by poorly defined wine tourism regulations, resulting in low tourist satisfaction and impairing the long-term sustainability of the wine tourism destination. A balanced approach to wine tourism development governed by a sound policy framework may minimize the adverse impact of tourism development in wine regions.

Ethical considerations and wine tourism

The question of ethics in tourism has been much discussed and debated among academicians in recent times. Tourism operates within the social and cultural settings of a place in order to generate economic value through the tourism offerings. This complex equation has invited academic discussion in the spheres of relations between ethical, economic, and aesthetic values in the context of modern tourism (Duffy and Smith 2003; Fennell 2006); public health, human rights regulation, and environmental protection (Donyadide 2010); and various other 'paradoxes which occur with its development and the role that ethics can play in the marketing of the tourism product' (Wheeler 1995).

In many countries, alcohol shares a rather complex relationship with social and cultural values due to the externalities associated with alcohol consumption (Baxter 2019). For instance, India, a multireligious and multicultural country with firm roots in its traditions and values, maintains mixed beliefs about alcohol consumption as an acceptable practice. Religious and ethical values discourage alcohol consumption in some sections of society (Viswani, Ambekar, and Ramadeep 2009). Therefore, some of the advanced economies use taxes and other regulations as a tool to 'offset negative externalities that alcoholic drinking imposes on society' (Griswold et al. 2018). Well-targeted wine tourism 'policy regime(s) guiding and managing the development' (Williams, Graham, and Mathias 2006) of wine tourism in wine regions may be able to avert any conflicts arising out of sociocultural complexities associated with alcohol consumption.

As research in wine tourism gains prominence among academicians, practitioners, and policymakers, much of the research focuses on wine tourism product development, experience creation, and wine tourism infrastructure development. Very few studies have focused on policy development, the regulatory environment of wine tourism development, and the role played by ethical concerns in policy development; and fewer so in the Indian context. Through this qualitative investigation, the author attempts to (a) review the existing policies and regulations imposed on the wine industry and wine tourism operations in Nashik, Maharashtra; and (b) explore if the ethical predicaments have any influence on the policymaking with respect to the wine industry in Nashik. The findings of this study may be useful for academicians, practitioners, and policymakers in reviewing wine tourism development policies through redefined interlinkages.

Methodology

This study is qualitative in nature and uses methodological triangulation for the purpose of data collection in order to enhance the validity and credibility of the research findings. The study uses a combination of both primary as well as secondary data sources due to the nature of the study. A combination of a systematic review of published resources and in-depth interviews with various stakeholders allowed the author to gain a deeper and comprehensive understanding of the phenomena. The study uses the grounded theory approach, which is usually employed in building 'theoretical explanation by specifying phenomena' (Corbin and Strauss 1990). The research used theoretical saturation or data saturation technique due to its usability in qualitative inquiry (Vasileiou et al. 2018). The primary data, in the form of in-depth interviews, were collected in 2019 and 2020. Through this research approach, the author intended to analyze information-rich narratives from multiple stakeholders, such as winery owners/managers, local tourism promotion boards, and wine industry experts, that resulted in a systematic and in-depth understanding of the area of research. Table 5.1 lists the participants of the in-depth interview. The author conducted the interviews using multiple means, including personal visits, telephonic, and email responses.

Table 5.1 List of interviewees

No.	*Organization*	*Designation*
Tourism Promotion Boards		
TB01	Shri Devasthan (Temple) Trust	Public Relations Officer
TB02	Maharashtra Tourism Development Corporation (MTDC), Government of Maharashtra	Regional Manager
TB03	All India Wine Producers' Association (AIWPA)	President
Wineries		
WO01	Soma Wine Village	Operations Manager
WO02	York Winery	Operations Manager
WO03	Vintage Wines Private Limited	Founder and Managing Director
Wine Industry and Wine Tourism Experts		
WE01	Independent	Sommelier and Wine Expert
WE02	Independent	Wine Maker

Source: Author

The inductive approach enabled the author to search and identify any underlying themes and patterns of meaning based on the data (Wilson and Little 2005). The research used the theoretical saturation or the data saturation technique. Saturation has emerged as the 'gold standard' in qualitative inquiry' (Vasileiou, Barnett, Thorpe, and Young 2018). The data saturation approach is based on the premise that when gathering fresh data no longer presents new theoretical insights nor reveals newer properties of the core theoretical categories (Charmaz 2006), the author collects no fresh responses.

A review of wine regulations in Maharashtra, India

The Indian wine story

The Indian wine industry and its development are almost synonymous with that of Maharashtra. Being the pioneer and front runner in the wine industry in India, Maharashtra State continues to lead in wine tourism development. India's commercial wine industry has witnessed a steady growth since the early 2000s (Kaddi 2013). Nashik, Maharashtra, has the largest number of wineries, with close to over 90% of the country's wines coming from this region (Sood 2019). Other important wine-producing states in India include Karnataka, Goa, and Himachal Pradesh, which together constitute 10% of the country's wine production. Nashik, over the years, has evolved as a wine tourism destination with a number of wine tourism attractions such as vineyard resorts, vineyard restaurants, wine festivals, wine tours, and various vineyard-based activities. Wines from Nashik are gaining prominence and popularity the world over with their unique varietals. Notable wine brands from Nashik include Sula, Zampa-Grover, Soma, York, Chandon, and Vallonne, among others (Csizmadia-Honigh 2015). With a young urban population that positions wine as a status choice, India has experienced a growing acceptance of wines among the upper middle classes. In 2018, the domestic wine industry had an estimated annual turnover of Rupees 600 crores (6 billion) with an average annual production of nearly 2 crores (20 million) liters and consumption of 1.5 crores (15 million) liters (Kasabe 2019). For 2019, the turnover increased to Rupees 700 crores (7 billion);

however, 2020 saw a dip of close to Rupees 150 crores due to disruption caused by the outbreak of the COVID-19 pandemic (Bhardwaj 2020). Wineries in Nashik have also expanded their export portfolio to major countries like the Netherlands, Hong Kong, Singapore, and Australia, among others. In 2018, India exported US$8,907,911 worth of wines to various countries (Sood, 2019). Given the steady growth of the wine industry, wine tourism in India is but a natural beneficiary, and Nashik has become synonymous with wine tourism among wine aficionados in India. The growing demand for experiential travel among Indian travelers can be seen as one of the key factors for the development of wine tourism in India (Kaddi 2013). In 2019, close to 55,000 wine tourists visited wineries in Nashik and Sula vineyards alone, recording a visitation of 35,000 wine tourists in December (Kasabe 2020). Nashik's SulaFest, one of the largest annual wine festivals in the country, recorded 40,000 visitors in 2019 (Sommelier India 2020).

Wine policy of Maharashtra

Regulations pertaining to alcoholic beverages – taxation and legislation – are controlled by the states in India, and therefore one sees variation in policies across India. While some states lean towards complete prohibition, others have been progressive in formulating wine-friendly policies in India. To give impetus to the growth of the wine industry in the state, the Government of Maharashtra implemented the Maharashtra Grape Processing Industrial Policy in 2001. This policy was essentially aimed at sustaining viticulture and wine production in the state. Some of the salient features that prevail to date were as follows:

a Easy and preferential credit facility through National Bank for Agriculture and Rural Development (NABARD).
b Declaration of the wine industry as a small scale industry.
c Concession in duties and taxes.
d Simplification of the licensing system and permission for wine production and sales.
e For the wineries, essential licenses for land, electricity supply, and infrastructure were made available under a single-window clearance system.
f Wine production units were accorded the status of food processing units in order to avail other benefits.
g Tourists were permitted to visit wine production units for wine tasting. These wine production units were also given licenses to sell wine on a retail basis in order to encourage cellar door sales.

Wine industry regulations in Maharashtra

State regulations on alcoholic beverages pertaining to control over selling, distribution, and pricing of the beverages are guided by Article 47 of the Constitution of India's Directive Principles. In India, the central government does not exercise direct control over the states' regulations over the production and sale of alcoholic beverages; therefore, this industry is one of the important tax revenue earners for the states. Every state and union territory has been authorized to formulate and implement its own set of rules and regulations for alcohol beverage control. This practice has led to a wide range of variations in wine regulations in India. States like Maharashtra, Karnataka, and Goa have adopted a liberal attitude towards the wine industry, while stringent rules prevail in most other states. States like Gujarat, Lakshadweep, Mizoram, Nagaland, and Manipur completely prohibit alcoholic beverages, including wine. Owing to

Table 5.2 Wine production, sale, and distribution related regulations for the state of Maharashtra

Winery activities and associated investments	*Policies and regulations (as of FY 2021)*
Setting up of wine production units, modernization and expansion of wine units	35% of the eligible project cost subject to a maximum of ₹5.00 crore in General Areas (MOFPI, 2019–2020)
Licensing for wineries	Licenses for the production of wines are granted under Maharashtra Manufacture of Beer and Wine Rules, 1966, through a simplified application process.
Foreign direct investment (FDI)	Up to 100% is allowed under the automatic route for brewing and distillation of liquor subject to the license by the appropriate authority.
Wine production	a Goods and services tax (GST) is not applicable to wines. Instead, a 20% value added tax (VAT) is levied by the state, and wineries can avail a reimbursement of 16% of the payment towards VAT upon filing tax returns. b The excise duty on wine manufactured from grapes produced in the state remitted 100% up to 2021.
Label registration fees	Waived fully
Cellar door sales	Regulated by State Excise Department. Wineries are permitted to sell wines on the premises through a simple application process
Sale and distribution outside Maharashtra	For sale of wines outside the state of Maharashtra, there is an export fee that the wineries have to pay, which differs from state to state. For instance, sales of wine in Karnataka are charged Rupees 300/- per liter in export fees, while sales in Delhi are charged only Rupees 5/- per liter.
Safety standards	a License required under the Food Safety and Standard Act 2006. b BIS Standards Certificate issued by Bureau of Indian Standards.

Source: Author

the proactive and progressive policies adopted by the state of Maharashtra, Nashik emerged as a leading example of wine industry development in the country. Table 5.2 illustrates the various wine regulations for the state of Maharashtra.

The Government of India's broadcast policy, which regulates advertising through mainstream media, strictly prohibits direct and surrogate (event sponsorship) advertising and any other promotion of consumption of alcoholic beverages, including wine. Therefore, wineries in the state resort to on-premises advertising in the vineyards, promotions through wine tastings, sponsorships, and point-of-sale promotions. Wine festivals have also emerged as a popular avenue for advertising wine brands.

Other wine industry promotion initiatives

There are a number of measures undertaken by Maharashtra State in order to promote wine industry in general and wine tourism specifically. Some of the notable initiatives are as follows:

a Under a joint initiative by the Government of India through the Agricultural and Processed Food Products Export Development Authority (APEDA) and the Government of Maharashtra through the Maharashtra Industrial Development Corporation (MIDC) for development of an Agri Export Zone (AEZ) to give impetus to the growth of the wine industry.

b Under the same agreement, the first and one of the successful wine parks was developed by MIDC called the Godavari Wine Park at Vinchur in Nashik. Some of the more notable wineries are situated within this wine park.

Wine tourism in Nashik and the ethical dimension

Religion and alcohol consumption in India – ideological underpinning

Ethical underpinnings in India are greatly shaped by the religious beliefs that have evolved in the country over the centuries. Belief systems, moral values, and the way of life in general have absorbed in themselves the best and otherwise from the various religions practiced. Nashik, much like the rest of the country, is a multireligious and multicultural society. Nashik has maintained mixed beliefs about alcohol consumption as an acceptable individual and social practice with its firm cultural and traditional roots. Hinduism is widely practiced in India and is usually considered by many as a far more diverse religion (Heath 2012). Some sections of Hindus do not consider alcohol consumption as a 'normal part of a social discourse or eating behavior, and there is a virtual absence of a normative pattern of drinking alcohol and attitude towards it' (Viswani, Ambekar, and Ramadeep 2009, p. 124) while some other sections consider it as an acceptable practice. Islam, one of the other widely followed religions in India, practices complete prohibition of alcohol consumption (Egan 2002). Mahatma Gandhi, a much revered and popular freedom fighter and social reformer of India, also advocated alcohol prohibition (Blocker, Fahey, and Tyrrell 2003). This complex yet interesting relationship that Indian society shares with alcohol consumption is perhaps underexplored in academic debates and researches. India as a country is highly nuanced with a multitude of complex attitudes shaped by its numerous beliefs, and it is rather challenging to present a definitive view on the acceptability or nonacceptability of alcohol.

Religious tourism in Nashik

Religious tourism comprises of journeys undertaken in search of truth and what is sacred or holy (Olsen and Timothy 2006, p. 3), and where the primary motivation of visiting holy sites and places of religious importance are often combined with visits to neighboring tourist attractions and vice versa (Vukonić 2002, in Simone-Charteris and Boyd 2011, p. 460). Nashik is one of the important religious tourism destinations for Hindus. Situated on the banks of the river Godavari (known as the Ganges of the South) (Babu 2008), Nashik finds mention in Indian mythological books such as the *Ramayana* (Eck 2012) and other ancient works of Indian literature, including some belonging to the pre-Christian era. Nashik is one of the four Hindu holy cities of India (others being Haridwar, Prayag, and Ujjain) that host the *Kumbh Mela*, a monthlong religious festival where pilgrims gather once every 12 years to take a holy dip in the rivers flowing through these cities (Bhardwaj 1983; Pletcher 2010; Yasuda, Raj, and Griffin 2018). It was estimated that close to 10 million pilgrims participated in *Kumbh Mela* held in 2015 (DNA India 2015). *Trimbakeshwar Jyotirlinga*, one of the 12 holy shrines dedicated to the Hindu deity *Shiva*, is situated in Nashik. *Trimbakeshwar* temple is one of the holiest sites for Hindus (Matsuo 2015; Pletcher 2010), attracting millions of devotees every year.

It is an interesting paradigm where two forms of tourism – religious and wine – are juxtaposed, making it a richly complex yet unique tourist destination where the contrast is but evident owing to a complex interrelationship between religion and alcohol consumption in India. In order to understand this interplay, relationship, and its influence on wine tourism policymaking better with that of the religious identity of Nashik, the author conducted in-depth interviews with important wine tourism stakeholders as well as experts. Using the inductive, grounded approach to analyze the data generated from interviews, the responses were largely confined to the following discussed factors:

a *Sociocultural acceptability of wines*: Societal values may be considered largely as a reflection of cultural and historic factors that shape their beliefs and attitudes. The highly nuanced fabric of society is made up of its people and their culture. Questions pertaining to typical travel itineraries and patterns of tourists in Nashik, respondents (WO01, WO02, WO03, TB02, TB03) observed that many wine tourists visiting vineyards expressed little or no inhibition or reservation in combining wine tours with visits to religious sites. So much so that some wineries actively included visits to the *Trimbakeshwar* temple as a part of their 'must visit' list for visiting wine tourists. The Ministry of Tourism, in association with the Maharashtra Tourism Development Board (MTDC), has been proactively promoting wine tourism globally. Wineries interviewed as a part of this study agreed that the Ministry of Tourism has continued to support wine tourism initiatives undertaken by the wineries. A respondent representing the Maharashtra Tourism Development Board (MTDC) (TB02) noted that

> [the] Ministry of Tourism has always been at the forefront of promoting wine tourism in the state, and there have been no instances of hesitation due to Nashik's religious identity. In fact, the India Grape Harvest Festival (a biennial wine festival) has been one of the very successful events that we co-host in association with AIWPA (All Indian Wine Producers' Association). We saw a footfall of over 20,000 visitors across 20 venues in Nashik in 2018. India Grape Harvest Festival, as a part of its program, also promoted other tourist attractions of Nashik, including places of religious importance such as Panchvati and Trimbakeshwar temple among others.

However, some voices of dissent were also recorded among respondents where some quarters expressed reservation towards the widespread growth of a culture where enjoyment of alcohol was promoted. (TB01), a respondent representing the Shri Devasthan Trust, highlighted that

> the temple has always maintained this stand and we have recently appealed to the district authorities that there should be no wine store or wine related activities within the 15 kilometers' radius of the temple. We believe this is important to maintain a spiritual atmosphere and sanctity of the place.

b *Economic dependency on both religious tourism and the wine industry*: The wine industry and religious tourism in Nashik are both important catalysts in the growth of the local economy. Besides job creation, both wine and religious tourism generate tax revenue to the local government. Questions pertaining to the acceptability of the wine industry and consequent wine tourism in Nashik, the responses indicated general acceptability owing to economic dependency on these sectors. One of the respondents (TB03) highlighted that

> this (wine tourism), in no way, disturbs the local population or the pilgrims. In fact, it acts a source of revenue for the local population. Wine tourism has contributed

> to the economy of the city, and has created ancillary jobs for the local population in and around Nashik. The hospitality industry has been the biggest beneficiary of wine tourism. Homestays and farm stays have also flourished, offering a perfect weekend getaway with local flavor. In fact, the wine industry has put Nashik on the global map, which would otherwise have been restricted to being a pilgrim destination.

Employment opportunities generated through wine tourism can be considered as one of the most important reasons for local communities to support wine tourism growth in Nashik. A respondent (WO02) from one of the most visited wineries in Nashik noted that

> I have been working with this winery for eight years now and have not come across any situation of resistance from the local community. It is because of wine tourism that Nashik has gained visibility among international tourists. More number of visitors are coming to Nashik due to wine tourism.

Identifying this opportunity, the Government of Maharashtra has taken several initiatives to promote wine tourism. The All India Wine Producers' Association (AIWPA), since its inception, has conceived and coordinated several wine tourism promotion initiatives along with the Government of Maharashtra. The respondent from AIWPA (TB03) indicated that

> [i]t [AIWPA] has coordinated with the Ministry of Tourism, and has taken major efforts in easing of policies, rules and regulations like permissions/licenses required, etc. It has also tied up with MTDC for marketing winery resorts. Under the current initiative of 'Ease of Doing Business' by the Government, there is a dedicated team in the tourism ministry which focuses on this issue. There are participations in the local fairs organized by MTDC, hosting of wine festivals/, etc. which further promotes the cause.

c *Geographical factors*: Nashik has been a grape-growing region for several decades. Unlike many wine regions in the world, India's climatic conditions allow two grape harvests a year: once during the wet summer months and once during the cooler winter months. In Nashik, too, this pattern has ensured that there is no dormancy for vines throughout the year, further increasing production.

> One cannot deny the fact that Nashik has also been the 'Grape Capital of India' since the 70s. It is only in the recent years that Nashik has also assumed the image of 'Wine Capital of India.' Wine, as a beverage, is a derivative of grapes, and the (wine) industry is a forward integration of viticulture.
>
> *(WO03)*

d Over the years, wine has come to be accepted as a natural by-product or derivative of grapes, and this has added to the general consensus, among locals and visitors alike, about Nashik being a hub of wine production and wine tourism.

e *Tourist segments*: There was a general census among respondents that both forms of tourism attracted district segments of tourists.

> There are broadly two categories of visitors to Nashik – pilgrim and tourism. Wine tourism forms part of the tourism subset and whereby the TG (target group, also

means target audience) for wine tourism is totally different from the pilgrims. The demographic profile of the people visiting temples/religious places and for wine tourism are mutually exclusive.

(TB02)

f Both forms of tourism continue to grow, attracting tourists from distinct demographic profiles.

There is a clear distinction between the profiles of tourists that visit Nashik for religious purposes and the ones that visit wineries. Wine tourists are a small segment who come, mostly, from tier 1 cities. They are mostly from upper middle-class families, are well educated and well-traveled. A matured audience who do not mind combining a winery visit with a temple visit.

(TO01)

Discussion

Ethical dimensions and impact on wine tourism policymaking

The findings of this study bring out several dimensions that highlight the unique interplay of factors allowing for an exploration of the relationship and interlinkages between ethical and regulatory influences. Firstly, Nashik's identity as a religious tourism destination continues to grow. Nashik has also continued to emerge as a popular destination for wine tourists due to its novel tourism offering and increasing interest in wines among travelers in India (Kaddi 2013). Religious tourism and wine tourism in Nashik have grown independent of one another owing to their distinct market segments and thus have come to be understood as mutually exclusive. Secondly, a common consensus was observed among stakeholders about the economic dependency of Nashik on the growth of the wine industry as well as the tourism industry due to religious tourism. This has led to general acceptability among stakeholders and the local population alike about the flourishing wine industry and its contribution. It was observed that some sensitivities and uneasiness continued to surround wine tourism promotion; however, a balanced approach in developing both forms of tourism would be a step in the right direction.

Thirdly, the findings suggest that there was no evident hesitancy or resistance on the part of the government, tourism promotion boards, tourism service providers, local communities, or other stakeholders in embracing the growing popularity of Nashik as a wine tourism destination given its religious identity. These identities are 'overlapping, yet (form) discrete elements of the larger sphere' (Burdick 2016, p. 178) of an organic image of the destination (Chon 1990). Similar tourism trends have been observed in the cities of Florence, Italy (Bambi et al. 2019; Maizza and Rosato 2008) and Ningxia, China (Hao, Li, and Cao 2015; Mu et al. 2007), where religious tourism and wine tourism have flourished together over the years. Further, religious attractions around the world are being commodified, and as a consequence, 'pilgrimages and other religious journeys are becoming tied to other types of tourism' (Simone-Charteris and Boyd 2011, p. 461). Such a trend, perhaps, has led to a softer divide between religious tourists and other kinds of tourists. Further, the findings also suggest that, although some quarters expressed concern towards growing wine tourism in Nashik, this did not have any visible policy implications.

Tourism operates within a complex environment encompassing sociocultural, economic, environmental, and legal factors. Many a times, given the unique socioreligious systems within a

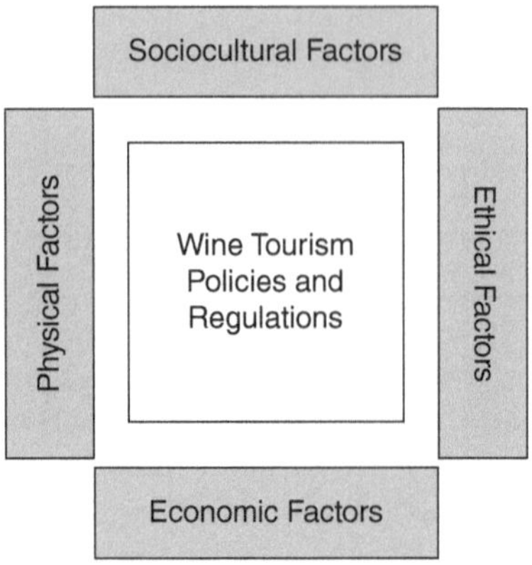

Figure 5.1 Suggested framework for external environmental influencers of wine tourism policy development in Maharashtra.

Source: Author

destination, ethical predicaments may also arise. Policymaking, therefore, for a destination must consider these factors in order to ensure a greater and long-term viability of the destination. Figure 5.1 illustrates the factors that may influence wine tourism policies and regulations in the case of Nashik, Maharashtra.

Impact of present policies and regulations on wine tourism

The progressive wine policy of the state, the Maharashtra Grape Processing Industrial Policy of 2001, no doubt laid down a general path for growth by giving the industry a preferential status. However, over time, the industry has evolved, and a need for newer policies to meet the challenges and complexities of the growing wine industry is now felt more than ever. Firstly, the tax structure on wines in India is rather complex, and a lack of uniformity in tax regulations has led to wine prices varying from state to state. This, to a large extent, has discouraged consumers from other states from buying wines originating from Maharashtra. This has also led to a lack of awareness about smaller, boutique, and upcoming wine brands of Nashik in other states. And, to some extent, it has affected the prospect of wine tourism for those wineries.

Secondly, export policy on wines is still at an evolving stage, as wine is not one among the most significant agricultural exports for India. Wine is treated no differently from any other processed foods by APEDA; therefore, no separate incentives are extended to wine exporters in India. Further, the absence of dedicated and stricter wine quality standards and regulations has also impaired the export of Indian wines. Reviewing regulating quality standards and its strict implementation may be able to give a further impetus to wine export. Greater visibility of Indian wines in international markets can be instrumental in attracting wine connoisseurs from around the world.

Finally, due to strict regulations that exist on advertising alcoholic beverages, including wines – directly or passively – wine brands have been resorting to event-based marketing strategies. The lack of a platform in popular media may affect the promotion of wine tourism to a great extent. Easing regulations in favor of sponsorship-based marketing activities can enable wine brands to reach a wider audience in the rest of India and globally.

Implications and conclusion

The study, in its attempt to critically evaluate the existing regulations under the current wine policy imposed on the wine industry in Nashik, highlights areas that have influenced the growth of wine tourism. On one hand, the Wine Policy of Maharashtra became an example of pro-growth initiatives by the state government in India; on the other hand, several aspects of the policy demand an immediate review in the light of evolving needs of the wine industry and the flourishing wine tourism industry. A uniform wine policy throughout the country with greater emphasis on quality, highlighting the regional uniqueness of wines, can be a way forward. Further, the study also investigates the ethical dimension and its influence on wine tourism development and wine tourism policymaking. The tourism industry operates in a complex sociocultural environment where multiple identities of the destination may sometimes lead to an uneasy relationship between tourism development and its multiple stakeholders. Using the findings of the study, the author proposes a framework interlinking the external environmental factors and wine tourism policy development in Maharashtra.

These findings may be particularly useful to policymakers, industry practitioners, and wine tourism stakeholders in the following ways:

a They may help in identifying the impact of the current regulations on wine tourism development.
b They may help in appreciating the interplay between ethical factors and wine tourism development. The policy framework must take into account not just the economic benefits that the growth might yield, but also the sociocultural and ethical sensitivities associated with it.
c Nashik presents a unique example of two forms of tourism – religious and wine – that are juxtaposed yet continuing to coexist, painting a richly complex destination imagery. Destination marketers can appreciate this identity of Nashik to promote it as a unique tourism destination.

Limitations of this investigation extend in two spheres. One pertains to the limited literature available with respect to ethical influences in wine tourism. The second is that the sample size is relatively limited and specific to Nashik, Maharashtra. To some extent, this may limit any generalized understanding related to ethical factors and their influence on wine tourism policy development. However, this study's findings open newer avenues of research in understanding the complexities associated with wine tourism development and its external environments encompassing sociocultural, economic, environmental, legal, and ethical factors.

References

Alebaki, M., and Ioannides, D. (2018). Threats and obstacles to resilience: insights from Greece's wine tourism. In Cheer, J. M., and Lew, A. A. (eds.), *Tourism, Resilience and Sustainability: Adapting to Social, Political and Economic Change* (pp. 132–148). London: Routledge.

Anderson, K. (2020). *Consumer Taxes on Alcohol: An International Comparison over Time*. CEPR Discussion Paper No. DP14388. https://ssrn.com/abstract=3535474

Anderson, K., Meloni, G., and Swinnen, J. (2018). Global alcohol markets: evolving consumption patterns, regulations, and industrial organizations. *Annual Review of Resource Economics*, 10: 105–132. https://doi.org/10.1146/annurev-resource-100517-023331

Babu, S. A. (2008). *Tourism Development in India: A Case Study*. New Delhi: APH Publishing.

Bambi, G., Iacobelli, S., Rossi, G., Pellegrini, P., and Barbar, M. (2019). Rural tourism to promote territories along the ancient roads of communication: case study of the rediscovery of the St. Francis's Ways between Florence and La Verna. *European Countryside*, 11(3): 462–474.

Baxter, R. A. (2019). *Wine and Health: Making Sense of the New Science, and What It Means for Wine Lovers*. San Francisco: Wine Appreciation Guild Press.

Bhardwaj, A. (2020). Wine prices could rise as Nashik industry looks to get going after lockdown. *The Print*, 17 July. https://theprint.in/india/wine-prices-could-riseas-nashik-industry-looks-to-get-going-after-lockdown/462250/

Bhardwaj, S. M. (1983). *Hindu Places of Pilgrimage in India: A Study in Cultural Geography*. Berkeley: University of California Press. https://books.google.co.in/books?id=D6XJFokSJzEC&dq=India+religious+sites&source=gbs_navlinks_s

Blocker, J. S., Fahey, D. M., and Tyrrell, I. R. (2003). *Alcohol and Temperance in Modern History: An International Encyclopaedia* (Vol. 1). Santa Barbara, CA: ABC-CLIO. https://books.google.co.in/books?id=BuzNzm-x0l8C&dq=mahatma+gandhi+and+alcohol&source=gbs_navlinks_s

Burdick, C. (2016). Branding the Alsatian oxymoron: The production of ambivalent identity. *Signs and Society*, 4(S1), 163–187.

Charmaz, K. (2006). *Constructing Grounded Theory: A Practical Guide through Qualitative Analysis*. London: Sage.

Corbin, J. M., and Strauss, A. (1990). Grounded theory research: procedures, canons, and evaluative criteria. *Qualitative Sociology*, 13(1):3–21.

Csizmadia-Honigh, P. (2015). *The Wines of India: A Concise Guide*. France: The Press Publishing Ltd.

DNA India (2015). Kumbh Mela begins in Nasik today: thousands take holy dip. DNA India. https://economictimes.indiatimes.com/news/politics-and-nation/kumbh-mela-begins-in-nashik-thousands-take-holy-dip/articleshow/48064499.cms?from=mdr

Donyadide, A. (2010). Ethics in tourism. *European Journal of Social Sciences*, 17(3): 426–433. https://citeseerx.ist.psu.edu/viewdoc/download?doi=10.1.1.1070.1164&rep=rep1&type=pdf

Duffy, R., and Smith, M. (2003). *The Ethics of Tourism Development*. Oxon: Routledge.

Eck, D. L. (2012). *India: A Sacred Geography*. New York: Potter/Ten Speed/Harmony/Rodale.

Egan, A. (2002). *Islam*. Oxford, UK: Heinemann. https://books.google.co.in/books?id=1svp1FuZNywC&source=gbs_navlinks_s

European Federation of Origin Wines. (2021). *What Are Origin Wines?* http://efow.eu/what-are-origin-wines/

Fennell, D. A. (2006). *Tourism Ethics* (Vol. 30). Bristol: Channel View Publications.

Gál-Czékus, I., and Nagy-Kovács, E. (2017). The multiplying effect of wine tourism. *Thinking Together The economy in practice*, 1 (1): 91–98. https://doi.org/10.12700/THTO.1.01.2017.1.7

Gore, J. (2008). *Comprehensive Study of Indian Wine Market*, Washington, DC: JCB International on behalf of the Wine Institute.

Griswold, M. G., Fullman, N., Hawley, C., Arian, N., Zimsen, S. R., Tymeson, H. D., Venkateswaran, V., Tapp, A. D., Forouzanfar, M. H., Salama, J. S., and Abate, K. H. (2018). Alcohol use and burden for 195 countries and territories, 1990–2016: a systematic analysis for the Global Burden of Disease Study 2016. *The Lancet*, 392(10152): 1015–1035.

Hall, C. M., Sharples, L., Cambourne, B., and Macionis, N. (2009). *Wine Tourism Around the World*. London: Routledge.

Hao, L., Li, X., and Cao, K. (2015). Toward sustainability: development of the Ningxia wine industry. *BIO Web of Conferences*, 5(3): 01021. EDP Sciences.

Heath, D. B. (2012). *Drinking Occasions: Comparative Perspectives on Alcohol and Culture*. London: Routledge. https://books.google.co.in/books?id=m1ARwTzLuYcC&dq=drinking+culture+india&source=gbs_navlinks_s

Kaddi, A. K. (2013). Event motivation study of wine festival visitors in Maharashtra. *Atna Journal of Tourism Studies*, 8(2): 1–18. https://doi.org/10.12727/ajts.10.1

Kasabe, N. (2019). Indian wine looks for stronger presence in global market. *Financial Express*, April 12. https://www.financialexpress.com/industry/indian-wine-looks-for-stronger-presence-in-global-market/1545754/

Kasabe, N., (2020). Nashik grape wine crushing season to begin from January 17. *Financial Express*. https://www.financialexpress.com/economy/nashik-grape-wine-crushing-season-to-begin-from-january-17/1824437/

Maizza, A., and Rosato, P. (2008). *Wine Tourism and Enhanced Value: A Comparison between Italian Successful Destinations*. Presented at the Fourth International Conference of the Academy of Wine Business Research, July 17–19, Siena, Italy.

Matsuo, M. (2015). The role of religious practices for the Indian middle class. In Bates, C., and Mio, M. (eds.), *Cities in South Asia*. London: Routledge.

Meloni, G., Anderson, K., Deconinck, K., and Swinnen, J., (2019). Wine regulations. *Applied Economic Perspectives and Policy*, 41(4): 620–649. https://doi.org/10.1093/aepp/ppz025

Ministry of Food Processing Industries (MOFPI). (2020). *Annual Report*. https://mofpi.nic.in/sites/default/files/english_2019-20_1.pdf

Montella, M. M. (2017). Wine tourism and sustainability: a review. *Sustainability*, 9(1): 113.

Mu, Z. L., Jian-Hong, W., Liu, J., Yan-Geng, J., and Xiting, L. (2007). Religious tourism and cultural pilgrimage: a Chinese Perspective. In Raj, R., and Morpeth, N. D. (eds.), *Religious Tourism and Pilgrimage Festivals Management: An International Perspective* (pp. 98–112). Wallingford: CABI.

Olsen, D. H., and Timothy, D. J. (2006). Tourism and religious journeys. In Timothy, D. H., and Olsen, D. J. (eds.), *Tourism, Religion and Spiritual Journeys* (pp. 17–38). Oxon: Routledge.

Pletcher, K. (2010). *The Geography of India: Sacred and Historic Places*. New York: Rosen Publishing Group. https://books.google.co.in/books?id=Mjr0X-8jrLAC&dq=Nashik+religious+sites&source=gbs_navlinks_s

Poitras, L., and Donald, G. (2006). Sustainable wine tourism: The host community perspective. *Journal of Sustainable Tourism*, 14(5): 425–448. https://doi.org/10.2167/jost587.0

Simone-Charteris, M. T., and Boyd, S. W. (2011). The potential for Northern Ireland to promote politico-religious tourism: An industry perspective. *Journal of Hospitality Marketing & Management*, 20(3): 457–483. https://doi.org/10.1080/19368623.2011.562438

Skinner, A. M. (2000). Napa Valley, California. Wine tourism around the world: development, management, and markets. In Macionis, N., Hall, C. M., Sharples, L., and Cambourne, B. (eds.), *Wine Tourism Around the World* (pp. 283–296). London: Routledge. https://doi.org/10.4324/9780080521145

Sommelier India. (2020). Don't miss SulaFest 2020 on the 1st & 2nd February 2020. *Sommelier India*. https://www.sommelierindia.com/dont-miss-sulafest-2020-on-the-1st-2nd-february-2020/

Sood, D. (2019). *Wine Production and Trade Update*. Mumbai: USDA foreign agricultural service. https://apps.fas.usda.gov/newgainapi/api/report/downloadreportbyfilename?filename=Wine%20Production%20and%20Trade%20Update_New%20Delhi_India_8-7-2019.pdf. Accessed on 24th January, 2021.

Tafel, M., and Szolnoki, G. (2020). Estimating the economic impact of tourism in German wine regions. *International Journal of Tourism Research*, 22(6): 788–799.

Trišić, I., Štetić, S., Privitera, D., and Nedelcu, A. (2020). Wine routes in Vojvodina province, Northern Serbia: A tool for sustainable tourism development. *Sustainability*, 12(1): 82.

Van Leeuwen, C., and Seguin, G. (2006). The concept of terroir in viticulture. *Journal of Wine Research*, 17(1): 1–10. https://doi.org/10.1080/09571260600633135

Vasileiou, K., Barnett, J., Thorpe, S., and Young, T. (2018). Characterising and justifying sample size sufficiency in interview-based studies: systematic analysis of qualitative health research over a 15-year period. *BMC Medical Research Methodology* 18(1): 1–18. https://doi.org/10.1186/s12874-018-0594-7

Viswani, M., Ambekar, A., and Ramadeep, P. (2009). Historical aspects of alcohol use in India: Role of gender and culture. In Browne-Miller, A. (ed.), *The Praeger International Collection on Addictions* (pp. 119–139). Santa Barbara, CA: ABC-CLIO. https://books.google.co.in/books?id=55NxDwAAQBAJ&dq=drinking+culture+india&source=gbs_navlinks_s

Wheeler, M. (1995). Tourism marketing ethics: an introduction. *International Marketing Review*. https://doi.org/10.1108/02651339510097720

Williams, P. W., Graham, K., and Mathias, L. (2006). Land use policy and wine tourism development in North America's Pacific Northwest. In Carlsen, J., and Charters, S. (eds.), *Global Wine Tourism: Research, Management, and Marketing* (pp. 27–46). Wallingford: CABI.

Wilson, E., and Little, D. E. (2005). A 'relative escape'? The impact of constraints on women who travel solo. *Tourism Review International* 9(2): 155–175. https://doi.org/10.3727/154427205774791672

Yasuda, S., Raj, R., and Griffin, K. (2018). *Religious Tourism in Asia: Tradition and Change Through Case Studies and Narratives*. Wallingford: CABI.

6

EXPLORING CO-CREATION PROCESS IN THE WINERIES

The relevance of social partner characteristics

Eduardo Moraes Sarmento, Sandra Maria Correia Loureiro, and Namércio Cunha

Introduction

In a global market, the exchange is regarded as one of the most important strategic resources that are critical for business company success (Hammervoll 2012; Johnson 1999; Powers and Reagan 2007). Hence, it is not a surprise that producers, distributors, retailers, customers, and even competitors join forces to co-create value solutions.

With the business environment becoming more competitive, a dyadic perspective of buyer–seller relations is increasing. Partners should be concerned about joining forces and working together to improve their joint performance and add value to their offers (Cannon et al. 2010). In sum, the success of a company can be influenced by the style of the relationship that is developed with other players (Arranz and de Arroyabe 2012; Cunha, Loureiro, and Rego 2015a; Monteiro, Guerreiro and Loureiro 2019; Terpend and Ashenbaum 2012).

Following this appeal for research, this chapter aims to explore how distributors and wine producers describe the personal and organizational social capabilities as facilitating facets of relationship quality (RQ) (it is an issue that depends on confidence, satisfaction, commitment, and trust). The wine sector is selected as a field of research because it has unique features which have not yet been deeply studied and could explain management specificities:

1 Portugal is an old wine country full of tradition and heritage, 'in which the wine culture has been flourishing for centuries' (Loureiro and Kaufmann 2012, p. 331).
2 The wine sector has not been deeply analyzed and studied in previous research.
3 This sector plays an important role in the national economy, and the recognition of the quality of the work carried out in this sector is being acclaimed internationally.

This chapter is organized as follows: after this introduction of our topic research, we present the theoretical background that supported this study, followed by a description of the methodology. The following sections discuss research findings from the empirical study, and the implications for managers of this study are discussed. Finally, limitations and new research directions are suggested.

 DOI: 10.4324/9781003143628-8

Theoretical background

Relationship marketing shifts the traditional concepts of short-term exchange and transactional marketing to social psychology and interpersonal relationship (Dixit 2017; Fournier 1998; Gummerus, Koskull, and Kowalkowski 2017; Tsai 2011). Service-dominant logic stresses that firms and customers jointly create value and build relationships. This approach also highlights that customer desires are ever-changing; consequently, firms must also change (Finney, Spake, and Finney 2011). Therefore, relationship marketing emerges as the core strategic element to be competitive in the 21st century. Athanasopoulou (2013) claims that within the field of relationship marketing, the concept of RQ promotes the development of a whole new stream that gains presence in the main marketing journals. This concept has been developed based on human relationship literature and relational marketing, and it is used to predict the dyadic consequences of established relationships by evaluating the partner's performance in its partnership role (Fournier 1998).

Wine producers and their relationships with distributors are engaged in understanding of how to achieve and maintain a high level of customer satisfaction. Understanding how to develop relationship strategies that go beyond traditional marketing and help to strengthen the RQ with the customer, and for that reason develop consistent loyalty and profits (Cunha, Rego, and Polónia 2015b; Drennan et al. 2015).

In the wine sector, the distributor is an important player who greatly influences the retail of bottled wine and, consequently, the consumer. Moreover, producers and distributors may work together by focusing on their customers and improving the relationship they have with them (Athanasopoulou 2013; Crosby, Evans, and Cowles 1990). The willingness and the ability to create effective long-term relationships with positive results for both partners is a core asset studied by researchers in order to better explore and understand the concept of RQ (Athanasopoulou 2013; Bobot 2011; Farooqi 2014; Loureiro and Cunha 2017).

A systematic literature review allows us to find several gaps in the literature. This process conducted used the database service Web of Science to search for the articles on RQ. The first selection shows a panel of 855 scientific articles. After applying various filters (read titles, abstract, consider journals in the ranking ABS), it was possible to focus on the observation of RQ in 84 final articles. As a result, we find three main gaps:

1. There is a lack of studies on the relationship dyadic. The previous studies tend to analyze the relationship only from the point of view of one part of the relationship, the seller or the buyer. Therefore, new studies are demanding that the perspective of all parties involved in a relationship should be considered to accurately grasp the true nature of the relationship.
2. In the few studies that address several different aspects of the relationship between two partners in a business context – for example, that of Athanasopoulou (2013), and other more recent studies analyzed in the systematic literature review – it was not possible to effectively find the major constructs related to RQ in a dyadic relationship;
3. There is a lack of identification of the relevance of the social capabilities to the RQ in a dyadic relationship.

In this vein, the current study explores how distributors and wine producers describe the organizational social capabilities and also which aspects they most value. This will be valuable in the wine tourism industry.

Methodology

We conducted in-depth interviews with wine industry professionals, taking into consideration all criteria of an exploratory study and a qualitative approach (Grounded Theory). In comparison to a quantitative approach, this one is more researcher-dependent, in which the data is interpreted in order to extract its meaning and is converted into information (Zikmund and Babin 2010). For Aaker et al. (2010, p. 162), 'The basic assumption behind qualitative methods is that an individual´s organization of a relatively unstructured stimulus indicates the person´s basic perceptions of the phenomenon and his or her reaction to it.'

A total of 11 relationships were analyzed, which were developed among top managers of wine-producing and distributing companies. These companies were contacted to get the final sample of 22 distributors and 22 producers interviewed.

All the in-depth interviews were transcribed by the interviewer to ensure its reliability (Malhotra 2019). After this procedure, we sifted the data through a cyclical process in which data interpretation, coding, and conceptualizing occurred simultaneously, 'albeit at different rates of progress' (Lindlof 1995, p. 215; Rego et al. 2016). We dedicated time and attention to analyzing the responses of each interview, which allowed us to identify the major themes/categories related to the research topic. Then, the emerging categories were used to code the data interviews. This process was finished when the 'dialogue' between data and the themes/categories became clear, and interpretations were stabilized. This process of category creation is based on the Grounded Theory (Batra, Ahuvia, and Bagozzi 2012). Data were processed through the WEBQDA software.

Case Study 6.1 The experience of wine producers in the north of Portugal

Results

Findings highlight Organizational Social Capabilities as a core aspect to develop an interactive relationship quality among stakeholders and fundamental to co-create wine experiences between market players and wine tourists. Relying on competent organizational and professional structures, knowing how to promote relevant social events that promote the producer's brand and increase the partner's focus, as well as the significance of the partner's history. Thus, Organizational Social Capabilities comprises Organisational Capability, Social Events promoted by the partners, and the history of the partners.

Organizational capability

This construct is not sufficiently developed in the RQ literature. The understanding that the partner operates with well-defined routines, according to an implemented organizational structure, with an able and strongly operational commercial and marketing structure, and also with competent professionals, conveys trust and facilitates the commercial relationship between both parties. If both partners have able organizational structures, the partnership will operate better with a greater likelihood of success for the commercial relationship between them.

Examples of sentences from interviews are as follows:

> When there is professionalism, from both organizations, and here we broaden it a little to what happens in the world of wine, to something that happens in the world of major consumption, when there is professionalism, there is clearly a good relationship…
>
> Terroir-Producer, among the producers we have, is one that has a more professional structure, more defined, with commercial and marketing areas and with the different areas defined and with professionals. And so when it's like that, and in the case of Wine Manager-Producer, it's easier to work, because there's more chance of things going well.

This subconstruct of the Organizational Social Capabilities construct is subdivided into three other subcategories: (1) organizational structure, (2) commercial/marketing structure, and (3) sales team quality.

Organizational structure

This subcategory is not sufficiently developed in the RQ literature. A company's organizational structure is a facilitator of the quality of the relationship formed between partners. This takes place because it makes processes smoother, simplifies solutions, and transmits security to the other party because they know what they can count on. The firm's organizational structure is a constant concern for those in charge to be able to respond to the needs of their partner and the market in general.

Examples of sentences from interviews are as follows:

> And this is the philosophy we have to follow. Well, it's like this, also in the producer, speaking about our case, in particular, there were a lot of internal changes that made this relationship easier.
>
> From the beginning of the year, everything is discussed, everything's in line, we're a machine, a perfectly oiled machine, very close to the producer, the market, the wholesalers and it's already a very big operation.

Commercial/marketing structure

This subcategory is not sufficiently developed in the RQ literature. Besides the firm's organizational structure, as a whole, we noticed an emphasis on the need for the partner to have a well-structured commercial and marketing team, able to give support to the other party and also to be a source of gathering information in the market. This market follow-up occurs with the producer's own teams supporting the distributor's commercial teams. The distributor also needs to have a commercial and marketing team able to ensure the brand's success in the market. This commercial capacity contributes to the high efficiency of commercial relations.

Examples of sentences from interviews are as follows:

> [W]e ourselves are very present in the market, we ourselves have a commercial structure to follow the commercial structure of the distributors. I think that's also very relevant. So as not to be a single voice, that feedback coming from the market is very important, various voices. That's an advantage we have.
>
> Because it wants to keep the brands always on top, and for that to happen it has very important back-up work. It's these sales promoters who help our salespeople to do business.

Sales team quality

The producer hopes his wine will be worked on and be given value as a brand in the market. For that to happen, commercial teams must be experienced, having thorough knowledge, and must be able to carry out differentiated work for the product. Therefore, the competencies of commercial teams contribute greatly to the success of the objectives aimed for by both parties, and in turn, to the quality of the relationship formed.

Examples of sentences from interviews are as follows:

> Consistency and quality. Consistency in the accompaniment, consistency in the salesperson's work, in terms of accompaniment, persistence, giving information to the buyer.
>
> Select-Producer benefits, because it has a strong distributor, obviously. It has an excellent team, focused, going back many years, many years, very uniform, some years ago, the newest one has been with the company for eight years, so there isn't a constant rotation of people, which is very good, which is very good, for us too it's very good, there's a culture.

Social events

The creation of social moments in which both business partners can share information about the product, between themselves or with other market players, is a facilitator of RQ. This is because it demonstrates the quality of the existing partner relationship and the commitment to the success of the goals aimed for. Those moments can be created through a wine tasting, through wine lunches or dinners, or, whenever possible, through visits to producers' vineyards and wineries. The objective is to create links between participants in the commercial relationship.

Examples of sentences from interviews are as follows:

> That's why I told you it's extremely important to receive the distributors and the salespeople at Quinta Douro, because the Quinta's involvement is the first sign of creating an emotional connection to the place, to the brand.
>
> Very often, the distributors themselves hold some events. Particularly wine-gastronomic dinners to promote the brand with the end consumers. But it's always work done jointly.

This subconstruct of the Organizational Social Capabilities construct is subdivided into three other subcategories: (1) wine tastings, (2) lunches and dinners, and (3) vineyard/winery visits. None of these constructs are developed in the RQ literature.

Wine tastings

This is a type of event that producers recognize as being important for bringing the brand's players together and thereby promoting their connection to it. This work, often carried out jointly, allows the two sides to respond to common needs and demonstrates their involvement in the commercial relationship formed.

Examples of sentences from interviews are as follows:

> [T]he company owner often goes to the Algarve to do tastings and, he's there all the time, because they like that.
>
> [T]hat is, if I'm asked to go to an event of that kind, I make sure internally that things go well. And probably I'll even go and get, if I can, oenologists from the production, because it's important to 'round up' 30, to 40 or 70 people, to make a big impact, with someone who really knows, knows what he's doing and often that relational is important here too.

Business lunches and dinners

The importance of these events is recognized by both partners in the business, be it lunches or dinners among themselves or with other market players. It is generally understood that the producer's participation is a bonus due to the technical sustainability they bring to the product and the history they represent with regard to the product. Holding these actions jointly is a way of demonstrating to the market the strength and quality of their commercial relationship.

Examples of sentences from interviews are as follows:

> Ana Almeida herself or the winemaker, very often, holds themed dinner parties or gastronomic dinners, to harmonize the food with the wine or with the salespeople that are present, so there's all this type of work done, to give training and at the same time create a connection between people, the product and the market.
>
> But we try, as much as possible, that the oenologists, now and again, appear together, as partners, with our distributors and they often go to the dinners organized by us at the clients' location, where there's the distributor or salesperson or even the owner of our distributors or of the restaurant, and then we take the oenologist to speak about the wine, to make a speech. We try, as much as possible, to involve our oenologists.

Vineyard/winery visits

When circumstances permit, visits to vineyards and wineries are events that provide memorable sensations for the participants. Knowing how to welcome and create memorable moments for visitors are opportunities to develop and deepen links with direct partners in the commercial relationship and also with other market players. These events are valued by both producers and distributors and are seen to be essential in working on the producer's brand and strengthening the partnership in the eyes of the market.

Examples of sentences from interviews are as follows:

> [A]nd our willingness to welcome people at Nova Quinta, as if they were in their own home. That ends up being some of the most positive factors that were focused on by the distributors. Because we also regularly welcome sales teams, because there's nothing better than sales teams visiting the estate, visiting the winery, tasting the wines with us, with the oenologist and understanding, really, where we are, what we do, and anyone coming to the Douro, will certainly never forget it.
>
> And so we also went to that estate, at the end of the day. It was two days, we were there for two or three days. And so, suddenly we arrive at the estate and there were 10 4×4s to pick us up. We get into the cars and they take us up the hill, to the top of Monte Oliveira, that's the estate, the whole hill was carved out of the slate, to plant the vines above the river. And we had, on the hilltop, we had a white table with wine glasses and with the wine served. It was the only thing there. He gave each of us a glass, we were about 40 people, he climbed up on some rocks, him and the family, and began speaking about the wine, there in that scenario. And I thought it was delightful, and after I commented on it to all my colleagues that the atmosphere he'd achieved in that scenario, with that performance, what it created with our sales team. That was worth more than 10 meetings of the normal kind.

Players' history

Family companies, with a family history going back generations, with consistent quality in their product, generate a reputation in the market that creates feelings of security and trust in the other party in the relationship. These characteristics are facilitators of RQ.

Examples of sentences from interviews are as follows:

> As I told you, these are companies that have been in the sector for a long time, they have a performance that has been consistent, always, since being more involved with distribution in the domestic market.
>
> They want to grow, they want to develop the business, but from a medium and long-term perspective, because all of them, as I said, are families and some are businesses going back generations.

This subconstruct of the Organizational Social Capabilities construct is subdivided into three other subcategories: (1) reputation, (2) family company, and (3) company history. These constructs are not developed enough in the RQ literature.

Reputation

Consistency of management and the respective reputation created in the market are facilitators of RQ because they transmit security to the other party and make them believe they are trustworthy partners to build a long-term commercial relationship with. The feeling that it is a project with growth potential, that it is here to stay for generations, with its own know-how, gives confidence to the other party in the commercial relationship.

Examples of sentences from interviews are as follows:

> They can sell with confidence. They can sell with confidence because they feel it's a project that's growing, it's a project that concentrates continuously on quality and not just on projection, on presenting products that don't have the due care in terms of presenting an image, in terms of product quality, in terms of price positioning.
>
> For us, our supplier is very important and we have to know if he's doing all right in the market, if it's a company with a future.

Family company

Companies with a family tradition covering generations transmit to the other partner security in the continuity of the relationship. This factor is valued, above all, by distributors as a facilitator of quality in the existing commercial relationship.

Examples of sentences from interviews are as follows:

> It's a very simple case, because, the history of this company is very old, it's lost in time, because it's a family business and so it goes back many years.
>
> Because we're speaking about a business that has a component that's very much family and tradition. There's an ancient house, there's a property with a fantastic view over the Douro valley, really, there's a whole history where, it makes the relationship and the people's presence important.

Company history

The history created and told around the producing brand is a factor greatly valued by distributors. This dimension facilitates the work of that same brand in the market and consequently helps to attain the goals sought by the two partners in the commercial relationship.

Examples of sentences from interviews are as follows:

> How many brands would love to have the history we have around here, to have these gardens, this house to transmit this history. And sometimes we forget to transmit this to the outside world.… Here there's an immensely rich history, that can be told.
>
> It's much easier, for me, to sell GOLDOR products, which is a brand from the Douro, it's the oldest house in terms of Port, it's 375 years old, and so it has a very long history.

Conclusion and implications

In a global world that easily replicates the similar resources that produce and offer goods and services, the mode how the players manage their business relationship could make the difference and consequently ensure the sustainability of a successful business (Bobot 2011; Chung et al. 2006; Nyaga et al. 2013; Van Bruggen, Kacker, and Nieuwlaat 2005).

This study allowed us to confront the theoretical concepts presented by RQ literature with what really happens in a dyadic exchange relationship between companies. The meaning of Organizational Social Capabilities highlights the importance of the companies that operate with competent professionals and a strongly operational commercial and marketing structure; the influence of a high reputation in the market with a history and/or a family heritage to tell, and the importance of the company organizes social events in order to develop relationships and share information about new wine products.

Events can be created through a wine tasting, wine dinners, or, whenever possible, through visits to producers' vineyards and wineries. These events may happen among distributors and producers, but also with tourists, particularly senior tourists. The wine social moments also allow presenting the heritage and the history of Portuguese wine, its properties, and how it has evolved. In this vein, wine tourism emerges as a mechanism that contributes to co-creating wine regions in general, and new wine products in particular. Managers of rural lodgings and wine producers should be more proactive in developing these social events to engage tourists and other players to the wine products.

However, if it is critical of the organizational competences, it is also essential to the personal social capabilities revealed by the professionals that intermediate the business relationship. It is fundamental that they reveal a true passion by this wine business, like real wine lovers that have deep knowledge about the sector and that have the capacity to transmit a big enthusiasm through the way that they promote the product and the brand. By the other side, the professionals must develop good relationships with the other players of the sector, institutions, producers, distributors, and consumers. This positive perspective can contribute to a healthy market share in which the players promote synergies that will benefit all. Moreover, they must sustain their behavior like a gentleman trade with distinct manners on the treatment of the business partner. This is a distinguishing factor that helps to make a big difference in professional businesses with other companies, but also with final consumers that could taste the wines or just visit the wine cellar, the vineyard, or the wine region.

Finally, professionals must have the capacity to understand and to communicate the wine intangibilities that also were bottled with the liquid product and will be served in a glass to the final consumer. This required the capacity to create and transmit the emotionality of the moments when the customers consume wine. These good moments could be in a celebration, in a wine tasting, or in a visit to the wine cellar.

Indeed, professionals involved in the wine sector could be characterized by the capacity to spend long hours by discussing the product nature that full fill a bottle of wine. Why? Because the product goes above the liquid and is rich in history and terroir particularities. The nicest countryside views and the vineyards have the capacity to make anyone completely 'overwhelmed.'

Limitations

This research was conducted keeping in mind the inquisitiveness and ethical principles of quality and accuracy, which must guide a scientific investigation. However, we are aware of the constraints involved, but also that it is an opportunity for further studies. Even so, every research design presents some limitations (Zikmund and Babin 2010). Below are the key limitations related to the current study:

1 The convenience sampling limits how well the study represents the intended population (Aaker et al. 2010; Zikmund and Babin 2010).
2 Our study is also focused on relationships in a specific industrial sector which restricts the generalization of the results.
3 In order to minimize self-assessment bias, we instructed the interviewees, who were representatives of the organization exchange relationship (mostly owners/managers), to answer from the perspective of the organization.

(Bloemer, Pluymaekers, and Odekerken 2013)

Further research opportunities

The current study analyzes the perspectives and the contexts used more repeatedly in previous studies regarding the type of market, relationships, and perspectives. When the exchange relationship is a B2B, the majority of the studies analyze organizations related to industrial goods and take into consideration the perspective of the buyer. There are few cases that analyze a dyadic perspective. In situations of B2C, companies related to the product market are analyzed, but only from the perspective of the buyer.

Regarding future research, several suggestions are put forward:

1 Future studies should analyze the antecedents and consequences in a dyadic perspective, and through a similar process of analysis, pursue a detailed model of constructs.
2 Expand upon our study for the analysis of international relationships between exporting and importing wine companies. It would also be interesting to replicate the study in other countries.
3 Identify wine marketing activities with greater relevance that promote wine brands to national and international consumers.
4 Additional research should explore the features of the Portuguese and international wine consumers and their expectations related to Portuguese wine.

References

Aaker, D. A., Kumar, V., Day, G. S., and Leone, P. (2010). *Marketing Research*. New York: John Wiley & Sons.

Arranz, N., and de Arroyabe, J. (2012). Effect of formal contracts, relational norms and trust on performance of joint research and development projects. *British Journal of Management*, 23(4): 575–588. https://doi.org/10.1111/j.1467-8551.2011.00791.x

Athanasopoulou, P. (2013). Relationship quality in services: Past, present, and future. In Kaufmann, H., and Panni, M. (eds.), *Customer-Centric Marketing Strategies: Tools for Building Organizational Performance* (pp. 171–190). Hershey, PA: IGI Global. https://doi.org/10.4018/978-1-4666-2524-2.ch009

Batra, R., Ahuvia, A., and Bagozzi, R. (2012). Brand love. *Journal of Marketing*, 76: 1–16. https://doi.org/10.1509/jm.09.0339

Bloemer, J., Pluymaekers, M., and Odekerken, A. (2013). Trust and affective commitment as energizing forces for export performance. *International Business Review*, 22(2): 363–380. https://doi.org/10.1016/j.ibusrev.2012.05.002

Bobot, L. (2011). Functional and dysfunctional conflicts in retailer-supplier relationships. *International Journal of Retail & Distribution Management*, 39(1): 25–50. https://doi.org/10.1108/09590551111104468

Cannon, J., Doney, P., Mullen, M., and Petersen, K. (2010). Building long-term orientation in buyer–supplier relationships: the moderating role of culture. *Journal of Operations Management*, 28(6): 506–521. https://doi.org/10.1016/j.jom.2010.02.002

Chung, J.-E., Sternquist, B., and Chen, Z. (2006). Retailer–buyer supplier relationships: the Japanese difference. *Journal of Retailing*, 82(4): 349–355. https://doi.org/10.1016/j.jretai.2006.08.008

Crosby, A., Evans, K., and Cowles, D. (1990). Relationship quality in services selling: an interpersonal influence perspective. *Journal of Marketing*, 54(3): 52–68. https://doi.org/10.2307/1251817

Cunha, P., Loureiro, S., and Rego, A. (2015a). Exploring the attitudes of bottled wine distributors toward wine producers in the Portuguese wine sector. *Journal of International Food and Agribusiness Marketing*, 27(1): 15–32. https://doi.org/10.1080/08974438.2013.805455

Cunha, P., Rego, A., and Polónia, D. (2015b). Corporate sustainability: a view from the top. *Journal of Business Ethics*, 143: 133–157. https://doi.org/10.1007/s10551-015-2760-8

Dixit, S. K. (ed.). (2017). *The Routledge Handbook of Consumer Behaviour in Hospitality and Tourism*. London: Routledge.

Drennan, J., Bianchi, C., Cacho-Elizondo, S., Loureiro, S., Guibert, N., and Proud, B. (2015). Examining the role of wine brand love on brand loyalty: A multi-country comparison. *International Journal of Hospitality Management*, 49: 47–55. https://doi.org/10.1016/j.ijhm.2015.04.012

Farooqi, S. (2014). The construct of relationship quality. *Journal of Relationships Research*, 5(2): 1–11. https://doi.org/10.1017/jrr.2014.2

Finney, R., Spake, D., and Finney, T. (2011). Lost in transition? The human influence on marketing's emerging service-dominant logic. *Journal of Management and Marketing Research*, 6: 1–16.

Fournier, S. (1998). Consumers and their brands: developing relationship theory in consumer research. *Journal of Consumer Research*, 24(4): 343–353. https://doi.org/10.1086/209515

Gummerus, J., Koskull, C., and Kowalkowski, C. (2017). Relationship marketing: past, present and future. *Journal of Services Marketing*, 31(1): 1–5. https://doi.org/10.1108/JSM-12-2016-0424

Hammervoll, T. (2012). Managing interaction for learning and value creation in exchange relationships. *Journal of Business Research*, 65(2): 128–136. https://doi.org/10.1016/j.jbusres.2011.05.011

Johnson, D. (1999). Social interdependence: Interrelationships among theory, research, and practice. *The American Psychologist*, 58(11): 934–945. https://doi.org/10.1037/0003-066X.58.11.934

Lindlof, T. R. (1995). *Qualitative Communication Research Methods*. Thousand Oaks, CA: Sage.

Loureiro, S., and Cunha, N. (2017). Wine prestige and experience in enhancing relationship quality and outcomes: wine tourism in Douro. *International Journal of Wine Business Research*, 29(4): 434–456. https://doi.org/10.1108/IJWBR-04-2017-0033

Loureiro, S., and Kaufmann, R. (2012). Explaining love of wine brands. *Journal of Promotion Management*, 18(3): 329–343. https://doi.org/10.1080/10496491.2012.696460

Malhotra, N. (2019). *Marketing Research: An Applied Orientation*. UK: Pearson.

Monteiro, P., Guerreiro, J., and Loureiro, S. (2019). Understanding the role of visual attention on wines' purchase intention: An eye-tracking study. *International Journal of Wine Business Research*, 32(2): 161–179. https://doi.org/10.1108/IJWBR-03-2019-0017

Nyaga, G., Lynch, D., Marshall, D., and Ambrose, E. (2013). Power asymmetry, adaptation and collaboration in dyadic relationships involving a powerful partner. *Journal of Supply Chain Management*, 49(3): 42–65. https://doi.org/10.1111/jscm.12011

Powers, T., and Reagan, W. (2007). Factors influencing successful buyer–seller relationships. *Journal of Business Research*, 60(12): 1234–1242. https://doi.org/10.1016/j.jbusres.2007.04.008

Rego, O., Cabrita, A., Rosa, H., Alves, S., Fonseca, V., Vouzela, C., Pires, F., and Bessa, R. (2016). Changes in milk production and milk fatty acid composition of cows switched from pasture to a total mixed ration diet and back to pasture. *Italian Journal of Animal Science*, 15(1): https://doi.org/10.1080/1828051X.2016.1141330

Terpend, R., and Ashenbaum, B. (2012). The intersection of power, trust and supplier network size: implications for supplier performance. *Journal of Supply Chain Management*, 48(3): 52–77. https://doi.org/10.1111/j.1745-493X.2011.03261.x

Tsai, S. (2011). Fostering international brand loyalty through committed and attached relationships. *International Business Review*, 20(5): 521–534. https://doi.org/10.1016/j.ibusrev.2010.10.001

Van Bruggen, G., Kacker, M., and Nieuwlaat, C. (2005). The impact of channel function performance on buyer–seller relationships in marketing channels. *International Journal of Research in Marketing*, 22(2): 141–158. https://doi.org/10.1016/j.ijresmar.2004.06.004

Zikmund, W., and Babin, B. (2010). *Essentials of Marketing Research*. Mason, OH: South-Western, Cengage Learning.

7
WINE TOURISM IN CANADA

Zainub Ibrahim, Rémy Tremblay, and Fiona McCarthy-Kennedy

Introduction

Canadian wine products satisfy the tastes of Canadians and individuals around the world, despite being a considerably small player in the international wine market. There are two principal wine regions in Canada, the Niagara Region and the Okanagan Valley, which produce the majority of grapes. Plus, three other emerging wine regions are in Prince Edward County, in Ontario; Nova Scotia; and Quebec.

Canada's wine history started in the 17th century but saw slow growth up until the late 1970s. The industry's delayed growth had to do with the challenges surrounding growing grapes in a Nordic climate. For regions like Quebec and Nova Scotia, the cold winters and humid summers were not ideal for grape growing. However, the wine industry and subsequent wine tourism industry have seen rapid growth in the last three decades. Much of this growth came with the introduction of the Canada-US Free Trade Agreement, policy changes to the General Agreement on Tariffs and Trade (GATT), and the introduction of the Vintners Quality Alliance. In an effort to keep the Canadian wine industry competitive with the introduction of Free Trade, the Ontario government incentivized grape growers to replace the native *labrusca* grapes with European *vinifera* varieties and other hybrids. Their introduction saw new wineries opening at a rapid pace, specifically in the Niagara Region and the Okanagan Valley, which led to further tourism developments across the regions.

Tourism development has often been a catalyst for economic growth. In many regions, tourism developments have created strategic linkages with each other as means to enhance the visitor experience. In the field of tourism can be established tourism clusters between suppliers, companies, leisure, industry, and other relevant organizations, working together but also in competition, with the principal aim to achieve competitive advantage (Fundeanu 2015). To enhance the success of wine tourism in Canada, many wineries have developed linkages with each other as well as other tourism and tourism-related business.

History of wine in Canada

Wine in the beginning

Throughout the colonial period, vines were brought from Europe to the New World, and attempts were made to grow grapes (Cattell 2014). Many of these early efforts failed, as the

DOI: 10.4324/9781003143628-9

European vines of the *Vitis vinifera* variety were not able to adapt to the harsh winter and were not resistant to diseases found in the east. For over a century, the Canadian wine industry relied on native *labrusca* grapes and hybrids such as *Niagara*, *duchess*, and *catawba* (Hope-Ross 2006). These grape varieties had tough skins, contained significantly less sugar, and were far more acidic than European grape varieties. They were often used to produce fortified wines, as they required large amounts of sweetening to be enjoyable.

The earliest documented history of grape growing and winemaking in Canada can be traced back to the 17th century in Nova Scotia. Historic documentation pointed to Louis Herbert as the first known planter of wine grapes in Bear River in 1611, but his absence from Nova Scotia after 1607 has invalidated these claims. In 1634, when French Governor Isaac de Razilly successfully planted his vineyard at Petite Riviere near LaHave in Nova Scotia, he was credited to be the first successful known planter of wine grapes in Canada. It is important to note that by 1713, when the British gained further control of Nova Scotia, wine was replaced by Caribbean rum and domestically produced ales.

The first commercial winery was founded in the early 19th century by Johann Schiller, an ex-soldier of German origin. Schiller moved to Upper Canada in 1798, and in 1811, he planted American hybrids from Pennsylvania on his land, lot 17, where he made wine for himself and his neighbors. Schiller is often cited as the 'father of commercial winemaking in Canada.' However, Alun Hughes (2010), former associate professor at Brock University, says there 'is little evidence to suggest that Schiller did nothing out of the ordinary.' However, it is likely his name lives on because grape-growing on lot 17 continued long after his death in 1816. In 1864 Schiller's winery was sold to Count Justin M. de Courtenay, who expanded the vineyards and later exhibited the wine he made in France (Cattell 2014, p. 6).

Prohibition

Prohibition ended the first part of grape-growing and winemaking history in Canada. Starting in the 19th century, Canada's Temperance Act gave local governments power to ban the sale of alcohol. It was first adopted by Prince Edward Island in 1907, and other provinces followed suit. When Ontario enacted the Temperance Act in 1916, grape growers had sufficient political influence to get an exemption for native wine. Ontario wine became the only legal alcoholic beverage that could be sold in the province (Cattell 2014). National prohibition came into effect in 1918 and remained in effect until December 31, 1919. Cattell (2014, p. 13) explains that this legislation prohibited the manufacture of intoxicating beverages, the importation and transportation of them into Canada, and the transportation of them into any part of Canada where their sale would be illegal. In 1921, even though national prohibition had ended, a law was passed making it illegal to import liquor or beer and export any kind of alcoholic beverage without a license. In Ontario, wine was the only alcoholic product that was available for purchase. With some restrictions in place, wine could only be purchased at a winery in quantities of a five-gallon keg or a case with 12 bottles. For a more comprehensive description of prohibition in Canada, refer to Hudson Cattells's *Wines of Eastern North America: From Prohibition to the Present—A History and Desk Reference*.

Wine quality: post-prohibition to present

With prohibition still underway in the United States, there was a market for Ontario wines (Cattell 2014). Wine production in Canada went from less than 500,000 gallons in 1921 to more than 6 million gallons in 1929 (Schreiner 1984, p. 204). Furthermore, the number of wineries in Ontario doubled, from 20 to 51, between 1925 and 1929. People opened wineries out of barns, basements, and garages – anywhere where a winery could be built. Demand for wine in Ontario increased exponentially, and 'anyone who could get a carload of liquor across the border could earn $75,000' (Hallowell 1972). With an increase in demand, there was a race to get the wine to market, and as a result, quality was ignored. There were no limits on the amount of sugar or water that could be added to wine products. If too much water was added, it was common to add colorants such as tar dyes and formaldehyde to make the watered-down wine product look more like pure wine.

Setting the standards for Ontario wine had already started after the end of prohibition. The Liquor Control Board of Ontario (LCBO) was introduced in 1926 as part of the Ontario Liquor Control Act – this officially ended provincial prohibition. The introduction of the LCBO led to more strict control of the quality of alcoholic beverages, and within three years, half of Ontario's wineries were out of business. Cattell (2014) explains that the prohibitionary experiences left lasting memories, and it would be decades before a new winery was licensed in Ontario.

Canada's wine industry went through a significant shift, specifically in Ontario, beginning in the 1970s. The introduction of new legislation like the Wine Content Act of 1976 set standards on volatile acidity, chemical use, preservatives, and the amount of time acceptable between grape harvest and grape delivery (Cattell 2014). However, it was not until a ruling under the GATT that the Ontario government revised the Wine Content Act. Ontario vintners were forced to produce higher-quality wines. Additionally, the price equalization policy under the Canada-US Free Trade Agreement in 1988 made the wine industry more competitive.

Throughout the later 1980s and 1990s, the Ontario wine and grape industry continued to grow remarkably fast. In the 1980s, Canada began the process of developing the Vintner's Quality Alliance (VQA), similar to the French appellation d'origin controlée, which would set the standard for Ontario's highest-quality wine. In 1982, Donald Ziraldo, former president of Inniskillin Wines, organized meetings with the Ontario wine industry to set up an appellation control system. The system would allow consumers to identify wines of the province based on the origin of the grapes from which they were produced (Cattell 2014, p. 170). Meetings concluded in 1986, and the VQA was officially introduced. The introduction of the VQA symbol on a bottle of Canadian wine assured consumers on standards of production, content, varietal percentage, appellation, and vintage (Hope-Ross 2006, p. 4).

To appeal to consumers' developing tastes, *labrusca* grapes were phased out and eventually banned from being sold as table wine, and *vinifera* was introduced. The *vinifera* harvest grew year over year – from 4,400 tons in the 1990s to 20,400 tons in 2000 (Cattell 2014). Icewine was one of the first wines in Canada to gain international acclaim and was included in the VQA's 1987 draft of rules and regulations (Cattell 2014). Canadian vintners realized that some *Vitis vinifera* grape varieties, like Reisling or Vidal, could withstand the cold climates and produce an impressive product. Cattell (2014, p. 262) explains that the freezing of grapes could increase concentration and improve flavor characteristics that came with letting the grapes hang on the vines for two additional months. Inniskillin winery in Ontario is currently one of the most notable producers of icewine in Canada and has won numerous awards for their product.

Wine tourism regions

Ontario

Niagara Region: the main region

The Niagara Region is Ontario's most extensive wine-growing region. The region attracts more than 13 million tourists per year (Niagara Canada 2017), and many come for the wineries, wine tours, and wine festivals. Many of the wineries are located between the Niagara Escarpment and Lake Ontario, making it an attractive and motivating factor for tourists looking to visit. Many of these wineries have developed linkages with each other and with other tourism-related businesses in the region to increase its appeal as a wine tourism destination.

The region produces 90% of the province's grapes, making it one of Canada's leading wine grape regions. It is the largest icewine producer in Canada, and many wineries specialize in Vidal Blanc and Riesling wine. In 2005, 14,270 acres of vineyard were harvested in the region from the 21,715 acres harvested in Canada (Hope-Ross 2006). However, this was not always the case. As Telfer (2001) points out, traditionally based on native *labrusca* grapes, the region was not noted for high-quality wines. It was not until the introduction of VQA and changes to the GATT that wine quality improved. The number of wineries in the region grew almost fivefold since 1988, with new wineries opening every year and continuous improvements made to existing ones (Carmichael and Senese 2012). Telfer (2001) argues that the rapid development of a large number of new wineries led to the establishment of wine tourism clusters throughout the region.

One of the most minor wine tourism clusters in the Niagara Region is found in the village of Jordan, located in the 'Bench' winery region. Cave Spring Cellars winery, along with its restaurant On the Twenty, opened in 1986 and 1993, respectively. The winery is located across from other heritage tourism buildings, including galleries, antique shops, and a museum. These businesses are often advertised together, and Cave Spring Cellars encourages tourists to visit the neighboring businesses. Telfer (2001) explains that it has become an out-of-the-way destination, which builds on the synergistic relationship between the tourism cluster and the winery. Beyond the village level, On the Twenty restaurant is part of a Taste of Niagara. Taste of Niagara brings together hotels, restaurants, wineries, and chefs, to name a few, to promote the use of local food in the tourism industry.

There is recognition of the value of tourism networking and cooperative partnerships in the town (Telfer 2001). Cooperative festivals and special events such as the 'Six Unforgettable Weekends of Summer' are an important way of promoting wine tourism in the area (Telfer 2001). Additionally, there are over 100 bed-and-breakfast establishments, and many have developed partnerships with the wineries to offer tourism packages to bring visitors to the area. Many of the wineries in Niagara-on-the-Lake are located within proximity to each other and can be easily visited in a day – bicycle tour operators have intentionally established their route to take tourists around the wineries. Niagara-on-the-Lake is home to over a dozen wineries, many of which are part of an association called Wineries of Niagara-on-the-Lake. The association markets products produced by the wineries and has developed partnerships with other tourism sectors from transportation to accommodation.

As the wineries become a significant force in the tourism industry, the level of synergy between the wineries and tourism continues to build (Telfer 2001). It is the collaboration of wineries and other tourism businesses that have contributed to wine tourism's growth and success in the region.

Prince Edward: the newcomer

Prince Edward County is a rural community located near Lake Ontario and south of Belleville. Agriculture has been the backbone of its economy, and the new and emerging wine industry is a reflection of a new generation of agritourism in the region. Prince Edward County is a newcomer to wine tourism in Canada, as the region did not receive its official appellation VQA until 2007. Just 20 acres of land were used for grape production in 2000, which grew to more than 600 acres in 2010. Together with Essex Pelee Island Coast, another relatively small wine region in Ontario, they are responsible for roughly 10% of the province's grape production. The region offers tourists a plethora of wine tour options and some festivals. The Vintner's Festival, founded in 2018, showcases winter art and offers food and wine pairings. It is common for culinary tourism and wine tourism to be linked. Wineries and restaurants are frequently advertised together. Some of the local wineries have even developed partnerships with local restaurants. Prince Edward County is home to a thriving tourism industry centered in one of Ontario's newest wine and culinary-making regions (Stolarick et al. 2010, p. 240).

The Okanagan Valley: the 'Napa' of Canada

The Okanagan Valley is located in the south end of the province of British Columbia. It is one of Canada's oldest and most significant wine tourism regions. The 124-mile-long valley begins at the town of Osoyoos in the south and runs north between the Cascade Mountains in the west and the Monashee Mountains in the east (Anderson, McGillivray, and Giberson 2007). For many years, tourism in the Okanagan Valley was dependent on its favorable climates, scenic landscapes, and orchard industries (Figure 7.1). Presently, the valley has a well-established

Figure 7.1 Nk'Mip Cellars and Cultural Centre, Osoyoos, Canada.

Source: Rémy Tremblay, 2020

wine industry that coincides with an equally remarkable shift in the region's tourism industry (Carmichael and Senese 2012).

With its diverse and unique landscapes, the valley has a long history of attracting tourists to the area. Much like the Niagara Region, the wine region and subsequent wine tourism industry have seen rapid growth in recent decades. Since the introduction of VQA quality standards, wineries grew from 14 in 1988 to 130 in 2009. The development of these wineries has caused a shift in Okanagan tourism's focus and the rural aesthetics that accompany it – such as integrated services and amenities. Wineries across the valley provide additional tourism-related products along with their core product, wine. Senese (2009) stated that 53 wineries provided tourism-related products and services in the Central Okanagan alone, 48% provided a gift (non-wine) shopping experience, 22% offered tours, 18% had full restaurant food service, and 12% provided accommodations.

Interestingly, the production of wine is secondary to the wine tourism experience in the Okanagan Valley. For example, Mission Hill, one of the region's largest wine producers, received a majority vote in 2011 from the West Kelowna District Council to take the next steps toward a $10 million expansion. This expansion had little to do with wine production, as it included a plan to develop a restaurant, cottages, a wellness center, a 120-room hotel, a museum, and a microbrewery.

The Okanagan Valley has seen significant growth in its wine industry since the first grapes were first planted 100 years prior, and its overall appeal as a wine tourism destination plays a critical role in maintaining overall competitiveness in the international marketplace. Festivals, tasting rooms, and accommodations are essential parts of the wine tourism experience in the region. It is a region that uses culture from its history and landscapes to remain competitive in the global market as a New World wine region.

Other Regions

Nova Scotia: growing and growing

Wine production in this province is significantly lesser when compared to the Okanagan Valley and the Niagara Region. It comprises several small producers, most of which produce fewer than 10,000 cases annually (Rimerman and Eyler 2017). Unlike those of Ontario or British Columbia, the province's climates make certain grape varieties more challenging to grow. Gillian Mainguy of the Winery Association of Nova Scotia says it was not until the 2000s, when winemakers were hired who had a better understanding of the *terroir* (Patten 2016), that wine quality and taste began to improve. Nova Scotia received its first appellation designation with Tidal Bay in 2012.

Quebec: thanks to hybridization

Quebec's wine industry was established in 1978, and it is the third-largest wine industry in Canada. However, it has yet to see the commercial success present in Ontario and British Columbia. Much of this is due to its climate, which has been a limiting factor for the province's wine industry. The intensely cold winters and short growing seasons limit the types of wine grapes that can be grown (Jones 2012). The province's wine industry is mostly focused on cold-tolerant interspecific hybrid grapes, such as Frontenac, Sabrevois, and St. Croix – but these hybrid varieties often produce low-quality wine (Slegers et al. 2015).

Case Study 7.1 The Nk'Mip Winery

Throughout Canada's history, Indigenous people have been stifled of their identity and undermined in their cultures. To counter the oppression brought on by colonialism, these communities have long used their culture and language to self-identify. The National Aboriginal Tourism Research Project indicated that one of the most significant barriers for Indigenous tourism development is the absence of coordinated tourism initiatives (O'Neil et al. 2015). The report identified the lack of partnerships and linkages with other Indigenous tourism businesses as one of the many critical economic growth barriers for these communities. The Osoyoos Indian Band (OIB), a First Nations government in the Okanagan Valley of British Columbia, is an interesting case of an Indigenous group using tourism business linkage through a unique tourism development project, the Nk'Mip Project, to benefit their economy, their culture and their surrounding environment.

The OIB, also known as Nk'Mip (pronounced 'in-Ka-meep'), was formed in 1877 and consisted of 540 members. For the OIB, their cultural identity is embedded in their lands, and it has laid the foundation for them to be economically self-sufficient. Through economic development activities, they have been able to rebuild their community, and their participation in the wine industry has played a critical role. Their lands stretch across 12,950 hectares and are located in the south end of the Okanagan Valley. They are one of two principal wine areas in the country, containing 25% of all vineyards in the valley (Anderson, McGillivray, and Giberson 2007; Aspler 2006). Their lands are also among the most environmentally diverse and home to one-third of Canada's endangered species (Anderson, McGillivray, and Giberson 2007). It is the land's dichotomous relationship that secures their economic independence while allowing them to preserve their cultural identity and traditions.

The OIB's involvement in the wine industry dates back to 1968, when members of the Band planted their first vineyard on 200 acres of land in the Okanagan Valley and called it Nk'Mip Vineyards. By 1969, they became the largest vineyard in the valley, supplying high-quality *vinifera* grapes to many wineries in the region. Through a strategic partnership with Vincor International (now Constellation Brands), Canada's largest wine producer, they are now one of the most extensive privately owned vineyards in western Canada. They grow Cabernet Sauvignon, Merlot, Pinot Noir, Pinot Blanc, Chardonnay, Cabernet Franc, Chancellor, Pinot Gris, Pinot Meunier, Scheurebe, Ehrenfelser, and Riesling. Constellation Brands leases approximately 800 acres for premium *vinifera* grapes, and the OIB manages 40 acres of the vineyard that supplies their winery, Nk'Mip Cellars.

Nk'Mip Cellars is part of a larger economic development project called the Nk'Mip. The project encompassed various planned developments to draw visitors to the area, including Nk'Mip Cellars – its associated vineyards and the Nk'Mip Desert Cultural Center. These two developments were expected to be crucial drivers of wine tourism and ecoculture tourism for the area and act as a vehicle for economic growth. Clarence Louie, Chief of the OIB, said that the most important aspect of their development plan involved incorporating and respecting heritage (Anderson et al. 2005). Incorporating culture and tradition into tourism developments is a customary practice amongst Indigenous populations, as validation of their culture is a way of survival. Bresner (2014) explains that Indigenous tourism destinations are often set on their people's history and culture as their primary means of attracting visitors to their destination.

The development of Nk'Mip Cellars made it the first Indigenous-owned and operated winery in North America. It opened in September 2002 as part of a joint venture with Constellation Brands. The $7 million development included 18,000 square feet of a winery, and a 20-acre vineyard. It is the first winery travelers to the area encounter, and it is the only winery directly accessible to drivers on the east–west route, known as the wine route. It was built to create small quantities of high-end wine; it has the capacity for only 18,000 cases. All their wine bottles don the VQA certification. They produce Chardonnay, Pinot Blanc, Pinot Noir, and Merlot, using grapes strictly from the Band's vineyards (Anderson, McGillivray, and Giberson 2007, p. 348). Most of their wine is sold at their properties, while the rest is sold at high-end restaurants, on specialty properties, and in some international markets.

Built directly adjacent to the winery is the Nk'Mip Desert Cultural Center. Anderson, McGillivray, and Giberson (2007) argue that the center acts as a value-added attraction for tourists visiting the winery – the experiences at the cultural center were meant to enhance the visitor's experience at the winery. The center was another way for the OIB to educate visitors on their heritage, culture, and the nature of their lands. The center offers a range of activities to visitors, from events to self-guided tours that showcase their culture and heritage. Their culture is an essential part of their identity, and visitors to the area will find their culture and traditions embedded in their businesses.

Viniculture has its roots in many aspects of history and has been used for centuries to emphasize cultural traditions and ancestral heritage. Wine consumption has often led individuals to reconnect to a particular region or build new connections to the land where it is made. For Indigenous communities in Canada, their land is an essential part of their culture. The OIB has used grape-growing and wine production to remedy their socioeconomic deficiencies and bring awareness to their heritage. Through the development of the Nk'Mip Project, they have created strategic linkages between their tourism businesses, which have allowed them to be economically autonomous while preserving, strengthening, and bringing awareness to their traditional values.

Conclusion

Canada's wine industry has been producing wine for over two centuries. Despite the crippling effect prohibition had on the Canadian wine industry, the industry has become a key factor in rural economic development. The wine industry is a considerably small player in the international market, but developing linkages between wineries and other tourism businesses has been critical for its economic growth. The strategic alliances between wineries and other tourism businesses present in the Okanagan Valley and the Niagara Region are two important examples of how strategic tourism development can be a catalyst for economic growth.

However, despite positive growth and favorable tourism visitation numbers to Canadian wine regions, the industry still faces many challenges. One of these challenges is inter-provincial exports, as market competitiveness remains high and leaves some smaller wineries vulnerable. Canadian vintners still face barriers that limit their abilities to bring their products across provincial borders – a stark contrast to other wine regions like the United States that have the ability to trade freely. Although the development of wine tourism in Canada has positively impacted the economy and has led to developments in more rural parts of the country, the gentrification of wine regions is another concern. The case study of Nk'Mip Cellars is an example of how an Indigenous community used strategic linkage in tourism development as means to be economically self-sufficient and protect their culture. However, for some Indigenous communities, gentrification could impact their social welfare if not carefully considered during any tourism development process.

Implications

The implications of this study extend to other marginalized and Indigenous communities with an interest in employing agritourism to complement and enhance their existing business practices. This case highlights the opportunity for sustainable community development as it encompasses a long-term strategy that considers cultural integrity, and social and economic needs. Relying solely on tourism as a means of economic sustainability is risky, as there are a variety of factors that can halt tourism operations altogether. In this case, through the Nk'Mip Winery, the OIB employs tourism clustering as an effective way to build connections between businesses, communities, and government. Consequently, by working with members of the private and public sectors, this marginalized community is able to effectively build a successful tourism destination while simultaneously protecting their lands and heritage. Like many Western developed countries, Canada is eager to reconcile its shameful history towards Indigenous groups, and in this case, tourism development allows them to strengthen their culture while promoting economic development in their community.

References

Anderson, R. B., Camp, R. D., Hindle, K. G., and Kayseas, R. (2005). *Corporate Aboriginal Alliances: A Case Study of the Osoyoos Indian Band.* Hawthorn: Swinburne University of Technology.

Anderson, R. B., McGillivray, S., and Giberson, R. J. (2007). The Nk'Mip Cellars: wine and tourism with an Aboriginal flavour. In Dana, L., and Anderson, R. B. (eds.), *International Handbook of Research on Indigenous Entrepreneurship* (pp. 336–351). Northampton, MA: Edward Elgar Publishing, Inc.

Aspler, T. (2006). *The Wine Atlas of Canada.* Toronto: Random House.

Bresner, K. (2014). Sharing identity through Indigenous tourism: Osoyoos Indian Band's Nk'Mip Desert Cultural Centre. *Anthropologica,* 56(1): 135–150.

Carmichael, B. A., and Senese, D. M. (2012). Competitiveness and sustainability in wine tourism regions: the application of a stage model of destination development to two Canadian wine regions. In P. H. Dougherty (ed.), *The Geography of Wine Regions, Terroir and Techniques* (pp. 159–178). New York: Springer Dordrecht Heidelberg.

Cattell, H. (2014). *Wines of Eastern North America: From Prohibition to the Present—A History and Desk Reference.* 1st Ed. Ithaca, NY: Cornell University Press.

Fundeanu, D. D. (2015). Innovative regional cluster, model of tourism development. *Procedia Economics and Finance,* 23: 744–749.

Hallowell, G. A. (1972). *Prohibition in Ontario, 1919–1923.* Ottawa: Ontario Historical Society.

Hope-Ross, P. (2006). *From the Vine to the Glass: Canada's Grape and Wine Industry.* Ottawa: Statistics Canada.

Hughes, A. (2010). *Looking back... with Alun Hughes: Johann Schiller: Father of Canadian Wine?* https://brocku.ca/social-sciences/geography/wp-content/uploads/sites/152/Johann-Schiller-Father-of-Canadian-Wine.pdf

Jones, N. (2012). The influence of recent climate change on wine regions in Quebec, Canada. *Journal of Wine Research*, 23(2): 103–113.

Niagara Canada (2017). *Niagara Tourism Profile*. Niagara: Niagara Canada.

O'Neil, B., Williams, P., Morten, K., Kunin, R., Gan, L., and Payer, B. (2015). *National Aboriginal Tourism Research Project 2015: Economic Impact of Aboriginal Tourism in Canada*. Vancouver: Aboriginal Tourism Association of Canada.

Patten, M. (2016). Canada's next great wine region is Nova Scotia. CBC. https://www.cbc.ca/news/canada/nova-scotia/nova-scotia-great-wine-1.3512006

Rimerman, F. A., and Eyler, R. (2017). *The Economic Impact of the Wine and Grape Industry in Canada 2015*. St. Helena, CA: Frank, Rimerman & Co. https://www.ontariocraftwineries.ca/wp-content/uploads/2017/03/Canada-Economic-Impact-Report-2015.pdf

Schreiner, J. (1984). *The World of Canadian Wine*. Vancouver: Douglas & MacIntyre.

Senese, D. (2009). *Agritourism, Amenity Resources and Organic Growth Poles*. Ottawa: Canadian Association of Geographers.

Slegers, A., Angers, P., Ouellet, É., Truchon, T., and Pedneault, K. (2015). Volatile compounds from grape skin, juice and wine from five interspecific hybrid grape cultivars grown in Québec (Canada) for wine production. *Molecules*, 20(6): https://doi.org/10.3390/molecules200610980

Stolarick, K., Denstedt, M., Donald, B., and Spencer, G. M. (2010). Creativity, tourism and economic development in a rural context: The case of Prince Edward County. *Journal of Rural and Community Development*, 5(1/2): 238–254.

Telfer, D. J. (2001). From a wine tourism village to a regional wine route: an investigation of the competitive advantage of embedded clusters in Niagara, Canada. *Tourism Recreation Research*, 26(2): 23–33.

8

KEY CHALLENGES IN GLOBAL WINE TOURISM

Vasco Santos, Paulo Ramos, and Bruno Sousa

Introduction

The growing global competitiveness, the need to innovate, and the desire to make a difference are at the origin of the progressive segmentation of the tourism market. This leads, therefore, to the diversification and multiplicity of tourist products. In this context, wine tourism as the intersection between tourist and winemaking activities has been a much-explored theme, nationally and internationally, by the scientific community, which has boosted scientific investigation about this issue. Globally, the phenomenon of wine tourism has become increasingly more important, as it is considered a trigger for the development of many rural areas, economically and socially (Presenza, Minguzzi, and Petrillo 2010). Getz and Brown (2006) state that the success of wine tourism depends on the good operationalization of the components of the wine tourism system, through the 'value chain of wine tourism': (stage 1) basic resources; (stage 2) viticulture; (stage 3) production of wine (wineries); and (stage 4) wine exports on one side, and wine tourism on the other.

A new definition of wine tourism has been proposed by Santos, Ramos, and Almeida (2018, p. 683):

> [W]ine tourism consists of an activity directly related to wine which provides a dynamic and versatile experience that integrates wine culture and heritage, to create emotions, sensations, attachment, and sensory impressions through the visit, allowing the wine tourist to become an advocate of that particular cellar, brand or wine region.

Therefore, wine tourism is a system composed of three subsystems: territory, tourism, and winemaking culture (Carta Europeia do Enoturismo 2006, p. 3). The experience and practice of wine tourism can be provided in a number of ways, from events and festivals, cultural and natural heritage, gastronomy, hospitality, wine tasting, visits to wineries and vineyard tours (Ramos, Santos, and Almeida 2018).

The most common resources and capabilities for developing wine tourism include incorporating wine tastings, services, and wine sales into winery visits, vineyard and wine cellar tours, and wine festivals. After the literature review of the special interest in wine tourism and

DOI: 10.4324/9781003143628-10

global characterization of wine and wine tourism sectors, the key challenges in global wine tourism are described. Finally, this chapter presents a case study that addresses the wine tourists' behavior in Porto wine cellars.

Special interest tourism in wine context

Tourism is a multifaceted and geographically complex activity. The evolution of tourism has promoted the emergence of new (and distinct) segments, niches, and even micro tourism niches. Tourism is progressively characterized by smaller market segments with different individual interests. A niche is a small market consisting of individual consumers or small groups of consumers with similar characteristics or needs. The growing global competitiveness, the need to innovate, and the desire to make a difference are at the origin of the progressive segmentation of the tourism market. This leads, therefore, to the diversification and multiplicity of tourist products (Braga and Sousa 2021). We easily find activities and/or products considered as specific interest tourism. Some special forms of tourism (SFT) are the result of technological developments (e.g., virtual tourism, e-tourism, or cyber tourism). SFT can be operated both on a first level, through a network of small companies (tour operators, hotels, etc.), making use of low barriers to entry, characteristics of the tourism industry, and employing information technologies as a form of 'home industry'; as at another level, on a larger scale, through transnational companies, using highly developed specialist systems that can provide a greater variety of tourism products and tourism services (e.g., dark tourism, pilgrimage, agritourism, military tourism, safari tourism, wedding tourism, wine tourism, accessible tourism, cruise tourism, film tourism, voluntourism, youth tourism, red-tape tourism, luxury tourism, wellness tourism).

Special interest tourism (SIT), interchangeably referred to as niche tourism, has become a noticeable phenomenon in the tourism industry and tourism literature since the 1980s. The concept of SIT is generally defined as travel for a specific interest or motivation with a provision for a customized experience (Douglas and Derrett 2001; Sundbo and Dixit 2020; Weiler and Hall 1992), which traditionally attracts a small number of tourists (Robinson and Novelli 2005; Sousa and Simões 2010). The rapid growth of SIT is largely due to the heterogeneity of the market products, as well as the increasing demand for more focused activities and interest-based travel experiences (Douglas and Derrett 2001; Ma, Kirilenko, and Stepchenkova 2020; Trauer 2006). As it is a type of tourism associated with different types of general tourism and even with niche tourism markets, we can consider wine tourism as a type of SIT, the production of which is a process guided by consumers' specific interests and motivations.

SIT has traditionally been regarded as the counterpoint alternative to mass tourism (Santos, Ramos, and Sousa 2021). However, there have been arguments that SIT and overall tourism (or mass tourism) are not necessarily mutually exclusive and often overlap (Ma, Kirilenko, and Stepchenkova 2020). According to Trauer (2006), the leisure specialization cycle demonstrates behavior, the cognitive system, and affective involvement. These three systems interact, where each facet of the involvement will end up influencing specific behaviors in different ways. The author highlights the facets of involvement of importance and fruition (attraction), identity and social cohesion (self-expression), lifestyle, and work (centrality) that underpin the concept of enduring involvement. From the demand perspective, niche tourists vary in motivations, subsegmenting on the continuum from GIT (general interest tourism, namely mass tourism) to SIT (Brotherton and Himmetoğlu 1997). According to Ma, Kirilenko, and Stepchenkova (2020), a 'dabbler' niche tourist (an unconvinced mass tourist who has a preliminary interest in SIT) is at the transitional stage from GIT to SIT and may still strongly resemble a mass tourist. From the supply perspective, destinations, tourism, and hospitality businesses cater to modern tourists' growing demand for excitement and personalized experiences (Robinson and Novelli 2005) and multi-motivational choices for destinations or smart tourism. According to Sousa and Simões (2010), tourists are increasingly looking for emotional stimuli, sensations with the purpose of 'buying feelings' rather than tangible or material products.

Specifically, the relationship between tourism and the wine industry reveals a number of positive synergies (Andrade-Suárez and Caamaño-Franco 2020). Wine tourism should be understood as visits to vineyards, wineries, wine festivals, and events in which tasting and/or experiencing the characteristics of winemaking regions are the principal pull factors for visitors. It can therefore be included in the subsectors of agricultural, rural, cultural, special interest, and, of course, industrial tourism (Duarte Alonso, Kok, and O'Brien 2020). Encouraging winemakers and tourist operators to collaborate in a very competitive market like tourism is no easy task. Cooperation, partnership, trust, and commitment to environmental preservation, customer satisfaction, destination loyalty, and quality are required by all participating members in order to make the cluster sustainable.

Even though wine production and tourism synergies have developed successfully in the mature wine regions of Europe (Italy, Spain, France, Portugal), nontraditional wine regions best follow the practices of synergetic economies in countries like the United States, New Zealand, Australia, Chile, and Argentina. Tourism and wine share a common developmental restriction, sustainability, and environmental quality (Karagiannis and Metaxas 2020).

The wine tourism industry has many forms and types regarding vineyards, cellars, wine houses, farms, wine hotels, services, structure, and dimensions. Wine tourism provides wine tourists with wine and wine tourism experiences. The special interest in wine tourism is related to the increasing role and impact that winemaking tourism has on different experiences, sensations, perspectives, motivations, and intentions. Santos et al. (2020) advocate that the wine and wine tourism experience plays a crucial role for wine companies to achieve a set of strategic and competitive advantages, as well as opportunities to continue growing, developing, and consolidating within wine tourism vacations contexts and wine trips that occur mainly among the wine tourism products and destinations. The most common resources and capabilities for developing wine tourism include incorporating wine tastings, services, wine sales into winery visits, vineyard and wine cellar tours, and wine festivals (Torres and Kunc 2016). These resources generate short-term revenue not only for wine firms but also for tour operators that offer a variety of complementary services related to the wine experience, such as dinners and excursions (Torres et al. 2020).

For instance, the wine sensory experience (WSE) has stood out as a very recognized and valid wine tourism and marketing construct by the scientific community, which has boosted scientific investigation about this issue. This phenomenon is due to the potential that wine products have as the core attraction in wine tourism destinations and products. WSE relates the intersection between wine tourists and winemaking activities (Sousa 2019). WSE involves wine tastings and cellar, wine house, and wine museum visits. Currently, and in the future, WSE has provided and boosted the creation of genuine and unique wine sensory appeal, more differentiated, tailor-made, mainly provided by wine tour guide/wine storyteller/winemaker/wine producer strictly about wine, e.g., wine bottle/wine brand/wine family/wine estate/winemaking (Santos, Ramos, and Sousa 2021).

Torres et al. (2020) argue that tourists usually visit multiple places along their travel routes rather than wineries alone. People engage in multidestination trips for four reasons: to provide variety, as they want to have other experiences; to decrease the probability of a completely unsatisfactory trip occurring; to please multiple decision-makers involved in the planning and execution of activities; and to leverage the lower marginal cost (in terms of time and money) of an additional activity as part of in a multidestination trip. In light of all this, the next section will discuss the global characterization of wine and wine tourism sectors. According to Festa et al. (2020), in developing wine tourism, a central role is clearly reserved for the territory, from the hardware point of view (landscape, cultural heritage, etc.) and from the software point of view (services, utilities, etc.). From this perspective, the contribution of the institutions in general and the local communities in particular, combined with the entrepreneurial energy of the wine tourism sector, could and should provide the impetus for the further development of wine tourism in different countries. Therefore, Karagiannis and Metaxas (2020) argue that, while global wine consumption is growing, environmental sustainability in the wine industry is receiving increased attention from retailers, governments, environmental groups, and consumers. Along with this growth, wineries are also facing issues on environmental sustainability, such as water consumption, waste management, and impact on the community. The local environment, community, and economy can all be negatively impacted by the wine-growing industry and vice versa. The pressure for sustainability practices from a niche marketing and wine tourism perspective, products, and services start from external drivers such as customer demands, investors, communities, government, and the need for compliance with regulations.

Global characterization of wine and wine tourism sectors

The relationship between wine and tourism is of a circular nature. The wine regions themselves attract wine and regular tourists that are a source of income for the wine industry of the region. The global wine market value in 2019 is estimated at around US$315 billion. The wine production in the world in 2019 was estimated at 260 mhl, presenting a decrease compared to the highest production numbers ever reached in 2018. The world wine consumption in 2019 is estimated at 244 mhl, marking a +0.1% increase compared to 2018. In 2019 the world wine export market grew compared with 2018, both in volume by +1.7% (1 by 05.8 mhl), and value by +0.9% (31.8 billion EUR) (OIV 2020). The global average per capita wine consumption is around 3.4 liters and US$45.79 per capita in 2020 (Statista 2020). One of its key contributions in terms of wine tourism is the way how it shapes the rural landscape of the main wine regions (the winescape). However, the global planted area of vineyards has not changed much since 2016 at around 7.4 mha. Even with the 2020 COVID-19 pandemic crisis pushing down the figures, it is expected that the global wine market can reach US$434.6 billion by 2027 (Reportlinker 2020).

The tourism sector amounted in 2019 to US$8.9 trillion, a share of 10.3% of the world's global GDP. It employs 330 million people (also a 10% share of the jobs in the world) and created revenue of US$1.7 trillion in visitor exports that are 6.8% of total exports and 28.3% of the world services exports (WTTC 2020). Due to the COVID-19 crisis, global tourist arrivals dropped dramatically, by 93% in June 2020, and the tourist arrivals also declined by 65% In the first six months of 2020 (ONWTO 2020), causing a loss of 440 million international arrivals and around US$460 billion in export revenues. This is five times more than the loss in the subprime financial crisis internationally felt by 2009. Wine tourism seems to be more exposed to the effects of travel restrictions than the wine sector by itself that, although the contraction in the on-trade consumption, can still sell its wines to the consumer off-trade (in-store and online). This may lead to potential disinvestment by the wineries in these activities. Therefore within this colossal crisis, both sectors need now, more than ever, to joint efforts to try to create a synergetic effect that can potentiate the recovery both in the wine sector and in the touristic sector, favoring wine tourism as a key.

Key challenges in global wine tourism

Table 8.1 shows the key challenges directly related to global wine tourism from the present to the future.

Table 8.1 Summary of key challenges and suggestions in global wine tourism

In spite of the pandemic crisis (COVID-19), wine tourism is considered one of the priority areas of government action and public policies, through the allocation of measures, support, and financing mechanisms, to make the value chain stronger.
The trend of the growth and development of supply and demand of wine tourism activities and experiences are widely known.
Worldwide, wine tourism has reached a level of excellence, and it's now more aligned with the New and Old World wine tourism image than ever before.
Improve physical and capacity conditions and also the requirements of health safety.
Integrate and further enhance the opinion makers and leaders: managers, stakeholders, players, marketers.
Develop sustainable and ecologic wine tourism practices.
Wine tourism is increasingly a product and wine tourism destination with high demand, which before the COVID-19 crisis has performed much better than expected.
Concerted action in digital promotion and marketing action plans to express the importance of wines in the promotion of wine tourism destinations.
Create virtual and augmented reality platforms in wine tourism: digital and hybrid wine events.
Reinforce strongly on digital channels and platforms: blogs, vlogs, websites, apps.
Encourage the implementation of smart wine tourism companies.
Significant importance of wine as a high involvement product in the continuous promotion of wine tourism in the world.
Opportunity to increase the quality and competitiveness of the wine tourism industry in relation to the main competitors and investors.
Investing and creation of new wine tourism projects, wine tourism units, wine routes and wine hotels, wine estates, cellar doors, among others).
Innovating in wine tourism creative activities for all: from kids to seniors.
Increase and foster more accessible and inclusive wine tourism companies.
Frequent application of the good practices' manual for wine tourism technical guide.

Case Study 8.1 Wine tourists' behavior in Porto wine cellars

In order to better illustrate some of the key issues that wine tourism faces, a case study based on a research study undertaken in the Porto wine cellars is presented. The Porto wine cellars are a relevant example of wine tourism due to the historical and cultural heritage that they represent, but also as an example of a concentrated area where many different wine cellars can be visited. The study was conducted in 2015 when Portugal was still recovering from the subprime financial crisis that led to a FMI/EU joint intervention. The Portuguese wine industry was one of the few sectors in the primary sector that continued to grow in a sustainable way, both internally and externally, with the ability to generate wealth and to help attract qualified workers to the interior of Portugal, valuing heritage and landscape. The production of wine states itself as a growing, global industry.

The case study aims to understand what makes the wine tourist move, namely relating to specific behavioral constructs such as involvement, destination emotions, place

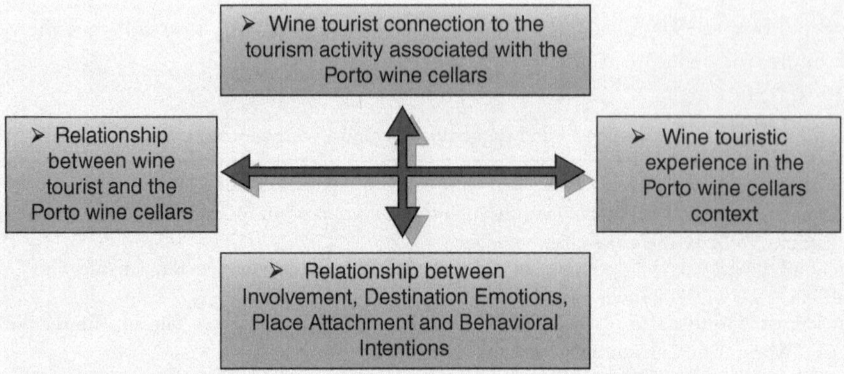

Figure 8.1 Case study framework.

Source: Authors

attachment, and behavioral intentions, in order to better understand wine tourist behavior (see Figure 8.1). It demonstrates how some of the key behavioral dimensions affect the relationship that wine tourists have on a particular experience.

The field study of wine tourist behavior there is still needed to better understand some of the drivers and outcomes of their visits, especially the factors and attributes that enhance future behaviors. This study focuses on the role and impact of the wine tourists' personal involvement (PI), wine product involvement (WPI), destination emotions (DE), and place attachment (PA) have on the wine tourist travel behavior (WTTB) whilst predicting behavioral intentions to visit Porto wine cellars. Ramos, Santos, and Almeida (2018) suggested that these constructs play a fundamental role and impact on behavioral tourists' intentions, including leisure, marketing, and tourism fields, in their content analysis of the relationship between involvement and emotions.

To better understand what was measured, a brief definition of these behavioral constructs is made.

- *Personal involvement*: Personal interest of the wine tourist with a product, object, situation, event, activity, or destination (Prayag and Ryan 2012; Yeh 2013).
- *Wine product involvement*: Personal involvement of the wine tourist with wine and their attributes/characteristics (Barber, Ismail, and Dodd 2008; Brown, Havitz, and Getz 2007).
- *Destination emotions*: Mental, physical, and psychological reaction expressed through negative or positive responses, depending on the experience that a wine tourist feels in a destination (Soscia 2013; White and Scandale 2005).
- *Place attachment*: The attachment of the wine tourist in relation to the place identity and place dependence developed in a wine tourism destination/activity/visit (Lewicka 2011; Yuksel, Yuksel, and Bilim 2010).
- *Behavioral intentions*: Personal intentions of the wine tourist to visit or revisit a wine tourism destination/activity/visit (Dixit 2017; Liu, Lee, and Kim 2015; Žabkar, Brenčič, and Dmitrović 2010).

A cross-sectional design was used, and the data gathering took place in four Porto wine cellars where samples of the similar and comparable population were obtained. The questionnaires were aimed at international wine tourists older than 18 years visiting Porto wine cellars. The top five nationalities were English, Spanish, French, Portuguese, German and Brazilian English, Spanish, French, Portuguese, and German.

In order to achieve a considerable sample size, when there is greater affluence and attraction of high visitation, data were collected in four Porto wine cellars during the visits for three weeks between late August and early September 2020. Only tourists who completed their visit were asked to participate in this study. A convenience sampling approach was employed in this study due to limited time and manpower. A total of 918 self-administrated questionnaires were considered for data analysis. The validation of the structural model was confirmed by multivariate statistical analysis with SEM using the PLS (v.3).

The wine tourists' profile

It was found that visits to Porto wine cellars are sought by middle-aged (almost 74% of respondents were between 31 and 60 years old) wine tourists, both men (51%) and women (49%), from various nationalities (the most respondents are of foreign origin), with a greater expression of France (24.5%) and, after, UK (12.2%) and Spain (10.8%), with similar values; then Germany (7.2%), Brazil (3.9%), and Portugal (2,6%) with less expression. Regarding academic qualifications, more than two-thirds have a university degree (see Table 8.2).

The majority sample of 79,6% stated the strong importance of the Porto wine tastings during the visits to the Porto wine cellars (Table 8.3).

The wine tourists stated that they spent an average of 1.87 (SD 1,09) days; 43.3% of the sample spent 1 day in the city, 37.9% spent 2 days and 11.8% spent 3 days (Table 8.4).

Considering the value of money spent on this visit, 46% agreed that it was a high effort of expense, while 35,3% disagreed with that statement (Table 8.5).

For 91.2% of the sample, this was their first time visiting the Porto wine cellars. Within the 8.8% that had already visited these wine cellars in the past, 50% were doing it for the second time, 19.2% for the third time, and 21.8% for the fourth time (Table 8.6).

When asked if they were visiting the wine cellars of Porto during the period of late August and early September because they were on holidays, 96.5% agreed with that statement, with only 3.5% of the sample having other reasons; 50.6% of the sample visited the cellars with family members, 40.9% with friends, 18.3% with a business colleague, and 3.2% went alone.

The descriptive statistics show that all items of the constructs present identical values in means. The 12 items of destination emotions (feelings) present a general mean of 6.72. The 15 items of destination emotions (pairs of emotions – positive and negative) present a general mean of 6.66. The 14 items of personal involvement present a general mean of 6.67. The five items of the consumer involvement profile (as part of personal involvement) present a general mean of 6.70. The five items of wine involvement present a general mean of 5.98. The eight items of place attachment present a general mean of 6.60. The four items of behavioral intentions present a general mean of 6.76. These results allow us to infer that wine tourists establish a strong involvement with Porto wine cellars. They also associate positive destination emotions to these cellars, and they feel connected and attached to culture, heritage environment, establishing a strong place identity and place attachment to Porto wine cellars.

After describing the general data, we verify the relationship between the relationships between the constructs using the partial least squares method (PLS) (as shown in Figure 8.2).

Table 8.2 Sociodemographic profile of the sample respondents

Gender (N = 914)	*Age (N = 912)*	*Education level (N = 914)*	*Country of origin (N = 916)*	*Job (N = 918)*
				Businessman/manager (12.6%)
	18 years old (2%)		Portugal (2.6%)	Freelancer/self-employed (16.4%)
	19–30 years old (17.2%)	Less than high school graduate (3.4%)	Spain (10.8%)	Middle/senior employed management (23.7%)
Male (51%)	31–40 years old (26.5%)	High school graduate (22.9%)	France (24/5%)	Civil servant (16.7%)
Female (49%)	41–50 years old (26.8%)	Degree (36.2%)	Germany (7.2%)	Worker (13.2%)
	51–60 years old (20.9%)	Master's degree (28.3%)	United Kingdom (12/2%)	Pensioner/retired (4.6%)
	61–70 years old (7.5%)	Doctorate degree (9.2%)	Brazil (3.9%)	Domestic/unemployed (1.9%)
	More than 70 years old (0.9%)		Other countries (27.8%)	Student (3.9%)
				Other (7%)
Gender (N = 914)	Age (N = 912)	Education level (N = 914)	Country of origin (N = 916)	Job (N = 918)
				Businessman/manager (12.6%)
	18 years old (2%)		Portugal (2.6%)	Freelancer/self-employed (16.4%)
	19–30 years old (17,2%)	Less than high school graduate (3,4%)	Spain (10.8%)	Middle/senior employed management (23.7%)
Male (51%)	31–40 years old (26,5%)	High school graduate (22,9%)	France (24.5%)	Civil servant (16.7%)
Female (49%)	41–50 years old (26,8%)	Degree (36,2%)	Germany (7.2%)	Worker (13.2%)
	51–60 years old (20,9%)	Masters degree (28,3%)	United Kingdom (12,2%)	Pensioner/retired (4,6%)
	61–70 years old (7,5%)	Doctorate degree (9.2%)	Brazil (3,9%)	Domestic/unemployed (1,9%)
	More than 70 years old (0,9%)		Other countries (27.8%)	Student (3,9%)
				Other (7%)

It is evident that personal involvement and wine product involvement had a vital role and impact in creating a destination emotion and a place attachment in the Porto wine cellars tourists. The destination emotions turned out to be more important than the Place Attachment to create behavioral intentions to visit and/or revisit Porto wine cellars. This study demonstrates that destination emotions are a strong antecedent of place attachment. This study also confirmed that there is a relationship between destination emotions and behavioral intentions.

Table 8.3 Tastings of Porto wine during the visits to the Porto wine cellars

Statement	*%* 1	2	3	4	5	6	7	*Mean*	*Std. Dev.*
It is very important for me the tastings of Porto wine during the visits in the Porto wine cellars (N = 901)	0.3	0.4	0.1	0.3	3.4	14.8	79.6	6.66	0,957

Table 8.4 Value of money spent during the visits to the Porto wine cellars

Euros spent on average during the visits (N = 904)							
%							
1 to 100€	*101 to 300*	*301 to 500*	*501 to 700*	*701 to 1000*	*1001 to 2000*	*2001 to 5000*	*More than 5000*
36.3	28.1	11.7	6.1	7.9	6.9	2.9	0.2

Table 8.5 The effort of the expense to visit the Porto wine cellars

Statement	*%* 1	2	3	4	5	6	7	*Mean*	*Std. Dev.*
The effort of this expense (visit Porto wine cellars) in your trip budget was high (N = 897)	1.4	16.9	27.7	11.4	4.7	9.7	20.9	6.66	0.957

Table 8.6 The first time visiting the Porto wine cellars

First time visiting the Porto wine cellars? (%) (N = 910)					
Yes	*No*				
91.2%	8.8%				
	Number of times previously visiting the Porto wine cellars				
	1	*2*	*3*	*4*	*5 or more*
	50%	19.2%	21.8%	2.6%	6.5%

These results demonstrate this direct link between place attachment and behaviour intentions in this particular case study of the Porto wine cellars. It was also found that destination emotions and place attachment have a mediator effect, favoring the relationship between personal involvement and place attachment and behavior intentions. This case demonstrated that involvement, destination emotions, and place attachment are definitely related to the behavioral intentions of the wine tourists.

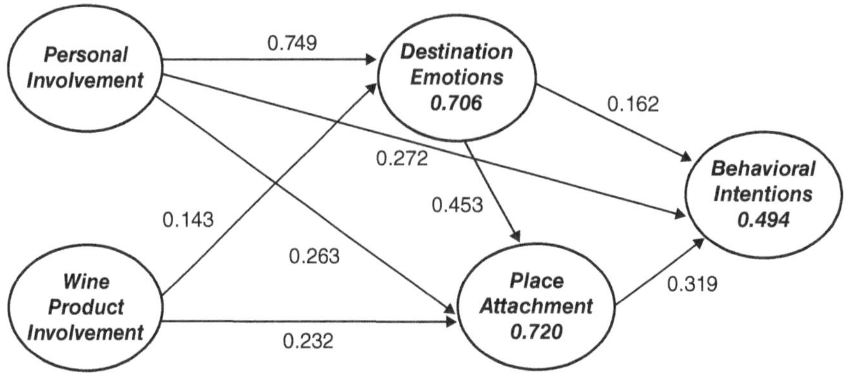

Figure 8.2 Structural model and trajectory.

Conclusions of case study

This research suggests that a stronger emotional connection between the visitors and the cellars should be emphasized, giving a more personal and customized experience, trying to understand the specific desires and expectations of the tourists in the detriment of the standardized visit and experience.

Wine tourism has undoubtedly been a trend that has become cemented by the quality and offer diversity. The Porto wine cellars are an example of a wine tourism destination. It has been systematically improved (new spaces and better and more integrated services). This led, until the COVID-19 pandemic crisis, to an increasing potential regarding quantity and quality, both in offer and demand. Regarding the role and impact of wine tourism in Portugal, namely the case of the Porto wine cellars, it is obvious that the characteristics, attributes, and critical factors of success, such as authenticity and cultural and territorial heritage, are critical for the majority of the wine tourists.

It is also important to segment the wine tourists and the different types of visits existing in the cellars. The managers of the Porto wine cellars should continuously monitor the socioeconomic impact of the visits: expenditure, number of visits, duration, and nationalities.

One of the concerns for the Porto wine cellars association should be raising awareness of the need for more and better information by wine tourism destination managers to reflect on better performance in guided tours. The evidence that forms the case study can contribute positively to developing and improving Porto wine cellars' positioning and marketing strategies. The differences and/or similarities of involvement on the destination emotions and place attachment by different segments of wine tourists should also be further explored and analyzed. Also, we suggest verifying whether international visitors with higher incomes have a significant influence on wine in longer stays and more visits to the cellars, and in their future behavior in terms of word-of-mouth and becoming true advocates for the region and their wines.

Implications

The findings may offer important implications for managing strategic marketing wine destination experiences, such as positioning and reinforcing the consolidation of the Porto wine cellars' destination branding. The results suggest new consequences on the behavioral intentions of the wine tourist. They also suggest new effects about the involvement, destination emotions, and place attachment and their effects, and predict some of the consumer behaviors in wine tourism. The findings are particularly valuable to wine tourism destination marketers with responsibilities in developing strategic planning, not only to make an effort to attract and retain more visitors, but also to engage in favorable recommendations from the wine tourists as a result of an increased level of involvement and emotion during their visit.

Suggestions to the wine tourism sector and future research

The COVID-19 outbreak is a sharp reminder that pandemics, like other rarely occurring catastrophes, have happened in the past and will continue to happen in the future. Even if we cannot prevent dangerous viruses from emerging, we should prepare to dampen their effects on society. The current outbreak has had severe (wine) economic consequences across the globe, and it does not look like any country will be unaffected (Donthu and Gustafsson 2020). So, in a disturbed global market due to the COVID-19 pandemic, the lessons from cooperation models in wine tourism regions should be strongly considered.

Future research should include new data collection methods, taking into account new side effects and forced operational changes due to COVID-19 on the financial and operational level for both tourism and the wine industry (e.g., the Portuguese case study). According to Haywood (2020), if there is an upside to crises like COVID-19, it's the ongoing, intense scrutiny of the past and exploration of more desirable futures. With communities-as-destinations now pulling together to better understand 'what is' through deep-dive wine destination assessments so as to determine 'what could and should be,' the whole process of figuring out how to transition toward transformation can proceed.

Pre-COVID, solidarity within communities-as-destinations had always been an uphill battle. Despite the prevalence of public/private partnerships in support of marketing and branding, the wine management of communities-as-destinations had remained an allusion, and few wine destination marketing organizations (DMOs) saw fit to adopt a much-needed managerial focus of wine tourism, particularly as a cluster. Thus, in an important wine country such as Portugal, it is crucial to understand the strengths and weaknesses of wine routes so as to (re)use them as further amplifiers for even greater expansion of wine tourism. In this respect, considering the natural connection between wine routes and related territory, a great contribution comes from the involvement/engagement of territorial stakeholders.

References

Andrade-Suárez, M., and Caamaño-Franco, I. (2020). The relationship between industrial heritage, wine tourism, and sustainability: a case of local community perspective. *Sustainability*, 12(18): 7453. https://doi.org/10.3390/su12187453

Barber, N., Ismail, J., and Dodd, T. (2008). Purchase attributes of wine consumers with low involvement. *Journal of Food Products Marketing*, 14(1): 69–86. https://doi.org/10.1300/J038v14n01_05

Braga J. L., and Sousa, B. (2021). The basic social process of 're-functionalising' and its implications for housing tourism: A niche tourism perspective. In Abreu, A., Liberato, D., González, E. A., and Garcia Ojeda, J. C. (eds.), *Advances in Tourism, Technology and Systems. ICOTTS 2020. Smart Innovation, Systems and Technologies* (vol. 209, pp. 125–141). Singapore: Springer. https://doi.org/10.1007/978-981-33-4260-6_12

Brotherton, B. O. B., and Himmetoğlu, B. (1997). Beyond destinations – special interest tourism. *Anatolia*, 8(3): 11–30. https://doi.org/10.1080/13032917.1997.9687118

Brown, G. P., Havitz, M. E., and Getz, D. (2007). Relationship between wine involvement and winerelated travel. *Journal of Travel & Tourism Marketing*, 21(1): 31–46. https://doi.org/10.1300/J073v21n01_03

Carta Europeia do Enoturismo. (2006). http://www.turismodeportugal.pt

Dixit, S. K. (ed.). (2017). *The Routledge Handbook of Consumer Behaviour in Hospitality and Tourism*. London: Routledge.

Donthu, N., and Gustafsson, A. (2020). Effects of COVID-19 on business and research. *Journal of Business Research*, 117: 284–289. https://doi.org/10.1016/j.jbusres.2020.06.008

Douglas, N., and Derrett, R. (2001). *Special Interest Tourism*. Brisbane, Australia: John Wiley & Sons.

Duarte Alonso, A., Kok, S., and O'Brien, S. (2020). Sustainable wine tourism development through the lens of dynamic capabilities and entrepreneurial action: an exploratory four-region perspective. *Tourism Recreation Research*, 45(3): 401–419. https://doi.org/10.1080/02508281.2020.1745000

Festa, G., Shams, S. R., Metallo, G., and Cuomo, M. T. (2020). Opportunities and challenges in the contribution of wine routes to wine tourism in Italy – a stakeholders' perspective of development. *Tourism Management Perspectives*, 33: 100585. https://doi.org/10.1016/j.tmp.2019.100585

Getz, D., and Brown, G. (2006). Critical success factors for wine tourism regions: a demand analysis. *Tourism Management*, 27(1): 146–158. https://doi.org/10.1016/j.tourman.2004.08.002

Haywood, K. M. (2020). A post COVID-19 future – tourism re-imagined and re-enabled. *Tourism Geographies*, 22(3): 599–609. https://doi.org/10.1080/14616688.2020.1762120

Karagiannis, D., and Metaxas, T. (2020). Sustainable wine tourism development: case studies from the Greek region of Peloponnese. *Sustainability*, 12(12): 5223. https://doi.org/10.3390/su12125223

Lewicka, M. (2011). Place attachment: how far have we come in the last 40 years? *Journal of Environmental Psychology*, 31(3): 207–230. https://doi.org/10.1016/j.jenvp.2010.10.001

Liu, X., Lee, J., and Kim, W. G. (2015). The role of the travel experience in the structural relationships among tourists' perceived image, satisfaction, and behavioral intentions. *Tourism & Hospitality Research*, 17(2): 1–12. https://doi.org/10.1177/1467358415610371

Ma, S. D., Kirilenko, A. P., and Stepchenkova, S. (2020). Special interest tourism is not so special after all: Big data evidence from the 2017 Great American Solar Eclipse. *Tourism Management*, 77: 104021. https://doi.org/10.1016/j.tourman.2019.104021

OIV. (2020). *Statistical Report on World Vitiviniculture 2019*. https://oiv.int/public/medias/6782/oiv-2019-statistical-report-on-world-vitiviniculture.pdf

Prayag, G., and Ryan, C. (2012). Antecedents of tourists' loyalty to Mauritius: the role and influence of destination image, place attachment, personal involvement and satisfaction. *Journal of Travel Research*, 51(3): 342–356. https://doi.org/10.1177/0047287511410321

Presenza, A., Minguzzi, A., and Petrillo, C. (2010). Managing wine tourism in Italy. *Journal of Tourism Consumption and Practice*, 2(1): 46–61.

Ramos, P., Santos, V., and Almeida, N. (2018). Main challenges, trends and opportunities for wine tourism in Portugal. *Worldwide Hospitality and Tourism Themes*, 10(6): 680. https://doi.org/10.1108/WHATT-08-2018-0055

Robinson, M., and Novelli, M. (2005). Niche tourism: An introduction. In M. Novelli (ed.), *Niche Tourism: Contemporary Issues, Trends and Cases* (pp. 1–11). Oxford: Elsevier.

Santos, V., Ramos, P., Almeida, N., and Santos-Pavón, E. (2018). Wine and wine tourism experience: a theoretical and conceptual view. *Worldwide Hospitality and Tourism Themes*, 21(3): 718–730. https://doi.org/10.1108/WHATT-09-2019-0053

Santos, V., Ramos, P., Almeida, N., and Santos-Pavón, E. (2020). Developing a wine experience scale: A new strategy to measure holistic behaviour of wine tourists. *Sustainability*, 12(19): 8055. https://doi.org/10.3390/su12198055

Santos, V. R., Ramos, P., and Sousa, B. B. (2021). Is wine tourism a niche tourism? Antecedents and consequences of wine as a niche tourism, a conceptual model. In de Carvalho, J. V., Rocha, Á., Liberato, P., and Peña, A. (eds.), *Advances in Tourism, Technology and Systems. ICOTTS 2020. Smart Innovation, Systems and Technologies* (vol. 208). Singapore: Springer. https://doi.org/10.1007/978-981-33-4256-9_32

Soscia, I. (2013). *Emotions and Consumption Behaviour*. Cheltenham: Edward Elgar Publishers.

Sousa, B., and Simões, C. (2010). Comportamento e perfil do consumidor de turismo de nichos. *Tékhne-Revista de Estudos Politécnicos*, 14: 137–146.

Sousa, B. M. (2019). A theoretical contribution from the perspective of innovation process in wine tourism contexts. *Marketing & Tourism Review*, 4(2): 1–18. https://doi.org/10.29149/mtr.v4i2.4744

Sundbo, J., and Dixit, S. K. (2020). Conceptualisations in tourism experience. In Dixit, S. K. (ed.), *The Routledge Handbook of Tourism Experience Management and Marketing* (pp. 15–26). London: Routledge.

Torres, J. P., Barrera, J. I., Kunc, M., and Charters, S. (2020). The dynamics of wine tourism adoption in Chile. *Journal of Business Research*, 127: 474–485. https://doi.org/10.1016/j.jbusres.2020.06.043

Torres, J. P., and Kunc, M. H. (2016). Market opportunity recognition in the Chilean wine industry: traditional versus relational marketing approaches. *Journal of Wine Research*, 27(1): 19–33. https://doi.org/10.1080/09571264.2016.1144584

Trauer, B. (2006). Conceptualizing special interest tourism – frameworks for analysis. *Tourism Management*, 27(2): 183–200. https://doi.org/10.1016/j.tourman.2004.10.004

Weiler, B., and Hall, C. M. (1992). *Special Interest Tourism*. London: Belhaven Press.

White, C. J., and Scandale, S. (2005). The role of emotions in destination visitation intentions: a cross-cultural perspective. *Journal of Hospitality & Tourism Management*, 12(1): 168–178.

Yeh, C. M. (2013). Tourism involvement, work engagement and job satisfaction among frontline hotel employees. *Annals of Tourism Research*, 42(10): 214–239. https://doi.org/10.1016/j.annals.2013.02.002

Yuksel, A., Yuksel, F., and Bilim, Y. (2010). Destination attachment: Effects on customer satisfaction and cognitive, affective and conative loyalty. *Tourism Management*, 31(2): 274–284. https://doi.org/10.1016/j.tourman.2009.03.007

Žabkar, V., Brenčič, M. M., and Dmitrović, T. (2010). Modelling perceived quality, visitor satisfaction and behavioural intentions at the destination level. *Tourism Management*, 31(4): 537–546. https://doi.org/10.1016/j.tourman.2009.06.005

PART II

Profiling and segmentation of the wine tourism market

9

UNDERSTANDING THE WINE TOURISM MARKETS

Brian Kee Mun Wong and Khairul Anuar Muhamad

Introduction

Wine tourism can be defined as 'a visitation to vineyards, wineries, wine festivals and wine shows for which grape wine tasting and/or experiencing the attributes of a grape wine region are the prime motivating factor for visitors' (Hall et al. 2000, p. 3). Instead of focusing solely on consumerism, wine tourists also desire to know about the history and culture of the wine regions, wine traditions and production processes, social encounters, and the terroir's attractions of the destination (Lee, McCole, and Holecek 2020). Travelers for wine tourism commonly seek unique wine experiences, and many may have the knowledge and rules to adhere to before deciding on a destination to visit (Vo-Thanh and Kirova 2017). Wine tourists are also attracted to private tours that give them the opportunity to have meaningful conversations with the winery owners or even the viticulture and oenology engineers. Besides, wine tastings often involve premium selections with a bold, distinctive character as part of their bucket list (Woldarsky and Geny-Denis 2019). In fact, some of the wine tourists or enthusiasts even brought their own wine bags to pack their preferred wines home.

In today's world, commercialized wine tourism requires effective marketing strategies to connect winery developers and wine producers with their target market. Traditional wine-producing nations are no exception. Wine tourism plays an essential role beyond just wine tasting, including bonding with consumers holistically from information search to visiting vineyards, harvesting grapes, wine production, and sales of products (Amarando, Assenov, and Visuthismajarn 2019). Wine tourism markets can be divided into three categories: 'Old World,' 'New World,' and 'Emerging World' (Chong 2017; Hall et al. 2009). Thus, this chapter will present some of the key wine tourism markets in accordance with these categorizations.

Old World wine tourism markets

Old World wine tourism, or wine tourism in Europe, is particularly popular in France, Italy, Spain, Portugal, and Germany (Gómez, Pratt, and Molina 2019; Hall et al. 2000). Despite being called 'Old World,' European wine tourism remains a focal destination among wine enthusiasts for its long history and reputation, coupled with beautiful vineyards and outskirt sceneries. While Old World wine tourism dates back to the 1980s, particularly in France (Hall et al. 2000),

DOI: 10.4324/9781003143628-12

the growth of the region remains positive year after year. According to a recent survey done by WineTourism.com, while most wineries around the world reported a loss of income due to the recent COVID-19 pandemic, surprisingly, four of the Old World destinations (i.e., France, Italy, Germany, Spain) reported revenue increases in 2020 (WineTourism.com 2019a).

France

Being among the pioneers in developing wine tourism in the 1980s, France has put in much effort with the expansion of tourist infrastructures and the opening of vineyards to the public and to public investment (Hall et al. 2009). Besides, making the vineyards and wine cellars accessible to the public for wine tasting further enhances France's popularity as a wine tourism destination (Hall et al. 2009). In addition, the diversity approach in France resulted in more instructive and culturally bound wine tourism products. In fact, the professionalization in the business market witnessed new offerings that combine vineyards and winery visits with corporate seminars and team-building activities (Bacchus 2019). The diversity of products is quite dynamic in the modern wine tourism sector, including master wine classes, cooking courses, creating personalized wine, overnight stays in vineyards or chateaux, easy hikes through the vineyards and their surroundings, courses in wine academies, or even well-planned wine routes such as La Grande Route des Vins (VisitFrenchWine.com n.d.).

In established wine-growing regions such as Burgundy, wineries are even promoting extremely personalized and exclusive arrangements such as 5-star accommodations, spa services, tastings with the winery owner, special private tours, and even helicopter flights. Burgundy has gained its reputation and importance as a wine-producing region and is among the first to introduce its own wine route. It must be noted that wine tourism in France has been supported by the success of Bordeaux wineries, as wine tourists respect and are curious about winery traditions, which Bordeaux wineries can offer (Bouzdine-Chameeva, Faugère, and Mora 2015). The INAO (Institut National des Appellations d'Origine) plays a role in attracting wine enthusiasts to visit Old World wineries, such as France, to dwell deeper into discovering their specificities.

Italy

Wine festivals such as the Open Cellars Festival (the largest wine festival worldwide) encourage visits to Italian wineries. Such a marketing effort by the Italian tourism board was effective in bringing wine tourists into Italy (Hall et al. 2009). In fact, Italy has been well-known for wine tourism in its Langhe vineyards, with Langhe's total wine tourism numbers increasing during the 2010s (Canovi 2017). Langhe vineyards have a long history of winemaking and attractive scenery, making it an interesting destination for wine tourists who want to learn about the making of Italian wine and, of course, to enjoy the beautiful vineyards and surrounding scenery (Canovi 2017). The Italian food and wine tourist is generally a cultured tourist, with greater capacity and a tendency to spend (Garibaldi 2018).

In today's world, most Italians organize their wine trips, relying heavily on the web, both to gather information and to book individual parts of the journey, but they also have a greater aptness than the generic tourist to book through intermediaries (Garibaldi 2018). It can also be highlighted that the use of food-and-wine-themed experiences has become more common, with the majority of Italian tourists who, regardless of whether they travel for seaside, mountain, or business tourism, have participated in at least one such activity related to food and wine during their trip (Garibaldi 2018). Italian wine tourism has steadily grown since 1993 when the foundation 'Wine Tourism Movement' encouraged the opening of about 100 wineries for public visitation in Tuscany (Oliviera and Filho 2016).

Spain

Spain has rich cultures, histories, languages, and cuisines to offer, and its wine offerings are quite diverse. The number of distinguishable wines available in Spain may offer different experiences to wine tourists, but it may also be confusing to new tourists who just embarked into wine tourism (Wacker 2019). Although Spain has the largest vineyard areas in the world (about 2.4 million acres) and has had fruitful production over the years, it was not until recently that Spanish wine tourism started to receive global recognition (Dao 2019). Different regions may offer different wine tourism products and services. However, commonly, the options cover winery visits, guided wine tours, wine tastings, winery accommodations, dedicated wine festivals, sightseeing, and cultural visits. Spanish wine tourism may also be combined with gastronomic tourism products, which is perceived as one of the best in the world for their taste, ingredient quality, and food variety. In fact, different regions have their own local cuisines, and often the cuisine is best appreciated when paired with the local wines produced in that region (Wacker 2019).

Wine selling is not the only aim of the Spanish wine tourism industry. The winery owners understand the importance of brand building and brand ambassadors to create the industry's competitiveness against its stronger neighbors, France and Italy. The distinctive positioning requires the public and private sectors to work hand in hand, developing exquisite wine tourism experiences that connect wine with history, culture, tradition, and cuisine of the local regions (Veseth 2017). Indeed, wine tourists in their previous encounters have complained that the wine tourism options in Spain were rather limited (CellarTours 2019). The history of Spain, formed through various nations that possess their own histories, cultures, and languages to a certain extent (Wacker 2019), should be an advantage to Spanish wine tourism development. The variety of cultures significantly influence the types of wine produced in each region. Hence, when viewed holistically, Spanish wine culture offers interesting varieties, yet also a melting-pot experience of many cultures within the country.

Portugal

Portugal received about 2.5 million wine tourists in 2019 (Haupt 2019). Despite being overshadowed by its more popular peers (i.e., France, Italy, Spain), Portugal has been promoting itself as a top wine tourism destination over the past few years. The nation has 14 key wine regions with 28 DOCs (*Denominação de Origem Controlada* – Protected Designation of Origin) (Santos, Vavdinos, and Martines 2020), where two of the regions are recognized as World Heritage by UNESCO: the Douro Valley Wine Region and the Pico Island Wine Region in the Azores. Douro Valley, among Portugal's many wine regions since 1756, is known to have the oldest wine appellation system in the world (Balch 2020). The facilities developed there include river cruises, scenic and historical train rides, multiple accommodation options, wine-related events, and tasting, and Port-dedicated wine routes (Correia et al. 2019). Porto, home of Port wine, recently upscaled its appeal by launching a €117 million museum, the World of Wine (WoW), to showcase the pride of Portuguese wine (Kang 2020). WoW opened for visitation on July 31, 2020, and it provides engaging educational experiences and celebrates the appreciation of oenology (Balch 2020; Kang 2020). It portrays Porto as a cultural wine destination and Portugal's confidence as a wine-producing nation at large.

In developing wine tourism, the domestic wine industry players need to synergize the elements of viniculture, food, and tourism. In that way, the destination will be able to provide a holistic gastronomic experience to ever-demanding tourists (Correia et al. 2019). The Portuguese Ministry of Tourism also has a plan that focuses on different angles of wine tourism: products (restaurants, accommodation, wine cellars, or wine producers), professional training, events, and campaigns to attract journalists and tour operators to Portugal, and it digitalizes the offer by creating a national online platform (Haupt 2019). The plan of developing Portugal into a wine tourism destination involved regional tourism entities, as it is also a way to articulate all the players involved in wine tourism. The Secretary of State for Tourism also declared that wine tourism already accounts for about 10% of those who want to visit Portugal and aims to turn Portugal into a 'must-see' wine tourism destination.

Germany

Germany's wine growing can be traced back to 2,000 years ago (Wines of Germany 2019) and has given rise to some of the wine regions as niche tourism destinations (Szolnoki 2018). The trend of wine tourism is also important as a means to achieve sustainable rural development since it includes the promotion of heritage and the local lifestyle of the destination. Germany is supported well by high domestic tourism traffic, provides a good and effective fundamental to wine tourism development. Despite its potential based on domestic tourism and wine production in Germany, wine tourism remained irrelevant until recently (Tafel and Szolnoki 2020). Six wine regions are located within a 1.5-hour drive from Frankfurt. Such proximity enables urbanites to visit any of the wine regions as a day or weekend trip, creating positive economic impacts to the regions (Tafel and Szolnoki 2020).

Franken, a German wine region, experienced a successful growth in wine tourism by having the aid of institutional guidance and training to wine tour guides (Dressler and Paunovic 2019). The wineries of the Saale-Unstrut and Sachsen wine regions enhance the images and opinions of wine tourism in the regions via website communication and online reviews (Dressler and Paunovic 2019). The Rheingau wine region offers not only wine but also history and nature, attracting first-time or rare tourists to the region and generating about €36 million in 2018 (Szolnoki 2018). Having said that, the wine tourism industry in Germany has shown huge potential, given the €3.2 billion revenue and 64,500 jobs created at the Franconia region and €1.3 billion revenue and 22,780 jobs generated from the Mosel region since 2012 (Tafel and Szolnoki 2020). Historic East German cities such as Weimar and Dresden are good candidates for strategic development in wine tourism.

New World wine tourism destinations

Despite being an emerging phenomenon, the New World wine tourism industry has proven itself to be more successful and competitive compared to the Old World ones. The New World wine tourism destinations include the United States, South Africa, New Zealand, Australia, Chile, Argentina, Israel, Canada, and Eastern Europe (Hall et al. 2000; Karagiannis and Metaxas 2020). The former five destinations received quite a substantial number of wine tourists (Hall et al. 2009), hence the focus of markets to discuss in this chapter.

United States

The United States generated about US$20 billion in revenue from wine tourism in 2016 (Becca 2018). In fact, California is the most famous American state for wine tourism due to its famous wineries (WineTourism.com 2019b). As one of the largest producers of wine, the United States eventually receives the highest number of wine tourists in the American continent (Oliviera and Filho 2016). The Napa Valley in San Francisco, California, has more than 500 wineries, receiving millions of tourists every year and contributing significantly to the region's economy. The 58-kilometer-long Napa Valley has experienced its glory in wine tourism due to several factors such as good production process, good wine quality, proximity to San Francisco, and the diversity of cultural and social activities available in the area.

Although wine tourism in Wisconsin and Minnesota is not as popular as in California, these areas have recently developed multiple tourist attractions, from winery visits to wine tasting opportunities and live music performances, to appeal to wine tourists (Lee, McCole, and Holecek 2020). The development moves, in fact, enhanced the number of wine tourists significantly, visiting Wisconsin's 118 wineries and Minnesota's 49 wineries in recent years. The growth improved both Wisconsin's and Minnesota's reputations for wine tourism. Wine tourism in the United States is known for the shorter duration of wine tours instead of longer ones and is often handled by tour operators (Garibaldi et al. 2017). The shorter tours are preferred by wine tourists in the United States to have an intensive feel of the New Word vineyards and a sumptuous taste of American wines, complementing the existing tours that tourists would have purchased when visiting the United States.

South Africa

In South Africa, wine tourism is most prominent in the Western Cape Province because of the concentration of high-quality South African vineyards (Ferreira and Hunter 2017). The best-known examples of high-quality South African vineyards are in the Stellenbosch region, which dominates wine production in South Africa with its historical heritage of wineries and wine tourism, providing Stellenbosch wineries with fame to be proud of (Grobbelaar 2020). To build a competitive advantage, the South African government launched the Wine Tourism strategy to enhance wine tourism experiences and produce better-quality wine. Tour operators who specialize in wine tourism are developed to showcase niche and off-the-beaten-track vineyards and boutique winery properties (Hawkins 2018). Wine tourists at Western Cape were reported to spend more than the average compared to other tourists, further enhancing economic growth in the region (News24 2018).

Customized and private wine tours were found to be the other key to the continuous growth of wine tourism in South Africa (News24 2018). The privacy allows the wine tourists to have a deeper and authentic experience, providing an immersive encounter for the tourists into the wine culture of South Africa. The tailor-made tours commonly include up-close interaction with the winemakers, personalized cellar tours, food and wine pairings, and tasting events (News24 2018). Such unique experiences are usually not found in the general package or scheduled wine tours. Perhaps that is the uniqueness of the South African wine tourism market – a way to distinguish itself from other contenders around the world.

New Zealand

New Zealand received over 1 million wine tourists between 2010 and 2015 (Baird and Hall 2016). The wineries are in several regions and open on a year-round basis, making it relevant to wine tourists visiting New Zealand in any season. The vineyards of Marlborough, a key wine tourism region in New Zealand, produce more than 75% of New Zealand's wine (Shavkatovich 2020). Coupled with its tourist-friendly facilities, the opportunities to taste high-quality Marlborough wine, attractive vineyard's landscapes, and architecture, the destination creates its niche attraction to the wine tourists. Hawkes Bay is also another prominent wine-producing region in New Zealand, where wine tourists flock into wineries such as Lime Rock Wines to experience first-hand high-quality wines, and often the tourists purchase a few bottles before leaving (Crick, Crick, and Tebbett 2020).

Wairarapa, in the southeastern corner of the North Island, offers exquisite rural life experience and high-quality wine within Martinborough's vineyards (Rainford 2017). Nevertheless, the wine production in Wairarapa's wineries and vineyards can be affected by the quantity of water available at times. Hence, adequate preparations are required to ensure vine growth and sustainability (Rainford 2017). Otago also welcomes wine tourists as part of its niche tourism product (Fountain 2016), offering high-quality wines such as Sauvignon Blanc, opportunities to experience wine production processes and wine tasting, calm weather, and beautiful sceneries. The government's effort to upscale its wine tourism offerings are essential for sustainable growth (Baird, Hall, and Castka 2018) and to further escalate New Zealand's potential to be among the top wine tourism destinations in the New World category.

Australia

Award-winning wineries in Australia not only offer wine but also gastronomic, food and beverage pairing, and adventurous activities (e.g., outdoor picnics, cooking classes, etc.) (WorldWideWineTours 2021). While wine quality from most of the Old World markets relies heavily on its soil and climate (terroir), Australia's wines focus on grape varieties (WorldWideWineTours 2021). As a result, not only does wine produced in Australia taste different from wine of the same variety produced overseas, but it can also vary enormously within Australia as well (Hough 2020). The flexibility regarding which grapes to grow and where to grow in Australia is perhaps the biggest advantage to the Australian wine tourism industry in managing the climate change issue (Hough 2020). Australia has always been very adaptable and has been introducing new varieties of grapes for a very long time. The nation has been using at least 160 grape varieties for wines. However, only about 15 to 20 of the varieties produce about 90% of Australian wines (Hough 2020).

As the seventh-largest wine-producing nation globally, Australia has about 170,000 hectares of vineyards that cut across four wine regions: New South Wales, South Australia, Victoria, and Western Australia (WineTourism.com 2019c). Australian wine is well received overseas, and 63% of its production is exported (Arthur 2020), with about 770 million liters of wine delivered to around 117 destinations globally over the years. China is the second-most valuable market, playing a big part in what has driven demand for red wine, in particular Shiraz, in recent years (Hough 2020). Experiential tourism, of which wine tourism is a subset, is itself the product of three things: online booking sites, cheap airfares, and Airbnb, all of which lowered the cost of travel significantly.

Chile

Chilean wine has its roots in the Colonial Era with the arrival of the conquistadors and Spanish missionaries who planted the first vines. It has since developed a great diversification of vines that has led to a strong global presence following decades of work (VinosdeChile n.d.). Chile, the eighth-largest wine-producing country globally (Figueroa and Rotarou 2018), has a twofer of extra-strength maritime and alpine advantages, complemented by suitable terroirs (VinosdeChile n.d.). The budget friendliness and variety of good-quality wines, coupled with picturesque landscapes, have further supported the nation in developing its wine tourism industry, offering a range of wine tours and wine festivals (Figueroa and Rotarou 2018). Despite reaching its full potential, Figueroa and Rotarou (2018) have already reported several sustainable challenges the market is facing: insufficient infrastructure, lack of public-private partnership, poor marketing and promotion strategies, lack of expertise, and, importantly, lack of coordination with other tourism sectors.

The most established and famous wine regions are in the Central Valley, which is close to the capital, Santiago. The closest to the city is Maipo Valley, which is mostly known for its renowned Cabernet Sauvignons. Of course, Carmenere and Merlot are also prominent in the Maipo Valley, a beautiful, peaceful location surrounded by the Andes Mountains, just outside of the bustling city (LBV 2019). Slightly further south is the Colchagua Valley (a few hours' drive from Santiago), with many luxury vineyards. The valley is known for its picturesque landscapes consisting of the beautiful vineyards growing up the hills with colonial estates. The Colchagua Valley produces wines similar to those of the Maipo Valley; Cabernet Sauvignon, Carmenere and Malbec. Finally, the Casablanca Valley is between Valparaiso and Santiago, so it is right between the Pacific and the Andes (LBV 2019). Wine tourists will also be able to behold the grandeur of the plantations from the lookout points, taste the varieties in centennial underground wine cellars, and tour the old manor houses (VinosdeChile n.d.).

Emerging World wine tourism destinations

Several emerging wine tourism destinations, particularly in Asia, started to gain momentum and to offer a new taste of wine tourism experience. Some of these popular destinations include: Shandong, Hebei, Ningxia, and Xinjiang in China; Pattaya, Hua Hin, and Kao Yai in Thailand; Nashik and Bengaluru in India; Yamanashi in Japan; Beqaa Valley in Lebanon; Bali in Indonesia; Inle Lake in Myanmar; and Battambang in Cambodia (Chin 2020; Eden 2019; Yan 2018). Among the many new wine tourism destinations in Asia, China and Thailand have received much attention recently.

China

China was reported as the fifth-biggest global wine consumption market in 2019, with approximately 1.78 billion liters of wine (Ma 2020). Despite being the fourth-biggest importer of wine globally, amounting to about €2.4 billion in 2018, China is also a producer of wine (Ma 2020). To date, China has as many as 13 wine regions: Beijing, Gansu, Hebei, Heilongjiang, Henan, Jilin, Liaoning, Ningxia, Shaanxi, Shandong, Tianjin, Xinjiang, and Yunnan (Duan et al. 2020; Li and Bardaji 2017, Wei 2015). However, not all wineries engage in wine tourism. The first purpose-built winery was established in Yantai, Shandong, by Changyu Pioneer in 1892 (Han 2011). Changyu, as a pioneer in the Chinese wine tourism scene, established the first professional chateau, 'Chateau Changyu-Castel,' and its own technical center in 2002 (Changyu 2014). Helan Mountain in Ningxia Hui autonomous region has recently been a focal point of the latest wine tourism in China (Wei 2015). The local government reported that the region had 86 established wineries, and that another 113 new wineries are currently under construction (Legal Network 2018). The wineries are organized into five clusters, namely, Hongsipu, Shizuishan, Qingtongxia, Yinchuan, and Yongning (Gu et al. 2021).

The Mi-Le wine region in Yunnan Province is reported to have received about half a million wine tourists yearly, creating a niche tourism product for the province (Duan et al. 2020; Wei 2015). As already a popular travel destination among holidaymakers to China, Yunnan also offers a holistic wine experience, ranging from dedicated wine tours and wine-themed restaurants to luxurious hotel accommodations, hot springs, European-bound architectures, historical sites, and local cultures (Duan et al. 2020; Wei 2015). To further capitalize on the higher spending power among the wine tourists visiting Mi-Le region, it is recommended that the local wineries improve their educational program and provide memorable experiences, while the local government focuses on developing personalized and well-designed marketing strategies (Duan et al. 2020). In general, the knowledge of Chinese wine culture among global wine enthusiasts is still low, and it requires greater effort from both the local and central tourism boards to promote and educate.

Thailand

Thailand is one of the more aggressive destinations in Southeast Asia that promotes wine tourism (Chong 2017). While Thailand has displayed its competency in this niche market, the nation still needs to improve its reputation and competitiveness (Chong 2017). Despite a newcomer, when wine tourists may not expect that a year-round warm country like this can offer good-quality wine products, the demand for wine tourism in Thailand has improved in recent years (Chong 2017; The Nation 2018). This has been aided by the popularity of wine importing to cater to tourists in Thailand. International events (e.g., the 11th Bangkok Grand Wine Tasting) and the presence of high-quality wineries further drive wine enthusiasts around the world to savor Thai wine first-hand (Amarando, Assenov, and Visuthismajarn 2019; The Nation 2018). Generally, there are three key wine regions in Thailand: Khao Yai, Pattaya, and Hua Hin (Tompkin 2019). High-quality wineries such as Siam Winery, PB Valley Khao Yai Winery, GranMonte Winery, Chateau de Loei Winery, and Monsoon Valley Winery have been productive in producing bottled wine for tourist consumption (Banks et al. 2013).

The Khao Yai wine region is just three hours away from Bangkok, making it a possible day trip or an overnight trip for any tourists to visit. The vineyards are located between 350 and 500 meters above sea level, backed by a stunning national park, beautiful natural limestone hills, rows of stone farmhouses, and historical Buddhist temples (Tompkin 2019). This wine region houses the winery that produces the internationally recognized Thai wine brand GranMonte. Monsoon Valley is renowned for its innovation and professionalism in wine production, having the tropical viticulture to keep planted vineyard vines alive and planted any time of the year at Hua Hin Hills Vineyard (Monsoon Valley 2020). Another key contributor to the Thailand wine tourism scene is the Silverlake Vineyard, which is located in the Pattaya wine region. It is known for adapting Australian viticulture to Thailand environments, producing branded wines such as Shiraz and Chenin Blanc (Renown Travel 2022). Given the continuous quality improvement of Thai wine, coupled with the support from the Tourism Authority of Thailand (TAT), the nation is believed to slowly pave its way to upscale its appeal to global wine tourists.

Case Study 9.1 Wine tourism development in China on the rise

Most vineyards in China are found throughout the northern region, usually set up in the middle of rustic villages and outskirt territories. Generally, Chinese wines are not globally known, as about 70% of the total wine production is consumed locally (Leong 2017). The infrastructure development somehow outspeeds the recognition China received in wine tourism. Construction projects from the coastal province of Shandong to the rural region of Ningxia commonly embed wine elements in their planning. Investments from local governments and international wine producers further escalate the promotion of wine tourism in China (Ly, Howson, and Begun 2014).

The economic impact on the wine regions can be impeccable when businesses and institutions from other sectors (e.g., banking, transportation, hospitality, and sports) are linked up together. In fact, the importance of sectorial linkages has seen its benefits in China so as the far-reaching wine regions, a similar phenomenon observed in Napa, California (Ly, Howson, and Begun 2014). The Ningxia region, which is just a stone's throw away from the Gobi Desert, has about 100 wineries producing 200 million bottles of wine by 2020, and it is positioned as the most productive wine region in China (Leong 2017). The development of wine tourism there is not restricted to just vineyards and wineries; it also has theme park-like attractions (e.g., wine castles, museums, recreational parks), purpose-built wine complexes, and international events. For example, the opening of Yantai Changyu International Wine City and the organization of the annual Yantai International Wine Festival further made Yantai a distinct wine tourism destination in Shandong, China (Chinadaily.com.cn 2015).

Smaller boutique wineries are also in abundance but harder for tourists to access because of their scattered locations in some of the most popular provinces, such as Shangdong, Shaanxi, and Ningxia. As wine culture is relatively new in China, it has the disadvantage of short production history as the Old World and New World markets may claim. Also, Chinese wine tourists differ in their knowledge and understanding of wine. To cater for this, larger and more established local wineries customize their wine tour packages and prepare relevant information for different types of tourists based on their knowledge level. For example, novices would first be introduced to the history of winemaking and its processes, while wine experts and connoisseurs are immediately fast-tracked to an introduction of grape varieties (Daxue Consulting China 2016; Leong 2017; Li 2020; Ly, Howson, and Begun 2014).

An example is the Co-organization of the World Wine Tourism Forum 2020 by the Ministry of Agriculture and Rural Affairs with the Ningxia government. The forum is a vital meeting of regional cooperation in the Ningxia wine industry, focusing on wine and culture, tourism, leisure, and art (Li 2020). The world's top 10 most promising wine tourism destinations or world's 'golden areas' was also revealed during the forum where Ningxia topped the list, alongside Yamanishi in Japan, Maipo in Chile, Quebec in Canada, and Wairarapa in New Zealand (Li 2020).

In conclusion, the rapid growth behind wine tourism participation in China forms the topic of discussion in this case study. The awareness of wine culture among the public remains low in China. Often the Chinese consumers perceive wine as a classy consumption from the West (Daxue Consulting China 2016), but as the wine starts to fill in the menu of Chinese gastronomy, the product's popularity will increase further. Vineyards and wineries may soon be filled with local and international tourists (Daxue Consulting China 2016) when wine consumption eventually becomes a norm. To secure a decent share of this potential market, it is really a matter of who first positions itself as a wine tourism destination that resonates well with the Chinese market (Daxue Consulting China 2016).

Conclusion

This chapter discussed the multiple wine tourism markets based on three key categorizations: Old World, New World, and Emerging World. The phenomenon of wine tourism not only combines the element of wine and tourism but often pulls in the other tourism products such as gastronomy, history, countryside, religious buildings, sun, sea and sand, and entertainment. Powerful economic influences were evident in most of the wine tourism markets highlighted in this chapter. It is no doubt that most of the tourism boards see the multiple possibilities of wine tourism, and hence, they invest in and grow this niche tourism product. While the Old World wine tourism destinations have a long history of the establishment, the growth of the markets remains positive.

At the same time, the New World markets are picking up momentum to create new segments and offer a good alternative to the wine tourists who prefer a new wine experience. Lastly, the Emerging World markets are indeed creating their own competitiveness, despite being the new kid on the block of the wine tourism industry. Regardless of Old, New, or Emerging World, wine tourists are now having an abundance of choices. While the New World and Emerging World wine tourism destinations are building up their competitiveness, the Old World destinations remain a preferred wine produce supplier among wine enthusiasts. As this chapter provides cues to the development of key wine tourism destinations, it is useful for both academics and the industry to dwell further into effective marketing strategies for a better growth opportunity in the 'new normal' environment of the tourism industry.

Implications

Despite the recent economic impact of travel lockdown on most industries, the growth of global wine consumption remains positive, particularly for wine produced from Old World destinations. The gradual enhancement of COVID-19 vaccination across the globe has slowly enabled certain nations to open up their borders (e.g., travel bubble agreement), welcoming tourists to visit selected wine tourism destinations.

For the tourism industry to return back to the robust condition it enjoyed before the pandemic remains a stretch target for the moment. However, the understanding of key wine tourism markets enables destination marketers to plan and strategize progressively, in preparation for a comeback once the borders are fully open for tourism in thes future. It is essential for any of the wine tourism markets to continue building a strong defined image, be it their brand, winery, or product, so that they may be able to tap into the real potential of this niche tourism product effectively.

References

Amarando, M., Assenov, I., and Visuthismajarn, P. (2019). Sustainable wine tourism and vineyards' environmental consciousness in Thailand. *African Journal of Hospitality, Tourism and Leisure*, 8(3): 1–13.

Arthur, R. (2020). Australian wine: how can the industry navigate its way through the pandemic? *BeveridgeDaily.com*, July 6. https://www.beveragedaily.com/Article/2020/07/06/Australian-wine-New-strategic-plan-to-navigate-the-coronavirus-pandemic?utm_source=copyright&utm_medium=OnSite&utm_campaign=copyright

Bacchus, A. (2019). How is French wine tourism organized? *WineTourBooking.com*, August 8. https://www.winetourbooking.com/en/wine-tourism-in-france-and-elsewhere/

Baird, T., and Hall, C. M. (2016). Competence-based innovation in New Zealand wine tourism: Partial strategies for partial industrialization. In Pechlaner, H., and Innerhofer, E. (eds.), *Competence-Based Innovation in Hospitality and Tourism* (pp. 197–224). Abingdon: Routledge.

Baird, T., Hall, M., and Castka, P. (2018). New Zealand winegrowers attitudes and behaviours towards wine tourism and sustainable winegrowing. *Sustainability*, 10(3): 797. https://doi.org/10.3390/su10030797

Balch, O. (2020). *The best wine experiences in Portugal*. https://www.theguardian.com/travel/2020/aug/25/best-wine-experiences-in-portugal-world-of-wine-porto

Banks, G., Klinsrisuk, R., Dilokwanich, S., and Stupples, P. (2013). Wines without Latitude: Global and local forces and the geography of the Thai wine industry. *EchoGéo*, 23: 1–18. https://doi.org/10.4000/echogeo.13368

Becca (2018). A Journey into the mind of a wine tourist: what keeps them coming back for more? *The Academic Wino*, November 1. http://www.academicwino.com/2018/11/wine-tourism.html/#:~:text=Specifically%2C%20this%20study%20found%20that,to%20a%20performance%20event%20in

Bouzdine-Chameeva, T., Faugère, C., and Mora, P. (2015). Wine tourism in Bordeaux, In Lee, K. (ed.), *Strategic Winery Tourism and Management* (pp. 89–115). Palm Bay, FL: Apple Academic Press.

Canovi, M. (2017). *Identity Construction and Its Influence on Wine Tourism Diversification Decisions: Case Study of Family Wineries in Langhe, Italy*. Doctoral thesis, University of Chester, UK.

CellarTours (2019). *Best Winery Hotels in Spain*. January 24. https://www.cellartours.com/blog/spain/ten-best-wine-hotels-in-spain

Changyu (2014). *Highlights in Pioneering*. http://www.changyu.com.cn:8189/about/before.html

Chin, P. (2020). Stunning vineyards in Southeast Asia for the best New Latitude wines. *Prestige*, August 3. https://www.prestigeonline.com/sg/travel/destinations/stunning-vineyards-in-southeast-asia-for-the-best-new-latitude-wines/

Chinadaily.com.cn (2015). Yantai, an international vine and wine city. *China Daily*, March 13. http://www.chinadaily.com.cn/m/shandong/yantai/2015-03/13/content_19806012.htm

Chong, K. L. (2017). Thailand wine tourism: a dream or a reality? *Asia Pacific Journal of Tourism Research*, 22(6): 604–614.

Correia, A. I., Cunha, R., de Matos, O. P., and Fernandes, C. (2019). Wine tourism experiences and marketing: The case of the Douro Valley in Portugal. In Sigala, M., and Robinson, R. N. S. (eds.), *Wine Tourism Destination Management and Marketing* (pp. 203–220). London: Palgrave Macmillan.

Crick, J. M., Crick, D., and Tebbett, N. (2020). Competitor orientation and value co-creation in sustaining rural New Zealand wine producers. *Journal of Rural Studies*, 73: 122–134. https://doi.org/10.1016/j.jrurstud.2019.10.019

Daxue Consulting China (2016). *Wine tourism in China is getting popular*. Daxue Consulting, July 1. https://daxueconsulting.com/wine-tourism-in-china-is-getting-popular/

Dressler, M., and Paunovic, I. (2019). *An exploration of digital innovation activity of German wineries in the regional tourism context: Differentiation and complementarity*. 1st International Research Workshop on Wine tourism: Challenges and futures Perspectives, EM Strasbourg Business School, Strasbourg, 46–59.

Duan, B., Arcodia, C., Ma, E., and Hsiao, A. (2020). The Xiao Zi effect on motivation to visit wineries from an identity perspective: The case of China. *Journal of Hospitality and Tourism Management*, 45: 359–369.

Eden, M. (2019). *Five of the Hottest Vineyards in Asia for Wine Lovers*. https://icstravelgroup.com/5-of-the-hottest-wineries-in-asia-for-wine-lovers/

Ferreira, S. L. A., and Hunter, C. A. (2017). Wine tourism development in South Africa: A geographical analysis. *Tourism Geographies*, 19(5): 676–698. https://doi.org/10.1080/14616688.2017.1298152

Figueroa, E., and Rotarou, E. (2018). Challenges and opportunities for the sustainable development of the wine tourism sector in Chile. *Journal of Wine Research*, 29(1): 1–22. https://doi.org/10.1080/09571264.2018.1532880

Fountain, J. (2016). *Just here for the scenery? Chinese holidaymakers and wine tourism in New Zealand. 9th Academy of Wine Business Research Conference* (pp. 582–591). Adelaide: University of South Australia.

Garibaldi, R. (2018). *First Report on Italian Food and Wine Tourism*. https://www.robertagaribaldi.it/first-report-food-and-wine-tourism/?lang=en

Garibaldi, R., Stone, M., Wolf, E., and Pozzi, A. (2017). Wine travel in the United States: a profile of wine travellers and wine tours. *Tourism Management Perspectives*, 23: 53–57. https://doi.org/10.1016/j.tmp.2017.04.004

Gómez, M., Pratt, M. A., and Molina, A. (2019). Wine tourism research: A systematic review of 20 vintages from 1995 to 2014. *Current Issues in Tourism*, 22(18): 2211–2249. https://doi.org/10.1080/13683500.2018.1441267

Grobbelaar, L. (2020). *Towards a conceptual framework for value co-creation in wine tourism*. Doctoral thesis, North-West University, Potchefstroom, 280 pp. http://repository.nwu.ac.za/bitstream/handle/10394/34907/Grobbelaar%20L%2020953674.pdf?sequence=1&isAllowed=y

Gu, Q., Zhang, H. P., Huang, S. S., and Zheng, F. (2021). Tourists' spatiotemporal behaviors in an emerging wine region: A time-geography perspective. *Journal of Destination Marketing and Management*, 19: 100513. https://doi.org/10.1016/j.jdmm.2020.100513

Hall, C. M., Johnson, G., Cambourne, B., Macionis, N., Mitchell, R., and Sharples, L. (2000). Wine tourism: an introduction. In: Hall, C. M., Sharples, L., Cambourne, B., and Macionis, N. (eds.), *Wine Tourism Around the World* (pp. 1–24). Oxford: Butterworth-Heinemann.

Hall, C. M., Sharples, L., Cambourne, B., and Macionis, N. (2009). *Wine Tourism around the World* (3rd ed.). New York: Routledge.

Han, L. (2011). The prospect and forecast of China's wine tourism in 2011. *Energy Procedia*, 5: 1616–1620.

Haupt, T. (2019). *Portuguese government plans to boost the wine tourism*. https://www.tourism-review.com/portugal-to-develop-into-wine-tourism-destination-news10986

Hawkins, B. (2018). *2nd Annual South African Food & Wine Tourism Report in Partnership with Wesgro*. https://exploresideways.com/blog/2nd-annual-south-african-food-wine-tourism-report-by-wesgro/

Hough, C. (2020). *Australia's grape and wine production has a history of innovation, revolutionising wine worldwide*. https://www.abc.net.au/news/rural/2020-12-02/grape-and-wine-production-in-australia-has-history-of-innovation/12921340

Kang, A. (2020). *Inside World of Wine, Portugal's Latest Tourist Destination*. https://www.forbes.com/sites/annakang/2020/08/30/inside-world-of-wine-portugals-latest-tourist-destination/?sh=b29b3925a6fb

Karagiannis, D., and Metaxas, T. (2020). Sustainable wine tourism development: Case studies from the Greek region of Peloponnese. *Sustainability*, 12(5223): 1–15. https://doi.org/10.3390/su12125223

LBV (2019). *Crash Course on Chilean wine*. https://www.labicicletaverde.com/crash-course-on-chilean-wine/

Lee, J. S.-H., McCole, D., and Holecek, D. (2020). Exploring winery visitors in the emerging wine regions of the North Central United States. *Sustainability*, 12(4): 1642. https://doi.org/10.3390/su12041642

Legal Network (2018). *By 2020, There Will Be More then 100 Wineries in Ningxia*. https://www.putaojiu.com/shangxun/186412.html

Leong, K. C. (2017). Wine tourism in China: how to explore emerging Disney-esque world *The Dallas Morning News*, September 27. https://www.dallasnews.com/arts-entertainment/travel/2017/09/27/wine-tourism-in-china-how-to-explore-emerging-disney-esque-world/

Li, Y. B., and Bardají, I. (2017). A new wine superpower? An analysis of the Chinese wine industry. *Cahiers Agricultures*, 26(6): 1–9. https://doi.org/10.1051/cagri/2017051

Li, Y. X. (2020). *Ningxia Named among World's 10 Most Promising Wine Destinations*. https://www.chinadaily.com.cn/a/202010/23/WS5f924640a31024ad0ba80897.html

Ly, P., Howson, C., and Begun, J. (2014). *Chinese Wine Tourism: Not Just about Wine*. https://soundideas.pugetsound.edu/faculty_pubs/3313/

Ma, Y. (2020). *Wine consumption in China 2000–2019*. Statista. https://www.statista.com/statistics/425098/china-wine-consumption/#:~:text=Wine%20consumption%20in%20China%20 2000%2D2019&text=In%202019%2C%20the%20Chinese%20population,country%20in%20the%20 past%20decade

Monsoon Valley (2020). *Our Story*. https://www.monsoonvalley.com/en/our-story

News24 (2018). *'Wine Tourism Grows by 16%': Emerging Routes Give Popular Franschhoek, Stellenbosch and Constantia a Run for Their Money*. https://www.news24.com/News24/wine-tourism-grows-by-16-emerging-routes-give-popular-franschhoek-stellenbosch-and-constantia-a-run-for-their-money-20180315

Oliviera, D. M., and Filho, A. F. C. (2016). City, architecture and wine: wine tourism in the contemporary society. *BIO Web of Conferences*, 7, 03026: 2–4. https://doi.org/10.1051/bioconf/20160703026

Rainford, E. J. (2017). *Ka Mua, Ka Muri – walking backwards into the future: an environmental history of the South Wairarapa Region 1984–2016*. Master of Arts thesis, Victoria University of Wellington, 133 pp. http://researcharchive.vuw.ac.nz/bitstream/handle/10063/6929/thesis_access.pdf?sequence=1

Renown Travel (2022). *Silverlake Vineyard*. https://www.renown-travel.com/daytripspattaya/silverlakevineyard.html

Santos, F. A., Vavdinos, N., and Martines, L.F. (2020). Progress and prospects for research of wine tourism in Portugal. Pasos. *Revista de Turismo y Patrimonio Cultural*, 18(1): 159–170. https://doi.org/10.25145/j.pasos.2020.18.010

Shavkatovich, R. S. (2020). Implementation of foreign experiences to develop wine tourism in Samarkand Region. *Indonesian Journal of Cultural and Community Development*, 5: 1–7. https://doi.org/10.21070/ijccd2020503

Szolnoki, G. (2018). New approach to segmenting tourists in a German wine region. *International Journal of Wine Business Research*, 30(2): 153–168. https://doi.org/10.1108/IJWBR-07-2017-0044

Tafel, M., and Szolnoki, G. (2020). Estimating the economic impact of tourism in German wine regions. *International Journal of Tourism Research*, 22: 788–799. https://doi.org/10.1002/jtr.2380

The Nation (2018). *Tourist demand drives wine industry to new heights in Thailand*. https://www.nationthailand.com/Corporate/30345134

Tompkin, J. (2019). *Bottle Shock: Thailand's Unlikely Emergence as a Wine Destination*. DW.com. https://www.dw.com/en/bottle-shock-thailands-unlikely-emergence-as-a-wine-destination/a-48147381

Veseth, M. (2017). *Sketches of Spain: State-of-the-Art Wine Tourism Does More than Sell Wine*. https://wineeconomist.com/category/spain/

VinosdeChile (n.d.). *Wine Diversity*. https://www.winesofchile.org/en/chile-a-wine-producing-country/wine-diversity/

VisitFrenchWine.com (n.d.). *Wine Tourism in France in Numbers*. https://www.visitfrenchwine.com/en/product/wine-tourism-france-numbers

Vo-Thanh, T., and Kirova, V. (2017). Wine tourism experience: A netnography study. *Journal of Business Research*, 83: 30–37. https://doi.org/10.1016/j.jbusres.2017.10.008

Wacker, J. P. (2019). *Everything you never wanted to know about Spanish wines (and a few things you did)*. Senior thesis, University of South Carolina, Columbia, 78 pp. https://scholarcommons.sc.edu/senior_theses/277

Wei, Q. (2015). Status of Ningxia Helan mountain wine tourism, market analysis and future development. *Tourism Overview*, 4: 189–190.

WineTourism.com (2019a). *Global Report of Covid-19 Impact on Wine Tourism*. https://www.winetourism.com/impact-covid-19-wine-tourism/

WineTourism.com (2019b). *California: The Most Well-Known Wine State in the USA*. https://www.winetourism.com/wine-region/california/

WineTourism.com (2019c). *Australian wine regions*. https://www.winetourism.com/wine-country/australia/

Woldarsky, V., and Geny-Denis, L. (2019). Development of a best practice manual in wine tourism in Portugal, *BIO Web of Conferences*, 12: 1–8. https://doi.org/10.1051/bioconf/20191203001

WorldWideWineTours (2021). *Popularity of Australian Wineries*. https://www.worldwidewinetours.com/australia/australia-wine-tours/

Yan, P. (2018, May 8). *Six Fascinating Travel Destinations in Asia for Wine Lovers*. Lifestyle Asia. https://www.lifestyleasia.com/bk/travel/destinations/6-fascinating-travel-destinations-asia-wine-lovers/

10
WINE TOURISM AND CONSUMER BEHAVIOR

Roberta Garibaldi, Matthew J. Stone, Erik Wolf, and Andrea Pozzi

Introduction

The consumption of wine tourism is a complex behavior. It deals with enjoying local wines and participation in wine activities, from consumptive (e.g., tastings) to more engaging (e.g., visits to wineries, wine events, and festivals), and it is influenced by internal factors that drive individuals' choices before and during their holiday. Personal attitudes and beliefs can stimulate the desire to understand and experience foreign culture through local wines (Chang, Kivela, and Mak 2010; Heldke 2003; Tse and Crotts 2005), while sociodemography, such as age, education, and income, can address the choice towards wine tourism activities. However, travelers rarely make their choices on their own. They collect information to evaluate and compare the offering available within destinations, then seek suggestions from family members and other relevant social groups to finalize their choice (Stone 2016).

Wine travelers' activities should be considered within the greater tourism landscape, rather than focusing on wine activities. Wine and food often go hand in hand: it is easy to think of a holiday that includes visits to wineries during the day, followed by a gourmet dinner. Stone, Garibaldi, and Pozzi (2019) found that wine travelers are more likely to participate in other food or drink activities while traveling, including gourmet meals, cooking classes, and beer activities. As for other segments, it is conceivable that wine tourism is rarely a discrete activity and is often undertaken in conjunction with some other forms of tourism (Charters and Ali-Knight 2002). A proper understanding of travelers' decisions in the context of wine tourism, therefore, suggests the adoption of a more holistic and inclusive approach.

The present chapter presents details of wine travelers' behavior in terms of factors that can influence their decisions. It, therefore, considers sociodemography, personal beliefs, attitude and values, social group influence and destination marketing mix, and attempts to evaluate their role in travelers' choices. Following Stone, Garibaldi, and Pozzi (2019), wine travelers are leisure travelers who have participated in wine activities (i.e., visit a winery, wine trail, or wine festival) while on a recent trip. Data come from the worldwide survey carried out by the World Food Travel Association for the *2020 Food Travel Monitor* (Stone, Migacz, and Garibaldi 2020). In particular, this chapter investigates: (1) what factors influence pre-trip decision making; (2) what information sources motivate wine travelers' prior to traveling; (3) what factors

 DOI: 10.4324/9781003143628-13

may influence wine travelers' onsite behavior; and (4) wine travelers' sharing behavior on social media. Wine travelers are statistically compared to other leisure travelers.

Review of the literature

Individual decision making in tourism

Following consumer behavior research (Engel 1968), traditional models of travelers' behavior describe how individuals narrow a choice from an awareness set to an evoked set to a final destination choice (e.g., Dixit 2017; Engel and Blackwell 1982; Van Raaij and Francken 1984). In these models, the first step is the 'generic decision' of whether (or not) to go on holiday, followed by the information acquisition, narrowing of options, and the final choice. Although decision-making does not always follow a pattern, it is widely accepted that individuals mainly follow their attitudes, beliefs, and values, evaluate alternatives and make their choices accordingly. It is conceivable that individuals who are more passionate about wine and food are more willing to search for novelty while traveling and try local wine and food products on holiday (Cohen and Avieli 2004; Hassan and Hall 2003; Kim, Eves, and Scarles 2009; Torres 2002).

In the alternative evaluation stage, individuals often pay attention to the destination marketing mix, including the availability of general activities and wine and food offerings, in addition to quality and price to help them to make a more conscious decision (Sirakaya and Woodside 2005). Empirical studies on wine and food tourism carried out in different countries (Dixit 2020; Garibaldi 2018, 2019; McKercher, Okumus, and Okumus 2008; Stone and Migacz 2016; Stone, Migacz, and Garibaldi 2020) noted that passionate travelers tend to choose destinations with a large variety of wine and food products and activities. But the evaluation process is not limited to comparing objective information. Other influences on decisions include online reviews, online and offline input from family members and relevant social groups (e.g., friends), as well as media and social media influences. Decrop (2006) acknowledged 'interpretive frameworks' in addition to the traditional process models of tourism decision-making. While the process of decision-making may seem straightforward in a model, it does not always operate this way in real life. For example, not all travelers make all of their own decisions. Stone (2016) found that about half of in-trip dining decisions were formally or informally delegated to other travel partners, who are considered to be 'social surrogates.'

Decision-making is ongoing during the trip itself, and not all choices are made before departing (Smallman and Moore 2010). Traditional models of travelers' behavior theorized that individuals constantly act upon their decision, which is always under evaluation (see Sirakaya and Woodside 2005). As travelers participate in a wide range of wine and food activities on holiday, they often check and evaluate alternatives even while traveling. In the pre-trip stage, the choice of the place to visit or where to eat can come from individuals' attitudes and are evaluated through offline and online suggestions and impressions (Garibaldi 2018, 2019; Stone and Migacz 2016; Stone, Migacz, and Garibaldi 2020). This process, however, seems not to follow a linear decision-making process. A number of purchase decisions (including dining out) are not often decided until on a trip (Woodside and King 2005). Additionally, travelers can visit an attraction (such as a winery) just to gratify the desire of only a single member of their party (a social surrogate), or because it is highly recommended by their reference group (Brown, Havitz, and Getz 2007). It appears that travelers can expressly delegate decisions, taking little or no role in the final choice. In this situation, they not only recommend choices, but also drive consumption of other people they are traveling with (Stone 2016).

Extending wine tourism theory

Academic studies have provided different definitions and conceptualizations of wine tourism through the years. However, there is a general consensus on conceiving wine tourism as travel mainly related to wine attractions and activities, such as visits to wineries, vineyards, events, festivals, and shows (Getz 2000; Hall 1996). The investigation of travelers' preference, despite its extensiveness, covered mainly visitors in wineries (Ali-Knight and Charters 2001; Bruwer and Alant 2009; Bruwer, Li, and Reid 2001; Corigliano 1996; Galloway et al. 2008; Marzo-Navarro and Pedraja-Iglesias 2012). These certainly succeeded in providing an exhaustive picture of these travelers while assuming that they all have a reasonable level of interest in wine.

However, it is conceivable that travelers who have a specific interest in wine often undertake wine tourism activities in conjunction with some other forms of tourism (i.e., food and beverage) (Charters and Ali-Knight 2002). Many predominantly North American empirical studies, using national samples of leisure travelers, investigated the likelihood of participating in wine, food, and beverage activities among the general population. The TAMS (2001) and World Food Travel Association (Stone and Migacz 2016) studies noticed that travelers who visit wineries and themed festivals often participate in other food and other beverage experiences while traveling. For example, 27.9% of American wine travelers and 31.8% of French wine travelers have also participated in a food or beverage tour while on a recent trip (Stone, Garibaldi, and Pozzi 2019). In studying individuals' behavior in wine tourism, this chapter adopts a broader perspective by examining wine tourism within the context of the range of tourism products available, primarily those related to food due to existing linkages between wine and food tourism. A more holistic approach was actually considered the most suitable for the purposes of the study.

Methodology

Survey design and data collection

As part of an international study for the World Food Travel Association, an online survey was distributed to a survey panel in the United States. Questions included past travel behavior, travel motivations, and attitudes toward food and drink travel. Respondents were at least 18 years of age and had taken at least one overnight trip (minimum 80 km/50 miles) in the past year. This chapter presents data from a sample of American leisure travelers (N = 1,491). The data was analyzed in SPSS 27.

The American survey was completed by 1,542 travelers in 2019. Because the age of alcohol consumption in the United States is 21, respondents under that age were removed from the study for a final sample of 1,491 leisure travelers. Respondents were proportionally balanced by gender (54% female/46% male) and generation (Boomers, Generation X, Millennials, and Generation Z). The average age of respondents was 40, the median age 42. As with leisure travel in general, education and income levels were higher for leisure travelers than for the general population. Table 10.1 shows the demographics of respondents.

Defining wine travelers

To qualify the sample, respondents were asked about their participation in a variety of food and beverage tourism activities. Following Stone, Garibaldi, and Pozzi (2019), wine travelers were defined as leisure travelers who have participated in a wine activity (i.e., visit to a winery,

Table 10.1 Demographics of the sample

Variable	*Category*	*Full sample (N = 1,491), %*
Generation	Boomers (born 1946–1964)	20
	Generation X (born 1965–1980)	35
	Millennials (born 1981–1996)	40
	Generation Z (born after 1996)	5
Gender	Male	46
	Female	54
Education	Have graduated from college	71
Annual Household Income	Under US$50,000	21
	US $50,000–99,999	37
	Over US$100,000	36
	Prefer not to answer	6
Children	Have children under 18 at home	38

Table 10.2 Percentage of travelers who are wine travelers based on demographics

Sample (N = 1,491)	*Wine travelers**
Full sample (US leisure travelers)	26.0% can be considered wine travelers
Boomers (born 1946–1964)	26.7% can be considered wine travelers
Generation X (born 1965–1979)	26.1% can be considered wine travelers
Millennials and Generation Z (born 1980–1998)**	25.9% can be considered wine travelers
Male	25.9% can be considered wine travelers
Female	26.2% can be considered wine travelers
Have graduated from college	28.0% can be considered wine travelers
Have not graduated from college	24.0% can be considered wine travelers
Under US$50,000	15.3% can be considered wine travelers
US $50,000–99,999	29.3% can be considered wine travelers
Over US$100,000	30.0% can be considered wine travelers

Notes:

* Wine travelers are here defined as leisure travelers who have visited a winery, wine trail, or wine festival while traveling in the past two years.

** Millennials were combined with Generation Z because Generation Z was a small cohort in this study due to their young age.

wine trail, or wine festival) while on a trip in the past two years. Over one-quarter of leisure travelers (26.0%) were considered wine travelers, slightly lower than, but comparable to, the percentage (31.4%) identified in a similar 2016 study (Stone, Garibaldi, and Pozzi 2019). Statistical comparisons are made between wine travelers and other leisure travelers to highlight similarities and differences.

Table 10.2 shows the percentage of wine travelers based on demographics. In general, the biggest differences tend to be on income and education level. Travelers with a household income of over $50,000 USD are about twice as likely to be wine travelers. Additionally, a slightly higher percentage of college graduates (28.0%) are wine travelers compared with non-college graduates (24.0%).

In this study, motivation was not used as the primary discerning factor for wine travelers. Early wine tourism researchers, such as Hall (1996), recognized that wine activities might be mainly for those whose primary travel motivations are related to wine (therefore suggesting that motivation can be used to define wine travelers). In the current study, only 18.5% of the sample (including 50.1% of wine travelers as defined in this chapter) have been motivated to take a trip or visit a destination for a winery or wine trail. This percentage likely underestimates the actual number of wine travelers because it did not consider those motivated by other wine activities, such as wine festivals and other activities, festivals, and events that may include or focus on wine. Additionally, it is often difficult to determine the primary or sole motivation for a trip. For example, a group of friends may take a weekend trip together to Napa Valley wine country. It is difficult to determine if the primary motivation for this trip was to spend time together, to get away from their daily lives, or to visit wineries. All work together to motivate the travelers. Even those traveling together on the same trip and participating in the same activities may express different primary travel motivations. In this study, they would be considered to be wine travelers because of their in-trip activity.

Results

Factors influencing wine travelers' decisions

Travel choices often follow individuals' attitudes and values. Individuals who participate in wine activities on holiday are more likely to be knowledgeable about wine: 63% of wine travelers define themselves as knowledgeable compared to 31% of other leisure travelers (see Table 10.3). Wine travelers are knowledgeable not just about wine, but also food and drink in general. This attitude can influence their willingness to search for and try other products and activities, suggesting that wine travelers are important to other culinary activities and attractions.

Table 10.3 Factors influencing the pre-trip decision-making process: differences between wine travelers and other leisure travelers

Variable	*Wine travelers*	*Other leisure travelers*	*Difference in means*	
	Mean (SD)	*Mean (SD)*	*t*	*Sig*
I consider myself knowledgeable about wine	3.66 (1.03)	2.76 (1.24)	−13.89	< .001**
I consider myself knowledgeable about beer	3.61 (1.11)	3.02 (1.30)	−8.56	< .001**
I consider myself knowledgeable about food and drink in general	3.97 (0.83)	3.68 (0.91)	−5.77	< .001**
When thinking about where I could visit, the availability of food experiences/ restaurants help me to choose between two different destinations	3.75 (0.84)	3.31 (1.05)	−8.37	< .001**
Prior to traveling to a destination, I seek out detailed information about where to eat and drink	3.98 (0.87)	3.64 (1.06)	−6.34	< .001**
Before I leave on a trip, I like to decide where to eat and drink at my destination.	3.58 (1.02)	3.36 (1.08)	−3.43	.001**

Notes:
Likert-type scale from 1 = strongly disagree to 5 = strongly agree
** Significant at $p \leq .001$

What may seem counterintuitive is that over one-third of wine tourists (37%) do not consider themselves well-informed about wine, suggesting that other factors may drive their decisions. This highlights the importance of wineries educating travelers rather than just providing an activity or diversion.

Destination marketing mix also plays a relevant role in the decision-making process. Among all travelers, 53.5% agree (or strongly agree) that the availability of food experiences and restaurants help them to choose between destinations to visit (see Table 10.3). The perceived importance of the types of food experiences and restaurants in destination choice is higher among wine travelers than the other group (69% of wine travelers compared to 48% of other leisure travelers). This states the importance of food and drink, not as an ancillary activity on a trip but as a key factor in destination choice.

The information search and decisions are often made before travel, with some spontaneous choices during the holiday. In this pattern, the main difference between wine travelers and the other group lies in pre-trip activities: Before traveling, wine travelers are more likely to both seek out detailed information about where to eat and drink (M = 3.98 on a 5-point Likert-type scale) and to decide where dining (M = 3.58). Both values are significantly greater ($p \leq .001$) than other leisure travelers.

Pre-trip decisions often follow online and offline suggestions and impressions from family members and friends. Table 10.4 shows what information sources have motivated travelers to visit a destination, restaurant/bar, or wine and food attraction. About two-thirds have been motivated by friends or family, but no significant difference (p = .165) can be noticed between wine travelers and the other group. However, when considering online sources (i.e., online review sites and videos), social media posts, and media (i.e., TV programs and magazine/newspaper articles), wine travelers are more likely to make decisions after reading advice and suggestions on these channels. Actually, nearly half of them have been motivated by online review sites, and nearly 40% were motivated by friend/family social media posts. It is important to note that these online sources and media influence destination choice, as well as food and drink decisions.

Table 10.4 Information sources that have motivated travelers to visit a specific destination, restaurant/bar, or culinary attraction: differences between wine travelers and other leisure travelers

	Wine travelers	*Other leisure travelers*	*Difference in means*	
	%	%	*Chi -Square*	*P*
My friends or family	68.5%	64.6%	1.92	.165
Online review sites (such as Yelp, TripAdvisor)	47.5%	39.4%	7.83	.005
Postings about food and drink from friends and relatives on social media	38.2%	26.8%	17.91	< .001**
TV programs or channels dedicated specifically to food or beverages	37.7%	24.8%	23.59	< .001**
Postings about food and drink from other people I follow on social media	31.0%	17.9%	29.19	< .001**
Articles in newspapers or magazines	32.0%	18.8%	28.85	< .001**
Online videos (such as YouTube or Vimeo)	27.1%	16.4%	21.31	< .001**

** Significant at $p \leq .001$

Factors influencing wine travelers' onsite behavior

The onsite behavior of wine travelers frequently reflects their personal attitudes. Wine travelers are more likely to make value-based decisions regarding dining while on a trip than other travelers. They more frequently make dining choices that reflect their own personal health and wellness values (M = 3.67), and they are more likely to try to eat at places that demonstrate social responsibility (M = 3.48). While this study did not investigate this in great detail, these findings imply that wine travelers' decision-making process for dining on a trip may consider additional factors that may be less important for other leisure travelers. Therefore, restaurants, wineries, and culinary attractions may be best served by communicating their health and sustainability values to travelers.

Social influences are important, especially to wine travelers. They recognized themselves as more influential regarding places to eat on a trip (M = 3.75). They, therefore, frequently act as social surrogates (Stone 2016) who make dining decisions for others (see Table 10.5). This implies that wine travelers may have a greater value to a destination than other leisure travelers. Additional data from the survey support this proposition. There is a significant difference ($p < .01$) between wine travelers and other travelers on in-trip expenditures on food and beverage, entertainment/cultural attractions, and shopping. Wine travelers report spending 19.8% more per day on food and beverage, 25.2% more on entertainment/cultural activities/recreation, and 28.2% more on shopping. While part of this difference may be due to higher income levels of wine travelers, even among the lower-income group (annual household income less than USD$50,000), wine travelers reported spending 40% more per day on food and drink ($p = .002$) than other travelers.

While many decisions are made before traveling, travelers are adaptable and flexible. Over 65% of both wine travelers and other leisure travelers agreed or strongly agreed that they made spontaneous decisions about where to eat while on a trip. Thus, marketing should not be limited to online media that require search but also a variety of point-of-purchase displays or other marketing at the destination. That wine travelers' behavior can also influence the decisions of other travelers can also be deducted by their use of social media on holiday. Not only do wine travelers rely more on social media for pre-trip information, but they are also more likely to share their travel experiences on different social media (including Facebook, Instagram,

Table 10.5 Factors influencing wine travelers' onsite behavior: differences between wine travelers and other leisure travelers

Variable	*Wine travelers*	*Other leisure travelers*	*Difference in means*	
	Mean (SD)	*Mean (SD)*	*t*	*Sig*
The food choices I make when traveling reflect my own personal health and wellness values.	3.67 (0.89)	3.44 (0.97)	−4.18	< .001**
When on a trip, I try to eat at places which demonstrate social responsibility.	3.48 (0.96)	3.22 (1.02)	−4.47	< .001**
I often influence my friends about where to eat while on a trip.	3.75 (.93)	3.43 (1.01)	−5.69	< .001**
My decisions regarding dining out are heavily influenced by my friends.	3.25 (1.08)	3.01 (1.07)	−3.67	< .001**

Notes:

Likert-type scale from 1 = strongly disagree to 5 = strongly agree

** Significant at $p \leq .001$

Table 10.6 Social media sharing activities: differences between wine travelers and other leisure travelers

Variable	*Wine travelers*	*Other leisure travelers*	*Difference in means*	
	Mean (SD)	*Mean (SD)*	*T*	*Sig*
I share my travel experiences on Facebook	3.22 (1.46)	2.73 (1.46)	−5.73	< .001**
I share my travel experiences on Instagram	2.84 (1.60)	2.35 (1.50)	−5.21	< .001**
I share my travel experiences on Twitter	2.16 (1.46)	1.86 (1.29)	−3.64	< .001**
I share my travel experiences on other social media (WhatsApp, Pinterest, Snapchat, etc.)	2.44 (1.49)	2.13 (1.42)	−3.56	< .001**

Notes:
Likert-type scale where: 1 = never; 2 = sometimes; 3 = about half of my trips; 4 = often; 5 = always
** Significant at p ≤ .001

and Twitter) than the other group (see Table 10.6). Among wine travelers, Facebook was the most frequently utilized (M = 3.22), followed by Instagram (M = 2.84). Not surprisingly, social media sharing was more frequent among younger wine travelers: among Millennials and younger individuals (born 1981–1998) (N = 154), the mean for Facebook and Instagram was respectively 3.60 and 3.45, indicating that these channels are used to share travel experiences on more than half of trips. This social media sharing is not limited to sharing of wine activities, suggesting that just having more wine travelers at a destination may be more likely to result in more social media posting activity by travelers.

Conclusions

This chapter investigated factors that influence travelers' decisions in wine tourism before and during their holidays by extending wine tourism theory and considering wine tourism within the context of other food products in the destination. Additionally, it compares wine tourists with other leisure travelers.

Being more knowledgeable in wine is positively correlated with being wine travelers, but these travelers also consider themselves more knowledgeable about both beer and food and drink in general. This suggests that wine travelers can be more willing to search and try other products and activities, making this segment attractive to non-wine-focused food and beverage tourism businesses (including bars, breweries, and restaurants). Findings are consistent with Stone, Garibaldi, and Pozzi (2019), who found that wine travelers participate in a greater variety of food and drink activities. The finding that over one-third of wine travelers consider themselves not well-informed about wine suggests that other factors may drive their decisions, such as social (being with friends), hedonic (just for fun), or educational motivations (to learn more about wine). It also indicates there is still an opportunity for wineries, wine bars, and wine tours to educate their visitors.

Wine travelers are more likely to be motivated to visit a destination, restaurant/bar, or culinary attraction by many media sources (including online reviews, social media postings, articles, TV programs, and online videos). Thus, destinations, attractions, and wineries should focus on a complete marketing mix to attract these valuable travelers. Wine travelers spend significantly more on food and drink, entertainment and cultural activities, and shopping per day than other leisure travelers.

Even personal values and beliefs drive wine travelers' decisions and onsite behavior. With regards to their dining choices, these travelers are more likely to choose places that reflect their own personal health and wellness as well as values, as emerged in other studies on food tourism (Cohen and Avieli 2004; Dixit 2019; Hassan and Hall 2003; Kim, Eves, and Scarles 2009; Torres 2002). This does not imply that all of their dining decisions are based on these criteria, but it shows that it is important for restaurants and other food and beverage providers to show their health and sustainability attributes to attract wine travelers. When faced with two seemingly equal choices, these criteria may make the difference for a restaurant. It is also important to recognize that not all decisions are made before traveling. Leisure travelers often make spontaneous choices, demonstrating the importance for the business to adopt a variety of strategies that may include, for instance, attractive storefronts, on-site advertising in hotels and tourist sites, and recommendations from locals.

Along with recognizing the influence of social groups (i.e., family and friends) in wine travelers' decisions, this chapter also found that this group can act as surrogates and can influence the choices of other individuals (Stone 2016). They are also more likely to share their travel experiences on social media. It seems, therefore, that not only do they influence others during a trip, but that their messages can also influence future friends or family members. This sharing creates a pattern beneficial to tourism businesses. Wine travelers post more on social media, and they are also more likely to use the same sources to discover motivations to travel. It can be argued, therefore, that the more wine travelers a destination attracts, the more social media posts will result and motivate future travelers to visit the place. Over time, this is likely to boost visitation.

It is important to remark that social media posts are related to travel experiences in general, not just about wine and food. This would suggest that wine travelers are especially important to destinations from a social media marketing perspective. While this research does not present a complete view of wine travelers' behaviors, it provides a glimpse at differences between the decision-making behavior of wine travelers when compared with others. It also finds many reasons that wine travelers are especially beneficial to destinations. Further research can continue to build upon the complexities of decision-making in wine tourism.

Implications

From a theoretical perspective, this chapter extends the wine tourism theory and considers this practice within the context of other food products in the destination. This approach allows for investigation of factors that influence travelers' decisions in wine tourism – and not merely for wine tourism activities – before and during their holidays.

Several practical implications can be drawn. Wine travelers are more knowledgeable about wine and food in general, and they are more likely to be motivated to travel by more sources, including social media, printed media, and TV shows and videos. Even personal health and wellness values and social responsibility are important. There is room for producers and destinations to exploit the traditional wine tourism experience, but particular attention must be paid to a complete marketing mix to attract these valuable travelers.

References

Ali-Knight, J., and Charters, S. (2001). The winery as educator: do wineries provide what the tourist needs? *Australian and New Zealand Wine Industry Journal*, 16(6): 79–86.

Brown, G. P., Havitz, M. E., and Getz, D. (2007). Relationship between wine involvement and wine-related travel. *Journal of Travel & Tourism Marketing*, 21(1): 31–46. https://doi.org/10.1300/J073v21n01_03

Bruwer, J., and Alant, K. (2009). The hedonic nature of wine tourism consumption: an experiential view. *International Journal of Wine Business Research*, 21(3): 235–257. https://doi.org/10.1108/17511060910985962

Bruwer, J., Li, E., and Reid, M. (2001). Wine-related life-style segmentation of the Australian domestic wine market. *Australian and New Zealand Wine Industry Journal*, 16(2): 104–108. https://hdl.handle.net/2440/12908

Chang, R. C. Y., Kivela, J., and Mak, A. H. N. (2010). Food preferences of Chinese tourists. *Annals of Tourism Research*, 37(4): 989–1011. https://doi.org/10.1016/j.annals.2010.03.007

Charters, S., and Ali-Knight, J. (2002). Who is the wine tourist? *Tourism Management*, 23(3): 311–319. https://doi.org/10.1016/S0261-5177(01)00079-6

Cohen, E., and Avieli, N. (2004). Food in tourism: attraction and impediment. *Annals of Tourism Research*, 31(4): 755–778. https://doi.org/10.1016/j.annals.2004.02.003

Corigliano, M. A. (1996). *Enoturismo. Caratteristiche della domanda strategie di offerta e aspetti territoriali e ambientali*. Milan: Franco Angeli.

Decrop, A. (2006). *Vacation Decision Making*. Wallingford, UK: CABI Publishing.

Dixit, S. K. (ed.). (2017). *The Routledge Handbook of Consumer Behaviour in Hospitality and Tourism*. London: Routledge.

Dixit, S. K. (ed.). (2019). *The Routledge Handbook of Gastronomic Tourism*. Oxon: Routledge.

Dixit, S. K. (2020). Marketing gastronomic tourism experiences. In Dixit, S. K. (ed.), *The Routledge Handbook of Tourism Experience Management and Marketing* (pp. 322–335). London: Routledge.

Engel, J. F. (1968). The study of human behavior. In Engel, J. F. (ed.), *Consumer Behavior: Selected Readings* (pp. 2–17). Homewood, IL: Richard D. Irwin Inc.

Engel, J. F., and Blackwell, R. D. (1982). *Consumer behavior* (4th ed.). Hinsdale, IL: The Dryden Press.

Galloway, G., Mitchell, R., Getz, D., Crouch, G., and Ong, B. (2008). Sensation seeking and the prediction of attitudes and behaviours of wine tourists. *Tourism Management*, 29: 950–966. https://doi.org/10.1016/j.tourman.2007.11.006

Garibaldi, R. (2018). *Primo Rapporto sul Turismo Enogastronomico Italiano 2018*. Bergamo: Centro Editoriale Librario Studium Bergomense (CELSB).

Garibaldi, R. (2019). *Rapporto sul Turismo Enogastronomico Italiano*. Self-Published.

Getz, D. (2000). *Explore Wine Tourism: Management, Development, Destinations*. New York: Cognizant.

Hall, C. M. (1996). Wine tourism in New Zealand. In Higham, J. (ed.), *Proceedings of the Tourism Down Under II: A Research Conference* (pp. 109–119). Dunedin: University of Otago.

Hassan, M. W., and Hall, C. M. (2003). The demand for halal food among Muslim travelers in New Zealand. In Hall, C. M., Sharples, L., Mitchell, R., Macionis, N., and Cambourne, B. (eds.), *Food Tourism around the World: Development, Management and Markets* (pp. 81–101). Oxford: Butterworth-Heinemann.

Heldke, L. (2003). *Exotic Appetites: Ruminations of a Food Adventurer*. New York: Routledge.

Kim, Y., Eves, A., and Scarles, C. (2009). Building a model of local food consumption on trips and holidays: A grounded theory approach. *International Journal of Hospitality Management*, 28(3): 423–431. https://doi.org/10.1016/j.ijhm.2008.11.005

Marzo-Navarro, M., and Pedraja-Iglesias, M. (2012). Critical factors of wine tourism: incentives and barriers from the potential tourist's perspective. *International Journal of Contemporary Hospitality Management*, 24(2): 312–334. https://doi.org/10.1108/09596111211206196

McKercher, B., Okumus, F., and Okumus, B. (2008). Food tourism as a viable market segment: it's all how you cook the numbers! *Journal of Travel & Tourism Marketing*, 25(2): 137–148. https://doi.org/10.1080/10548400802402404

Sirakaya, E., and Woodside, A. G. (2005). Building and testing theories of decision making by travellers. *Tourism Management*, 26: 815–832. https://doi.org/10.1016/j.tourman.2004.05.004

Smallman, C., and Moore, K. (2010). Process studies of tourists' decision-making. *Annals of Tourism Research*, 37(2): 397–422. https://doi.org/10.1016/j.annals.2009.10.014

Stone, M. J. (2016). Deciding not to choose: Delegation to social surrogates in tourism decisions. *Tourism Management*, 57: 168–179. https://doi.org/10.1016/j.tourman.2016.06.002

Stone, M. J., Garibaldi, R., and Pozzi, A. (2019). Exploring additional food and beverage activities of wine travelers. In Dixit, S. K. (ed.), *The Routledge Handbook of Gastronomic Tourism* (pp. 89–96). London: Routledge.

Stone, M. J., and Migacz, S. (2016). *The American Culinary Traveler: Profiles, Behaviors, & Attitudes*. Portland, OR: World Food Travel Association.

Stone, M. J., Migacz, S., and Garibaldi, R. (2020). *2020 Food Travel Monitor*. Portland, OR: World Food Travel Association.

TAMS (Travel Activities and Motivation Survey) (2001). *Wine and Cuisine Profile Report*. Toronto: Lang Research.

Torres, R. (2002). Toward a better understanding of tourism and agriculture linkages in the Yucatan: Tourist food consumption and preferences. *Tourism Geographies*, 4(3): 282–306. https://doi.org/10.1080/14616680210147436

Tse, P., and Crotts, J.C. (2005). Antecedents of novelty seeking: International visitors' propensity to experiment across Hong Kong's culinary traditions. *Tourism Management*, 26: 965–968. https://doi.org/10.1016/j.tourman.2004.07.002

Van Raaij, W. F., and Francken, D. A. (1984). Vacation destinations, activities and satisfactions. *Annals of Tourism Research*, 11(1): 101–112. https://doi.org/10.1016/0160-7383(84)90098-7

Woodside, A. G., and King, R. L. (2005). Qualitative comparative analysis of travel and tourism purchase-consumption systems. In March, R., and Woodside, A. G. (eds.), *Tourism Behaviour: Travelers' Decisions and Actions* (pp. 66–72). Wallingford, UK: CABI Publishing.

11

PREFERENCES AND CHARACTERISTICS OF WINE TOURISTS

Kuan-Huei Lee

Wine was first used as a form of medicine in ancient history back in 1500 BC. It was considered first as a sacred item and a symbol of civilization before it became part of our diet and an essential element in most food and beverage establishments (Civitello 2011). Tourists travel for different motives, and wine tourists are a special interest group that is attracted by the consumption of wine and travel to different destinations with the objective of searching for wine-related products. This chapter introduces consumer behavior, consumer needs, and preferences, and profiles of wine consumers from involvement, personality, and lifestyle. Classification of wine tourists are presented as well as a brief description of wine events and festivals. The case of how bar businesses overcame the downturn affected by the pandemic is also discussed.

Understanding consumer behavior

There is decision-making behind every consumption choice. The process usually starts with the recognition of a need or problem, followed by information search, evaluation of alternatives, a decision on the chosen alternative, and post-purchase evaluation. During this process, the consumers are influenced by their own cognitive, habitual, and affective constraints. One influential factor that affects this process is the degree of involvement that the consumer possesses toward a specific product, message, or situation. The definition of involvement is 'a person's perceived relevance of the object based on inherent needs, values, and interests' (Zaichkowsky 1985, p. 342). Havitz and Dimanche (1990, p. 184) proposed the following definition of involvement:

> a psychological state of motivation, arousal, or interest between an individual and recreational activities, tourist destinations, or related equipment, at one point in time, characterized by the perception of the following elements: importance, pleasure value, sign value, risk probability, and risk consequences.

Mitchell (1979, p. 195) viewed involvement as 'an individual level, internal state variable that indicates the amount of arousal, interest or drives evoked by particular stimulus or situation.'

Antecedents of involvement include personal factors such as needs, importance, interest, and value; object or stimulus factors such as differentiation of alternatives, source of communication, and content of communication; and situational factors such as purchase and occasion. One commonly used instrument to measure antecedents of involvement is the Consumer

DOI: 10.4324/9781003143628-14

Involvement Profile (CIP) that measures involvement using five antecedents (Laurent and Kapferer 1985, p. 43):

- The perceived significance of the product (its personal meaning).
- The perceived risk related to the product is divided into two different facets.
- The perceived significance of negative implications for bad choices and the perceived probability of making this error.
- The symbolic value the customer attributes to the merchandise, its acquisition, or its consumption.
- The emotional or hedonic value of the product and its capacity to offer pleasure and affects.

Another widely used involvement scale is the Personal Involvement Inventory (PII) developed by Zaichkowsky that assess involvement in three facets (Zaichkowsky 1985, p. 342):

- *Personal*: Awareness, beliefs, or desires that stimulate a person to obtain an object.
- *Physical*: Features of the object differentiate and increase awareness.
- *Situational*: Where relevance or awareness toward the object temporarily increases.

Consumer needs and preferences

The consumption of items is strongly related to personal needs, the requirement required to be fulfilled for different reasons. One way of classifying consumer's needs is the Hierarchy of Needs by Abraham Maslow. Maslow classified five different levels of needs, from lower- to higher-order (Figure 11.1):

- *Physiological needs*: This is the first and lowest need on Maslow's hierarchy of needs, which includes all basic needs required for human survival such as food, water, clean air, clothing, and shelter. Once this level of need is fulfilled, it will move to the second level on the hierarchy.

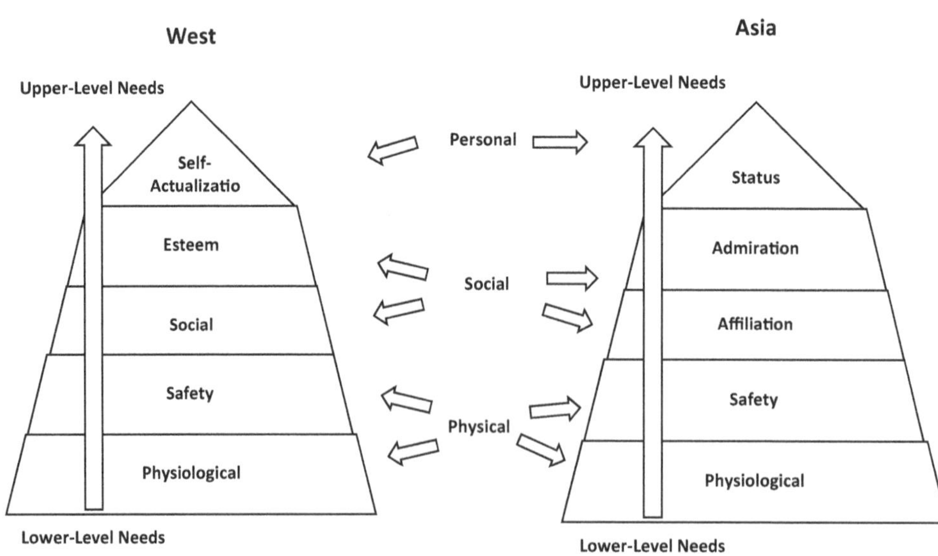

Figure 11.1 Maslow's Hierarchy of Needs and Schütte and Ciarlante Asian Equivalent Model.

- *Safety needs*: The second level of needs is the need for safety and security, which includes not only physical safety but also psychological stability that one can control in the environment and lifestyle.
- *Social needs*: The third level of Maslow's hierarchy is social needs, which include affection, love, belongingness, and acceptance.
- *Esteem needs*: This need has two orientations; the inward-directed needs refer to self-esteem, personal success, independence, and satisfaction; the outward-directed needs include the need for reputation, status, prestige, and recognition from others.
- *Self-actualization needs*: This is the highest level in Maslow's hierarchy, which includes the need for personal growth and self-fulfillment.

Other scholars have suggested a slightly different perspective from an Asian perspective when it comes to the order and priorities of Maslow's hierarchy needs. Ap (2004) compared the top personal and societal values between Asians and Americans and found that personal achievement and individualism are characteristics of Western society; on the other hand, Asians tend to be more collectivist and present behaviors such as respect to seniors and authority, filial piety, loyalty, and reciprocity, values shared by Confucianism. The highest level of need for Western cultures is related to a personal need, the need for self-actualization. In the case of Asian cultures, the highest level of need is still a social need, the need of status, which is the awareness of social image and dignity or face in a group setting.

Personality traits and consumer behavior

Personality traits also influence consumer's choices and preferences. Personality traits reflect one's characteristics, thought patterns, feelings, and behaviors. The Big Five personality traits were developed in the 1980s and classified personality into five general factors, also known as the five-factor or OCEAN model: (1) openness to experience; (2) conscientiousness; (3) extraversion; (4) agreeableness; and (5) neuroticism (McShane and Von Glinow 2019).

People who have high scores in *openness to experience* tend to be more imaginative, curious, and more willing to try new things and receive new ideas. People who score high in *conscientiousness* tend to be more organized, goal-driven, methodical, and proactive. People with more *extroversion* are more talkative, outgoing, and sociable; *extroverts* like to mingle with others more than introverts. People with higher *agreeableness* tend to be more willing to help others, more considerate, more generous, and more flexible. People with a high score in *neuroticism* are more anxious, insecure, and temperamental compared to low neuroticism scorers, who tend to be calm and secure.

Other factors that influence consumer behavior studies by academics include innovativeness, need for uniqueness, need for recognition, sensation seeking, optimum stimulation level, variety and novelty seeking, compulsive consumption, and consumer ethnocentrism (Schiffman and Wisenblit 2015).

- *Need for uniqueness*: People are seeking to be different related to others through acquiring different products and services. The need for uniqueness is related to individual social status; people with a high need for uniqueness usually adopt new products and brands faster than others.
- *Optimum stimulation level (OSL)*: This is the degree through which to measure an individual affinity to complex, unusual, and novel experience against the preference of uncluttered, calm, and simple experience. People with high OSL are more willing to take risks, try

new products and services, and search for special experiences while shopping. OSL scores are related to personal lifestyle; people with a lifestyle equal to their OSL level are satisfied with their lives; if their OSL score is higher than their lifestyle, people tend to be bored and seek activities that can bring excitement. On the other hand, people with lower OSL levels compared to their lifestyle are overwhelmed and tend to seek quieter activities and relaxation.

- *Sensation seeking*: It is the need that one has for complex, unusual, novel sensations and experience, as well as the willingness to take risks in order to experiment with the desired experience. People with a higher sensation-seeking score are likely to include volunteerism in their holidays.
- *Variety and novelty seeking*: This is the behavior to switch previously used brands to another new alternative or adopt a product in a novel way.
- *Need for cognition (NFC)*: NFC measures a person's desire or enjoyment of thinking. People with high NFC are more likely to respond to ads and will seek product information online; however, people with low NFC find it more difficult to be focused and tend to be attracted by peripheral information of an ad.
- *Compulsive consumption*: It is an addicted and out-of-control buying action. This behavior might cause problems to the shopper and sounded family members. Studies show the majority of compulsive buyers are females who often purchase things that they do not need or do not use. Severe compulsive buyers will require psychological intervention to control this behavior.

Consumer profile of wine tourists

Tourists travel for different reasons. For wine tourists, the main travel motivation is the experience of wine-related products. Wine tourism can be defined as 'a visitation to vineyards, wineries, wine festivals and wine shows for which grape-wine tasting and/or experiencing the attributes of a grape-wine region are the prime motivating factor for visitors' (Hall and Mitchell 2000, p. 447). According to Brown and Getz (2005), wine tourists form a special interest group that travels based on their desire to visit regions of wine production and wineries even when they are traveling for other reasons. The wine tourist's main travel reason might not always be wine, but they would visit wine-related establishments in the destination when there is a chance.

Wine tourists have been classified according to their consumer behavior using different traits and tools. Galloway et al. (2008) found that sensation seeking was closely related to spending on wine and visits to wineries by wine tourists. Barber, Taylor, and Deale (2010) discovered that environmental personality characteristics influence the purchase intention of wine tourists. The involvement with wine also influences the motivation and patterns of wine tourists to visit a winery (Nella and Christou 2014). The study identified four segments of wine tourist: (1) wine lovers, (2) neophytes, (3) occasional wine tourists, and (4) hangers-on. The segment of wine lovers has the most income and motivation to visit a specific winery, the neophytes are low-income students with an interest in wine, the occasional tourists are attracted more by the local gastronomy than by wine, and the hangers-on are not wine drinkers, and they consider the winery a pure tourist attraction.

Byrd et al. (2016) listed nine wine destination attributes that might attract visitors; these attributes form the core and supplementary services: (1) wine product and consumption experience, (2) general winescape features such as winery production facility and testing room, (3) educational opportunities, (4) entertainment and events, (5) natural environment and rural landscape, (6) relaxation and recreation, (7) customer service, (8) socializing and wine clubs, and (9) other tourism and hospitality services. The study found among these nine attributes that customer service was the primary predictor of repeat visitation intentions.

Adopting the classification from lifestyle studies, wine tourists can be classified according to the intensity and affinity towards wine into the following four categories (Figure 11.2):

- *Amateurs*: People who are attracted to wine but with only limited knowledge about it.
- *Newcomers*: People who like wine and see it is a good opportunity for relationship building; they are also keen to learn more about wine.
- *Professionists*: People with substantial knowledge about wine, who know the origin and the process, and have the capability to discuss the organoleptic properties of different varieties.
- *Wine experts*: People who know about wine and buy large quantities of wine for home consumption and visit wineries with organized groups.

In their study, Ignatov and Smith (2006) segmented a group of gastronomic tourists in Canada into 'wine tourists,' 'food tourists,' and 'food and wine tourists,' based on psychographics, demographics, trip characteristics, and media habits. Brown, Havitz, and Getz (2006) used a Wine Involvement Scale (WIS), developed based on Consumer Involvement Profile (CIP) scales, to test wine clubs and wine events in Western Canada. Pratt (2010) also used the WIS to study the moderating factor of wine involvement on tourists' behaviors; the result suggests that the behavior of participants with a high wine involvement scale was more consistent with wine tourists compared with those with a low wine involvement scale.

Another study used the scale of Zaichkowsky's Personal Involvement Inventory to study a group of wine tourists. These tourists were segmented into high, medium, and low involvement in wine according to their behavior, self-image, attitude, and intention to visit wine festivals (Yuan et al. 2008).

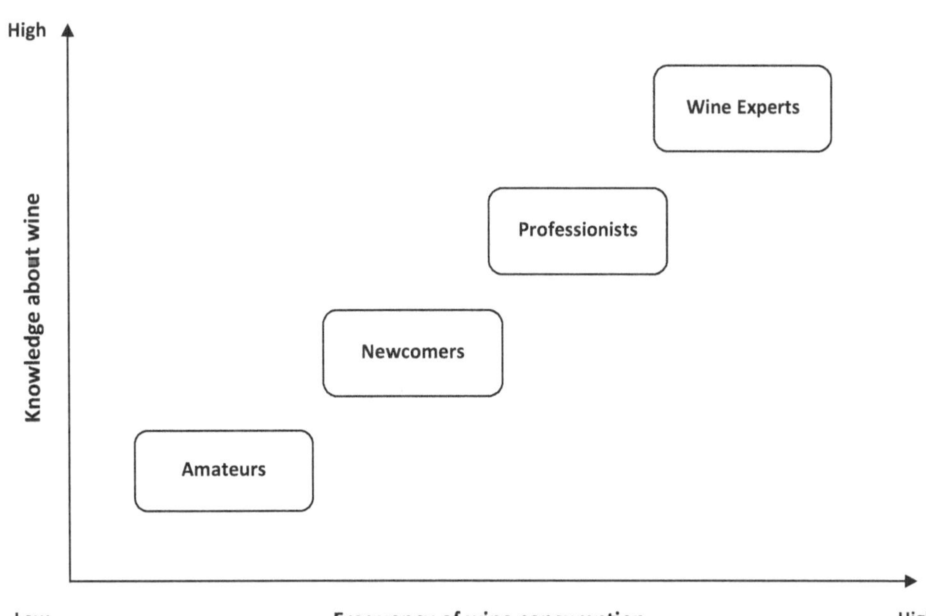

Figure 11.2 Classification of wine tourists according to lifestyle.

The wine-related lifestyle (WRL) research instrument was developed based on Grunert's food-related lifestyle (Bruwer, Li, and Reid 2002). The WRL instrument has five dimensions: (1) wine consumption situations; (2) ways of shopping; (3) quality/attributes; (4) drinking rituals; and (5) consequences of wine consumption. The study interviewed over 300 households in Adelaide, South Australia, and classified these households into five different segments: (1) purposeful inconspicuous premium wine drinkers, (2) ritual-oriented conspicuous wine enthusiasts, (3) enjoyment-seeking social wine drinkers, (4) fashion/image-oriented wine drinkers, and (5) basic wine drinkers.

The study of Tassiopoulos, Nuntsu, and Haydam (2004) employed a structured questionnaire adapted from a past study to investigate a group of wine tourists in South Africa. The study found the existence of psychographic motivations among the wine tourists studied. Another study targeted tourists to wineries in Matakana, New Zealand, using the list of values (LOW) as an instrument; the study found most winery visitors were well-educated professionals with a well-developed set of expectations related to the quality of wine tourism experience. The study identified two main segments in this group of winery tourists; the 'achiever' and the 'fun-lover' (Simpson, Bretherton, and de Vere 2004).

Participants in wine fairs and festivals

One of the most frequent destinations for wine tourists is wine fairs and festivals. Most wine producer countries organize wine festivals periodically; countries such as Australia, the United States, Germany, France, Brazil, and Argentina organize food and wine fairs and festivals that include wine tasting, grape pressing, regional music, and food. The success of each of these festivals requires support from different stakeholders that directly or indirectly influence the festival. The government is an important stakeholder in that the expectation and provisions associated with the festival will influence the destination's tourism activities in general. Other stakeholders include the hospitality, food, and wine industries, residents, media, sponsors, volunteers, and attendees, among others (Figure 11.3).

The organization of wine festivals will bring economic, social, and environmental short-term and long-term benefits to the local community; some benefits for the short term and long term are (Hede 2008, p. 94):

Short-term benefits are:

- Increased demand for local food and wine produce
- Increased trade at participating restaurants, hotels, and cafes
- Positive economic impact
- Improved opportunities for leisure
- Improved opportunities for cultural exchanges through wine
- Reduced wine miles traveled
- Increased awareness of the Slow Wine movement
- Reduced diversion rates from landfill

Long-term benefits are:

- Increased consumer preference for locally produced food and wine comparable to imported produce
- Increased investment in local food and wine industries

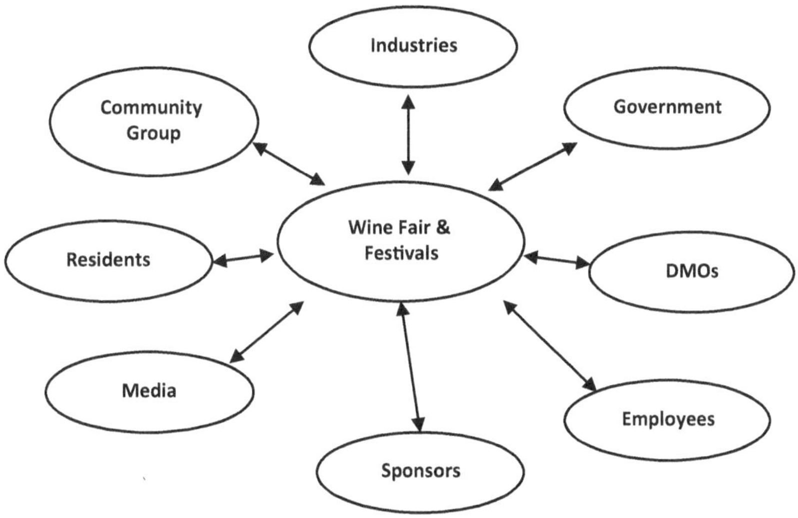

Figure 11.3 Stakeholders in wine fairs and festivals.

- Increased opportunities for long-term employment in the hospitality, food and wine industries
- Increased exporting of local food and wine produce
- Improved social and intellectual capital in the host community
- Increased recognition and tolerance of cultural diversity in the host community
- Reduction in the carbon footprint of festival and wine industries
- Greater support for the Slow Wine movement

Case Study 11.1 Wine festival, event, and business

Slow Wine

Part of Slow Food International, Slow Food is a nonprofit organization founded by Carlo Petrini in 1989 in Italy. The movement spread into 160 countries worldwide, with around 100,000 members joining local chapters called convivia. The central philosophy of Slow Food is good, clean, and fair food. According to Schneider (2008, p. 390):

> Good food is defined as being tasty and diverse and produced in such a way as to maximize its flavor and connections to a geographic and cultural region. Clean food is sustainable, helping to preserve rather than destroy the environment, while fair food is produced in socially sustainable ways that emphasize social justice and fair wages.

Food is an inevitable part of our lifestyle, and we are responsible for the choice we make regarding cultivation, production, distribution, and consumption of food-related products.

Slow Wine, together with Slow Cheese, Slow Fish, Slow Meat, and others, is part of the Slow Food movement that follows the same good, clean, and fair values in wine. The movement supports small-scale Italian winemakers that continue using the traditional way of production that respects the environment and terroir. Italy is the leading wine producer in the world, together with France and Spain; wine production in Italy in 2019 was 47.5 million hectoliters (Statista 2019). Wine is part of Italy's identify and heritage; Slow Wine promotes those Italian winemakers that safeguard the biodiversity of grape varieties.

ProWein/ProWine

ProWein, one of the largest industry meetings and trade fairs of wines and spirits, started in the late 1990s. Up to 2021, five locations hosted the trade show worldwide: ProWein Düsseldorf, ProWine São Paulo, ProWine Shanghai, and ProWine Asia (Hong Kong and Singapore). A local trade show is held yearly in each of these locations; ProWine Asia rotates annually the venue between Hong Kong and Singapore. Due to the COVID-19 pandemic, some locations needed to postpone their exhibition time to later dates during 2020 and 2021, e.g., the originally scheduled ProWein 2021 in Düsseldorf was postponed to 27–29 March 2022.

ProWine Singapore co-hosts with Food and Hotel Asia FHA-Food and Beverage, happening every two years in Singapore, It is the largest food and beverage trade event and exhibition in Asia with 2,000 exhibitors from over 100 countries worldwide. There is an exhibition of cutting-edge food technology, a culinary competition, and professional seminars using 80,000 square meters of exhibition area, attracting 43,000 attendees.

Bar business in the World and in Singapore

The bar and lounge businesses have been affected by the COVID-19 pandemic in most parts of the world, with lockdown and closure of businesses for a long period of time. New initiatives can be found from consumers to assist their favorite bars to overcome this downturn; in Belgium, the 'helpy hour' initiative was launched when customers are paying the price of two to get one drink, as opposed to the 'happy hour' when you get two drinks paying for one. In Denmark, Carlsberg's launched the 'adopt a keg' initiative, where customers buy a virtually filled keg at home during the lockdown period; the virtual keg would be exchanged for real beer in a real bar after lifting of restrictions.

There are about 4,700 restaurants and bars in Singapore, over 660 are bars (Statista 2020). The Singapore Nightlife Business Association (SNBA) is a nonprofit organization that groups bar runners in Singapore. The COVID-19 pandemic hit hardly the business of nightclubs and bars in Singapore. At least 45 bars permanently closed, and many obtained food licenses to convert into restaurants or bistrós since 2020. Consumers started ordering food and beverage deliveries to the home. The 1887 Virtual Bar launched online on April 9, 2020, using Instagram Live and Zoom to the community of consumers; this is a weekly session in which guests and feature local bartender icons meet online. Consumers can purchase gift card offers, remote house or corporate party packages, online masterclasses, and delivery options for wine and cocktails. The Old Man, ranked 38th in the World's 50 Best Bars, started offering their signature cocktails in sealed bags for their customers to take away during the lockdown period in mid-2020.

Conclusion

Wine tourists are consumers that consider the consumption of wine as a serious interest; most of them are highly involved in wine-related activities. Academics found through using different measurement scales the consumption behavior of this special interest group. Studies found different categorizations on consumer's involvement with wine and the relationship of their lifestyle activities with wine. Those consumers that possess a strong interest in wine are highly motivated to do activities related to wine in their leisure times, such as the visitation of wineries and the enrollment into wine tasting workshops.

There are several wine fairs and festivals around the world that these consumers visit frequently. Success of these fairs and festivals is determined by how well their stakeholders work with one another, aiming to provide quality service and experience to wine tourists. With a better understanding of wine tourists, destination marketers and event organizers can plan activities that best suit the need of this special interest group.

Many fairs, festivals, and trade shows related to wine have been postponed since 2020 due to the outbreak of the COVID-19 pandemic in most parts of the world. Cases of wine fairs and festivals have been introduced as well as some innovative ways organized by businesses to overcome the shortfall in sales since 2020.

Implications

This chapter presented consumer behavior and preferences using the Consumer Involvement Profile (CIP), Maslow Hierarchy of Needs, and personality traits. Many studies explored the consumer profile of wine tourists; for example, the study of Nella and Christou (2014) found four different segments of wine tourists; Ignatov and Smith (2006) presented three groups of wine and food tourists; Pratt (2010) used the Wine Involvement Scale (WIS) to identify high-involvement participants and low-involvement participants to wine. The study proposed the classification of wine tourists according to their lifestyle into four different categories depending on the importance of wine in their daily lives. The practical aspect of wine tourism was presented in the chapter by mentioning the importance of wine fairs and festivals around the world and described two major wine events in the world: Slow Wine, and ProWein or Pro Wine. The case of how bar operators around the world confronted the outbreak of the COVID-19 pandemic was also stated with some details focusing on the bar business operators in Singapore.

References

Ap, J. (2004). Intercultural behavior: glimpses of leisure from an Asian perspective. In Weiermair, K., and Mathies, C. (eds.), *The Tourism and Leisure Industry: Shaping the Future* (pp. 123–134). Binghamton: Haworth Hospitality Press.

Barber, N., Taylor, D. C., and Deale, C. S. (2010). Wine tourism, environmental concerns, and purchase intention. *Journal of Travel & Tourism Marketing*, 27(2): 146–165. https://doi.org/10.1080/10548400903579746

Brown, G., and Getz, D. (2005). Linking wine preferences to the choice of wine tourism destinations. *Journal of Travel Research*, 43(3): 266–276. https://doi.org/10.1177/0047287504272027

Brown, G. P., Havitz, M. E., and Getz, D. (2006). Relationship between wine involvement and wine-related travel. *Journal of Travel & Tourism Marketing*, 21(1): 31–46. https://doi.org/10.1300/J073v21n01_03

Bruwer, J., Li, E., and Reid, M. (2002). Segmentation of the Australian wine market using a wine-related lifestyle approach. *Journal of Wine Research*, 13(3): 217–242. https://doi.org/10.1080/0957126022000046510

Byrd, E. T., Canziani, B., Hsieh, Y.-C. J., Debbage, K., and Sonmez, S. (2016). Wine tourism: Motivating visitors through core and supplementary services. *Tourism Management*, 52: 19–29. https://doi.org/10.1016/j.tourman.2015.06.009

Civitello, L. (2011). *Cuisine and Culture: A History of Food and People*. Hoboken, NJ: John Wiley & Sons.

Galloway, G., Mitchell, R., Getz, D., Crouch, G., and Ong, B. (2008). Sensation seeking and the prediction of attitudes and behaviors of wine tourists. *Tourism Management (1982)*, 29(5): 950–966. https://doi.org/10.1016/j.tourman.2007.11.006

Hall, C. M., and Mitchell, R. (2000). Wine tourism in the Mediterranean: A tool for restructuring and development. *Thunderbird International Business Review*, 42(4): 445–465. https://doi.org/10.1002/1520-6874(200007/08)42:4%3C445::AID-TIE6%3E3.0.CO;2-H

Havitz, M. E., and Dimanche, F. (1990). Propositions for testing the involvement construct in recreation and tourism context. *Leisure Sciences*, 12(3): 179–195. https://doi.org/10.1080/01490409009513099

Hede, A.-M. (2008). Food and wine festivals: stakeholders, long-term outcomes and strategies for success. In Hall, C. M., and Sharples, L. (eds.), *Food and Wine Festivals and Events around the World: Development, Management and Markets* (pp. 85–100). Burlington, MA: Butterworth-Heinemann.

Ignatov, E., and Smith, S. (2006). Segmenting Canadian culinary tourists. *Current Issues in Tourism*, 9(3): 235–255. https://doi.org/10.2167/cit/229.0

Laurent, G., and Kapferer, J.-N. (1985). Measuring consumer involvement profiles. *Journal of Marketing Research*, 22(1): 41–53. https://doi.org/10.1177/002224378502200104

McShane, S. L., and Von Glinow, M. A. Y. (2019). *Organizational Behavior* (4th ed.). New York: McGraw-Hill Education.

Mitchell, A. A. (1979). Involvement: a potentially important mediator of consumer behavior. *Advances in Consumer Research*, 6(1): 191–196.

Nella, A., and Christou, E. (2014). Segmenting wine tourists on the basis of involvement with wine. *Journal of Travel & Tourism Marketing*, 31(7), 783–798. https://doi.org/10.1080/10548408.2014.889639

Pratt, M. (2010). *Moderating Effects of Wine Involvement in Wine Tourism*. Paper presented at the *5th. International Academy of Wine Business Research Conference*, Auckland, NZ.

Schiffman, L. G., and Wisenblit, J. (2015). *Consumer Behavior* (11th ed.). England: Pearson.

Schneider, S. (2008). Good, clean, fair: the rhetoric of the Slow Food movement. *College English*, 70(4): 384–402. https://www.jstor.org/stable/25472277

Simpson, K., Bretherton, P., and de Vere, G. (2004). Lifestyle market segmentation, small business entrepreneurs, and the New Zealand wine tourism industry. *Journal of Quality Assurance in Hospitality & Tourism*, 5(2–4): 157–188. https://doi.org/10.1300/J162v05n02_09

Statista. (2019). Wine production worldwide in 2019. https://www.statista.com/statistics/240638/wine-production-in-selected-countries-and-regions

Statista. (2020). Number of restaurants in Singapore from 2010 to 2019. https://www.statista.com/statistics/723478/number-of-restaurants-in-singapore/

Tassiopoulos, D., Nuntsu, N., and Haydam, N. (2004). Wine tourists in South Africa: a demographic and psychographic study. *Journal of Wine Research*, 15(1): 51–63. https://doi.org/10.1080/0957126042000300326

Yuan, J., Morrison, A. M., Cai, L. A., Dodd, T. H., and Linton, S. (2008). Segmenting wine festival visitors using their personal involvement with wine. In Hall, C. M., and Sharples, L. (eds.), *Food and Wine Festivals and Events around the World* (pp. 146–165). Burlington, MA: Butterworth-Heinemann.

Zaichkowsky, J. L. (1985). Measuring the involvement construct. *The Journal of Consumer Research*, 12(3): 341–352. https://doi.org/10.1086/208520

12

WINE TOURISM'S INSTITUTIONAL FRAMEWORK AND GOVERNING SYSTEM

Evidence from Portugal and six international reference markets as benchmarks

Alexandre Guedes

Introduction

The world's largest wine-producing countries have been witnessing an investment boom in wine tourism facilities as it is assumed to be the best way to sell wine directly to consumers and a way to build a destination brand image and a reputation, increasing visitors' flow, local spending, and contributing to boosting regional development.

However, as domestic and international competition increases, wineries and regional wine destinations need to innovate in their product and processes. In wine tourism innovation processes and tourism's institutional framework and governing system play an important role in enhancing the industry's competitiveness.

Moreover, wine tourism is a systemic and functional phenomenon that extends its boundary beyond the vineyard and/or winery facility (property) and embraces a more complex notion of *winescape* (Hall et al. 2000; Nowak and Newton 2006; Peters 1997), which conceptualizes the 'wine region as a tourist destination' (Byrd et al. 2016, p. 20). Thus, the concept of winescape is a perceivable outcome of a complex form of multifunctional land use (Antrop 2005) based on mutualistic cooperation between wine and tourism (Guedes and Rebelo 2019). Therefore, it depends on a collective commitment that goes beyond the strict limits of the farm economy and grape and wine production.

Accordingly, the level of development of wine tourism is reliant on the quality of the organizational structures and the institutional ecosystem, which 'affects the environmental behavior of companies in the tourism sector' (Lavandoski et al. 2016, p. 267). Thus, wine tourism calls for an integrated steering approach that focuses on governing the winescape's system, allowing the allocation of resources, and exercising control and coordination (Bramwell and Lane 2011; Rhodes 1996) of different governing systems involved, namely in the wine and tourism sectors, at the national, regional, and local levels.

The institutionalization process and its influence on the development of wine tourism attain significant importance due to its symbiotic nature. This is particularly critical regarding the

DOI: 10.4324/9781003143628-15

winescape's dimensions (natural scenery and landscape, the winery structures and vineyards, and the tourist facilities) (Lavandoski et al. 2016, p. 277) and the need to integrate the planning, promotion, and marketing strategies of the destination as a whole.

Therefore, the goal of this chapter is to analyze wine tourism's institutional framework and governing system based on wine and tourism organizations that provide websites. Of particular significance for wine tourism institutional and governance, assessment is the number of links within and between wine and tourism institutional websites, used as a proxy for institutional relations. These links showcase the level of online engagement between institutions involved in both sectors and display the level of internal and external multisectoral integration of a wine tourism destination.

To fulfill this goal, this chapter examines Portugal's wine tourism institutional framework and governance system by explicitly integrating at the regional level the Douro Wine Region (DWR), one of the oldest demarcated and regulated wine regions in the world (established in 1756), where the renowned Port wine is produced (Caldas and Rebelo 2013). Moreover, to strengthen the analysis and conclusions, we additionally examined the online institutional structure and governance model of six reference markets in wine tourism (Italy, France, the United States, Australia, South Africa, and New Zealand) by focusing on specific wine regions as a benchmark, following the same criteria applied to the DWR. This will allow a cross-institutional environment and governance comparison between countries and provide a framework that contributes to organizational improvement by observing comparable structures in some of the most relevant wine tourism destinations in the world.

Literature review

Governance systems

Governance systems are based mainly on bureaucratic hierarchies, networks, or markets (Pahl-Wostl 2009; Thompson et al. 1991), which vary according to 'degree of formality of institutions and the importance of state and non-state actors' (Pahl-Wostl 2009, p. 358), i.e., the intensity of self-organizing practices and collaboration among various players in the system.

Despite the capacity of developed destinations to self-organize, McLennan et al. (2014) argue that governments can also encourage and help the 'learning process' and promote inter-institutional cooperation and collaboration through strategic planning (McLennan et al. 2014, p. 116). Therefore, cooperation is vital to ensure the governance system's purpose due to the vulnerable and limited capacity of a particular institution to deliver results or tackle challenges individually. Cooperation procedures may improve their common competence. Despite the importance of cooperation and collective commitment, institutions – particularly innovative ones – are also essential to change practices and procedures to successfully guide governance systems on the right path (McLennan et al. 2014).

At present, policy development and implementation are becoming increasingly dependent on multilevel and polycentric governance systems (Ostrom, Tiebout, and Warren 1961), involving a wide array of actors that contribute through cooperation. The nation-state's gradual decline of influence and dominance (Pahl-Wostl et al. 2007) due to the expansion of globalization as well as civil society's enhanced involvement is giving rise to important societal changes that encourage integrated and modular governing regimes and promote social learning processes.

However, the gradual waning of formal governments involvement in destination governance in support of more participation of other destination stakeholders' networks (Pastras and Bramwell 2013) as well as a convergence towards public-private partnerships (PPPs), the so-called 'QUANGO model' (Pike and Page 2014, p. 210), has lessened central government's capacity to coordinate and guide tourism policymaking and activities. The downside of hybrid governance regimes is the rise of a series of ineffective results such as replication of functions and initiatives, promiscuous business partnerships (e.g., public-private contracts, policymaking), and economic competitiveness support rather than encouraging policies addressing sustainability issues (Augustyn and Knowles 2000; Bramwell and Rawding 1994; Lutz and Ryan 1997). Also, monocentric and bureaucratic governance systems, which fully control and decide all aspects of society, tend to be less prone to encourage fluid and systematic (social) learning processes (Mostert et al. 2007; Tippet et al. 2005) through information and knowledge sharing, which are central in fostering innovation (McLennan et al. 2014) and promoting destination development.

A destination's political environment encompasses a complex ecosystem formed by a diverse range of institutions and actors with varying degrees of influence and participation in a multi-faceted range of relationships (Pike and Page 2014). New institutionalization explores relations between institutions at distinct spatial scales and focuses on how institutions evolve through collaborative interactions between individual and collective actors, which encompasses 'actors' shared and conflicting values and perceptions, and the structural context' (Pastras and Bramwell 2013, p. 392). This approach (new institutionalization) underlines the governmental, institutional sphere as a relational system, with institutions cooperating with distinct areas (economic, sociopolitical, and cultural), including principles and values that are internal and external to institutions.

Therefore, the successful coordination of tourism by the state is contingent on it ensuring 'mutual cooperation and priorities about actions among different government organizations and between the state and actors beyond government' (Pastras and Bramwell 2013, p. 390). Consideration should also be pointed to the geographical scale of the activities and authority of government in tourism destinations – whether they operate at national or subnational scales, as well as on the relations between public and private institutions, and the broader socioeconomic and political framework.

Overall, tourism destination development must involve an integrated multilevel strategy that needs to negotiate organizational limits and create value for the individual players (Haugland, Grønseth, and Aarstad 2011).

Multilevel destination strategies and destination capabilities

Destinations encompass a plethora of stakeholders offering various products and services, which requires a certain level of coordination and a multilevel organization and commitment. Destinations are also part of an ecosystem of other destinations and external players with whom they establish relationships and links. Likewise, individual destinations tend to follow complementary strategies (Haugland, Grønseth, and Aarstad 2011) within a larger geographical framework or national destination structure, establishing inter-destination bridge ties to assist the development of these larger regions or destinations.

Destination capabilities are also dependent on resources and skills that are provided by individual players within the destination, as well as external actors. Also, destination capabilities are dependent on the destination stakeholder's combined capacity to 'integrate, build, and reconfigure internal and external competencies to address rapidly changing environments' (Teece, Pisano, and Shuen 1997, p. 516). In the case of wine tourism destinations, this is particularly important since it depends on an interconnected effort that integrates resources and competencies distributed across several different stakeholders within two main institutional ecosystems, i.e., wine and tourism. Wine tourism destinations may consequently gain from their capabilities in two fundamental areas, i.e., marketing and communicating products and services as well as adjusting the production of products and services according to the needs of the market. This can be achieved through cooperative branding based on shared attributes, which provides a regional organization (and smaller ones) with a coherent cognitive image (Cai 2002), generating effective use of (wine) tourism resources and marketing methods.

Destination capabilities are also contingent on coordination at the destination level and are associated with interorganizational bridge ties, either through direct links among players or indirect links to external actors, a notion that Haugland, Grønseth, and Aarstad (2011, p. 281) applied to inter-destination relations. The notion of bridging network ties or bridging social capital – i.e., 'a quality created between people' (Burt 1997, p. 339) – underlines relationships among otherwise independent systems. Distinct local interorganizational forms affect both how and the extent to which external information and knowledge will be apprehended and lead to imitation and innovation as well as improve the destination's competence to build integrated multilevel strategies. As such, more integrated destinations can more easily acknowledge the benefits of ties to other players in the system (Haugland, Grønseth, and Aarstad 2011, p. 281), namely inter-sectorial players (wine and tourism) and intra-sectoral ties between different levels or scales of tourism entities (e.g., DMO and national authorities). Hence, destinations that can connect destination capabilities and establish coordination at the destination level will be more qualified to build multilevel policies to generate value for destination stakeholders, as opposed to disengaged and incoherent strategies (Haugland, Grønseth, and Aarstad 2011).

Summing up, a destination's success is reliant on a competitive strategy that pursues a local actor's 'profitable and sustainable position against the forces that determine industry competition' (Porter 1998, p. 1), needing to build competitive advantages to endure in the future. In that sense, micro and macro perspectives of destination competitiveness (Dwyer and Kim 2003) are interrelated since firm-specific behaviors determine competitiveness, which is also influenced by an environmental condition provided by the destination's interorganizational bridge ties to accelerate innovation and imitation, which consequently raises market awareness.

Hence, local stakeholder's existence succeeds mainly on a destination's ability to persuade tourists and capacity to 'ensure that its overall 'appeal,' and the tourist experience offered, must be superior to that of the alternative destinations open to potential visitors' (Dwyer and Kim 2003, p. 369). It is a complex effort that requires an agreement on shared strategic ambitions that are dependent on effective inter-institutional relations and coordination between actors, which requires a clear knowledge of all of a single stakeholder's resources and a sound understanding of the destination's capabilities. In the case of wine tourism destination organizations, there must be full awareness of all actors (e.g., wineries) and each one's capabilities. Coordination is essential to structure relationships between the destination's players and develop shared destination goals as well as execute actor-level strategies. Developing innovation activities can be accomplished by institutionalizing ties to external players, namely public wine authorities that can interchange information and knowledge, which is critical for innovation at tourism destinations and individual firms.

The theories presented here, which help explain the broader context of governance systems, new institutionalization, and integrated multilevel destination strategies, serve as foundations for the analysis and interpretation of the empirical study.

Methodology

The analysis of the Portuguese wine tourism governance system is accomplished by acknowledging inter- and intra-institutional bridge ties between national and subnational wine and tourism organizations and integrating the DWR at the regional level. The Portuguese wine tourism institutional framework is identified by acknowledging all national public organizations (national wine and tourism institutions) and determining all edges linking to public and private organizations at regional and local tier levels. Wine tourism institutional and governance assessment is based on the number of links within and between wine and tourism institutional websites analyzed in 2018, used as a proxy for institutional interactions. These ties demonstrate the level of online cooperation between institutions involved in both sectors. Moreover, cooperative actions among actors and organizations encompass several topics of cooperation (e.g., providing information, document sharing) and assume various levels of intensity (e.g., number and density of links between institutions, the number of wineries). Cooperative actions can occur between and among public and private organizations.

To broaden our knowledge in this area, we additionally examined online institutional structures and governance models of six international reference destinations,[1] which have different interorganizational structures. By comparing destinations, it is possible to identify functional and organic gaps between destinations and develop multilevel policies and actions as well as improve cooperation. Therefore, benchmarking is a valuable method to identify good practice standards (Kozak 2002).

Results and discussion

The Portuguese institutional ecosystem (Figure 12.1) is comprised of a wide range of entities exhibiting an atomized pattern and consisting predominantly of hermetic organizations. The institutional network shows a minimal level of bridge ties, revealing weak institutional interaction and a low level of information and data disclosure within or between wine and tourism institutions. The wine tourism's institutional framework provides a feeble representation of the DWR, which is not formally recognized as a destination and therefore does not have a dedicated website. Moreover, there is limited or virtually no information about the DWR's wineries, besides an incomplete list provided by the national wine authority ('Vini Portugal').

The Portuguese official websites lack a linking out (external linking) strategy, which could allow to scale up resources and coordination (e.g., databases) as well as establishing trust and encouraging positive participation and contribution among institutional actors. Oppositely, the Portuguese institutional pattern displays a fragmented network of links exhibiting just one connection between the national tourism authority's website ('Turismo de Portugal') and the subnational level representing the Douro wine sector, with one link to the association of Port wine producers ('Port Wine Cellars').

The national tourism authority does not provide any connection to the national wine organization ('Vini Portugal') that has marketing responsibilities in the global markets by using the 'Wines of Portugal' website and brand. Yet, this relation is asymmetric since 'Vini Portugal' includes a link within its website dedicated to 'Wine Tourism,' which provides access to a technical guide on wine tourism hosted on 'Turismo de Portugal's' website. The disconnection shown here between two institutional hubs that have responsibilities in promoting tourism and wines in foreign markets limits the marketing effort that is required by wine tourism, which calls for the cooperation and articulation of both institutional frameworks. This vulnerability has a negative cascade effect on the business sector, namely on wine-producing companies which operate based on the assumption of mutualistic cooperation between wine and tourism (Guedes and Rebelo 2019) to expand the range of wine consumers as well as strengthen its perceived value (Gill, Byslma, and Ouschan 2007). This could deter the destination's competitiveness due to the lack of interorganizational bridge ties to accelerate innovation and imitation, which

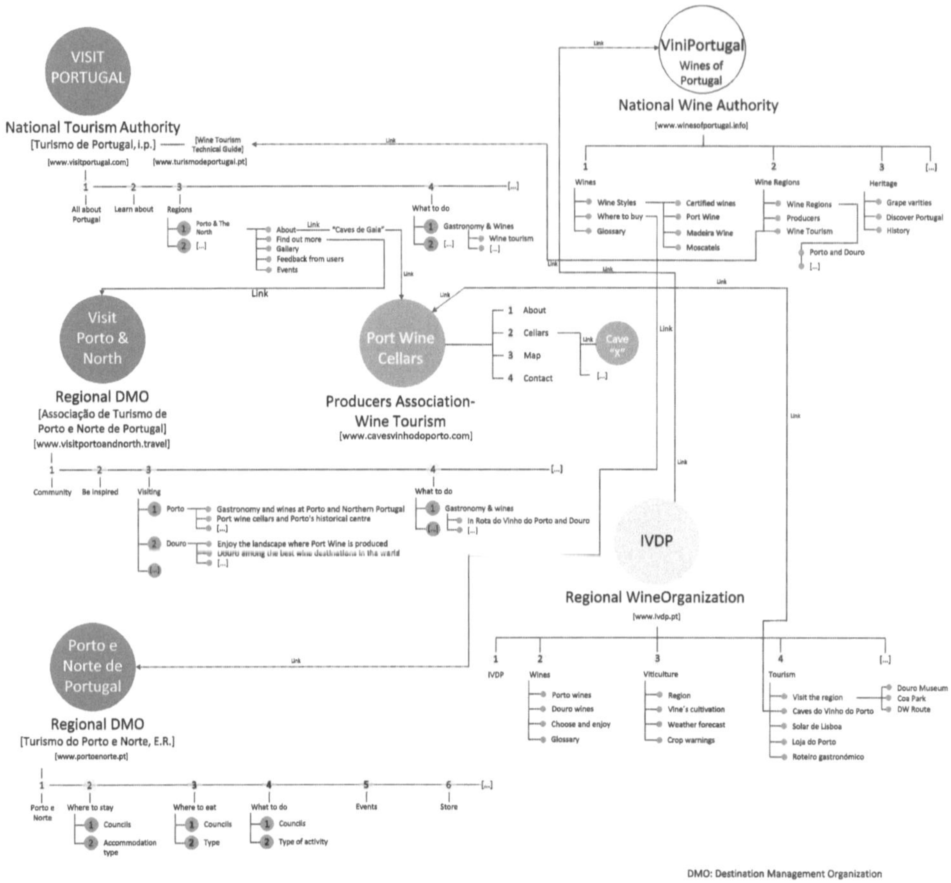

Figure 12.1 Portuguese wine tourism's governing system.

consequently raises market recognition. Moreover, wine tourism's development purpose of bringing the point of consumption closer to the point of production is hampered.

Within the tourism institutional framework, two regional-tier DMOs represent the North of Portugal region where the Douro is located. This suggests a strong organizational inefficiency due to overlapping and fragmented functions associated with the North of Portugal's destination marketing, despite one having assigned roles within the external market and the other within the domestic market. There is also no evidence of supporting links of shared information and data between the two organizations, which again exposes ineffectiveness. Still, the national wine organization, 'Vini Portugal,' which is responsible for marketing wines in the global market, links out to the regional DMO's website ('Turismo do Porto and North of Portugal') that has duties assigned to the domestic market. It would be expected that both national- and regional-level organizations directly involved in marketing wine and the Douro destination abroad would display formal levels of interaction as an elemental order rather than two entities with different marketing scopes.

The analysis of inter-institutional relations also reveals that the regional wine organization responsible for the wine sector with responsibilities in the Douro, the 'Instituto dos Vinhos do Douro e Porto' (IVDP), provides no referral to the DWR's wineries, nor to the tourism organizations – namely, the regional DMOs – despite including a link to an internal webpage that relates general tourism information. There is a void that can be filled and improved through adequate inter-institutional cooperation with the regional DMO (e.g., 'Turismo do Porto and Norte de Portugal') by providing a referral that allows users to be more aware of the Douro's destination.

The overall governing system reveals a monocentric structure and a high level of top-down governance based on central leadership government due to the low level of communicating vessels to promote fluid and systematic learning processes through information and knowledge sharing. This system is somehow in contradiction with a strategy towards public-private partnerships at the regional DMO level, which has led to the replication of functions and initiatives. There is also a minimal level of engagement and interaction between institutions in different-scale tiers as well as between national-tier organizations involved in wine and tourism, which is demonstrated through the lack of bridge ties and information diffusion. The limited cooperation between institutions within the Portuguese (and DWR) governance system is responsible for limiting the level of destination capabilities, which corresponds, namely, to the destination stakeholder's collective ability to build and integrate marketing and communicating products and services.

Moreover, regional- and local-tier organizations tend to mimic the central government's behavior of command over lower layers of decision. This behavior goes as far as not including a clear and thorough list of the Douro's wineries, or specific information on the wine region's characteristics (e.g., terroir attributes). Despite this weakness, there are political signs that show greater recognition of wine tourism in Portugal, proclaimed in a recent national action plan for wine tourism that calls for participatory mobilization.

Despite the limited assessment of the Portuguese institutional network that sustains wine tourism, it is important to look at some of the world's most important wine tourism destinations in six international markets (Figure 12.2) to recognize good practice standards.

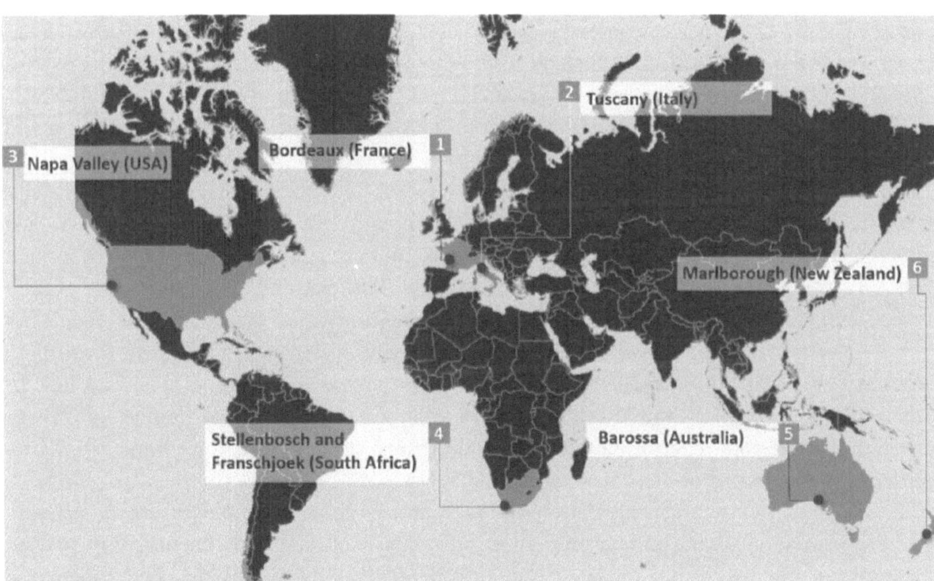

Figure 12.2 Wine tourism destinations benchmarks.

In general, the benchmark models (Appendices 12.1–12.6) demonstrate vertical and hierarchical coherence, starting from the central government national tier down to the regional level, securing interorganizational links that allow information and data to be shared within the same institutional framework. In some cases, the organizational structure provides hybrid systems that vertically connect two organizations located in the two institutional environments, wine and tourism. This model is superimposed on a pattern of inter-institutional relations between the sectors of tourism and wine on the same scale of power, i.e., through links between national organizations in the two sectors.

The observed governance systems reveal an active coordination system that ensures an adequate flow of information between both wine and tourism institutional environments. For instance, the national Italian tourism organization (Appendix 12.2) communicates directly with the national ('Citá del Vino' and 'Italian Wine Movement') and regional associations within the wine sector (case analyzed: the 'Chianti Classico Wine Consortium'). This structure, in turn, allows the dissemination of a detailed list of all wineries, which can be accessed selectively with the use of filters combining tourist services and equipment. In the Italian case, the national tourism authority provides a direct link to the regional DMO (in our case, to 'Visit Tuscany'), which makes available important bridge ties to wine tourism programs.

The benchmark cases indicate that it is a common practice to internalize links that have, in many cases, a reciprocal correspondence, i.e., symmetrical institutional relation (tie) (e.g., the South African model – Appendix 12.5). Also, when there is no direct link between both national wine and tourism governmental organizations, as the New Zealand case shows (Appendix 12.6), inter-institutional ties are established between the national tourism authority ('Tourism New Zealand') and all wine tourism destinations (regional level, which in our case correspond to 'Destination Marlborough Trust'), that sequentially provide a link to the wine region's website ('Wine Marlborough').

Another feature that characterizes our benchmark cases is that they provide either a national or/and regional website dedicated to the theme of wine tourism, being common at the regional-tier level to identify DMOs that overlap with the wine region. In the case of the United States (Appendix 12.3), as it is a federal republic, a bridge tie is provided between both tourism and wine state tier organizations, which, according to our example, is established between 'Visit California' and 'Discover California Wines' (California Wine Institute), which also provides a connection to the Napa Valley wine destination. This shows a clear integration of wine and tourism in either case.

The French example (Appendix 12.1) exhibits a dense governance model with numerous inter-institutional relations where wine tourism obtains strategic recognition through a robust and effective online platform owned by the national tourism authority, entitled 'Visit French Wine,' which is exclusively dedicated to wine tourism. This platform serves as a strategic hub that secures links to wine regions and regional wine tourism entities. In this case, the 'Bordeaux Wine Trip' was analyzed, providing a plethora of detailed information on wine tourism in Bordeaux (e.g., farms, routes, accommodation, wine tourism programs) as well as a link to Bordeaux's wine organization. Also, the national wine authority, 'Vins de France,' provides a close inter-institutional articulation with all wine regions, conveying relevant information and ensuring an institutional online connection.

The observed international institutional models also reveal a common pattern of strategic links between regional wine and/or tourism organizations and wine routes managed by private associations or integrated into the wine and/or tourism regional institution. The Italian case illustrates the most intricate and dense model, which provides a regional route in addition to subregional routes managed by private associations. This is a common denominator that was recognized on all models, which do not have correspondence in the case of the DWR.

Another relevant piece of information that was captured, which enables great visibility to the wineries as well as transparency in the sector, is the availability of a complete inventory of wineries, in most cases published in a dedicated way, with detailed information, and in some cases (e.g., Australia, France, the United States) with separate landing pages. Some of the models (e.g., France) provide web-based information systems, which allow for segmented data search depending on the tourist's motivations.

Conclusion

This chapter has contributed to providing an outlook on wine tourism's institutional framework and governing system in Portugal (and the DWR) and six international reference markets as benchmarks. Intra- and inter-institutional cooperation between wine and tourism organizations at different governmental tier levels and external players are examined by probing website links as a measure of interaction.

The Portuguese case displays a negligible engagement between wine and tourism national authorities, a minimal level of integration between national and regional governmental tier levels, and almost no association with single stakeholders. These findings demonstrate limited devolution of information and resources from national to regional and local tiers of decision and interest. The lack of inter-institutional engagement is responsible for limiting the destination's capabilities leveraged by the stakeholder's collective ability to integrate and build a common strategy to market and communicate wine tourism destination products and services. The overall structure of the Portuguese institutional network seemingly discloses a monocentric and dominant governmental leadership form that constrains access to and flow of information. Accordingly, it limits the possibility of developing the DWR as a wine tourism destination and inhibits its competitive advantage, which requires extensive coordination between actors as opposed to an atomized system. Moreover, the Portuguese governance model lacks an aggregating reference framework at the regional level that embodies the DWR as a wine tourism destination, which could be established through a DMO structure or a wine route that establishes a geographical limit, connects, and allows the coordination of a network of players (e.g., wineries).

The observed international wine tourism destination governance systems exhibit less centralized structures alongside a multilevel approach and cooperation between wine and tourism institutional environments. The level of bridge ties between and within institutions supports the disclosure and access to more information and data concerning each wine destination (e.g., wineries, experiences, wine routes, terroir). The internalization of links that have reciprocal correspondence – i.e., symmetrical institutional relation (tie) – is common among these international benchmarks. Also, the investigated governance systems include organizations at the national and/or regional tier level that embody either the wine tourism product or the wine tourism destination, which in each circumstance reveals a more advanced level of wine tourism destination development as well as a more dynamic involvement of each actor and widespread coordination between actors.

This study contributes to developing a research stream on the governance of wine tourism destination models based on the number of links between institutional websites and provides evidence on the quality of the organizational structures and institutional ecosystem, which influences the environmental behavior of companies in the wine tourism sector. This research also has policy implications by encouraging critical analysis and comparison between different wine tourism destination governance models and identifying relevant inter-institutional constraints and weaknesses in the Portuguese case that can be detected in other cases.

Further research in the field of wine tourism destination governance should also extend knowledge on operational and internal organizational effectiveness related to wine destination governance by developing in-depth interviews with board members and a panel of experts.

Implications

This study contributes to the body of knowledge on governance and institutional theory by providing an integrated analysis of wine tourism. This research also delivers practical managerial contributions by drawing attention to the relative paucity of coordination mechanisms within Portuguese wine tourism's institutional framework and governing system. The study further highlights the need for a sounder articulation between the entrepreneurial ecosystem and institutional players to strengthen the DWR wine destination's global awareness. Additionally, this investigation provides an international institutional benchmark and governing system that can be used to steer wine tourism development in Portugal.

Appendices

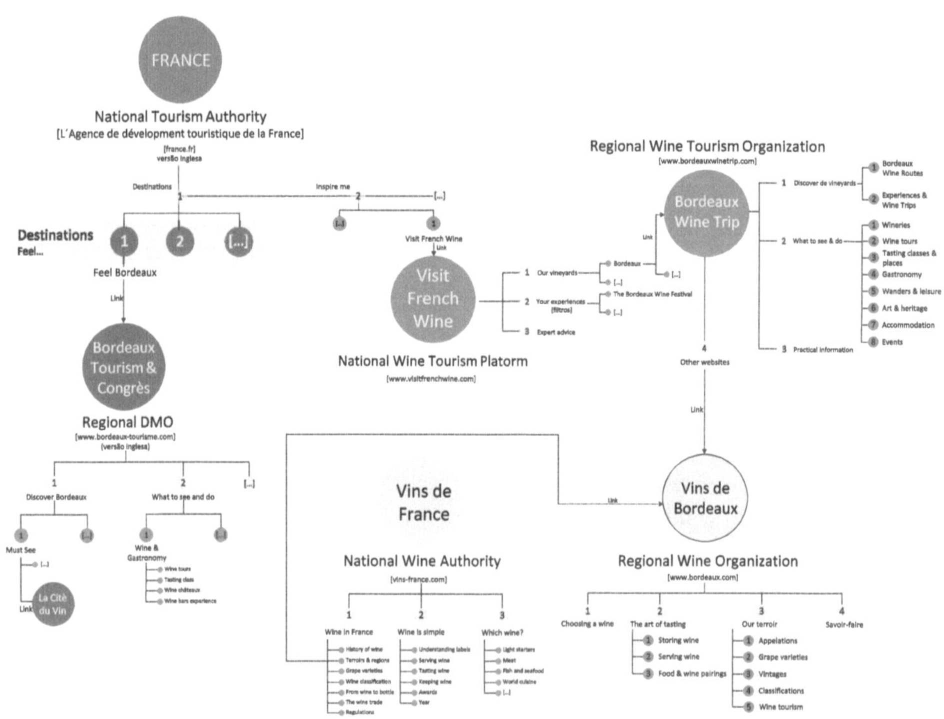

Appendix 12.1 French wine tourism's governing system.

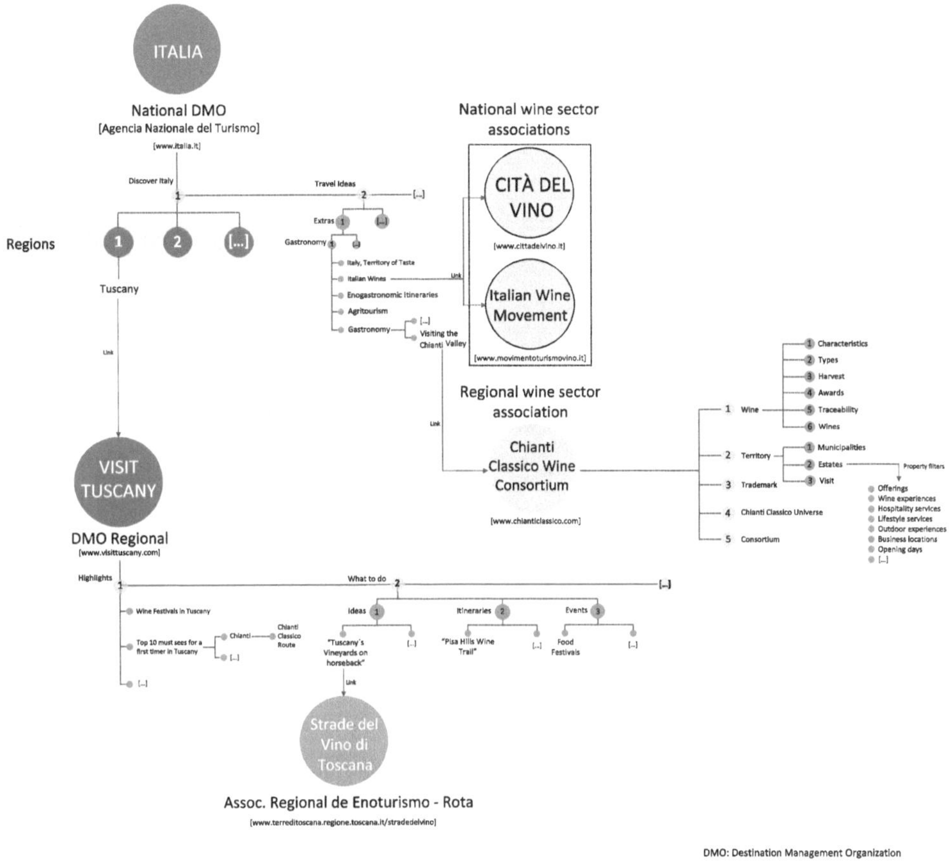

Appendix 12.2 Italian wine tourism's governing system.

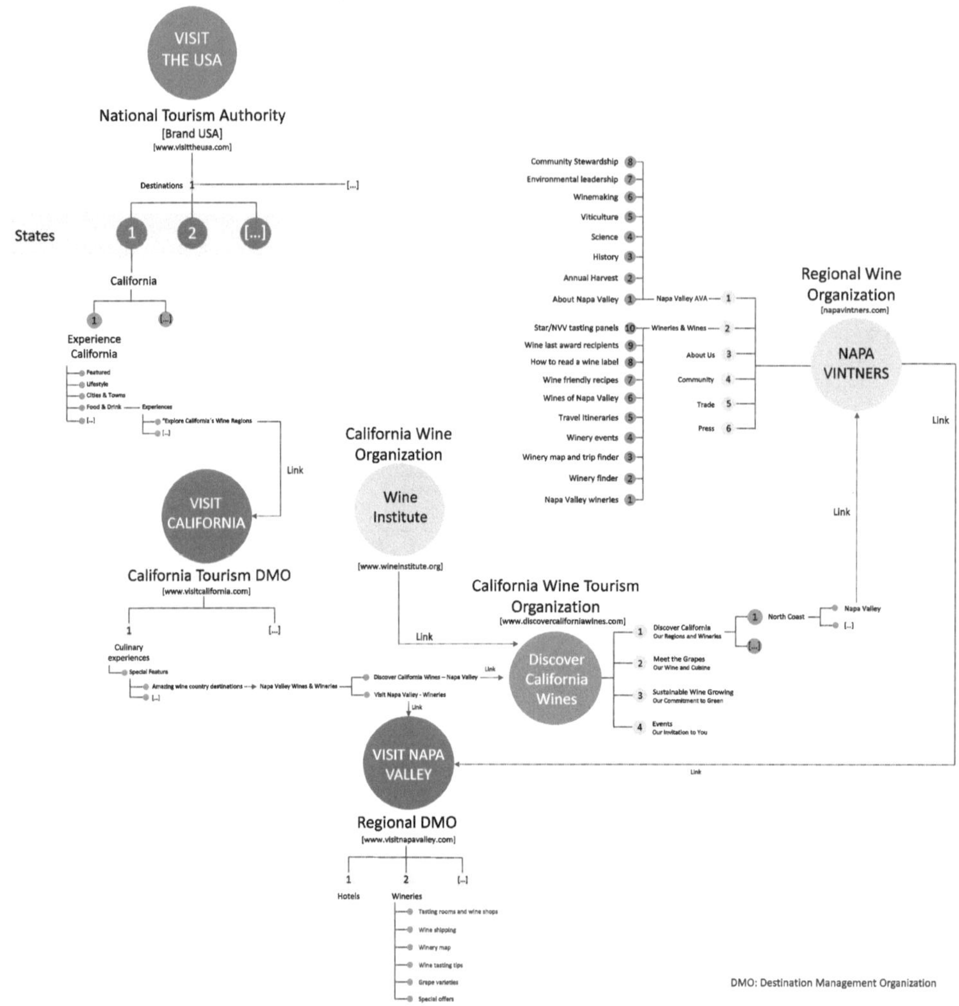

Appendix 12.3 American wine tourism's governing system.

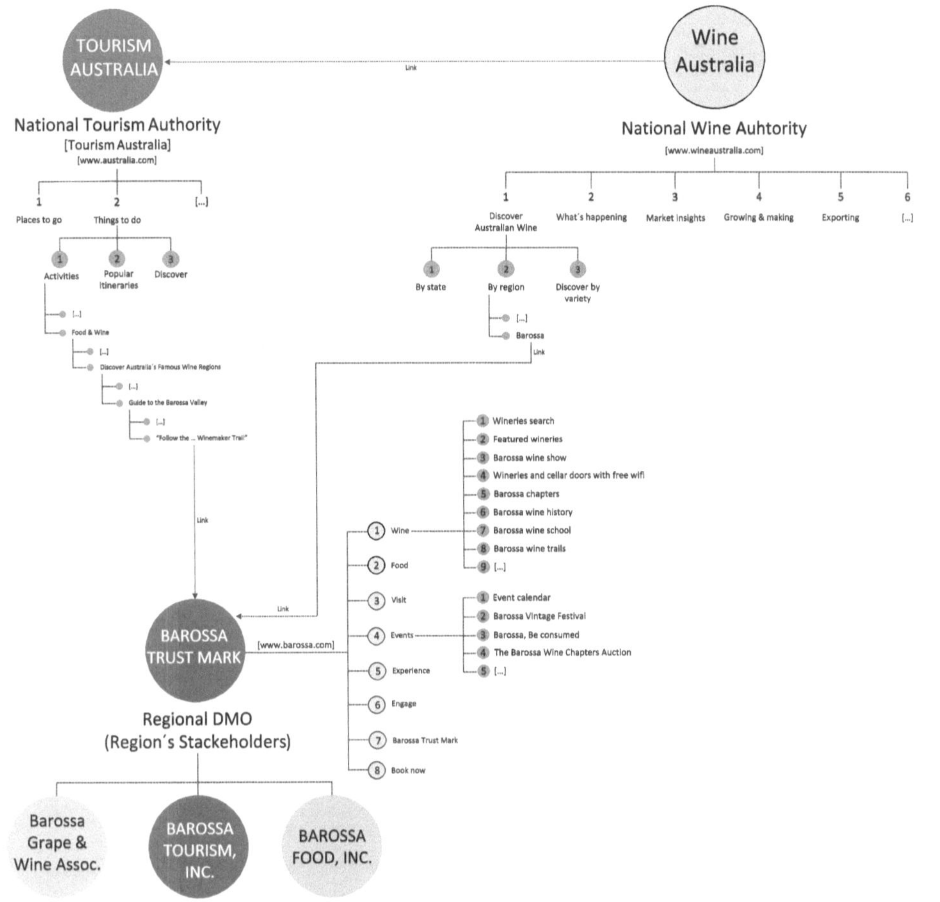

DMO: Destination Management Organization

Appendix 12.4 Australian wine tourism's governing system.

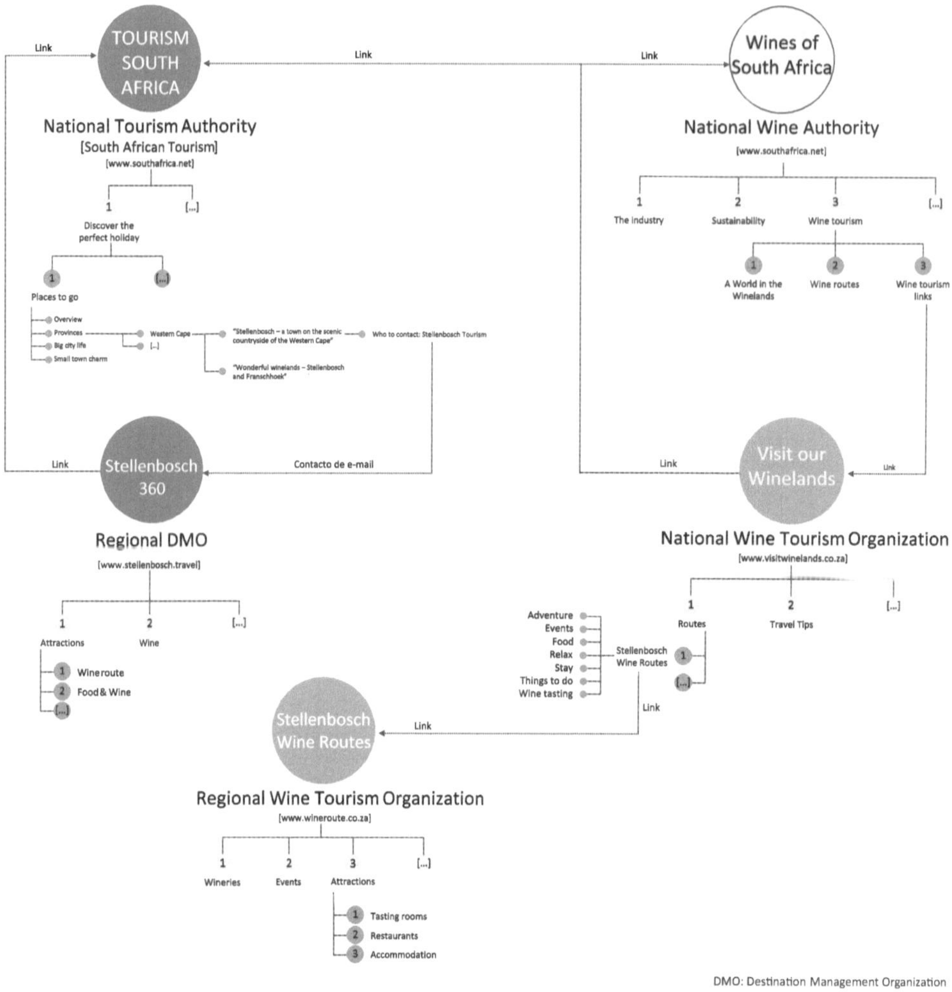

Appendix 12.5 South African wine tourism's governing system.

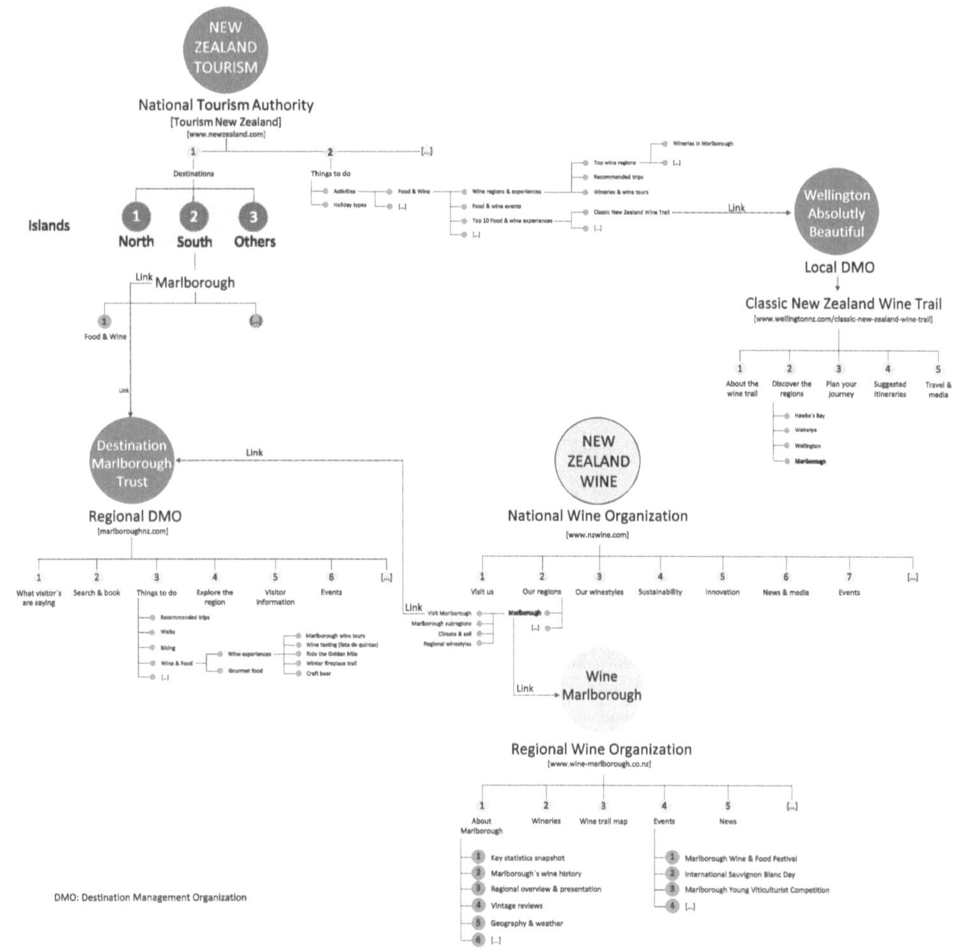

Appendix 12.6 New Zealand wine tourism's governing system.

Acknowledgements

This study has received support from the FCT – Portuguese Foundation for Science and Technology, project UIDB/04011/2020.

Note

1 France, Italy, United States, South Africa, Australia, and New Zealand.

References

Antrop, M. (2005). Why landscapes of the past are important for the future. *Landscape Urban Plan*, 70(1): 21–34. https://doi.org/10.1016/j.landurbplan.2003.10.002

Augustyn, M. M., and Knowles, T. (2000). Performance of tourism partnerships: a focus on York. *Tourism Management*, 21: 341–351. https://doi.org/10.1016/S0261-5177(99)00068-0

Bramwell, B., and Lane, B. (2011). Critical research on the governance of tourism and sustainability. *Journal of Sustainable Tourism*, 19(4–5): 411–421. https://doi.org/10.1080/09669582.2011.580586

Bramwell, B., and Rawding, L. (1994). Tourism marketing organizations in industrial cities: Organizations, objectives and urban governance. *Tourism Management*, 15(6): 425–434. https://doi.org/10.1016/0261-5177(94)90063-9

Burt, R. (1997). The contingent value of social capital. *Administrative Science Quarterly*, 42(2): 339–365. https://doi.org/10.2307/2393923

Byrd, E., Canziani, B., Hsieh, Y.-C., Debbage, K., and Sonmez, S. (2016). Wine tourism: Motivating visitors through core and supplementary services. *Tourism Management*, 52: 19–29. https://doi.org/10.1016/j.tourman.2015.06.009

Cai, L. (2002). Cooperative branding for rural destinations. *Annals of Tourism Research*, 29(3): 720–742. https://doi.org/10.1016/S0160-7383(01)00080-9

Caldas, J., and Rebelo, J. (2013). Portuguese wine rating: an old product a new assessment. *Wine Economics and Policy*, 2: 102–110. https://doi.org/10.1016/j.wep.2013.11.004

Dwyer, L., and Kim, C. (2003). Destination competitiveness: Determinants and indicators. *Current Issues in Tourism*, 6(5): 369–414. https://doi.org/10.1080/13683500308667962

Gill, D., Byslma, B., and Ouschan, R. (2007). Customer perceived value in a cellar door visit: the impact on behavioural intentions. *International Journal of Wine Business Research*, 19(4): 257–275. https://doi.org/10.1108/17511060710837418

Guedes, A., and Rebelo, J. (2019). Merging wine and tourism-related services: evidence from the Douro (Portugal) Wine Region. *Journal of Wine Research*, 30(4): 259–274. https://doi.org/10.1080/09571264.2019.1652152

Hall, C., Johnson, G., Cambourne, B., Macionis, N., Mitchell, R., and Sharples, L. (2000). Wine tourism: an introduction. In Hall, C. M., Sharples, L., Cambourne, B., and Macionis, N. (eds.), *Wine Tourism around the World: Development, Management and Markets* (pp. 1–23). Oxford: Butterworth-Heinemann.

Haugland, S., Grønseth, H., and Aarstad, J. (2011). Development of tourism destinations: An integrated multi-level perspective. *Annals of Tourism Research*, 38(1): 268–290. https://doi.org/10.1016/j.annals.2010.08.008

Kozak, M. (2002). Destination benchmarking. *Annals of Tourism Research*, 29(2): 497–519. https://doi.org/10.1016/S0160-7383(01)00072-X

Lavandoski, J., Pinto, P., Silva, J., and Vargas-Sánchez, A. (2016). Causes and effects of wine tourism development in wineries. *International Journal of Wine Business Research*, 28(3): 266–284. https://doi.org/10.1108/IJWBR-08-2015-0032

Lutz, J. R., and Ryan, C. (1997). Impacts of inner city tourism projects: the case of the International Convention Centre, Birmingham, UK. In Murphy, P. E. (ed.), *Quality Management in Urban Tourism* (pp. 41–53). Chichester: Wiley.

McLennan, C., Ritchie, B., Ruhanen, L., and Moyle, B. (2014). An institutional assessment of three local government-level tourism destinations at different stages of the transformation process. *Tourism Management*, 41: 107–118. https://doi.org/10.1016/j.tourman.2013.09.007

Mostert, E., Pahl-Wostl, C., Rees, Y., Searle, B., Ta'bara, D., and Tippet, J. (2007). Social learning in European river basin management: barriers and fostering mechanisms from 10 river basins. *Ecology and Society* 12(1): 19. https://www.ecologyandsociety.org/vol12/iss1/art19/

Nowak, L., and Newton, S. (2006). Using the tasting room experience to create loyal customers. *International Journal of Wine Marketing*, 18(3): 157–165. https://doi.org/10.1108/09547540610704738

Ostrom, V., Tiebout, C. H., and Warren, R. (1961). The organization of government in metropolitan areas: a theoretical inquiry. *American Political Science Review*, 55: 831–842. https://doi.org/10.2307/1952530

Pahl-Wostl, C. (2009). A conceptual framework for analysing adaptive capacity and multi-level learning processes in resource governance regimes. *Global Environmental Change*, 19: 354–365. https://doi.org/10.1016/j.gloenvcha.2009.06.001

Pahl-Wostl, C., Craps, M., Dewulf, A., Mostert, E., Tabara, D., and Taillieu, T. (2007). Social learning andwater resources management. *Ecology and Society*, 12(2): 5. https://www.ecologyandsociety.org/vol12/iss2/art5/

Pastras, P., and Bramwell, B. (2013). A strategic-relational approach to tourism policy. *Annals of Tourism Research*, 43: 390–414. https://doi.org/10.1016/j.annals.2013.06.009

Peters, G. (1997). *American Winescapes: The Cultural Landscapes of America's Wine Country*. Boulder, CO: Westview Press.

Pike, S., and Page, S. (2014). Destination marketing organizations and destination marketing: a narrative analysis of the literature. *Tourism Management*, 41: 202–227. https://doi.org/10.1016/j.tourman.2013.09.009

Porter, M. (1998). Competitive strategy: the core concepts. In Porter, M. (ed.), *Competitive Advantage: Creating and Sustaining Superior Performance* (pp. 1–30). New York: The Free Press.

Rhodes, R. (1996). The new governance: governing without government. *Political Studies*, 44: 652–667. https://doi.org/10.1111/j.1467-9248.1996.tb01747.x

Teece, D., Pisano, G., and Shuen, A. (1997). Dynamic capabilities and strategic management. *Strategic Management Journal*, 18(7), 509–533. https://doi.org/10.1002/(SICI)1097-0266(199708)18:7<509::AID-SMJ882>3.0.CO;2-Z

Thompson, G., Frances, J., Levacic, R., and Mitchell, J. (eds.) (1991). *Markets, Hierarchies and Networks: The Coordination of Social Life*. London, UK: Sage Publishers.

Tippet, J., Searle, B., Pahl-Wostl, C., and Rees, Y. (2005). Social learning in public participation in River Basin Management – early findings from HarmoniCOP European case studies. *Environmental Science & Policy*, 8(3): 287–299. https://doi.org/10.1016/j.envsci.2005.03.003

13

WINE TOURIST MOTIVATIONS AND PERCEPTIONS OF DESTINATION ATTRIBUTES

Brian Kee Mun Wong and Christy Yen Nee Ng

Introduction

Wine tourism product combines wine products with travel and tourism services. It is not a new phenomenon, but not a stagnant one either. Certain regions that have been developed into wine tourism destinations see significant tourism growth, such as the Baja California region in Mexico (Meraz Ruiz and Ruiz Vega 2016), California and Oregon in the United States (Karlsson and Karlsson 2017), La Rioja region in Spain (Karlsson and Karlsson 2017; Meraz Ruiz and Ruiz Vega 2016), the Peloponnese region in Greece (Karagiannis and Metaxas 2020), and the Shandong and Yantai regions in China (Han 2011). The emergence of wine tourism products provides tourists with an alternative perspective to experience a destination that is usually off the beaten track. The leisure mode of visiting these new tourism destinations enables tourists to truly appreciate the picturesque landscapes, the local people's hospitality, and memorable sensory encounters. Thus, each wine tourism destination developed its unique character and winery products, taking advantage of its unique and natural attractions.

Motivation is a crucial element in tourism study (Wong, Musa, and Taha 2017) and a significant determinant of wine tourists' behavior. The tourists' inner state drives and leads their behavior to satisfy physiological and psychological needs. Wine tourists may not be the usual traveler; we understand who they are. Even though we believe most wine tourists should be wine enthusiasts themselves, it may not be accurate for every one of them. It is also worth noting that wine enthusiasts and wine lovers may differ significantly. The former are generally wine specialists, connoisseurs, or any in between, who have high expectations and knowledge on the wineries they visit and the wine products they consume and purchase. Often, the wine trip or tour requirements differ significantly from non-wine enthusiasts and wine lovers. Thus, exploring wine tourists' motivations enables an in-depth understanding of their actual expectations, aims, and requirements when involved in wine tourism.

Motivation theories

The essential concept of motivation is 'need,' where it is crucial to understand human motivating behavior (Mansfeld and Pizam 1999). Travel motivation occurs when a traveler realizes

DOI: 10.4324/9781003143628-16

the existence of a need deficiency. It is essential to understand why people travel and what factors influence their intention to travel, and their destination choice. Curiosity and the urge for new experience and knowledge in an unfamiliar situation may also motivate an individual into action (Podoshen 2012). Though there is no universal definition of motivation, this chapter proposes the operational definition of motivation as 'a set of psychological and physiological needs that are triggered by curiosity and the urge to gain new experiences and knowledge in a less familiar destination' (Wong 2015).

The understanding of travel motivation can be referred to several theoretical papers such as the hierarchy of needs theory (Maslow 1943; Tikkanen 2007), the travel career ladder theory (e.g., Moscardo and Pearce 1986; Wong and Musa 2014), expectancy theory (e.g., Deci 1975; Vroom 1964), drive theory (e.g., Gnoth 1997), Crompton's theory (e.g., Crompton 1979), escaping/seeking theory (e.g., Iso-Ahola 1980; Snepenger et al. 2006), the means-end theory (e.g., Klenosky 2002; Uysal, Li, and Sirakaya-Turk 2008), and the push and pull theory (e.g., Dann 1976; Uysal and Hagan 1993; Uysal, Li, and Sirakaya-Turk 2008). These theories are widely used as the basis of many tourism studies.

According to Maslow's Hierarchy of Needs, the physiological level of needs, which forms the lowest level of needs, deals with the essential needs of life to a person (Bishop 2016). However, in wine tourism, the wine tourists' physiological needs (hunger, rest, thirst) do not significantly influence their participation in such a tourism niche (Deng 2013; Coughlan 2017). The safety needs of wine tourists need to be accommodated by the wine tourism providers. A lack of fulfillment in wine tourists' perceived safety needs in the wine tours they plan to partake in can reduce their likelihood of participating in the same wine tours, negatively affecting the wine tourism industry (Ferreira 2015).

Next, social needs refer to requests for being loved and cared for by society (Coughlan 2017). In wine tourism, social needs are interconnected with participation, as wine tourists want to build social groups and gain acceptance in society over wine consumption (Deng 2013). On the other hand, esteem needs can be achieved in wine tourism when a wine enthusiast is often referred to during wine tourism activities. The enthusiast is often a good and reliable source of information, advising other wine tourists on wine product purchase decision-making (Taylor and Barber 2016).

Lastly, the highest level of Maslow's Hierarchy of Needs is self-actualization, where it constitutes needs that make a person capable of doing anything to their full potential (Bishop 2016). Wine tourism can fulfill self-actualization needs because wine tours can provide various degrees of psychological and emotional involvement for individual wine tourists with close tourist engagement and the importance of attraction to satisfy wine tourists' self-actualization needs (Puryova 2013). As displayed in Maslow's Hierarchy of Needs, the five levels of needs provide a good view of how and why wine tourists are involved in the wine tourism phenomena (Coughlan 2017).

Moscardo and Pearce (1986) and Pearce (1991) adopted the concept of Maslow's Hierarchy of Needs in tourists' motivations studies and developed the Travel Career Ladder (TCL). It is a multi-motive model, proposing to understand tourists' motivations through five different needs and the motivation changes based on the accumulated travel experiences (Ryan 1998) throughout their life span. The developmental and dynamic TCL framework's five needs are physiological, safety or security, relationship, self-esteem or development, and fulfillment. As in Maslow's theory, TCL suggests that lower levels of the ladder typically have to be satisfied before moving on to higher levels. However, the total pattern of tourists' motives is more meaningful in describing the phenomenon instead of just focusing on a single motive. Through a more comprehensive understanding of the specific needs in each level of the ladder, a broader spectrum of different psychological needs and motives can be cataloged. Though it is usual for tourists to have fulfilled the rest of the needs before achieving the fulfillment need, it is not always the case.

Pearce and Lee (2005) argued that some tourists might ascend the TCL, probably due to the higher needs requirement upon reaching each stage from previous experiences; while some may stay at a certain level, restricted to advance higher by certain limitations, such as financial and health issues (e.g., Loker-Murphy 1995). This will particularly be true in the wine tourism context, and wine tours are not available in any other place except for wine producers' destinations. Moreover, to reach the vineyards requires slightly larger expense on transportation and special tours. Wine tourists may start from any level within TCL, ascending or descending depending on their previous experiences, knowledge of the activity, and the specific activity's investment level (Wong and Musa 2014). Besides that, travelers may also shift along the self-directed side or the other-directed side, or along both sides of the ladder (Wong 2015). An example of a modified TCL was presented by Wong and Musa (2014) in their study of travel motivations among retirement tourists to Malaysia through the well-known Malaysia My Second Home (MM2H) program (as shown in Figure 13.1).

Another motivational theory that is commonly adopted in tourism studies is the push and pull theory. Based on Tolman's work (1959), Dann (1977) proposed pull-push motivation in

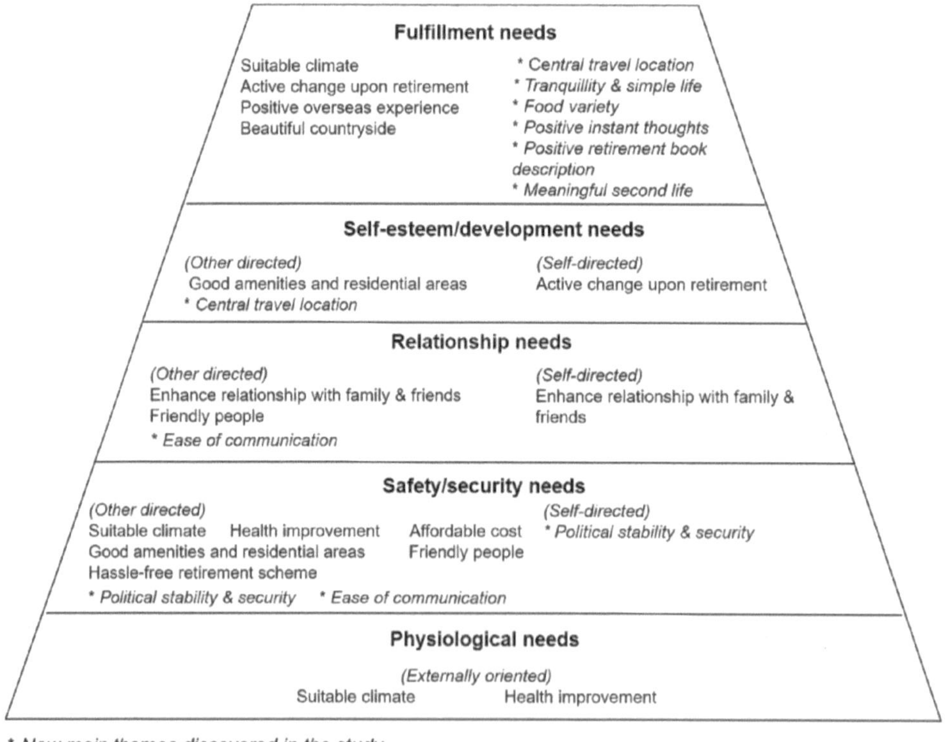

Figure 13.1 Modified Travel Career Ladder (TCL).

Source: Wong and Musa (2014)

tourism studies. The theory plays a paramount role in understanding when and where an individual would travel. It assumes that individuals travel as they are being pushed by their internal desire while at the same time being pulled by external forces (Jeftovic 2015; Uysal and Hagan 1993; Uysal, Li, and Sirakaya-Turk 2008). The attributes are general and not destination-specific. The explanation of travel motivation based on the push and pull motivation theory has been established in most tourism kinds of literature. Different push and pull motivators will drive individuals to travel and select their destinations accordingly.

Though most push factors are insubstantial desires of travelers, the literature suggests that travelers are pushed by internal desires initially, which may include escape, rest and relaxation, social interaction (family and friends), health and fitness, knowledge improvement, adventure, and prestige (Uysal and Hagan 1993). Instead of just establishing travel desires, push factors are also used to determine when and where to travel (Crompton 1979) and complement pull factors (Dann 1977). The domain of the push factors explains 'why' travelers travel. It is about the socio-psychological predisposition to travel (Dann 1981). Thus, push factors are referred to as 'the desire to travel,' associating with the decision 'whether to go' (Kim, Ick-Keun, and Giri 2007). An individual may alter his/her intention to utilize or not to utilize the full range of products. For example, in a tourism scenario, the tourists may decide whether to go on a wine trip or to have other leisure activities as an alternative. Furthermore, the decision can be spontaneous (Wong and Musa 2014).

Realizing the push motives, wine tourism destination marketers may improve their competitiveness by solving the problems that cause push motives through essential and appropriate activities and attractions (Correia, Oom do Valle, and Moço 2007). In return, this will help the destination create influential pull factors and eventually create the destination's attractiveness and specific features as perceived by individual travelers (Uysal and Jurowski 1994). The pull factors influence tourist's destination selection, where a mixture of facilities and services play an essential role (Hu and Ritchie 1993). Once tourists decide to travel, they will then consider the pull factors of a destination that attracts them (Oh, Uysal, and Weaver 1995). Therefore, pull factors generally are the destination attributes that correspond adequately to the push motivators (Dann 1981).

Tourists may not consider only a single pull factor, but various factors simultaneously as long as they respond well with the push factors (Dann 1981). It is worth noting that any of the pull factors of a specific destination may be driven by one or more push factors (Klenosky 2002). In general, activities commonly and easily accessible to travelers in their home destination (e.g., entertainment and sports) will be the least essential pull factors (Jamrozy and Uysal 1994). The understanding of travel motivation through push and pull theory enables wine tourism destination marketers to distinguish or forecast travelers' demand and supply of their offerings to attract the tourists. Pull factors are claimed to be the maintenance attributes by Uysal, Li, and Sirakaya-Turk (2008), where their absence may not ensure travelers' satisfaction. The examples of push and pull factors that motivate travelers to travel are illustrated in Figure 13.2.

Wine tourists usually consider several variables before making a destination selection. The variables can be categorized into four groups: (a) internal variables (i.e., push motivation, values, images, lifestyles, tourists' personality); (b) external variables (i.e., destination pull factors, hindrances, family, friends, reference group influences, marketing mix, social class, household-related factors such as power level, lifestyle, and style of decision making); (c) proposed wine tour's features (group size, time, distance, and duration of the trip); and (d) experiences of the vineyard trip (feeling and mood during vacation, post-purchase assessment). The interlinks among the variables upon consideration will influence the final selection of a specific destination.

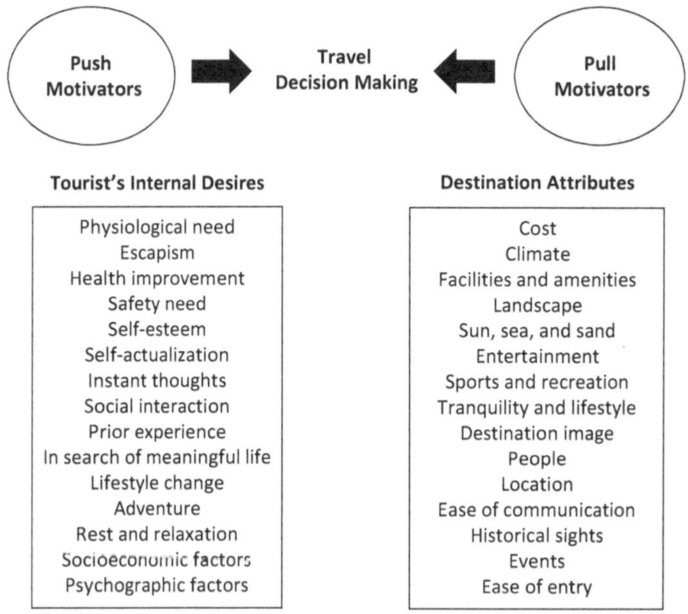

Figure 13.2 Examples of push and pull travel motivation factors.

Source: Authors

Push motivators in wine tourism

Push factors are generally origin-related, intangible, and sociopsychological variables that drive individuals to travel and explain their travel desire (Crompton 1979; Klenosky 2002; Yoon and Uysal 2005). Therefore, it makes sense that most push factors are primary motivators that represent the tourist's intrinsic desires and their satisfaction with a travel need (Uysal and Jurowski 1994).

Educational experience

The first push factor that will affect tourist motivations for participating in wine tourism is the educational experience. Vo-Thanh and Kirova (2017) stated that educational experience is one of the main motivations in the wine tourism industry, where personal development was recognized as a critical attribute desired by wine tourists. This is because of the satisfaction and good memories wine tourists will experience through recreational, cultural, and culinary activities and accommodations, all of which may offer an educational element. Thus, wine tasting and wine-food pairing events at local restaurants are the most common educational activities (Mihailescu 2015).

The level of wine knowledge has been associated with the monthly wine expenditures as well as the frequency of wine consumption. Alebaki, Menexes, and Koutsouris (2015) also mentioned that most research efforts have focused on subjective knowledge (i.e., how much a consumer perceives or thinks that s/he knows) rather than objective knowledge because the consumer has a stronger connection with experience-based knowledge. It is also easier to be assessed, not by some sort of test but rather by standardized scales.

The importance of educational experiences is further highlighted mainly by the tourists' decision to leave their homes and participate in wine tourism (Vibeke and Vibeke 2020). The opportunities to learn about wine in general, the winemaking processes, and the wineries' products drive tourists towards engaging in wine tourism (Grobbelaar 2020; Hazard et al. 2016). Thus, among the outcomes of such educational experiences from wine tourism is enhancing personal growth among the tourists (Grobbelaar 2020).

Escapism

Often wine tourists think about wine tourism as an opportunity to relax, given wine tourism's holiday opportunities (Ngan 2019). Among the activities perceived as an opportunity to relax during the wine tourism encounter include wine drinking, visiting vineyards and wineries, and observing rural scenery (Peršurić, Damijanić, and Šergo 2019). Afonso et al. (2018) believe that the dimensions of motivations are not consensual. This is where motivations of 'escape' and 'socialization' in wine tourism can be combined. This is because it encompasses the desire for relaxation in a rural scenario, enrichment through a unique esthetic experience, and an escapist retreat from daily routines while enjoying a group's companionship, such as family or friends. While wine tourists may gaze at a winescape, gaining an esthetic experience, they may also engage in various recreational options typically available in wine tourism destinations (Vo-Thanh and Kirova 2017). Besides, there are numerous activities that tourists may participate at the destinations, such as hot air ballooning, bicycle tours, vineyard hiking, vineyard tours by horse and carriage, riding a grape picker, and grape stomping and harvesting. Thus, the activities complement the need for escapism among wine tourists.

Social need

The opportunity to meet with other people, such as winemakers, other tourists, and perhaps wine enthusiasts, is another push motivator in wine tourism (Jeftovic 2015). The minimal risks involved in such activity drive wine tourists to meet and interact with other tourists at the wine tourism destination (Jeftovic 2015). With a common aim to participate in wine tours, the tourists find the opportunity to meet and learn from the winemakers, together with their family, friends, and others whom they may not know before the tour. The social linkage among the tourists somehow provides a better condition for making wine purchase decisions, since tourists are generally reluctant to take too many risks about wine purchases (Jeftovic 2015).

Adventure

Adventure is another push factor motivating wine tourism (Uphill 2017). An adventurous experience can help reduce boredom among the tourists, provide change from the routine tourism agendas, and create novelty experiences for wine tourists (Dyer 2015; Uphill 2017). The adventure derived from participation in wine tourism activities also creates an opportunity for the tourists to sense achievement, challenge, and competency (Vorobiova 2017), which can be described as the self-actualization need in Maslow's Hierarchy of Needs. Although adventure generated from wine tourism may be risky (Uphill 2017), it can still be a decisive push factor in wine tourism.

Product involvement

The concept of product involvement is often used in studying consumer behavior as it affects their decision-making (Afonso et al. 2018). In the context of wine product involvement, Bruwer and Huang (2012, p. 463) explained wine product involvement as 'a motivational state of mind of a person with wine or wine-related activity which reflects the extent of personal relevance of the-wine related decision to the individual in terms of one's basic values, goals, and self-concept.' Instead, involvement with wine reflects the interest, enthusiasm, and excitement consumers show toward wine. This push motivation factor can best complement a strategic marketing plan by the respective wine tourism destination's marketers, highlighted in the next section.

Pull motivators in wine tourism

Pull factors represent the external attributes that attract travelers and help them visit a particular destination (Klenosky 2002; Uysal and Jurowski 1994). According to Uysal and Hagan (1993), the pull factors cover both tangible resources (e.g., recreation, beaches, facilities, and cultural attractions) and tourists' perceptions and expectations (e.g., benefit expectation, marketing image, and novelty). This shows that pull factors are related to the decision 'where to go' (Kim, Ick-Keun, and Giri 2007), depending on their perceptions of the destination's attributes.

Wine tasting opportunity

By the nature of wine tourism itself, the opportunity for wine drinking is a crucial pull motivator at the wine tourism destinations (Dyer 2015; Lee, McCole, and Holecek 2020; Picard, Moreira, and Loloum 2018). The firsthand tasting, without payments or distribution by intermediaries, is an opportunity that wine tourists may not get outside the wine tourism phenomenon. While such an opportunity dated back to the 1930s, the tradition remains until now. Besides, with the improved production quality of table wines, a continuous increase in opportunities to the new world of wine tourism destinations such as California, United States, has been observed since the 1960s (Dyer 2015). Similar empirical evidence has been reported in South Africa, where wine drinking is the most crucial factor in wine tourists' intentions to visit South Africa as a wine tourism destination (Goosen 2014).

Wine quality

In Austria, wine quality has been considered a pull factor that motivates the niche wine tourists to visit the nation (Stockl et al. 2016). Also, assurances of wine quality through system usage have been considered necessary if wine tourism destination marketer wants to attract more wine tourists, thus further escalating the importance of wine quality as a pull motivator (Khanna 2016). While tourists may purchase wine due to good customer service, the wine quality's reputation will often influence their purchase decision. Hence, wine tourists will likely pick a destination with good wine quality perception (Grobbelaar 2020).

Product marketing

The product involvement behavior among wine tourists drives them to engage in several wine tourism activities, such as tasting wine, experiencing the tasting room services, enjoying the area's natural resources, and many more (Byrd et al. 2016). Thus, several studies suggest that the level of involvement with wine affects consumer behavior and wine tourism pursuits. López-Guzmán, Vieira-Rodríguez, and Rodríguez-García (2014) further suggested that food and wine involvement have become one of the most important reasons for wine tourists to visit Spain in recent years. Between 2008 and 2012, the number of wine tourists in Spain grew by 27.5% in primarily two regions, namely Sherry and Penedés. This is due to the success of a project that Spain's government launched called 'Tasting Spain,' in which the mission is to develop wine tourism linked to culinary tourism. This new tourism product aimed to promote Spanish gastronomy on an international scale, create new destinations specializing in gastronomic experiences, and develop strategies for promoting and marketing these places.

Winescape

Another pull motivator to wine tourism is the attractiveness of vineyards' scenery (Goosen 2014; Lee, McCole, and Holecek, 2020; Picard, Moreira, and Loloum 2018; Woldarsky and Geny-Denis 2019). Byrd et al. (2016) highlighted this factor as 'winescape.' Most vineyards set their own visitation time, depending on the season and accessibility. Wine tourists will then choose which vineyard(s) to visit based on the availability, perceived attractiveness, and perceived quality of the vineyards (Goosen 2014; Woldarsky and Geny-Denis 2019). A study by Sorensen (2013) reported the "attractiveness of scenery" as a vital motivator to encourage wine tourists to participate in wine tourism in Bulgaria. The study also revealed that this factor plays an essential role in influencing tourists' choice of visiting one vineyard. The same notion is also echoed by Gmelch and Gmelch (2011) in their study of wine tourism in California, United States.

Destination image

Byrd et al. (2016) stated that the winescape as a destination image has progressed over time. It has developed from the core vineyard and winery facility to a wine region's more significant conceptualization as a tourist destination. This is where the three distinct components of an advanced winescape must consist of vineyards, winemaking activity, and tasting rooms. Thus, wine tourists will be motivated to find a rural environment supportive of grape growing and related activities such as winemaking, a picturesque vineyard landscape, and a tasting room where they can sample the wine (Byrd et al. 2016; Lee, McCole, and Holecek 2020; Picard, Moreira, and Loloum 2018; Vo-Thanh and Kirova 2017). Furthermore, the wine tourism destination image plays an essential role in attracting tourists, where some researchers believe that wine tourists perceive a more comprehensive range of benefits before purchase and during winery visits. This includes the surrounding community's tourism aspects such as outdoor recreational areas, regional heritage, cultural attractions, and local dining and lodging (Byrd et al. 2016; Vo-Thanh and Kirova 2017).

Entertainment

The following destination attribute of wine tourism is entertainment (Ngan 2019). Besides the natural beauty of the countryside and the vineyards, tourists are also attracted by the entertainment opportunities available for their leisure and pleasant experiences (Sorensen 2013). However, it is worth noting that entertainment is generally considered less critical to wineries when catering to wine tourists than wine production for tasting and consumer perceptions

Table 13.1 Summary of push and pull motivators in wine tourism

Push factors motivating wine tourism	*Pull factors motivating wine tourism*
Educational experiences	Wine tasting opportunity
Alebaki, Menexes, and Koutsouris (2015); Grobbelaar (2020); Hazard et al. (2016); Vo-Thanh and Kirova (2017); Vibeke and Vibeke (2020)	Dyer (2015); Goosen (2014)
Escapism	Wine quality
Ngan (2019); Peršurić, Damijanić, and Šergo (2019); Afonso et al. (2018); Vo-Thanh and Kirova (2017)	Grobbelaar (2020); Khanna (2016); Stockl et al. (2016)
Social need	Product marketing
Jeftovic (2015)	Byrd et al. (2016); Guzmán, Rodríguez, and García (2014)
Adventure	Winescape
Dyer (2015); Uphill (2017); Vorobiova (2017)	Goosen (2014); Gmelch and Gmelch (2011); Sorensen (2013)
Product involvement	Destination image
Afonso et al. (2018); Bruwer and Huang (2012)	Byrd et al. (2016); Vo-Thanh and Kirova (2017)
	Entertainment
	Ngan (2019); Sorensen (2013); Vo-Thanh and Kirova (2017)

(Sorensen 2013). Nevertheless, wine tourists generally will feel excited when a special event occurs in wine destinations (Vo-Thanh and Kirova 2017). Therefore, wine tourists will be more attracted to wine destinations, especially if there is an event such as cultural attractions to experience different cultures.

The primary elements to entertain wine tourists are wine, food festivals, and shows. The interrelation of art, music, and cultural heritage in entertaining tourists is also necessary. Numerous specialty shops such as wine shops within wine destinations are also an excellent entertainment element. As wine tourists are increasingly motivated by entertainment while engaging in wine tourism, perhaps wine tourism developers need to put in more effort to provide alternative entertainment experiences to wine tourists compared to before (Sorensen 2013). The summary of the push and pull motivators in wine tourism is listed in Table 13.1.

Importance of understanding motivations in wine tourism

As discussed in this chapter, the wine tourism market has expanded beyond the Old World and New World wine tourist markets to emerging wine markets (e.g., Thailand and China) in recent years (Chong 2017; Hall et al. 2009; Keene 2018). Therefore, understanding wine tourism motivations is crucial to fill the academic research gaps in this field, particularly among the emerging wine markets. Besides, helpful information about why wine tourists engage in wine tourism and what factors influence them to choose a specific wine tourism destination (Goosen 2014; Keene 2018) will be helpful not just for the academics to understand the phenomenon for the practitioners better. Equipped with the tourists' proper understanding, marketers can craft specific marketing strategies to create an emotional attachment with potential wine tourists. Personal emotions can be established when the created appeals satisfy the potential wine tourists' needs and wants (Jeftovic 2015).

The recognition of wine tourism motivations and destination attributes' perception is vital for the wine industry's development. Such an understanding can help destination marketers create targeted promotions for tours and perhaps organize virtual tours at wine-producing regions to satisfy wine tourists who cannot physically visit the destination (Garibaldi et al. 2017; Jeftovic 2015). The motivators coupled with the appropriate psychographic data may help the marketers predict customers' loyalty to wine brands, wine consumption forecast, potential tourists' knowledge of wine, and their willingness to visit vineyards and learn about wine. Even the preparation of tours in the distant future can benefit from well-anticipated wine tourists' motivations (Jeftovic 2015).

Case Study 13.1 Thailand – the making of an emerging wine tourism destination

Thailand is a more recent wine tourism destination than the Old World (European) and New World (American, South African, and New Zealander) destinations. Thailand's potential to be a rising destination for wine tourism started to receive attention from other industry players (Chong 2017; Hall et al. 2009). In recent years, Thailand has been seen as an exciting place for wine tourism because of its touristic appeal and natural environment (Amarando, Assenov, and Visuthismajarn 2019). Even so, studies about Thailand's wine tourism industry are still at the infant stage since the industry itself is still in its early stage of development (Amarando, Assenov, and Visuthismajarn 2019).

Compared to Italy and France, Thailand's wine tourism industry has received little attention from tourists (Amarando, Assenov, and Visuthismajarn 2019). Generally, awareness of wine tourism and wineries in Thailand has been rare overseas, resulting in the unfamiliarity of its wine tourism products among wine tourists (Chong 2017). Due to the little awareness, the pull motivations are inadequate. Hence wine tourists are not motivated to engage in Thailand's wine tourism offerings.

The Thailand Authority of Tourism's (TAT) aggressiveness and strong presence in the global tourism arena can offer a strong value proposition for the nation to portray its destination attributes within the wine tourism industry. It is a new niche segment from which Thailand perhaps can benefit by leveraging its global tourism campaign. Such attempts will help the nation improve tourists' perceptions of Thailand as a wine tourism destination (Amarando, Assenov, and Visuthismajarn 2019).

Another destination attribute of Thailand is its lower wine prices due to the Thai government's tax incentive offerings for wine production (Amarando, Assenov, and Visuthismajarn 2019). Lower prices for Thai wines can make Thai wines more affordable to potential wine tourists, encouraging potential wine tourists to visit Thailand. Hence, tourists who may not visit Thailand specifically for wine tourism may take advantage of the opportunity to taste and purchase Thai wines (Amarando, Assenov, and Visuthismajarn 2019).

Despite the current low awareness of wine tourism in Thailand, it must be noted that tourist demand for wine tourism in Thailand has improved with the rise of wine tourism in Thailand in recent years (Chong 2017; The Nation 2018). Furthermore, the popularity of wine importing to the 37 million tourists who visited Thailand in 2018, and wine tourism events such as the 11th Bangkok Grand Wine Tasting, make Thailand an exciting destination for wine tourists to savor Thai wine first-hand (The Nation 2018). Therefore, it is evident that the pull factor of 'wine tasting opportunity,' especially among the first to taste it, drives wine tourists to visit Thailand for wine tourism purposes.

Customized wine events and the presence of high-quality wineries in Thailand have been noted as having the potential to encourage wine tourists further to choose Thailand as a wine tourism destination (Amarando, Assenov, and Visuthismajarn 2019; Mihailescu 2015). Examples of high-quality wineries in Thailand include Siam winery, PB Valley Khao Yai Winery, GranMonte winery, and Chateau de Loei Winery, proving themselves in the high production of quality bottled wine for tourist consumption in 2011 (Banks et al. 2013).

In general, Thailand's potential to be an upcoming wine tourism destination in the region will probably become reality in a matter of time. The government support and involvement in building and growing this new niche tourism product will further escalate the effort. Thailand is blessed with many attractive destination attributes that can easily complement its effort to build its reputation as a new wine tourism destination. Thus, the nation needs to understand wine tourists' motivations to prepare and implement more strategic marketing plans to achieve the objective. Perhaps Thailand's high-quality wineries should have more exposure in the global wine scene to attract wine tourists and try out this 'new kid on the block' of wine tourism.

Conclusion

This chapter discussed the motivations of wine tourism, which include the perception of wine tourism destination attributes. While we may explain the variables through conventional theories such as Maslow's Hierarchy of Needs and Travel Career Ladder, this chapter also described the push and pull theory's motivation factors. As a widely adopted theory in the study of tourism motivations, five push factors (i.e., educational experience, escapism, social need, adventure, product involvement) and six pull factors (i.e., wine tasting opportunity, wine quality, product marketing, winescape, destination image, entertainment) are explained and summarized in Table 13.1. This chapter acknowledges the importance of understanding wine tourists' motivations from both academic and practical perspectives. Several combinations of motivations can assist wine tourism destination marketers and operators in designing differentiated packages that address the specific needs of wine tourists. As the wine tourism industry needs to satisfy wine tourists' needs and wants, the proper understanding of their motivations with an appropriate set of psychographic data will enable wine tourism providers to plan better wine production, prepare more personalized tours, and hence build the capability to cater to the wine tourism market more effectively (Garibaldi et al. 2017; Jeftovic 2015).

Implications

This study yields important managerial and marketing implications because they contribute to stakeholders' understanding of what motivates wine tourists to visit a wine region and increase their involvement. In order to increase the intention to return, wine tourism destination marketers should promote activities that encourage involvement with wine (i.e., winemaking, farming, and harvestings) and participation in wine-related events (i.e., wine festivals, seminars, and trade shows). Furthermore, to boost involvement, wine tourism developers, owners, and marketers are best advised to develop their wine tourism products with combined elements of escapism, socialization opportunities, accommodation, and local gastronomic experiences. Hence, this chapter facilitates the marketing of vineyard tours and winery products in a more informed manner. It provides valuable guidance to wine tourism destination marketers by adding value to wine tourists' experiences, promoting positive word-of-mouth, and enhancing their repeat visits.

References

Afonso, C., Silva, G., Gonçalves, H., and Duarte, M. (2018). The role of motivations and involvement in wine tourists' intention to return: SEM and fsQCA findings. *Journal of Business Research*, 89: 313–321. https://doi.org/10.1016/j.jbusres.2017.11.042

Alebaki, M., Menexes, G., and Koutsouris, A. (2015). Developing a multidimensional framework for wine tourist behavior: Evidence from Greece. *Wine Economics and Policy*, 4(2): 98–109. https://doi.org/10.1016/j.wep.2015.11.002

Amarando, M., Assenov, I., and Visuthismajarn, P. (2019). Sustainable wine tourism and vineyards' environmental consciousness in Thailand. *African Journal of Hospitality, Tourism and Leisure*, 8(3): 1–13. https://www.ajhtl.com/uploads/7/1/6/3/7163688/article_35_vol_8_3__2019.pdf

Banks, G., Klinsrisuk, R., Dilokwanich, S., and Stupples, P. (2013). Wines without latitude: global and local forces and the geography of the Thai wine industry, *EchoGéo*, 23: 1–18. https://journals.openedition.org/echogeo/13368?lang=en

Bishop, J. (2016). An analysis of the implications of Maslow's Hierarchy of Needs for networked learning design and delivery. *Int'l Conf. Information and Knowledge Engineering*, IKE'16, 49–54. https://www.researchgate.net/profile/Jonathan_Bishop4/publication/307533183_An_analysis_of_the_implications_of_Maslow's_Hierarchy_of_Needs_for_networked_learning_design_and_delivery/links/580bf60e08ae74852b5a7363.pdf

Bruwer, J., and Huang, J. (2012). Product involvement and consumers' BYOB behaviour in the South Australian on-trade wine market. *Asia Pacific Journal of Marketing and Logistics*, 24(3): 1–34. https://doi.org/10.1108/13555851211237911

Byrd, E., Canziani, B., Hsieh, Y.-C., Debbage, K., and Sonmez, S. (2016). Wine tourism: motivating visitors through core and supplementary services. *Tourism Management*, 52: 19–29. https://doi.org/10.1016/j.tourman.2015.06.009

Chong, K. L. (2017). Thailand wine tourism: a dream or a reality? *Asia Pacific Journal of Tourism Research*, 22(6): 604–614. https://doi.org/10.1080/10941665.2017.1308389

Correia, A., Oom do Valle, P., and Moço, C. (2007). Modeling motivations and perceptions of Portuguese tourists. *Journal of Business Research*, 60(1): 76–80. https://doi.org/10.1016/j.jbusres.2005.10.013

Coughlan, L. (2017). *A critical assessment of the culinary preferences of international tourists to South Africa* [Doctoral thesis]. North-West University, Potchefstroom, 252 pp. https://repository.nwu.ac.za/bitstream/handle/10394/26315/L%20Coughlan%20-%2026256096.pdf?sequence=1&isAllowed=y

Crompton, J. L. (1979). Motivations for pleasure vacation. *Annals of Tourism Research*, 6(4), 408–424. https://doi.org/10.1016/0160-7383(79)90004-5

Dann, G. (1976). 'The holiday was simply fantastic.' *Tourism Review*, 31(3), 19–23. https://doi.org/10.1108/eb057732

Dann, G. (1977). Anomie, ego-enhancement and tourism. *Annals of Tourism Research*, 4(4), 184–194. https://doi.org/10.1016/0160-7383(77)90037-8

Dann, G. (1981). Tourism motivation: an appraisal. *Annals of Tourism Research*, 8(2), 189–219. https://doi.org/10.1016/0160-7383(81)90082-7

Deci, E. L. (1975). *Intrinsic Motivation*. New York: Plenum.

Deng, S. (2013). *'I Prefer a Dry Red Thanks': A Consumer Behavioural Study of Resident Auckland Chinese Wine Consumption and Wine-Related Tourism* [Master's thesis]. Auckland University of Technology, Auckland, 159 pp. https://openrepository.aut.ac.nz/bitstream/handle/10292/5515/DengS.pdf?sequence=3&isAllowed=y

Dyer, S. (2015). Democratizing visions of luxury and the good life in California Wine Country: Wine tourism from repeal to the eve of the 'Wine Revolution.' *Business and Economic History On-Line*, 13: 1–12. https://thebhc.org/sites/default/files/Dyer_BEHO_Final%20Draft_0.pdf

Ferreira, D. P. (2015). *Push-and Pull Forces within Outbound Destination Choice* [Master's thesis]. Nelson Mandela Metropolitan University, Port Elizabeth, South Africa, 267 pp. https://core.ac.uk/download/pdf/145036214.pdf

Garibaldi, R., Stone, M., Wolf, E., and Pozzi, A. (2017). Wine travel in the United States: a profile of wine travellers and wine tours, *Tourism Management Perspectives*, 23: 53–57. https://doi.org/10.1016/j.tmp.2017.04.004

Gmelch, G., and Gmelch, S. B. (2011). *Tasting the Good Life: Wine Tourism in the Napa Valley*. Bloomington: Indiana University Press.

Gnoth, J. (1997). Tourism motivation and expectation formation. *Annals of Tourism Research*, 24(2): 283–304. https://doi.org/10.1016/S0160-7383(97)80002-3

Goosen, W. (2014). *Wellington as a Developing South African Wine Tourism Destination* [Doctoral thesis]. Stellenbosch University, Stellenbosch, 118 pp. http://scholar.sun.ac.za/handle/10019.1/96032

Grobbelaar, L. (2020). *Towards a Conceptual Framework for Value Co-Creation in Wine Tourism* [Doctoral thesis]. North-West University, Potchefstroom, 280 pp. http://repository.nwu.ac.za/bitstream/handle/10394/34907/Grobbelaar%20L%2020953674.pdf?sequence=1&isAllowed=y

Guzmán, T. L., Rodríguez, A. V., and García, J. R. (2014). Profile and motivations of European tourists on the Sherry wine route of Spain. Tourism Management Perspectives, 11, 63–68. https://doi.org/10.1016/j.tmp.2014.04.003

Hall, C. M., Sharples, L., Cambourne, B., and Macionis, N. (2009). *Wine Tourism around the World* (3rd ed.). New York: Routledge.

Han, L. (2011). The prospect and forecast of China's wine tourism in 2011. *Energy Procedia*, 5: 1616–1620. https://doi.org/10.1016/j.egypro.2011.03.275

Hazard, W., Magrath, C., Mohanty, A., and Nogueira, T. (2016). *Wine Tourism Development in Northern Greece* [Bachelor of Science thesis]. Worchester Polytechnic Institute, Worchester, MA, 64 pp. https://web.wpi.edu/Pubs/E-project/Available/E-project-042816-042630/unrestricted/Wine_Tourism_Development_in_Northern_Greece.pdf

Hu, Y., and Ritchie, J. R. B. (1993). Measuring destination attractiveness: a contextual approach. *Journal of Travel Research*, 32(2): 25–34. https://doi.org/10.1177/004728759303200204

Iso-Ahola, S. E. (1980). *The Social Psychology of Leisure and Recreation*. Dubuque, IA: William C. Brown Co. Publishers.

Jamrozy, U., and Uysal, M. (1994). Travel motivation vacations of overseas German visitors. *Journal of International Consumer Marketing*, 6(3–4): 135–160.

Jeftovic, A. (2015). *Opportunities for Development of Wine Tourism in the Region of Fruska Gora* [Master's thesis]. Singidunum University, Belgrade, 77 pp. https://singipedia.singidunum.ac.rs/preuzmi/41716-opportunities-for-development-of-wine-tourism-in-the-region-of-fruska-gora/1795

Karagiannis, D., and Metaxas, T. (2020). Sustainable wine tourism development: Case studies from the Greek region of Peloponnese. *Sustainability*, 12: 5223. https://doi.org/10.3390/su12125223

Karlsson, P., and Karlsson, B. (2017). The four successful types of wine tourism. *Forbes*. https://www.forbes.com/sites/karlsson/2017/07/21/the-four-successful-types-of-wine-tourism/?sh=6e531a2c36fa

Keene, M. M. (2018). *Modelling Luxury Wine Preference: A Study of Business Travellers from China* [Doctoral dissertation]. Purdue University, West Lafayette, IN, 134 pp. https://hammer.purdue.edu/articles/thesis/Modeling_Luxury_Wine_Preference_A_Study_of_Business_Travelers_from_China/7439003

Khanna, R. (2016). *Evaluating Push and Pull Factors in for Gastronomy Tourism Destination Decisions – The Case of Amish Country* [Master's thesis]. Kent State University College of Education, Health, and Human Services, New Delhi, 37 pp. https://etd.ohiolink.edu/apexprod/rws_olink/r/1501/10?p10_etd_subid=119449&clear=10

Kim, K., Ick-Keun, O., and Giri, J. (2007). College student travel: a revised model of push motives. *Journal of Vacation Marketing*, 13(1): 73. https://doi.org/10.1177/1356766706071201

Klenosky, D. B. (2002). The 'pull' of tourism destinations: a means-end investigation. *Journal of Travel Research*, 40(4), 385(311). https://doi.org/10.1177/004728750204000405

Lee, J. S.-H., McCole, D., and Holecek, D. (2020). Exploring winery visitors in the emerging wine regions of the North Central United States. *Sustainability*, 12(4), 1642. https://doi.org/10.3390/su12041642

Loker-Murphy, L. E. (1995). Backpackers in Australia: A motivation-based segmentation study. In *Proceedings from the Tourism Research and Education Conference* (pp. 115–128). Canberra, Australia: Bureau of Tourism Research.

López-Guzmán, T., Vieira-Rodríguez, A., and Rodríguez-García, J. (2014). Profile and motivations of European tourists on the Sherry wine route of Spain. *Tourism Management Perspectives*, 1: 63–68. https://doi.org/10.1016/j.tmp.2014.04.003

Meraz Ruiz, L., and Ruiz Vega, A. V. (2016). Wine tourism in Baja California, Mexico: an analysis of its supply and a comparison with the wine region of La Rioja, Spain. *Investigaciones Turísticas*, 12: 73–98. https://doi.org/10.14198/INTURI2016.12.04

Mansfeld, Y., and Pizam, A. (1999). *Consumer Behaviour in Travel and Tourism*. New York: Haworth Hospitality Press, an imprint of Haworth Press.

Maslow, A. H. (1943). A theory of human motivation. *Psychological Review*, 50: 370–396.

Mihailescu, R. (2015). Is there a scope for organic wine tourism development? A focus on South African wine industry. *Rivista di Scienze del Turismo*, 6(1–2): 11–21. https://doi.org/10.7358/rst-2015-02-miha

Moscardo, G. M., and Pearce, P. L. (1986). Historical theme parks: an Australian experience in authenticity. *Annals of Tourism Research*, 13(3): 467–479. https://doi.org/10.1016/0160-7383(86)90031-9

Ngan, Y. M. (2019). *Port Wine Experience for the Chinese Market*. [Master's thesis]. Marketing Department, Quinta & Vineyard Bottlers, Vinhos, 32 pp. https://repositorio-aberto.up.pt/bitstream/10216/126696/2/390246.pdf

Oh, C. H., Uysal, M., and Weaver, P. A. (1995). Product bundles and market segments based on travel motivations: a canonical correlation approach. *International Journal of Hospitality Management*, 14(2), 123–137. https://doi.org/10.1016/0278-4319(95)00010-A

Pearce, P. L. (1991). Dreamworld: A report on public reactions to dreamworld and proposed developments at dreamworld. In *A Report to Ernst and Young on Behalf of the IOOF in Conjunction with Brian Dermott and Associates*. Townsville, Australia: Department of Tourism, James Cook University.

Pearce, P. L., and Lee, U. I. (2005). Developing the travel career approach to tourist motivation. *Journal of Travel Research*, 43(3): 226–237. https://doi.org/10.1177/0047287504272020

Peršurić, A. S. I., Damijanić, A. T., and Šergo, Z. (2019). The wine tourism terroir: Experiences from Istria. *ToSEE – Tourism in Southern and Eastern Europe*, 5: 319–333. https://bib.irb.hr/datoteka/1023640.22_IlakPersuric_TezakDamijanic_Sergo.pdf

Picard, D., Moreira, C. N., and Loloum, T. (2018). Wine magic consumer culture, tourism and terroir. *Journal of Anthropological Research* (74)(4): 1–15. https://doi.org/10.1086/699943

Podoshen, J. S. (2012). Dark tourism motivations: simulation, emotional contagion and topographic comparison. *Tourism Management*, 35: 263–271. https://doi.org/10.1016/j.tourman.2012.08.002

Puryova, R. (2013). *Examining Main Motivational Factors in Dark Tourism – Gallipoli, Turkey* [Master's thesis]. Institute of Graduate Studies and Research, Eastern Mediterranean University, Gazimağusa, 116 pp. http://i-rep.emu.edu.tr:8080/xmlui/handle/11129/1999

Ryan, C. (1998). The travel career ladder: an appraisal. *Annals of Tourism Research*, 25(1): 936–957. https://doi.org/10.1016/S0160-7383(98)00044-9

Snepenger, D., King, J., Marshall, E., and Uysal, M. (2006). Modeling Iso-Aloha's motivation theory in the tourism context. *Journal of Travel Research*, 45(2): 140–149. https://doi.org/10.1177/0047287506291592

Sorensen, Z. (2013). *Exploring Wine Tourism in Bulgaria: A Pathway to Sustainable Rural and Tourism Development?* [IIIEE Master's thesis]. Lund University, Lund, 50 pp. https://lup.lub.lu.se/student-papers/search/publication/4091904

Stockl, A. F., Tischler, S., Bauer-Krosbacher, C., Koja, T., and Mathis, J. K. (2016). Does wine enhance the attractiveness of a destination? *Proceedings of the 9th Academy of Wine Business Research Conference* (pp. 598–607). University of South Australia, Adelaide.

Taylor, C., and Barber, N. A. (2016). How will my wine purchase decision be viewed by others? *Journal of Wine Research*, 27(3): 202–225. https://doi.org/10.1080/09571264.2016.1173533

Tikkanen, I. (2007). Maslow's hierarchy and food tourism in Finland: Five cases. *British Food Journal*, 109(9): 721–734. https://doi.org/10.1108/00070700710780698

The Nation (2018). *Tourist demand drives wine industry to new heights in Thailand*. https://www.nationthailand.com/Corporate/30345134

Tolman, E. C. (1959). *Psychology: A Study of a Science*. New York: McGraw Hill.

Uphill, R. (2017). *A Contemporary Overview of Visitor Motivations and Expectations at Naracoorte Caves National Park, South Australia* [Master's thesis]. College of Humanities, Arts and Social Sciences, Flinders University, Adelaide, 181 pp. https://flex.flinders.edu.au/file/60b18ffb-d566-4b99-81f9-ee40e58041f9/1/ThesisUphill2017OA.pdf

Uysal, M., and Hagan, L. A. R. (1993). Motivation of pleasure travel and tourism. In Kahan, M., Olsen, M., and Var, T. (eds.), *Encyclopedia of Hospitality and Tourism* (pp. 798–810). New York: Van Nostrand Reinhold.

Uysal, M., and Jurowski, C. (1994). Testing the push and pull factors. *Annals of Tourism Research*, *21*(4), 844–846. https://doi.org/10.1016/0160-7383(94)90091-4

Uysal, M., Li, X., and Sirakaya-Turk, E. (2008). Push-pull dynamics in travel decisions. In Oh, H. (ed.), *Handbook of Hospitality Marketing Management* (pp. 413–439). Oxford: Butterworth-Heinemann.
Vibeke, C., and Vibeke, J. (2020). *Uncorking Wine Tourism Prospects of Regional Tourism Sustainability in Spain* [Bachelor's thesis]. Linnaeus University, Sweden, 58 p. https://www.diva-portal.org/smash/get/diva2:1393775/FULLTEXT01.pdf
Vo-Thanh, T., and Kirova, V. (2017). Wine tourism experience: A netnography study. *Journal of Business Research*, 83: 30–37. https://doi.org/10.1016/j.jbusres.2017.10.008
Vorobiova, N. (2017). *Motivations of Tourists in Wine Regions: The Case of La Rioja* [Master's thesis]. University of Algarve, Faro, Portugal, 1–53 pp. https://sapientia.ualg.pt/bitstream/10400.1/10116/1/Dissertation%20%20Natalia%20Vorobiova%281%29.pdf
Vroom, V. H. (1964). *Work and Motivation*. New York: John Wiley & Sons.
Woldarsky, V., and Geny-Denis, L. (2019). Development of a best practice manual in wine tourism in Portugal. *BIO Web of Conferences*, 12, 03001. https://doi.org/10.1051/bioconf/20191203001
Wong, K. M. (2015). *'Malaysia My Second Home' (MM2H): Retirees' Motivations, Satisfaction, and Post-Satisfaction Intentions* [Doctoral thesis]. University of Malaya, Malaysia, 403 pp. http://studentsrepo.um.edu.my/4741/1/%E2%80%98MALAYSIA_MY_SECOND_HOME%E2%80%99_(MM2H)_-_RETIREES%E2%80%99_MOTIVATIONS,_SATISFACTION,_AND_POST-SATISFACTION_INTENTIONS.pdf
Wong, K. M., and Musa, G. (2014). Retirement motivation among 'Malaysia My Second Home' participants. *Tourism Management*, 40: 141–154. https://doi.org/10.1016/j.tourman.2013.06.002
Wong, B. K. M., Musa, G., and Taha, A. Z. (2017). Malaysia My Second Home: the influence of push and pull motivations on satisfaction. *Tourism Management, 61C*: 394–410. https://doi.org/10.1016/j.tourman.2017.03.003
Yoon, Y., and Uysal, M. (2005). An examination of the effects of motivation and satisfaction on destination loyalty: a structural model. *Tourism Management*, 26(1): 45–56. https://doi.org/10.1016/j.tourman.2003.08.016

14

WINE CONSUMERS' KNOWLEDGE OF WINE AND THEIR WINE SELF-CONFIDENCE IN DIFFERENT SALES CONTEXTS

Alet C. Erasmus and Renée van Wyk

Introduction

The wine industry is very important for the South African economy. The country prides itself in having 17 official wine routes that are established in terms of ideal geographical wine-growing locations, plus a multitude of other wine routes, including the Cape 62 wine route that is the world's longest, and the Green Mountain Eco wine route that is the world's first biodiverse wine route. Amid fierce competition among restauranteurs across the world, the global award in 2018 for the 'Best Hotel Wine List in the World' was awarded to the South African chef Chantel Dartnell, from the Restaurant Mosaic – one of many other awards that she had received over time. Unavoidably, therefore, wine is an important product in South Africa's agriculture, cuisine, as well as its tourism industry. Before the COVID-19 pandemic, the export value of wine in South Africa was the second largest of all agricultural commodities in the country. Compared to other countries, South Africa is categorized as one of the New World wine-producing countries that employ different winemaking philosophies from those of the so-called Old World wine-producing countries such as France and Italy (Swart and Smit 2009).

The earliest evidence of wine production can be found in the Bible's Old Testament, concerning Noah's wine-drinking habits. Records also note wine consumption among the Greek, Roman, and Egyptian gods, with indications that wine had served as a status symbol in Egypt, as well in as early Mesopotamia, where Iraq, Syria, and Turkey are located today. Today, Christians still use wine to celebrate the holy sacraments, using red wine to symbolize the blood of Christ. Since the industrial revolution and with the emergence of a consumer society, it has become customary for wine to represent wealth rather than nobility. Nowadays, wine is often given as a gift and used to celebrate memorable occasions, pairing wine with selected dishes to complement meals. As a commodity, wine is a complex product with indisputable symbolic significance due to assigned meaning that relates to status, religion, and even sexuality (Devonish 2011). The emphasis on status, and the more contemporary sexual connotations to wine, are derived from reinforcement through advertising in the media (Vlachvei, Notta, and Ananiadis 2009).

DOI: 10.4324/9781003143628-17

Recent statistics indicate that significant shifts are expected in the global wine industry in the future, for example, in the profile of wine drinkers. Younger consumers are steadily shifting from other alcoholic beverages such as beer, and more females nowadays prefer wine. New wine markets are also entering the scene, for example, Japan, China, India, and Russia (Anderson, Nelgen and Pinilla 2017; Resnick 2008). Another trend is for the wine industry to adopt more sustainable wine production practices to reduce its carbon footprint and divert to organic, biodynamic, and sustainable vini- and viticulture practices (Schäufele and Hamm 2017; Sogari, Mora, and Menozzi 2016). Reports indicate significant progress in the total ethical accredited production volumes of wine, which stood at 48% of the total harvest of 1.4 million tonnes in South Africa in 2017 (Loots 2017). Even health-related issues concerning wine consumption are attracting more attention (Chang, Thach, and Olsen 2016). Recent reports suggest that young adult females will become a prominent, new wine drinking segment in South Africa. They particularly prefer lighter wines, even sweet wines, such as a sweeter rosé. An emphasis will also be placed on low-alcohol wines with a premium appeal (Wine in South Africa, 2020).

The prevailing issue for consumers is that the range of wines to choose from, and the characteristics of wine, are continuously changing and expanding. This can be very intimidating – whether selecting wines in a retail store or wine cellar, or when confronted with a wine list in a restaurant. Over time, wine has become very specialized in terms of price, exclusivity, and novelty – all characteristics that are associated with the prestige of wine. Undeniably, therefore, consumers' knowledge of wine is crucial in terms of their self-confidence to choose wine. Consumers' wine self-confidence (WSC) is a rather intricate matter in that it could differ from one context to another. For example, it is an entirely different matter to choose wine in a retail store than to choose wine in a social context, such as a restaurant, where the people who are present might judge one's choices. This apparent sensitivity about wine choices dates back to the early Roman Empire, where a wine drinker was judged on issues such as the origin of the wine, vintage, variety of grapes used for the wine, and even a person's capacity to drink! ('The Symbolism of Wine through the Ages' 2011).

Segmentation of wine consumers

Marketers generally segment wine consumers to design relevant marketing strategies (Bruwer, Roediger, and Herbst 2017b). Typically, people's interest in wine, their involvement, their knowledge, and even their consumption of wine are considered (Resnick 2008; Yuan et al. 2008). Notwithstanding consumers' involvement with wine, price is generally used by most people as an indication of the quality of the wine (Bruwer and Buller 2012). Therefore, consumers are inclined to choose a more expensive wine when they want to impress somebody.

Over time, scholars have categorized wine consumers in various ways, assigning different names to describe them, mostly to indicate people's involvement with wine (Neuninger et al. 2016), and their subjective and/or objective knowledge of wine (Kelley, Hyde, and Bruwer 2015; Hussain, Cholette, and Castaldi 2007), their level of attention to luxury brands (Hall 2016) and lifestyle differences (Kelley, Hyde, and Bruwer 2015; Bruwer and Li 2007).

The complexity associated with wine choices

The characteristics of wine can be judged only after the purchase unless an opportunity for wine tasting exists. It is rather intimidating to select wine because of the variety of wines that are available on the market and because so much depends on the intrinsic properties of wine that the naked eye cannot see. Therefore, wine is described as 'an information-intensive experience product' (Bruwer, Saliba, and Miller 2011), where consumers have to make trade-offs between selected desirable attributes of different wines (Novemsky et al. 2007). Any doubt about the suitability of wine may increase consumers' risk perception. This could constitute a combination of functional, social, financial, and physical risk caused by the likelihood that one's choice of wine may cause social embarrassment, create concern about affordability, and/or hesitation concerning the alcohol content or chemical composition of the wine (Outreville and Desrochers 2016). The best way to counteract any of these concerns would be to learn more about the different types and attributes of wine.

Consumers' wine knowledge

Elements of wine knowledge

Although wine preference is very personal, pertinent quality indicators are used to evaluate and compare wine. Wine is firstly distinguished in terms of the type of wine, namely red, white, rosé, sparkling, and liquored wine (Kallas, Escobar, and Gil 2013). Thereafter, intrinsic and extrinsic product-related characteristics of the wine are distinguished (Charters and Pettigrew 2007).

The intrinsic attributes of wine represent all the objective quality indicators that culminate as a consequence of the wine production processes, such as the grape variety, the vintage, the alcohol content by volume, and the style and taste of the wine (Atkin and Johnson 2010). This excludes any external cues on the bottle, such as the closure, brand name, and labeling. The extrinsic attributes are more tangible and easier to judge and compare but are subjective quality indicators, such as the brand name, country, or region of origin, price, packaging, labeling, any awards or accolades that a wine may have received in the past, promotional material, the history and reputation of the winemaker, and even the shelf position and aesthetics (Barber and Almanza 2007; Kallas, Escobar, and Gil 2013). When faced with a complex wine purchase decision, the extrinsic cues – for example, the brand name, evidence of the wine region, label design, or price – mostly serve as shortcuts to identify the preferred wine (Atkin and Johnson 2010; Bruwer and Buller 2012; Velikova, Charters, and Cogan-Marie 2016).

Types of wine knowledge

Knowledge, per se, is multidimensional and is defined in several ways (Alba and Hutchinson 2000). Knowledge is acquired through exposure to factual information and personal experience, whereby consumers learn how to use and apply what they know; for example, how to correctly serve different types of wine. *Explicit knowledge* refers to the factual information that someone possesses and that can be communicated verbally. *Tacit knowledge*, on the contrary, is more difficult to communicate and refers to a person's 'know-how' in a specific situation or concerning a specific product. It is demonstrated through a person's behavior, use, and consumption practices (Marx-Pienaar 2015), such as to confidently choose a wine that would be appropriate for the occasion and how to serve the wine. Consumers can further their knowledge about wine through increased exposure and experience with it, conducting actual information searches, and exposure to promotional material, such as wine promotions, and by attending wine tastings. Increased knowledge about wine will enable a person to differentiate between the nuances of different wines and will reduce confusion when selecting wine (Bruwer and Buller 2012; Taylor, Dodd, and Barber 2008).

A person's wine knowledge can also be objective or subjective in nature (Dodd et al. 2005). *Objective wine knowledge* refers to the amount of measurable factual information that a person possesses (Dodd et al. 2005; Thomas and Pickering 2003). This type of knowledge usually increases over time with increased exposure and experience (Vigar-Ellis, Pitt, and Caruana 2015; Robson et al. 2014). Generally, consumers with higher levels of objective knowledge are more inclined to try out new wines (Vigar-Ellis, Pitt, and Caruana 2015), typically consult more written information sources, and rely on intrinsic cues when selecting wine (Barber, Taylor, and Strick 2010; Dodd et al. 2005; Phau and Suntornnond 2006) due to an ability to confidently filter and distinguish relevant product attributes (Kardes, Kim, and Lim 2001; Wirtz and Mattila 2003). *Subjective wine knowledge* refers to a person's 'self-assessed' or 'self-perceived' knowledge, which indicates what consumers think they know about wine (Alba and Hutchinson, 2000; Teagle, Mueller, and Lockshin 2010). Increased experience with wine is more likely to boost consumers' subjective knowledge, rather than their objective wine knowledge (Barber, Almanza, and Dodd 2008). Consumers' subjective wine knowledge largely determines their wine purchase behavior (Lockshin, Jarvis, d'Hauteville, and Perrouty 2006; Mueller et al. 2010) and even frequency of wine consumption (Barber, Almanza, and Dodd 2008; Vigar-Ellis, Pitt, and Caruana 2015). Consumers are inclined to overstate their subjective wine knowledge, thinking that they know more than what they do (objective wine knowledge) (Veale 2008).

In reality, consumers' subjective and objective knowledge are entangled. A consumer with limited objective knowledge about wine may think that he/she knows more (higher subjective knowledge) (Dodd et al. 2005). A consumer may even have high levels of both subjective and objective knowledge or any combination of the two (Veale 2008). Both subjective and objective knowledge influence the information sources that consumers will consult for personal or social consumption (Barber and Taylor 2013; Barber, Taylor, and Strick 2010).

Choosing wine at off-premise and on-premise sales locations

The context, location, or situation in which wine is purchased and/or consumed influences consumers' risk perception and may complicate their decision behavior (Bruwer, Saliba, and Miller 2011). Two distinct wine purchasing locations can be distinguished, namely off-premise- and on-premise locations.

Off-premise locations refer to places such as retailers and supermarkets, liquor stores, and wine specialty stores. In South Africa, the largest percentage of wine is sold at off-premise locations, specifically in supermarkets (Wine in South Africa 2020), indicating that wine has become part of many households' routine grocery shopping. This is, however, less prestigious than to shop at a specialty store that stocks a more impressive variety of wines and where sales personnel are better trained to assist. Psychologically, it is less intimidating to select and purchase wine at off-premise locations where consumers are free to consult tangible product cues such as labels, brand names, and the available marketing material in their own time, and where sales personnel can be consulted (Lockshin and Corsi 2012).

On-premise locations refer to contexts where wine is purchased and consumed on site, such as in restaurants (Bruwer, Saliba, and Miller 2011), where the presence of others during the wine selection process could be intimidating for the less experienced (Lockshin and Corsi 2012). What makes matters worse is that available product information in the locations is generally limited to whatever is printed on the menu (Bruwer, Chrysochou, and Lesschaeve 2017a; Bruwer, Saliba, and Miller 2011) unless the waiter or sommelier makes a recommendation (Lockshin and Corsi 2012). An easy solution is mostly to opt for well-known brand names rather than to experiment (Olsen, Thompson, and Clarke 2003). Wine is often consumed on-premise, as part of a learning experience, and is then purchased off-premise (Lockshin and Corsi 2012).

Consumers' wine self-confidence (WSC)

An understanding of consumers' WSC is crucial in terms of an understanding of consumers' wine purchase behavior and wine consumption (Barber, Ismail, and Taylor 2007). Any feeling of insecurity when selecting wine signifies a lack of WSC. This is generally due to lower levels of subjective knowledge; thus, the perception that one lacks the relevant knowledge to deal with the alternatives that are available to choose from. Increased WSC decreases anxiety when selecting wine and inspires consumers to try out new and different wines (Atkin and Johnson 2010; Olsen, Thompson, and Clarke 2003).

An explication of consumer self-confidence (CSC)

Theoretically, consumers' self-confidence (CSC) indicates an individual's subjective evaluation of his/her own ability to perform a particular task (Clark, Goldsmith, and Goldsmith 2008) and to act with authority in the marketplace (Bearden, Hardesty, and Rose 2001). The need to understand consumers' self-confidence in the marketplace was initially explored by Bearden, Hardesty, and Rose (2001), who defined the construct and developed a validated measurement scale that would be fitting to measure the construct. Their encompassing study eventually proposed a six-dimensional scale that contained 31 scale items that assimilated under two higher-order dimensions, namely (1) *decision-making* (DM) and (2) *protection* (PROT), that were further delineated:

Decision-making confidence (DM): Because a consumer's confidence is challenged during different stages of the decision process, five underlying dimensions of DM are distinguished (Bearden, Hardesty, and Rose 2001; Olsen, Thompson, and Clarke 2003).

- Information acquisition (IA) entails a consumer's ability to confidently evaluate the suitability of a particular product and to acquire additional information if needed (Bearden, Hardesty, and Rose 2001; Loibl et al. 2009) to reduce uncertainty (Barber, Ismail, and Taylor 2007). When an abundance of information exists, as with wine, a decision-making process can be simplified by reverting to a few selected cues (Chaney 2000), such as a trusted brand name, or personal information and recommendations obtained from acquaintances or sales personnel (Mueller et al. 2010; Resnick 2008). Experienced consumers inevitably rely on personal experiences (Barber 2009; Bearden, Hardesty, and Rose 2001; Lockshin and Hall 2003). Wine consumers are generally interested in primary product attributes such as the origin of the wine, price, quality indicators such as previous accolades, taste, vintage, brand name, and grape variety. This information can mostly be accessed from wine labels (Mueller et al. 2010; Tang, Tchetchik, and Cohen 2015), which often represent the only information source that wine consumers rely on. This elevates the importance of wine labels beyond their aesthetic value and distinctiveness (Barber, Ismail, and Taylor 2007).
- Information processing (IP) encompasses a consumer's ability to appropriately apply an array of information (Bearden, Hardesty, and Rose 2001). This dimension of CSC was omitted in the adapted scale of Loibl et al. (2009), because they argued that consumers would anyway restrict information sources to those that they are capable of using.

- Consideration-set formation (CSF) describes a consumer's ability to distinguish the most preferred product alternatives and knows where to procure them (Bearden, Hardesty, and Rose 2001). Because many wines are so similar and consumers do not always have clear preferences, it may be rather difficult to make trade-offs between the attributes of the different wines (Novemsky et al. 2007). Wine experts are generally better equipped to do so (Bearden, Hardesty, and Rose 2001). In the final adapted WSC scale of McClung, Freeman, and Malone (2015), this dimension was excluded because they argued that the associated complexity of a continuously expanding range of wine brands is bound to exceed consumers' comprehension in the future.
- Personal outcomes (PO) reflect the influence of consumers' purchase decisions on their personal experience, for example, the feeling of personal satisfaction, regret, and even risk perception (Bearden, Hardesty, and Rose 2001; Cho, Bonn, and Kang 2014; Erasmus, Donoghue, and Fletcher 2015).
- Social outcomes (SO) refer to a person's self-belief that a particular product decision will result in favorable outcomes, for example, generating compliments (Bearden, Hardesty, and Rose 2001; Erasmus, Donoghue, and Fletcher 2015). Negative feedback may, however, cause social embarrassment.

Protection (PROT): Two dimensions are distinguished to indicate that CSC also entails the ability to confidently protect oneself during the purchase situation (Bearden, Hardesty, and Rose 2001; Olsen, Thompson, and Clarke 2003).

- Persuasion knowledge (PK) reflects a consumer's perceived ability to apply marketing and promotional tactics to his/her personal advantage and to oppose those they are not interested in (Bearden, Hardesty, and Rose 2001; Erasmus, Donoghue, and Fletcher 2015).
- Marketplace interfaces (MI) may be more intricate, in that it indicates a consumer's ability to take a stand, to voice concerns when needed, and to take action in a situation where one's consumer rights have been infringed (Bearden, Hardesty, and Rose 2001; Erasmus, Donoghue, and Fletcher 2015).

Consumers' wine self-confidence (WSC)

The CSC scale of Bearden, Hardesty, and Rose (2001) was adapted by Olsen, Thompson, and Clarke (2003) to explore consumers' WSC. They reformulated Bearden's scale items to refer specifically to wine purchases and designed a WSC scale including 27 scale items. In 2015, McClung, Freeman, and Malone adapted this scale to focus on wine purchasing strategies. The scale was slightly adapted in 2018 (Van Wyk 2018) for a South African context to compare consumers' WSC in on- and off-premise locations. This endeavor is briefly described in the following section.

An exploration of South African consumers' WSC in different sales locations

Methodology

A survey that involved 690 regular wine consumers was conducted in South Africa in 2017 (Van Wyk 2018). The research aimed to explore wine consumers' wine purchasing habits, their knowledge of wine, but particularly their WSC in different locations. Respondents were 21 years of age and older (57.4% males; 42.6% females), and the income levels within the sample were well distributed (29.5% lower-, 38.2%, middle-, and 32.3% upper-income consumers), which made it possible to explore possible income level differences.

Results

Consumers' wine purchases: Due to the study's focus on wine consumers' WSC at on- and off-premise locations, respondents also had to describe their wine purchases. The majority of the sample (64.2%) either purchased wine monthly or occasionally; 22.2% purchased wine once or twice per week; while daily purchases were the exception (0.7%). The 2.9% who indicated that they had never purchased wine themselves, were indeed wine drinkers. Similar to other studies that had indicated that wine purchasing had become part of households' grocery shopping (Bruwer and Huang 2012; Forbes 2012; Velikova, Murova, and Dodd 2013), the majority of the sample purchased wine at a retailer, such as supermarkets and retail wholesalers, while approximately one-third of the sample patronized liquor stores. Understandably, almost 90% seldom purchased wine at specialty stores or wineries, the preferred shopping destination of wine connoisseurs (Ndanga, Louw, and Van Rooyen 2008), as these outlets are not always accessible. This study was conducted before the COVID-19 pandemic, which might explain why more than 80% of the respondents had never purchased wine online before.

Objective knowledge test: The self-developed test, which was limited to wine terminology and relevant skills such as the serving of wine, required true/false responses as used previously (Bruwer 2014; Bruwer and Buller 2012; Mueller, Francis, and Lockshin 2008). The sample's mean score across the ten items for this test (M = 6.89) revealed that their objective knowledge was average, which concurred with the findings of Barber (2009) and Bruwer and Buller (2012). Many of the respondents were unable to discriminate between white- and red wine cultivars and were uncertain about wine's alcohol content. As expected, higher-income respondents, who are more likely to choose from a wider range of wines, performed significantly better in the objective knowledge test.

Subjective knowledge test: Questions on food and wine pairing, wine service, wine quality, and wine selection were included. Results indicated respondents' subjective judgment of their wine knowledge, and thus, their perception of their wine knowledge compared to that of others they know. This revealed that 33.04% viewed their knowledge as limited; almost half of the sample (47.54%) perceived their knowledge of wine to be average, while near one in five (19.42%) thought they were well informed. A pitfall of any subjective measurement is that respondents have to compare themselves to others they know. One's own judgment may therefore be inflated if your peers are poorly informed, and vice versa.

A statistical comparison of respondents' scores for the objective and subjective knowledge tests revealed that respondents underestimated their knowledge of wine.

Consumers' wine self-confidence (WSC): The WSC test for off-premise and on-premise locations was based on research by Olsen, Thompson, and Clarke (2003) that was adapted from the original CSC scale of Bearden, Hardesty, and Rose (2001), for use in a wine context (WSC). Bearden's adapted CSC-scale had been used successfully in a South African context in 2015 by Erasmus, Donoghue, and Fletcher (2015), and by McClung, Freeman, and Malone (2015) in an American context, to measure consumers' WSC. The following items were included to measure two primary dimensions of consumers' WSC, namely decision-making confidence (comprising four underlying dimensions), and protection (comprising two underlying dimensions).

Decision-Making
Information acquisition (IA)

1 I know exactly where to find the information I need before making a wine purchase.
2 I know how to search for information I might need before making a wine purchase.
3 I am confident in my ability to do an investigation about wine before a purchase.
4 I know the right questions to ask when shopping for wine/ordering wine.
5 I have the skills required to obtain needed information before making a wine purchase.

Consideration-Set Formation (CSF)

6 I am confident in my ability to recognize a good brand of wine worth considering.
7 I can tell which brands of wine will meet my expectations.
8 I trust my own judgment when deciding which brands of wine to consider.
9 I know where to find the best stores to purchase wine.
10 I can easily limit my focus on a few good brands of wine when making a decision.

Personal Outcomes (PO)

11 I often have doubts about the wine purchase decisions I make.
12 I am frequently concerned about which wine to buy.
13 I often wonder if I have made the right wine selection.
14 I never seem to buy the right wine for me.
15 Too often, the wine I buy is not satisfying.

Social Outcomes (SO)

16 My friends are impressed with my ability to pair wine and food.
17 I impress people with the wine purchases I make.
18 I can choose a good wine for an occasion.
19 I receive compliments on my ability to distinguish between wines.

Protection
Persuasion Knowledge (PK)

20 I know when a wine salesperson or waiter is trying to pressure me to buy a particular wine.
21 I can see through the sales gimmicks used to get people to buy wine.
22 I can use sales gimmicks to my advantage.

Marketplace Interfaces (MI)

23 I am unwilling to complain to a manager/waiter about poor service when purchasing wine.
24 I am hesitant to ask for the advice of an expert in a store or restaurant about wine.

Consumers' WSC at different locations: The original Wine Self-Confidence Scale (Olsen, Thompson, and Clarke 2003) was reformulated to refer to specific purchase locations, which were investigated in two separate sections, with 24 similar questions each, framed to specify the off-premise or on-premise context, using five-increment Likert-type agreement scales. Because the original scale was adapted, exploratory factor analysis was used to extract the relevant dimensions of WSC in the South African context, for each purchase location. Three factors, rather than the five proposed in the original scale of Bearden, Hardesty, and Rose (2001) and Olsen, Thompson, and Clarke (2003), were extracted for each location, although only one of the factors (dimensions) concurred across the two locations.

The dimensions for consumers' WSC at *off-premise locations*, are:

- *Persuasion Knowledge*, indicating wine consumers' persuasion confidence when selecting wine, merged items from the original scale's Persuasion Knowledge and Consideration-Set Formation dimensions. Despite being the strongest off-premise WSC dimension (M = 3.69; Maximum = 5), consumers' persuasive competency was still only above average.
- *Decision Confidence* merged items related to the original dimensions, Social Outcomes, Information Acquisition, and Consideration-Set Formation, and accrued a mean score (M = 3.34; Maximum = 5), suggesting that wine consumers' decision confidence in this context is slightly above average.
- *Apprehension* puts the focus on wine consumers' personal limitations when selecting wine, assimilating items relating to Personal Outcomes and Marketplace Interfaces from the original scale. The mean (M = 2.13; Maximum = 5) suggests that wine consumers are only slightly deterred by the challenge faced in off-premise locations.

The dimensions for consumers' WSC at *on-premise locations* are:

- *Proficiency* merged items from the original dimensions Persuasion Knowledge (PROT), Consideration-Set Formation, Information Acquisition, and Social Outcomes (DM). Although the strongest of the three dimensions in this context (M = 3.68; Maximum = 5), wine consumers' confidence to choose wine on-premise and knowing where to look for information are merely above average.
- *Social Outcome* reflects the social implications of selecting wine on-premise, admitting related social implications when choosing wine in the presence of others. This is the second-strongest dimension (M = 3.12; Maximum = 5) and moderately strong.
- *Apprehension* combined items from Personal Outcomes (DM) and Marketplace Interface (PROT), indicating slight hesitation (M = 2.08; Maximum = 5) when selecting wine on-premise. This is the only dimension that concurs with those at off-premise locations.

The strongest dimension of consumers' WSC at off-premise locations, is Persuasion Knowledge. Although Decision Confidence is slightly weaker, both dimensions are moderately strong in terms of their contribution towards consumers' WSC. The relevance of Apprehension in influencing consumers' overall WSC at off-site locations should not be negated. Based on its relative strength compared to the other two dimensions, it indisputably dampens consumers' WSC at off-site locations.

The relative strength of the dimension Proficiency at on-site locations is highly likely attributed to consumers' subjective knowledge of wine. However, their eventual WSC when choosing wine in a public domain is jeopardized by Social Outcomes, which is only moderately strong and is weakened by Apprehension. The strength of this dimension of WSC is slightly below average, but its relative strength, compared to the other dimensions that are supposed to reveal consumers' competence, indicates the relevance of anxiety, uneasiness, and hesitation.

The research also explored possible income differences in consumers' WSC, anticipating that perceptions of affordability would influence consumers' exposure to different wines and their wine consumption behavior that are crucial in learning about wine. On off-premise locations, the highest income consumers' Decision Confidence and Persuasion were significantly higher than that of middle-income consumers, who were again significantly more confident than low-income consumers. Apprehension (doubtfulness) was significantly higher among middle- than upper-income consumers concerning wine choices. For on-premise locations, all three dimensions of WSC, the two higher income categories seem significantly more confident compared to the low-income group concerning Proficiency, Social Outcomes, and Apprehension. Therefore, income level exerts a significant influence on consumers' WSC, in boosting their ability to negotiate wine choices, to contemplate it, and to reduce hesitation. In the hospitality and tourism industries, these might be detrimental in terms of the attraction or dissuasion of their target markets.

Conclusion

The study revealed pertinent differences in how consumers' WSC is structured at off- and on-premise locations. This is worth noting in our understanding of consumers' behavior in different sales contexts and how wine consumers can be supported to more confidently conclude wine purchase decisions. In both locations, an element of Apprehension, signifying consumers' hesitation to decide, exists. Even though this feeling of intimidation and anxiety seems only moderately strong, its strength nevertheless indicates that it is a stumbling block in both locations. Consumers' uncertainty could be explained in terms of their performance in the subjective knowledge test: Around 52% perceived their knowledge as being average, and 17% considered themselves very knowledgeable about wine. It is therefore not surprising that wine consumers are not at ease in any of the contexts. In addition, wines and brands are continuously expanding, which further complicates matters. Consumers' subjective knowledge of wine may also explain their above-average Decision Confidence and Proficiency at off-premise and on-premise locations. The outcomes of the very basic objective knowledge test ($M = 6.89$) indicates that wine consumers still have much to learn about wine. It also became clear that the social implications of a wine decision may indeed jeopardize consumers' WSC at more intimidating on-premise locations. It is worthy to note that the social dimension of WSC did not emerge as a separate entity for off-premise locations, while it was indisputably identified for on-premise locations, amid indication that feelings of Apprehension are a reality. The latter was not very strong in any of the locations, which makes sense as consumers seem under the impression that their wine knowledge is reasonably good. Therefore, despite respondents' thinking that they know a fair amount about wine and that they scored reasonably well in the objective knowledge test, their social confidence is merely moderate, which negatively influences their overall WSC score. The research also indicated that higher-income consumers are indeed significantly more confident concerning wine choices, irrespective of the purchase location, and feelings of doubt and anxiety (apprehension) are significantly weaker, although present.

Academic implications

Empirical evidence is provided that consumers' WSC differs in terms of how it is structured in different locations, with confirmation that on-premise locations undeniably exert an amount of social pressure. This research indicates that consumers' decision confidence is product specific.

Therefore, CSC could be explored in alternative purchase situations as well, to gain a better understanding of consumers' decision processes and what inputs are needed to guide more satisfactory purchase outcomes.

Marketing implications

Concerning wine sales and wine tourism, which is a very important element of the hospitality and tourism industry, it should be noted that consumers' WSC can be boosted only through increased objective knowledge about wine. With expectations that many exciting changes and additions to existing wine ranges are still to come, the wine industry should pre-empt wine consumers' underlying uncertainties to ensure that they are attracted to new developments rather than be overwhelmed by what is in the pipeline for them to enjoy. The question, however, is how to present wine information without overloading consumers with facts and figures. The complicated issue of communicating information about wine, broadly yet informally, is a challenge that marketers should pursue. On-premise locations might benefit from offering wine tastings, even in a restaurant, while specific information at the point of sale will reduce perceived risks and possible discomfort when making a wine purchase decision. It will be extremely counterproductive for the wine industry if wine consumers simply remain brand loyal to overcome their own uncertainties, as wine industries' enthusiastic initiatives for the future will be in vain.

Case Study 14.1 A wine tour: pleasure or agony?

James and Jill are young professionals who decided to explore the wine route in Stellenbosch, South Africa, to celebrate the opportunity to go on vacation after the very strict COVID-19 lockdown period. They join a small tour group for a day trip to different wine farms. The first stop entails a wine tasting and chocolate pairing that Jill is particularly excited about. Considering themselves to be fairly inexperienced with wine, this is an opportunity they want to optimize to learn more about wine. Seated on their own at the first winery, the couple enjoys the **beautiful** view and the wine-tasting experience, discussing their preferences. They purchase two bottles of wine, one white and one red, planning to purchase more wine from other wineries on the route. At the second stop, another tour bus joins their company and they are guided through the cellars before being seated around a large rectangular table in a tasting room. An enthusiastic gentleman across the table starts making conversation and tells them about his love of wine, his love for the Cape Winelands, and about his world travels. Another guest joins the conversation and explains that he is a chef who specifically joined the bus tour to select wines for his restaurant. The conversation between the other two guests becomes rather noisy, interrupted by laughter, while they share their experiences and preferences about wine. Although the young couple forms part of the conversation, they mostly listen to the gentlemen's stories. The groups have the opportunity to purchase wine from the winery shop. While the world traveler and the chef place orders for cases of wine that need to be delivered, the young couple decides to restrict their purchases to two single-bottle gift packs, explaining to their travel companions that they have already purchased wine at their previous stop. The bus departs with James and Jill to their third stop to enjoy a cheese-and-wine pairing for lunch, leaving the other group behind. At the third wine tasting, the couple rushes towards a two-seater table. After lunch, they have a brief chat with the salesperson in the retail section, and they purchase six bottles of the cellar's award-winning Chenin Blanc.

Questions:

1 Which stop was probably the most enjoyable for the couple, and why?
2 At which stop did they learn more about wine? Why?
3 How was the couple's wine self-confidence affected at the three wine tastings? Explain.
4 Why was the young couple more willing to purchase wine at the first and the third stop than the second?
5 Why did the young couple probably prefer to sit alone at the third wine cellar rather than join another conversation?
6 To what would you attribute the wine choices of the chef and the world traveler at the second cellar?
7 What could the young couple have done at the second stop to optimize the situation for their own benefit?
8 What does the young couple's choice of wine at the third stop indicate?

References

Alba, J. W., and Hutchinson, J. W. (2000). Knowledge calibration: What consumers know and what they think they know. *Journal of Consumer Research*, 27(2): 123–156.

Anderson, K., Nelgen, S., and Pinilla, V. (2017). *Global Wine Markets, 1860 to 2016: A Statistical Compendium*. Adelaide: University of Adelaide Press. https://doi.org/10.20851/global-wine-markets

Atkin, T., and Johnson, R. (2010). Appellation as an indicator of quality. *International Journal of Wine Business Research*, 22(1): 42–61.

Barber, N. (2009). Wine consumers information search: Gender differences and implications for the hospitality industry. *Tourism and Hospitality Research*, 9(3): 250–269.

Barber, N., Almanza, B., and Dodd, T. (2008). Relationship of wine consumers' self-confidence, product involvement, and packaging cues. *Journal of Foodservice Business Research*, 11(1): 45–64.

Barber, N., and Almanza, B. A. (2007). Influence of wine packaging on consumers' decision to purchase. *Journal of Foodservice Business Research*, 9(4): 83–98.

Barber, N., Ismail, J., and Taylor, D. C. (2007). Label fluency and consumer self-confidence. *Journal of Wine Research*, 18(2): 73–85.

Barber, N., and Taylor, D. C. (2013). Experimental approach to assessing actual wine purchase behavior. *International Journal of Wine Business Research*, 25(3): 203–226.

Barber, N., Taylor, D. C., and Strick, S. (2010). Selective marketing to environmentally concerned wine consumers: a case for location, gender, and age. *Journal of Consumer Marketing*, 27(1): 64–75.

Bearden, W. O., Hardesty, D. M., and Rose, R. L. (2001). Consumer self-confidence: Refinements in conceptualization and measurement. *Journal of Consumer Research*, 28(1): 121–134.

Bruwer, J., and Buller, C. (2012). Country-of-origin (COO) brand preferences and associated knowledge levels of Japanese wine consumers. *Journal of Product & Brand Management*, 21(5): 307–316.

Bruwer, J., Chrysochou, P., and Lesschaeve, I. (2017a). Consumer involvement and knowledge influence on wine choice cue utilisation. *British Food Journal*, 119(4): 830–844.

Bruwer, J., and Huang, J. (2012). Wine product involvement and consumers' BYOB behaviour in the South Australian on-premise market. *Asia Pacific Journal of Marketing and Logistics*, 24(3): 461–481.

Bruwer, J., and Li, E. (2007). Wine-related lifestyle (WRL) market segmentation: demographic and behavioural factors. *Journal of Wine Research*, 18(1): 19–34.

Bruwer, J., Roediger, B., and Herbst, F. (2017b). Domain-specific market segmentation: a wine-related lifestyle (WRL) approach. *Asia Pacific Journal of Marketing and Logistics*, 29(1): 4–26.
Bruwer, J., Saliba, A., and Miller, B. (2011). Consumer behaviour and sensory preference differences: Implications for wine product marketing. *Journal of Consumer Marketing*, 28(1): 5–18.
Bruwer, N. M. (2014). *Perceptions of Wine*. Diploma of Cape Wine Master. Cape Wine Academy.
Chaney, I. M. (2000). External search effort for wine. *International Journal of Wine Marketing*, 12(2): 5–21.
Chang, K. J., Thach, M. W. L., and Olsen, J. (2016). Wine and health perceptions: exploring the impact of gender, age, and ethnicity on consumer perceptions of wine and health. *Wine Economics and Policy*, 5(2): 105–113.
Charters, S., and Pettigrew, S. (2007). The dimensions of wine quality. *Food Quality and Preference*, 18(7): 997–1007.
Cho, M., Bonn, M. A., and Kang, S. (2014). Wine attributes, perceived risk, and online wine repurchase intention: the cross-level interaction effects of website quality. *International Journal of Hospitality Management*, 43: 108–120.
Clark, R. A., Goldsmith, R. E., and Goldsmith, E. B. (2008). Market mavenism and consumer self-confidence. *Journal of Consumer Behaviour*, 7(3): 239–248.
Devonish, H. (2011). 'Wine,' Women and Song: The More Things Change ... *Sexuality & Culture*, 15, 332–344. https://doi.org/10.1007/s12119-011-9108-z
Dodd, T. H., Laverie, D. A., Wilcox, J. F., and Duhan, D. F. (2005). Differential effects of experience, subjective knowledge, and objective knowledge on sources of information used in consumer wine purchasing. *Journal of Hospitality and Tourism Research*, 29(1): 3–19.
Erasmus, A. C., Donoghue, S., and Fletcher, L. (2015). Proposing persuasion knowledge as the antecedent of consumers' self-confidence during a complex purchase decision. *Journal of Marketing Management*, 3(2): 27–38.
Forbes, S. L. (2012). The influence of gender on wine purchasing and consumption: an exploratory study across four nations. *International Journal of Wine Business Research*, 24(2): 146–159.
Hall, D. (2016). Exploring wine knowledge, aesthetics, and ephemerality: clustering consumers. *International Journal of Wine Business Research*, 28(2): 134–153.
Hussain, M., Cholette, S., and Castaldi, R. M. (2007). Determinants of wine consumption of US consumers: an econometric analysis. *International Journal of Wine Business Research*, 19(1): 49–62.
Kallas, Z., Escobar, C., and Gil, J. M. (2013). Analysis of consumers' preferences for a special-occasion red wine: a dual response choice experiment approach. *Food Quality and Preference*, 30(2): 156–168.
Kardes, F. R., Kim, J., and Lim, J.-S. (2001). Consumer expertise and the perceived diagnosticity of inference. *Advances in Consumer Research*, 19: 409–410.
Kelley, K., Hyde, J., and Bruwer, J. (2015). US wine consumer preferences for bottle characteristics, back label extrinsic cues and wine composition: a conjoint analysis. *Asia Pacific Journal of Marketing and Logistics*, 27(4): 516–534.
Lockshin, L., and Corsi, A. M. (2012). Consumer behaviour for wine 2.0: a review since 2003 and future directions. *Wine Economics and Policy*, 1(1): 2–23.
Lockshin, L., and Halstead, L. (2005). *A Comparison of Australian and Canadian Wine Buyers Using Discrete Choice Analysis*. Sonoma State University-School of Business & Economics Wine Business Program.
Lockshin, L., Jarvis, W., d'Hauteville, F., and Perrouty, J.-P. (2006). Using simulations from discrete choice experiments to measure consumer sensitivity to brand, region, price, and awards in wine choice. *Food Quality and Preference* 17(3): 166–178.
Loibl, C., Cho, S. H., Diekmann, F., and Batte, M. T. (2009). Consumer self-confidence in searching for information. *Journal of Consumer Affairs*, 43(1): 26–55.
Loots, J. (2017). *State of the SA Wine Industry – 2017*. http://vinpro.co.za/state-sa-wine-industry/
Marx-Pienaar, N. J. M. M. (2015). *Urban South African Consumers' Ability to Discursively Reflect on the Sustainability of Their Households' Fresh Produce Consumption and Waste Practices amidst Global Concerns about Climate Change* [Doctoral thesis]. University of Pretoria, Pretoria, South Africa. http://repository.up.ac.za/dspace/handle/2263/46277?show=full
McClung, S. R., Freeman, K. A., and Malone, D. (2015). Consumer self-confidence in wine consumers: the role of knowledge-based factors. *Journal of Promotion Management*, 21(5): 601–613.
Mueller, S., Francis, L., and Lockshin, L. (2008). *The Relationship Between Wine Liking, Subjective, and Objective Wine Knowledge: Does It Matter Who Is Your 'Consumer' Sample?* AWBR Academy of Wine Business Research.
Mueller, S., Lockshin, L., Saltman, Y., and Blanford, J. (2010). Message on a bottle: the relative influence of wine back label information on wine choice. *Food Quality and Preference*, 21(1): 22–32.

Ndanga, L. Z. B., Louw, A., and Van Rooyen, J. (2008). *Increasing Domestic Consumption of South African Wines: Exploring the market potential of the 'Black Diamonds.'* Working Paper. Pretoria, South Africa: Department of Agricultural Economics, Extension and Rural Development, University of Pretoria.

Neuninger, R., Mather, D., Duncan, T., and Aitken, R. (2016). Questioning the way that we measure consumers' product involvement levels: how wine awards exposed differing involvement levels. *Universal Journal of Management*, 4(11): 615–620.

Novemsky, N., Dhar, R., Schwarz, N., and Simonson, I. (2007). Preference fluency in choice. *Journal of Marketing Research*, 44(3): 347–356.

Olsen, J. E., Thompson, K. J., and Clarke, T. (2003). Consumer self-confidence in wine purchases. *International Journal of Wine Marketing*, 15(3): 40–51.

Outreville, J. F., and Desrochers, J. (2016). Perceived risk: an experimental investigation of consumer behavior when buying wine. *Journal of Consumer Behaviour*, 15(6): 549–559.

Phau, I., and Suntornnond, V. (2006). Dimensions of consumer knowledge and its impacts on country of origin effects among Australian consumers: a case of fast-consuming product. *Journal of Consumer Marketing*, 23(1): 34–42.

Resnick, E. (2008). *Wine Brands Success Strategies for New Markets, New Consumers, and New Trends*. New York: Palgrave Macmillan.

Robson, K., Plangger, K., Campbell, C., and Pitt, L. (2014). Objective and subjective wine knowledge: evidence from an online study. *Proceedings of the 8th International Conference of the Academy of Wine Business Research*, Geisenheim, June 28–30.

Schäufele, I., and Hamm, U. (2017). Consumers'perceptions, preferences, and willingness-to-pay for wine with sustainability characteristics: a review. *Journal of Cleaner Production*, 147: 379–394.

Sogari, G., Mora, C., and Menozzi, D. (2016). Factors driving sustainable choice: the case of wine. *British Food Journal*, 118(3): 632–646.

Swart, E., and Smit, I. (2009). *The Essential Guide to South African Wines: Terroir and Travel*. Green Point, South Africa: Cheviot Publishing.

Tang, V. C. M., Tchetchik, A., and Cohen, E. (2015). Perception of wine labels by Hong Kong Chinese consumers. *Wine Economics and Policy*, 4(1): 12–21.

Taylor, D. C., Dodd, T. H., and Barber, N. (2008). Impact of wine education on developing knowledge and preferences: an exploratory study. *Journal of Wine Research*, 19(3): 193–207.

Teagle, J., Mueller, S., and Lockshin, L. (2010). *How Do Millennials' Wine Attitudes and Behaviour Differ from Other Generations?* Paper presented at the *5th International Academy of Wine Business Research Conference*, Auckland, New Zealand, February 8–10.

The symbolism of wine through the ages. (2011). http://www.whitewine.co.uk/news/the-symbolism-of-wine-through-the-ages/

Thomas, A., and Pickering, G. (2003). The importance of wine label information. *International Journal of Wine Marketing*, 15(2): 58–74.

Van Wyk, R. (2018). *Wine Consumers' Knowledge of Wine and Their Self-Confidence When Selecting Wine in Different Sales Contexts* [Master's dissertaion]. University of Pretoria. http://hdl.handle.net/2263/68636

Veale, R. (2008). Sensing or knowing? Investigating the influence of knowledge and self-confidence on consumer beliefs regarding the effect of extrinsic cues on wine quality. *International Journal of Wine Business Research*, 20(4): 352–366.

Velikova, N., Charters, S., and Cogan-Marie, L. (2016). *Consumer Knowledge and Preferences for Wine amongst Visitors to Burgundy*. Paper presented at *9th Academy of Wine Business Research Conference*, Adelaide, Australia: 154.

Velikova, N., Murova, O., and Dodd, T. H. (2013). Emerging wine market in the Dominican Republic: consumer market analysis. *Wine Economics and Policy*, 2(2): 76–84.

Vigar-Ellis, D., Pitt, L., and Caruana, A. (2015). Knowledge effects on the exploratory acquisition of wine. *International Journal of Wine Business Research*, 27(2): 84–102.

Vlachvei, A., Notta, O., and Ananiadis, I. (2009). Does advertising matter? An application to the Greek wine industry. *British Food Journal*, 111(7): 686–698. https://doi.org/10.1108/00070700910972378

Wine in South Africa (2020). https://www.euromonitor.com/wine-in-south-africa/report

Wirtz, J., and Mattila, A. S. (2003). The effects of consumer expertise on evoked set size and service loyalty. *Journal of Services Marketing*, 17(7): 649–665.

Yuan, J., Morrison, A., Cai, L. A., Dodd, T., and Linton, S. (2008). Segmenting wine festival visitors using their personal involvement with wine. In Hall, C. M., and Sharples, L. (eds.), *Food and Wine Festivals and Events around the World* (pp. 146–165). Oxon: Routledge.

15

THE EMERGING WINE TOURIST

Perspectives of multicultural, first-time winery visitors

Karen Goodlad and Susan Phillip

Introduction

The Committee on Tourism and Competitiveness (CTC) of the United Nations World Tourism Organization (UNWTO n.d.), 'the United Nations agency responsible for the promotion of responsible, sustainable and universally accessible tourism,' defines wine tourism as 'tourism whose purpose is visiting vineyards, wineries, tasting, consuming and/or purchasing wine, often at or near the source' and categorizes it as a subset of gastronomy tourism. CTC defines gastronomic tourism as 'a type of tourism activity which is characterized by the visitor's experience linked with food and related products and activities while traveling.' According to UNWTO, both gastronomy and wine tourism 'contribute to branding destinations, maintaining and preserving local traditions and diversities, and harnessing and rewarding authenticity' (UNWTO).

Wine tourism marries the production of wine at wineries with activities for tourism, including tasting and purchasing wines, participating in wine-related activities, and enjoying the scenery. In varying degrees, wineries, especially smaller ones, depend on visitors for sales and distribution (Dodd and Bigotte 1997; Yue, Govindasamy, and Kelley 2019). In New York State, the focus of this study and a state made up predominantly of small, family-owned wineries, vineyards attracted 1.43 million visitors for a total of 4.71 million visits, with an economic impact of $1.33 billion in 2019 (Hudson Valley 2020).

The profile of the wine consumer as white, affluent, educated, and mostly male has endured for decades (Folwell and Baritelle 1978). Wine tourism research offers similar findings about wine tourists (Charters and Ali-Knight 2002; Dodd 1995; Dodd and Bigotte 1997; Mintel 2017; Yue, Govindasamy, and Kelley 2019; Gunasekar et al. 2021). Researchers have also acknowledged the importance of younger cohorts in the sustainability of wine tourism for some time (Fountain and Charters 2009). More recent studies on African American and Hispanic wine consumption present information to support the inclusion of younger cohorts in the marketing strategies of the wine industry (Hammond, Sydnor, and Kang 2014; Velikova, Wilkinson, and Harp 2016). Missing in the research is the perspectives of young multiculturals.

Despite a growing interest among African Americans and Hispanics (Hammond, Sydnor, and Kang 2014; Velikova, Wilkinson, and Harp 2016), the wine industry has not shifted its market segmentation to include this market beyond its traditional products to them of mostly

DOI: 10.4324/9781003143628-18

sweet and fortified wines (Gaiter 2020; Worobiec 2020). This study evaluates the perspectives of the first-time winery visit of young, multicultural, urban undergraduates in the Department of Hospitality Management of New York City College of Technology (2021a, 2021b), better known as City Tech. Part of the City University of New York (CUNY), City Tech serves a large, culturally diverse urban population (http://www.citytech.cuny.edu/about-us/) and is credited with propelling its students into the middle class (Leonhardt 2017).

Despite data to support diversifying its markets, the US wine industry continues to exclude people of color in its marketing and has long been criticized for racial bias and persistently operating as an 'exclusive club' (McIntyre 2020). The study aims to fill the gap in the existing body of knowledge on the expectation and behavior intentions of an emerging market that has been ignored by the wine industry. Evidence of the intention of culturally diverse undergraduate hospitality students visiting a winery for the first time is presented and bolsters why the wine industry should take notice of this demographic. The reflections, amassed over seven years, are the outcomes of an assignment that required the students to visit a vineyard and winery in a wine region of their choice. Considering the location of the college, most participants visited wine regions around New York City, with a few choosing to visit urban wineries.

Background

Wine tourism is a 'fast growing and financially lucrative segment across the United States,' attracting affluent and well-traveled visitors (Garibaldi et al. 2017, p. 56). According to the National Association of American Wineries, the major US wine-producing states are: California, primarily the American Viticultural Areas (AVAs) of Napa Valley and Sonoma County; Washington; New York; and Oregon (Wine America n.d.); all have wine tourism offerings. The United States has the world's largest domestic market for wine consumption, and in 2014 domestic visitors made up the majority of its 15 million wine tourists, who tended to be older, affluent couples visiting wineries near to their homes (Mintel 2017).

Mintel's information mirrors the profile of the US wine tourist revealed in other studies in the United States and abroad (Dodd 1995; Mitchell and Hall 2001). Mintel's 2017 study also found that Millennials (born between 1981 and 1996) are drinking more wine than any other age group; that they have overtaken Baby Boomers (born between 1946 and 1964) as the largest generation in the United States (Fry 2020). This presents opportunities for the US wine industry.

The dynamics of the market revealed in previously cited studies are playing out in the US wine industry and, by extension, wine tourism. Boomers have been the wine industry's most important market to date, both maturing in the 1990s. However, returns will decrease as Baby Boomers age and face health concerns. Millennials have been identified as the emerging market to replace the Boomers, but their earning power, lifestyle characteristics, and engagement with wine do not align with the wine industry, as did the Boomers (McMillan 2020).

Notwithstanding the importance of the Baby Boomers, recent data reveals that '43% of the 75 million Millennials in the U.S. identify as African American, Hispanic or Asian' (Nielsen 2018), and that even if slowed, growth in the Hispanic population 'outpaces' that of other groups (Noe-Bustamante, Lopez, and Krogstad 2020). Added to this, a 2019 survey by the Wine Market Council found that wine drinkers in the 21–29 age cohort of its study 'were significantly more ethnically/racially diverse' (Wine Market Council 2019, p. 26). In the United States, the consumer buying power for minorities grew to $2.4 trillion in 2018 (Weeks 2019). These statistics point to the importance of learning about the experiences of young, multicultural wine tourists.

Multicultural consumers' perspective of the wine industry

Demographics alone do not tell the story of a multicultural consumer base; current events lend information as well. In the racial reckoning that followed George Floyd's killing by police in May 2020, wineries and professional associations responded with statements, pledges, and the appointment of diversity directors (McIntyre 2020). The impetus for the wine industry's remedial steps has come from Black wine professionals and enthusiasts who are increasingly vocal about the lack of representation and inclusion and the indignities of racism in the workplaces and at wine events. There are reports of Black professionals being insulted and their knowledge slighted (Asimov 2020). Consumers who expressed knowledge of wines and discernment in making selections beyond what was stereotyped were subject to microaggression and directed to cheaper brands that, presumably, they could better afford. People of color are followed in wine shops, and their expressions of wine knowledge are often met with incredulity (Asimov 2020).

Even without widespread support of the wine industry, minority-owned wine businesses and organizations are dismantling stereotypes, broadening markets and representation in the wine industry. Their numbers as owners, oenologists, directors, and managers are growing, as are organizations formed to market, support, and inspire, among them the Association of African American Vintners and the Mexican American Vintners Association. Latinx and Hispanics, who have provided much of the skilled and unskilled labor in wineries, have been integral to the viticultural history of the United States and are now moving to the forefront of the industry (Asimov 2004).

References and images in the media of multicultural persons drinking wine have helped to popularize wine among a broader audience. For example, the main character in the TV show *Scandal* was often seen drinking red wine, and the Netflix movie *Uncorked* documents a young Black man's journey to become a Master Sommelier while facing adversity both in the workplace and in his personal life (Asimov 2014; Gyarkye 2020). Wine has also been popularized by Black athletes (Mobley 2018) and clubs such as Divas Uncorked Collaborative Consortium, whose events have been featured on Oprah.com (Powell n.d.). The interest of wine enthusiasts is a probable indicator that the demographic group in which they are included would be responsive to marketing for wine and wine tourism. Whatever the basis for the wine industry's reluctance to embrace people of color, buying power cannot be the excuse. Market research and other data provide ample information to support broadening the market outreach to ethnic minorities beyond selling stereotypical products (Gaiter 2020).

Despite these indicators, the image of the wine industry as white, affluent, and male persists, owing to strong marketing associations and the industry's reluctance to embrace multicultural consumers. The study of culturally diverse wine students' first-time visit to a winery takes place against this backdrop.

Literature review

According to Carlsen (2004, p. 6), Australia's first wine tourism conference held in 1998 'set the parameters of wine tourism research' and explored the synergies of wine and tourism, including 'the promotion of wine regions and events through tourism, increased cellar door sales to tourists, added value to regional production and new business opportunities in wine tourism.'

Wine tourists are mainly drawn from general tourists visiting a wine region; consequently, there is 'synergy' between wine and tourism (Carlsen and Dowling 2001, p. 5; Dodd 1995) that involves attractions beyond the winery (Charters and Ali-Knight 2002). Multiple definitions of wine tourism have been offered over the years, stressing either consumer, touristic, or marketing elements (Carlsen and Dowling 2001). Hall et al. (1997, p. 5) offers among the earliest and most cited definitions of wine tourism: 'visitation to vineyards, wineries, wine festivals and wine shows for which grape wine tasting and/or experiencing the attributes of a grape wine region are the prime motivating factors for visitors.' Research by Getz (cited in Getz and Brown 2006, p. 147) proposed three perspectives on wine tourism: 'wine producers, tourism agencies and consumer.' '"Wine tourism" is, simultaneously a form of consumer behavior, a strategy by which destinations develop and market wine-related attractions and imagery, and a marketing opportunity for wineries to educate and to sell their products directly to customers.'

Foundational research focused on 'visitors to the wineries, not wine consumers' (Getz and Brown 2006, p. 148). Wine tourism in Australia and New Zealand, where innovations in wine production and growth in tourism converged in the 1990s and was the subject of early research (Beverland 1998; Bruwer 2003; Carlsen and Dowling 2001; Hall et al. 1997; Mitchell and Hall 2001), as was the United States (Dodd 1995; Dodd and Bigotte 1997; O'Neill and Palmer 2004).

Wine tourism research evolved from focusing on cellar door experiences (Dodd and Bigotte 1997; Mitchell and Hall 2001) to exploring motivation – differentiating the generational experiences and preferences (Fountain and Charters 2009). Alonso, Fraser, and Cohen (2007) found that a younger educated cohort, wanting to learn about wine, have the potential to be loyal wine tourists. Additional studies appeared, and they included research on wine tourism motivations and preference (Charters and Ali-Knight 2002; Getz and Brown 2006) and on service (Carlsen 2011).

Research by Bruwer (2003, p. 433) indicates that wine tasting and wine purchasing were primary motivators for wine tourism and that the 'beauty' and 'setting' of the 'vineyards, wineries, and wines' – the 'winescape' – were also motivations for wine tourism. The significance of the 'winescape' and its pleasurable or hedonic aspects in the wine tourism experience are explored by Rueger-Muck and Bruwer (2019) and by Bruwer and Alant (2009). Quadri-Felitti and Fiore (2012) examine Pine and Gilmore's (1999) Experience Economy construct in relation to wine tourism. Research on wine tourists has been developing along with research on wine tourism (Alebaki and Iakovidou 2010). Wine tourism is a 'marriage' with tourism that includes socializing, relaxing, and learning (Dodd 1995, p. 5) and that involves attractions beyond the winery (Charters and Ali-Knight 2002; Getz and Brown 2006). Research by Charters and Ali-Knight (2002) and by Fountain and Charters (2009) highlights personal development as a motivation of wine tourists.

Wine tourists 'are not a single homogeneous group' (Ali-Knight and Charters 2001, p. 79; Charters and Ali-Knight 2002; Velikova, Wilkinson, and Harp 2016). However, the wine industry has operated under the premise that its market owes its success solely to affluent, educated (Dodd 1995; O'Neill and Palmer 2004), and male consumers (Campbell 2009). Even as this population ages, there is little evidence that the wine industry is seizing opportunities to tap multicultural markets, especially African Americans, leading Hammond, Sydnor, and Kang (2014, p. 137) to state that 'wine may be an inadvertent segregating beverage.' Hammond, Sydnor, and Kang (2014) and Velikova, Wilkinson, and Harp (2016) are among the few academic studies focusing on the opinion, attitudes, and interest related to wine consumption among African Americans and Hispanics, respectively. The profiles of the US wine consumer that emerged from both studies were similar to those of others (Dodd and Bigotte 1997; Wine Market Council 2019). Hammond, Sydnor, and Kang (2014, p. 124) found that African Americans shared the 'demographic characteristics as the target population' of white wine consumers, absent race.

The scope of wine tourism research has been broadened by studies of young winery visitors in Alonso, Fraser, and Cohen (2007) and Stergiou, Airey, and Apostolakis (2018). While researchers have found that the majority of winery visitors are from urban areas (Carlsen 2004), research lacks analysis of the wine tourism experience of multicultural undergraduates, the subjects of this study.

Subjects

Study participants are students in the Department of Hospitality Management, whose mission is to educate 'students for careers in the hospitality industry through foundational knowledge of hospitality operations and experiences that cultivate diverse perspectives, lifelong learning, collaboration, and community engagement' (https://www.citytech.cuny.edu/hospitality/).

Course and assignment

Wine and Beverage Management is a required course in which students are expected to develop a basic understanding of the geographic, historic, economic, and cultural aspects of fermented beverages. Students discuss winemaking and learn professional tasting procedures, including vocabulary particular to beverage management.

The study of wine is a multisensorial experience, involving sight, smell, taste, and touch to make complex connections between viticulture, vinification, and geography (King, Donaldson, and Harry 2012; Parr 2019). Based on studies of educational theorists (Dewey 1938; Kolb and Kolb 2012), place-based learning became a valued aspect of the course (Goodlad and Leonard 2018). Each semester, students are required to visit a winery, brewery, or distillery to identify and analyze the unique qualities of the experience. The objective is for students to communicate their understanding of the factors that affect the taste of wine and their experience at a vineyard and winery.

Methodology

The data are drawn from reflective writing of 179 students between 2013 and 2019. The writings were posted on the OpenLab, the college's proprietary open-source platform for teaching and learning. The writing samples were evaluated, and 59 samples were discarded because they were based on visits to distilleries, breweries, and roasters. One sample included information about both a winery and a brewery. The IBM Watson Natural Language Processing Tool and Tone Analyzer and Grounded Theory methods were used to interpret student perspective on visiting a winery, including transportation, evaluation of customer service, knowledge of the personnel, the visitor experience, and intention to return.

The IBM Watson Natural Language Processing Tool and Tone Analyzer was used for linguistic analysis to detect language associated with emotion in 119 writing samples of winery visits (IBM, Tone Analyzer, Watson 2017). The research process similarly described by High (2012) involved gathering and compiling student writing, sorting each submission by semester, deidentifying the data, then sorting by type of location visited (winery, distillery, brewery, or roaster). The IBM Watson program identified the words, sentences, and paragraphs for representation of five emotions – joy, anger, disgust, sadness, and fear – and produced a value for each. The results demonstrate the overall emotion and the different emotions students expressed.

The Grounded Theory method, introduced in 1967 by Glaser and Strauss (as cited by Strauss and Corbin 1994) to analyze qualitative data, was used in this research to identify patterns in writing samples. Specifically, the authors looked at how student writing revealed the student's expectations, impressions, intent to return, and challenges faced. The author's notes were shared and discussed. The initial findings were categorized into four themes: knowledge of the tasting room staff's contribution to the winescape; the significance of traveling to the region; the students' technical knowledge of the factors that affect the taste of wine; and students' likelihood to recommend or return to a winery. Deidentified quotes were then selected as examples of each theme.

Analysis

Watson tone analyzer, overall emotion, and distinct emotions

The results of the Watson Natural Language Processing Tool and Tone Analyzer revealed a positive sentiment 100% of the time and was consistent whether it was viewed as a whole, by semester, or by individual student. The aggregated value of the emotions was 0.70, indicating emotion is evident. The overall value of emotion for all semesters ranged from 0.63 to 0.75. Figure 15.1 shows the semester-by-semester results. Two semesters' values indicate overall emotion in the 'high possibility' range (0.75–1), while all other semesters provided evidence of emotion (0.5–0.75) (IBM, Tone Analyzer, Watson 2017).

The results of each of the distinct emotions – joy, anger, disgust, sadness, and fear – were aggregated for all student writing from spring 2013 to spring 2019 (see Figure 15.2). Joy is the emotion with the highest result, 0.67. Anger, disgust, sadness, and fear were all detected at low levels, below 0.5 (IBM, Tone Analyzer, Watson 2017).

Wine tourism is strongly associated with emotion (Rueger-Muck and Bruwer 2019; Quadri-Felitti and Fiore 2012), especially for first-time visitors (Bruwer and Alant 2009), and when the visitor anticipates a learning experience (Leri and Theodoridis 2019). Considering the instructions for the assignment, the amount of emotion associated with the student writing

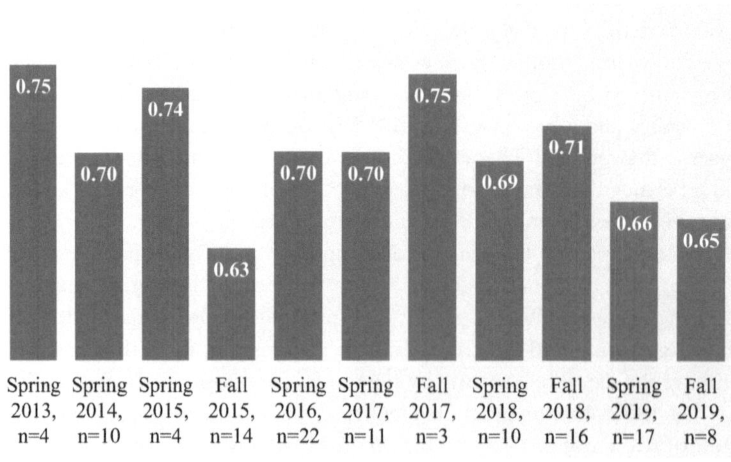

Figure 15.1 IBM Watson Natural Language Processing Tool and Tone Analyzer: overall emotion value.

Source: IBM, Tone Analyzer, Watson (2017)

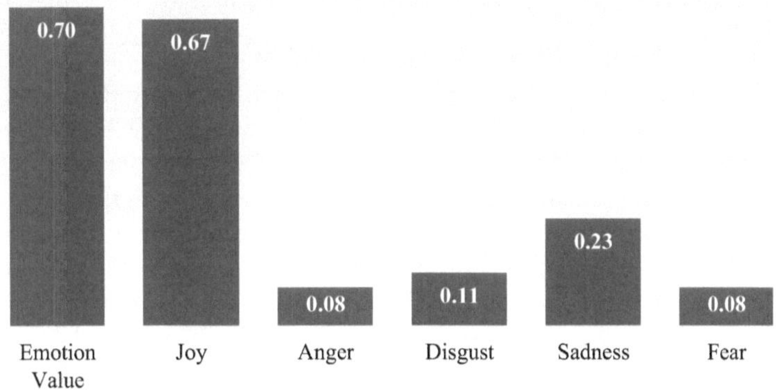

Figure 15.2 IBM Watson Natural Language and Tone Analyzer: overall emotion and five distinct emotions score, winery visits.

Source: IBM, Tone Analyzer, Watson (2017)

is expected; in keeping with those findings, overall, emotion was most evident when students described their expectations. Emotion, mostly those associated with joy, was apparent when describing the winescape, specifically, the knowledge of the tasting room personnel, the physical location of the vineyard and winery, wine tasting, and their recommendation to visit a winery. The emotions of anger, disgust, fear, and sadness were most closely associated with expectations not met.

Writing about expectations was not a requirement of the assignment, yet students expressed what they were feeling as they prepared for their visit. As Fountain and Charters (2009) present, the education component lends itself to expressions of emotion by young first-time winery visitors: 'As I was approaching the winery, I felt eager to learn all about the winery and excited because it was my first time being at one.' 'It was my first time to go to a vineyard in the USA. … I had no experience of tasting wines.… I was excited to go to a vineyard with some knowledge of wine tasting.'

The wine tasting was often associated with emotion, specifically the word 'excited.' 'We were excited for the rosé but this was not even close to the complexity and flavor of the first sparkling rosé we tried.' Another student expressed familiarity with wine and eagerness to taste something unique: 'Most of the wines smell, taste, and look familiar, but the one that I get the most excited about is the orange wine or sauvignon blanc reserve.' While the word 'excite' was not used explicitly by this student, their word choice offered heightened emotion: '[the winery] is truly a great place where you can not only carry out a tasting of unique and unusual wine collections, but also purchase innovative and delightfully tasty wine varietals, as well as have lots of fun too.' The student's written reflection of tasting wine parallels the findings of King, Donaldson, and Harry (2012: 13), who wrote, 'the real-time embodied process of tasting wine and experiencing perceptions shifting in response to social cues was a highly valuable learning experience for students.'

The positive emotions the students felt as first-time wine tourists made them want to share the experience with others. This desire reflects other research, but of different demographics (Fountain and Charters 2009; Leri and Theodoridis 2019; Rueger-Muck and Bruwer 2019). Emotions were notable in the summaries of wine tourism experience, especially regarding the winescape. A student addressed the knowledge and customer service of staff: 'the tasting room personnel was exceptional. They knew exactly what they were talking about.' The student continued, stating the staff created a 'great atmosphere for the rest of the wine connoisseurs that were being attended to while I was there.' Another student expressed that the winescape was the reason to participate in wine tourism in the future: 'The warm welcoming hospitality and customer service displayed by the staff, allowed this tasting to feel more like a one day destination getaway as opposed to a trip to a vineyard.' The student continued, 'it was such an amazing experience that I scheduled a larger group tasting for later on this month.' As seen in this and other examples, the word 'excited' was used by a student when considering returning to a vineyard: 'I am very excited to perhaps visit the vineyard again when the grapes start to grow.' Another wrote in a reflective tone: 'I felt accomplished and a feeling of coming full circle.'

Grounded theory analysis

Four themes emerged from the writing samples: knowledge of the tasting room staff's contribution to the winescape, significance of traveling to the region, the students' technical knowledge of the factors that affect the taste of wine, and the students' likelihood to recommend or return to the winery.

Knowledge of the tasting room staff's contribution to the winescape

Students often wrote about the knowledge and attitude of the staff in conjunction with the tasting room itself. 'At the end of our guided tour, we arrived at the tasting room. It was simple and bright, with a beautiful unobstructed view of the vines. The staff was attentive and very knowledgeable on all the wines in the portfolio.' Another student's connection is evident here: 'The staffs [*sic*] are very friendly and knowledgeable. You will learn a lot from them.' Here, the attitude is professional: 'The knowledge of tasting room personnel was very good and very professional. … I learned about some great dishes that could be served with different types of wine.'

Not only did students find value in the knowledge of the tasting room staff, but some purposely sought out specific learning experiences: 'I choose this particular Vineyard because they not only were very knowledgeable in the wine viniculture [*sic*] but I was able to get a private tour and interview the Winemaker.'

Significance of traveling to the region

The dependence on public transportation presented a challenge. 'We thought about taking the LIRR … but we realized that it would be too much because after the LIRR we had to take a cab. So we decided on having my father drive us.' Yet, some did use public transportation: 'It was a long ride on the LIRR but the scenery of Long Island and traveling over the Long Island sound was so peaceful and gave me a chance to just relax and unwind.'

The long travel time impacted/affected choice. One student changed plans because of time constraints associated with traveling to a wine region: 'the drive was about 2 hours give or take, so my brother and I just canceled that plan.' Location drove another student's choice: 'I decided to choose that vineyard … because of the time it took to travel there.' For some, the long train ride was an opportunity to experience the changing views of the landscape, stating, 'from the concrete jungle of New York City … to the natural beauty of Long Island.'

Students' technical knowledge of the factors that affect the taste of wine

What becomes apparent is students' knowledge of wine, which is likely to be more advanced than non-hospitality students or other young, first-time wine tourists (Stergiou, Airey, and Apostolakis 2018). Students' writing reveals technical knowledge of the attributes of the AVA, grape varieties, viticulture, or vinification. An example includes a student's accurate explanation of the challenges of fermentation when relying on wild yeast and observed a 'majority of the red grapes are fermented in barrels and a small portion in stainless steel.' There is also evidence of students comparing their own knowledge with that of the tasting room staff: 'the host was very knowledgeable. I was able to trace back what she mentioned to what we learned in wine class.' Additionally, their knowledge is revealed when presenting information about AVA: 'the microclimate, makes it suitable for growing grapes of a European variety since it helps the soil retain heat in the winter' and viticulture 'vertical canes that grew between six and a half to seven feet tall and the removal of the leaves to allow air to reach the grapes to prevent mold.'

At times, the students' wine knowledge drives their desire to repeat the wine tourism experience. Student's youthful tone and use of wine vocabulary demonstrates this: 'I totally recommend the vineyards as a destination to visit, and I would like to come back again during the harvest season because the bud had just come up when we visited.'

Students included eloquent tasting notes, providing examples of the tactile experience of taste and the smell which stimulate cognitive activity (King, Donaldson, and Harry 2012; Parr 2019): 'The color of the wine was clear and light, fruity with pear and apple scents that are present but not overwhelming. The smell of vanilla was present but only lightly' and 'it coated my mouth like a blanket it had notes of dark cherries, blackberries and vanilla from fermenting in old French barrels. It left my mouth savoring for more' are examples of students' personal tasting notes.

Students' likelihood to recommend or return to a winery

When students wrote about participating in wine tourism again, they often associated their engagement with the winescape as a motivation, this mirrors findings by Leri and Theodoridis (2019). 'The environment and experience make me want it to come back [*sic*].' The time of year influenced a reason for another visit: 'Unfortunately, we didn't go there at the right time. I think it will be much better to have a tour in summer and take a look at the growing vines.'

Wine tourism has been tied to the student's internal desires to improve and reflects the research findings of Charters and Ali-Knight (2002) and Fountain and Charters (2009), who identified personal development as a motivation for wine tourism. Two students expressed this desire: 'I didn't know what to expect ... I was able to see the country, network, and add another trip to my long list of places I want to see.' and 'I felt like I was a tourist in my own city and I will try to plan a trip to [a winery] again so I can have a tour.'

Visits to urban wineries ignited the desire in students to engage with wine tourism in traditional wine regions. This student's first visit triggered curiosity and a desire to develop tasting skills: 'My next visit will be to a Vineyard [*sic*] where I will be trying to compare the similarities and differences between a winery and a vineyard – just to see for myself if there are any.' 'It was amazing to see all the behind the scenes, and we got to meet [the wine maker]! My mission would be going to an actual vineyard in Long Island or Upstate in the future.' Echoing this sentiment, a student from the same cohort wrote: 'I would say that my overall experience was great, and I hope to visit a winery with actual vineyard there in the future.'

It is significant to mention that most students did not prioritize the taste of wine as a reason to participate in wine tourism as reflected in other studies (Fountain and Charters 2009; Leri and Theodoridis 2019; Rueger-Muck and Bruwer 2019). Students rarely stated that the taste of the wine was a reason to recommend or participate in wine tourism again. One student did mention 'good wine' but also included a reference to the winescape for a reason to recommend wine tourism: 'I would recommend this winery to anyone who is looking to try good wines and also enjoy wonderful atmosphere and [a] spacious tasting room.'

Results

The study reveals that the emotions, perspectives, experiences, and behavior intentions of multicultural, undergraduate hospitality students visiting a winery for the first time reflect similar findings of wine tourists in other research: For example, how wanting to learn, the winescape, and behavior of others stimulate positive emotions during the winery visit (Leri and Theodoridis 2019); the importance of winescape to first-time (and repeat) visitors (Bruwer and Alant 2009), and awareness of the winescape's connection to well-informed tasting room personnel (Rueger-Muck and Bruwer 2019; Quadri-Felitti and Fiore 2012). The importance of tasting wine, tasting room staff knowledge, developing their knowledge of wine, and touring the winery are represented in their writing and matches that of Gen Y wine tourists (Fountain and Charters 2009). However, not found in other studies and revealed in this research is how students' technical knowledge contributes to their decision-making associated with wine tourism, such as their ability to connect attributes of the AVA, viticulture, and vinification to the overall experience as well as their intention to participate in wine tourism again. Despite research reflecting the strength of multicultural markets (Hammond, Sydnor, and Kang 2014; Velikova, Wilkinson, and Harp 2016), wineries and wine tourism organizations have been slow or even reluctant to embrace them. The study reinforces the potential market of younger multiculturals for wine tourism.

Case Study 15.1 A circle of knowledge

The rigor associated with the introductory wine course in the Department of Hospitality Management at New York City College of Technology provides the foundation for inspired students to pursue a career in beverage management, which, like the subjects of this case study, many pursue while completing their degrees. Now the beverage director of a prominent New York City restaurant group, Roberto is one such student whose experience with wine tourism helped him develop his wine knowledge and identify his career goals. He started off uncertain as to which path in the hospitality industry he would take and became continually inspired, both in the classroom and in the vineyard.

Like many of the participants in this study, Roberto's first visit to a winery was part of the introductory course, which is the subject of this study. He shared that his visit to a winery was after harvest, yet 'that one experience opened up [my] interest past the textbooks.' His experience in the course led him to enroll in three elective wine courses at the college, including participation in a Julia Child Foundation–funded wine blending project in which students work under the guidance of Christopher Nicolson, resident winemaker of Red Hook Winery, an urban winery in Brooklyn, New York. The first visit to a winery not only triggered Roberto's desire to visit other wine regions but led to more learning and career opportunities. He declared that the visit to the winery 'really completed a circle of knowledge I didn't even know I was missing.' The physical act of completing the steps associated with winemaking made everything I had studied make sense. Being able to speak with the winemaker is certainly something that I will never forget.' About this winery and wine blending experience, Roberto states: 'To be able to touch, smell, squeeze, taste and use all of my bodily senses to understand wine making was an experience I will always remember.'

These experiences led to more wine tourism, but farther afield, and included overnight stays and other tourism activities. 'Visiting the winery, with the added personal "back-stage" experience, inspired me to want to visit other wineries and learn as much as I could about different styles of winemaking in different climates around the world.' Immediately after his first visit, he began to explore wine tourism with the detailed planning of each visit. 'I was privileged to be able to take a trip the following year to Napa Valley and visit four wineries, speaking with several winemakers about their process and mission and their wines.' He intends to travel to France, Italy, and South Africa and include overnight stays in wine regions. Each trip will be designed to explore the region and to expand his knowledge of wine tasting, viticulture, and vinification.

In this case study, a young, multicultural hospitality student and wine tourist identified not only an enjoyable tourism activity, but a career direction after visiting a winery for the first time. Focusing on the winescape, which included unique experiences such as discussions with winemakers and tastings that embodied a multisensorial immersion into the winery's unique offerings, presented him with reasons to continue to engage in wine tourism.

To attract young multicultural, educated wine tourists, wineries and wine tourism promoters must seek to create, as Roberto states, 'unforgettable experiences,' engaging the visitor in their entirety.

Conclusion

This study illustrated the experience of culturally diverse young people, who are informed about wine and wine tourism, as they engage in wine tourism for the first time. A student illustrates the gap between current marketing in wine tourism and demand from a culturally diverse market:

> The friendly, easy-going, very scenic, easy to navigate, hard to miss environment is a great change from the city and suits those who look to get away from city life very well. The only weakness I see is that with Generation X aging and the Millennials picking up on the trends of their predecessors, the vineyard may have a problem with what I believe will be an explosion of wine drinkers from New York, if just one local celebrity picks up one of [the winery's] great wines. I can easily see Shawn 'Jay-Z' Carter supporting this winery, and with his influence and maturity, [a] tsunami-like wave of people would swing by the vineyard. As these two generations age, the

vineyard is going to seriously have to think about expanding its visiting grounds and having access control measures in place, something I would enjoy being a part of maybe someday in the future.

Implications of the study

The perspective on wine tourism of multicultural undergraduate hospitality students is important to consider in the market segmentation of wine tourism, because they reflect many of the preferences and values that have contributed to the industry's growth as revealed in decades of research. Winery owners, the wine industry, and wine marketers must seek and evaluate the views of new and emerging wine tourists in order to improve service, increase purchases at the cellar door, enhance the tasting room experience and the overall guest experience, and build loyalty. Moreover, wineries and wine tourism professionals need to broaden the content and context of the winery experience to build inclusive environments, fostering a sustainable wine tourism industry.

References

Alebaki, M., and Iakovidou, O. (2010). Segmenting the Greek wine tourism market using a motivational approach. *New Medit*, 9: 31–40.

Ali-Knight, J., and Charters, S. (2001). 'The winery as educator: Do wineries provide what the wine tourist needs? *Australian and New Zealand Wine Industry Journal*, 16(6): 79–86.

Alonso, A. D., Fraser, R. A., and Cohen, D. A. (2007). Does age matter? How age influences the winery experience. *International Journal of Culture, Tourism and Hospitality Research*, 1(2): 131–139. https://doi.org/10.1108/17506180710751678

Asimov, E. (2004). Pickers to vintners: a Mexican-American saga. *New York Times*, October 17.

Asimov, E. (2014). Red wine is the drink of choice on 'Scandal' and 'The Good Wife'. *New York Times*, October 28.

Asimov, E. (2020). Black wine professionals demand to be seen. *New York Times*, June 29.

Beverland, M. (1998). Wine tourism in New Zealand – maybe the industry has got it right. *International Journal of Wine Marketing*, 10(2): 24–33. https://doi.org/10.1108/eb008679

Bruwer, J. (2003). South African wine routes: some perspectives on the wine tourism industry's structural dimensions and wine tourism product. *Tourism Management*, 24(4): 423–435. https://doi.org/10.1016/S0261-5177(02)00105-X

Bruwer, J., and Alant, K. (2009). The hedonic nature of wine tourism consumption: an experiential view. *International Journal of Wine Business Research*, 21(3): 235–257. https://doi.org/10.1108/17511060910985962

Campbell, K. (2009). Wine clubs and the wine industry – adjusting the stereotypes of wine enthusiasts in America. *World Wine Guide*. https://ezinearticles.com/?Wine-Clubs-and-the-Wine-Industry---Adjusting-the-Stereotypes-of-Wine-Enthusiasts-in-America&id=1768681

Carlsen, J. (2004). A review of global wine tourism research. *Journal of Wine Research*, 15(1): 5–13. https://doi.org/10.1080/0957126042000300281

Carlsen, J. (2011). Assessing service quality at wineries and cellar doors through service mapping. *International Journal of Wine Business Research*, 23(3): 271–290. https://doi.org/10.1108/17511061111163087

Carlsen, J., and Dowling, R. (2001). Regional wine tourism: a plan of development for Western Australia. *Tourism Recreation Research*, 26(2): 45–52. https://doi.org/10.1080/02508281.2001.11081342

Charters, S., and Ali-Knight, J. (2002). Who is the wine tourist? *Tourism Management*, 23(3): 311–319. https://doi.org/10.1016/S0261-5177(01)00079-6

Dewey, J. (1938). *Experience and Education*. New York: Simon and Schuster.

Dodd, T., and Bigotte, V. (1997). Perceptual differences among visitor groups to wineries. *Journal of Travel Research*, 35(3): 46–51. https://doi.org/10.1177/004728759703500307

Dodd, T. H. (1995). Opportunities and pitfalls of tourism in a developing wine industry. *International Journal of Wine Marketing*, 7(1): 5–16. https://doi.org/10.1108/eb008636

Folwell, R. J., and Baritelle, J. L. (1978). United States. Department of Agriculture. Economics, S. *The U.S. wine market*. Washington, DC: Dept. of Agriculture, Economics, Statistics, and Cooperatives Service: For sale by the Supt. of Docs., U.S. Govt. Print. Off.

Fountain, J., and Charters, S. (2009). Generation Y as wine tourists: their expectations and experiences at the winery-cellar door. In Benckendorfff, P., Moscardo, G., and Pendergast, D., *Tourism and Generation Y* (1st ed., pp. 47–57). Wallingford: CABI.

Fry, R. (2020). Millennials overtake Baby Boomers as America's largest generation. *Pew Research Center*. https://www.pewresearch.org/fact-tank/2020/04/28/millennials-overtake-baby-boomers-as-americas-largest-generation/

Gaiter, D. J. (2020). Being Black in the white world of wine. *SevenFifty Daily*. https://daily.sevenfifty.com/being-black-in-the-white-world-of-wine/

Garibaldi, R., Stone, M. J., Wolf, E., and Pozzi, A. (2017). Wine travel in the United States: a profile of wine travelers and wine tours. *Tourism Management Perspectives*, 23: 53–57. https://doi.org/10.1016/j.tmp.2017.04.004

Getz, D., and Brown, G. (2006). Critical success factors for wine tourism regions: a demand analysis. *Tourism Management*, 27(1): 146–158. https://doi.org/10.1016/j.tourman.2004.08.002

Goodlad, K., and Leonard, A. E. (2018). Place-based learning across the disciplines: a living laboratory approach to pedagogy. *InSight: A Journal of Scholarly Teaching*, 13: 150–164. https://doi.org/10.46504/14201808go

Gunasekar, S., Das, P., Dixit, S. K., Mandal, S., and Mehta, S. R. (2021). Wine-experienscape and tourist satisfaction: through the lens of online reviews. *Journal of Foodservice Business Research*. https://doi.org/10.1080/15378020.2021.2006039

Gyarkye, L. (2020). 'Uncorked' review: a fresh take on the father-son drama: streaming movie review. *New York Times*, March 27, 2020.

Hall, C. M., Cambourne, B., Macionis, N., and Johnson, G. (1997). Wine tourism and network development in Australia and New Zealand: review, establishment and prospects. *International Journal of Wine Marketing*, 9(2): 5–31.

Hammond, R., Sydnor, S., and Kang, E. (2014). Reaching an underserved wine customer: connecting with the African American wine consumer. *FIU Hospitality Review*, 31(3): 118–144.

High, R. (2012). The era of cognitive systems: an inside look at IBM Watson and how it works. *IBM Corporation, Redbooks*.

Hudson Valley. (2020). *Economic Impact of New York Wine and Grapes. Wine Magazine*1 [online] Available at: https://hvwinemag.com/economic-impact-of-new-york-wines-and-grapes/

IBM, Tone Analyzer, Watson. (2017). https://tone-analyzerdemo.mybluemix.net/

King, T. J., Donaldson, J. A., and Harry, E. (2012). Tasting wine: a learning experience. *Journal of Experiential Education*, 35(3): 464–479. https://doi.org/10.1177/105382591203500306

Kolb, A. Y., and Kolb, D. A. (2012). *Experiential Learning Spiral*. Boston: Springer.

Leonhardt, D. (2017). America great working-class colleges. *New York Times*, January 18, 2017.

Leri, I., and Theodoridis, P. (2019). The effects of the winery visitor experience on emotions, satisfaction and on post-visit behaviour intentions. *Tourism Review*, 74(3): 480–502. https://doi.org/10.1108/TR-07-2018-0092

McIntyre, D. (2020). The wine industry is overwhelmingly white. Now, the push for inclusivity is gaining momentum. *The Washington Post*, June 26, 2020.

McMillan, R. (2020). The state of the wine industry report 2020. *Silicon Valley Bank*. https://www.svb.com/wine-report

Mintel. (2017). *Wine Tourism*. TTA No. 3 March 2017.

Mitchell, R., and Hall, C. M. (2001). The influence of gender and region on the New Zealand winery visit. *Tourism Recreation Research*, 26(2): 63–75. https://doi.org/10.1080/02508281.2001.11081344

Mobley, E. (2018). How fine wine became the NBA's new status symbol. *San Francisco Chronicle*, December 2, p. M4.

New York City College of Technology. (2021a). About City Tech. *New York City College of Technology*. https://www.citytech.cuny.edu/about-us/mission.aspx

New York City College of Technology. (2021b). Department of Hospitality Management. *New York City College of Technology*. https://www.citytech.cuny.edu/hospitality/

Nielsen. (2018). Black impact: consumer categories where African Americans move markets. *Nielsen Insights*. https://www.nielsen.com/us/en/insights/article/2018/black-impact-consumer-categories-where-african-americans-move-markets/

Noe-Bustamante, L., Lopez, M. H., and Krogstad, J. M. (2020). U.S. Hispanic population surpassed 60 million in 2019, but growth has slowed. *Pew Research Center.* https://www.pewresearch.org/fact-tank/2020/07/07/u-s-hispanic-population-surpassed-60-million-in-2019-but-growth-has-slowed/

O'Neill, M. A., and Palmer, A. (2004). Wine production and tourism: adding service to a perfect partnership. *Cornell Hospitality Quarterly*, 45(3): 269–284. https://doi.org/10.1177/0010880404263075

Parr, W. (2019). Demystifying wine tasting: cognitive psychology's contribution. *Food Research International*, 124: 230–233. https://doi.org/10.1016/j.foodres.2018.03.050

Pine, B. J., II, and Gilmore, J. H. (1999). *The Experience Economy: Work Is Theatre and Every Business a Stage.* Boston, MA: Harvard Business School Press.

Powell, J. (n.d.). *The Wine Divas.* https://www.oprah.com/omagazine/julie-powell-meets-the-wine-divas/2

Quadri-Felitti, D., and Fiore, A. M. (2012). Experience economy constructs as a framework for understanding wine tourism. *Journal of Vacation Marketing*, 18(1): 3–15. https://doi.org/10.1177/1356766711432222

Rueger-Muck, E., and Bruwer, J. (2019). Wine tourism and hedonic experience: a motivation-based experiential view. *Tourism & Hospitality Research*, 19(4): 488–502. https://doi.org/10.1177/1467358418781444

Stergiou, D. P., Airey, D., and Apostolakis, A. (2018). The winery experience from the perspective of Generation Z. *International Journal of Wine Business Research*, 30(2): 169–184. https://doi.org/10.1108/IJWBR-03-2017-0018

Strauss, A., and Corbin, J. (1994). Grounded theory methodology: an overview. In Denzin, N., and Lincoln, Y. (eds.), *Handbook of Qualitative Research* (1st ed., pp. 273–285). Thousand Oaks, CA: Sage.

United Nations World Tourism Organization. (n.d.). Gastronomy and wine tourism. *Product Development.* https://www.unwto.org/gastronomy-wine-tourism

Velikova, N., Wilkinson, K. T., and Harp, S. S. (2016). Reaching the U.S. Hispanic wine market: highlighting the opportunities. *Journal of Food Products Marketing*, 22(1): 43–64. https://doi.org/10.1080/10454446.2014.949991

Weeks, M. (2019). Minority markets see economic growth. *UGA Today.* https://news.uga.edu/multicultural-economy/

Wine America. (n.d.). United States wine and grape industry FAQs. *Wine America.* https://wineamerica.org/policy/by-the-numbers

Wine Market Council. (2019). Wine and other alcohol consumption among 21–29 year old U.S. wine consumers. *Research.* https://winemarketcouncil.com/research/

Worobiec, M. (2020). A voice for Black winemakers. *Wine Spectator.* https://www.winespectator.com/articles/leaders-association-african-american-vintners-discuss-building-diverse-wine-industry

Yue, C., Govindasamy, R., and Kelley, K. (2019). Mid-Atlantic wine tourism consumer preference: an econometric approach. *International Journal of Wine Business Research*, 31(3): 327–343. https://doi.org/10.1108/IJWBR-03-2018-0011

16

SENIOR TRAVELERS AND WINE TOURISM EXPERIENCE

Opportunities of an evolving market segment

Izaskun Zurbitu-Aldama and Adela Balderas-Cejudo

Introduction

Population aging is a well-known phenomenon in most developed countries, where the proportion of older people has been steadily growing over the past century (Harper and Leeson 2008). As the global senior population expands, there is increasing scholarly interest in the intersection between tourism and the aging population (Otoo, Kim and Choi 2020). The aging of populations in most Western countries and many non-Western countries, as well as changes in older adults' sociodemographics and travel patterns, have made the older segment an appealing target population for the global tourism and the travel industry (Patterson 2017).

Senior travelers are gradually embracing the desire for memorable experiences in their tourist activities (Patterson and Pegg 2009), that is, memorable, personal, and unique experiences where seniors can learn and broaden their minds through their travels. It is becoming increasingly important in an aging world that tourism providers offer a greater understanding and deeper insights about what makes an experience 'memorable' for senior tourists. Wine and food have long been regarded as an important component of the tourist experience. According to Carlsen (2004), many wine regions and tourism destinations have realized that the benefits of wine tourism extend well beyond the cellar door to virtually all areas of the regional economy and into the urban areas that generate the majority of wine tourists. Nevertheless, wineries also face an increasing need to add value to wine tourism experiences and also a need to deeply understand their visitors' motivations and expectations in a changing world. Experience seekers are moving away from standardization and material goods and are looking for 'authenticity' in the places they visit and select customized travel (Rachão et al. 2020).

Our world has been brought to a halt by COVID-19. More than ever, knowing the preferences and needs of customers in general and the senior market segment, in particular, will be key so as to be able not only to market services and products correctly but also to accomplish the expectations of a cohort that does not want to do leisure travel as previous generations did. Therefore, it is important that tourism providers respond to the changing needs of senior travelers seeking memorable experiences but putting safety first in this unprecedented situation linked to unprecedented challenges. Understanding what type of tourism and experiences senior travelers want to do in the future, understanding their motivations, expectations, needs, and behavior, and accounting for these changes in a proactive way will not only provide exciting

DOI: 10.4324/9781003143628-19

opportunities for research, but it will also have practical implications for all stakeholders. Therefore, this chapter throws light on the memorable wine tourism experiences that seniors are willing to undertake. It shows that there are specific tools that may be used to design and create unique experiences for this evolving market segment.

Senior travelers: an attractive segment for wine tourism

According to *World Population Prospects 2019* (United Nations 2019), by 2050, 1 in 6 people in the world will be over the age of 65, up from 1 in 11 in 2019. All societies in the world are amidst this longevity revolution – some are at their early stages, and some are more advanced. Virtually every country in the world is experiencing growth in the size and proportion of older persons in their population (United Nations 2019). In this aging world, the older market segment has become an attractive target. Major shifts in society, such as globalization, technology, climate change, and desire for well-being and safety, will culminate in new travel attitudes and consumer demands affecting all targets, and seniors will be no exception. The tourism industry has recognized the market potential of older people, and tourism research has tended to focus on developing competitive business and marketing strategies to target these consumers (Dixit 2017). Seniors are wealthier, healthier, more tech-friendly, more mobile, and more sophisticated travelers than previous generations. Tourism and travel is now an attractive option for many seniors who have the time and discretionary income to spend on enjoying their retirement years (Patterson and Balderas 2020).

The senior segment is growing and evolving, and their consumption patterns are also changing. Nowadays, it is considered a fact that the aging population is expected to have several marketing implications for all industries (Nella and Christou 2016), and the tourism industry is no exception. The market segment of senior or older tourists can be easily considered as an ideal target for tourism marketers, not only because of its current size and purchasing power, but also because of demographic and social forecasts that it will continue to grow in numbers (Schröder and Widmann 2007). Additionally, it is a promising target due to the time flexibility of seniors after retirement (Jang and Wu 2006), and because it can constitute a marketable to attenuate seasonal imbalances of tourist destinations.

There has been a bigger demand for tourism services not typically associated with older adult tourists of previous generations. Consistent with Wei and Milman (2002), it is particularly relevant to look at senior tourists 'satisfaction with trip experiences because they are likely to be experienced tourists with greater knowledge in vacation evaluation (Ruys and Wei 2001). Senior tourists are increasingly seeking out memorable experiences (Vigolo 2017). A memorable experience occurs when an individual feels that there is some emotional, intellectual, physical, and/or spiritual stimulation that absorbs them, and becomes closer and more attracted to the atmosphere of a specific setting such as a restaurant or hotel resort (Tung and Ritchie 2011). The more senses that an experience engages, the more effective and memorable the experience will be. This is because the greater the number of images that visitors can obtain and retain from an environment that is not part of their routine life, the longer they will remember them for a long time afterwards (Andereck et al. 2006).

Pine and Gilmore proposed in their 1999 work *The Experience Economy: Work Is Theatre and Every Business a Stage* that the advent of the experience economy era, pointing out that it is the fourth stage of development after the agricultural, industrial, and service economies, and that experiences are unforgettable or memorable occurrences created by consumers. Tourism studies have emphasized the significance of memorable experiences because a person's memory is the single most important information source for an individual when making a decision to revisit or to spread word-of-mouth comments about a particular destination (Oh, Fiore, and Jeoung 2007).

Following Sotiriadis (2015), tourism destinations, businesses, and different stakeholders are constantly looking for ways to improve, differentiate, and promote their offering aiming to achieve a better positioning and gain competitive advantage in an increasingly competitive sector. The advent of COVID-19 is necessitating the revision of the design of the tourist experience, and in the future there will be the need to adapt to the health measures associated with social distancing while entering the 'new normal' (Gössling, Scott, and Hall 2021) and the post-COVID-19 era.

All these factors should be considered for the design of appealing tourism products and services. Changes in the characteristics of tourism demand in general, and senior travelers in particular, can create clear market advantages for the destinations and wineries that are willing to adapt their offerings to their needs. Rethinking, innovating, and rebooting a seriously damaged sector in an attempt to see tourism beyond COVID-19 is essential. Wineries' marketers and the tourism industry, in general, need to respond to this changing market sector so as to be able to offer memorable experiences to their guests.

Applying the experience economy model to wine tourism

The experiential view of consumers' behavior is critical to understanding wine tourism. The relevance of the hedonic theories to explain the experiences desired by customers is increasing. In the last decade's wine tourism researchers have investigated the experiential concepts behind the tourism activities (Brown and Getz 2005; Charters, Fountain, and Fish 2009; Dixit 2020; Galloway et al. 2008; Getz and Carlsen 2008; Pikkemaat et al. 2009; Quadri-Felitti and Fiore 2012) among others.

The seminal hedonic-experiential model proposed by Pine and Gilmore (1999) analyzes the customer experience based on four realms – educational, entertainment, escapist, and aesthetic experiences – known as 'the four Es.' The study on consumer experiences based on Pine's framework has been widely acknowledged in different tourist activities. Regarding cultural activities, Bideci and Albayrak (2018) indicated the underlying dimensions of museum experiences – that is, edutainment, comfort, escape, and aesthetic. Radder and Han (2015) confirmed that edutainment also made the strongest contribution to the prediction of overall satisfaction and behavioral intentions in museums. Oh et al. (2007) provided a starting point for quantitatively evaluating experience economy concepts and practices within the leisure industry, specifically in bed and breakfast, followed by other activities such as cruises lines (Hosany and Witham 2010) and heritage trails researched by Hayes and MacLeod (2007). Other touristic sites such as religious sites, music festivals, or zoos have also qualitatively been studied based on Pine and Gilmore's experience economy realms (Mehmetoglu and Engen 2011; Musa et al. 2015, 2017).

The four underlying dimensions of the experiences may also be a logical framework to understand wine tourism activities, but such a framework has been implemented by few researchers: Ali-Knight and Carlsen (2003) in the case of Australian wineries; Pikkemaat et al. (2009) in the South Tyrol Wine route; and Quadri-Felitti and Fiore (2012) in the case of Lake Eire wineries in the United States. Consumer experiences incorporating the four elements of the experience model lead to stronger memories and consequent positive consumer valuations (Gilmore and Pine 2002). Memorable experiences are created when a company uses services as the stage and goods as props to engage the users in an inherently individual and personal way (Gilmore and Pine 2002).

The four Es dimensions are positioned along with two matrices of experiences: consumer participation (active or passive) horizontally and consumer connection (absorption or immersion) vertically. These matrices are divided into four quadrants: education, entertainment, escapism, and aesthetics.

Most of the wine tourism authors have narrowed their research to one or two dimensions. For example, authors based their research only on: education and aesthetics (Bruwer and Alant 2009);, aesthetics and escapism (Cohen and Ben-Nun 2009; Fountain and Charters 2010), with a special interest in the combination of education and entertainment (Getz and Carlsen 2008; Park et al. 2008); or only entertainment (Axelsen and Swan 2009). Yet, a holistic approach to wine tourism experiences is needed (Quadri-Felitti and Fiore 2016). From the quantitative approach, only Quadri-Felitti and Fiore (2016) have adapted the original Oh et al. (2007) operationalized scale to measure the 4Es framework in wine tourism finding significant differences in the experience economy priorities between different wine tourism actors, such as winery owners and grape farmers; lodging operators; restaurateurs and retailers; and recreation and attraction operators.

The element of education (active participation + absorption quadrant)

Learning and education are constant top tourist motivators (Ballantyne and Packer 2011) no matter the age, gender, or lifestyle cycle. Education as learning facts can play an essential role in creating memorable experiences, not only in wine tours (Pikkemaat et al. 2009; Pine and Gilmore 1999; Quadri-Felitti and Fiore 2012; Saayman and van der Merwe 2015) but also in wineries (Kim et al. 2010) and in other cultural and recreational activities.

The element of entertainment (passive participation + absorption quadrant)

In contrast with the educational element, entertainment entails a passive participation from the consumers' point of view. That means the consumers passively observe others' activities and/or performances (Manthiou et al. 2014), such as cellar music concerts, wine-blending demonstrations, or museums and heritage site visits. Tourists may come to a winery for a socially enjoyable experience in many cases, having no special interest in wine, as researched in Generation Z (Stergiou, Airey, and Apostolakis 2018).

Originally coined by Pine and Gilmore (1999), 'edutainment' is the overlap between both absorptive elements – entertainment and education. It can be seen as the mixture of informative, social, and fun aspects of wine tourism activities (Fernandes and Cruz 2016; Getz and Carlsen 2008). Actually, the wine itself may also be embedded as an additional element in many other different cultural and entertainment activities providing 'edutainment.'

The element of escapism (active participation + immersion)

Similar to education, tourists are active in escapist experiences, but immersed in the wine tourism experience itself. Numerous recreational activities in wineries and wine tours engage tourists in rural settings which are very different from their urban or semiurban way of living. These activities by themselves are top motivators to visit wineries in many cases (Anker et al. 2015; Fountain and Charters 2010).

Wine events should offer active participation through winemaking workshops, harvesting activities, and other events that may foster the educational and escapism realms of the experiences (Quadri-Felitti and Fiore 2013; Vo Thanh and Kirova 2018; Rachão et al. 2020). The wine aesthetic dimension is paramount following Quadri-Felitti and Fiore, 2013. Concurring with Rachão et al. (2020), further empirical research on co-creative activities in wine tourism is needed due to their significant participative role in adding extra value to food and wine tourism experiences.

The element of aesthetics (passive participation and immersion)

In contrast with escapist experiences, consumers are passive receivers of aesthetic experiences but also are immersed in the wine tourism experience in a different time and/or place. The aesthetic element embodies the so-called 'winescape' concept or the aesthetic motivation that is central in wine tourism. Being a concept thoroughly researched in wine journals, 'winescape' is, according to Bruwer and Johnson (2010), 'encapsulating the interplay of natural landscape and setting; heritage, architecture, and artifacts within a winery; a winery's vineyard, cellar door, and wines; complementary products and services; signage and layout; and people at a winery.' The aesthetic experience dominates the likelihood of positive memories and destination loyalty in the wine tourism context, according to Quadri-Felitti and Fiore (2013).

Case Study 16.1 Vivanco Winery: 'Sharing wine culture' with seniors

This case study presents the Vivanco Winery (www.vivancoculturadelvino.es) in Rioja, Spain (see Figure 16.1). Selected as one of the top 50 world wineries to visit in 2020, this family-owned winery is considered one of the main touristic destinations for seniors in Rioja.

Figure 16.1 Vivanco Winery, Rioja (Spain).

Vivanco attracted over 70,000 visitors in 2019 (70% national). It is a magnificent 9,000-sqm facility divided into vineyards, a modern underground winery, five permanent museum exhibition halls, tasting rooms, a restaurant, a wine shop, and meeting rooms. Named among the most impressive wine museums in the world, the Vivanco Museum was created in 2004 thanks to the family's passion for wine culture. The collection of artistic and historic pieces is truly unique, representing years of work in recovering and safeguarding wine-related heritage by the Vivanco family (currently, the fourth generation of winemakers, Rafael and Santiago). The museum includes ethnographic pieces such as centuries-old wine presses and remarkable pieces of art inspired by wine from world master painters, such as Dürer, Picasso, Miró, or Warhol. The exhibition is rounded off with an impressive garden, The Garden of Bachus features every vine in the world, and a documentation center has more than 8,000 monographs and objects. A varied range of wine-focused activities is centered on the museum and winery visits, offered by 20 employees dedicated to wine tourism.

Senior tourism provides an interesting opportunity to break the seasonality of the peak demand in the late summer and fall months as well as festivities and weekends. Preseniors (50–55 years) and seniors (above 55 years old) play an essential role as the main tourist segments the whole year. Targeting this niche market that is less influenced by vacations, weather, or short breaks is key to increasing the scalability of the wine tourism business unit. According to their own data, seniors account for more than 30% of the total tourists at Vivanco in 2019.

Personalization and authenticity are key drivers in the experience creation at Vivanco for senior tourism. The wine-related activities may be defined within the four realms of experiences: aesthetics, entertainment, education, and escapism. Seniors can enjoy Vivanco in their timeframe and own taste, building their own experience blending and choosing from the large variety of activities detailed in Table 16.1.

Vivanco is an example of wine tourism for seniors mainly based on passive immersion activities with a strong educational purpose and 'edutainment' concept. From the educational point of view, original authenticity is especially keynoted by seniors since many of the heritage artifacts collected at the museum are first-of-their-kind: winemaking equipment, pieces of art, and others. The magnificent selection of pieces of art and ethnography show sensitivity and care about wine culture that drives Vivanco into an exceptional level of authenticity. Seniors also highlight the importance of the local winery guides, passionate for their work, and true believers in the wine and region.

Three segments of seniors are defined based on the choice of experiences: local culture seekers, independent food travelers, and wine geeks. The 'culture seekers' are mainly Spanish seniors traveling on guided tours from nearby cities. Visiting on weekdays, these seniors are interested in Vivanco as a cultural landmark. Not interested in the wine itself, they are looking for gastronomic pleasures as a good complement to their cultural visits. Their ideal experience contains absorptive elements, entertainment and educational – that is, visiting the wine museum and winery in groups with a local guide and ending the journey by eating local food at the restaurant. The other two segments, 'independent food travelers' and 'wine geeks,' are interested in combining educational and escapist activities. Therefore, the winery designed the exclusively holistic Vivanco Experience. The 2.5-hour museum and winery visit with a local guide in their own language is blended with wine tastings held in extraordinary locations (such as the Garden of Bachus or the Octagonal Room). 'Food travelers' are mainly couples traveling by car independently, mainly from France and other European countries. They usually spend two or three days in the region. Their average sales ticket is higher because of the wine shopping. The 'independent wine geeks' mainly come from the United States with a high interest in wine. Vivanco is considered to be one of their 'must' experiences in Spain due to its wine

Table 16.1 Wine tourism activities at Vivanco

	Passive participation	*Active participation*
Absorption	Entertainment	Educational
	Museum free guided visits	Underground winemaking facility tour guided by local experts**
	Museum guided tours with local guides**	Visiting heritage and ethnographic pieces' collection**
	Visiting art gallery**	Wine tastings during visits
	Retail shopping	Wine seminars
	Wine bar tasting	Culinary-wine pairing menu at the restaurant**
	Visiting the Corkscrew Collection	Culinary menu designed with 'wine as an ingredient' at the restaurant
	Vivanco Kids experiences	Local food meals with wine at the restaurant**
Immersion	Esthetic	Escapist
	Visit of the building and surroundings**	Enjoying the Vivanco Experience (2.5-hour long holistic experience)**
	Discovering unique pieces of art	Harvesting experience in fall
	Enjoying the 'wow' walking through the Art Gallery in the museum**	External events with local partners (hot air ballooning, cycling tours, visiting medieval villages, etc.)
	Participating in the 'winescape' (integrated in all winery visits and at least 25% time of the whole visits)**	
	Enjoying the 'wow effect' of the Octagonal Aging Room at the winery**	

Source: Authors' own analysis and elaboration on Vivanco data.

Note:

** Popular wine tourism-related activities for seniors.

Source: Analysis framework by Haller, Hess-Misslin, and Mereaux (2020), following Pine's original framework

heritage collection. They are looking for a holistic experience around the wine, including pairings with food and wine shopping.

Vivanco embodies an example of a wine-related cultural space with a strong educational purpose, designed with a mass media approach and based on the cultural authenticity of wine heritage. Focused on pre-senior and senior segments, digital technology is not widely used in the experiences, in contrast with other hybrid wine-culture venues, such as the Cité du Vin, France. Authentic experiences based on heritage, wine culture, and personalization seem to be very successful for the senior market in this case.

Conclusions

Changes in the characteristics of tourism demand, needs, and expectations can create clear market advantages for the destinations that are willing to adapt their offerings to the tourism needs of each generation group (Nella and Christou 2016). The potential of the seniors' market for stable or even increasing growth is a very promising element (Horneman et al. 2002; Patterson and Pegg 2009; and Patterson 2017), and tourism businesses targeting the aging population with an effective marketing strategy are going to be monetarily rewarded.

Since each winery tourist visits the winery with a different motivation and level of wine involvement, they will perceive different types of values. Accordingly, it would be critical to investigate winery visitors' values toward winery attributes (Lee, McCole, and Holecek 2022).

Distinctive wine tourism experiences for seniors may be related to touristic 'hybrid venues' in 'destination wineries.' Vivanco shows a great example of so-called 'hybrid venue' embedded in a winery. Defined by Passebois Ducros and Euzéby (2020), this term refers to 'hybrid structures that are emerging in the leisure area dedicated to the exploration of a cultural theme.' Concerning the aforementioned case study, Vivanco also turned into a so-called 'landmark winery or destination winery' (Ferreira and Hunter 2017; Getz and Brown (2006) – that is, a winery with a great visual impression and symbolic significance that has become a monument to wine and the main attraction in one of the most important wine regions in the world, Rioja.

Senior tourism may be especially attracted to places and experiences like Rioja and Vivanco in the near future, since they may prefer to avoid the hustle and bustle associated with larger, crowded tourism areas or other wine and food regions in the current COVID-19 situation and post-COVID time, in which safety will come first. Despite its world reputation as one of the top Spanish wine regions, the bucolic and rural Rioja is still underdeveloped in terms of wine tourism, though it enjoys an increasing number of visitors – 860,000 in 2019, enjoying a total of 119 wineries (ACEVIN 2020).

In this time for change, resilience, and reset, COVID-19 has provided a unique opportunity to rethinking aging, businesses and wine tourism in an attempt to reimagine and envision a better future capable of deeply understand needs and wants. Concurring with Horneman et al. (2002), in order to cater to the diversity of needs of the seniors' market, a more comprehensive understanding of travel behavior is necessary in order for tourism products and services to be tailored.

Implications

Becoming aware of the potential of senior travelers and being able to adapt to the changing trends of this evolving market segment is essential for the industry. Understanding seniors' motivations and the kind of experiences they value, and accounting for these changes in a proactive way, will not only provide exciting opportunities for research, but it will also have practical implications for public policymakers, industry practitioners, wineries, and travel marketers.

This chapter presented a successful example of the experience economy model theory applied to the seniors' segment's wine experiences. Furthermore, the model suggested aims to provide an appropriate and simple framework to design and adapt post-COVID wine tourism memorable experiences for seniors. This approach can be useful, especially for the wineries' marketers and tourism companies, in terms of systematically improving the seniors' experiences and evaluating visits more effectively.

Acknowledgements

We thank Mr. Eduardo Diez, Wine Tourism Director at Vivanco, as well as the Vivanco Foundation for their valuable time and information on this case study.

References

ACEVIN (2020). *Informe de visitantes a bodegas y museos del vino asociados a las Rutas del Vino de España 2019.* https://wineroutesofspain.com/wp-content/uploads/2020/10/informe-de-visitantes.pdf

Ali-Knight, J., and Carlsen, J. (2003). An exploration of the use of 'extraordinary' experiences in wine tourism. http://academyofwinebusiness.com/wp-content/uploads/2010/05/File-019.pdf

Andereck, K., Bricker, K. S., Kerstetter, D., and Nickerson, N. P. (2006). Connecting experiences to quality: understanding the meanings behind visitors' experiences. In Jennings, G., and Nickerson, N. P. (eds.), *Quality Tourism Experiences* (pp. 81–98). London: Routledge.

Anker, T. B., Sparks, L., Moutinho, L., and Grönroos, C. (2015). Consumer dominant value creation. *European Journal of Marketing*, 49(3–4): 532–560. https://doi.org/10.1108/EJM-09-2013-0518

Axelsen, M., and Swan, T. (2009). Designing festival experiences to influence visitor perceptions: the case of a food and wine festival. *Journal of Travel Research*, 49: 436–450.

Ballantyne, R., and Packer, J. (2011). Using tourism free-choice learning experiences to promote environmentally sustainable behaviour: the role of post-visit action resources.' *Environmental Education Research*, 17(2): 201–215. https://doi.org/10.1080/13504622.2010.530645

Bideci, M., and Albayrak, T. (2018). An investigation of the domestic and foreign tourists' museum visit experiences. *International Journal of Culture, Tourism, and Hospitality Research*, 12(3): 366–377. https://doi.org/10.1108/IJCTHR-02-2018-0029

Brown, G., and Getz, D. (2005). Linking wine preferences to the choice of wine tourism destinations. *Journal of Travel Research*, 43(3): 266–276. https://doi.org/10.1177/0047287504272027

Bruwer, J., and Alant, K. (2009). The hedonic nature of wine tourism consumption: an experiential view. *International Journal of Wine Business Research*, 21(3): 235–257. https://doi.org/10.1108/17511060910985962

Bruwer, J., and Johnson, R. (2010). Place-based marketing and regional branding strategy perspectives in the California wine industry. *Journal of Consumer Marketing*, 27(1): 5–16. https://doi.org/10.1108/07363761011012903

Carlsen, P. J. (2004). A review of global wine tourism research. *Journal of Wine Research*, 15(1): 5–13. https://doi.org/10.1080/0957126042000300281

Charters, S., Fountain, J., and Fish, N. (2009). You felt like lingering: experiencing 'real' service at the winery tasting room. *Journal of Travel Research*, 48(1): 122–134.

Cohen, E., and Ben-Nun, L. (2009). The important dimensions of wine tourism experience from potential visitors' perception. *Tourism and Hospitality Research*, 9(1): 20–31.

Dixit, S. K. (ed.). (2017). *The Routledge Handbook of Consumer Behaviour in Hospitality and Tourism*. London: Routledge.

Dixit, S. K. (ed.). (2020). *The Routledge Handbook of Tourism Experience Management and Marketing*. London: Routledge.

Fernandes, T., and Cruz, M. (2016). Dimensions and outcomes of experience quality in tourism: The case of Port wine cellars. *Journal of Retailing and Consumer Services*, 31: 371–379. https://doi.org/10.1016/j.jretconser.2016.05.002

Ferreira, S. L., and Hunter, C. A. (2017). Wine tourism development in South Africa: a geographical analysis. *Tourism Geographies*, 19(5), 676–698. https://doi.org/10.1080/14616688.2017.1298152

Fountain, J., and Charters, S. (2010). Generation Y as wine tourists: their expectations and experiences at the winery-cellar door. In Benckendorff, P., Moscardo, G., and Pendergast, D. (eds.), *Tourism and Generation Y* (pp. 47 57). Wallingford: CAB International.

Galloway, G., Mitchell, R., Getz, D., Crouch, G., and Ong, B. (2008). Sensation seeking and the prediction of attitudes and behaviours of wine tourists. *Tourism Management*, 29: 950–966. http://doi.org/10.1016/j.tourman.2007.11.006

Getz, D., and Brown, G. (2006). Benchmarking wine tourism development: the case of the Okanagan Valley, British Columbia, Canada. *International Journal of Wine Marketing*, 18(2): 78–97. https://doi.org/10.1108/09547540610681077

Getz, D., and Carlsen, J. (2008). Wine tourism among generations X and Y. *Tourism: An International Interdisciplinary Journal*, 56(3): 257–269.

Gilmore, J. H., and Pine, B. J. (2002). Differentiating hospitality operations via experiences: why selling services is not enough. *Cornell Hotel and Restaurant Administration Quarterly*, 43(3): 87–96. https://doi.org/10.1016/S0010-8804(02)80022-2

Gössling, S., Scott, D., and Hall, M.C. (2021). Pandemics, tourism and global change: a rapid assessment of COVID-19. *Journal of Sustainable Tourism*, 29(1): 1–20. https://doi.org/10.1080/09669582.2020.1758708

Green, B. (2005). *Marketing to Leading-Edge Baby Boomers*. Ithaca, NY: Paramount Publishing.

Haller, C., Hess-Misslin, I., and Mereaux, J.-P. (2020). Aesthetics and conviviality as key factors in a successful wine tourism experience. *International Journal of Wine Business Research*, 33(2): 176–196. https://doi.org/10.1108/IJWBR-12-2019-0063

Harper, S., and Leeson, G. (2008). Introducing the *Journal of Population Ageing*. *Journal of Population Ageing*, 1(1): 1–5. https://doi.org/10.1007/s12062-009-9012-6

Hayes, D., and MacLeod, N. (2007). Packaging places: designing heritage trails using an experience economy perspective to maximize visitor engagement. *Journal of Vacation Marketing*, 13(1): 45–58. https://doi.org/10.1177/1356766706071205

Horneman, L., Carter, R. W., Wei, S., and Ruys, H. (2002). Profiling the senior traveler: an Australian perspective. *Journal of Travel Research*, 41(1): 23–37. https://doi.org/10.1177/004728750204100104

Hosany, S., and Witham, M. (2010). Dimensions of cruisers' experiences, satisfaction, and intention to recommend. *Journal of Travel Research*, 49(3): 351–364. https://doi.org/10.1177/0047287509346859

Jang, S., and Wu, C.-M. (2006). Seniors' travel motivation and the influential factors: an examination of Taiwanese seniors. *Tourism Management*, 27: 306–316. https://doi.org/10.1016/j.tourman.2004.11.006

Kim, J.-H., Ritchie, J., and Tung, V. (2010). The effect of memorable experience on behavioral intentions in tourism: a structural equation modeling approach. *Tourism Analysis*, 15(6): 637–648. https://doi.org/10.3727/108354210X12904412049776

Lee, K., Kladou, S., Usakli, A., and Shi, Y. (2022). Inspiring winery experiences to benefit destination branding? Insights from wine tourists at Yantai, China. *Journal of Hospitality and Tourism Insights*, 5(1): 116–137. https://doi.org/10.1108/JHTI-06-2020-0109

Manthiou, A., Lee, S., Tang, L., and Chiang, L. (2014). The experience economy approach to festival marketing: vivid memory and attendee loyalty. *Journal of Service Marketing*, 28(1): 22–35. https://doi.org/10.1108/JSM-06-2012-0105

Mehmetoglu, M., and Engen, M. (2011). Pine and Gilmore's concept of experience economy and its dimensions: an empirical examination in tourism. *Journal of Quality Assurance in Hospitality and Tourism*, 12(4): 237–255. https://doi.org/10.1080/1528008X.2011.541847

Musa, G., Mohammad, I., Thirumoorthi, T., Moghavvemi, S., and Kasim, A. (2015). Exploring visitors' experience using strategic experiential modules (SEMs): the case of Zoo Negara, Kuala Lumpur. *International Journal of Tourism Cities*, 1(3): 234–253. https://doi.org/10.1108/IJTC-08-2014-0012

Musa, G., Najmin, S., Thirumoorthi, T., and Taja, A. Z. (2017). Examining visitors' experience with Batu Cave, using the four realm experiential theory. *International Journal of Tourism Cities*, 3(2): 105–120. https://doi.org/10.1108/IJTC-10-2016-0037

Nella, A., and Christou, E. (2016). Extending tourism marketing: implications for targeting the senior tourists' segment. *Journal of Tourism, Heritage & Services Marketing*, 2(1): 36–42. https://doi.org/10.5281/zenodo.376336

Oh, H., Fiore, A. M., and Jeoung, M. (2007). Measuring experience economy concepts: tourism applications. *Journal of Travel Research*, 46(2): 119–132. https://doi.org/10.1177/0047287507304039

Otoo, F. E., Kim, S., and Choi, Y. (2020). Understanding senior tourists' preferences and characteristics based on their overseas travel motivation clusters. *Journal of Travel & Tourism Marketing*, 37(2): 246–257. https://doi.org/10.1080/10548408.2020.1740136

Park, K. S., Reisinger, Y., and Kang, H. J. (2008). Visitors' motivation for attending the South Beach Wine and Food Festival, Miami Beach, Florida. *Journal of Travel & Tourism Marketing*, 25(2): 161–181.

Passebois Ducros, J., and Euzéby, F. (2020). Investigating consumer experience in hybrid museums: a netnographic study. *Qualitative Market Research*, 24(2): 180–199. https://doi.org/10.1108/QMR-07-2018-0077

Patterson, I. (2017). *Tourism and Leisure Behaviour in an Ageing World*. Oxfordshire: CAB International.

Patterson, I., and Pegg, S. (2009). Serious leisure and people with intellectual disabilities: benefits and opportunities. *Leisure Studies*, 28(4): 387–402. https://doi.org/10.1080/02614360903071688

Patterson, I., and Balderas, A. (2020). Continuing and emerging trends of senior tourism: a review of the literature. *Journal of Population Ageing*, 13: 385–399. https://doi.org/10.1007/s12062-018-9228-4

Pikkemaat, B., Peters, M., Boksberger, P., and Secco, M. (2009). The staging of experiences in wine tourism. *Journal of Hospitality Marketing & Management*, 18(3): 237–253. https://doi.org/10.1080/19368620802594110

Pine, B. J., and Gilmore, H. J. (1999). *The Experience Economy: Work Is Theatre and Every Business a Stage*. Boston: Harvard Business School Press.

Quadri-Felitti, D., and Fiore, A. M. (2012). Experience economy constructs as a framework for understanding wine tourism. *Journal of Vacation Marketing*, 18(1): 3–15. https://doi.org/10.1177/1356766711432222

Quadri-Felitti, D., and Fiore, A. M. (2013). Destination loyalty: effects of wine tourists' experiences, memories, and satisfaction on intentions. *Tourism and Hospitality Research*, 13(1): 47–62. https://doi.org/10.1177/1467358413510017

Quadri-Felitti, D., and Fiore, A. M. (2016). Wine tourism suppliers' and visitors' experiential priorities. *International Journal of Contemporary Hospitality Management*, 28(2): 397–417. https://doi.org/10.1108/IJCHM-05-2014-0224

Rachão, S. A. S., Breda, Z., Fernandes, C., and Joukes, V. (2020). Food-and-wine experiences towards co-creation in tourism. *Tourism Review*, 76(5): 1050–1066. https://doi.org/10.1108/TR-01-2019-0026

Radder, L., and Han, X. (2015). An examination of the museum experience based on Pine and Gilmore's experience economy realms. *Journal of Applied Business Research*, 31(2): 455–470. https://doi.org/10.19030/jabr.v31i2.9129

Ruys, H., and Wei, S. (2001). Senior tourism. In Douglas, N., Douglas, N., and Derrett, R. (eds.), *Special Interest Tourism*. Australia: John Wiley and Sons.

Saayman, M., and van der Merwe, A. (2015). Factors determining visitors' memorable wine-tasting experience at wineries. *Anatolia*, 26(3): 372–383. https://doi.org/10.1080/13032917.2014.968793

Sotiriadis, M. D. (2015). Culinary tourism assets and events: suggesting a strategic planning tool. *International Journal of Contemporary Hospitality Management*, 27(6): 1214–1232. https://doi.org/10.1108/IJCHM-11-2013-0519

Stergiou, D. P., Airey, D., and Apostolakis, A. (2018). The winery experience from the perspective of Generation Z. *International Journal of Wine Business Research*, 30(2): 169–184. https://doi.org/10.1108/IJWBR-03-2017-0018

Tung, V. W. S., and Ritchie, J. B. (2011). Exploring the essence of memorable tourism greenexperiences. *Annals of Tourism Research*, 38(4): 1367–1386. https://doi.org/10.1016/J.ANNALS.2011.03.009

United Nations, Department of Economic and Social Affairs, Population Division. (2019). *World Population Ageing 2019: Highlights* (ST/ESA/SER.A/430). https://www.un.org/en/development/desa/population/publications/pdf/ageing/WorldPopulationAgeing2019-Highlights.pdf

Vigolo, V. (2017). *Older Tourist Behavior And Marketing Tools*. Cham, Switzerland: Springer. https://doi.org/10.1007/978-3-319-47735-0

Vo Thanh, T., and Kirova, V. (2018). Wine tourism experience: a netnography study. *Journal of Business Research*, 83: 30–37. https://doi.org/10.1016/j.jbusres.2017.10.008

Wei, S., and Milman, A. (2002). The impact of participation in activities while on vacation on seniors' psychological well-being: a path model application. *Journal of Hospitality & Tourism Research*, 26(2): 175–185. https://doi.org/10.1177/1096348002026002006

PART III

Wine tourism as an instrument for regional development

17
WINE TOURISM
A story rooted in place

Caroline Morrow

Introduction

This chapter explains how wine tourism destinations build resilience when their destination narratives connect wine to place. Wine tourism destinations can strengthen and diversify their destination identities by using stories and related visitor experiences to highlight links between local wine, landscapes, traditions, and people. A region's wine heritage does not need to be centuries old to develop wine tourism, but the story of wine does need to be rooted in place for a wine tourism destination's identity to have meaning and integrity.

At the Culinary Tourism Alliance, food tourism is considered a subset of cultural tourism that includes any experience where a visitor interacts with food *and/or drink* that reflects the local history, heritage, and cultures of a place (Culinary Tourism Alliance 2020). Wine tourism is thereby categorized as a subset of cultural tourism and also a subset of culinary tourism. The terms food tourism, gastronomy tourism, and culinary tourism are considered synonymous, whereas agritourism is seen as a subset of culinary tourism focusing on food and drink production. Importantly, wine tourists at their core are cultural tourists – a point that sits at the foundation of this chapter. Depending on the motivations of the visitor, wine tourism can also be a subset of agritourism. This is the case when a wine tourism experience focuses on the processes behind grape growing and wine production.

Understanding links between wine tourism and other forms of tourism combats tendencies to oversimplify the offerings of a destination – a behavior that does a disservice to destination identity and weakens resilience. Nuance adds richness to a destination narrative and builds longevity into the appeal of a destination identity. For the curious traveler, a deeper level of exploration is possible from repeat visits to the same destination. When visitors are aware of the intricate connections between a destination's cultures, traditions, landscapes, and people, they gain additional meaning from their trips and a stronger sense of belonging to the destination. Destination developers and tourism businesses both have a role to play in connecting wine tourism destination identities and narratives to place.

This chapter starts by exploring connections between wine and place, examining how these relate to building strong wine tourism destinations with supportive narratives. Next, the chapter dives into the role of foodways in wine tourism development, and an argument is made against developing wine tourism in isolation. Then, recommendations are presented

DOI: 10.4324/9781003143628-21

on how to increase touchpoints and weave wine tourism into other types of tourism as a tool for broadening potential markets. A case study from Ontario, Canada, is used to support this point. Afterwards, the chapter discusses how a narrative tied to place, along with related experiences, elevates the visitor journey. This is followed by practical tips in the form of checklists for destination developers and tourism businesses to strengthen connections to place. Finally, the chapter wraps up with a focus on future thinking around building resilience into wine tourism development.

Wine and place

A natural pairing

Wine is inextricably linked to place. So much so that the concept of terroir is associated more with wine than any other food or drink product. Even the novice wine drinker knows that soil and climate influence the character of the final bottled product. Wine's local connections, however, reach deeper still. They transcend regional histories, traditions, and landscapes. Although wine regions have long touted their growing conditions and specialized production techniques as indicators of quality, these points of differentiation are not enough to attract a strong visitor base. In the eyes of a prospective visitor, what sets one wine region apart from another is often everything that it offers *beyond* the wine itself.

Wine tourism destinations

A wine tourism destination is one where the position of wine in the destination's identity is undeniable. Typically, this means that the visitor has diverse opportunities to interact with wines that reflect the history, heritage, and cultures of the place. Increasingly, wine tourism offerings are moving beyond tasting and purchasing to include immersive experiences that enable visitors to develop a better appreciation of wine's connection to the place. For example, visitors can develop a deeper understanding of a wine tourism destination when they learn how wine production relates to local identity. From traditional wine tourism businesses, such as producers, to complementary tourism assets like grocery stores, diverse visitor touchpoints can be leveraged to make these connections clear. Stocking local wines and sharing stories that tie wine to place are key strategies that can be applied across the tourism value chain.

There can be both urban and rural aspects to wine tourism destinations. Businesses all along the urban-rural transect combine to form a destination. Hub-and-spoke models are common. In these models, visitors may choose to interact with local wine products in an urban area, perhaps at a museum or tasting room, as well as travel outside of town to see where the grapes are grown and how they are transformed into wine. The stories that are told about the role of wine in the destination and the frequency with which the visitor is exposed to related touchpoints contribute to the overall destination narrative. It should be noted that a wine tourism destination label does not necessarily exclude other descriptors. Tuscany, for example, is a wine tourism destination, a rural tourism destination, an agritourism destination, a walking tourism destination, and a cultural tourism destination, amongst many other identities.

Destination narratives

Positioning wine tourism within a broad context is an important strategy for destination developers. When a wine tourism destination's offering reflects local landscapes, cultures, and traditions, it transforms into something unique that cannot be replicated elsewhere. Regional characteristics become important points of differentiation that add value to the visitor experience, making it feel one of a kind. By highlighting connections between wine and place, strategic storytelling adds to a destination's appeal, entrenching wine into the region's identity. Showcasing these connections tells a destination narrative that is richer and more meaningful than one focusing on wine alone. Fortunately for wine tourism developers and business owners, there is no shortage of stories linking wine to local cultures, foods, traditions, landscapes, and agriculture – especially when local foodways are leveraged.

Furthermore, complementary urban-rural narratives can strengthen a destination image and contribute to a well-rounded wine tourism destination. Such destinations may see visitors enjoying local wines at a bustling city restaurant one evening and the next day reveling in the tranquility of rural landscapes, touring the facilities that produced the wine they drank the night before. Diverse touchpoints make for a stronger destination, one with complementary activities and draws. It also counteracts the concentration of visitors to city centers by dispersing them across a wider area that can more effectively absorb visitor numbers, spreading out both the positive and negative externalities of tourism.

Robust destination narratives attract visitors, and robust destination identities encourage repeat visitation. Visitors encounter destination narratives in the trip planning process but also during their trips. Destination narratives are therefore important platforms for celebrating points of differentiation between one wine tourism destination and another. Of course, a destination narrative can be upheld only if the visitor experience advertised matches the one received. There cannot be a gap between expectations and experience. Without integrity, even the best-crafted destination narrative cannot be upheld for long. Destination identities and narratives must align.

Foodways and wine tourism

Understanding foodways

Foodways describe how culture, environment, economics, and history connect. They explain how ingredients and dishes relate to the heritage and traditions of an area or people. Foodways tell the story of how food and drink are woven into the fabric of a community – for example, how specific foods and wines developed side by side within a given geographic area. For visitors, the stories behind regional drinks and dishes are more than interesting anecdotes. They explain how and when these products are enjoyed by locals, providing a glimpse into local cultures and, through this, making eating and drinking experiences more meaningful.

Foodways are dynamic. They evolve in relation to social, economic, and environmental conditions. Italy's relationship with the tomato, for example, is a classic foodways story. Tomatoes today are often associated with Italian cuisine, through dishes like Caprese salad and *tagliatelle al ragù*. However, when tomatoes were first introduced to Europe from the Americas, they were considered poisonous, and many people refused to eat them. Over time the tomato became a staple ingredient for Italian cooks, and through waves of emigration, Italy's tomato-based recipes have traveled across the world and been adapted many times over, birthing, amongst other dishes, Argentinian *Milanesa napolitana* and Chicago deep-dish pizza.

Another foodways story is that of wine in Argentina. Today, wine plays a key role in the country's food and drink culture. Yet, wine grapes are not native to the region. Vines and production techniques were brought over by the Spanish in the 1500s. With ingenuity, the Argentinean wine industry evolved based on the local context. Argentina's now-renowned wine culture reflects many factors, including immigration patterns, local innovation, and global demand. As the world developed a taste for Argentinean wines, their exports grew, and a complementary wine tourism industry emerged, attracting visitors from near and far.

For travelers interested in foodways, food and drink are about more than sustenance. They are ways of understanding place. Showcasing foodways through storytelling reinforces for visitors the unique connections between cuisine and place. Foodways exist at the destination level and also at the business level. If the food and drink products offered by a business do not neatly fit into a broader regional narrative, personal foodways can also add value to the visitor experience. When a menu explains a restaurant owner's family connection to the products they serve, it illustrates personal foodways and makes the visitor experience more meaningful.

Imagine a restaurant owned by a newcomer family that serves traditional food and drink made using family recipes from their homeland. Over time, these recipes may or may not be influenced by their new surroundings, potentially through the inclusion of new ingredients or production techniques. Stories behind the recipes, the families these belong to, and how the recipes relate to their new context are of interest to a culinary tourist. The relationship between business- and destination-level foodways should not be overlooked. When a visitor develops a personal connection to local foodways through a wine tourism activity at a business, this tangibly adds to the visitor experience of the destination.

Foodways as expressions of destination identity

Understanding foodways allows wine tourism developers and business owners to tell stories that add context to local traditions and thereby greater value to the visitor experience. Foodways offer genuine points of differentiation that combat the homogenization of food and drink offerings, something that is all too easy in a globalized world. In a time when, for many, the disconnect between agricultural production and gastronomic consumption is widening, foodways provide an urban-rural link. Through the context that foodways bring, a meal or a drink becomes more meaningful.

Foodways are useful tools in building a destination narrative. They help to connect wine tourism experiences to place by linking these to a broader context. Foodways reflect the interactions of the many pieces that add context to a destination, from its cultures to its landscapes, and they show off the richness of the destination's identity – something that is essential to crafting a strong destination narrative. Importantly, the intricacies of the stories behind local foodways appeal to cultural tourists because they allow visitors to get to know a place better. Since wine tourists are at their core cultural tourists, businesses and destination developers can leverage foodways to enhance wine tourism experiences and showcase connections between wine and place.

Foodways as connections to place

Foodways add meaning to travel by drawing connections to place, highlighting why a visitor's experience in one destination cannot be replicated elsewhere. Foodways are based on local characteristics, making them unique points of differentiation. Foodways stories are valued by cultural tourists and anyone looking to relate more deeply to the places they visit. When visitors understand foodways, connections between wine and place become richer and more memorable. A glass of Loire Valley Cabernet Franc is more meaningful when enjoyed with a traditional meal of *roulade d'agneau* overlooking the vineyards where the grapes were grown.

Tastes of place

Tastes of place are expressions of foodways. They are manifestations of local culture and identity that visitors and locals can interact with. These interactions bring to life the food and drink of an area, resulting in more memorable visitor experiences. Sampling local wines in a cellar, for example, is a taste of place experience that directly connects participants to a destination's foodways. However, not all tastes of place are necessarily consumable. They can also include intangible assets. Rural hospitality, agricultural landscapes, and good company, for instance, are tastes of place that, through wine tourism activities, allow visitors to connect with a destination.

No wine tourism destination has one overarching taste of place. No single dish, drink, or agricultural technique can fully define a destination's identity. Instead, a destination is made up of an array of tastes of place that each add value to the visitor experience. Food, wine, climate, architecture, and hospitality, among others, all intermingle to create a complete story of regional foodways. For example, enjoying local food and wine at a cultural festival as opposed to a sterile laboratory setting provides additional context that adds to the visitor experience, allowing the visitor to connect to the essence of the destination and better understand its identity.

The more often a visitor is exposed to local tastes of place, the more clearly a destination's identity, including its foodways, is understood. Local tourism businesses can strengthen a destination narrative through the tastes of place that they include in their service offering. With thoughtful storytelling, both online and in-person, destination developers and tourism businesses can strategically tie diverse tastes of place together, making these resonate more effectively with visitors. Menus, chalkboards, websites, and social media are just a few of the platforms that can be used to share stories about foodways and tastes of place with visitors.

More than a wine tourism destination

Imagine that a pair of young professionals are planning a weekend away. They want to relax in a rural area where they can connect with nature, eat great food, and stay in a small town. High on their list of must-do activities are hiking and experiencing local culture. They decide to visit a wine region because of the beautiful vineyard landscapes, walking trails, and rural gastronomy. For these travelers, wine is a key part of any vacation dining experience. They are also curious to learn about wine production processes and to try local specialties. They plan their walking route specifically to have lunch at a winery where they can admire beautiful views. In this example, wine adds to the destination's appeal, but it is only part of it. For these visitors, wine is a way to dive into local cultural traditions and imbue deeper meaning and personal connection to the landscapes across which they hike. Wine is not, however, their primary motivation for travel. Travelers with broad motivations make up a large part of the visitor market and should not be overlooked by wine tourism destinations.

It's easy to attract the serious wine enthusiast. They will travel great distances to see the birthplace of their favorite varietal or to sample bottles unavailable in their home country. Great wine alone, however, does not automatically equate to a great wine tourism destination. To create a resilient destination, the offering must *also* appeal to the travel companion of the wine enthusiast and attract the wine curious, cultural tourists, and travelers seeking rural getaways. When debating between two wine regions, the one with the more holistic offering is, for many travelers, the obvious choice. A well-rounded destination offers diverse touchpoints, allowing visitors (regardless of their travel motivations) to interact with the destination in multiple ways via its cultures and landscapes. Such destinations allow visitors to connect with the essence of the place on more than one level.

When wine tourism is combined with other forms of tourism, such as food tourism, rural tourism, or outdoor tourism, it appeals to broader markets. Although many visitors would happily spend an afternoon at a vineyard, only a niche segment will want to make it the raison d'être for their trip. Many more think of wine as *part* of an itinerary. Destination developers can combine wine tourism with other forms of tourism to maximize potential markets and reach the full spectrum of wine tourists. Everyone from a visitor checking must-try vintages off a list, to someone casually ordering a bottle of local wine with dinner, to the rural tourist seeking vineyard landscapes can be considered a potential visitor market. For visitor segments with less experience in wine tourism, the prospect of participating in wine tastings and winery tours can be intimidating. However, when wine tourism is combined with other activities, like hiking or picnic lunches, it can become more welcoming and appealing. By developing a layered, multifaceted visitor offering, wine tourism destinations become approachable and attractive to a larger group of potential visitors.

Case Study 17.1 Prince Edward County, Ontario

Prince Edward County (PEC) is a Canadian wine region that is home to around 40 wineries. It is located on a peninsula on the north shore of Lake Ontario, a two-hour drive east of Toronto. Prince Edward County is a Designated Viticulture Area characterized by limestone-rich soils, a lake-moderated climate, and cold winter temperatures that result in the need to seasonally bury vines underground. Pinot Noir and Chardonnay are two particularly popular grapes in the region; however, Riesling, Gamay, Cabernet Franc, Baco Noir, and many other varieties are also grown.

Prince Edward County is a popular weekend getaway destination, particularly for urbanites looking to escape the city. Located in proximity to one of Canada's busiest motorways, it is accessible from large population centers, including Toronto, Ottawa, and Montreal. Sandbanks Provincial Park is a keystone attraction for the area, boasting an impressive freshwater sand dune system that appeals to naturalists and beachgoers alike. Wineries, as seen in Figure 17.1, are a key part of the visitor experience, but only part of it. Prince Edward County's rural landscapes are home to a large number of small boutique businesses that contribute to the overall destination narrative. These include restaurants, farms, cideries, meaderies, breweries, bed and breakfasts, and artist studios.

A defining character trait of the region is the bounty of food and drink experiences on offer. PEC's diverse visitor offering means that the destination appeals to broad visitor segments. Wine tourists will find plenty to do, but so too will those seeking a rural getaway or entertainment. Agro-alimentary offerings, from roadside farm stands, to cider tastings, to photoshoots in sunflower fields, are defining features of the destination. Wine is woven into other offerings, whether that be through the vineyard landscapes or local bottles featured at one of PEC's trendy accommodations. Wine becomes an essential part of the visitor experience, but not in a way that feels overbearing. Visitors can incorporate as few or as many wine tourism activities into their itineraries as they choose. Wine tourism offerings vary in time commitment and price, catering to different visitor preferences.

Figure 17.1 Prince Edward County, Ontario.

Source: Culinary Tourism Alliance/Agatha Podgorski

Prince Edward County's destination marketing and management organization celebrates place through complementary tourism assets and ties wine tourism into its other visitor offerings. For many years, the Taste Trail was a primary trip planning tool for visitors to PEC. Distinctive blue road signs accompanied by print and digital maps made it easy for visitors to design an itinerary. The trail played a key role in building the region's identity as a place for quality food and drink experiences. The 2018–2019 trail guide lists 32 stops ranging from wineries to restaurants (Prince Edward County 2018). Although the Prince Edward County Taste Trail is frequently used as a case study for culinary tourism development tools in Ontario, it was recently dissolved. Food and wine tourists planning a trip today can consult the Prince Edward County Official Tourism site for information about the local food and drink experiences on offer. In addition, Wine Country Ontario and the Prince Edward County Wine Growers Association provide maps and trip planning tools listing the region's wineries.

Wine is firmly ingrained in the overall visitor experience, but with the wide variety of complementary experiences on offer, the destination can meet the needs of diverse visitor profiles – from rural tourists, to food tourists, to outdoor enthusiasts. PEC can thereby capitalize on a wide array of visitor segments while directing visitor spending to its local food and drink businesses. Unfortunately, the rapid growth in popularity of Prince Edward County and its limited geographic size have combined to make overtourism, especially during the summer months, an increasing concern for many residents of (and visitors to) the region. Local destination developers are currently exploring ways to combat the environmental and social-cultural externalities that can come with popularity as a tourism destination.

Prince Edward County provides an interesting case study for wine tourism destination development. With its diverse visitor offering and a narrative linked to place, farm-to-table dining, craft alcohol, and an array of locally owned businesses are key to the destination's appeal. Regardless of a visitor's primary motivation for travel, integrating wine tourism into an itinerary is simple and approachable. Visitors interact with the local wine industry in many ways, including sampling a local bottle at dinner, participating in a guided wine tour, or stocking up on supplies before heading home.

Prince Edward County has successfully linked wine tourism to its overall destination image while still appealing to broad visitor segments through its complementary activities. The intermingling of wine tourism with other types of tourism fits the destination narrative. It adds enough diversity to build resilience into the destination's appeal, but not so much that it dilutes the destination's identity. Although the destination applies many good practices, such as diversifying its food and drink narrative and offering, it is suffering from overtourism in the high season, and there are lessons to be learned from this as well. When using diverse and complementary activities to appeal to a wide range of target markets, wine tourism destination developers must keep the carrying capacity of the destination top of mind in their planning processes.

Elevating the visitor experience

A growing number of visitors want to connect deeply with the places they visit. Not only are they looking to feel touched (even transformed) by their trips, they are also hoping to extract meaning from their travels and deliver benefits to the communities they visit. Itinerary decisions can help visitors achieve these goals. For instance, those seeking purposeful travel experiences may focus their spending choices on local businesses and strategically pick destinations that do not suffer from overtourism. Often, this kind of thoughtful travel also benefits the visitors themselves, whether this is because of a thought-provoking conversation with a local business owner, or from learning a new skill such as how to stomp grapes or when to decant a wine.

Immersive experiences led by local guides are an effective tool for connecting visitors to place. This is reflected in the growth of local-led tourism activities through platforms like Airbnb Experiences. Visitors want to learn about the places they visit. Connecting directly to those in the food and wine industry is a meaningful way to achieve this. Local guides can tell participants about the region's wines but also introduce them to the vintners, landscapes, and history of the local area. These kinds of experiences would be incredibly difficult for the independent traveler to replicate, making the local guide a key tool for connecting more deeply to place.

Regardless of whether visitors are motivated by food, wine, or something else entirely, people need to eat and drink, and this is an advantage for wine tourism destinations. Food and wine tend to evolve together in situ, and as such, their combination comes with inherent storytelling opportunities around local culture and traditions. Eating experiences can be elevated when connected to wine and vice versa. On the part of the visitor, there is also growing expectation that travel experiences be immersive and multisensory. Of course, by its very nature, wine tourism is at an advantage to meet this demand, since eating and drinking are two of the only activities that allow visitors to activate all five senses. It is no wonder that food and drink are being leveraged by everyone from ecotourism providers to conference center operators, to elevate visitor experiences. To have integrity and meaning, however, food and drink experiences must connect to broader destination identities.

Connecting wine tourism to a broader destination identity

Checklist for wine tourism destination developers

By leveraging connections between wine and place and integrating elements of regional identity into the destination brand and products, destination developers can strengthen ties between visitors and the destination. The more frequently visitors are exposed to regional points of differentiation, the stronger this bond will feel. Of course, consistency between the destination narrative and the offering on the ground is essential. This is true for regional products, like wine trails, as well as for the wine tourism experiences offered by individual businesses. The following is a checklist for wine tourism destination developers seeking to connect wine tourism to place through a broader destination identity.

- Have a clear destination narrative that is echoed across the visitor journey.
- Connect wine to other cultural and natural assets, including agricultural landscapes, food products, events, traditions, etc.
- Consistently celebrate foodways, helping visitors understand how wine relates to the broader picture. (For example, if a specific wine is consumed on feast days, explain this to visitors. It will help them to better understand the product and the destination identity.)
- Empower local businesses to echo the regional narrative and add to it with their own stories. (For example, build capacity in a broad range of tourism business owners so that they understand the regional narrative and can explain the role of wine in the destination, regardless of whether or not wine is a focus of their business. At the same time, encourage creativity so that they can provide complementary stories, as this adds integrity and richness to the overall destination narrative.)
- Connect local businesses to each other so that they are aware of each other's offerings and stories, empowering them to be effective ambassadors for the region. (For example, virtual meet-and-greets, industry events, online forums, and mailing lists are some ways that destination developers can help to connect businesses to each other.)
- Add volume to business stories by sharing these through social media, marketing materials, etc., because these stories add richness and detail to the overall destination narrative.

- Celebrate the people behind local wines, adding a human element to the regional narrative, connecting visitors to place, and fostering in them a sense of belonging within the destination.
- Align regional product development with the overall destination narrative. (In designing wine routes, certification programs, and events, it is important to base design decisions on the regional narrative.)
- Work with local businesses to develop immersive, multisensory experiences that bring the destination narrative to life. (For example, empower businesses to design offerings that reflect their own business in alignment with the destination narrative, as this ensures consistency in what is advertised and what visitors actually experience. Of course, doing so also helps visitors to connect more deeply with the destination.)

Checklist for wine tourism businesses

Wine tourism businesses can help visitors connect to place by offering experiences that highlight links between wine and the destination while providing opportunities for visitors to learn about the destination's other assets. A wine tasting paired with local food would be one way to do this. The following is a checklist for wine tourism businesses seeking to connect wine to a broader destination narrative through the visitor experiences that they offer. The checklist targets wine tourism businesses; however, other tourism businesses, like food retailers and tour agencies, can also provide visitors with wine tourism touchpoints, thereby adding integrity and consistency to the destination narrative.

- Work with local destination developers to explore how your business story fits into the regional tourism narrative.
- Regularly share stories about your business's connection to place online and in-person.
- For all online channels, select high-quality photos of your wine tourism products and experiences, including images of people enjoying these products. (Visitors, especially those motivated by food and drink, are resourceful. They will go online in advance to decide where, and often what, to eat or drink. Including high-quality photos of people enjoying your tourism offerings helps prospective visitors understand what you offer and inspires visitation.)
- Regularly share compelling content through your social channels and encourage your customers to do the same. (Be sure to reshare user-generated content, as it often appears to potential visitors as more authentic than content produced by a business.)
- Strive to have all visitors leave your business with a clear understanding of your story and its connection to the region. (For example, wine products, complementary food products, and related stories can all celebrate connections to place.)
- Empower your staff to be ambassadors for the region. (For example, offering opportunities for your staff to visit other local businesses allows them to provide visitors with personal reflections and recommendations for other experiences to have in the region.)
- Offer unique tastes of place to help visitors connect to local foodways.
- Highlight local ingredients, suppliers, and partners. (For example, collaborating with other local businesses adds value to your offering. It creates links between your business and the destination and strengthens your unique selling points by making them impossible to replicate. Celebrating partnerships might mean naming them on your menu or posting about them on social media.)
- Empower your staff to share engaging stories about local ingredients, suppliers, and partners. (For instance, consider inviting your suppliers to meet your staff or organizing

fieldtrips so that your staff can visit suppliers, like coopers or farmers. This personal connection enriches the stories your staff share with visitors.)

- Celebrate your ties to local community. (For example, if you participate in events, tours, routes, etc., highlight these! Including your business in regional tourism products by becoming a member of your local destination marketing and management organization will also tie your business to place.)
- Offer immersive, multisensory wine tourism experiences. (Help visitors learn about and connect with your business and region in a memorable, hands-on way.)
- Offer multiple opportunities for visitors to interact with your products and take these home with them. (When you provide a compelling story, visitors will want to support your business, bring your products home with them, and share these with family and friends. Don't miss this opportunity! Portable tastes of place can inspire repeat visitation and act as a referral for the gift recipients as well.)

Moving forward with resilience

Shocks to the tourism industry can come at unexpected times and in unpredictable ways. Building resiliency to disturbances is essential to successful wine tourism destination management. Travel restrictions, visitor hesitancy, and other legacies of the COVID-19 pandemic mean that destinations will need to diversify their visitor base and show flexibility in the markets that they pursue.

As the world emerges from the pandemic, patrons will, at least in the short and medium term, travel closer to home. Wine tourism destinations that target international visitors at the expense of domestic markets will need to rethink their offering to align with the needs of domestic visitors. Innovation in meeting local visitor needs in wine tourism destinations is already underway around the world. These include developing loyalty programs, focusing on outdoor events, offering online food and wine catering services, and communicating with local markets via social media (International Organisation of Vine and Wine and United Nations World Tourism Organization 2020). Connecting wine tourism to other tourism activities, like gastronomy tourism or recreation tourism, broadens potential visitor markets. Similarly, diversity in tourism product and experience offerings makes destinations more resilient.

Wine tourism is one of a few types of tourism that is innately multisensory. When designed thoughtfully, wine tourism experiences have the potential to connect visitors to place and inspire a sense of belonging and attachment to the destination. This can prove valuable in encouraging repeat visitation. Aligning experience development and product development with a strong destination narrative is key to ensuring that visitor expectations are met (and exceeded). This is essential to the longevity and resilience of the destination. At the same time, complementary tourism activities reinforce parallel connections to place, yielding a destination that supports longer trip itineraries. This can encourage overnight stays, attract diverse visitors, and foster regional development through a larger and broader visitor spend.

A destination narrative is an essential tool for transmitting the cultural value of wine and building the appeal of the overall destination. At the same time, to maximize the value of wine in destination development, its inextricable connection to place must stay at the forefront of the visitor experience. Behind every bottle is a unique set of people, landscapes, and techniques. Understanding these foodways makes it possible for everyone from the wine afficionado to the wine novice to more meaningfully connect with a wine tourism destination. By providing diverse opportunities for visitors to interact with wine, tourism businesses and destination developers can foster a sense of connection to the destination. Linking wine to (and integrating it within) other forms of tourism increases the frequency of visitor interactions with local wine and improves the likelihood that visitors will grasp the key role of wine in the destination's identity. The story of wine truly comes to life for visitors when both the destination narrative and the visitor experience are rooted in place.

References

Culinary Tourism Alliance. (2020). *Food Tourism 101*. https://www.culinarytourismalliance.com/food-tourism-101

International Organisation of Vine and Wine and United Nations World Tourism Organization. (2020). *Recreating Bonds Is Essential for Innovation in Wine Tourism*. https://www.oiv.int/en/oiv-life/recreating-bonds-is-essential-for-innovation-in-wine-tourismnbsp

Prince Edward County. (2018). Taste Trail 2018–2019. *Picton*. https://www.visitpec.ca/wp-content/uploads/2018/05/TasteTrail_Extracted.pdf

18

STRATEGIES FOR DEVELOPING WINE TOURISM DESTINATIONS

The case of the Douro Valley (Portugal)

Eduardo Cândido Cordeiro Gonçalves and Ricardo Jorge da Costa Guerra

Introduction

Today's ideal planning and management of a tourism destination implies the participation of multiple public and private actors and needs a framework that embodies the 'negotiation, discussion and cooperation' between them. Thus, taking the Douro Valley (Portugal) as a territorial subject, a cluster that currently crosses the wine and tourism sectors, we propose to present the optimal design of the planning and strategy for the sustainable and responsible development of tourism in the region (Aguiar 2002; Aguiar et al. 2001).

The vineyards and wine are recognized as the region's main attraction, the basis of a landscape shaped by humans for centuries, with which the Port Wine brand is associated. Currently, there is a clear redefinition of the agricultural areas in the Douro, giving them a complementary tourist character and enhancing their heritage resources, while increasing and modernizing the specificities of agricultural technologies around viticulture. The logic of the integrated development of this tourism destination is based on sustainable development based on a territorialist paradigm, where wine tourism requires a robust economic and agricultural base. In addition, the 'identity culture' (Covas and Covas 2007) is important in a rural context, as it is largely due to the endogenous and exogenous valorization of these territories. The development models of these oenological destinations are often the result of the approaches in which they are thought; namely, they are seen in terms of supply or demand or are approached in terms of the motivation of the tourist or the design of the destination (Getz 2000; Hall et al. 2000).

The study of wine tourism has been pursued by several authors, whose study focuses on several countries with wine regions, of which the following stand out: Canada (Hackett 1998; Telfer 2001; Williams and Dossa 2003), Australia (Dowling and Carlsen 1999), New Zealand (Beverland 1998; Hall 2003), Chile (Sharpley 2002), France (Frochot 2000), Hungary (Szivas 1999), Italy (Pavan 1994), Spain (Gilbert 1992), South Africa (Bruwer 2003), United States (Dodd 1995; Skinner 2000), and Portugal (Brás 2010; Costa and Dolgner 2003; Guedes 2006; Mitchell, Hall, and McIntosh 2000; Vaz 2008).

DOI: 10.4324/9781003143628-22

Douro Valley: a wine tourism destination

Considered the oldest demarcated and regulated wine region in the world, the Douro Demarcated Region (DDR), in the North of Portugal, includes an area of 24,000 hectares which, in 2001, was classified by UNESCO as a World Heritage Cultural Landscape (CCDRN 2006).

Considered an 'area of exceptional tourist aptitude' (RCM 139/2003), the Douro deserved the elaboration of a timely development strategy – the *Tourism Development Plan for the Douro Valley* (2004). Despite having constituted 'a strategic reference for tourism development' for this territory, as highlighted in the *Regional Tourism Agenda – Action Plan for Tourism in Northern Portugal* (2015), that plan was consequently updated with the creation of the *Structure Mission for the Demarcated Region of the Douro* (RCM 116/2006), which worked to update the *Tourism Development Plan for the Douro Valley* (2013–2015). In fact, already with the *National Strategic Tourism Plan* (2007), the Portuguese government had considered the Douro a 'Priority Tourist Pole. (Portugal, Conselho de Ministros 2003, 2006; Turismo de Portugal I, P 2007).

Methodology

Considering the epistemological framework of the social sciences, we fundamentally adopt qualitative/deductive and/or mixed methodologies. Therefore, starting from the collection of pre-existing information through the collection of data from various sources of information, we follow an exploratory perspective of a case study. We use both a systematic consultation and analysis of documentary sources to substantiate the objectives proposed for the study and serial data that characterize the tourist destination Douro. The empirical component involves the use of open techniques, namely the interview from the perspective of supply (stakeholders and local trade), complemented by the questionnaire survey from the perspective of the demand in order to enable the design of the tourist profile. Also, using benchmarking to other regions that are wine tourism references, the data collection focused on reading, statistical testing, as well as using the content analysis technique to ensure a treatment that substantiates and validates the problem and further interpretation.

In this investigation, we use the case study method that allows us to examine the object of study considering its 'natural context,' with the referred collection of data in multiple sources, with the question of research as 'Why?' or 'How?' with a focus on current issues and in which the results depend strongly on the capacity of 'integration of the researcher' (Quivy and Campenhoudt 2008).

SWOT analysis

The SWOT analysis, a very popular management tool in marketing, serves to make the strategic planning of companies, organizations, and tourism destinations. This instrument allows the possibility of analyzing scenarios and supporting decision-making. The analysis provides a diagnosis of the situation of the company itself and the environments that surrounds it (Kotler 2013; Kotler and Armstrong 2018; Tocquer and Zins 2004).

Considering the opportunities and threats the Douro region faces as a wine tourism destination, its strengths, and weaknesses, a SWOT analysis is highlighted in Table 18.1.

According to the various studies analyzed (CCDRN 2008; Comunidade Internacional do Douro 2014; Turismo do Porto e Norte de Portugal 2015) and the strategic assumptions of the Portuguese *Tourism Strategy 2027* (Turismo de Portugal 2017) that point to an offer based on endogenous resources, the Douro Valley has distinctive elements, as shown in Figure 18.1.

Table 18.1 SWOT analysis of the Douro Valley as a tourism destination

Strengths	*Weaknesses*
- Climate conducive to wine production - Quality of wines, their diversity, as well as their reputation - Geographic location and unique landscape - Good accessibility - Typical and unique local gastronomy - Culture, history, heritage - Classification of World Heritage Cultural Landscape by UNESCO - Existing tourist activities - Investors who wish to develop wine tourism - The population – welcoming, friendly, and willing to receive - Symbiosis between wine and the typicity of the region - Tasting as motivation for wine tourism - Awards are given to the region from international recognition	- General lack of offer of accommodation - Lack of tourism packages that combine wine tourism with other activities - Lack of a preponderant brand image, which marks the market and allows creation of the desired image as a place of visit or destination - Poor directional signage as well as for the wine routes themselves - Lack of networking between regions, tour operators, other tourism agents and the local economy - Difficulty in mobilizing the sector around collective communication - Aged population - Opening hours of establishments dedicated to wine tourism are not very flexible, when they exist
Opportunities	*Threats*
- Bet on the qualitative and not the quantitative - Emergence of new markets - The growth of the media as a promotional element - International fairs and events as a promotional strategy for the region and their respective products - Continued increase in tourist demand worldwide and in the region - Use of innovations that are implemented in the region - Use of concepts such as emotion, sensation and learning - Consider the family, meetings, and incentive tourism segments - Bet on professionalism and training	- Climate changes - International economic crisis - Pandemics and other health disasters - Competition from Portuguese wine regions such as competition from international regions and the development of wine tourism in new countries - Aging of the population in general - Decline in wine consumption - Political and economic crises that lead to discard the creation of development funds

Source: Adapted from Guerra et al. (2020).

It is clear that wine is a central element of this territory. For each of these elements, differentiating factors that support their importance can also be associated.

Definition of the objectives

The Douro Valley region faces several challenges due to lack of sustainable tourism development, and in this sense, the following general and specific objectives have been defined:

1 Globally improve the indicators of tourist demand (overnight stays, average stay, occupancy rate, tourism receipts, etc.)
2 Valorization of the territory, the communities, and improving the quality of the tourist offer (attracting investment, improving the quality of infrastructures, offering endogenous

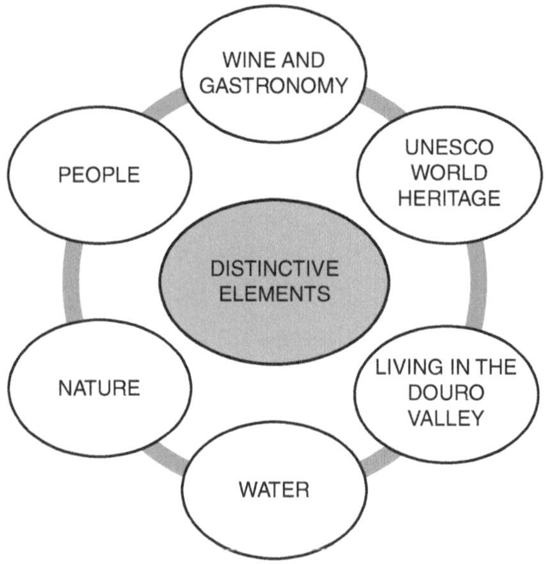

Figure 18.1 Distinctive elements of the Douro Valley as a tourism destination.

Source: Elaborated by the authors

products, strengthening professional training, improving tourism signage, preserving cultural heritage, etc.)

3 Strengthen governance and promote a spirit of cooperation among stakeholders (territorial cohesion and partnerships, integration in national and international networks, foster public and private investment, implementation of public projects, etc.)

4 Raise the levels of awareness and attractiveness of the Douro brand in the national and international market (positioning based on differentiating elements, promotion for potential target market segments, promoting events with projection, fam trips, and press trips, encouraging living in the Douro Valley, etc.)

Achieving these objectives presupposes a strategy aligned to different levels, namely: the definition of strategic tourism products, the identification of target markets with a greater probability of success, and the intended positioning.

Strategic options: tourism products, markets, and positioning

This strategy proposal for tourism development in the Douro Valley is based on a set of guidelines divided into three areas: tourism products, source markets, and strategic positioning.

The diversity and distinctiveness of existing resources allow the Douro Valley region to structure its tourism offer by defining different categories of tourism products that meet different motivations (Table 18.2).

In each of these products, the region offers a more active or passive experience, allowing the markets to live an experience according to their preferences and needs.

Throughout the territory, tourism products gain a differentiated prominence, showing different levels of priority in investment and development. Despite the prominence of one or more tourism products, the capital of the resources offers the region a set of strategic products

Table 18.2 Strategic tourism products to be developed in the Douro Valley

Tourism products	
Strategic products	Gastronomy and wine tourism (Douro Cellars and Estates) Nautical tourism (cruises and Douro Tours) Cultural and landscape touring Nature tourism
Complementary products	Religious tourism Living in the Douro Health and wellness tourism Business tourism

Source: Elaborated by the authors from primary data; Turismo do Porto e Norte de Portugal (2015); Turismo de Portugal (2017).

Table 18.3 Strategic source markets for the development of tourism in the Douro Valley

Source markets	
Priority markets	Portugal Spain France
Strong betting markets	Brazil Germany United Kingdom United States
Moderate betting markets	Italy Netherlands Belgium Switzerland China

Source: Elaborated by the authors from primary data, INE (2019), Turismo do Porto e Norte de Portugal (2015), and Turismo de Portugal (2017).

(core products based on the experience of the destination, resources, infrastructure, and the offer of local companies) and complementary ones (those which are not the main ones but are important in enriching the tourist experience) that allow involvement in different experiences considering the uniqueness of the territory.

Each market has unique characteristics, and should therefore be approached, in a particular and creative way, according to a specific and duly substantiated strategy. In this sense, strategic source markets are divided into current priority markets, strong betting markets, and moderate betting markets (Table 18.3). According to more recent sources (INE 2019; Turismo do Porto and Norte de Portugal 2015; Turismo de Portugal 2017), the current priority markets, with both strong and moderate bets, were categorized according to criteria of the evolution of tourism demand of the Douro and globally, and data on the economic and demographic situation of each.

The current priority markets demonstrate a good knowledge of the destination, are markets with greater tourism demand, and have greater proximity and excellent accessibility. Strong betting markets have a significant tourist demand for the destination, have a margin of growth (recent high growth rates), have strong and potentially growing economies, and have seen a recent increase in accessibility. Finally, there are markets with a moderate bet that also have high potential in terms of tourism, and tourist demand for the destination is also moderate today; they have geographical proximity (except for China) and require a considerable investment effort.

The definition of the positioning of the Douro Valley region is based on the differentiating aspects of its territory, which, taken as a whole, make it a unique destination. The ambition must be to position Douro as a high-quality tourism destination. The central elements of the future positioning are clearly gastronomy and wines (Porto, DOC Douro, Sparkling Wines, and *Moscatel*, from Wine Cellars and Estates in the Douro), history, culture and identity, the Douro River and exuberant landscape surroundings, the classification as World Heritage of Humanity by UNESCO, people and their hospitality, the diversity of natural resources, security, tranquility and well-being, quality and modernity, and the proximity and accessibility to emerging markets.

With the main objective of attracting greater tourism flow and increasing the economic benefits for the local population, the Douro Valley must have a positioning strategy of high and distinctive quality. The pillars of the Douro Valley region are its identity matrix, composed of vineyards, wine, a river with strong tourist potential, landscape, heritage, and culture, thus being distinctive factors and of recognized world value.

Marketing mix policies

The marketing policies for the municipalities of the Douro Valley converge with the general guidelines of Turismo de Portugal and the Regional Tourism Authority of Porto and North of Portugal. However, this is not reflected in all the analyses presented, due to the specificity of the Douro territory, considering the challenges and unique resources existing in this region. The chosen marketing strategy, and taking into account the specificity of tourism, is based on the definition of seven elements: *product*, *price*, *communication*, *distribution*, *people*, *processes*, and *physical evidence*.

Product

The product policy for this region includes the adequacy of the tourist offer to specific market segments. The tourism product in the municipalities of the Douro Valley has great potential for development due to the unique conditions that this region presents in terms of tourism. With the main themes linked to gastronomy and wines, nature tourism, cultural and landscape touring, and also nautical tourism on the Douro River, this region should be responsible for attracting investment with the objective to creating support units for a high-quality and innovative tourism. The development of cultural routes, support structures for nature and adventure tourism, and various events are some examples that can serve to attract visitors throughout the year.

The product strategy focuses on the complementarity of several tourist products, both strategic and complementary. Despite this multi-product approach already presented in Table 18.2, this work will focus on the product of gastronomy and wines, with the clear objective of presenting solutions for the development of wine tourism in the region. Table 18.4 presents the main guidelines of the product policy, which include the objectives that relate to gastronomy and wine tourism products.

Price

Price is one of the most important components that influence the decision to travel. However, it is not an easily controllable element by the destination management entities, as it is the

Table 18.4 Gastronomy and wine tourism product politics to be developed in the Douro Valley

Tourism product	*Core attractions*	*Priority segments*	*Objectives*
Gastronomy and Wine Tourism	• Quality wines recognized in the world (Port Wine, DOC Douro, *Moscatel* and Sparkling Wine from Vale Varosa) • Rich Douro cuisine • Emblematic wine cellars and estates with a strong tradition • UNESCO World Heritage Site – Alto Douro Wine Region • Douro Wine Villages	• Young couples and small groups (between 25 and 60 years old) with high purchasing power/ sociocultural level • Professionals in the catering and wine sector	• Attract visitors to a more frequent, more intense and differentiating consumption, promoting their return. • Attract visitors from competing destinations through a different positioning, considering the uniqueness of the region. • Capture new segments of source markets by calling on unique experiences and specific motivations. • Attract visitors who are already at the destination by presenting complementary experiences. • Foster the attraction of visitors in times of less demand, considering the sustainability of the destination.

Source: Elaborated by the authors.

competence of the sector agents that make up the value chain. In this sense, at the destination level, the strategy involves associating the value with the intended positioning, which is recognized by the markets as a value adjusted to the quality and projected tourist experience. The definition of the price also aims to gain market share and increase the competitiveness of the destination.

For the definition of a price choice strategy, two levels of coherence must be respected: internal coherence, regarding the agents' cost and profitability conditions; and external coherence, considering the purchasing capacity of the markets and the prices of the competitors. In addition, this choice has to be consistent in relation to destination positioning decisions and also in relation to the distribution strategy.

In this sense, as can be seen in Table 18.5, the policy to be followed in the medium and long term should proceed to the adoption of a *premium* price and, immediately, the price standard should aim at gaining market share to the competitors improving its competitiveness. In both cases, a value-added strategy, with a quality/price ratio sought by the visitor, will be essential to encourage the choice of destination in relation to its competitors.

Each local tourism agent in the sector should, on a regular basis, look at benchmarking strategies, taking note of other offers in competing destinations, at the same time that they align with the desired positioning for the Douro Valley region. The price sensitivity of the source markets may be higher or lower depending on several factors already mentioned.

Communication

In terms of communication, the attitude of communicating can be understood as a set of efforts to make a destination that wants to be 'sold' perceived by potentials and current visitors.

Table 18.5 Price strategies to be developed in the Douro Valley

Pricing strategies	*Features*	*Relationship with future positioning*	*Decision*
Competition	Prices defined based on competition can choose between cost strategies (lower value), differentiation (high value), or focus strategies (specific segments).	According to the diagnosis, the municipalities covered by the Douro Wine Villages project demonstrate a lower price than the competitors, conferring a remarkable quality. The destination tourism agents must choose to gain a competitive advantage over the competitors through decisions that lead to a differentiation with the price adjustment according to the increase in the destination's qualification.	To follow in the immediate
Premium	Higher prices accompanied by the quality and exclusivity of the destination and its products of distinct value. The destinations agents should attract market segments with high economic power.	Currently, this should not be the strategy to follow, as it will have to be built and based on an increase in the quality of service and the tourist experience that is intended for the destination in a medium- or long-term view. It is intended, therefore, that the destination aspires to the practice of a higher price associated with a high-quality product, taking advantage of the distinctiveness and recognition of the Douro Region.	To follow in the medium, long term

Source: Elaborated by the authors.

In addition, it is through communication that the destination is able to transmit the desired positioning.

Pointed out as one of the main gaps for the development of tourism in the Douro region by entities, agents, and visitors, this strategy considers six specific objectives in terms of communication: promoting Douro in market segments with a potential appetite for its consumption, increasing its notoriety; captivate consumer attention for Douro's culture and identity; arouse the desire to visit a unique region; encourage a longer stay in the region; induce the desire to return; and increase online presence.

Table 18.6 Gastronomy and wine tourism communication politics to be developed in the Douro Valley

Tourism product	*Markets with largest predisp. consumption*	*Priority segments*	*Message to communicate*	*Communication channels*
Gastronomy and Wine Tourism	Portugal, Spain, France, Brazil, United States, Italy, United Kingdom, Germany, Netherlands, Switzerland	• Young couples and small groups (between 25 and 60 years old) with high purchasing power/ sociocultural level • Professionals in the catering and wine sector	• High value of gastronomic and wine heritage and unique sensory experiences • UNESCO World Heritage Site – Alto Douro Vinhateiro • First demarcated and recognized wine region in the world – Douro • Production of quality notorious wines (Port Wine, DOC Douro, *Moscatel*, and Sparkling Wines from Vale Varosa) • Emblematic estates with a strong tradition and new innovative projects in the wine and catering sector.	• Wine and gastronomy social clubs • Specialized press • Recommendation from family and friends • Internet (websites and social networks) • Brochures/ catalogs • Tourism fairs (general and by product) • Mobile and apps • Institutional newsletter and emails • Family trips and press trips • Magazines and specialized publications

Source: Elaborated by the authors.

As can be seen in Table 18.6, these objectives will translate into communication actions to be carried out in different periods for equally different markets and segments, with the purpose of informing and promoting the destination – the variety and uniqueness of the offer. Taking into account an internationalization strategy, it is necessary to develop multilingual promotional materials.

Distribution

Regarding the destination's marketing and *distribution* strategy and its products and services, and considering the importance of intermediaries in the process and the increasing relevance of the internet for travel planning and acquisition, the guideline to be adopted should follow a mixed approach of traditional and online channels (direct and indirect channels).

This document advocates the need to maximize the main access channels used by different markets for each tourism product. Despite the growing trend towards direct booking of services, especially flights and accommodation, it appears that the intermediary sector

Table 18.7 Distribution channels to be considered for the development of gastronomy and wine tourism in the Douro Valley

Tourism product	*Markets with largest predisp. consumption*	*Priority segments*	*Distribution channels*
Gastronomy and Wine Tourism	Portugal, Spain, France, Brazil, United States, Italy, United Kingdom, Germany, Netherlands, Switzerland	• Young couples and small groups (between 25 and 60 years old) with high purchasing power/sociocultural level • Professionals in the catering and wine sector	• Specialized travel agencies • Internet (website and others) • Specialized sites • Social networks • Mobile applications (apps) • DMC (destination management companies) • Other professionals in the sector

Source: Elaborated by the authors.

remains important. These data are in accordance with the current visitor profile of the Douro region.

The selection of distribution channels should take place according to the purchasing profile of current and potential tourist consumers in the region. In addition to the simultaneous use of direct and indirect channels to place tourism products/services, the Douro region and its tourism agents must take into account that there is a consumption pattern for the gastronomic and wine tourism product (Table 18.7).

People

In the tourism sector, people play an essential role in the success of any strategy. In this element of the marketing mix, any person involved, directly or indirectly, in the production and/or delivery of the tourism product or service can be considered.

It is not enough to put an excellent product on the market. The level of development of the people who provide the accompaniment from the beginning of the choice of destination, during the trip, the stay, and on the return, is a fundamental factor of differentiation. Employees must be as prepared as possible. Increasing their motivation and measuring their level of satisfaction is also necessary. In Table 18.8, it is possible to observe the distinctive characteristics of the people that live/work in the Douro Valley, the intervention audiences, and the proposed actions.

Governance should make every effort so that the relationship of cooperation and mutual assistance with all actors of tourism is growing in the development of joint work in order to achieve success.

Processes

One can define the process as the way to do and how to achieve the highest degree of effectiveness and efficiency possible, thus also looking for the best possible result. They are the procedures of public and private operation in the provision of services. All details are important

Table 18.8 Actions proposed for the element people to be developed in the Douro Valley

Tourism product	*Distinctive features*	*Intervention audiences*	*Proposed actions*
Gastronomy and Wine Tourism	- Hospitality - Douro culture and identity - Qualification and training - Valuing human relationships	- General population - Local, regional, national and international agents and actors - Public and private institutions - Scientific community	- Promote employment, qualification, and valuing of the people - Involvement and awareness of the population - Policy information and transparency - Preservation of authenticity and ways of living for local communities - Motivation of the population and local agents - Governance

Source: Elaborated by the authors.

to diagnose the points to improve in the destination, consequently increasing the degree of visitor satisfaction.

The proposed actions for this element of the marketing mix include the requalification/innovation/modernization of the tourism sector and the destination, the improvement of accessibility, the continuous qualification of human resources and the improvement of the services provided, and the creation of a management structure for the destination, for the development of an observatory project for the tourism sector, and also for networking with the various partners.

Physical evidence

Physical evidence is considered to be everything a customer perceives from the destination from the moment the customer comes into contact with it. It is the environment where the service is delivered/provided. The appearance of the destination, the communication strategy, accessibility, accommodation, catering, signage, and support infrastructure, among others, influence the customer's perception of the quality of the service. Because of the intangibility of the services, it is difficult for visitors to objectively assess their quality.

In this context, the following actions can be considered: the requalification/enhancement of the destination and infrastructures, the improvement and standardization of signage/interpretative information, the creation of high-quality infrastructures/modernity/innovation, and support and incentives for requalification and investment.

Action plan

The action plan for the development of the tourism sector in the Douro Valley aims to guide the entire strategy in a logical sequence and linked by phases that are all interconnected, thus unfolding in seven main actions to be developed, subdivided into complementary actions, such as in Table 18.9.

The strategy defined for the destination must be updated and readjusted whenever there is evidence that demonstrates the need to adapt the model defined in the action plan.

Table 18.9 Action plan proposed for the development of the tourism sector in the Douro Valley (2019–2023)

Actions to develop	*2019*	*2020*	*2021*	*2022*	*2023*
A.I. MANAGEMENT STRUCTURE					
A.I.1 Creation of the management structure and reinforcement of governance					
A.I.2 Promotion of a cooperation spirit among stakeholders					
A.I.3 Maximizing synergies and integration in national and international networks					
A.I.4 Creation of an observatory for the tourism sector and the investor support					
A.II. STRUCTURING AND QUALIFYING THE OFFER					
A.II.1 Inventory of tourism resources					
A.II.2 Strengthening of professional training and qualification of human resources					
A.II.3 Improvement of road and rail mobility and navigability and tourism signage					
A.II.4 Development and improvement of infrastructure in tourist areas					
A.II.5 Development of strategic tourism products					
A.II.6 Definition and organization of thematic routes					
A.II.7 Structuring inclusive tourism and developing accessibility for people with reduced mobility					
A.III. INVESTMENT ATTRACTION AND INNOVATION					
A.III.1 Promotion and attraction of innovative investment in hotels, restaurants, wine tourism, and entertainment					
A.III.2 Presentation of the destination to potential investors					
A.III.3 Modernization and enhancement of the network of heritage and territory interpretation centers					
A.IV. 'PRO-TOURISM' EDUCATION					
A.IV.1 Raising awareness of the importance of tourism to residents					

(Continued)

Table 18.9 (Continued)

Actions to develop	*2019*	*2020*	*2021*	*2022*	*2023*
A.IV.2 Raising awareness among tourism entrepreneurs about the valorization of quality local products					
A.IV.3 Disclosure of the actions developed					
A.V. COMMUNICATION AND PROMOTION					
A.V.1 Development of notorious corporate image					
A.V.2 Creation of promotional material					
A.V.3 Creation of an official tourism portal for the Douro Valley					
A.V.4 Strengthening the positioning and reputation of the Douro destination in the national and international market					
A.V.5 Raising and holding reference events					
A.V.6 Educational visits, family trips, and press trips					
A.V.7 Participation in tourism promotion events in Portugal and other countries					
A.VI. SUSTAINABILITY AND QUALITY					
A.VI.1 Adoption of the Guide of Good Practices developed for the Douro Valley					
A.VI.2 Development of a certification model					
A.VI.3 Sustainable Tourism Douro Program					
A.VI.4 Protection of natural and cultural heritage					
A.VI.5 Promotion of efficient waste management					
A.VII. DESTINATION AND MARKET MONITORING					
A.VII.1 Monitoring the satisfaction of residents and tourism agents (regional, national, and international)					
A.VII.2 Monitoring visitor satisfaction with the experience					
A.VII.3 Study of the visitor's profile					
A.VII.4 Monitoring of the preferences and habits of the travel markets					
A.VII.5 Monitoring of official tourism data					

Source: Elaborated by the authors.

Conclusion

The present study makes it possible to present a set of findings that allow us to highlight new knowledge about the object of analysis: update of an internationalization marketing strategy for tourism in the Douro Valley region. Based on a set of guidelines set out in tourism products, sending markets, and strategic positioning, it is possible to validate an action plan for the development of the tourism sector in the Douro Valley. Countless challenges face the region, especially for institutions, local businesses, and the host community. It is up to the agents who are responsible for the economic and social development of this unique territory to be able to apply best practices with a view to sustainable development, which will bring more benefits to all those interested in this process. Therefore, as a result of pre-existing documentary research and the reference bibliography and validity through the empirical component of the investigation, certain strategic paths and guidelines should be followed to improve the attractiveness and competitiveness of the Douro Valley region with regard to the combination of wine and tourism as the core product.

With a view to the sustainable development of wine tourism, it is necessary to ensure several requirements, which can be analyzed in two aspects: at the level of the destination/region and at the individual level of a wine tourism unit, firstly presenting the requirements and recommendations for the region of the Douro Valley can be confirmed in terms of wine tourism. This must be understood as an opportunity to relate with the visitor, as an activity, simultaneously, of investment in relations with the visitor (creation of a brand and of the firm and profitable relationships), but also as an additional source of income. The most relevant requirements include, from the outset, the necessary existence of several tourist agents in the territory, such as wine tourism units, accommodations, restaurants, travel agencies, tour operators and entertainment companies, museums, wine interpretation centers, wine shops, and tasting services and courses.

It is important that this offer contain both quantity and quality in order to give dimension to it. It also differentiates the fact that Douro Valley is inserted in a demarcated region (DOC) that is an identification of the territory, thus obtaining an administrative body, an appropriate brand, and tourism signage. There is also a need to consider encouraging the qualification of infrastructures and services to achieve levels of excellence. Another aspect to be taken into account is improving the signage on road, rail, and river access. The definition of means and tools that enable greater proximity and relationship with the customer should be encouraged, as well as the development of a post-experience service and other loyalty strategies.

It is also important to have a database with updated customer contacts, present or potential. The training and specialization of entrepreneurs and wine tourism collaborators is essential, as is the good understanding of local wines by employees of other tourist units, such as restaurants, hotels, and local businesses. The pricing policy must be adequate and competitive, with a clear focus on a quality/price ratio. The existence of wine stores and other regional products that must have specific areas for wine tasting, with the appropriate equipment, a shopping area, and guided tours in several languages, should be promoted. The stores should also have a diversified offer that goes beyond the simple offer of wines from the producer, such as gourmet products, to homemade productions of sweets, chocolates, handicrafts, crockery, clothing, merchandising, among others.

In wine regions, it is essential to link wine to tourism. Furthermore, wine may even be the most important differentiating and motivating factor in visiting the territory. In this regard, the Douro Wine Villages project should continue to be a bet for dynamizing the territory (Gonçalves et al. 2018). The dynamism of tourism entertainment in this network of villages will offer the visitor more reasons for attractiveness and greater levels of satisfaction. It is important to value heritage, art, and architecture directly related to wine production. Some estates manage to attract more visitors by combining their products with elements of art and architecture (Vaz 2008).

Furthermore, wine should have a very close relationship with Douro gastronomy, providing harmonious care for the type(s) of wine that are produced and commercialized. For this, wine tourism units, restaurants, and other food and beverage units should be encouraged and targeted for specialization actions in order to explore and enhance this perfect symbiosis. Endogenous products in gastronomy should be valued. It is also advisable to cross-sell products (wines, cheeses, olive oil, and sweets, among others).

The ecotourism and sustainability aspect in the production of wines and in the provision of educational services and programs for visitors is certainly a differentiating element, which meets current trends in tourist demand. Environmental practices and sustainability are factors that tourists currently value the most. In the Douro Valley, it is essential to organize and promote activities and entertainment available to visitors. It is advisable that these activities can meet the visitor's senses, as well as provide a special and unique atmosphere, with activities in rural areas and the offer of meals and/or accommodations on the wine grower's properties. Education actions on the type of wine, its production, and vinification secrets are also good proposals to attract tourists. It is important to mention that the perception of wine regions on the part of tourists has changed over time, and currently, the 'aesthetics of regions' is emphasized – that is, the whole experience – instead of focusing only on wine and its production. The promotion of cross-selling with other offers in the territory should also be a good practice to follow.

References

Aguiar, F. (2002). O Alto Douro Vinhateiro, uma paisagem cultural, evolutiva e viva. *Douro-Estudos & Documentos*, 7(13): 143–152.

Aguiar, F., Andresen, T., Dias, J., and Silva, P. S. (Coord.) (2001). *Plano intermunicipal de ordenamento do território do Alto Douro Vinhateiro-Relatório*. Vila Real: Universidade de Trás-os-Montes e Alto Douro.

Beverland, M. (1998). Wine tourism in New Zealand – maybe the industry has got it right. *International Journal of Wine Marketing*, 10(2): 24–33. https://doi.org/10.1108/eb008679

Brás, J. M. (2010). *As rotas de vinho como elemento de desenvolvimento económico*. Aveiro: UA.

Bruwer, J. (2003). South African wine routes: some perspectives on the wine tourism industry's structural dimensions and wine tourism product. *Tourism Management*, 24(4): 423–435. https://doi.org/10.1016/S0261-5177(02)00105-X

CCDRN. (2006). *Alto Douro Wine Region – World Heritage*. Porto: CCDRN.

CCDRN. (2008). *Plano de Desenvolvimento Turístico do Vale Do Douro 2007–2013*. Porto: CCDRN.

Comunidade Intermunicipal do Douro. (2014). *Estratégia Integrada de Desenvolvimento Territorial da Região do Douro 2014–2020*. Vila Real: CIM Douro e augusto Mateus Associados.

Costa, A., and Dolgner, R. (2003). *O enquadramento legal do enoturismo. In 1° Encontro do Turismo em Espaço Rurais e Naturais*. Coimbra: Escola Superior Agrária de Coimbra.

Covas, A., and Covas, M. M. (2007). *A ecosocioeconomia das áreas rurais de baixa densidade: um ensaio teórico-metodológico*. In *XIII Congresso I da APDR – Recriar e Valorizar o Território*. Angra do Heroísmo.

Dodd, T. (1995). Opportunities and pitfalls of tourism in a developing wine industry. *International Journal of Wine Marketing*, 7(1): 5–16. https://doi.org/10.1108/eb008636

Dowling, R., and Carlsen, J. (1999). *Wine Tourism: Perfect Partners. Proceedings of the First Australian Wine Tourism Conference*. Canberra: Bureau of Tourism Research.

Frochot, I. (2000). Wine tourism in France: a paradox? In Hall, C. M., Sharples, L., Cambourne, B., and Macionis, N. (eds.), *Wine Tourism around the World: Development, Management and Markets* (pp. 67–80). Oxford: Butterworth-Heinemann.

Getz, D. (2000). *Explore Wine Tourism: Management, Development & Destinations*. New York: Cognizant Communication Corporation. https://doi.org/10.1016/S0261-5177(02)00033-x

Gilbert, D. (1992). Touristic development of a viticultural region of Spain. *International Journal of Wine Marketing*, 4 (2): 25–32. https://doi.org/10.1108/eb008597

Gonçalves, E., Guerra, R., Mendonça, J. P., and Pereira, V. (2018). *Douro em movimento, aldeias com vida – Carta de boas praticas*. Maia: Ed. ISMAI; CEDTUR; Douro Generation.

Guedes, C. P. S. (2006). *O comportamento do visitante das caves do vinho do Porto: Da hipérbole de Baco à exploração Venusiana*. Aveiro: Universidade de Aveiro.

Guerra, R., Gonçalves, E., Martins, H., and Mendonça, J. P. (2020). Enoturismo e 'Boas Práticas.' Uma aplicação ao Vale do Douro. In Guerreiro, A., Maduro, A., Gonçalves, E., and Custódio, J. (eds.), *Enomemórias e Enoturismo – O património secular do vinho (História, tradição, identidade)*. Maia: ISMAI; CEDTUR; e MVA.

Hackett, N. (1998). *Vines, Wines and Visitors: A Case Study of Agricultural Diversification into Winery Tourism*. Burnaby: Simon Fraser University.

Hall, C. M. (2003). Politics and place: an analysis of power in tourism communities. In Singh, S., Timothy, D. J., and Dowling, R. K. (eds.), *Tourism in Destination Communities* (pp. 99–114). Wallington: CABI.

Hall, C. M., Sharples, L., Cambourne, B., and Macionis, N. (eds.) (2000). *Wine Tourism around the World: Development, Management and Markets*. Oxford: Butterworth Heinemann.

INE. (2019). *Estatísticas do Turismo 2018*. Lisboa: INE.

Kotler, P. (2013). Marketing para o Século XXI (7.ª Edição). Lisboa: Editorial Presença.

Kotler, P., and Armstrong, G. (2018). *Principles of Marketing*. (Global ed.). London: Pearson Education Limited.

Mitchell, R., Hall, C. M., and McIntosh, A. (2000). Wine tourism and consumer behaviour. In Hall, C. M., Sharples, L., Cambourne, B., and Macionis, N. (eds.), *Wine Tourism around the World: Development, Management and Markets* (pp. 115–135). Oxford: Butterworth-Heinemann.

Pavan, D. (1994). L'Enoturismo tra fantasia e método. *Vignevini*, 21 (5): 18–19.

Portugal. Conselho de Ministros. (2003). Resolução do Conselho de Ministros, No. 139/2003 *Diário da República* no. 199 Série I Parte B de 29/08/2003.

Portugal. Conselho de Ministros. (2006). Resolução do Conselho de Ministros, No. 116/2006 *Diário da República* no. 182/2006, Série I de 2006-09-20.

Quivy, R., and Campenhoudt, L. V. (2008). *Manual de investigação em ciências sociais* (5th ed.). Lisbon: Gradiva.

Sharples, L. (2002). Wine tourism in Chile: a brave new step for a brave new world. *International Journal of Wine Marketing*, 14(2): 43–53. https://doi.org/10.1108/eb008742

Sharpley, R. (2002). Introduction. In Sharpley, R., and Telfer, D. (eds.), *Tourism and Development: Concepts and Issues* (pp. 1–7). Clevedon: Channel View Publications.

Skinner, E. (2000). Napa Valley, California: a model of wine region development. In Hall, C. M., Sharples, L., Cambourne, B., and Macionis, N. (eds.), *Wine Tourism around the World: Development, Management and Markets* (pp. 283–296). Oxford: Butterworth Heinemann.

Szivas, E. (1999). The development of wine tourism in Hungary. *International Journal of Wine Marketing*, 11(2): 7–17. https://doi.org/10.1108/eb008692

Telfer, D. (2001). From a wine tourism village to a regional wine routes: an investigation of the competitive advantage of embedded clusters in Niagara. Canada. *Tourism Recreation Research*, 22(2): 23–33. https://doi.org/10.1080/02508281.2001.11081340

Tocquer, G., and Zins, M. (2004). *Marketing do Turismo*. Portugal: Instituto Piaget.

Turismo de Portugal (2007). *Plano estratégico nacional do turismo: Para o desenvolvimento do turismo em Portugal*. Lisboa: TP.

Turismo de Portugal. (2017). *Estratégia Turismo 2027*. Lisboa.

Turismo do Porto e Norte de Portugal. (2015). *Estratégia de marketing turístico do Porto e Norte de Portugal*. Porto: TPNP.

Vaz, L. (2008). *O enoturismo em Portugal: da cultura do vinho ao vinho como cultura*. Lisbon: Universidade de Lisboa.

Williams, P., and Dossa, K. (2003). Non-resident wine tourist markets: implications for British Columbia's emerging wine tourism industry. *Journal of Travel and Tourism Marketing*, 21(19): 31–46. https://doi.org/10.1300/J073v14n03_01

19

WINE TOURISM DEVELOPMENT

From the host community's perspective

Shuangyu Xu and Carla Barbieri

Introduction

The past three decades experienced a rapid growth of wine tourism throughout traditional wine regions (e.g., Spain, Italy) as well as emerging ones in the 'New World' (e.g., Australia, South Africa, the United States of America [USA]). Abundant research has examined different aspects of wine tourism, including tourists' motivations and preferences (Charters and Ali-Knight 2002; Quintal et al. 2017), contributions to regional development (Carmichael and Senese 2012; Job and Murphy 2006), wine festivals, and events (Mason and Paggiaro 2012; Taylor and Shanka 2002), synergies between wine culture and heritage (Mitchell, Charters, and Albrecht 2012), and marketing and promotion efforts (Barber, Donovan, and Dodd 2008), to name a few.

The literature documents that wine tourism delivers a mix of benefits, which include boosting economic growth, conserving rural lifestyle and landscape, and facilitating regional development and promotion (Mitchell and Hall 2006). However, this set of studies has predominantly adopted either the wine tourist perspective, to understand their consumer behavior, or a macro-regional economic perspective, to uncover issues associated with its development (Gomez, Pratt, and Molina 2019). Yet, the host community's perspective is an overlooked piece from the existing literature. Understanding and involving the host community is fundamental to gaining community residents' support for and, ultimately, the sustainability of tourism development (Eusebio, Vieira, and Lima 2018; Holladay et al. 2020; Latkova and Vogt 2012). It is especially important to assess the level of social capital, defined as the cornerstone of social relations, ties, and structure within a group, to foster the development and success of tourism (Adler and Kwon 2002; McGehee et al. 2010). Thus, a better understanding of residents' perspectives as well as their social capital in wine tourism destinations, especially emerging ones, is critical to fostering the development and success of tourism (Franken, Gomez, and Ross 2018; Xu, Barbieri, and Seekamp 2020).

Despite the importance of residents' attitudes and communities' social capital to support and sustain the success of tourism development, such a line of research focused on wine tourism is limited. The importance of understanding and involving the host community for tourism success, together with the still-limited knowledge pertaining to wine tourism, urges for a closer examination of wine tourism development from the host community's perspective with an integrated approach of residents' attitudes and communities' social capital. This chapter does so by summarizing a study

DOI: 10.4324/9781003143628-23

conducted to examine host communities of a wine tourism destination in terms of residents' attitudes toward wine tourism development and wine tourism–related social capital.

Literature review

Wine tourism

Wine tourism, defined as 'travel related to the appeal of wineries and wine country, a form of niche marketing and destination development, and an opportunity for direct sales and marketing on the part of the wine industry' (Getz 2000, p. 4), uniquely stands at the crossroads of the wine and tourism industry. Early wine tourism research predominantly focused on wine tourism products and their development, with most research being descriptive in nature and confined to Australia and New Zealand (Mitchell and Hall 2006). The continuous growth of wine tourists and cellar door sales and the significance of wine tourism to regional development led to a significant expansion of wine tourism research since the 2000s (Gomez et al. 2019). Thus far, the plethora of wine tourism research has examined wine tourism development (Carmichael and Senese 2012), winery and cellar door (Bruwer et al. 2013), wine tourist behavior (Quintal et al. 2017), wine events and festivals (Mason and Paggiaro 2012), marketing and promotion (Qiu et al. 2013), critical success factors for wine regions (Jones, Singh, and Hsiung 2013), wine tourism models (e.g., testing relationships of wine tourism behavior constructs; Shapiro and Gomez 2014), and wine-related education (Alonso et al. 2014).

Gomez, Pratt, and Molina (2019) further sorted previous wine tourism research into two areas. The first one focused on wine tourism regional development, either investigating the role and contribution of wine tourism to regional development in rural areas, or examining wine tourism networks, clusters, and development policies (Mitchell and Hall 2006; Telfer 2001). The second one centered on the experiential wine tourist, looking into their psychographic profiles (e.g., satisfaction, motivation) and as associated with marketing and promotion practice (Grybovych, Lankford, and Lankford 2013; Kolyesnikova and Dodd 2008). With the aforementioned foci on either the winery/vineyard owners and managers or wine tourists, little is known about the host community's perspective toward wine tourism development. This is especially problematic considering that developing sustainable wine tourism is closely related with local community and understanding the specific impacts caused by wine-related tourism (Montella 2017; Poitras and Getz 2006).

Resident attitudes toward tourism development

A robust research line has examined residents' attitudes toward general tourism development for decades, primarily to evaluate tourism impacts. Attitudes and perceptions are used interchangeably for using similar measurement items and scales (Andereck and Nyaupane 2011; Xu et al. 2016). Residents' attitudes toward tourism were found to be directly associated with their support for additional development and the revitalization of communities (Lankford and Howard 1994; Latkova and Vogt 2012). Different variables (e.g., demographics, job dependency on tourism, community satisfaction, community attachment) influence residents' attitudes toward tourism development (Deccio and Baloglu 2002; McGehee and Andereck 2004; Nunkoo and Ramkissoon 2011). Despite the ample research on residents' attitudes toward tourism developments, the recent quest for sustainability has triggered a renewed interest in residents' attitudes toward tourism especially associated with environmental and social impacts (Gursoy, Chi, and Dyer 2009; Xu et al. 2016).

The Social Exchange Theory (SET) has been predominantly employed to assess residents' attitudes toward tourism development (Eusebio, Vieira, and Lima 2018; Park, Nunkoo, and Yoon 2015; Vargas-Sanchez, Plaza-Mejia, and Porras-Bueno 2009). In brief, it suggests that people make decisions to engage in exchanges based on their evaluations of both costs and benefits that occurred during the process (Choi and Murray 2010; Wang and Chen 2015). Accordingly, local residents are more likely to engage in tourism development if the perceived benefits from tourism outweigh the costs (Chen and Chen 2010). Some commonly perceived benefits associated with tourism include increased job opportunities, enhanced quality of life, and cultural exchanges with tourists, while commonly perceived costs include increases in prices for goods and services, traffic congestion, and damages to the natural environment and landscape (Ko and Stewart 2002; Latkova and Vogt 2012; Rasoolimanesh et al. 2015). Based on SET, residents' attitudes toward tourism development have been measured in terms of perceived positive and negative impacts within the three dimensions – economic, social, and environmental – of sustainability (Chen and Chen 2010; Vargas-Sanchez, Plaza-Mejia, and Porras-Bueno 2009).

In addition to residents' perceptions about tourism impacts on host communities (Andereck and Nyaupane 2011), perceived tourism impacts at the individual level (i.e., personal benefits) have also been examined (McGehee and Andereck 2004; Wang and Pfister 2008). However, previous studies failed to validate the scale developed to measure personal benefits (Xu et al. 2016). That is, a measure of personal benefit to be incorporated in the examination of residents' attitudes toward tourism development is still needed (Andereck and Nyaupane 2011).

Social capital

Social capital is defined as features of social organization (e.g., networks, trust) that facilitate coordination and cooperation for mutual benefit (Putnam 1995). Social capital has long been recognized for its economic significance (Moscardo 2012), importance to community ties (Woolcock and Narayan 2000), emphasis on collective actions by gluing people together via social networks (Shucksmith 2000), and its incorporation of trust in economic development (Falk and Kilpatrick 2000; Moscardo 2012).

Social capital has been measured differently based on the theoretical frameworks and perspectives adopted. From the network perspective, measurements distinguishing between bonding (i.e., internal relationships that occur horizontally within a community) and bridging (i.e., ties and relations between individuals or communities across distinct social networks) social capital (Putnam 1993). Some considered trust, defined as the expectation community members have in other members' good intentions and actions (Dirks 2000), a critical component of social capital (Leahy and Anderson 2010) and the foundation for enabling community actions (Falk and Kilpatrick 2000). Others believed that information sharing between members in communities captures the reciprocity of information flow in networks (Adler and Kwon 2002), facilitates access to broader sources of information (Coleman 1988), and thus is an important component of social capital (Li and Barbieri 2020). Some researchers found that through collective action – actions taken by members within a group to further their common interest (Xu, Barbieri, and Seekamp 2020), small businesses are capable of mobilizing social relations, which results in improved economic performance and opportunities for growth (Brunori and Rossi 2000). Long-term collective action, especially related to tourism development, led to community collaboration (Hwang, Stewart, and Ko 2012). Based on various measurements, social capital was found to be closely associated with the success of tourism development and essential to the evaluation of a region's readiness to undertake tourism development (Macbeth,

Carson, and Northcote 2004). Several attributes, such as length of residence (McGehee et al. 2010), income (Kim and Won 2003), and socioeconomic status (Ryan et al. 2005) influence residents' perceptions of social capital.

Given its significance in community development (Flora 2004), social capital has been examined in both the general tourism and niche tourism settings, focusing on exploring either its synergistic relationship with tourism or association with tourism entrepreneurship (Gil Arroyo, Knollenberg, and Barbieri 2021; Zhou, Chan, and Song 2017). Previous research has investigated the synergistic relationships between tourism and social capital from two perspectives. The first perspective is social capital within a community as instrumental to developing community-based niche tourism, such as ecotourism (Jones 2005), rural tourism (Park et al. 2012), and craft-beverage tourism (Gil Arroyo, Knollenberg, and Barbieri 2021). A second perspective states that tourism positively impacts social capital through bridging with networks outside the community, creating public spaces for social interaction, and even establishing local tourism organizations (Moscardo 2012). More specifically, some studies have identified how components (e.g., trust and social network) of social capital positively impact the success of tourism businesses and entrepreneurs (KC et al. 2019; Franken, Gomez, and Ross 2018; Zhao, Ritchie, and Echtner 2011) while others studied how tourism creates and strengthens social ties between families, friends, and others (Moscardo 2012; Mura and Tavakoli 2014).

Compared to the broader tourism literature, research of social capital in wine tourism is limited. Among those, social capital has been mainly examined from winery businesses' perspective, emphasizing its significance to regional development in rural areas (Hall 2005), to support new wine-entrepreneurship in emerging regions (Franken, Gomez, and Ross 2018), or to improve competitive advantages of a destination (Jackson and Murphy 2006). There is a scarcity of research looking into the level of social capital in communities associated with wine tourism (Macbeth, Carson, and Northcote 2004; McGehee et al. 2010; Xu et al. 2016).

Case Study 19.1 Wine tourism in the Piedmont region of North Carolina: Community's Perspective

Wine trails, an emerging phenomenon of wine tourism, have been widely developed worldwide. The USA alone has at least one wine trail in 48 states, with a total number of 2,898 wine trails nationwide (Xu, Barbieri, and Seekamp 2020). Despite such rapid growth, wine trails are understudied in general wine tourism research, as most studies have focused on the entire wine region. As such, the scant research on wine tourism from the host community's perspective is even more acute related to wine trails. Yet, the intrinsic cooperation network among wine-related businesses and community services needed for the development of wine trails (Del Chiappa, Bregoli, and Kim 2019) deems wine trails the perfect setting to investigate the host community's perspective.

We chose wine trails in the Piedmont Triad, North Carolina (NC), in the USA as a case study to investigate the host community's attitudes toward wine tourism development. NC is the home of the first wine grape in the USA and the only place in the world where every major type of grape is grown (Visit NC 2020). NC hosts 200 wineries, 23 wine trails, and five American Viticultural Areas (Visit NC 2020). The Piedmont region is the heart of NC's wine country, covering 12 counties of 5,875.2 square miles, with 646,333 households (U.S.

Census Bureau 2013) whose average median income is $41,873. Approximately 17% of its population falls below the poverty level, and only 17.9% have a bachelor's or advanced degree (U.S. Census Bureau 2010). In the past few decades, the region has been shifting away from the traditional textile, tobacco, and timber (furniture) industries to develop its wine industry, and it now accommodates 11 wine trails.

This study included two wine trails (Haw River, Surry County), as they are the most representative of the Piedmont region based on their spatial and tourism characteristics. Haw River has a superior characterization indicating easy accessibility and high tourism service comprehensiveness; Surry County has a marginal characterization with easy accessibility but moderate tourism service comprehensiveness (Xu et al. 2016). Both trails share similarities in the number of composing wineries (n = 4), length (Haw River = 43 miles, Surry County = 32 miles), and proximity to highway (Haw River = .08 miles, Surry County = .02 miles). Yet, they differ in tourism service offerings and trail spatial patterns. Haw River, which has a cluster of wineries in the middle of the trail, has a very comprehensive tourism offering (e.g., private events, food services, art-related components, outdoor recreation options). Surry County has an even greater distribution of wineries along the trail, but their offerings are less diverse.

Methods

A survey questionnaire designed for this study collected information on residents' sociodemographic characteristics, involvement with the wine industry and wine tourism, their attitudes and perceived personal benefits associated with local wineries, and the perceived relationship between Piedmont wineries and local communities. Residents' overall perceptions (attitudes, perceived personal benefits, social capital) associated with local wineries were assessed using three modified scales. The modified attitudes scale (Ko and Stewart 2002; Smith et al. 2013) comprised of 24 items equally representing three dimensions of perceived positive and negative *Economic*, *Sociocultural*, and *Environmental* impacts. The modified personal benefits scale (Andereck and Nyaupane 2011; McGehee and Andereck 2004; Wang and Pfister 2008) included seven items comprising two dimensions, *Personal Enhancement* and *Community Sentience*. The modified scale used to assess a community's level of social capital associated with local wineries (Grootaert et al. 2004; McGehee et al. 2010) included 18 items representing five dimensions of social capital: *Trust*, *Information Sharing*, *Collective Action*, *Bonding*, and *Bridging*.

Using a stratified random sampling approach, 663 households residing in communities within a 10-mile zone around each wine trail was selected. A drop-off/pick-up method was used to hand-deliver questionnaires to households; the field researcher explained study purposes, dropped off the surveys, and asked households to hang the completed surveys in a provided bag on the doorknob on scheduled date (two to three days after the drop-off) for pick-up, if they agreed to participate. If no surveys were hung on the doorknob on the scheduled day for pick-up, the researcher left a stamped envelope along with a brief note asking residents to mail back the surveys once completed. A total of 344 (Haw River: 164; Surry County: 155) surveys were collected representing an overall response rate of 51.9%. Altogether, 300 complete surveys were included for analysis.

The SPSS statistical package was used to analyze responses through a series of descriptive and multivariate statistics. Descriptive statistics were used to outline respondents' demographic characteristics and wine involvement, their attitudes and perceived personal benefits toward local wineries, and communities' wine tourism–related social capital. Multivariate regressions ($p < 0.05$) were used to examine the influence of residents' demographics, their wine involvement, and wine trail types on residents' attitudes toward wine tourism development and on the communities' wine tourism–related social capital.

Findings

Most respondents were female (58.4%) and averaged 52 years old. About one half (52.0%) did not have a four-year college degree. Almost one-third of the respondents (31.5%) were retired. Nearly half of the residents reported a pre-tax annual household income of less than $50,000 and were rooted in their neighborhood for more than 20 years (Table 19.1). Most residents

Table 19.1 Residents' overall characteristics, level of wine involvement, and perceived relationship with local wineries

Sociodemographic Indicators	*Percentage of Respondents (%)*
Pre-tax household income (n = 237)	
$25,000 or less	20.3
$25,000–$49,999	28.3
$50,000–$74,999	22.8
$75,000–$99,999	16.4
$100,000 or more	12.2
Home ownership (n = 294)	
Home owners	84.0
Home renters	16.0
Number of years living in current neighborhood (n = 221)	
Less than 3 years	9.2
3–5 years	13.4
6–10 years	15.5
11–20 years	18.4
21–34 years	17.7
35–50 years	18.0
51 years or more	7.8
Household involvement with wine-related activities[1] *(n = 295)*	
Subscriber to a wine-related magazine	2.0
Member or friend of a wine club	4.7
Member of a wine-related organization	0.7
Follower of online wine-related social media	4.7
Respondent of any informal wine social groups	3.7
No connection with any wine-related activity	89.8
Wine involvement index[2] *(n = 295)*	
None	89.8
One	6.4
Two	2.7
Three or more	1.1
Perceived Piedmont wineries and community relationship (n = 294)	
Excellent	8.6
Good	36.3
Average	16.5
Fair	7.3
Poor	3.6
I don't know	27.7

1 Percentages sum to more than 100%, as respondents were able to select multiple categories.
2 Index constructed by adding the five indicators of wine-related activities (1 = Yes; 0 = No).

Table 19.2 Residents' attitudes toward wine tourism development

Dimensions (n = 284)	*Mean*[1]
Economic dimensions	*3.22*
Tourists' spending	3.69
Variety of local businesses	3.60
Number of local businesses	3.43
Number of jobs	3.40
Real estate and property tax	3.28
Prices of goods and services	3.20
Community economic stability	3.17
Economic inequality among	3.06
Sociocultural dimensions	*3.24*
Variety of cultural activities	3.54
Conservation of local heritage	3.31
Sense of community identity	3.26
Residents' quality of life	3.23
Number of recreational facilities	3.19
Communities' small-town feeling	3.16
Quality of public services	3.08
Crime in the community	2.99
Environmental dimensions	*3.17*
Beauty of local landscapes	3.54
Environmental consciousness	3.27
Health of local ecosystems	3.23
Tranquility of community	3.17
Quality of public infrastructure	3.13
Traffic congestion	3.08
Overcrowding in public areas	3.07
Littering	2.96
Overall attitudes toward wine tourism development	*3.21*

1 Measured using a 5-point scale from 1 = significantly decreased to 5 = significantly increased.

owned their current residence and neither got involved in wine-related activities nor visited the study wine trails in the past three years, but believed that an overall good to excellent relationship existed between wineries and local communities.

Overall, respondents had neutral attitudes toward local wineries (Table 19.2) and did not see any major negative impacts associated with the establishment of wineries in their communities. The *Sociocultural* dimension was most positively rated, closely followed by *Economic* and *Environmental* dimensions. Major benefits of wine tourism development include increases in tourists' spending, in the variety of cultural activities, and in the beautification of local landscapes. Residents' demographic characteristics and their level of wine involvement were significantly associated with their overall attitudes (Table 19.3) toward local wine tourism development and all three comprising dimensions. Respondents' visit frequency to the study wine trails were positively associated, but their length of residence in the neighborhood were negatively associated with their overall attitudes (and all three comprising dimensions) toward local wine tourism development. In addition, residents' age was positively associated with their overall attitudes and the *Sociocultural* and *Environmental* dimensions, while their residence ownership was negatively associated with the *Economic* dimension.

Similarly, respondents perceived moderate levels of personal benefits associated with local wine tourism development (Table 19.4). Respondents agreed that they cared more about the

Table 19.3 Residents' demographics and wine involvement as associated with their attitudes

Independent Variables	*Residents' Attitudes*			
	Overall	*Economic*	*Sociocultural*	*Environmental*
Age	.168*	.014	.179**	.220**
Education level	.106	.119	.103	.056
Residence ownership (Yes, otherwise)	−.095	−.213**	−.044	.004
Length of residence in current neighborhood	−.179**	−.215**	−.136*	−.166*
Visit frequency to study wine trails	.236**	.169**	.233**	.194**
Household wine involvement	.046	−.036	.078	.082
Model statistics				
R	.347	.347	.336	.317
R^2	.121	.120	.113	.101
p−value	< .001	< .001	< .001	< .001

*p < .05, **p < .001

Table 19.4 Residents' perceived impacts of wine tourism development at the personal level

Personal Benefits (n = 291)	*Mean*[1]
Personal enhancement	*3.05*
My understanding of other culture has increased	3.18
I have more opportunities to participate in recreational activities	3.16
The quality of my personal life has improved	2.96
My property value has increased	2.92
Community sentience	*3.26*
I care more about my community's cultural resources	3.31
I care more about my community's natural resources	3.31
I feel my community is a better place to live	3.16
Overall perceived personal benefits (mean)	*3.14*

1 Measured using a 5-point scale from (1) strongly disagree to (5) strongly agree.

community's cultural and natural resources because of local wineries but did not observe significant increases in their quality of life or property value. Respondents rated higher in their sentience toward the community than their personal enhancement. Residents' visit frequency to local wine trails as well as their level of wine involvement were significantly associated with their overall perceived personal benefits and its two comprising dimensions (Table 19.5).

Respondents also perceived a moderate level of social capital associated with local wineries (Table 19.6). *Collective Action* was the most highly rated dimension, closely followed by *Trust* and *Information Sharing*, while *Bonding* and *Bridging* were the least highly rated. Most respondents agreed that local wineries participated in community events and that they could easily find information of local winery-related events and knew where to find information of local winery-related events. Conversely, a third of respondents would neither hang out with friends at local wineries nor volunteer in local winery-related events if needed.

Residents' demographic characteristics and their relationship with local wineries were significantly associated with the overall social capital and all six comprising dimensions (Table 19.7). Residents' visit frequency to study wine trails was positively associated with the overall social

Table 19.5 Residents' demographics and wine involvement as associated with their personal benefits

Independent Variables	*Personal Benefits*		
	Overall	*Personal enhancement*	*Community Sentience*
Age	−.050	−.061	−.025
Education level	.066	.084	.036
Residence ownership (Yes, otherwise)	.004	.026	−.018
Length of residence in current neighborhood	−.068	−.058	−.071
Visit frequency to study wine trails	.280**	.264**	.264**
Wine involvement	.183**	.191**	.155*
Model statistics			
R	.399	.400	.354
R^2	.160	.160	.125
p-value	< .001	< .001	< .001

$^*p < .05$, $^{**}p < .001$

Table 19.6 The association of community's social capital with local wineries

Items by Dimensions (n = 290)	*Mean*[1]
Collective action	*3.18*
Local winery supports community need	3.32
People in my community participate in local winery-related events	3.23
I will volunteer in local winery-related events if needed	2.71
Local winery participates in community events	3.52
Trust	*3.07*
I trust winery to make decisions that protect community interests	3.18
Local residents trust winery to make decisions that protect community interests	3.09
My community has a shared vision for local winery development	2.96
Information sharing	*3.07*
I can easily find information of local winery-related events	3.40
I know where to find information on local winery-related events	3.35
My community has a forum to discuss winery-related concerns	2.76
My community has a forum to discuss winery-related opportunities	2.75
Bridging	*2.91*
I have acquaintances who own or manage a local wine-related business	2.93
I attend events hosted by local wineries and other businesses	2.93
I have acquaintances who own or manage a local winery	2.88
I attend events co-hosted by several local wineries	2.86
Bonding	*2.77*
I attend private parties at local winery	2.83
I talk to my neighbors about visiting local winery	2.81
I hang out with friends at local winery	2.70
Overall Community's Social Capital	*3.00*

1 Measured on a 5-point Likert scale ranging from 'Strongly disagree' (1) to 'Strongly agree' (5).

Table 19.7 Residents' demographic characteristics and relationship with local wineries as associated with community's social capital

Independent Variables	Social Capital (standardized β and significance)					
	Overall	Bonding	Bridging	Trust	Collective Action	Information Sharing
Age	.074	−.015	.022	.093	.047	.168*
Education level	−.076	−.086	−.131*	−.057	−.048	.031
Annual household income	.082	.042	.133*	.046	.065	.062
Length of residence in current neighborhood	−.070	−.042	−.015	−.100	.001	−.132
Visit frequency to study wine trails	.463**	.482**	.518**	.268**	.335**	.312**
Relationship with Piedmont wineries	.078	.073	.044	.015	.098	.106
Model statistics						
R	.473	.489	.534	.281	.351	.372
R^2	.223	.239	.286	.079	.123	.139
p-value	< .001	< .001	< .001	.007	< .001	< .001

* $p < .05$, ** $p < .001$

capital and its six comprising dimensions. Residents' education level was negatively associated, but their annual household income was positively associated with the *Bridging* dimension. In addition, residents' age was positively associated with the *Information Sharing* dimension.

Discussion and implications

Contrary to previous studies in which local residents had an overall positive attitude toward tourism development and rated economic benefits the highest (McGehee and Andereck 2004; Vargas-Sanchez, Plaza-Mejia, and Porras-Bueno 2009), local communities in this study were relatively neutral toward wine tourism development and rated most favorably on *Social-cultural* benefits. These findings may be related to the inception stage of local wine tourism development, whereas previous research studied well-established tourism destinations with a mature tourism industry (Harrill and Potts 2003; Williams and Lawson 2001). Considering that most residents were rooted in the communities, with a good portion being retirees, they may have a higher appreciation for new sociocultural activities and opportunities created by local wineries, which help diversify their leisure time. The finding that residence ownership was negatively associated with the *Economic* dimension may be explained by the results from personal benefits – that residents did not experience significant increases in their property value or the quality of life and thus failed to rate favorably in the *Economic* dimension toward local wine tourism development.

Considering that visit frequency to wine trails was positively associated with residents' overall attitudes and perceived personal benefits, and that those who stayed longer in their current neighborhood tend to rate more negatively, wineries and tourism authorities may consider placing more efforts on communicating the benefits of wine tourism to local communities through activities and events or marketing campaigns (e.g., flyers, social media), especially targeting on those who have long been in the region and community. In such promotion efforts, wine tourism's role in increasing tourists' spending, diversifying local businesses and cultural activities, and beautifying local landscapes should be highlighted. To trigger residents' interests in wine tourism and increase their visit frequency to wine trails, a combination of both wine-related (e.g., wine tasting, wine social gatherings) and non-wine-related activities (e.g., private celebration parties, concerts, art exhibitions) should be promoted (Lockshin and Spawton 2001).

The overall moderate level of wine tourism–related social capital is likely also related to the inception stage of local wine tourism, and the resulted limited residential interests in wine-related activities. Similar to residents' attitudes, results on social capital also found that their visit frequency to the study wine trails was positively associated with the overall social capital associated with wineries, which suggest that encouraging visits among residents to wineries is key to forging relationships with local communities and to increasing positive attitudes toward their businesses (Adler and Kwon 2002). Given that higher levels of social capital are associated with more desired socioeconomic outcomes (Moscardo et al. 2017; Ellison, Steinfield, and Lampe 2007), it is imperative to build wine-related social capital in local communities, especially the least favorably rated *Bonding* and *Bridging* dimensions. Possible strategies for bonding with host communities and increasing awareness include hosting local meetings and events and providing on-site recreation opportunities.

Considering that the *Bridging* dimension was positively associated with residents' household income and negatively associated with their education level, wineries and tourism authorities may consider paying special attention to involving and including households with a lower level of income or design activities and events that may cater to the interests of residents with a higher education background. This study also found that older residents were more resourceful in finding information and identifying opportunities for local winery-related events, making sense as they are more connected given their age and familiarity with the region.

Conclusion

This study examined residents' attitudes toward wine tourism development and local communities' wine tourism-related social capital. In doing so, this study enhances understanding of wine tourism development from the host community's perspective, which contributes to the sustainable development of the local community with regards to genuinely involving residents and maximizing their benefits. Study results also provide management and marketing implications of wine tourism development and illuminate future wine tourism development policies.

References

Adler, P. S., and Kwon, S. W. (2002). Social capital: prospects for a new concept. *Academy of Management Review*, 27(1): 17–40. https://doi.org/10.5465/amr.2002.5922314

Alonso, A. D., Bressan, A., O'Shea, M., and Vlad, K. (2014). Educating winery visitors and consumers: an international perspective. *Current Issues in Tourism*, 17(6): 539–556.

Andereck, K. L., and Nyaupane, G. P. (2011). Exploring the nature of tourism and quality of life perceptions among residents. *Journal of Travel Research*, 50(3): 248–260. https://doi.org/10.1177/0047287510362918

Barber, N. A., Donovan, J. R., and Dodd, T. H. (2008). Differences in tourism marketing strategies between wineries based on size or location. *Journal of Travel & Tourism Marketing*, 25(1): 43–57. https://doi.org/10.1080/10548400802164889

Brunori, G., and Rossi, A. (2000). Synergy and coherence through collective action: some insights from wine routes in Tuscany. *Sociologia Ruralis*, 40(4): 409–423. https://doi.org/10.1111/1467-9523.00157

Bruwer, J., Coode, M., Saliba, A., and Herbst, F. (2013). Wine tourism experience effects of the tasting room on consumer brand loyalty. *Tourism Analysis*, 18: 399–414. https://doi.org/10.3727/108354213X13736372325957

Carmichael, B. A., and Senese, D. M. (2012). Competitiveness and sustainability in wine tourism regions: the application of a stage model of destination development to two Canadian wine regions. In Dougherty, P. H. (ed.), *The Geography of Wine: Regions, Terroir and Techniques* (pp. 159–178). New York: Springer.

Charters, S., and Ali-Knight, J. (2002). Who is the wine tourist? *Tourism Management*, 23(3): 311–319. https://doi.org/10.1016/S0261-5177(01)00079-6

Chen, C. F., and Chen, F. S. (2010). Experience quality, perceived value, satisfaction and behavioral intentions for heritage tourists. *Tourism Management*, 31(1): 29–35. https://doi.org/10.1016/j.tourman.2009.02.008

Choi, H. C., and Murray, I. (2010). Resident attitudes toward sustainable community tourism. *Journal of Sustainable Tourism*, 18(4): 575–594. https://doi.org/10.1080/09669580903524852

Coleman, J. (1988). Social capital in the creation of human capital. *American Journal of Sociology*, 94: 95–120.

Deccio, C., and Baloglu, S. (2002). Nonhost community resident reactions to the 2002 Winter Olympics: the spillover impacts. *Journal of Travel Research*, 41(1): 46–56. https://doi.org/10.1177/0047287502041001006

Del Chiappa, G., Bregoli, I., and Kim, A. K. (2019). Inter-sectorial collaboration in networks: a boundary object approach to wine regions. *Tourism Planning & Development*, 16(6): 591–611. https://doi.org/10.1080/21568316.2018.1543727

Dirks, K. T. (2000). Trust in leadership and team performance: evidence from NCAA basketball. *Journal of Applied Psychology*, 85(6): 1004–1012. https://doi.org/10.1037/0021-9010.85.6.1004

Ellison, N. B., Steinfield, C., and Lampe, C. (2007). The benefits of Facebook 'friends': social capital and college students' use of online social network sites. *Journal of Computer-Mediated Communication*, 12 (4): 1143–1168. https://doi.org/10.1111/j.1083-6101.2007.00367.x

Eusebio, C., Vieira, A. L., and Lima, A. (2018). Place attachment, host-tourist interactions, and residents' attitudes towards tourism development: the case of Boa Vista Island in Cape Verde. *Journal of Sustainable Tourism*, 26(6): 890–909. https://doi.org/10.1080/09669582.2018.1425695

Falk, I., and Kilpatrick, S. (2000). What is social capital? A study of interaction in a rural community. *Sociologia Ruralis*, 40(1): 87–110. https://doi.org/10.1111/1467-9523.00133

Flora, C. B. (2004). Community dynamics and social capital. In Rickerl, D., and Francis, C. (eds.), *Agroecosystems Analysis* (pp. 93–107). Madison, WI: American Society of Agronomy.

Franken, J., Gomez, M., and Ross, R. (2018). Social capital and entrepreneurship in emerging wine regions. *Journal of Wine Economics*, 13(4): 419–428. https://doi.org/10.1017/jwe.2018.37

Getz, D. (eds.) (2000). *Explore Wine Tourism: Management, Development and Destinations*. New York: Cognizant Communication Corporation.

Gil Arroyo, C., Knollenberg, W., and Barbieri, C. (2021). Inputs and outputs of craft-beverage tourism: the destination resource acceleration framework. *Annals of Tourism Research*, 86, 103102. https://doi.org/10.1016/j.annals.2020.103102

Gomez, M., Pratt, M. A., and Molina, A. (2019). Wine tourism research: a systematic review of 20 vintages from 1995 to 2014. *Current Issues in Tourism*, 22(18): 2211–2249. https://doi.org/10.1080/13683500.2018.1441267

Grootaert, C., Narayan, D., Jones, V. N., and Woolcock, M. (2004). Measuring social capital: an integrated questionnaire. https://elibrary.worldbank.org/doi/abs/10.1596/0-8213-5661-5

Grybovych, O., Lankford, J., and Lankford, S. (2013). Motivations of wine travels in rural Northeast Iowa. *International Journal of Wine Business Research*, 25(4): 285–309. https://doi.org/10.1108/IJWBR-07-2012-0023

Gursoy, D., Chi, C. G., and Dyer, P. (2009). Locals' attitudes toward mass and alternative tourism: the case of Sunshine Coast, Australia. *Journal of Travel Research*, 49(3): 381–394. https://doi.org/10.1177/0047287509346853

Hall, C. M. (2005). Rural wine and food tourism cluster and network development. In Hall, D., Kirkpatrick, I., and Mitchell, M. (eds.), *Rural Tourism and Sustainable Business* (pp. 149–164). Clevendon: Channelview Press.

Harrill, R., and Potts, T. D. (2003). Tourism planning in historic districts: Attitudes toward tourism development in Charleston. *Journal of the American Planning Association*, 69(3): 233–244. https://doi.org/10.1080/01944360308978017

Holladay, P. J., Dixon, A. W., Nguyen, M. C., Nguyen, B. L., Xu, S., and Price-Howard, K. (2020). Stakeholder perceptions of Da Nang, Vietnam as a tourism gateway city. *Journal of Tourism Insights*, 10 (1): 8. https://doi.org/10.9707/2328-0824.1089

Hwang, D., Stewart, W., and Ko, D. (2012). Community behavior and sustainable rural tourism development. *Journal of Travel Research*, 51(3): 328–341. https://doi.org/10.1177/0047287511410350

Jackson, J., and Murphy, P. (2006). Clusters in regional tourism: an Australian case. *Annals of Tourism Research*, 33(4): 1018–1035. https://doi.org/10.1016/j.annals.2006.04.005

Job, H., and Murphy, A. (2006). Germany's Mosel Valley: can tourism help preserve its cultural heritage? *Tourism Review International*, 9: 333–347. https://doi.org/10.3727/154427206776330526

Jones, M. F., Singh, N., and Hsiung, Y. (2013). Determining the critical success factors of the wine tourism region of Napa from a supply perspective. *International Journal of Tourism Research*, 7: 261–271. https://doi.org/10.1002/jtr.1984

Jones, S. (2005). Community-based ecotourism: the significance of social capital. *Annals of Tourism Research*, 32(2): 303–324. https://doi.org/10.1016/j.annals.2004.06.007

Kc, B., Morais, D. B., Smith, J. W., Peterson, M. N., and Seekamp, E. (2019). Using social network analysis to understand trust, reciprocity, and togetherness in wildlife tourism microentrepreneurship. *Journal of Hospitality & Tourism Research*, 43(8): 1176–1198. https://doi.org/10.1177/1096348019840794

Kim, N. S., and Won, Y. S. (2003). Relationship between social capital and the leadership of private organization. *The Journal of Rural Society*, 13(1): 171–207.

Ko, D. W., and Stewart, W. P. (2002). A structural equation model of residents' attitudes for tourism development. *Tourism Management*, 23(5): 521–530. https://doi.org/10.1016/S0261-5177(02)00006-7

Kolyesnikova, N., and Dodd, T. H. (2008). Effects of winery visitor group size on gratitude and obligation. *Journal of Travel Research*, 47: 104–112. https://doi.org/10.1177/0047287507312411

Lankford, S., and Howard, D. (1994). Developing a tourism impact attitude scale. *Annals of Tourism Research*, 21(1): 121–139. https://doi.org/10.1016/0160-7383(94)90008-6

Latkova, P., and Vogt, C. A. (2012). Residents' attitudes toward existing and future tourism development in rural communities. *Journal of Travel Research*, 51(1): 50–67. https://doi.org/10.1177/0047287510394193

Leahy, J. E., and Anderson, D. H. (2010). Cooperation gets it done: social capital in natural resources management along the Kaskaskia River. *Society and Natural Resources*, 23(3): 224–239. https://doi.org/10.1080/08941920802378897

Li, J., and Barbieri, C. (2020). Demystifying members' social capital and networks within an agritourism association: a social network analysis. *Tourism and Hospitality*, 1(1): 41–58. https://doi.org/10.3390/tourhosp1010004

Lockshin, L., and Spawton, T. (2001). Using involvement and brand equity to develop a wine tourism strategy. *International Journal of Wine Marketing*, 13(1): 72–81. https://doi.org/10.1108/eb043371

Macbeth, J., Carson, D., and Northcote, J. (2004). Social capital, tourism and regional development: SPCC as a basis for innovation and sustainability. *Current Issues in Tourism*, 7(6): 502–522. https://doi.org/10.1080/13683500508668200

Mason, M. C., and Paggiaro, A. (2012). Investigating the role of festivalscape in culinary tourism: the case of food and wine events. *Tourism Management*, 33(6): 1329–1336. https://doi.org/10.1016/j.tourman.2011.12.016

McGehee, N., and Andereck, K. (2004). Factors predicting rural residents' support of tourism. *Journal of Travel Research*, 43(2): 131–140. https://doi.org/10.1177/0047287504268234

McGehee, N., Lee, S., O'Bannon, T. L., and Perdue, R. (2010). Tourism-related social capital and its relationship with other forms of capital: an exploratory study. *Journal of Travel Research*, 49(4): 486–500. https://doi.org/10.1177/0047287509349271

Mitchell, R., Charters, S., and Albrecht, J. N. (2012). Cultural systems and the wine tourism product. *Annals of Tourism Research*, 39(1): 311–335. https://doi.org/10.1016/j.annals.2011.05.002

Mitchell, R., and Hall, C. M. (2006). Wine tourism research: the state of play. *Tourism Review International*, 9(4): 307–332. https://doi.org/10.3727/154427206776330535

Montella, M. M. (2017). Wine tourism and sustainability: a review. *Sustainability*, 9: 113. https://doi.org/10.3390/su9010113

Moscardo, G. (2012). Building social capital to enhance the quality-of-life of destination residents. In Uysal, M., Perdue, R., and Sirgy, M. J. (eds.), *Handbook of Tourism and Quality-of-Life Research: Enhancing the Lives of Tourists and Residents of Host Communities* (pp. 403–421). Dordrecht: Springer Netherlands.

Moscardo, G., Konovalov, E., Murphy, L., McGehee, N., and Schurmann, A. (2017). Linking tourism to social capital in destination communities. *Journal of Destination Marketing & Management*, 6(4): 286–295. https://doi.org/10.1016/j.jdmm.2017.10.001

Mura, P., and Tavakoli, R. (2014). Tourism and social capital in Malaysia. *Current Issues in Tourism*, 17(1): 28–45. https://doi.org/10.1080/13683500.2012.718320

Nunkoo, R., and Ramkissoon, H. (2011). Developing a community support model for tourism. *Annals of Tourism Research*, 38(3): 964–988. https://doi.org/10.1016/j.annals.2011.01.017

Park, D. B., Lee, K. W., Choi, H. S., and Yoon, Y. (2012). Factors influencing social capital in rural tourism communities in South Korea. *Tourism Management*, 33(6): 1511–1520. https://doi.org/10.1016/j.tourman.2012.02.005

Park, D. B., Nunkoo, R., and Yoon, Y. S. (2015). Rural residents' attitudes to tourism and the moderating effects of social capital. *Tourism Geographies*, 17(1): 112–133. https://doi.org/10.1080/14616688.2014.959993

Poitras, L., and Getz, D. (2006). Sustainable wine tourism: the host community perspective. *Journal of Sustainable Tourism*, 14(5): 425–448. https://doi.org/10.2167/jost587.0

Putnam, R. (1993). The prosperous community: Social capital and public life. *The American Prospect*, 13(4): 35–42.

Putnam, R. (1995). Bowling alone: America's declining social capital. *Journal of Democracy*, 6(1): 65–78.

Qiu, H., Yuan, J., Ye, B., and Hung, K. (2013). Wine tourism phenomena in China: an emerging market. *International Journal of Contemporary Hospitality Management*, 25(7), 1115–1134. https://doi.org/10.1108/IJCHM-06-2012-0087

Quintal, V., Thomas, B., Phau, I., and Soldat, Z. (2017). Using push-pull winescape attributes to model Australian wine tourist segmentation. *International Journal of Wine Business Research*, 29(4): 346–372. https://doi.org/10.1108/IJWBR-01-2017-0007

Rasoolimanesh, S. M., Jaafar, M., Kock, N., and Ramayah, T. (2015). A revised framework of social exchange theory to investigate the factors influencing residents' perceptions. *Tourism Management Perspectives*, 16: 335–345. https://doi.org/10.1016/j.tmp.2015.10.001

Ryan, V. D., Agnitsch, K. A., Zhao, L., and Mullick, R. (2005). Making sense of voluntary participation: a theoretical synthesis. *Rural Sociology*, 70(3): 287–313. https://doi.org/10.1526/0036011054831198

Shapiro, M., and Gomez, M. I. (2014). Customer satisfaction and sales performance in wine tasting rooms. *International Journal of Tourism Research*, 11: 451–463. https://doi.org/10.1108/IJWBR-09-2012-0026

Shucksmith, M. (2000). Endogenous development, social capital and social inclusion: perspectives from LEADER in the UK. *Sociologia Ruralis*, 40(2): 208–218. https://doi.org/10.1111/1467-9523.00143

Smith, J., Anderson, D., Davenport, M., and Leahy, J. (2013). Community benefits from managed resource areas: an analysis of construct validity. *Journal of Leisure Research*, 45(2): 192–213. https://doi.org/10.18666/jlr-2013-v45-i2-3011

Taylor, R., and Shanka, T. (2002). Attributes for staging successful wine festivals. *Event Management*, 7(3): 165–175. https://doi.org/10.3727/152599502108751569

Telfer, D. J. (2001). Strategic alliances along the Niagara wine route. *Tourism Management*, 22(1): 21–30. https://doi.org/10.1016/S0261-5177(00)00033-9

U.S. Census Bureau. (2010). *State and county quick facts*. https://www.census.gov/quickfacts/qfd/index.html

Vargas-Sanchez, A., Plaza-Mejia, M. D. L. A., and Porras-Bueno, N. (2009). Understanding residents' attitudes toward the development of industrial tourism in a former mining community. *Journal of Travel Research*, 47(3): 373–387. https://doi.org/10.1177/0047287508322783

Visit NC. (2020). *Wine: A wine history well aged*. https://www.visitnc.com/wine

Wang, S., and Chen, J. (2015). The influence of place identity on perceived tourism impacts. *Annals of Tourism Research*, 52: 16–28. https://doi.org/10.1016/j.annals.2015.02.016

Wang, Y., and Pfister, R. E. (2008). Residents' attitudes toward tourism and perceived personal benefits in a rural community. *Journal of Travel Research*, 47(1): 84–93. https://doi.org/10.1177/0047287507312402

Williams, J., and Lawson, R. (2001). Community issues and resident opinions of tourism. *Annals of Tourism Research*, 28(2): 269–290. https://doi.org/10.1016/S0160-7383(00)00030-X

Woolcock, M., and Narayan, D. (2000). Social capital: implications for development theory, research, and policy. *World Bank Research Observer*, 15: 225–249. https://doi.org/10.1093/wbro/15.2.225

Xu, S., Barbieri, C., Anderson, D., Leung, Y. F., and Rozier-Rich, S. (2016). Residents' perceptions of wine tourism development. *Tourism Management*, 55: 276–286. https://doi.org/10.1016/j.tourman.2016.02.016

Xu, S., Barbieri, C., and Seekamp, E. (2020). Social capital along wine trails: spilling the wine to residents. *Sustainability*, 12(4): 1592. https://doi.org/10.3390/su12041592

Zhao, W., Ritchie, J. R. B., and Echtner, C. M. (2011). Social capital and tourism entrepreneurship. *Annals of Tourism Research*, 38(4): 1570–1593. https://doi.org/10.1016/j.annals.2011.02.006

Zhou, L., Chan, E., and Song, H. (2017). Social capital and entrepreneurial mobility in early-stage tourism development: a case from rural China. *Tourism Management*, 63: 338–350. https://doi.org/10.1016/j.tourman.2017.06.027

20

WINERY OWNERS' PERCEPTIONS AND MOTIVATIONS TOWARDS WINE TOURISM

Discourses of winery owners in Langhe, North Italy

Magali Canovi and Andy Lyon

Introduction

Residents' perceptions and motivations towards tourism and future tourism development have received considerable attention from academics since the 1960s and is still a topic that is one of the most studied in the tourism literature (Xu et al. 2016). Such attention is hardly surprising given the crucial role of the host community in the success of tourism planning and development. Tourism development projects and activities are likely to be threatened if designed and implemented without the support of local communities (Gursoy, Jurowski, and Uysal 2002). Scholars largely agree that understanding local communities' perceptions and motivations is beneficial for tourism planners and policymakers in order to identify and address major concerns, maximize benefits, and minimize potential resistance (Ribeiro et al. 2017). From the local community perspective, tourism studies have primarily focused not only on investigating residents' attitudes towards tourism in general but also on their perceptions of tourism impacts in particular (Chen, Hsu, and Li 2018; Vargas-Sanchez et al. 2015), while other stakeholder groups have been largely overlooked, particularly business owners (Sanchez Cañizares, Castillo Canalejo, and Núñez Tabales 2016). In part, this study addresses that general need while focusing on winery owners' perceptions and motivations towards tourism development in a particular geographical context.

Numerous stakeholders operate within the tourism development system and have various positions, influence, and power. Not all stakeholders are equal in the development process; some have more influence and power over development than others. Stakeholders can be split into two main groups, namely active and passive. Active stakeholders are those who affect decisions or actions, while passive stakeholders are those who are affected (either positively or negatively) by those decisions (Grimble and Wellard 1997). If stakeholders can affect decisions or actions regarding development concerns, then that makes them active in development outcomes. Winery owners are active stakeholders in that they can affect outcomes relating to

 DOI: 10.4324/9781003143628-24

economic, social, and environmental concerns in the region. Understanding and analyzing the discourses of these active stakeholders can therefore help to further the debate around tourism development both conceptually and empirically (Lyon, Hunter-Jones, and Warnaby 2017).

Using qualitative evidence from a sample of 20 winery owners, this chapter examines their perceptions of and attitudes towards tourism in the North Italian wine region of Langhe, a region long renowned for its wine and which is increasingly popular as a tourist destination. Langhe is characterized by traditional 'Old World' winemaking practices and traditions in which its wine produce and its producers are rooted in and synonymous with the place itself. With this, coupled with the fact that World Heritage status was awarded to Langhe by UNESCO in 2014, it is no surprise that over the past decade, Langhe has developed as a wine tourism destination that is increasingly popular internationally (DMO Piemonte 2018). Moreover, the steady rise in tourist numbers to the area has led to an increased awareness amongst wine business owners about the potential benefits of tourism development, and a number of them have introduced various tourism-related activities at their wineries, notably wine tastings, winery visits, cellar door sales, B&B accommodations, and restaurants. Others, however, have resisted wine tourism as being antithetical to their core activity and identity.

This chapter examines winery owners' perceptions of and motivations towards wine tourism as a business and a regional rural tourism diversification option. The research objectives are:

- To understand the reasons for diversification into wine tourism in the Langhe region.
- To analyze the discourses of winery owners to better understand their business motivations in wine tourism.
- To make recommendations to planners and marketers of wine tourism regions regarding sense of place development.

Literature

Active stakeholders

One of the principal themes of the sustainable development of tourism destinations is the involvement of the local community in decisions about and the implementation of social and economic change (Sharpley 2014). There has been a growing body of literature on this subject since the early works of Cohen (1984) and Murphy (1985). Local communities are key stakeholders in tourism development and can affect or are affected by their involvement in the planning of the industry (Simpson 2001; Sook et al. 2014). Residents within a community (including business owners) will demonstrate behaviors that either support or do not support tourism depending on their perceptions of the positive and negative impacts of tourism on themselves and on the communities in which they reside (Ribeiro et al. 2017). Focusing specifically on rural tourism development, Wilson et al. (2001) conclude that success cannot be achieved without the participation and collaboration of the business community both directly and indirectly in tourism development.

Rural tourism development/diversification

Rural tourism, as a particular diversification strategy, originated during the early twentieth century in the Alps and the UK and progressively expanded across Europe (Canoves et al. 2004). During the agricultural crisis in the 1980s, national governments and municipalities promoted tourism as an efficient rural development strategy (Farmaki 2012). While the initial intention of promoting tourism was to develop a side activity that allowed rural business owners to strengthen their economic position, tourism in rural areas rapidly became a sector of its own (Barlybaev, Akhmetov, and Nasyrov 2009). There is a general belief that tourism, as a strategy of economic growth and diversification, can act as a potential economic panacea in rural areas, both at the local and individual farm levels (Sharpley and Vass 2006).

Initially, the main research theme in rural tourism development focused on examining and revealing the economic motives and benefits driving business owners to engage in tourism. Key economic reasons for engaging in diversification center on the generation of additional income (Barbieri and Mahoney 2009), long-term security in farming (Sharpley and Vass 2006), and new employment opportunities (Yang 2012). Within a wine-producing context, it is suggested that tourism is a short-term, beneficial strategy to increase cellar door sales (Charters and Menival 2011). However, researchers found that after having engaged in tourism diversification, personal and social goals tend to be highly accomplished, whereas economic drivers (e.g., additional income) showed much lower levels of accomplishment, even though these motives were initially ranked as most significant (Barbieri 2010). Hence, tourism diversification has increasingly been recognized as consisting of a complex web of both economic and noneconomic goals (Flanigan, Blackstock, and Hunter 2015; Hansson et al. 2013).

Limited academic attention, however, has been paid to the family unit and to how family involvement in the business affects rural tourism diversification (Hansson et al. 2013). This is quite surprising given that most rural businesses are family-owned and managed (Fitz-Koch et al. 2018). Neglecting the direct role of the family on diversification in the rural sector might limit our understanding of the motives driving rural family businesses to exploit entrepreneurial opportunities (Fitz-Koch et al. 2018). One of the first studies to criticize this approach was Hansson and colleagues' (2013) research on farmers' motives for diversifying their farm business in Sweden. They highlight the context-dependency of farmers' motivations for diversification and emphasize the need to consider the social context, notably the context of the farm family. Similarly, Tew and Barbieri (2012) note that some motivations are family-centered, as respondents in their study indicated that agritourism diversification was important for the farm family, particularly for enhancing their quality of life and for keeping the farm in the family.

Family-centered motivations

Within the family business literature, it is now well understood that the involvement of the family in the firm leads to a complex set of goals and motivations (Sharma, Chrisman, and Chua 1997). Family businesses tend to predominantly value noneconomic goals and affective endowments, which have been grouped together under the concept of socioemotional wealth (SEW). SEW has been defined as the 'non-financial aspects of the firm that meet the family's affective needs, such as identity, the ability to exercise family influence, and the perpetuation of the family dynasty' (Gomez-Mejia et al. 2007, p. 106). The SEW model implies that family firms are predominantly motivated to sustain their socioemotional wealth. Hence, family firms are believed to be fundamentally different from nonfamily firms, due to the fact that they are predominantly concerned about family-centered noneconomic goals (Berrone, Cruz, and Gomez-Mejia 2012).

The SEW model encompasses five dimensions, which are likely to influence family firms' decision-making process, notably: family control and influence; family members' identification with the firm; binding social ties; emotional attachment; and renewal of family bonds to the firm through dynastic succession (Berrone, Cruz, and Gomez-Mejia 2012). The first dimension is concerned with family members' desire to keep control and have influence over the firm. The family needs to be in control of the firm to preserve their socioemotional wealth.

Engaging into diversification, as in this case, agritourism diversification is likely to reduce SEW 'by having to appoint nonfamily members to various business units, reducing family influence over the units, decreasing centralization of decision making, and the like' (Berrone, Cruz, and Gomez-Mejia 2012, p. 260). It might require family firms to hire external employees and/or managers who might be more educated and possess enhanced managerial knowledge and skills and could threaten family control (Schmid et al. 2015). The second dimension relates to the family's identification with the firm. Organizational identification is especially prominent if family members are intimately tied to the firm. It is believed that a family's identity is closely linked to the family firm if it carries the family name (Berrone, Cruz, and Gomez-Mejia 2012). The third dimension of binding social ties is concerned with family firms' social relationships, whereas the fourth dimension expresses the importance of emotions linked and attached to the family business. The final dimension of SEW – the renewal of family bonds to the firm through dynastic succession – relates to the family's intention to transfer the firm to the next generation (Berrone, Cruz, and Gomez-Mejia 2012).

The importance of these five dimensions is, however, likely to vary for each family. While some families may put increased emphasis on the continuity of the family business by transferring business control to the next generation, other families might stress the emotional attachment to the business (Cennamo et al. 2012). Thus, family firms' decisions (as in this case, wine tourism diversification decisions) are not driven by an economic logic but are taken in the light of their SEW (Berrone, Cruz, and Gomez-Mejia 2012).

Place identity

Furthermore, within a rural context, research has revealed that people living and/or working in a rural environment tend to have stronger place bonds than people living in urban environments (Anton and Lawrence 2014). It is believed that 'people self-define through places, thereby developing a place identity. Such an identity arises from particular values, attitudes, and beliefs about the physical world as well as direct experiences with this environment' (Strzelecka, Boley, and Woosnam 2017, p. 63). As a consequence, individuals' motivations, attitudes, and perceptions are likely to be influenced and shaped by their place identity (Hallak, Brown, and Lindsay 2012). Bonaiuto et al. (2002, p. 636) refer to place identity as 'that part of people's personal identity which is based on or built upon the physical and symbolic features of the places in which people live.' Thus, as individuals start to develop an emotional attachment and become socially and psychologically invested in a particular place, that place becomes part of their sense of self (Anton and Lawrence 2014) and is, therefore, a fundamental element to the construction of self-identity.

To summarize, business owners (in this case, winery owners) are active stakeholders in tourism development and, as such, are key members of the community. Diversification into tourism is done for a variety of reasons, involving family and SEW motivations. Place attachment and self-identity are also important factors in determining the motives and actions regarding tourism development. The following case study explores these aspects in relation to wine tourism in Langhe, Italy.

Methodology

Case study

The North Italian wine region of Langhe is used as a case study to explore the motivations underlying winery owners' decisions to engage in tourism. Langhe, situated in the southern part of the Piedmont region, has been recognized for its long-standing tradition in winemaking, its interaction with the environment, and the aesthetic qualities of the area. The region secured its reputation during the early 1990s when Barolo was recognized as 'one of the world's great wines' (Rosso 2014), which led to an increased interest by tourists to visit these places of wine production. The inscription on the UNESCO World Heritage List in 2014 generated increased international visibility and further enhanced Langhe's reputation as an internationally renowned tourism destination. The region has continued to attract wine tourists from around the world – as demonstrated by an 81% growth of tourist arrivals to the region between 2006 and 2016 (DMO Piemonte 2018).

For the Langhe region, however, it could be argued that this significant and unexpected increase in tourist numbers meant that residents and winery owners were not prepared, or even unwilling, to accommodate tourists. One reason is that wineries in Langhe generate their income from wine/grape sales and wine export and do not depend on tourism for their economic survival. While this situation initially led to opposition and resistance on the part of winery owners, over the years, the region witnessed a steady rise in agritourism facilities as well as in the number of wineries developing tourism-related activities on site. These include wine tastings, winery visits, cellar door sales, B&B accommodations, and restaurants. Others, however, have resisted wine tourism, seeing it as antithetical to their core activity and identity as Langhe winemakers.

Methods chosen

A case study approach is adopted to examine winery owners' perceptions of and motivations towards wine tourism. This qualitative approach is the best way to respond to *what*, *how*, and *why* questions when investigating a contemporary phenomenon in depth and within its real-life context. The main objective of this research is to gain an in-depth understanding of the fundamental assumptions influencing winery owners' perceptions to engage in and pursue wine tourism activities.

The winemaking system in Langhe is largely based on small-sized, family-owned wineries (exclusively run and managed by family members), with an annual wine production ranging between 30,000 and 100,000 liters. The criteria for the sampling strategy were as follows. First, winery owners were contacted who have recently engaged in wine tourism activities. A snowballing sampling approach was adopted in which a small number of winery owners led the researchers to others through their personal contacts. This allowed the researchers to meet and interact with numerous winery owners, as well as reach owners who opposed and/or limited wine tourism activities at their winery, which would not have been possible otherwise. Family owners, who hold the decision-making power and actively run the family business on a daily basis, were selected for this research (see Table 20.1). Qualitative, semi-structured interviews were conducted with family owners from 20 family wine firms, which have diversified to a greater or lesser extent into tourism (see Table 20.2).

Data analysis

Thematic and discourse analysis was used to analyze the data. While thematic analysis revealed the content of participants' discourses and was considered the first step of the analysis process, discourse analysis allowed for a deeper understanding of what was said as it is 'the study of language in use' (Gee 2014, p.17). This approach is also sensitive to reflexivity and takes into account the fact that the interaction between researchers and participants plays an important

Table 20.1 Interview sample

Case number	*Location*	*Foundation of winery*	*Current generation*	*Annual wine production (in liters)*
1	Barolo	1919	3rd	30,000
2	Barbaresco	1958	2nd	30,000
3	Barolo	1885	4th	80,000
4	Mondovì	1990	2nd	50,000
5	Barbaresco	1978	2nd	70,000
6	Dogliani	1924	3rd	65,000
7	Neive	1950	2nd	21,000
8	Neive	1964	5th	120,000
9	Barbaresco	1948	4th	20,000
10	Monforte d'Alba	1982	2nd	150,000
11	Neive	1965	3rd	65,000
12	La Morra	1941	3rd	45,000
13	Serralunga d'Alba	1957	3rd	110,000
14	Monforte d'Alba	1878	2nd	90,000
15	Novello	1991	4th	85,000
16	Serralunga d'Alba	1953	3rd	85,000
17	Diano d'Alba	1927	3rd	30,000
18	Castiglione Falletto	1979	2nd	180,000
19	Barbaresco	1971	3rd	50,000
20	La Morra	1959	6th	120,000

Table 20.2 Diversification activities

Case number(s): 1–20	*Wine tastings*	*Winery visits*	*Cellar door sales*	*B&B*	*Other*
3	✔	✔	✔	✔	Restaurant
9, 16, 19, 20	✔	✔	✔	✔	
12, 13	✔	✔	✔		Restaurant
4, 11	✔	✔	✔		Sightseeing tours
6	✔	✔	✔		Camping
5, 7, 8, 10, 14, 15, 17, 18	✔	✔	✔		
2	✔		✔	✔	
1	✔	✔			

part in the construction of discourses (Phillips and Hardy 2002). Considering the sociocultural, historical, and geographical context of winery owners allows researchers to interpret discourses differently and analyze why certain things have been said. Winery owners' discourses are strongly embedded in their particular contexts.

Findings

Resistance to wine tourism and place identity

Although all participating winery owners had engaged to some extent in wine tourism, the majority of winery owners (65%) revealed their reluctance to pursue diversification and invest in additional tourism-related activities. The empirical data show that due to wineries' high export level, which can reach up to 85% for some wineries, winery owners are economically stable. Diversification into tourism is not recognized as a necessity, and therefore, it is not primarily linked to financial motives. In addition, one can observe that these winery owners strongly identify with their occupation/profession. Winery owners' lifestyles, as well as their passion for the job and their attachment to the local place of production (Groth and Curtis 2017), result in a strong identification with their occupation and the construction of a *producer* identity. Interestingly, the results show that these owners engaged in tourism diversification solely by providing winery visits, wine tastings, and cellar door sales. Larger tourism-related investments, such as B&Bs and restaurants, were not considered.

The analysis shows that communities who live and work on the land develop a strong attachment and identification to that land and place. Winery owners in Langhe have generally grown up there, and their passion for making wine typically plays a central role in their everyday lives. It is also typical that children help their parents in the vineyards and cellars and listen to stories about the family traditions and histories. These include tales of ancestors' successes and sacrifices made to drive their wine business forward, or the contribution of each generation to the business, as well as the family's secret winemaking traditions and processes. Wine tourism, however, provides a new dimension to these ongoing (hi)stories, and our empirical data show that the development of the tourism industry in Langhe has led to conflicting views and perceptions amongst business owners regarding their support for and involvement in tourism.

> **Our primary activity**, and it is important never to forget that is winemaking.
>
> *(Case 10)*

> **Our primary activity**, and it is important never to forget that is winemaking.
>
> *(Case 17)*

> This is **not a public institution**, but it is a business.
>
> *(Case 20)*

They strongly identify with their winemaking profession, whereas wine tourism is not regarded as a profession and is likely to threaten owners' sense of self. In some cases, owners display their commitment to being a conventional wine producer. They focus on their agricultural role (e.g., 'you are what you produce'), the place attachment (e.g., 'we've grown up in these surroundings'), as well as their passion for the job (e.g., 'we are in love with what we are doing'). In this instance, winery owners narrate stories about the family, the place, and their passion for the winemaking profession:

> We don't sleep at night because we're concerned about the harvest; is it going to be compromised by the rain, the season, hail storm. You are doing everything possible so that you have the highest quality wine because it is **our ambition**, **our passion**.
>
> *(Case 4)*

Winery owners have generally grown up in this rural environment and have been exposed to winemaking from a young age. The winery owner in Case 4 emphasizes her commitment to the winemaking profession. She aspires to produce 'the highest quality wine' and continuously strives to live up to her expectations. She does not consider wine tourism as a viable option and concentrates on her core winemaking activity, as she desires to be recognized as a 'real' winemaker.

> This is a winery, **not a tourism business**. I wouldn't be able to also run an agritourism business, because that would be **too much**. I believe that you have to make choices, if you want your work to be done correctly. It would **disturb the other work.**
>
> *(Case 8)*

This winery owner indicates his reluctance to pursue wine tourism activities. His statement 'you have to make choices, if you want your work to be done correctly' implies his priority for the winemaking profession. He assumes that engaging in tourism diversification requires a considerable amount of time and would interfere with being a professional winemaker – meaning that he would not be able to practice his profession 'correctly.' In this instance, business decisions are taken in the light of the winemaking profession.

> It [tourism diversification] might work as we have the facilities, but … opening a restaurant, not at all. So that's **not our profession**, so we are **not interested** in doing that.
>
> *(Case 18)*

This owner admits to having adequate facilities to either invest in the construction of a wine tourism business or the development of a restaurant; however, his statement 'this is not our profession' implies his priority for the winemaking profession, indicating that they are 'not interested' in engaging in wine tourism. Accordingly, these findings show that owners strongly identify with the winemaking profession and actively construct a producer identity when highlighting their reluctance to engage in wine tourism.

Family-centered motivations

We also found evidence that winery owners' motivations for engaging in wine tourism are driven by family-centered goals and SEW. Family owners' accounts are underpinned by the importance of the family unit. The majority of participating family wineries are managed and controlled by the nuclear family. Family owners express their desire to exercise and maintain full control of the family winery, manage the day-to-day activities and hold the decision-making power.

> This is a **family business** we are only 5 people, and **we** take care of all the business, also with the vineyards … **My wife** helps out if she has time, she does also take care of the B&B … Also [name] is **my brother-in-law**, my wife's brother; **he** is the winemaker; so he is more in the vineyards. When I'm not here, **he** takes care of some visits. So **we** try to arrange **ourselves**. It is only **family**. There is also an older woman, that is my **mother-in-law** and [name], my **father-in-law** is working in the vineyards.
>
> *(Case 19)*

This quote from Case 19 and the one that follows from Case 9 highlight pertinent pronouns and winery owners' reference to family business in bold. This shows that the family is perceived as the dominant coalition, influencing the decision-making process, and demonstrates the importance of the family in managing and keeping control of the winery. Sometimes the dominant coalition of the firm includes the involvement of the extended family, as in this case, the in-laws. The extended family manages every aspect of the business, including the wine production, vineyards, winery visits, tastings, and the B&B accommodations as well as the commercial and marketing aspects of the winery.

The notion of control plays a major role in the construction of these discourses by expressing their reluctance to employ outside personnel and delegate authority to nonfamily members. The appointment of nonfamily members to various business units, as in this case wine tourism-related areas of the business, is likely to reduce family influence over these units and decrease centralization of decision-making (Berrone, Cruz, and Gomez-Mejia 2012), and thus affect SEW. The inclusion of family members in the winery, rather than nonfamily employees, is emphasized on multiple occasions as a way of safeguarding the tradition and keeping family control.

> We **don't have the intention to grow** bigger. I think as we are now with the agritourism business, the winery, the wine tastings and visits, that is **the maximum we can do** for it to **stay a family business**. We run from the morning to the evening though. But it is very satisfying as we still have **direct contact** with guests and clients from the winery.
>
> *(Case 9)*

Families generally express their desire to stay in control of the various tasks in the winery, including the tourism-related activities. In this particular discourse, statements, such as 'we don't have the intention to grow,' 'this is the maximum we can do,' and 'stay a family business' relate to family owners' lack of entrepreneurial spirit. In this instance, pursuing agritourism diversification is likely to result in the loss of family control over wine tourism-related activities. Families tend to preserve personal control over the business and avoid delegation and decentralization of authority and decision-making. The level of families' engagement in agritourism diversification is thus influenced and guided by the need to keep family control, linking into SEW motivations. In this case, the majority of wine-producing families have taken the decision to limit wine tourism activities at their winery to a level where they are still able to remain in control.

Furthermore, we found that emotions are likely to influence family firms' activities.

> **My grandfather** built this winery in 1964, he decided straight away to create a space for hospitality, for receiving clients/tourists; which was inaugurated in 1967. So for us it was normal to receive tourists.
>
> *(Case 8)*

Highlighting the importance of the family history as well as the family tradition reveals family members' emotional attachment to the firm. Every generation has contributed to the success of the family firm. Ancestors' success stories play an important role in the construction of winery owners' accounts. The fact of running the family wine firm at the present time means that ancestors have taken the right decisions and have managed the winery successfully.

> I'm **proud** to be able to drive the company forward and follow in the footsteps of my ancestors. So I'm feeling **proud**.
>
> *(Case 20)*

> Knowing that this company has been passed down from father to son for such a long time is something wonderful. It's a great **privilege**.
>
> *(Case 26)*

The positive emotion of *pride* is evident in winery owners' accounts. Emotions pervade every aspect of people's life and more often than not dominate over reason and guide their decision-making process. Positive emotions, as in this case pride, have been found to encourage flexibility and openness in decision-making (Baron et al. 2008). Winery owners' sense of pride and respect demonstrate a strong emotional attachment to the family firm.

> I have 2 sisters so we are 3 from the new generation. So for them [the parents] it was important whoever decided to join the company that we would have our specific role in order to avoid any fights between my 2 sisters and me. So everyone has his or her own space to be able to express him- or herself… So the business aims to grow, we're currently investing, for example in the construction of the restaurant, we're looking for new vineyards, we're looking for new countries to sell our wines. We're always looking to grow.
>
> *(Case 20)*

In this particular case, the family took the decision to engage in wine tourism by opening a wine shop next to the winery and an agritourism business on the outskirts, to include the new generation in the family firm. Each family member would thus occupy a distinct position within the business. The importance of succession resulted in the expansion of the business through developing wine tourism activities. These findings are consistent with other studies' empirical findings that show how engagement in wine tourism provides employment opportunities for family members (e.g., Barbieri and Mahoney, 2009; Schilling, Sullivan, and Komar 2012). This shows the family's priorities are to keep the family together and avoid internal conflicts to guarantee the continuity of the family wine firm and again linking into Family SEW, which is developed further below.

Likewise, the importance of succession and the involvement of the subsequent generation lead to the adoption of long-term projects (Sirmon and Hitt 2003).

> **In 10 years** I would like to finish the extension, finish the B&B, finish the restaurant and that would be already a big project … and frankly this would be what I set for myself to do **in the future** … and then leave it for my son what I will be constructing **in the next 10 years**. Already this project is very ambitious. It's not a small project for us.
>
> *(Case 13)*

Winery owners tend to adopt a long-term orientation (in bold in the previous quote) with the desire to pass on the family firm to the next generation. As is evident in the Case 13 excerpt, the family's future projects are ambitious and require a significant amount of investment. The family plans to increase the wine production, restructure and enlarge the winery, and develop an agritourism business as well as a restaurant. The subsequent generation plays an important part in the current development plans. Current investments in the family wine business serve as future opportunities for subsequent generations, thus highlighting the importance of a generational change. The majority of winery owners express their desire to engage in agritourism diversification to guarantee employment for new family members. Family wine firms diversify into agritourism through investing in the construction of wine tasting facilities, B&Bs, and restaurants to offer the subsequent generation distinct areas of responsibilities within the firm. The subsequent generation thus plays an important role in family wine firms' development plans.

Conclusions and research implications

This research set out to examine the perceptions and motivations underlying winery owners' decisions to engage in wine tourism. The chapter has established that the motivations for engaging in wine tourism are considerably more complex when the family context, winery owners' identification with their profession, and their affective attachment to the place of production are taken into account.

Our findings have potential practical implications for rural development policy (e.g. Common Agricultural Policy) at the European level, as well as for governments and destination management organizations (DMOs) at the regional level. First, tourism diversification is recognized as a central objective of the rural development policy across European Union member states, as a means for enhancing the quality of life of rural communities (Hansson et al., 2013). In this instance, our findings provide policymakers with valuable insights into the motivations underlying rural business owners' decisions to engage or resist diversification. Particularly, we show how their motivations for diversification are inextricably linked to their place attachment/place identity and the situation of the family.

Second, at the regional level, the findings of this study are likely to have implications for governments and DMOs when promoting and presenting diversification initiatives for rural regeneration. When pursuing economic regeneration, regional governments primarily aim for social cohesion, sustainability, and economic growth. As active stakeholders, the business community has an important role to play in developing the tourism industry, and their views should be sought as key community players. Regional government officials need to take into consideration the complexity of rural families' motivations for diversification and position their message accordingly to ensure receptivity. While government officials might highlight the economic benefits of diversification at the individual, societal, and regional level, including increased profits, generation of additional income, as well as employment creation and retention, the findings of this study clearly indicate that a more refined approach is needed, emphasizing the family benefits of tourism diversification.

Third, the findings of this study have shown that some rural business owners still limit or resist tourism diversification on the basis of an anticipated loss of identity. In this instance, tourism planners and DMOs need to take into consideration rural business owners' strong identification with their profession, as well as their affective attachment to the place of production.

Finally, in order to achieve the right balance between tourism prosperity and rural communities' well-being, local authorities and DMOs should offer continuous support and guidance in terms of training and education programs. These programs should provide rural business owners with an in-depth understanding of tourism planning and development at the local level and highlight not only the economic benefits of tourism, but also the enjoyment and satisfaction this particular type of diversification can bring, without threatening their core activities and occupational identity.

References

Anton, C. E., and Lawrence, C. (2014). Home is where the heart is: the effect of place of residence on place attachment and community participation. *Journal of Environmental Psychology*, 40: 451–461. https://doi.org/10.1016/j.jenvp.2014.10.007

Barbieri, C. (2010). An importance-performance analysis of the motivations behind agritourism and other farm enterprise developments in Canada. *Journal of Rural and Community Development*, 5: 1–20.

Barbieri, C., and Mahoney, E. (2009). Why is diversification an attractive farm adjustment strategy? Insights from Texas farmers and ranchers. *Journal of Rural Studies*, 25: 58–66. https://doi.org/10.1016/j.jrurstud.2008.06.001

Barlybaev, A. A., Akhmetov, V. Y., and Nasyrov, G. M. (2009). Tourism as a factor of rural economy diversification. *Studies on Russian Economic Development*, 20: 639–643. https://doi.org/10.1134/S1075700709060094

Baron, J., Somogyi, S., Hall, D., and Carrington, A. (2008). Wine marketing online. In *Proceedings of the BbWorld Asia Pacific 2008 Conference*, Melbourne, Australia.

Berrone, P., Cruz, C., and Gomez-Mejia, L. R. (2012). Socioemotional wealth in family firms theoretical dimensions, assessment approaches, and agenda for future research. *Family Business Review*, 25: 258–279. https://doi.org/10.1177/0894486511435355

Bonaiuto, M., Carrus, G., Martorella, H., and Bonnes, M. (2002). Local identity processes and environmental attitudes in land use changes: the case of natural protected areas. *Journal of Economic Psychology*, *23*(5): 631–653. https://doi.org/10.1016/S0167-4870(02)00121-6

Canoves, G. Villarino, M., Priestley, G. K., and Blanco, A. (2004). Rural tourism in Spain: an analysis of recent evolution. *Geoforum*, 35: 755–769. https://doi.org/10.1016/j.geoforum.2004.03.005

Cennamo, C., Berrone, P., Cruz, C., and Gomez-Mejia, L. R. (2012). Socioemotional wealth and proactive stakeholder engagement: why family-controlled firms care more about their stakeholders. *Entrepreneurship Theory and Practice*, 36: 1153–1173. https://doi.org/10.1111/j.1540-6520.2012.00543.x

Charters, S., and Menival, D. (2011). Wine tourism in Champagne. *Journal of Hospitality and Tourism Research*, 35: 102–118. https://doi.org/10.1177/1096348010384597

Chen, N., Hsu, C. H., and Li, X. R. (2018). Feeling superior or deprived? Attitudes and underlying mentalities of residents towards mainland Chinese tourists. *Tourism Management*, 66: 94–107. https://doi.org/10.1016/j.tourman.2017.11.007

Cohen, E. (1984). The sociology of tourism: approaches, issues, findings, *Annual Review of Sociology*, 10: 373–392. https://doi.org/10.1146/annurev.so.10.080184.002105

DMO Piemonte (2018). Rapporti Statistici dei flussi turistici in Piemonte. https://www.regione.piemonte.it/web/temi/cultura-turismo-sport/turismo/osservatorio-turistico/rapporti-statistici-dei-flussi-turistici

Farmaki, A. (2012). An exploration of tourist motivation in rural settings: the case of Troodos, Cyprus. *Tourism Management Perspectives*, 2: 72–78. https://doi.org/10.1016/j.tmp.2012.03.007

Fitz-Koch, S., Nordqvist, M., Carter, S., and Hunter, E. (2018). Entrepreneurship in the agricultural sector: a literature review and future research opportunities. *Entrepreneurship Theory and Practice*, 42: 129–166. https://doi.org/10.1177/1042258717732958

Flanigan, S., Blackstock, K., and Hunter, C. (2015). Generating public and private benefits through understanding what drives different types of agritourism. *Journal of Rural Studies*, 41: 129–141. https://doi.org/10.1016/j.jrurstud.2015.08.002

Gee, J. P. (2014). *An Introduction to Discourse Analysis: Theory and Method* (4th ed.). London: Routledge.

Gomez-Mejia, L. R., Haynes, K. T., Núñez-Nickel, M., Jacobson, K. J., and Moyano-Fuentes, J. (2007). Socioemotional wealth and business risks in family-controlled firms: evidence from Spanish olive oil mills. *Administrative Science Quarterly*, 52: 106–137. https://doi.org/10.2189/asqu.52.1.106

Grimble, R., and Wellard, K. (1997). Stakeholder methodologies in natural resource management: a review of principles, contexts, experiences and opportunities, *Agricultural Systems*, 55(2), 173–193. https://doi.org/10.1016/S0308-521X(97)00006-1

Groth, T. M., and Curtis, A. (2017). Mapping farmer identity: why, how, and what does it tell us? *Australian Geographer*, 48: 365–383. https://doi.org/10.1080/00049182.2016.1265881

Gursoy, D., Jurowski, C., and Uysal, M. (2002). Resident attitudes: a structural modeling approach. *Annals of Tourism Research*, 29(1): 79–105. https://doi.org/10.1016/S0160-7383(01)00028-7

Hallak, R., Brown, G., and Lindsay, N. J. (2012). The Place Identity–Performance relationship among tourism entrepreneurs: a structural equation modelling analysis. *Tourism Management*, *33*(1): 143–154. https://doi.org/10.1016/j.tourman.2011.02.013

Hansson, H., Ferguson, R., Olofsson, C., and Rantamäki-Lahtinen, L. (2013). Farmers' motives for diversifying their farm business – the influence of family. *Journal of Rural Studies*, 32: 240–250. https://doi.org/10.1016/J.JRURSTUD.2013.07.002

Lyon, A., Hunter-Jones, P., and Warnaby, G. (2017). Are we any closer to sustainable development? Listening to active stakeholder discourses of tourism development in the Waterberg Biosphere Reserve, South Africa. *Tourism Management*, 61: 234–247.

Murphy, P. (1985). *Tourism: A Community Approach*. London: Routledge.

Phillips, N., and Hardy, C. (2002). *Discourse Analysis: Investigating Processes of Social Construction* (Vol. 50). Thousand Oaks, CA: Sage.

Ribeiro, M. A., Pinto, P., Silva, J. A., and Woosnam, K. M. (2017). Residents' attitudes and the adoption of pro-tourism behaviours: the case of developing island countries. *Tourism Management*, 61: 523–537. https://doi.org/10.1016/j.jhtm.2020.06.017

Rosso, M. (2014). *Barolo – The Jewel of the Langa*. Turin, Italy: Omega Edizioni

Sanchez Cañizares, S. M., Castillo Canalejo, A. M., and Núñez Tabales, J. M. (2016). Stakeholders' perceptions of tourism development in Cape Verde, Africa. *Current Issues in Tourism*, 19(10): 966–980. https://doi.org/10.1080/13683500.2015.1008428

Schilling, B. J., Sullivan, K. P., and Komar, S. J. (2012). Examining the economic benefits of agritourism: the case of New Jersey. *Journal of Agriculture, Food Systems, and Community Development*, 3: 199–214. https://doi.org/10.5304/jafscd.2012.031.011

Schmid, T., Ampenberger, M., Kaserer, C., and Achleitner, A. K. (2015). Family firm heterogeneity and corporate policy: evidence from diversification decisions. *Corporate Governance: An International Review*, 23: 285–302. https://doi.org/10.1111/corg.12091

Sharma, P., Chrisman, J. J., and Chua, J. H. (1997). Strategic management of the family business: past research and future challenges. *Family Business Review*, 10: 1–35. https://doi.org/10.1111/j.1741-6248.1997.00001.x

Sharpley, R. (2014). Host perceptions of tourism: a review of the research. *Tourism Management* 42: 37–49.

Sharpley, R., and Vass, A. (2006). Tourism, farming and diversification: an attitudinal study. *Tourism Management*, 27, 1040–1052. https://doi.org/10.1016/j.tourman.2005.10.025

Simpson, K. (2001). Strategic planning and community involvement as contributors to sustainable tourism development. *Current Issues in Tourism*, *4*(1): 3–41.

Sirmon, D. G., and Hitt, M. A. (2003). Managing resources: linking unique resources, management, and wealth creation in family firms. *Entrepreneurship Theory and Practice*, 27: 339–358. https://doi.org/10.1111/1540-8520.t01-1-00013

Sook, F., May, L., Songan, P., and Nair, V. (2014). The impact of local communities' involvement and relationship quality on sustainable tourism in rural. *Procedia – Social and Behavioral Sciences*, 144: 60–65. https://doi.org/10.1016/j.sbspro.2014.07.274

Strzelecka, M., Boley, B. B., and Woosnam, K. M. (2017). Place attachment and empowerment: do residents need to be attached to be empowered? *Annals of Tourism Research*, 66: 61–73. https://doi.org/10.1016/j.annals.2017.06.002

Tew, C., and Barbieri, C. (2012). The perceived benefits of agritourism: the provider's perspective. *Tourism Management*, 33: 215–224. https://doi.org/10.1016/j.tourman.2011.02.005

Vargas-Sanchez, A., do Valle, P. O., da Costa Mendes, J., and Silva, J. A. (2015). Residents' attitude and level of destination development: an international comparison. *Tourism Management*, 48: 199–210. https://doi.org/10.1016/j.tourman.2014.11.005

Wilson, S., Fesenmaier, D. R., Fesenmaier, J., and Van Es, J. C. (2001). Factors for success in rural tourism development. *Journal of Travel Research*, 40(2): 132–138.

Xu, S., Barbieri, C., Anderson, D., Leung, Y. F., and Rozier-Rich, S. (2016). Residents' perceptions of wine tourism development. *Tourism Management*, 55: 276–286.

Yang, L. (2012). Impacts and challenges in agritourism development in Yunnan, China. *Tourism Planning & Development*, 9(4): 369–381. https://doi.org/10.1080/21568316.2012.726257

21

WINE TOURISM IN RURAL JAPAN

An integrated development instrument

Chuanfei Wang

Introduction: wine tourism and rural development in Japan

The Japanese winemaking industry has steadily grown in terms of both quantity and quality since the late 1990s. At the turn of the century, Japan started to make its name in the wine world, and it has been considered a wine-producing region by international professionals. It was first introduced in the second edition of *The Oxford Companion to Wine* (Robinson and Harding 1999) in 1999 and in the seventh edition of *The World Atlas of Wine* (Johnson and Robinson 2013). At the same time, there have been great efforts in building those winery areas into tourism destinations. The newly emerged wine tourism is not only a business phenomenon but also an economic strategy currently taken by both public and private actors to cope with the economic challenges in rural Japan (Kodama 2019; Naitō 2009; Takahashi 2015).

This chapter examines wine tourism in Japan as an instrument, integrating urban and rural social resources to stimulate rural economic development. It addresses the question of how wine tourism has been established by entrepreneurs in their local rural communities, where developing the tourism industry is taken as a national agenda to stimulate the economy. Japan's case highlights that the collaborative approach between entrepreneurs and government, instead of the one of top-down or bottom-up, is key in making wine tourism possible as an integrated approach to rural development. It also illustrates that the rural and urban should be considered as two interrelated social and economic entities. This indicates that solutions to rural challenges need to integrate social resources in both urban and rural settings.

The challenges that the Japanese have been experiencing in the rural areas can be summarized as economic decline and community shrinking caused by long-term depopulation, specifically, the low fertility rate, aging population, and younger generations' moving to the cities. However, they are not unique to Japan. An OECD (Organisation for Economic Co-operation and Development 2006) report titled 'The New Rural Paradigm: Policies and Governance' pointed out the global issue of increasing urbanization. It thus called for countries to develop policy initiatives with an integrated approach, focusing on local advantages and integrating various sectoral policies at regional and local levels and to improve coordination of sectoral policies at the central government level to meet the needs of their rural regions. In the past decade, tourism was taken as a strong integrated approach in some countries (Correia and Brito 2016).

DOI: 10.4324/9781003143628-25

Among various forms of tourism, wine tourism has increasingly expanded in both well-established or newly emerged winemaking areas. While it was initially a marketing tool for increasing wine sales, it has been adopted as a strategic means to integrate resources both inside and outside the rural community and to commodify the rural community as a whole as tourism products (Gammack 2006). Studies have shown that successful commodification of rural tourism resources through wine tourism offers tourists a holistic experience of a territory instead of an isolated activity related only to wineries. This suggests that if wine tourism serves as an instrument of integrating rural tourism resources, it should be a form of tourism providing a vast array of attractions and culture from multiple industries in a region where wine is produced, such as restaurants, local lifestyle, arts, crafts, nature exploration, and local history (Etcheverria 2016; Getz and Brown 2006; Gmelch and Gmelch 2011).

Therefore, wine tourism requires a wide array of actors. Getz (2000) pointed out three major perspectives from the marketing viewpoint, including wine producers, tourism agencies, and consumers. Similarly, from a sociological perspective, I proposed the 'wine world' approach (Wang 2019), suggesting that wine tourism be considered as a construct by actors including business entrepreneurs, government, volunteers, and consumers. This chapter aims to show the significance of collaboration between the top policymakers and grassroots entrepreneurs in making wine tourism an integrated instrument for rural development.

My own long-term ethnographic research in Japan has discovered these various levels of promotion in expanding wine tourism and, more importantly, employing it as a means for rural revitalization. Since 2015, my fieldwork has included wine tourism in Yamanashi Prefecture, the case study that this chapter focuses on, and the prefectures of Kumamoto, Yamagata, Iwate, Nagano, Kanazawa, and Hokkaido. I interviewed various local actors involved in wine tourism, including wine tourism event designers and organizers, winery owners and sommeliers, local government officials in charge of tourism, and volunteer wine tourism guides. I also participated in its wine tourism events regularly for fieldwork observations. Therefore, this chapter provides a chronicle overview of how Japanese wine tourism has been established and how it has been employed as a tool for rural development. Since this chapter was composed at the end of 2020, during the COVID-19 pandemic, it also covers the impacts that this pandemic has made on Japan's wine tourism and, more importantly, how those actors involved have tried to make the business survive. This part is based on an interview I conducted on November 10, 2020, with one of the core wine tourism organizers.

Wine tourism in Japan: its definition, brief history, and instrumental roles

Although Japan has been making grape wine since the late 19th century (Wang 2017), tourism themed by wine has only recently been established. The Japanese language of wine tourism, 'wain tsūrizumu,' is a loanword from English, first used at the Wine Tourism Yamanashi event in 2008. According to the event's official web page, wine tourism is defined as follows.

> Wine Tourism Yamanashi is an event in which tourists take a tour of the wine-producing region and enjoy the whole region. The purpose of organizing this event was to hope a lot of people can enjoy the region. On the days of the event, circulation buses in the region involved in the event are provided to help tourists visit wineries conveniently. After application to the event, tourists receive winery information, map, bus timetable, and tickets. Based on the information provided by these materials and

> by the homepage of the event on the internet, tourists set up their travel plans before coming to the event, including which winery to visit, which areas to go, where to eat, and so on. Then, on the days of the event, tourists enjoy the region of grape growing and wine producing according to the travel plans they made on their own.

It is worth noting that the Japanese term *wain tsūrizumu* is registered by the organizer of the Wine Tourism Yamanashi, General Incorporated Association Wine Tourism, as a trademark in Japan to differentiate wine tourism from other forms of tourism themed by wine. In other words, any form of wine-related tourism without the organizations and activities described earlier cannot use the term *wine tourism*. The trademark protects the model of Wine Tourism Yamanashi and has helped expand it recently to other wine-producing regions, such as Hokkaido, Nagano, and Yamagata in Japan.

Wine Tourism Yamanashi is commonly regarded as the most well-established regional wine tourism in Japan, which successfully allocates wine tourism resources in and outside of the region and attracts urban tourists to come and consume in this rural region. Tourist participation is easy; entrance tickets are purchased online for 4,000 yen, and tourists are mailed a participation card, a map, a bus schedule, and a list of wineries. By 2019, it had been an annual event held twice a year on a weekend during spring and autumn, collaboratively involving grape growers, wineries, restaurants, shops, morning markets, and the local government in Yamanashi Prefecture. In 2018, this model of wine tourism was also being copied by other domestic winemaking regions using the same name and tourist products involved. Yamagata Prefecture in the northeast, for example, held its first wine tourism in summer 2018, called 'wine tourism Yamagata,' adopting the same pattern as to how wine tourism was organized in Yamanashi (Fieldwork, June 9, 2019).

Beyond the rural community, wine tourism is also regarded as significant to the development of the entire wine producing industry in Japan. Domestic wines began to face stronger price challenges from imported European wines in the domestic market in 2019 when the Economic Partnership Agreement with the European Union came into effect. Wine tourism is one of the key sales channels to Japanese winemakers, who have expected wine tourism to help attract the Japanese public to know and consume domestic wines (The Japan Agricultural News 2019; Nihon Keizai Shimbun 2018).

The following sections take a closer look at how wine tourism was established as an integrated approach for rural development with the collaboration of the top and the grassroots actors. As the 'world approach' suggests, national and local governments at the top provide and carry out the policies of establishing and promoting wine tourism. The grassroots actors include three groups – tourism organizers, winemakers, and volunteers. Tourism organizers are the entrepreneurs initiating and maintaining wine tourism by networking all of the other actors involved to integrate wine tourism resources of the nature and culture, such as grape farming and its landscape as well as winemaking skills and concepts. Winemakers in wineries provide winery tours to spread local winemaking culture and wine tasting. And local individual residents or groups volunteer to provide free tour guide services to the tourists in the local area for non-winery but wine-history-and-culture-related tourist locations, such as shrines and museums.

While every actor is significant in sustaining wine tourism, this chapter focuses on the entrepreneurial tourism organizers as well as on national and local governments, because both of them, from private and governmental sectors of the society, play a leading role in establishing wine tourism in Japan as an integrated effort for increasing rural development.

The actors from the top: national and local governments

Rural revitalization has long been on the national agenda in Japan. From governments to grassroots levels, some of the policy initiatives and strategies have been undertaken since the 1990s, including the national policy of merging cities, towns, and villages; the campaign of town building; the approach of volunteerism; and the recent attempt to participate in the Trans-Pacific Partnership Agreement (Assmann 2016; Delaney 2016; Matanle and Rausch 2011; Rausch 2016). Tourism has been taken as a significant instrument to achieve this goal by integrating the rural and the urban resources. In particular, wine tourism has been seen as a new and promising form of tourism to revive and sustain rural communities (Kodama 2017).

Japan's national government has issued policies supporting building wine tourism in rural areas. At the highest level, a new policy called *kankōrikkoku*, or 'national revitalization through tourism,' serves to establish the tourism industry as a key instrument for national economic development in the 21st century, and especially in the years before and after the 2020 Tokyo Olympic Games. It also supports an overall national economic shift in Japan from manufacturing to service industries. Currently, the domestic policy of rural revitalization places the importance on rural economic development by relating local regions to urban consumers of various forms of tourism, focusing thematically on natural and cultural resources, in regional and local areas in contemporary Japan (Graburn 2009; Hendry 2009; Jones 2016; Oedewald 2009).

This policy has been supplemented by a number of national policy-directed programs, including developing new tourism themes, which ended in 2018, and developing a themed tourism scheme since 2019, which is a new program built upon the previous one (Japan Tourism Agency 2020). Among them is the Japan Heritage program, initiated in 2015, which includes a major winemaking region of Japan and its winemaking history in 2018 (see Wang 2020 for details). This scheme includes the Tokyo Olympics as the goal to designate a total of 100 Japanese regional spots as national heritage sites for attracting international tourists to visit. However, by November 2020, new spots had continued to be designated as Japan Heritage locations, despite postponement of the Olympic Games to 2021 because of the COVID-19 pandemic.

This scheme has played an important institutional role in legitimizing the local wine history and thereby facilitating the boost of wine tourism in a region without a long-standing tradition of wine. Practically, it has been effectively used as a catching brand to develop wine tourism products at the local level. Since 2018, there have been new wine tourism products designed as Japan Heritage routes. More recently, in June 2020, another new designation from the Japan Heritage program was given to Ushikushi city in Tsukuba Prefecture and Kofu city in Yamanashi Prefecture for their winemaking history, which is named as '140 Years of Japanese Wine: Crystals of Japanese Culture Brewed with Domestic Grape.' The histories related to wine include the first government-owned and private-owned wineries and the first 'chateau' with large-scale production and a division-of-labor system in Japan. The historical narratives help boost wine tourism in the way that they are embodied in the specific wineries and facilities.

In the meantime, local governments with wine-related resources have positively responded to the national policy of *kankōrikkoku* since its inception by building and expanding wine tourism as an important industry for developing local economies. The government of Yamanashi Prefecture has worked directly on wine tourism events, providing financial and policy support (Fieldwork, November 23, 2018). Located to the east of Tokyo near Mount Fuji, Yamanashi Prefecture can be said to be one of the most developed wine regions in Japan because of its long history of grape growing and winemaking. It is said to be the cradle of Japan's winemaking, the first company there having been established in 1877, marking the beginning of Japan's grape winemaking at the national level. Since then, it has been the leading region for experimentation, with local and Western methods of winemaking. It is currently Japan's largest winemaking area. Of the nation's 303 wineries, 81 are located in Yamanashi (National Tax Agency 2019).

Unlike Yamanashi, most of the prefectures with wineries are newly developed wine-producing regions. The local government of Nagano Prefecture in the west of Japan, for example, has been supporting the establishment of new wineries to help sustain the long-run economic and social development of the local community by further developing wine tourism. In the Ueda city of the same prefecture, the local government welcomed Mercian, one of Japan's biggest winemaking corporations, to open a winery there in 2018. At the same time, Ueda city's neighbor city, Tomi city, has been the major wine-producing area of the prefecture. For the local government, construction of the winery was meant to establish the city as a wine-producing region. Mercian, which brought its institutional knowledge of grape growing and wine producing, was also expected to exchange that knowledge with wineries in the neighboring cities of the same prefecture so that the quality of wine produced in the entire prefecture would be improved. More importantly, the local government expected to establish wine tourism in the prefecture, connecting all of the tourism resources, including the wineries, restaurants, and hot spas, and attract tourists from the outside (Interview, Mr. Matsuo from Mercian Mariko on April 29, 2018).

Case Study 21.1 The actors from the grassroots: entrepreneurial tourism organizers

Individual entrepreneurs who themselves are self-employed in the tourism industry are the important actors in building and maintaining wine tourism. In such companies, the number of full-time employees is usually between one and three; when there are wine tourism events, part-timers are hired. They function as planners, designers, and operators, connecting wine tourism resources from wineries to transportation companies in the local.

The Wine Tourism Yamanashi event was held for the first time in Yamanashi Prefecture in 2008. It is currently run by two companies – LOCALSTANDARD Ltd., which designs and plans the event, and Tabizen Ltd., which is responsible for marketing directly to customers and operating services such as reservations, tickets, and customer service before and on the day of the event.

Mr. Oki Takayuki, owner of LOCALSTANDARD, and Mr. Yomoto Takayuki cofounded the event. Although both grew up in the region and coincidentally have the same first name, they did not know each other before coming to work together on wine tourism. In 2007, Oki established LOCALSTANDARD, specializing in designing wine tourism and organizing the event every year.

Before engaging in wine tourism, neither of them had ever heard the term wine tourism before. Oki ran a I bar, where he sold Western food modeled on popular dishes in New York, Italy, and Spain and various imported drinks, including wine from countries well known for winemaking. After selling foreign food and drinks for years, he turned his attention to local products and switched all of the wines at his shop. Meanwhile, Yomoto had worked in Tokyo and came back to his hometown to take over his family company making metal sheet painting. The experience of working at the family company inspired him to become an initiator of the Wine Tourism Yamanashi. At his family company, he learned that local companies relied heavily on financial support from the local governments. He realized that government support would not lead to a bright economic future for the entire local community, and he hoped to make some changes. In 2003, he created a business study group called KOFU Pride, aiming to find specific solutions for helping Kofu, the capital city of Yamanashi Prefecture, to be independent of the local government's subsidies. The group held monthly meetings at Oki's shop with people who were also local business owners to look for local uniqueness and discuss strategies. In a meeting, the owner of a local winery, who had studied winemaking in Australia, talked about wine tourism in Australia. The story inspired both Yomoto and Oki to think about the wine produced in their hometown and be willing to organize something similar in Yamanashi (Nikkei Style 2015).

To both men, wine tourism was seen more as a strategy than as a business opportunity, attracting people from outside the area and creating new opportunities for the local economy. How wine tourism could achieve such a goal was well explained by Oki in an interview as follows.

> We are doing this not only to support wineries in the region. We are also thinking about how we can survive. If we don't make a market in the region, we cannot survive. So, we thought about what would be the best strategy to attract more people around the nation and increase the number of visitors. To do this, we considered using wine as a weapon. Yamanashi has wineries trying to improve the quality of grapes to make good wine. But good quality doesn't guarantee a good market. So, a space where consumers could access those wines was needed. This is the bottleneck in all of Japan. If more and more wine lovers come, the tourist population will increase. They drink wine and some of them will want to stay for a night. Then, this opens a market for the hotel industry. By doing this, we think the region can survive. Wine tourism is a strategy for solving daily problems.
>
> *(Nikkei Style 2015)*

The local entrepreneurs share the idea with the government that wine tourism functions to integrate local tourism resources and consumers from the outside of the region. This was seen in a map (see Figure 21.1) displayed at the entrance of the Katsunuma Library in Yamanashi Prefecture, where wine tourism is carried out every year. It marks the locations of wineries and other wine-related facilities. It also offers information on places where tourists can experience local food and culture other than wine – for example, noodles and porcelain crafting. Thus, this map represents the idea of wine tourism as an instrument of integrating and branding the entire region as a whole. This whole set of cultural and natural resources are constantly delivered to the tourists, mostly from the outside of the region, through wine tourism events every year.

Discussion: contributions and challenges in wine tourism

In only a decade, Yamanashi Prefecture has become a well-established wine tourism industry based on its existing winemaking industry. From the beginning, wine tourism was designed for multiple purposes, including marketing effects for the wineries, economic stimulation, and social revitalization to the local community. To put it another way, wine tourism is envisioned

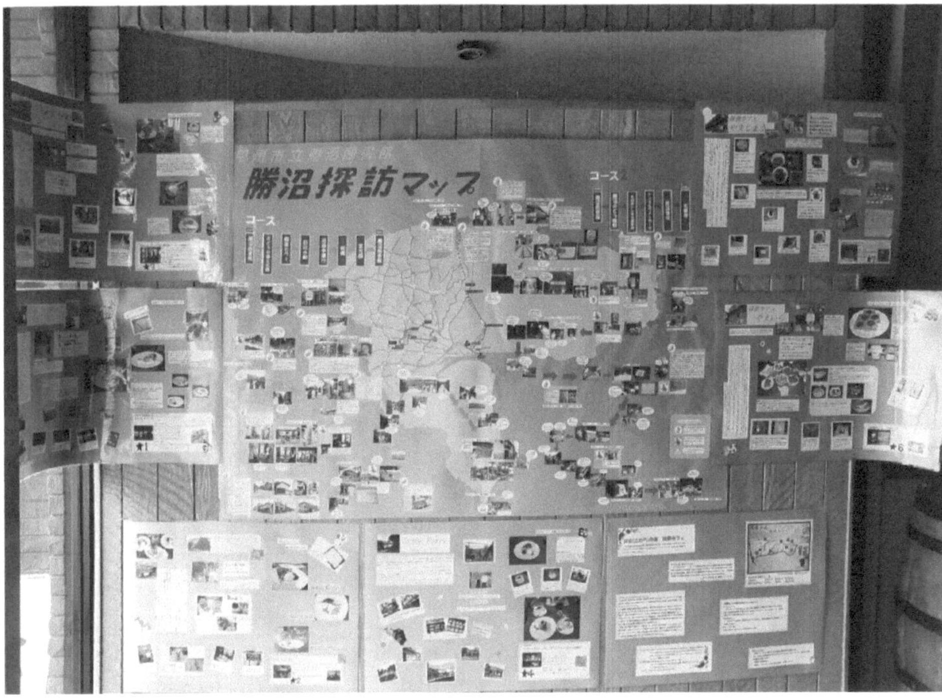

Figure 21.1 A map offers a full picture of the local places where tourists can visit. Two sample walking routes themed on wine are specially marked.

Source: Photograph by author, November 23, 2019

as an integrated instrument to address rural challenges. This collaborative approach between local entrepreneurs and government has played a key role in making it possible for national programs supporting tourism development in general, as mentioned earlier in this chapter. It is worth noting that the event received subsidies from the local government in the first year and in following years. Moreover, several programs have encouraged young people to engage in the winery business, such as providing financial support and free winemaking training to entrepreneurs who are willing to build a winery. This includes the offering of agricultural land for farming wine grapes at a low price, and renting residential houses at a rate lower than the average market price to young families who move to the local community in order to establish wineries. There are also policies supporting female winery founders who can receive a low-interest loan from the bank to start a winery. And these policies are not limited to Yamanashi.

When it comes to whether wine tourism has fulfilled its roles, it could be evaluated in several terms. Firstly, visitors to the Wine Tourism Yamanashi festival had been steadily growing every year by the outbreak of COVID-19 in 2020, from 1,284 visitors in 2008 to 2,123 visitors in 2011. Since 2012, to maintain the quality of tourist experience, the number of visitors has been limited to no more than 1,500. At the same time, festival organizers raised the price of the entry ticket to double that of the previous year's festival price. Surprisingly, the increased price did not cause a drop in visitors. Since then, the number of visitors has been about 1,300 on average (1,393, 1.244, and 1.368 in 2016, 2017, and 2018). The majority of visitors are from outside the region. For the 2018 festival, 90% of the visitors were not locals. Secondly, day-trippers were decreasing, whereas visitors staying over two nights increased by about 4.2%

between 2016 and 2018. This indicates that wine tourism was able to create some economic opportunities for the hotels in the region. Thirdly, wine tourism helps increase immediate wine purchases. The number of bottles purchased by wine tourism visitors was increased in both the cheap (less than 1,500 yen) and expensive (between 1,501 and 2,000 yen) price ranges, but decreased in the middle (more than 2,000 yen) range (Ippan shadanhōjin waintsūrizumu 2019).

Wine tourism could be considered successful based on the aforementioned statistics, for the goals of attracting tourists from the outside to the region and stimulating the regional economy through wine tourism were achieved. Moreover, wine tourism contributed to the growing reputation of local wine producers as the number of regular winery visitors increases. Veteran winery visitors who had visited more than 11 wineries in the region increased by 12.4% from 2016 to 2018 (Ippan shadanhōjin waintsūrizumu 2019). Furthermore, Wine Tourism Yamanashi has become a model for the local actors to design other tourism products with a similar concept of integrating local tourism resources. Some of them are not themed by wine but include local wineries as a part of tourist destinations. Within the tourism of local crafts, for example, tourists tried their hand at making porcelain with the local craftsman of wine glasses. In short, in the past decade, wine tourism achieved the expected goal of attracting urban tourists and made an economic contribution to the locals.

However, this does not mean there are no new challenges. Wine tourism has not been successful in attracting young tourists in their 20s, who account for only 7% of the tourist population to the wine tourism event (Ippan shadanhōjin waintsūrizumu 2019). Moreover, wine tourism profitability is questionable. According to a core member, Mr. Sudo, whom I interviewed on November 10, 2020, their experience of the past years showed that wine tourism has four characteristics: it has low profits; it takes time and effort; the cost goes up (due to labor demand for preparing and providing service before and on the day); and it is difficult to hold a large number of tourists on the event day for maintaining service quality. This is to say that it might not be sustainable as a business to make a profit and survive if the government subsidies ran out.

The global health crisis in 2020 raised another challenge, as all of the wine tourism was terminated. The organizers tried online wine tourism by organizing online tours of wineries, including the online 'tasting' of the wine from those wineries with a professional introducing the wines and describing their features. In the hope that online wine tourism could contribute to winery sales, those wineries provided discounted prices to the tourists who attended the online wine tourism and purchased the wine for tasting at home during the online tour. Such online wine tourism experiences were not comparable to those held on-site in terms of the scale. They were organized in small groups, about 10 to 20 people at a time. However, each experience was fully attended every time. By the time I interviewed Mr. Sudo in early November 2020, it seemed to be only a temporary strategy for people involved in the wine tourism business to try to survive through the COVID-19 crisis.

Conclusion: wine tourism as an integrating instrument of rural development

This chapter has demonstrated the Japanese case of wine tourism being employed as an integrated instrument to revitalize rural society. It has focused on two key actors, the government from the top and the local entrepreneurs from the grassroots, both of whom collaborate to build the wine tourism industry in a winemaking region without a long-standing reputation. Two points are highlighted. Firstly, rural challenges should be addressed by resources in both urban and rural areas. This means wine tourism should be regarded as a form of tourism that attracts tourists from beyond the local territory, in many cases rural areas. Secondly, the collaboration between the governmental and private actors is key in establishing wine tourism. The policy support from the government is significant in initiating wine tourism at the regional level, especially if it is expected to play a role in integrating various resources in and outside of the region.

This wine tourism case from Japan shows a Japanese approach to their rural challenges, and it cannot be considered a refined model. Questions such as how to make sustainable wine tourism remain the major theme that local actors consider in their daily business operations. Moreover, to a large degree, it may not be able to represent the complicated issues that many other wine tourism destinations have faced. However, it does imply some conceptual directions especially to those wine tourism destinations without long-standing wine-producing traditions or whose wine-producing history has not yet been discovered as a source for building tourism at the local level. The effectiveness of wine tourism as a tool for local development relies on a framework that can mobilize and integrate resources in and outside of the winemaking region. This framework requires collaborations of actors from the government to business, and even to each resident. The local and national governments' role in supporting wine tourism builders by mobilizing resources outside the region is important, especially in the initial phase. Endeavor by a single group of actors to develop wine tourism, integrating sources inside and outside of a region, is not easy. Wine tourism is practiced in a region with wine and its related natural and cultural resources, but it requires economic, knowledge, and social sources beyond the region to make it possible and, at best, sustainable in the long run.

Acknowledgement

This work was supported by the Japan Society for the Promotion of Science under Grant-in-Aid for Young Scientists (B) 17K13853, 'The Cultural Construction of the Wine Tourism Experience in Japan, China, and Australia.'

References

Assmann, S. (2016). Agriculture in Japan: free trade and the need for reform. In Assmann, S. (ed.), *Sustainability in Contemporary Rural Japan: Challenges and Opportunities* (pp. 49–61). London and New York: Routledge.

Correia, R., and Brito, C. (2016). Wine tourism and regional development. In Peris-Ortiz, M., Rama, M., and Rueda-Armengot, C. (eds.), *Wine and Tourism: A Strategic Segment for Sustainable Economic Development* (pp. 27–39). Switzerland: Springer International Publishing.

Delaney, A. E. (2016). Social sustainability in post-3.11 coastal Japan: the significance of social capital. In Assmann, S. (ed.), *Sustainability in Contemporary Rural Japan: Challenges and Opportunities* (pp. 3–17). London and New York: Routledge.

Etcheverria, O. (2016). Wine tourism and gastronomy. In Peris-Ortiz, M., Rama, M., and Rueda-Armengot, C. (eds.), *Wine and Tourism: A Strategic Segment for Sustainable Economic Development* (pp. 161–177). Switzerland: Springer International Publishing.

Gammack, J. G. (2006). Wine tourism and sustainable development in regional Australia. In Carlsen, J., and Charters, S. (eds.), *Global Wine Tourism: Research, Management and Marketing*. Oxon: CAB International.

Getz, D. (2000). *Explore Wine Tourism: Management, Development and Destinations*. New York: Cognizant Communication Corporation.

Getz, D., and Brown, G. (2006). Critical success factors for wine tourism regions: a demand analysis. *Tourism Management*, 27: 146–158. https://doi.org/10.1016/j.tourman.2004.08.002

Gmelch, G., and Gmelch, S. B. (2011). *Tasting the Good Life: Wine Tourism in the Napa Valley*. Bloomington: Indiana University Press.

Graburn, N. (2009). The past and the other in the present: kokunai kokusaika kanko – domestic international tourism. In Guichard-Anguis, S., and Moon, O. (eds.), *Japanese Tourism and Travel Culture* (pp. 21–36). London and New York: Routledge.

Hendry, J. (2009). Fantasy travel in time and space: a new Japanese phenomenon? In Guichard-Anguis, S., and Moon, O. (ed.), *Japanese Tourism and Travel Culture* (pp. 129–145). London and New York: Routledge.

Ippan shadanhōjin waintsūrizumu [General Incorporated Association Wine Tourism] (2019). 調査報告書 「ワインツーリズム参加者に見る山梨県峡東地域の現状と課題」 (Chōsa hōkoku-'ho 'waintsūrizumu sankasha ni miru Yamanashi ken kōtō chiiki no genjō to k'dai') [Survey report, 'Current situation and issues in the eastern part of Yamanashi Prefecture as seen by wine tourism participants'] (p. 5). https://wineresort.jp/wp-content/uploads/2020/06/日本遺産山梨県峡東地域観光客実態調査分析業務（平成30年度).pdf

Japan Tourism Agency. (2020). テーマ別観光による地方誘客事業 (Tēma-betsu kankō ni yoru chihō yūkyaku jigyō) [Local tourist attraction business by thematic tourism]. https://www.mlit.go.jp/kankocho/shisaku/kankochi/theme_betsu.html

Johnson, H., and Robinson, J. (2013). *The World Atlas of Wine* (7th ed). London: Mitchell Beazley.

Jones, T. (2016). Hokkaido's overlapping protected areas and regional revitalization: the case study of Toya-Usu geopark in Shinkotsu-Toya national park. In Assmann, S. (ed.), *Sustainability in Contemporary Rural Japan: Challenges and Opportunities* (pp. 151–166). London and New York: Routledge.

Kodama, T. (2017). 世界で活発化するワインツーリズム (Sekai de kappatsu-ka suru waintsūrizumu) [Wine tourism becoming more active in the world]. *国際貿易と投資 [Kokusai bōeki to tōshi]*, 29(1): 191–199.

——— (2019). 地理的表示制度の活用を通した日本ワインの地域ブランド推進 ワインツーリズム推進策の一要素として(Chiri-teki hyōji seido no katsuyō o tōshita Nihon wain no chiiki burando suishin waintsūrizumu suishin-saku no ichi yōso to shite) [Through the utilization of geographical indication system to promotion of regional brands of Japanese wine as an element of wine tourism promotion measures]. 日本国際観光学会論文集 (Nihon kokusai kankō gakkai ronbun-shū)[*Proceedings of the Japan Society for International Tourism*], 26: 161–173. https://www.jafit.jp/thesis/pdf/19_n04.pdf

Matanle, P., and Rausch, A. S. (2011). *Japan's Shrinking Regions in the 21st Century: Contemporary Responses to Depopulation and Socioeconomic Decline*. Amherst, NY: Cambria.

Naitō, N. (2009). ワインツーリズムの可能性と示唆 (Wain tsūrizumu no kanōsei to shisa) [Possibility and suggestion of wine tourism]. *Space and Environment Institute Report*.

National Tax Agency (2019). 製造業者の概況 (Seizōgyōsha no gaikyō) [Overview of the wineries]. http://www.nta.go.jp/taxes/sake/shiori-gaikyo/seizogaikyo/kajitsu/pdf/h29/29wine01.pdf

Nihon Keizai Shimbun (2018). 日欧EPA、恩恵まずワインから 2019年2月発効(Nichiō EPA, onkei mazu wain kara 2019-nen 2 tsuki hakkō) [Japan-EU EPA, first of all benefit from wine, effective February 2019]. https://www.nikkei.com/article/DGXMZO39014290W8A211C1TJC000/

Nikkei Style (2015). 参加者8年目で倍増 「ワインツーリズムやまなし」(Sankasha 8-nen-me de baizō 'wain tsūrizumu yamanashi') [Doubled in the 8th year of participants 'Wine Tourism Yamanashi']. November 6. http://style.nikkei.com/article/DGXMZO93398550Z21C15A0I00000

Oedewald, M. (2009). Meanings of tradition in contemporary Japanese domestic tourism. In Guichard-Anguis, S., and Moon, O. (eds.), *Japanese Tourism and Travel Culture* (pp. 105–128). London and New York: Routledge.

Organisation for Economic Co-operation and Development (2006). *The New Rural Paradigm: Policies and Governance*. http://www.oecd.org/regional/regional-policy/thenewruralparadigmpoliciesandgovernance.htm

Rausch, A. S. (2016). The Heisei Municipal Mergers: measures of sustainability, equality and identity. In S. Assmann (ed.), *Sustainability in Contemporary Rural Japan: Challenges and Opportunities* (pp. 1229–1541). London and New York: Routledge. Kindle.

Robinson, J., and Harding, J. (1999). *The Oxford Companion to Wine* (2nd ed.). Oxford: Oxford University Press.

Takahashi, T. (2015). *農林水産物・飲食品の地理的表示:地域の産物の価値を高める制度 利用の手引(Nōrin suisanbutsu inshoku-hin no chiri-teki hyōji: chiiki no sanbutsu no kachi o takameru seido riyō no tebiki) [Geographical Indications of Agriculture, Forestry and Fishery Products, Food and Drink: Guide for Using the System to Increase the Value of Local Products]*. 農山漁村文化協会(Nōsan gyoson bunka kyōkai).

The Japan Agricultural News (2019). ワイン=観光 パスタ=商品力 連携強め対抗 日欧EPAで北海道(Wain = kankō pasuta = shōhin-ryoku renkei tsuyome taikō nichiō EPA de hokkaidō) [Wine = tourism pasta = product power, strengthening cooperation, Japan-EU EPA, Hokkaido]. https://www.agrinews.co.jp/p48345.html

Wang, C. (2017). Joining the global wine world: Japan's wine making industry. In Niehaus, A., and Walravens, T. (eds.), *Feeding Japan: The Cultural and Political Issues of Dependency and Risk* (pp. 225–250). New York: Palgrave Macmillan.

——— (2019). The 'Worlds Approach' to gastronomic tourism: The case of wine tourism in Japan. In Dixit, S. K.(ed.), *The Routledge Handbook of Gastronomic Tourism* (pp. 395–402). London: Routledge.

——— (2020). The making of wine heritage in Japan. *The Special Issue of Asian Food and Cultural Governance: Food Heritage, Culinary Soft Power and Globalization. Asian Anthropology*, 20: 61–76. https://doi.org/10.1080/1683478X.2020.1773622

22
WINE TOURISM GENTRIFICATION
Transforming rural agricultural regions in the USA

Shenrui (Demi) Deng, Byron Marlowe, and Robert J. Harrington

Introduction

Over the last half-century, wine regions in the United States have evolved. These regions tend to have traditional agricultural areas with growing populations. Characterized by a high-quality lifestyle, those wine regions now have evolved into destinations attracting wine tourists and residents. This dynamic change fits the definition of gentrification and modern colonialism due to the 'usurpation of the land' through an influx of outside operators, new businesses, and tourists themselves (Smith 2012) to wine regions throughout the United States. This process also impacts residents, workers, and farmers being pushed out, creating a feeling of displacement, elitism, and social exclusion. The wine industry has grown substantially throughout all 50 states in the United States. With projected growth expected to continue above that of the overall economy, wine tourism becomes increasingly essential. The most recognized states for quality wines, including California, Washington, and Oregon, are under pressure to resolve the good, the bad, and the ugly of gentrification. The wine regions in these states are traditionally rural, agricultural areas; with the growth of wineries, tasting rooms, lodging services, and visitors, these areas have been transformed from locations with limited activities and services to areas with high-end restaurants, events, travel services, and luxury lodging operations. The dynamic process can best be described as gentrification. Specifically, population growth and real estate appreciation of those wine regions started as early as the 1980s but have significantly increased since the 2000s. These wine regions have embraced tourism due to benefits in the forms of income and employment. While tourism development is frequently touted in a positive light, it can also produce detrimental effects on a region, thereby influencing the quality of life positively and negatively.

Glass (2010) pioneered classical gentrification when she attempted to describe the complicated urban process that occurred in central London. With the replacement of an existing population by the incoming middle classes, the tenurial transformation of property renting to property owning, the rehabilitation of shabby cottages to expensive, stylish flats, and the shift of the whole social characters, the development activity evolves in both beautiful and ugly ways, accelerated by the role of private or government-aided factors or a combination of both (Lees et al. 2008). Glass (2010) captured the core agent of this dynamic change in that the new gentrifiers transformed everything. For this study, an investigation into the transformation

DOI: 10.4324/9781003143628-26

of rural wine tourism destination areas in the United States through the lens of gentrification is conducted through three cases: Napa Valley, Walla Walla, and Willamette Valley American Viticultural Areas (AVAs). The transformation has built up a relatively affluent and exclusive enclave marked by the proliferation of wineries, population increase, and new residential offerings in each. Gotham (2005) conceptualized tourism gentrification to depict revitalized areas known for the influence of tourism. He also conducted an empirical analysis specific to the Vieux Carre, titled the Crown Jewel of New Orleans, and highlighted the dual process of globalization and localization in modernization through redevelopment and tourism growth in a broader context under macroeconomics.

Recently, local authorities have grappled with the winery and hospitality businesses that desire to market their wines to visitors, but that runs up against the anticipation of residents who have moved to these wine regions to experience a quiet agricultural countryside. A series of winery marketing event restrictions in the Napa Valley AVA in California, the Willamette Valley AVA in Oregon, and the Walla Walla AVA in Washington are local authorities' interventions to stop the transformation of rural wine regions into event-driven destinations. These governments attempt to alleviate the stakeholders' conflict of interest in generating revenue by not attracting the traffic and crowds typical of large festivals, and events are controversial. However, this conflict is not going away, and elevated applications for new wineries being protested and applications for use and event permits for existing wineries being denied are often a source of conflict between private businesses and citizens of the wine region. For wineries, the direct-to-consumer model is essential for survival, particularly under such a challenging circumstance of recovery from COVID-19; therefore, promoting wineries and inviting guests to these rural destinations is vital for the economic concern. However, we must question ourselves: At what expense and balance can we assume these wine tourists bring to these wine regions? Do they negatively or positively influence these areas, especially when the community population increases, tourism infrastructure numbers escalate, and home prices upsurge?

For those regional government and city planners, sustainable strategies are anticipated to lessen the potential/existing problems listed and pave a viable way to support the development of wine regions in the long turn. Key issues include:

- Gentrification and exponential growth in real estate prices and living cost
- The disparity of tourism effects on the economy and the environment
- Agricultural and service industry worker wages lagging behind region price increases
- Intensive relationships among the winery businesses, the local authority, and the community
- Negative environmental influences (e.g., pollution, habitat loss)
- Adverse social impacts (e.g., loss of cultural identity, growth of aggression)

The cases will concentrate on the perspective of tourism gentrification to analyze the Napa Valley AVA (California), the Willamette Valley AVA (Oregon), and the Walla Walla Valley AVA (Washington). The initiations of tourism gentrification focusing on wine regions stemmed from two reasons.

First, US wine regions have evolved from traditional agricultural areas in grape growing to produce quality wine products and foster wine-related tourism experience by elite gentrifiers who rely on their capital to boost the economy and gain profits. Those gentrifiers appear to have formed micro-networking that centers on residence and leisure spaces. The expansion of wine regions along with the capital and cultural circulation, an outcome of gentrification, tightly fits the role of the Experience Economy. However, wine tourism research from the lens of gentrification is lacking despite ongoing imposed restrictions and implicit issues identified in the literature.

Second, the dynamics of production and consumption of the land, rooted in the patterns of rural and tourism gentrification, heavily impact the previous residents, workers, and farmers who are challenged to adapt to the change. The gap thus creates spatial fragmentation, social exclusion, racial dimensions, and even elite colonialism. The adverse consequences may lead the wine regions to a deadlock in sustainability. Finding solutions to lessen the potential hazards caused by tourism gentrification and navigating an ecological blueprint are keys to continuing wine tourism success.

Over the last five decades, gentrification has had several important phases and influences on US wine regions. More specifically, tourism gentrification influenced the grape-growing regions of Napa Valley, Willamette Valley, and Walla Walla Valley AVAs in different manners. Negative consequences of tourism gentrification include an unequal exchange between residents and visitors, physical-spatial variation, and unexpected sociocultural transformation amplified along with the development of those wine regions. Key indicators (e.g., population growth and median house value trend) were investigated to identify the status quo of each wine region under the spectrum of gentrification, and a reasonable strategic development plan was drafted.

Five waves of gentrification

Nelson, Oberg, and Nelson (2010) listed an array of influencing factors to rationalize the occurrence of comprehensive gentrification grounded in the supply – e.g., rent gaps by Smith (1987), gentrifiable properties such as land and farms – and demand – e.g., expanding pool of new potential gentrifiers – model. Thus, gentrification has evolved over time with five distinct waves (see Figure 22.1) and integration of gentrification patterns (from Aalbers 2019; Hackworth and Smith 2001; Hyra et al. 2020). These identified waves and patterns provide guidance for the current and future growth of wine tourism regions.

The first wave started around the 1970s, mainly in the form of sporadic housing reconstruction. This wave was led by upper-income individuals who were willing to move to urban areas and rebuild the old housing units. The anchoring phase (second wave) carried on in the 1980s when gentrifiers took advantage of the downturn in property values to deepen privatization and expand the upgrading process to more neighborhoods. In New York City, homesteading Lower East Side and SoHo mirrored the continuum of gentrification in 1980. This movement was initiated by the US Department of Housing and Urban Development to support urban rebuilding through a positive lens and highlight its role of genesis rather than genocide. It is acknowledged that bureaucratic departments, to some degree, give a hand to the advancement of gentrification. The second wave gradually faded due to an economic recession that temporarily halted the expansion of gentrification.

As the reinvestment ensued, fancy retail and professional services were developed, and chic boutiques and upscale recreation complexes were built up to conform to the lifestyle of higher-income categories. In this wave, physical improvements become pretty prominent when the magnitude of change reaches a certain level – a series of revitalizations result in asset appreciation and social mix reshaping. Benefiting from the increased taxes and numerous government aids, subsidies, and public utility initiatives, these attributes facilitated the advancement of the dynamic process of gentrification. By the early 1990s, the support of the state became a more weighted medium to reactivate the transition of gentrification from the second wave to the third wave. Government, corporate, and the partnerships between those two parties have become the agents of consistent urban renewals. Neoliberal urbanism expanded to the concerns the social reproduction, social control, and even political effect, led by the states

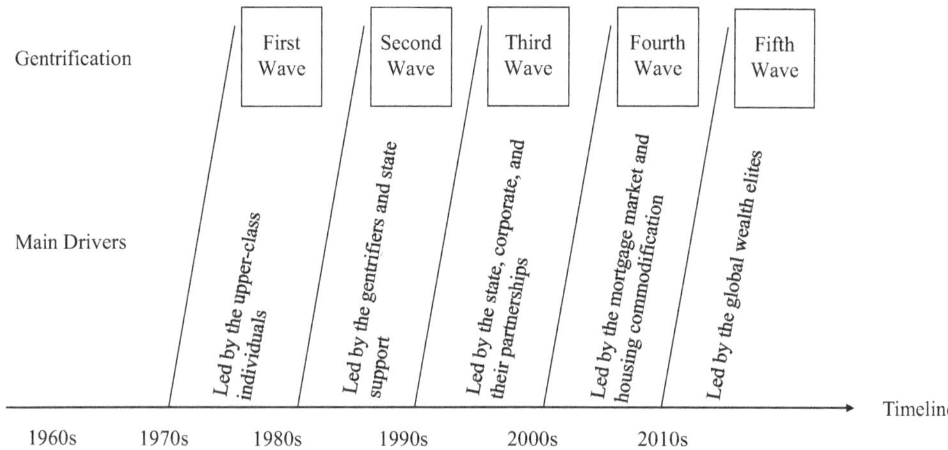

Figure 22.1 Five waves of gentrification.

Source: Adapted from Aalbers (2019); Hackworth and Smith (2001); Hyra et al. (2020)

(Smith 2012). Meanwhile, gentrification is not confined to economic and social shifts and political and cultural elements reform.

Twenty-first-century gentrification has been marked by the prominent role of state support, which has been complemented by the return of investment (financial drive) in the mortgage market. The flows of international capital in the mortgage market pushed real estate commodification and fostered another round of appreciation and gentrification. Not until the exposure of the financial crisis in 2007 was the growth of the fourth wave paused. By the 2010s, global wealth elites' arrivals brought in vigor and vitality, stimulating the fifth wave of gentrification. Platform capitalism (e.g., Airbnb) has become a new channel to transform capital into neighborhoods close to arty or cultural areas (Aalbers 2019). In other words, tourism gentrification again turns out to be a hit due to the flexibility of incorporating artistic and cultural components. Tourism gentrification in wine regions is one of the most delicate cases requiring a delicate balance between gentrification drivers and the need to maintain a sustainable rural living environment (see Figure 22.1).

Tourism gentrification

Tourism gentrification refers to 'the transformation of a middle-class area into a relatively affluent and exclusive enclave marked by a proliferation of corporate entertainment and tourism venues' (Gotham 2005). In other words, the development of these destinations depends heavily on the tourism industry. The gain of a relatively higher annual household income than the national median, the guarantee of employment rate in the tourism-related business, and the contribution of accumulative tourism revenues for local regeneration are the products of exemplary tourism gentrification. For rural agricultural areas lacking viable developing strategies, the coexistence of tourism and gentrification is seemingly a positive way out (Cocola-Gant 2018). Instead of prioritizing residents' expectations for community renewal, tourists' demand becomes the main driver to invigorate the development of those agricultural regions.

The requirements based on tourists (e.g., upgrading accommodation and hedonic consumption) are amassed and promoted with the support of influential capitalists interested in creating an environment for a new round of profit maximization. The proliferation of a tourism-oriented zone creates substantial tourist-friendly spaces while endorsing a precursor's role for destination image. The constant support of tourism infrastructures, local travel and tourism authorities, travel services, and other related stakeholders secured destination visitors' snowballing effects and tourism revenue every year. Many stakeholders who advocate tourism gentrification also benefit from rejuvenation. For many traditional wine-producing regions such as Beaujolais, France, and Central Otago, New Zealand, who have experienced a downturn or constraint on rural area development, wine-related tourism contributes to the rural reconstruction and various socioeconomic benefits such as sustaining the wine production business and employment with subsequent involvement and support by local authorities (Alonso et al. 2015; Hall 2005). Undoubtedly, the development of tourism gentrification in wine regions also increases the awareness of the cultural significance of local food and wine to a broader audience, thereby enhancing the destination's competitiveness (Hall 2005).

Negative consequences of tourism gentrification

As tourism gentrification has expanded, so has the rebellion against it. The resistance against tourism gentrification is reflected in the economic aspect, unbalanced exchange between residents and visitors, physical–spatial variation, and sociocultural transformation. Direct displacement caused by economic dynamics is one result (Cocola-Gant, Gago, and Jover 2020). Tourism consistently enhances the value of properties, stimulating the construction of temporary lodging or travel services such as hotels, motels, resorts, transportation services, and travel agencies by replacing the old residential properties. The lack of available housing, upgrading costs, and a social networking shortage for the other residents has forced the lower-income tenants to move out. As such, the regeneration sacrifices equality and even forms discrimination.

Apart from the economic influence, the cohabitation of tourists and residents may raise another issue given the undesirable consequence of such and the impact on the residents' well-being perceptions. The most frequent issue is pollution, such as noise pollution, resulting in difficulty sleeping and even mental health issues. Tenants were confronted with the situation in which tourists heavily intervened in their lifestyle. The substitution of unknown housing users may also evoke concerns about community security. Eventually, more and more middle-class families may lose faith in living in these tourism-oriented areas. Many times, the stakeholders tried to disguise the ugly side of gentrification by using words of regeneration.

The gentrifiers who settled down in rural landscapes desired to consume nature and wilderness since rural areas like wine regions are the representations of tranquility, family, nature, and stability (Lees et al. 2008). The increasing significance of service did open up opportunities for footloose entrepreneurs and workers lured to rural living (Nelson, Oberg, and Nelson (2010)). However, the overdevelopment of tourism gentrification (e.g., overcrowding) may end up pushing out those gentrifiers in addition to the previous tenants from the destination to relocate. Further, it is acknowledged that the winery visitors and wine consumers are becoming more demanding (Alonso et al. 2015). How far should the business owners and travel authorities in a wine region adapt to the constant demands and needs from the visitors' perspective? The impact of leveraging the communities' benefits is worth discussing further.

On top of physical–spatial related influence, social and cultural identity loss is also a severe consequence. The attractiveness of destination tourism will convert many local sites into a complete tourism zone, erasing the unique cultural identities step by step. Plog's (2001) destination life cycle clearly illustrated how a popular destination heavily dependent on tourism lost its appeal and beauty as a home for local communities.

While we have outlined the intricate tensions rooted in various contemporary gentrification in this chapter, when reviewing the literature, we were a bit surprised at how little attention was raised to the wine regions in the United States over the trajectory of gentrification. All wine regions and their neighborhoods in the United States are changing and evolving due to the impact of wine tourism and related effects on these traditionally rural settings. Despite the continuous argument on land-use management in numerous wine regions, most geographical areas have transformed from rustic agriculture-focused landscapes to emerging tourism-oriented destinations with many complexes to support the development of wine tourism. Meanwhile, the relative vacant or abandoned residential zones in the wilderness have shifted into middle-class, centered living places. The transformation of wine regions has revealed the mutually entwined key features of green gentrification and tourism gentrification. However, the studies of gentrification in the wine region contexts are undervalued, and no data, to the best of our knowledge, have been sourced empirically to measure gentrification across wine regions. The scare research motivates us to explore the wine regions through the gentrification lens and reexamine its regeneration strategies in the long run.

Case Study 22.1 Napa Valley, California

The Napa Valley, California, AVA has signs of gentrification in 1861, when Charles Krug was credited with establishing the first commercial winery. His success and leadership sparked a wave of new growth, and by 1889 more than 140 wineries were in operation, including Schramsberg (founded in 1862), Beringer (1876), and Inglenook (1879). The early 1940s marked a crucial point in Napa Valley's history when these early vintners realized they would be more successful working together than on their own. In 1944, seven vintners signed the agreement of association that formed the Napa Valley Vintners trade association, now 550 wineries strong. The prominence of Napa Valley wine on the world stage is primarily due to their vintners' efforts since the 1970s. People like Robert Mondavi, one of Napa Valley's most renowned winemakers, fully embodied the collective spirit and camaraderie that gave rise to Napa wines' success and premium quality. If a single event can be credited with putting Napa Valley on the map, it was the Paris Tasting of 1976. This blind, comparative tasting pitted Cabernet Sauvignon and Chardonnay from California against the best wines of Bordeaux and Burgundy in a blind tasting. When the tasting was done, the judges had given top honors to Chateau Montelena Chardonnay and Stag's Leap Wine Cellars Cabernet Sauvignon. Napa Valley would never be the same. The number of wineries would flourish from a few dozen to several hundred today, with impacts of gentrification on each community in the AVA.

Napa Valley has been a site of intense conflicts over commercial revitalization, agricultural preserve, and neighborhood integrity. Although Napa Valley was well documented as a place of wild grapes growing in history, the expansion of *Vitis vinifera* grapes growing started in the middle of the 1800s. Long before, the first commercial winery was sparked in 1861. The success of the winery ignited a wave of wineries and business growth. Afterward, the introduction of the first agricultural preserve paved the way for wineries' regeneration, substituting the rural landscape and other crop growing and production. By 2020, over

550 wineries resided in the Napa Valley AVA. Though just 4% of California's wine grapes are harvested here annually, Napa Valley is big in stature. Relying on the picturesque landscape, unique geologic history, and renowned table wines, more and more recreational infrastructures such as trendy restaurants, hip wine tasting rooms, high-scale shopping centers, chic boutiques, public art museums, and various travel services were built to cater to the increasing demand of tourists from all over the world.

Napa Valley is regenerated as a paradise for wine tourists. Numerous destination marketing organizations (e.g., Napa Tourism Improvement District, Downtown Napa Association) put efforts into increasing Napa Valley's reputation as an ideal tourism destination. In recent decades, winery owners and local DMOs have teamed up to boost the Experience Economy by initiating a series of wine-related festivals and activities. More and more middle-class families who desire to reside in such a unique environment moved into the areas and kept capitalizing on the local properties, tourism infrastructures, and auxiliary facilities. Tourism has become a fundamental strategy for Napa Valley to import spending and bolster revenue. With the growth of consumption-based tourism, property values have escalated dramatically, along with the cost of living for residents. Those original residents who fail to survive with the community's change are discriminated against and eventually pushed out of this region. The whole demographic structure has witnessed a significant change over the decades. For example, median household income in 2018 was $84,753, up 40.2% since 2000; but median real estate prices are up 120.89% since 2000 (see Table 22.1).

Although the Napa Valley region has attempted to confront the potential adverse effects caused by wine tourism development, the stakeholders have tried to remain transparent and committed to sharing a mutual ground on the mission of 'sustaining agriculture and the environment and to providing leadership and services to advance the health, safety, and economic wellbeing of current and future generations' (Napa County 2019, p. 3). The adoption of the winery definition ordinance is an outcome. The local authorities attempted to be clear on the land use regulations by clarifying whether a winery can have its own on-site restaurant, the process of wedding and musical festivals, and the limited construction of wine tourism facilities designed to protect agriculture development. The initiative of 'Napa Green Land' conveys their decision on the environmental leadership for wine region development.

Table 22.1 Indicators to underlie rural and tourism gentrification cases in Napa Valley

Year	*Population*	*Number of wineries*	*# Overnight visitors (thousand)*	*Median household income*	*Unemployment rate*	*Median house value*
1980	99,199	60		$18,887.00		
1990	110,765	120		$36,733.00		$182,320.00
2000	124,279		22,300	$51,738.00	2.70%	$251,300.00
2010	134,051	400	28,000	$67,389.00	4.80%	$571,500.00
2018	140,530	550	38,500	$84,753.00	3.30%	$603,700.00
2019	139,623		40,000	$92,768.00	2.90%	$635,900.00

Source: US Census Bureau (2020)

Walla Walla, Washington

California usurps much of the fame the wine world has to offer when it comes to Western US wines. It gets the movies, it gets the name recognition, and it gets the Parker points. However, a wine region with an echoing name is giving California wines a run for their money: Walla Walla. The Walla Walla AVA (meaning 'many waters' in the indigenous Native American language) is mainly in the southeastern region of Washington State, but around 43% of planted area in the AVA also creeps into part of Oregon. The statement 'the finest wines reflect nature, and Mother Nature does not recognize political borders' by the Walla Walla Valley Wine Alliance is the best clarification of the Walla Walla AVA geographical belongings. One of the most defining characteristics of Walla Walla is the ground floor. It is made of loess – loosely compacted deposits of windblown sediment that give the soil a yellow-gray color. Loess is sandy, windblown silt that can measure anywhere from two to four feet deep in the valley. The annual rainfall can be as low as seven inches to the west, all the way up to 22 inches near the base of the Blue Mountains on the east. The elevation range is just as dramatic, ranging anywhere from 400 feet to 2,000 feet above sea level. The latitude is at 46°, equivalent to the midway point of Bordeaux and Burgundy in France.

Started in the 1970s, Walla Walla commercial wineries sprung out. By 2002, over 50 wineries were licensed in Walla Walla. Due to the continued growth of renowned wineries, the unique artistic landscape, and the prominence of the finest wines, Walla Walla becomes a vibrant hub for tourism. More and more chic restaurants, lodging, and boutiques have built up to cater to the increasing demand. These characteristics have made Walla Walla a desirable destination for relocation and investment of gentrifiers.

The wine region has grown along with Walla Walla's town, and now 145 wineries call the Walla Walla AVA in Washington home. The population in the Walla Walla AVA has increased – up 52% since 1970 – and median real estate prices are up over 220% since 2000 (see Table 22.2).

As Walla Walla wine gained more popularity, the Walla Walla Valley Wine Alliance was formed in 2001, aiming to 'support and promote quality and excellence in the Walla Walla Valley wine industry and further national and international recognition of the region' (Walla Walla Valley Wine Alliance 2021). To bring recognition for strict environmental standards and high-quality farming practices, Walla Walla entered into the team to implement sustainable vineyard management in 2008, which has developed more sustainable grape farmers. Assuredly, Walla Walla is centering on how to stand out regionally, nationally, and internationally in the name of quality wine and wine tourism. Besides winery owners, destination marketing

Table 22.2 Indicators to underlie rural and tourism gentrification cases in Walla Walla

Year	*Population*	*Number of wineries*	*Number of overnight visitors (thousand)*	*Median household income*	*Unemployment rate*	*Median home value*
1980	47,435					
1990	48,439			25,808		
2000	55,180	16		36,800		$149,300
2010	58,781		273	43,119	7.50%	$178,000
2018	60,615	120	1,066	48,678	4.80%	$286,500
2019	62,083	145		61,202	4.90%	$324,800

Source: US Census Bureau (2020)

organizations (DMOs) have tried to leverage greater resources for marketing Walla Walla as a premier wine tourism destination and to draw more visitors to this area by incorporating outdoor recreation, special events, historical tourism, and agritourism with wine. Wine tourism appears to be a natural fit for local development from the DMO's perspective.

Willamette Valley, Oregon

Starting in the 1930s, the Willamette Valley's prairie lands were utilized to grow seed crops. In 1965, winegrowers who were disillusioned with California Pinot Noir began to move north and started the Oregon Pinot Noir planting journey along with a small amount of related varietals such as Pinot Gris. Sitting on the same latitude as the premier Pinot Noir regions (Burgundy, France) in the world, the Pinot Noir growing in this area is not easy because this grape varietal is extremely sensitive to the microclimate. With the protection from Oregon's Coastal Range mountains, winemakers, oenologists, and vineyard workers take care of the grape with a passion to create quality wine. The wine pioneers in this region, full of enthusiasm and diligence, paved the way for quality wine development in this area. Willamette Valley's flagship Pinot Noir is a leader in the global market and wins a well-deserved reputation as one of the globe's premier Pinot Noir growing and producing areas.

As the valley was authorized as an AVA in 1983, more and more investors were conducive to the terroir – old volcanic, sedimentary seabed, and loess soils. Starting from five bonded wineries with 35 vineyard acres in the 1970s, Willamette Valley is home to nearly 700 distinguished wineries. Nine appellations are nested in this valley: Chehalem Mountains, Dundee Hills, Eola-Amity Hills, Laurelwood District, McMinnville, Ribbon Ridge, Tualatin Hills, Van Duzer Corridor, and Yamhill-Carlton. Although not every single winery employs sustainable, organic, or biodynamic certification, most wineries in Willamette Valley apply sustainable practices to address environmental stewardship and social responsibility. The commitment to safeguarding the ecology system and rich environmental heritage has continued for more than 100 years since the Oregon wine growers were aware of the potential adverse effects of grape growing to the land they love.

With the growth of numerous trendy restaurants, bars and lounges, tasting rooms, wine shops, and delightful accommodations (e.g., B&B, scenic wine region resorts), Willamette Valley turns out to be a popular tourist destination and is geared to provide a variety of customized wine tourism experiences (e.g., private winery/vineyard tour, cellar tasting, harvesting experience). Along with the guided wine tour, tourists have the flexibility to attend wine-related events, festivals, harvesting, and other activities (e.g., live entertainment, balloon adventures) with family-friendly or dog-friendly options. Kayaking, paddleboarding, birding, and cycling are also available. Willamette Valley seems to rely extensively on tourism, and observers may be hard-pressed to trace back to the original outlook.

To promote the destination awareness of Willamette Valley, six visitor associations were fostered to initiate marketing strategies in all forms. Vineyards and wine are the iconic assets for this area. To sell wine tourism products and services, the wine tours are mingled with an array of inclusive experiences such as cultural experiences (e.g., historic carousel museum, iconic speed museum, Grammy award-wining Bach festival), farm experiences (e.g., black-and-white truffles forage and feast), culinary experiences (e.g., ale and cider trails, wine-and-cheese events), and landscape experiences (e.g., enchanted forest hiking and national scenic byway walking) around all the appellations areas. In 2018, the tourism industry provided 22,600 direct jobs, a 3.1% increase compared to 2010. Average employee earnings from the tourism industry

Table 22.3 Indicators to underlie rural and tourism gentrification cases in Willamette Valley

Year	*Population McMinnville*	*Number of wineries*	*Number of overnight visitors (thousand)*	*Median household income*	*Unemployment rate*	*Median home value*
1980	14,080	34				
1990	17,928	70		$29,281.00		
2000	26,417	135	34,000	$40,919.00	4.30%	$149,300.00
2010	32,187	418	50,000	$44,013.00	10.90%	$178,000.00
2018	33,662	493	86,000	$51,820.00		$182,000.00
2019	34,347	676	88,000	$54,254.00	3.50%	$325,700.00

Source: US Census Bureau (2020)

increased by 6.2% compared to 2010. Visitor spending in 2018 increased by 3.7% compared to 2010 (see Table 22.3).

Both residents and stakeholders are gaining from the expansion of tourism gentrification at the current phase in Willamette Valley. They appear to be quite pleased with the market fit and neglect the potential adverse consequences of tourism gentrification as it goes on, since no intended strategy of controlling the ugly side is planned.

Discussion

Tourism impacts every corner of the wine regions (Napa Valley, Walla Walla, and Willamette Valley) in our research and is the key catalyst for local economic development. There is no better way to experience a wine region than by indulging in culture, gastronomy, nature, and heritage-related experience based on earlier research in wine and food tourism (Ottenbacher et al. 2016). From small local businesses to large hospitality corporations in those wine regions, they all benefit from tourism development. However, as specified earlier, the negative influences of tourism are also prominent. The term *tourism gentrification* conveys a variety of existing or potential undesirable consequences. For Napa Valley, the declining population, the extraordinary home value median, rocketing visitors, and number of wineries highlight the opposing positions between tourism gentrification and residents' quality of life. The disparity of tourism effects on the economy and community environment has likely persuaded considerable numbers of former residents (including gentrifiers) to relocate to other living areas. Local authorities and other stakeholders are not immune from observing the impact of this phenomenon. Hence, additional regulations that subject operators to more limited construction of tourism-related infrastructures and hospitality services (e.g., winery restaurants) are commonly enacted to relieve the tension between community well-being and tourism gentrification. The utilization of such emergent strategies is after the fact in wine regions and are less effective in reducing the current imbalance status in the absence of well-defined goals. Thus, emergent anti-gentrification strategies become inadequate and even generate counterproductive effects in times of change for both the community and operators. To minimize negative effects and facilitate continued success in rural wine regions, deliberate planning upfront and prompt reviews bringing together a variety of regional stakeholders are helpful to adjust strategies to conform to the continuously changing situations and needs of the community.

In the same vein, the exponential growth in real estate prices in Walla Walla and Willamette Valley has been a phenomenon with no noticeable decrease in residents' populations to date.

A growing impact of many unconstructive environmental and social influences are likely to occur without any proper intervention. The reliance on the market determination is not a great long-term strategy for these two wine regions, as past experiences in Napa Valley and other locations indicate that tourism gentrification waves will eventually lead to additional challenges for the local community development in a sustainable pattern.

Unlike the different tourism gentrification strategies, all three wine regions are dedicated to embracing a mutual agreement with environmental and social sustainability concurrent with wine production in the long run. This deliberate strategy sets a clear direction for all the wineries within the wine regions to follow and cooperate with the stakeholders involved.

Strategic development plan

- *Intended strategy.* The concept of determining a clear strategy on what a wine region intends to achieve is a core element of the strategic process. To lessen the potential for detrimental effects of tourism gentrification, this strategy should not only be based on market and fit relationships but also proactively consider the social processes involved to achieve buy-in from a variety of stakeholders as well as identified (e.g., cost-of-living challenges, congestion) and unidentified issues (which may be specific to the local environment). Ensuring plans for a wine tourism campaign and overall strategy are developed to balance the desire for community renewal, tourism demand, and sustainable resident quality of life that are also vital to continued success, as demonstrated by the cases presented in this chapter.
- *Deliberate strategy.* As suggested by Harrington and Ottenbacher (2015), these are deliberate and planned activities that are identifiable in advance. While these efforts may or may not become fully actualized, leadership should ensure that cooperation with stakeholders is obtained, political support is secured, infrastructure needs are included upfront, and activities are designed to fulfill both residents' and visitors' requirements. Further, deliberate planning is integrated to address future housing, pollution, congestion, cost of living, and other quality-of-life issues. Finally, impacted stakeholders should be brought into the planning process early on and consulted within a deliberate fashion.
- *Emergent strategy.* Any business environment can be described as complex, dynamic, and uncertain. Thus, while strategic planning is promoted as deliberate in nature, unexpected opportunities are likely to emerge, and new options may become viable as time goes on. This phenomenon may include new products, additional services, and unique experiences, as well as local traditions becoming economically feasible as authentic additions to the touristic business landscape. These emergent strategies should be implemented to provide residents with entrepreneurial opportunities rather than concentrating on external providers when possible. Unique experiences, services, or products that are authentic to the region can become 'hidden gems' that enhance visitors' overall experience and unique skillsets, which can also alleviate drivers of gentrification.
- *Realized strategy challenges.* From intended strategy to deliberate and emergent elements, a key to lessening the potentially detrimental effects of wine tourism gentrification is to measure the success of what strategies are realized using a balanced approach for both business and quality-of-life success residents. In addition to traditional visitor satisfaction, community residents' and other stakeholders' satisfaction should be regularly estimated. Finding best practices for communicating positive impacts and activities to improve the quality of life of those in the community is an additional role of the strategic process in wine tourism.

Conclusion

There is more to creating a fantastic destination for wine regions than just having a stellar cellar door and scenic winery. Tourism is a must medium to promote those wine regions and contribute to economic development and local reconstruction. Besides factors like wine quality, excellence of the wine-related experience, tourism and hospitality infrastructures, and facilities, other areas of the infrastructure are required for success, such as transportation, travel services, hotels, and recreation services. Every wine region has created its way of differentiating itself and attracting visitors. Good infrastructure and facilities in a destination pull demand for its tourism products and experiences.

For a thriving wine destination, a sufficient population of wineries and other activities is required. As wine tourism and surrounding regions mature, a variety of challenges undoubtedly arise, such as noise and air pollution, loss of land use, loss of cultural identity, displacement, high living costs, the tension between residents and visitors, and growing aggression against further development. How to draw a strategy based on the market, fit, and social processes calls for serious consideration. How to balance the interests of different stakeholders and set up a grounded goal for the sustainable development of wine regions, in the long run, is also a test for local authorities as well as the stakeholders involved.

In an earlier study of wine region cases (Germany and France), six key strategic planning attributes impacted wine tourism strategy success. These included (1) the strategy itself, (2) cooperation among stakeholders, (3) leadership issues, (4) region promotion, (5) regional products/services, (6) and quality standards (Harrington and Ottenbacher, 2015). Further, they used strategic planning concepts of intended, deliberate, emergent, and realized strategy applied to the wine tourism context. Table 22.4 provides a summary of these descriptions with additional questions that wine region stakeholders should assess to facilitate prosperous wine tourism and minimize the adverse effects of gentrification.

In summary, when facing the rapid change of wine regions' progress, the local authorities should anticipate the potential challenges and be open-minded to embracing innovative wine tourism destination strategies that can address concerns of wine region gentrification and other problems associated with an increased population, average household real estate value, and tourism visitors.

As a result of gentrification, the good, the bad, and the ugly of wine tourism products and services will emerge. While these products and services will likely meet the needs and values of the higher-household-income gentrifiers who are visitors or have relocated their home to a rural wine region, developing more premium wine experiences to meet the expectation of the lifestyle that comes with higher discretionary spending of those new residents will be needed. Allowing for the past to be present makes sure that the re-establishing of local wine traditions incorporates the wine tourism product and offerings. Annual reminders of the spring barrel release or harvest parties are marked by history for winemakers. These staples of the vintage year connect the current growing year to the past and remind the wine tour of the agricultural cycle of grape growing and winemaking.

Innovations and sustainable methods of managing events and the tourism ecosystem with the local government must also emerge as visitations grow in the wine region. Embracing locals' needs and having residents take part in the political process of growing wine tourism is a critical aspect of regional growth and success. For example, regions that find ways or signage to direct visitors to wine tourism routes specific to wine country can assist in alleviating rural road congestion created by increased population from gentrification. As professionals in the wine industry strive for higher quality levels and higher standards to assessing the wine they

Table 22.4 Wine tourism planning and gentrification challenges

Strategic planning concepts	*Examples planning practices in the wine tourism context*	*Questions to consider*
Intended Strategy: What do we intend to achieve?	– Integrative tourism campaign plan – Develop a clear strategy based on market and fit as well as social processes and issues	– How has the campaign and overall strategy been developed to balance the desire for community renewal, tourism demand, and sustainable resident quality of life?
Deliberate Strategy: These are deliberate and planned efforts and activities to move goals from intended to realized where possible.	– Efforts to achieve cooperation with stakeholders – Maintain political support and leadership – Ensure infrastructure needs are part of the process – Tie plan with other quality activities for both residents and visitors	– Are efforts to address future housing, pollution, congestion, cost of living, and other quality-of-life issues part of the region's deliberate strategies? – Are impacted stakeholders included in the deliberate strategic planning?
Emergent Strategy: Given a complex and uncertain business environment, new opportunities may emerge or possible options may become more or less viable as time goes on.	– New products/services/ experiences become viable – Local traditions are identified and economically viable – Innovations and sustainable methods become available – Quality levels and communication	– What new experiences can provide residents with entrepreneurial opportunities? – What 'hidden gems' can be part of an authentic touristic terroir for the region? – How can new business opportunities provide pride and success for both local operators and residents in general?
Realized Strategy Challenges: One aspect is measuring outcomes, and a second is to implementing goals and objectives that minimize potential detrimental effects	– Measure of outcomes – Balanced scorecard approach for both business and quality-of-life success – Measuring satisfaction for both visitors and community – Assessing rural community sustainability	– How to assess whether regional success is perceived by all stakeholders? – How to communicate positive impacts – economic and community – to stakeholders?

Source: 'Wine Tourism: Transforming US Rural Agricultural Regions' (Harrington and Marlowe 2019)

produce, they must also maintain a balanced development of their wine tourism products and experiences.

The cases on wine country gentrification shared in this chapter show that gentrification in rural wine regions is a critical yet understudied area of the field. A greater understanding of the local attitudes towards the continued growth of residents, tourists, and the average value of home real estate needs to be explored. Increased number of residents and higher home values for each of the wine regions highlighted in this chapter are present. The result is higher-income households within the wine region, and gentrification in rural wine regions is concluded.

References

Aalbers, M. B. (2019). Introduction to the forum: from third to fifth-wave gentrification. *Tijdschrift voor Economische en Sociale Geografie*, 110(1): 1–11. https://doi.org/10.1111/tesg.12332

Alonso, A. D., Bressan, A., O'Shea, M., and Krajsic, V. (2015). Perceived benefits and challenges to wine tourism involvement: an international perspective. *International Journal of Tourism Research*, 17(1): 66–81. https://doi.org/10.1002/jtr

Cocola-Gant, A. (2018). Tourism gentrification. In Lees, L., with Phillips, M. (eds.) *Handbook of gentrification studies* (pp. 281–293). Cheltenham: Edward Elgar Publishing.

Cocola-Gant, A., Gago, A., and Jover, J. (2020). Tourism, gentrification and neighbourhood change: an analytical framework – reflections from Southern European cities. In Oskam, J. (ed.), *The overtourism debate* (pp. 121–135). Bingly: Emerald.

Glass, R. (2010). Aspects of change. In Brown-Saracino, J. (ed.) *The Gentrification Debates: A Reader* (pp. 19–26). New York: Routledge.

Gotham, K. F. (2005). Tourism gentrification: the case of New Orleans' vieux carre (French Quarter). *Urban Studies*, 42(7): 145–176. https://www.jstor.org/stable/pdf/43197316.pdf

Hackworth, J., and Smith, N. (2001). The changing state of gentrification. *Tijdschrift voor Economische en Sociale Geografie*, 92(4): 464–477. https://doi.org/10.1111/1467-9663.00172

Hall, M. (2005). Rural wine and food tourism cluster and network development. In Hall, D., Kirkpatrick, I., and Mitchell, M. (eds.), *Rural Tourism as Sustainable Business* (pp. 149–164). New York: Channel View Publications.

Harrington, R. J., and Marlowe, B. (2019). Wine tourism: Transforming US rural agricultural regions. In *FUA Summer Session*. Florence, Italy.

Harrington, R. J., and Ottenbacher, M. C. (2015). Wine tourism strategy-making: a model for planning and implementation. In Lee, K. (ed.), *Strategic Winery Tourism and Management: Building Competitive Winery Tourism and Winery Management Strategy* (pp. 9–30). Florida: Apple Academic Press.

Hyra, D., Fullilove, M., Moulden, D., and Silva, K. (2020). *Contextualizing Gentrification Chaos: The Rise of the Fifth Wave*. https://www.american.edu/spa/metro-policy/upload/contextualizing-gentrification-chaos.pdf

Lees, L., Slater, T., and Wyly, E. (2008). *Gentrification*. New York: Taylor & Francis Group.

Napa County (2019). *Napa County Strategic Plan 2019–2022*.

Nelson, P. B., Oberg, A., and Nelson, L. (2010). Rural gentrification and linked migration in the United States. *Journal of Rural Studies*, 26(4): 343–352. https://doi.org/10.1016/j.jrurstud.2010.06.003

Ottenbacher, M. C., Harrington, R. J., Fauser, S., and Loewenhagen, N. (2016). Should culinary tourism and hospitality service attributes be defined as primary tourism drivers? An expectancy-fulfillment grid approach. *Journal of Foodservice Business Research*, 19(5): 425–440. https://doi.org/10.1080/15378020.2016.1185879

Plog, S. (2001). Why destination areas rise and fall in popularity: an update of a Cornell Quarterly Classic. *Cornell Hotel and Restaurant Administration Quarterly*, 42(3): 13–24.

Smith, N. (1987). Gentrification and the rent gap. *Annals of the Association of American Geographers*, 77(3): 462–465.

Smith, N. (2012). New globalism, new urbanism: gentrification as global urban strategy. In Brenner, N., and Theodore, N. (eds.), *Spaces of Neoliberalism: Urban Restructuring in North America and Western Europe* (pp. 80–103). https://doi.org/10.1002/9781444397499.ch4

US Census Bureau (2020). *The U.S. census 2020*. https://www.census.gov/data.html

Walla Walla Valley Wine Alliance (2021). *Walla Walla Valley Wine Alliance – Winegrowers sustainable trust, Walla Walla Uncovered*. https://www.wallawallauncovered.com

23

ROLE OF GREAT WINE CAPITALS NETWORK IN MAKING WINE TOURISM A COMPETITIVE FACTOR OF A TERRITORY

Adela Balderas-Cejudo, Jorge García Martín, and Lázaro Echegaray Eizaguirre

Introduction

Academic studies on wine tourism have rapidly grown over recent decades. The wine tourism business has increased, and it represents today a key leverage element to improve wineries' performances (Montella 2017), that is, to determine the weight that wine may have as a competitive factor of territory in wine-producing areas. Wine tourism is not only viewed as an entrepreneurial opportunity for winemakers, but also considered to represent a vehicle of employment generation and socioeconomic development of less favored peripheral regions (Alebaki and Koutsouris 2019; Getz 2000). According to Sotiriadis (2015), tourism destinations, businesses, and different stakeholders are constantly looking for ways to improve, differentiate, and promote their offering, aiming to achieve better positioning and gain a competitive advantage in an increasingly competitive sector. Wineries also face an increasing need to add value to wine tourism experiences (Faugère et al. 2013).

Undoubtedly, the COVID-19 pandemic constitutes an exceptional shock event posing tourism's greatest challenge since the 2008 global financial crisis (Rogerson and Rogerson 2020). Gössling, Scott, and Hall (2021) emphasize that the worldwide spread of the pandemic not only caused a global crisis for hospitality and tourism but also dramatically changed consumers' wants and market demands that were satisfied by existing tourism value chains and business models during the pre-COVID-19 era. COVID-19 has brought the world to a standstill. Innovation, new strategies, new visions, stronger bonds, and much better communication should be new ways to work and build strength.

Since the mid-1990s, the field of wine tourism has developed (Brochado et al. 2021; Carlsen 2004; Gunasekar et al. 2021), and it had become an important form of tourism when a steady growth of tourists interested in wine who displayed relatively high spending patterns could be observed (Getz and Brown 2006; Pikkemaat et al. 2009). The competitive positioning of wine tourism regions is an important strategic issue, as many regions are now aggressively marketing themselves to attract high-yield tourists (Bruwer et al. 2017; Getz and Brown 2006; Williams 2001).

DOI: 10.4324/9781003143628-27

The increasing volume and value of wine tourism experiences have been documented in many countries (Barney 1991). A nation's or region's competitiveness depends on the capacity of its industry to investigate and upgrade (Porter 1985). According to Williams (2001), there is a need to focus greater attention on identifying those experiential attributes which bind the various physical components of the wine tourism experience together.

This chapter is focused on positioning wine tourism as a competitive factor of a territory and the case of Great Wine Capitals (GWC). GWC is a network of 11 major global cities in both the Northern and Southern Hemispheres, which share a key economic and cultural asset: their internationally renowned wine regions. It is the only such network to encompass the so-called 'Old' and 'New' Worlds of wine, and it aims to encourage travel, education, and business exchanges. The wine tourism industry needs research to further study specific areas of this market (Stewart, Bramble, and Ziraldo 2008). Associations such as GWC insisting on the need to know the market and seeking new opportunities for improvement and growth is not only necessary, but also positive for the industry.

Wine tourism as a competitive factor of a destination

It has long been recognized that the tourism sector is highly vulnerable to disruption by hazard events, with localized phenomena such as earthquakes, bushfires, volcanic explosions, tsunamis, or floods, as well as global events such as disease pandemics (Laws, Prideaux, and Chon 2007; Lenggogeni, Ritchie, and Slaughter 2019). The need to rebuild the sector, building a more resilient tourism economy, and understanding a new customer in the post-COVID era will be crucial.

Getz and Brown (2006, p. 79) define wine tourism destinations as 'regions which base some or all of their appeal on wineries and wine-related benefits.' Following Bruwer et al. (2017), wine tourism as a special interest element of tourism is not yet greatly endowed with in-depth theory, and definitions relating to the main research field(s) in which it is rooted, namely tourism and leisure, have generated many still ongoing and much-deliberated issues (Alant and Bruwer 2004). Wine tourism as a discrete field of research is inchoate, yet there has been rapid development in this field since the mid-1990s (Carlsen 2004). The definition and conceptualization of wine tourism has not resulted in a uniform approach and continues to evolve (Bruwer 2003).

What seems clear is that the importance of wine tourism is rising rapidly all over the globe, not only because of its affordance to address the ever-growing requirements of tourists to experience creative tourism activities, but also for its ability to support sustainable destination development (Sigala 2014). Despite the links between wine and tourism, formal attempts to create a greater economic advantage out of the relationship between the two industries are only a relatively recent development (Hall and Mitchell 2000).

To different degrees, five drivers affect wine tourists: word-of-mouth recommendations, product attractiveness, tourism services, information available on the internet, and tour operators' recommendations (Torres et al. 2020). When viewed from a marketing perspective, wine tourism is likely to determine the experiences sought by potential and actual wine tourists. In fact, most definitions of wine tourism relate to the traveler's motivation and experiences (Getz and Brown 2006). Certain authors defined wine tourism as visits to vineyards, wineries, wine festivals, and wine shows for which wine tasting and/or experiencing the attributes of a grape-wine region are the prime motivating factors for visitors (Hall 1996; Hall and Mitchell 2000). Getz (2000) stressed that there are at least three major perspectives on wine tourism: wine producers, tourism agencies (representing the destinations), and consumers. Thus, 'wine tourism' is, simultaneously, a form of consumer behavior, a strategy by which destinations develop and market wine-related attractions and imagery, and a marketing opportunity for wineries to educate and to sell their products directly to consumers (Getz and Brown 2006). Barth and Salazar (2011) state that wine tourism can be defined as travel to experience wineries and wine regions and their links to lifestyles. According to Alebaki and Lakovidou (2011), authenticity, regional culture, and gastronomy are closely linked to wine tourism.

Wine tourism, or winery visitation, is emerging as an increasingly significant sector of both the wine and tourism industries (Barth and Salazar 2011; Sparks 2007). A fairly broad consensus contends that the wine region of origin adds value in consumers' eyes as it represents a significant choice criterion (Carlsen and Boksberger 2013; Perrouty, d'Hauteville, and Lockshin 2006; Tustin and Lockshin 2001), for which consumers are prepared to pay the price (Schamel and Anderson 2003). The demand for wine tourism is based on 'the motivations, perceptions, previous experiences and expectations of the wine tourist' (Hall, Johnson, and Mitchell 2000, p. 6). Nowadays, experience seekers are moving away from standardization and purely material goods and are looking for 'authenticity' in the places they visit and select customized travel (Rachão et al. 2020). The factors underpinning motivations are complex due to the convoluted nature of the wine consumer–tourist relationship (Bruwer et al. 2017).

Wine tourism is thus an area of growing interest because of its potential to contribute to regional development and employment at times of rural restructuring, particularly through the development of inter- and intra-industry networks (Hall et al. 1997). Hall and Mitchell (2000) discuss the concept of tourist terroir, which they define in terms of the 'unique combination of the physical, cultural and natural environment that gives each region its distinctive tourist appeal.' This concept expands the notion of winescapes to encompass more of the feeling of the region, which is a culmination of all of its physical and cultural parts (Sparks 2007).

While it is true that the aim of visitors to wine regions is to taste and even to buy wine (Alant and Bruwer 2004), the secondary experiences and motivations for such experiences are crucial to understanding what primarily motivates a person to visit a wine region and a winery cellar door. The characteristics and motives of wine tourists need to be better understood (Charters and Ali-Knight 2000).

The activity of tourists visiting winery cellar doors in a wine region is the essence of wine tourism. The motivations of wine tourists are considered an important aspect in the overall understanding of wine regions and wineries/cellars of the needs and expectations of their customers (Alant and Bruwer 2004; Hall, Johnson, and Mitchell 2000). Core destination appeal includes features such as attractive scenery, pleasant climate, moderately priced accommodations, easy-to-obtain information, well-signposted wine trails, and a variety of things to see and do (Getz and Brown 2006).

Wine tourism has the capacity to play a significant role in national and regional sustainable tourism-development plans through its contribution to sustaining the economic and social bases of regions, as well as environmental dimensions (Hall and Mitchell 2000). Achieving competitive advantage in times of rapid change requires that tourism stakeholders have a clear understanding of the direction of change and its implications for business or destination management (Dwyer et al. 2009).

Great wine capitals (GWC)

Since its creation in 1999, the GWC has been the international reference network that brings together some of the most emblematic regions of the wine industry, both in the so-called 'Old World' and in the 'New World.' Its main objective, as the GWC itself points out on its website, is the promotion of tourism, culture, and commercial exchange among its 11 current members: Adelaide (South Australia), Bilbao–Rioja (Spain), Cape Town–Cape Winelands (South Africa), Bordeaux (France), Lausanne (Switzerland), Mainz-Rheinhessen (Germany), Mendoza (Argentina), Porto (Portugal), San Francisco–Napa Valley (USA), Valparaíso–Casablanca Valley (Chile), and Verona (Italy).

The GWC directs its resources to develop competitive advantages among its members, having culture, heritage, and geographical location as fundamental resources. Within its strategic plan, the GWC develops various initiatives to improve the competitiveness of its members, such as the annual contest aimed at wine tourism sustainability, art, or culture; the annual symposiums with international experts in the wine industry; or the creation of a network of travel agencies to facilitate exchanges between the 11 regions.

The first approach to wine tourism leads us to analyze promotional activities and events used to meet the needs and desires of consumers. Among the more common tools used by members include visits to wineries, wine tastings, or a wide variety of culinary events. In fact, gastronomy is one of the drivers most used by the GWC to attract tourism to its regions. Two clear examples are Adelaide and Bilbao–Rioja, promoting gastronomy based on kilometer zero.

Other widely shared activities are travels through the vineyards – on foot, by bicycle, or on horseback. It is interesting to mention the case of Adelaide, where the offer combines walks through vineyards with the contemplation of the wild fauna represented by kangaroos or koalas, among other animal species. The wine festivals related to the different milestones in the winemaking process are well-established activities by practically all wine capitals. Within the cultural chapter, the activities most cited by members (GWC website) are those related to seminars of different contents, to the terroir history, to music (both classical and pop-rock music), or to art. At this point, it is worth noting the success of the exhibition at the d'Arenberg winery (Adelaide) of an important collection of works by Salvador Dalí, including 25 sculptures in brocade by the brilliant surrealist author, perfectly combined with the tasting of the wines from the prestigious winery (D'Arenberg 2020). Likewise, other regions such as San Francisco–Napa Valley, possibly the most developed wine tourism destination in either the Old or the New World according to Barber, Taylor, and Deale (2010), has also touched on some activities or tools to promote wine tourism, among which stands out a significant offer of art (in galleries, museums, and elsewhere) that this territory has known to combine with the world of wine.

On the other hand, some territories have developed some type of differentiated and unique promotional activities or events that improve their value proposition. Thus, Cape Town–Cape Winelands (Franschhoek Wine Tram 2020) counts among its attractions impressive tram tours through the extensive vineyards of the Franschhoek Valley in the heart of the Cape Winelands. The tracks on which the trams circulate were originally built in 1904, and today they serve as a connecting element between wine and tourism in the area, a clear example of the integration of historical elements in the henoturistics offer. The Argentine wine region of Mendoza, which currently receives about a million tourists, includes in its portfolio a wide range of activities: from horseback riding through vineyards, to participation in harvesting activities, to complete classes of cooking taught in some of its most prestigious wineries (Gregori, María Teresa 2017). Lausanne for its part, through a group of 12 wineries, has been developing a show since 2017 that includes a sample of theater and literature combined with the tasting of their wines (GWC website).

Regarding the geographical composition of the GWC wine consumer, there is an increase in Chinese consumers who are interested not only in the wines but also in the whole range of wine tourism, mainly in Napa Valley, Verona, and Adelaide. While their quantitative importance is not very significant, the relevant issue to highlight is their presence and future evolution, which is expected to rise. According to the Napa Valley Visitor Profile (2019), after the Americans and the Canadians, the Chinese visitor segment is in the third position in this region. Something similar happens in Adelaide, where the Chinese tourist also reaches the third position. A gradual increase in this segment of the demand in all regions, driven by a growing Chinese middle class willing to travel, is expected as soon as the COVID-19 pandemic passes.

Another factor of interest connected with the previous one is the degree of internationalization of demand. In this sense, capitals such as Verona in first place given its great artistic attraction, Bourdeaux, or even Bilbao–Rioja with 37% of foreign visitors and increasing every year (Consejo Regulador DOCA Rioja 2020), are regions where the importance of the international visitor is more significant. On the other hand, only 19% of Napa Valley's 4 million visitors come from abroad (The Napa Valley Visitor Profile 2019).

From a demographic point of view, the most prominent part of the demand is quantitatively made up of men, albeit by a small margin, as a result of the progressive participation of women in the world of wine tourism. Likewise, another important population segment, although with less purchasing power than that of the over-45s, is that composed of young people. However, this segment is having an unequal impact on the GWC territories. While in San Francisco–Napa Valley, it is one of the most nourished (The Napa Valley Visitor Profile 2019), this is not the case in Bourdeaux, where the most numerous segments are those of middle-aged people (Observatorio Vitivinícola Argentino 2020).

We will conclude this brief description of the GWC with a look at the socioeconomic profile of the visitor. This profile is made up of upper-middle-class or higher, with significant purchasing power and higher education. The wine tourist who visits the different places of the Great Wine Capitals values the quality of the wines as well as the gastronomy of the region. According to Garibaldi (2020), in general terms, this is a consumer who is looking for experiences around tasting and buying wines as well as guided tours through the vineyards or inside the wineries – although except for buying, the rest of the experiences have been reduced by the effect of COVID-19.

Case Study 23.1 Bilbao-Rioja wine region, Spain

La Rioja (DOC La Rioja) is one of the most important wine areas in the north of Spain and has had the necessary conditions for growing grapes since the mid-19th century (Ledesma Silva and Cobos 2017). Due to the advent of oidion and phylloxera in French vineyards, the area of La Rioja began to be used by the French, who reached note buyers in the most important towns of the region (Gobierno de la Rioja 2012). Thereon, great wineries were born in the last century, and native strains were rediscovered, such as the Tempranillo variety. The area obtained its first Denomination of Origen in 1925, and in 1991 the Rioja Denomination of Origen Degree (Arimany-Serrat, Ferreras Noger, and Rabaseda I Tarres 2016).

The Rioja wine district is located in the provinces of La Rioja, Alava, and Navarra. It is divided into three regions: The Rioja Alta and Rioja Alavesa are in the west and influenced by the Atlantic, and the Rioja Baja is in the east and influenced by the Mediterranean. The vineyard area covers almost 64,000 hectares (158,000 acres), with an average annual production of 360 million bottles. Rioja was granted Denomination of Origin status in 1926 by the Spanish Ministry of Agriculture and in 1991 became the first Qualified Denomination of Origin (Denominación de Origen Calificada) in recognition of the high quality of its wines and the strict quality control imposed on grapes and wine in the region (Great Wine Capitals 2020a, 2020b, 2020c). The sector is considered as one of the best in the wine world due to several factors: the prime location for the type of soil, the climate, and the river sources (Ledesma Silva and Cobos 2017, p. 47).

Regarding wine tourism, the case study of the Bilbao–Rioja wine area is an outstanding one so far as it includes a rural wine region such as La Rioja, a geopolitical region itself, and a city like Bilbao, a capital which belongs to the geographical area of Basque Country. Bilbao has undergone a big change during the last years, as it was an industrial city which grew thanks to the mining industry and the navigation, among other things. It is remarkable that the port of Bilbao is one of the largest in Spain, having a high transit of passengers.

Both regions are highly interconnected, and some of Rioja´s areas even belong to the Basque Country region. Therefore, there has been a very good and solid relationship between both Bilbao and La Rioja since ancient times, and together they form a very important industrial center. Bilbao's airport is connected with many other European airports, and both cities are well linked through a highway. This area is well-known for its innovation through technology development centers and its tourism destinations. This is thanks to its wide range of cultural sites, gastronomy, wine cellars, and diverse landscapes.

The relationship with tourism is an essential factor in the Bilbo-Rioja wine area. The area offers a wide range of activities to visitors, from the most affordable country homes to luxury hotels, mountains areas to seaside locations, summer festivals in every village, sensational gastronomy, and many other cultural offers, including the world of wine, which is always present. For example, there are many tourism tours in the most representative cities of La Rioja, such as Haro, a city full of historic palaces and contemporary art.

In addition to this, there are many different tours starting in Bilbao's airport, from where visitors are picked up and transferred to Haro, where they can choose among a variety of hotels and city tours to breathe wine or tour classical and interesting wineries. Private seated tasting of the most exclusive wine collections is also included in these tours, and normally the taste is accompanied by Iberian ham, cheese, and olive oil from La Rioja. Tourists also have the opportunity to enjoy the traditional restaurants of the area.

There are very important cellars in Haro's surroundings, like Viña Real, which are glad to host people. This is also the case of Marques de Riscal, one of the most significant cellars in the area, which ranked sixth among the World's Best Vineyards in 2020. The jewel in the crown is the Marriott International hotel chain: Marqués de Riscal, a Luxury Collection Hotel designed by Frank O. Gehry (Figure 23.1).

This hotel combines the most deeply rooted winemaking tradition with an avant-garde design, luxury, and the most advanced technology of the 21st century. In addition to the scenery and the gastronomy, it's possible to enjoy a new experience unique to Spain: vinotherapy. The visitor who comes to the region will find a wide range of tourist attractions: the towns and villages of Alava and La Rioja, the monumental old part of Victoria-Gazteiz, the capital of the Basque Country, and many other excursions (marquesderiscal.com).

Figure 23.1 Marqués de Riscal Hotel.

Source: Marqués de Riscal Website (n.d.)

From the mountains of Sierra de Cantabria the wineries extend down the slope towards the Ebro river basin. Here you can find the winery of CUNE (La Compañía Vinícola del Norte de España), which launched the Viña Real brand in the 1920s and became one of the pioneering wineries to make crianza wines in oak barrels in the area. The grapes of La Rioja Alavesa are at the heart of this winery. The Viña Real winery in Laguardia, Alava, designed by the French architect Phillippe Mazieres, is composed of three buildings. Both wineries provide the opportunity to visit their cellars and offer an unforgettable walk through its wine village. The tour provides a combination of tradition and innovation and includes the winemaking cellars, an impressive barrel aging cellar designed by Gustave Eiffel, and a cemetery of wines with historic bottles (cune.com).

Conclusions

Looking at the GWC as a whole, creating competitive advantages based on differentiation is seen as a *sine qua non* factor to compete on international markets with greater assurance, and in this context, wine could be used as a resource with which to generate a flow of tourists. However, to build these competitive advantages, it is also necessary that the differentiation has a strong connection with the land and terroir.

In relation to demand, it is vital for Bilbao–Rioja and for the rest of the members to understand consumer behavior and also be able to connect with them in order to establish brand loyalty. According to Dawson, Fountain, and Cohen (2011), those brand-loyal consumers not only continue to purchase wine after they have returned home, but also propagate the images, stories, and experiences with other potential consumers. Word-of-mouth has been identified as the most important source of information used by wine tourists prior to their visit to a winery cellar door (Bruwer and Reilly 2006).

In accordance with all this, Bilbao–Rioja has been able to develop important competitive factors, as shown in this chapter. Thus, the internal connection between the wine region of La Rioja and the metropolitan area of Bilbao is guaranteed by an important road network as well as by the motorway. As far as the external connection is concerned, the international airport on the one hand and the port of Bilbao on the other, a stopover for some of the most important cruise lines, facilitate access to the wine region from the international shipping markets.

Having said this, the development of its competitive strategies has not been concentrated in one single direction. Thus, the wine sector has created over the last few years a new wine architecture, turning many of its wineries into impressive 'wine temples,' such as Marqués de Riscal or Viña Real. A strategy that guarantees every year an important attraction on demand for wine tourism.

On the other hand, the gastronomic richness of the region where Basque and Rioja cuisines meet is another competitive factor with an attractive force over demand. In addition to the centuries-old cuisine of La Rioja, new Basque cuisine has a wide and exquisite offer. Not for nothing does the Basque Country have the highest concentration of Michelin-starred restaurants in Spain (The Gourmet Journal 2020).

Wine therapy is another of the strategic tools successfully implemented in Bilbao–Rioja. The therapeutic benefits derived from wine are spread throughout the region through its extensive network of spas and thermal health centers, in which the tourist lives a unique experience discovering a new form of relationship with wine.

In short, Bilbao–Rioja competes at an international level not only for its excellent wines but also because it has been able to develop its own differentiation in international markets. The main competitive factors on which it has based its differentiation, mentioned in this chapter, are capable of generating not only a major attraction on demand but also significant profitability for the wine tourism business as a whole.

Implications

The contents developed so far have important implications, both for the wine industry, mainly producers and marketers, and for the tourism sector. As we have seen, the creation of an international network of wine-producing territories can contribute decisively to finding and developing an essential competitive factor to fight successfully in international markets. In the not-too-distant future, new international networks will be formed for the same purpose, to compete with an advantage in a global market.

When the time comes, it will be their own members who will be in charge of contributing to the network, through their knowledge, employing the best strategies to compete successfully. Clear examples of this, developed in the Great Wine Capitals, have been explained in this chapter. Thus, to the zero kilometer gastronomic strategies, the artistic exhibitions of the highest level, or the streetcars transporting tourists through the vineyards of the Cape Winelands, other, more concrete strategies have been developed in Bilbao–Rioja.

Indeed, this historic and renowned Spanish wine-growing region has been able to develop its own competitive advantage over time. Its model, of proven success, consists of the union of three important factors: a great knowledge of the tourist market treasured in this part of Spain (including gastronomy, culture, and luxury hotels), a monumental architecture created by a significant part of its nearly 600 wineries, and an efficient network of communications by land, air, and sea, capable of connecting in record time with the main international hubs.

If we add to all this the excellence of its wines – one of the great economic engines of the area, according to the Consejo Regulador de la Denominación de Origen (2020) – the result is an interesting wine tourism model with implications for other wine regions, mainly from the GWC network.

In addition, other implications that arise from the new market segments, detected in some of the aforementioned wine regions, will undoubtedly expand through the international network. Therefore, it will be convenient to closely analyze the consequences of the incorporation into wine tourism of a large segment of the young population, as is the case of San Francisco–Napa Valley, or the increasingly extensive incorporation of a very large Chinese middle class, eager to know the international wine tourism offer.

These circumstances will also have new implications for Bilbao–Rioja. In the first place, it will be interesting to monitor the commercial margins derived from the more-than-probable increase in prices as demand increases. Secondly, it will also be interesting to know the degree of adaptation of the wine tourism offer to the tastes and needs of the new consumers.

For all these reasons, both Bilbao–Rioja and the other competitive wine regions will have to create their own 'superior value offer,' aimed not only at the new target, but also at their shareholders, employees, and collaborators. However, as of this writing, most of these implications are in a latent state, waiting to emerge with force once the consequences of the severe pandemic plaguing the tourism industry in general and wine tourism in particular have passed.

References

Alant, K., and Bruwer, J. (2004). Wine tourism behaviour in the context of a motivational framework for wine regions and cellar doors. *Journal of Wine Research*, 15(1): 27–37. https://doi.org/10.1080/0957126042000300308

Alebaki, M., and Koutsouris, A. (2019). Wine tourism destinations across the life-cycle: A comparison of Northern Greece, Peloponnese and Crete. In Sigala, M., and Robinson, R. N. S. (eds.), *Wine Tourism Destination Management and Marketing* (pp. 463–482). Cham: Palgrave Macmillan.

Alebaki, M., and Lakovidou, L. O. (2011). Market segmentation in wine tourism: a comparison of approaches. *Tourism: An International Multidisciplinary Journal*, 6(1): 123–140.

Arimany-Serrat, N., Ferreras Noger, A., and Rabaseda I Tarres, J. (2016). Análisis economic financiero del sector vinícola de La Rioja en un entorno de crisis. *Intangible Capital*, 12(1): 268–294.

Barber, N., Taylor, D. C., and Deale, C. (2010). Wine tourism, environmental concerns, and purchase intention. *Journal of Travel & Tourism Marketing*, 27: 146–165. https://doi.org/10.1080/10548400903579746

Barney, J. (1991). Firm resources and sustained competitive advantage. *Journal of Management*, 17(1): 99–120. https://doi.org/10.1177/014920639101700108

Barth, S., and Salazar, J. (2011). Wine tourism and consumer behaviors related to wine purchases. *Journal of Tourism Insights*, 1(1). Article 2. https://doi.org/10.9707/2328-0824.1001

Brochado, A., Stoleriu, O., and Lupu, C. (2021). Wine tourism: a multisensory experience. *Current Issues in Tourism*, 24(5): 597–615. https://doi.org/10.1080/13683500.2019.1649373

Bruwer, J. (2003). South African wine routes: some perspectives on the wine tourism industry's structural dimensions and wine tourism product. *Tourism Management*, 24(4): 423–435. https://doi.org/10.1016/S0261-5177(02)00105-X

Bruwer, J., Pratt, M. A., Saliba, A., and Hirche, M. (2017). Regional destination image perception of tourists within a winescape context. *Current Issues in Tourism*, 20(2): 157–177. https://doi.org/10.1080/13683500.2014.904846

Bruwer, J., and Reilly, M. (2006). The power of word-of-mouth communication as an information source for winery cellar door visitors, *Australian & New Zealand Wine Industry Journal*, 21(3):43–52.

Carlsen, J. (2004). A review of global wine tourism research. *Journal of Wine Research*, 15(1): 5–13. https://doi.org/10.1080/0957126042000300281

Carlsen, J., and Boksberger, P. (2013). Enhancing consumer value in wine tourism. *Journal of Hospitality & Tourism Research*, 39(1): 132–144. https://doi.org/10.1177/1096348012471379

Charters, S., and Ali-Knight, J. (2000). Wine tourism – a thirst for knowledge? *International Journal of Wine Marketing*, 12(3): 70–80.

Consejo Regulador de la Denominación de Origen Calificada Rioja. (2020). *Rioja, Memoria Anual 2019*, La Rioja: Consejo Regulador de la Denominación de Origen Calificada Rioja.

Consejo Regulador DOCA Rioja. (2020). *Los turistas internaconales eligen Rioja*. https://www.riojawine.com/noticias/los-turistas-internacionales-eligen-rioja/

D'Arenberg. (2020). *Dali at d'Arenberg*. https://www.darenberg.com.au/dali-at-darenberg

Dawson, D., Fountain, J., and Cohen, D. A. (2011). Seasonality and the lifestyle 'conundrum': an analysis of lifestyle entrepreneurship in wine tourism regions. *Asia Pacific Journal of Tourism Research*, 16(5): 551–572. https://doi.org/10.1080/10941665.2011.597580

Dwyer, L., Edwards, D., Mistilis, N., Roman, C., and Scott, N. (2009). Destination and enterprise management for a tourism future. *Tourism Management*, 30(1): 63–74. https://doi.org/10.1016/j.tourman.2008.04.002

Faugère, C., Bouzdine-Chameeva, T., Pesme, J. O., and Durrieu, F. (2013). The impact of tourism strategies and regional factors on wine tourism performance: Bordeaux vs. Mendoza, Mainz, Florence, Porto and Cape Town. *Mendoza, Mainz, Florence, Porto and Cape Town (January 16, 2013)*. https://doi.org/10.2139/ssrn.2201563

Franschhoek Wine Tram. (2020). *Our Story*. https://winetram.co.za/

Garibaldi, R. (2020). *Roberta Garibaldi: The Impact of COVID-19 on Wine Tourism*. https://www.greatwinecapitals.com/industry-resources/roberta-garibaldi-the-impact-of-covid-19-on-wine-tourism/

Getz, D. (2000). *Explore Wine Tourism: Management, Development and Destinations*. New York: Cognizant Communication Corp.

Getz, D., and Brown, G. (2006). Benchmarking wine tourism development: The case of the Okanagan Valley, British Columbia. Canada. *International Journal of Wine Marketing*, 18(2): 78–97.

Gobierno de La Rioja (2012). *Zonas y regiones vinícolas. ICVV (2017)*. La Rioja, Spain: Instituto de Ciencias de la Vid y del Vino.

Gössling, S., Scott, D., and Hall, C. M. (2021). Pandemics, tourism and global change: a rapid assessment of COVID-19. *Journal of Sustainable Tourism*, 29(1): 1–20. https://doi.org/10.1080/09669582.2020.1758708

Great Wine Capitals. (2020a). *The Great Wine Capitals Global Network – A World of Excellence*. https://www.greatwinecapitals.com/about/

Great Wine Capitals. (2020b). *Bouche à Oreille*. https://www.greatwinecapitals.com/best-of-wine-tourism-awards/bouche-a-oreille/

Great Wine Capitals. (2020c). *Adelaide, South Australia*. https://www.greatwinecapitals.com/capitals/adelaide-south-australia

Gregori, María Teresa. (2017). Mendoza y el agroturismo: un espectro de colores detrás de la luz. *Revista de Divulgación Científica. Facultad de Ciencias Agrarias–Uncuyo*, 7.

Gunasekar, S., Das, P., Dixit, S. K., Mandal, S., and Mehta, S. R. (2021). Wine-experienscape and Tourist satisfaction: through the lens of online reviews. *Journal of Foodservice Business Research*. https://doi.org/10.1080/15378020.2021.2006039

Hall, C. M. (1996). Wine tourism. In Higham, J. (Ed.), *Proceedings of the Tourism Down Under II: A research conference* (pp. 109–119). New Zealand: The University of Otago.

Hall, C. M., Cambourne, B., Macionis, N., and Johnson, G. (1997). Wine tourism and network development in Australia and New Zealand: review, establishment and prospects. *International Journal of Wine Marketing*, 9(2): 5–31. https://doi.org/10.1108/eb008668

Hall, C. M., Johnson, G. R., and Mitchell, R. D. (2000). Wine tourism and regional development. In Hall, C. M., Sharples, E., Cambourne, B., and Macionis, N. (eds.), *Wine Tourism around the World: Development, Management and Markets* (pp. 196–225). Oxford: Butterworth Heinemann.

Hall, C. M., and Mitchell, R. (2000). Wine tourism in the Mediterranean: a tool for restructuring and development. *Thunderbird International Business Review*, 42(4): 445–465. https://doi.org/10.1002/1520-6874(200007/08)42:4%3C445::AID-TIE6%3E3.0.CO;2-H

Laws, E., Prideaux, B., and Chon, K. (2007). Crisis management in tourism: challenges for managers and researches. In Laws, E., Prideaux, B., and Chon, K. (eds.), *Crisis Management in Tourism* (pp. 1–12). Cambridge: Cab International.

Ledesma Silva, E., and Cobos, A. R. (2017). Innovation in the wines of Rioja. *Journal of Science and Research*, 2(8): 44–49. https://doi.org/10.26910/issn.2528-8083vol2iss8.2017pp44-49

Lenggogeni, S., Ritchie, B. W., and Slaughter, L. (2019). Understanding travel risks in a developing country: a bottom up approach. *Journal of Travel and Tourism Marketing*, 36(8): 941–955.

Marqués de Riscal Website. (n.d.). https://www.marquesderiscal.com/

Montella, M. M. (2017). Wine tourism and sustainability: a review. *Sustainability*, 9(1): 113. https://doi.org/10.3390/su9010113

Observatorio Vitivinícola Argentino, (2020). *Una mirada a las tres regiones vitivinícolas líderes: Napa, Rioja y Burdeos.* https://observatoriova.com/2017/10/una-mirada-a-las-tres-regiones-vitivinicolas-lideres-napa-rioja-y-burdeos/

Perrouty, J. P., d'Hauteville, F., and Lockshin, L. (2006). The influence of wine attributes on region of origin equity: an analysis of the moderating effect of consumer's perceived expertise. *Agribusiness*, 22(3): 323–341. https://doi.org/10.1002/agr.20089

Pikkemaat, B., Peters, M., Boksberger, P., and Secco, M. (2009). The staging of experiences in wine tourism. *Journal of Hospitality Marketing & Management*, 18(2–3), 237–253. https://doi.org/10.1080/19368620802594110

Porter, M. E. (1985). *Competitive Advantage: Creating and Sustaining Superior Performance*. New York: Free Press.

Rachão, S. A. S., Breda, Z., Fernandes, C., and Joukes, V. (2020). Food-and-wine experiences towards co-creation in tourism. *Tourism Review*, 76(5): 1083–1091. https://doi.org/10.1108/TR-01-2019-0026

Rogerson, C. M., and Rogerson, J. M. (2020). COVID-19 tourism impacts in South Africa: government and industry responses. *GeoJournal of Tourism and Geosites*, 31(3): 1083–1091. https://doi.org/10.30892/gtg.31321-544

Schamel, G., and Anderson, K. (2003). Wine quality and varietal, regional and winery reputations: hedonic prices for Australia and New Zealand. *Economic Record, Garry Barrett publisher September*, 79(246): 357–369. https://doi.org/10.1111/1475-4932.00109

Sigala, M. (2014). Wine tourism around the world: development, management and markets. *Journal of Wine Research*, 25(2): 133–134. https://doi.org/10.1080/09571264.2014.907136

Sotiriadis, M. D. (2015). Culinary tourism assets and events: suggesting a strategic planning tool. *International Journal of Contemporary Hospitality Management*, 27(6): 1214–1232.

Sparks, B. (2007). Planning a wine tourism vacation? Factors that help to predict tourist behavioral intentions. *Tourism Management*, 28: 1180–1192. https://doi.org/10.1016/j.tourman.2006.11.003

Stewart, J. W. Bramble, L., and Ziraldo, D. (2008). Key challenges in wine and culinary tourism with practical recommendations. *International Journal of Contemporary Hospitality Management*, 20(3): 303–312.

The Gourmet Journal. (2020). *Estrellas Michelin 2020: listado completo.* https://www.thegourmetjournal.com/noticias/estrellas-michelin-2020-listado-completo/

The Napa Valley Visitor Profile. (2019). *Destination Analysts.* https://assets.simpleviewinc.com/simpleview/image/upload/v1/clients/napavalley/2018_Napa_Valley_Visitor_Profile_Final_for_Web_46ac1a26-e3eb-454e-869f-c01219e8a363.pdf

Torres, J. P., Barrera, J. I., Kunc, M., and Charters, S. (2020). The dynamics of wine tourism adoption in Chile. *Journal of Business Research*, 127: 474–485. https://doi.org/10.1016/j.jbusres.2020.06.043

Tustin, M., and Lockshin, L. (2001). Region of origin: does it really count? *Australian and New Zealand Wine Industry Journal*, 16(5): 139–143.

Williams, P. (2001). Positioning wine tourism destinations: an image analysis. *International Journal of Wine Marketing*, 13(3): 42–58.

24
FROM WINE TO VERMOUTH
Events as guardians of the territory

Francesc Fusté-Forné

Introduction

Food and wine events contribute to the valorization of regional cultural and natural landscapes and the socialization of culinary heritages. Wine events develop strong relationships between agriculture and tourism, and a growing number of international events have promoted the creation of territorial brands – because of the pairing of 'food' and 'place.' As a result, gastronomy destinations are developed, with management and marketing implications that have spread a bidirectional relationship between rural and urban environments.

This chapter is focused on the event 'Reus viu el vi' (Reus lives wine), which is celebrated in Reus, southern Catalonia, northeastern Spain. The event includes stands of wine and vermouth producers. Both wine and vermouth, which is a drink also made of grapes, are identity markers of the region. Departing from a case study approach, this research analyzes 'how' and 'why' a wine event contributes to the preservation of wine heritage. Drawing on a conversation with the founder and organizer of the event and supplemented with secondary data, this chapter discusses the meanings attached to the event as a protector and promoter of landscape values. Implications for research and practice are described.

Food events and tourism

The relationship between food and tourism has rapidly increased in recent decades (Ellis et al. 2018; Okumus 2021), and food and beverages are constantly used as tourism attractions (Dixit 2019; Everett 2019). Food communicates a sense of place (Sims 2009) and has a growing impact on destination planning and management (Knollenberg et al. 2020; Rachão et al. 2019). This is especially manifested in wine tourism (Sigala and Robinson 2019). Food and wine tourism widely engage visitors with local landscapes through food (Long 2004), and food events are regarded as experiences where food heritage is protected and promoted (Mason and Paggiaro 2012).

Hoeller (2001) affirms that event tourism is one of the fastest-growing economic activities in contemporary tourism. Getz and Page (2016) add that event tourism has specially boomed since 2008. Getz (1991) worked on the understanding of event tourism, and he underlined that 'events comprise a key element in both the origin area (i.e., events are an important

DOI: 10.4324/9781003143628-28

motivator of tourism) as well as within the destination area (i.e., events feature prominently in the development and marketing plans of most destinations)' (Getz and Page 2016, p. 593). In this sense, event tourism generates large economic and social impacts on communities. In particular, Virginio et al. (2019) state that events are less affected in times of crisis, because they are not influenced by the seasonality of tourism, and in turn, generate benefits for other segments of tourism (see also Seraphin 2020) – for example, rural or urban tourism. Events make places more livable and engage locals and visitors with the valorization of *our* shared legacy. In this sense, festivals and cultural celebrations

> are links for the preservation of traditions, the construction and legitimation of group identity, the communitas between subcultural groups, the generation of social and cultural capital and the value of celebrations to establish the identity of the place and civic pride.
>
> *(Getz and Page 2016, p. 606)*

Food festivals are increasingly used to identify destinations (Lee and Arcodia 2011) and are a source of visitor satisfaction (Carneiro et al. 2019). In this sense, Yang et al. (2020) point out that 'when an event is paired or marketed jointly with a destination, a synthesis between the two occurs' (p. 2). As a consequence, a *unique* knowledge is generated about local communities, local landscapes, and local heritage. The organization of events is crucial for destination branding (Campillo 2012) where events are powerful tools to transform a particular food or beverage into an identity symbol: for example, cheese in Bra (Italy), tomato in Buñol (Spain), or beer in Munich (Germany), to cite only a few examples. There are also many regions around the world that celebrate wine festivals (see Yeoman et al., 2012), such as the Melbourne Food and Wine Festival, which is one of the most popular wine events (Hall and Sharples 2008).

Food events effectively communicate tangible and intangible heritages. Why? Food events protect the artisanal agriculture, and they sustain regional communities (Berno and Fusté-Forné 2020; Star, Rolfe, and Brown 2020), especially in rural areas where food events have become significant drivers for tourism (Getz and Robinson 2014). Also, events are experiences which, based on the preservation of local culture and traditions, are identified as attraction factors that improve the image of destinations (Campillo-Alhama and Martínez-Sala 2019). Previous research states that 'wine festivals and events not only attract visitors but also play a vital role in creating exposure to regional wines and wineries, increasing regional wine sales and stimulating wine tourism' (Kruger and Viljoen 2019, p. 1). This chapter adds texture to this understanding of wine events focused on the case of a local festival celebrated in the city of Reus, southern Catalonia, in northeastern Spain. The wine experience is not only a food experience but also a celebration of agriculture and industrial heritage in the context of tourism (see Yen and Wang 2020).

Study method

This research is built on a case study (see, for example, Fusté-Forné and Berno 2016; Xiao and Smith 2006), which is useful when a study addresses questions prefaced with 'how' and 'why' (Yin 2009). This research used a case study approach (see also Getz 2013) to analyze 'how' and 'why' a wine event contributes to the protection and promotion of wine heritage. Based on a conversation with the founder and organizer of the event, and supplemented with secondary data (official reports and news), this research aims to understand the significance of the festival and to scrutinize the event as a gatekeeper of winescapes.

Results

Reus viu el vi, translated as 'Reus lives wine,' is a wine festival organized in the city of Reus since 2009. The event is organized by the Chamber of Commerce of Reus, which represents five counties in Catalonia (Baix Camp, Conca de Barberà, Priorat, Ribera d'Ebre, and Terra Alta). In Spain, a chamber of commerce is a public law corporation that provides services to businesses in a region. The Chamber of Commerce of Reus manages six wines with a Protected Designation of Origin (PDO), more than any other body in the same territorial scale. Four of these six PDO belong uniquely to the Reus Chamber: PDO Conca de Barberà, PDO Montsant, PDO Priorat, and PDO Terra Alta. In addition, PDO Tarragona and PDO Catalonia are managed together with other Chambers of Commerce.

The analyzed festival includes the participation of wineries from the six different PDOs, which offer their products both for tasting and selling. In this sense, the fair is held in the Llibertat Square during the first weekend of June (Friday to Sunday) and includes the 'Fair of the PDOs of the demarcation' and, since 2018, the 'Reus Vermouth Show,' which was added to celebrate the tenth edition of the festival. The fair offers several packs for visitors, such as the pack 'Reus lives wine,' which includes a glass, the catalog, a cup holder, five tastings, and the hat of the fair at a price of €13. The last edition held in 2019 included the participation of 40 wineries, which presented more than 160 wine references (García Jardí 2019). In relation to its origin, the event was created due to several factors, as highlighted by an interview with the representative of the Chamber of Commerce. On the one hand, Reus is a shopping and urban destination in Catalonia (Reus Promoció 2020), with more than 100,000 inhabitants (Idescat 2020). Reus is located in a territory with a large agricultural tradition where the production of wine and vermouth (see Figure 24.1) have been a symbol of its identity for centuries (Fusté-Forné 2019). The fair was born as an initiative to protect and promote wine as a local product.

A celebration with alcohol-based beverages

With the arrival of the *Festa Major de Sant Pere* (the biggest annual festival in Reus, which celebrates Saint Peter, patron saint of the city, together with *Mare de Déu de Misericòrdia*) on June 29, the streets of Reus are full of people. Both bakeries and pastry shops prepare their traditional cherry cake (*coca de cireres*), restaurants offer the traditional dish of beef and rice, people walk down the streets with a bag of Laurie frites in their hands, and all the terraces are ready to sit and drink a vermouth or a *masclet* (a local beverage). In Reus, *Festa Major* has been documented since the 12th century, when Saint Peter became the city's patron. From a historical point of view, it is the most important festival in the city in a pre-industrial society, and until the beginning of the 19th century, when Carnival started to receive growing attention. However, despite being in the midst of a crisis at the beginning of the 20th century, the *Festa Major* never lost its ritual, represented by the *tronada* (a local firework) and the performance of the traditional festive elements, such as the Giants, the Dwarves, and the Mule. This context also meant that the cuisine associated with the festival was below minimums, as seen in an article in the newspaper *Foment*, published on June 29, 1909, which stated that the *Festa Major de Reus* was in the end stages, because even the famous beef and rice have nothing left but a memory.

The relationship between Reus and alcohol production is both historical and contemporary, and it is illustrated with beverages such as *aguardiente*, vermouth, and *masclet*. Thus, a strong linkage between food and drink is manifested in local festivals celebrated in the city. Ancient documentation says that political and religious authorities were offered a soda, as a refreshment, during the celebration of *Festa Major*. Earlier, the concept of the procession was

Figure 24.1 Vineyard for wine and vermouth production in Reus.

Source: Author

longer than it is now, and an official snack became necessary, including wine, *aguardiente*, and food. In this sense, we can observe a significant feature of Reus's 'food and drink': alcohol. Reus is located in the middle of an agricultural area of wine growing and production, which resulted in a very important pre-industrial *aguardiente* industry, with many producers and artisan workshops, which allowed to export *aguardiente* by sea everywhere. However, with the arrival of phylloxera, and the impact that this had on wine production in the region during the late 19th century, growers started to popularize aromatic wines, such as vermouth, which does not apply any processes of distillation. Therefore, vermouth is the current product in the line of alcohol production in Reus, and this shows that there has always been a relationship between Reus and alcohol production, making famous expressions such as '*Reus, Paris, London*'.

Nowadays, the role of vermouth in the *Festa Major* is obvious. What used to be 'let's have an *aguardiente*' is now substituted by the vermouth, which has created a meeting point to socialize with family and friends in the terraces of the old quarter. Vermouth has an extensive presence in the official program of the *Festa Major* (for example, with a number of vermouth tastings), but also throughout the year with permanent spaces such as the Vermouth Museum, or the recently opened Vermouth and Wine House, in the Enological Station. Also, in recent years, vermouth has been recovered and consolidated its cultural and social space: 'let's share a vermouth' makes more sense than ever. Vermouth is part of the city landscape. In the same way, there is another drink, *masclet*, which is made from vermouth and *plim* and has been promoted as the drink of the Reus festival. If beef and rice is the dish of the *Festa Major*, the *masclet* is the beverage. Thus, 'Masclet Parade' and 'Masclet Contest' are organized with the participation of various bars and restaurants in the city. These establishments are involved in the promotion of *masclet*, and they add other ingredients to the mixture of vermouth and *plim* in order to make each combination unique, while respecting the original recipe. Every year, after the fireworks celebrated in June 24, and once the official speaker has made the first tasting of *masclet*, barrels are thrown in Mercadal Square, and 'Masclet Parade' runs through the city streets.

In this sense, while culinary and food traditions may have changed, and the origins of gastronomy as part of *Festa Major* are old, the relationship between food and beverages and *Festa Major* is a landmark of the city. Culture is dynamic and evolves over time, but festive food remains well established as part of local lifestyles. In this sense, cherry cake, beef and rice, Laurie frites, *masclet*, and vermouth are elements of the city's food identity. Both social and business daily life in Reus revolves around food and beverages, which influence traditions that reflect how we were and how we are.

In relation to the origin of the festival 'Reus lives wine,' the Chamber representative states that there were many wine celebrations in Catalonia and Spain; however, there was no fair that brought together several PDOs. Also, a decade ago, when you went to restaurants in Reus, you found that local wines were not included in the menus. 'Can you imagine a restaurant in Logroño that is not full of Rioja?' Reus is the commercial capital of all the PDOs represented by the Chamber, and according to the Chamber representative, 'we wanted to raise awareness about the possibilities of wines that are close to home, to make them known for people and restaurants in the city.' In 2022 the 12th edition of the fair was celebrated (which should have been held in 2020 and 2021). 'Reus is a point of reference' for regional wineries, and some winemakers negotiate up to three-quarters of their sales in Reus. The Chamber representative underlines that there were 'too many fairs, and that meant we had to make our fair unique, for example, with more than one PDO and many different wineries.' Reus has engaged with the festival, which is part of its identity. People and restaurants have also widely participated. In this sense, another added value of the festival 'Reus lives wine' is that it takes place during the first weekend of June. The fair represents the beginning of the program of the *Festa Major*. This has contributed to the active participation of both locals and visitors, year after year. This fair is a 'must' in the city's annual activities. It is an experience around wine and vermouth that allows people to know the soul of the producers and socialize with them.

Food and wine festivals are spaces for the promotion of food culture and, especially, of wine culture. Thus, there are wineries that participate every year while others change – people discover the territory through its winegrowers and makers – which represents a wide range of wine references available. 'We all like a French or California wine, but we have a lot of quality wines next to our homes that we don't know about it.' In this sense, the fair also plays a role as protector and communicator of the territorial values of wine, and the associated production and consumption processes, and makes them available to the general public. In relation to recent trends such as creative and participatory tourism, the Chamber representative said, 'the fair organizes activities to engage visitors as a prize for the first person to taste a wine from each PDO.'

Previous news and reports confirm 'the full consolidation of the event as the great showcase of the region's wineries and the work that, from the organization, is done to project this event internationally' (Reus Digital 2015). The participation of other PDO wines in Catalonia and Spain, and some international wineries, make the fair a way of communicating to the world the Reus city brand (Ajuntament de Reus 2015). The fair transforms Reus as a Catalan capital of wine every June, based on a leitmotif: to promote regional wines. The event acknowledges the 'traditional agri-food capital of Reus and places the fair as an example of the ability to support and act as an umbrella for the promotion of local products' (Siré 2017). In addition, Reus was the capital of Catalan Culture in 2017 (Ajuntament de Reus 2017), which further consolidated its projection. The 2017 edition was the most crowded to date, as 60,000 wine tastings and 33,000 food tastings – also based on traditional products of the territory, such as hazelnuts and olive oil – were served during the event (Martínez 2017). At the 2018 fair, which celebrated its tenth anniversary, more than 80 wineries participated (Díaz Bages 2018).

The fair has expanded its implications over the years. Even today, the PDOs have their monographic fairs (for example, Priorat wine is celebrated in Falset), and Reus is the place where all the PDOs that communicate territorial identity come together. Both event formats coexist with a single or multiple presence of regional wine PDOs. The aim of the Chamber of Commerce as a public body, which includes the business fabric of the territory, is to help its companies, and in particular wine companies. They are able to promote wine products directly at the event, and restaurants are ideal partners to co-create experiences of production and consumption focused on local products. Restaurants are prescribers of local foods and drinks. This impact does not only rely on wine; it extends to vermouth, which is the flag of Reus, and which has been incorporated as part of the fair. This collaboration has also led to a solid complementary offer of gastronomy, to pair with wine and vermouth, and has also expanded the influence of the event to other towns in the territory such as Cambrils, which has recently started to organize a similar fair with a focus on coastal gastronomy.

Discussion and conclusions

This chapter has reviewed the crucial role of a wine event for the protection and promotion of wine landscapes and the meanings attached to the event from the perspective of its founder and organizer, and supplemented with data obtained in official reports and regional news. In this sense, results show the impact of the event in the city, the region, and the people who participate. 'Reus lives wine' has emerged as a city event and an essential must-go in the food and festive calendar of Reus. It represents a well-established celebration in its social agenda. The fair brings together wineries of up to six PDOs of territorial proximity, four of them under the direct area of influence of the Chamber of Commerce of Reus (Conca de Barberà, Montsant, Priorat, and Terra Alta). The participation of several wineries also facilitates the experience towards new producers and opens the door to the consolidation of rural entrepreneurship through a festival. 'Reus lives wine' is a fair of wine and vermouth, and it is a space for meeting and socializing based on wine culture and tradition. In this sense, wine growing and making are awarded a tourist value to protect and promote wine heritage and landscapes.

Implications of this wine experience expand destination stakeholders' knowledge about the communication of a sense of place and community. Figure 24.2 shows these relationships, where wine experts participate in wine growing and winemaking. Wine people are very close to the land because they know the historical and contemporary dynamics of wine tradition. In this sense, wine culture encapsulates both the wine landscapes (for example, environmental values) and the wine heritages (for example, wine pairing), which are part of wine production. Also, a wine is valorized when it enters the realms of wine consumption. It is obvious that the production and consumption of wine are intertwined, and a wine event is the place where the co-creation of the wine experience is shared between hosts and guests. A wine event engages producers with the development of a wine storytelling, which expands the

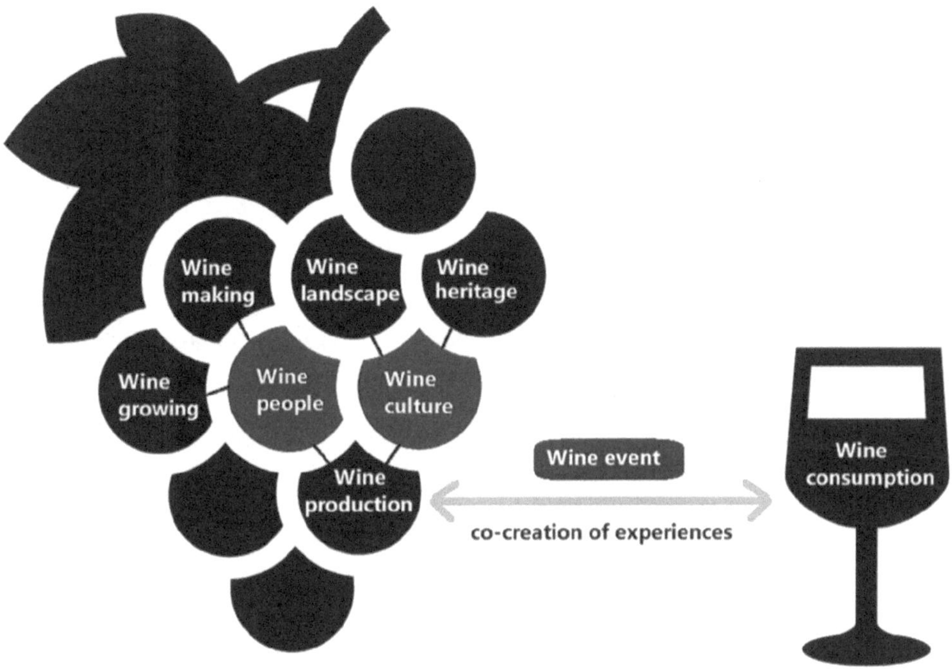

Figure 24.2 Wine values in 'Reus lives wine.'

Source: Author

wine-based knowledge of visitors. At the same time, this sustains the preservation of 'wine' and the promotion of the narratives attached to the product – drawing on a local event that is based on local foods and beverages.

In relation to the theoretical implications of this study, the research highlights the power of food to contribute economically and socially to regional development and to individual and social well-being (Batat et al. 2019), which is especially relevant when food is communicated via small, sustainable, and slow tourism projects (Fusté-Forné and Jamal 2020; Hatipoglu, Ertuna, and Salman 2020) such as a food fair. Wine events foster solid linkages between agriculture and tourism and elevate a form of rural lifestyle (wine growing and making) into a tourism attraction. This chapter has analyzed 'how' and 'why' 'Reus lives wine' became a crucial event focused on the celebration of wine and vermouth. Food and beverage tourism contributes towards the comprehension of a local identity (see, for example, Bessière 1998; Dixit 2020; Enzenbacher 2019; Kim and Ellis 2015; Kim and Iwashita 2016) – Reus means alcohol-based beverages.

Event tourism contributes to 'combatting seasonality of demand, spreading tourism geographically, and assisting in other forms of urban and economic development' (Getz and Page 2016, p. 597). Further research must focus on the perspectives of wine producers that participate in the event. Also, the opinion of visitors must be analyzed to explore the ways they experience the event, their motivation, and satisfaction. In this sense, 'festival and event attendees are continually seeking unique, authentic, experientially-oriented festivals and events with a greater focus on attendees actively participating in the experience' (Brennan, Yiheng, and Song 2020, p. 1). Drawing on the understanding of event tourism as a driver for sustainable development, future research must also explore the implications of food and beverages events towards 'a more harmonic utilization of resources [that] can be achieved guarantying, therefore, [that] their long-term viability and a better environment for locals to live in, work and grow is assured' (Boukas 2019, p. 82).

Wine events and festivals engage visitors with the celebration of *authentic* products, which rely on *unique* landscapes (see Fusté-Forné 2016, 2020). The event has promoted the city as a capital of wine (and vermouth) in Catalonia and served as a branding strategy for the city of Reus (Cartanyà 2018), paired with wine (and vermouth). The festival has promoted an effective integration of wine (and vermouth) in local trade and restaurants, and with a large participation of residents and tourists who are able to discover the territory close to where the wine is sold. This is the main practical implication of the research. In this sense, 'events have become a large focus for economic development in rural and regional areas. Tourism has been a key driver of food festivals, where they play a role in attracting visitors and generating expenditure in the local economy' (Star, Rolfe, and Brown 2020, p. 325). However, the fair must search for new ways to interact with participants and visitors (for example, vermouth shows) and further understand the meaningful relationships between food and events within tourism systems.

References

Ajuntament de Reus. (2015). *Del 5 al 7 de juny es onceptua la setenaoncept (segona internacional) de la Reus Viu el Vi.* http://www.reus.cat/sites/reus/files/150601_dossier_reus_viu_vi.pdf

Ajuntament de Reus. (2017). *Reus 2017 – Capital de la Cultura Catalana.* https://capitalcultura.reus.cat/

Batat, W., Peter, P. C., Moscato, E. M., Castro, I. A., Chan, S., Chugani, S., and Muldrow, A. (2019). The experiential pleasure of food: a savoring journey to food well-being. *Journal of Business Research*, 100: 392–399. https://doi.org/10.1016/j.jbusres.2018.12.024

Berno, T., and Fusté-Forné, F. (2020). Imaginaries of cheese: revisiting narratives of local produce in the contemporary world. *Annals of Leisure Research*, 23(5): 608–626. https://doi.org/10.1080/11745398.2019.1603113

Bessière, J. (1998). Local development and heritage: traditional food and cuisine as tourist attractions in rural areas. *Sociologia Ruralis*, 38 (1): 21–34. https://doi.org/10.1111/1467-9523.00061

Boukas, N. (2019). Rural tourism and residents' well-being in Cyprus: towards a onceptualized framework of the appreciation of rural tourism for islands' sustainable development and competitiveness. *International Journal of Tourism Anthropology*, 7(1): 60–86. https://doi.org/10.1504/IJTA.2019.098105

Brennan, B. S., Yiheng, L., and Song, H. (2020). Online marketing and electronic commerce integration: a case study examining an international wine festival. *The Journal of Internet Electronic Commerce Research*, 20(2): 1–20. https://doi.org/10.37272/JIECR.2020.04.20.2.1

Campillo, C. (2012). El desarrollo de políticas estratégicas turísticas a través de la marca acontecimiento en el municipio de Elche (2000–2010). *PASOS, Revista de Turismo y Patrimonio Cultural*, 10(1): 119–129. https://doi.org/10.25145/j.pasos.2012.10.011

Campillo-Alhama, C., and Martínez-Sala, A. M. (2019). La estrategia de marketing turístico de los Sitios Patrimonio Mundial a través de los eventos 2.0. *PASOS. Revista de Turismo y Patrimonio Cultural*, 17(2): 425–452. https://doi.org/10.25145/j.pasos.2019.17.029

Carneiro, M. J., Eusébio, C., Caldeira, A., and Santos, A. C. (2019). The influence of events cape on emotions, satisfaction and loyalty: the case of re-enactment events. *International Journal of Hospitality Management*, 82: 112–124. https://doi.org/10.1016/j.ijhm.2019.03.025

Cartanyà, A. (2018). Vermut al matí I vi a la tarda. *Reusdigital.cat*, June 3. https://reusdigital.cat/noticies/economia/vermut-al-mati-i-vi-la-tarda

Díaz Bages, A. (2018). La 10a edició de la fira Reus Viu el Vi presenta una mostra dedicada al vermut com a novetat. *Reusdigital.cat*, May 28. https://reusdigital.cat/noticies/reus/la-10a-edicio-de-la-fira-reus-viu-el-vi-presenta-una-mostra-dedica-al-vermut-com

Dixit, S. K. (ed.). (2019). *The Routledge Handbook of Gastronomic Tourism.* New York: Routledge.

Dixit, S. K. (ed.). (2020). *The Routledge Handbook of Tourism Experience Management and Marketing.* London: Routledge.

Ellis, A., Park, E., Kim, S., and Yeoman, I. (2018). What is food tourism? *Tourism Management*, 68: 250–263. https://doi.org/10.1016/j.tourman.2018.03.025

Enzenbacher, D. J. (2019). Exploring the food tourism landscape and sustainable economic development goals in Dhofar Governorate, Oman: Maximising stakeholder benefits in the destination. *British Food Journal*, 122(6): 1897–1918. https://doi.org/10.1108/BFJ-09-2018-0613

Everett, S. (2019). Theoretical turns through tourism taste-scapes: the evolution of food tourism research. *Research in Hospitality Management*, 9(1): 3–12. https://doi.org/10.1080/22243534.2019.1653589

Fusté-Forné, F. (2016). Tasting cheesescapes in Canterbury (New Zealand). *New Zealand Geographer*, 72(1): 41–50. https://doi.org/10.1111/nzg.1211

Fusté-Forné, F. (2019). La Gastronomia i la Festa Major. *NW La Revista de Reus*, 83: 9–15.

Fusté-Forné, F. (2020). Say Gouda, say cheese: travel narratives of a food identity. *International Journal of Gastronomy and Food Science*, 22: 100252. https://doi.org/10.1016/j.ijgfs.2020.100252

Fusté-Forné, F., and Berno, T. (2016). Food tourism in New Zealand: Canterbury's foodscapes. *Journal of Gastronomy and Tourism*, 2(2): 71–86. https://doi.org/10.3727/216929716X14720551277763

Fusté-Forné, F., and Jamal, T. (2020). Slow food tourism: an ethical microtrend for the Anthropocene. *Journal of Tourism Futures*, 6(3): 227–232. https://doi.org/10.1108/JTF-10-2019-0120

García Jardí, E. (2019). L'11a edició de Reus Viu el Vi comptarà amb una quarantena de cellers. *Reusdigital.cat*, June 5. https://reusdigital.cat/noticies/economia/l11a-edicio-de-reus-viu-el-vi-comptara-amb-una-quarantena-de-cellers

Getz, D. (1991). *Festivals, Special Events, and Tourism*. New York: Van Nostrand Reinhold.

Getz, D. (2013). *Event Tourism: Concepts, International Case Studies, and Research*. Elmsford, NY: Cognizant Communication Corporation.

Getz, D., and Page, S. J. (2016). Progress and prospects for event tourism research. *Tourism Management*, 52: 593–631. https://doi.org/10.1016/j.tourman.2015.03.007

Getz, D., and Robinson, R. N. (2014). Foodies and food events. *Scandinavian Journal of Hospitality and Tourism*, 14(3): 315–330. https://doi.org/10.1080/15022250.2014.946227

Hall, C. M., and Sharples, L. (2008). *Food and Wine Festivals and Events around the World: Development, Management and Markets*. Oxford: Routledge.

Hatipoglu, B., Ertuna, B., and Salman, D. (2020). Small-sized tourism projects in rural areas: the compounding effects on societal wellbeing. *Journal of Sustainable Tourism*: 1–23. https://doi.org/10.1080/09669582.2020.1784909

Hoeller, E. H. (2001). Turismo de Eventos: Centreventos CauHansen de Joinville/SC. In M. G. R. Ansarah (ed.), *Turismo: Segmentação de Mercado* (pp. 62–80). São Paulo: Futura.

Idescat (2020). Població. https://www.idescat.cat/tema/xifpo

Kim, S., and Ellis, A. (2015). Noodle production and consumption: from agriculture to food tourism in Japan. *Tourism Geographies*, 17(1): 151–167. https://doi.org/10.1080/14616688.2014.978812

Kim, S., and Iwashita, C. (2016). Cooking identity and food tourism: the case of Japanese udon noodles. *Tourism Recreation Research*, 41(1): 89–100. https://doi.org/10.1080/02508281.2016.1111976

Knollenberg, W., Duffy, L. N., Kline, C., and Kim, G. (2020). Creating competitive advantage for food tourism destinations through food and beverage experiences. *Tourism Planning and Development*: 1–19. https://doi.org/10.1080/21568316.2020.1798687

Kruger, M., and Viljoen, A. (2019). Terroir wine festival visitors: uncorking the origin of behavioural intentions. *Current Issues in Tourism*, 24(5) 616–636. https://doi.org/10.1080/13683500.2019.1667310

Lee, I., and Arcodia, C. (2011). The role of regional food festivals for destination branding. *International Journal of Tourism Research*, 13(4): 355–367. https://doi.org/10.1002/jtr.852

Long, L. M. (2004). *Culinary Tourism*. Lexington: University Press of Kentucky.

Martínez, I. (2017). La Reus Viu el Vi serveix 60.000 tastos i prop de 33.000 degustacions gastronòmiques. *Reusdigital.cat*, June 11. https://reusdigital.cat/noticies/reus/la-reus-viu-el-vi-serveix-60000-tastos-de-vi-i-prop-de-33000-degustacions

Mason, M. C., and Paggiaro, A. (2012). Investigating the role of festivalscape in culinary tourism: the case of food and wine events. *Tourism Management*, 33(6): 1329–1336. https://doi.org/10.1016/j.tourman.2011.12.016

Okumus, B. (2021). Food tourism research: a perspective article. *Tourism Review*, 76(1): 38–42. https://doi.org/10.1108/TR-11-2019-0450

Rachão, S., Breda, Z., Fernandes, C., and Joukes, V. (2019). Food tourism and regional development: a systematic literature review. *European Journal of Tourism Research*, 21: 33–49.

Reus Digital (2015). Les xifres avalen un any més la Reus Viu el Vi. *Reusdigital.cat*, June 7. https://reusdigital.cat/noticies/economia/les-xifres-avalen-un-any-mes-la-reus-viu-el-vi

Reus Promoció (2020). Activitats recomanades. https://www.reusturisme.cat

Seraphin, H. (2020). COVID-19: an opportunity to review existing grounded theories in event studies. *Journal of Convention and Event Tourism*, 22(1): 3–35. https://doi.org/10.1080/15470148.2020.1776657

Sigala, M., and Robinson, R. (eds.). (2019). *Wine Tourism Destination Management and Marketing*. Cham, Switzerland: Springer.

Sims, R. (2009). Food, place and authenticity: local food and the sustainable tourism experience. *Journal of Sustainable Tourism*, 17(3): 321–336. https://doi.org/10.1080/09669580802359293

Siré, J. (2017). La Cambra reuneix el millor tast de vi per consolidar la reactivació del sector. *Reusdigital.cat*, June 10. https://reusdigital.cat/noticies/reus/la-cambra-reuneix-el-millor-tast-de-vi-consolidar-la-reactivacio-del-sector

Star, M., Rolfe, J., and Brown, J. (2020). From farm to fork: is food tourism a sustainable form of economic development? *Economic Analysis and Policy*, 66: 325–334. https://doi.org/10.1016/j.eap.2020.04.009

Virginio, D. F., Moreira, Y. S., Lima, P. M. P., and Chacon, L. S. (2019). Interface entre municípios, Turismo e eventos: Um olhar para a Microrregião do Litoral Sul Potiguar e impactos possíveis. *Revista de Turismo Contemporâneo*, 7(1): 49–70. https://doi.org/10.21680/2357-8211.2019v7n1ID16931

Xiao, H., and Smith, S. L. (2006). Case studies in tourism research: a state-of-the-art analysis. *Tourism Management*, 27(5): 738–749. https://doi.org/10.1016/j.tourman.2005.11.002

Yang, F. X., Wong, I. A., Tan, X. S., and Wu, D. C. W. (2020). The role of food festivals in branding culinary destinations. *Tourism Management Perspectives*, 34: 1–11. https://doi.org/10.1016/j.tmp.2020.100671

Yen, T. F. T., and Wang, M. H. (2020). Wine cultural event as a growing phenomenon: Role of novelty, value and satisfaction in developing WOM. *Asian Journal of Education and Social Studies*: 29–41. https://doi.org/10.9734/ajess/2020/v10i230264

Yeoman, I., Robertson, M., Ali-Knight, J., Drummond, S., and McMahon-Beattie, U. (2012). *Festival and Events Management*. London: Routledge.

Yin, R. K. (2009). *Case Study Research: Design and Methods*. Thousand Oaks, CA: Sage.

PART IV

New approaches and practices in wine tourism marketing

25

MARKETING DESTINATIONS THROUGH WINE TOURISM

Using the example of developing wine tourism in Austria

Stephanie Tischler and Albert F. Stöckl

Introduction

While competition in the wine business environment intensified globally, many tourism destinations around the world started to position themselves as wine tourism regions in recent years (Getz and Brown 2006; Williams 2001a). Wine tourism has become a popular leisure activity internationally, which is appealing to various target groups (Sigala and Robinson 2019, p. 1).

Retrospectively, the Austrian 'glycol wine scandal' in 1985 can be seen as a blessing in disguise and laid the foundation for an exceptional rise of the Austrian wine industry (Steiner and Seiler 2005) that finally resulted in many well-organized and successful wine tourism initiatives and strategies of national and regional wine marketing bodies as well as individual businesses.

This chapter aims at (1) discussing strategic approaches in wine tourism destination marketing, (2) identifying relevant wine tourism destination attributes and experiences that can be marketed in the wine tourism context, and (3) presenting the case of wine tourism in Austria using current examples of successful wine tourism destinations.

Theoretical background

The development of wine tourism in Austria

In 1985, the 'glycol wine scandal' hit Austria. Diethylene-glycol – an automobile antifreeze – was considered safe and very difficult to detect by the authorities. It was added to thin, sour wines to make them smoother and more palatable. Tanker loads of doctored wine were shipped off to Germany, masquerading as quality wine with forged documentation to support its alleged origin. Besides such 'upgrading' of simple wines, some 'inventive' wineries even created wines without grapes (Stöckl 2006). When the fraud was detected, hundreds of (mostly big) wineries went bankrupt; owners or managers involved went to jail (Austrian Wine Marketing Board 2016).

The German Ministry of Health warned of the dangers of drinking Austrian wines, and the US Bureau of Alcohol, Tobacco, and Firearms banned the import of Austrian wine. Poland, France, and Great Britain followed suit and removed Austrian wines from their shelves. Greece

DOI: 10.4324/9781003143628-30

banned all imports of Austrian wines. Denmark blacklisted 50 varieties (Steiner and Seiler 2005). The scandal destroyed consumer trust across the world and resulted in almost zero Austrian wine exports. The market for Austrian wine at home and abroad was devastated for years (Austrian Wine Marketing Board 2016).

Integrity in the wine business demanded a completely fresh start. To respond, a new and strict wine law was introduced, which required a complete examination of all wine stocks (Austrian Wine Marketing Board 2016). This implementation of stringent quality standards, combined with consumer consciousness of estate bottling and wine purity, led to a rebuilding of the Austrian wine culture in ways nobody could have foreseen before the scandal. Through *Vinea Wachau*, the Wachau's growers established quality benchmarks that went beyond Austria's new wine laws and set the stage for international prestige. South Styria (*Südsteiermark*) emerged from a century of obscurity to national stardom. By the late 1990s, Austrian wine enjoyed unprecedented export success and prestige that has continued to this day (Robinson 2015).

The wine scandal is often seen as a blessing in disguise, as it laid the foundation for an exceptional rebirth and rise of the industry (Knoll 2000). This development is often referred to as 'the Austrian *Weinwunder*' (wine 'miracle'). The *Weinwunder* was carried by a new, young generation of wine growers and makers and lead to a break with old customs: grapes were selected and thinned in order to minimize and optimize the harvested material. Vineyards were tended to more carefully in order to harvest only healthy, perfect grapes. Cellar technology was updated to a modern standard. In the footsteps of their pioneer parents, the new generation of vintners caters better and better to the tastes of new consumers. The quality has spiraled upwards. Besides, the machinery of commercial marketing has discovered the winemaking industry. The best vintners, just like the best chefs and hosts, have become the stars of the Austrian culinary revolution (Steiner and Seiler 2005). Simultaneously, this area marked the beginning of a flourishing wine tourism sector in Austria. Following the scandal, it was a must for all wineries to be open to the public (Stöckl 2006). From the mid-1990s on, the general attitude towards wine underwent a change, and Austrian wine, wineries, and wine architecture were booming; wine became a 'lifestyle product,' and direct sales ex cellar door started to increase. This marked the starting point of developing wine tourism in Austria, which finally became very popular (Steiner and Seiler 2005).

At the same time, competition in the wine business environment intensified globally, and positioning as a wine tourism region became a strategic issue of destinations around the world (Getz and Brown 2006; Williams 2001a). Wine tourism finally became a popular leisure activity appealing to various target groups (Sigala and Robinson 2019, p. 1).

Marketing wine tourism destinations

Based on mentioned international developments, Williams (2001a, p. 42) suggested a shift from 'an emphasis on wine production processes and related facilities to aesthetic and experiential values associated with more leisurely recreational and tourist pursuits.' Consequently, a clear focus on wine tourism experiences based on high-quality wines is needed, accompanied by aspects such as the landscape as well as culinary, educational, event hosting, and cultural dimensions. Similarly, Getz and Brown (2006, p. 156) claim that the experience of a wine destination is more important than specific attractions and products.

Many definitions of wine tourism focus exactly on these wine-related experiences, which are based on the interplay of the wine and winery experience as well as the destinations attributes (e.g., Williams 2001b; Saayman and van der Merwe 2014). Wine tourism is, therefore, focused not solely on visiting wineries and tasting wines, but on providing a much broader range of experiences (Okech 2016, p. 72).

Similarly, Bruwer and Alant (2009, p. 242) argue that 'wine tourism involves more than just visiting wineries and vineyards but is instead the culmination of a number of unique experiences that include the surrounding environment, ambiance, atmosphere, regional culture, and local wine and food.' Consequently, they propose that winescape elements – specifically those related to the natural landscape as a brand element – shall be marketed by wine regions. The wineries themselves should develop promotional messages that emphasize the core wine tourism product elements such as wine tasting and/or buying as well as other hedonic experience elements. Furthermore, Bruwer and Alant (2009) suggest that marketing efforts targeted at repeat visitors should focus strongly on wine-related attributes, while marketing activities addressed at first-time visitors should emphasize the elements of the natural environment.

Sigala and Robinson (2019, pp. 1–2) argue that regardless of the wine tourist type, all wine tourists wish to immerse themselves into the winescape in order to experience and learn about the landscape, scenery, traditions, culture, and heritage. Therefore, the attractiveness and authenticity of the winescape become crucial factors besides the provision of high-quality wine. Developing wine tourism experiences based on the physical, cultural, and natural resources of a wine destination and its wineries have therefore become a strategic issue in wine destination marketing.

Gomez, Lopez, and Molina (2015, p. 213) conclude that a wine destination comprises natural and cultural resources, infrastructure, socioeconomic environment, climate, atmosphere, social conditions, and gastronomy (Gomez, Lopez, and Molina 2015, p. 213). Croce and Perri (2010, p. 61) argue that using such local resources of a destination and turning them into identifiable and appealing tourist products gives a terroir a distinct identity that helps to differentiate from competitors. Analyzing the area's resources is, therefore, the starting point before a terroir can be transformed into a tourist destination.

Summing up, it can be said that the critical features of the wine tourism experience cover the core wine product (e.g., wineries, winery staff, festivals), the core destination appeal (e.g., scenery, climate, accommodation, wine trails) as well as the cultural product (e.g., traditional wine villages, accommodation with regional character, fine dining and gourmet restaurants) (Getz and Brown 2006, p. 156).

When marketing wine tourism, destinations should therefore include regional and cultural activities in the region besides winery visits. Furthermore, marketing efforts of wine tourism destinations have to be target group-oriented and consequently emphasize the core offer for each segment (e.g., wineries, culinary activities, children's activities). By bundling various activities for different segments, destination marketers can develop tailor-made marketing strategies which are more efficient to potential visitors, wineries, and other providers in the region. Building alliances between tourist attractions, hotels, restaurants, and other facilities can help to attract potential wine consumers of different segments and with different needs and interests. This can eventually lead to higher consumer loyalty (Cohen and Ben-Nun 2008).

Improving a destination's marketing effectiveness and efficiency is a central aim in destination marketing and requires adequate tools and approaches (Soteriades 2012). Besides value-chain, strategic marketing, e-marketing and clustering approaches that have been proven to improve the effectiveness of destination marketing (Soteriades 2012), and creating and delivering unique, extraordinary, and memorable tourism experiences are seen as essential in order to maintain a sustainable competitive advantage of tourism destinations (Hudson and Ritchie 2009; Ritchie and Crouch 2003). Culinary and wine experiences are especially capable of fostering positive and memorable moments in life as they are able to appeal to all senses (Everett 2016;

Morgan 2006). Pleasureful and memorable experiences demonstrably lead to consumer emotional attachments to (regional) brands, an increased willingness to pay, and consumer loyalty (Stöckl, Rinke, and Eisingerich 2016). Sigala (2019b, p. 139) concludes that 'effective wine tourism experiences are those that can engage wine tourists behaviorally, emotionally, mentality/congnitively but also spiritually.'

As a result, customer experiences can create long-term competitive advantages (Shaw and Ivens 2002, p. 210). Destination management and marketing efforts should therefore target at creating and delivering unique, extraordinary, and memorable tourism experiences to maintain a sustainable competitive advantage (Dixit 2020; Hudson and Ritchie 2009; Ritchie and Crouch 2003).

In the same vein, Byrd et al. (2016) argue that researching which elements of a winescape motivate a tourist to visit, revisit, and recommend the destination to others is a fundamental element of developing successful destination marketing strategies for wineries and wine regions. Based on this, they developed a model covering core and supplementary services in a wine region. The authors propose to maintain focus on the core consumer benefits of tasting and buying wine, but also to extend the experience by supplementary services that facilitate enjoyment of the wine product. As a result, they suggest higher associations with future intentions to visit, buy wine, and promote via word-of-mouth (Byrd et al. 2016).

Despite the mentioned findings and recommendations regarding 'supplementary services that facilitate the enjoyment of wine' (Byrd et al. 2016, p. 28), one must consider that the 'average consumer' is far from being a wine connoisseur. The average consumer usually devotes little time to the purchasing of most goods he or she selects in major retail chains, and his/her decisions are influenced predominantly by price. However, changing consumer purchasing behavior in major retail chains leads to the assumption that wine 'culture' in general is evolving (Bernetti, Casini, and Marinelli 2006). Therefore, it becomes evident that strategies, activities, or measures that are taken to create bonds between consumer and company or brand gain increasingly in importance, incorporate strategies, and receive growing attention in research (Fournier 1998). Empirical evidence shows that pleasureful and memorable experiences demonstrably lead to consumer emotional attachments to (regional) brands, increased willingness to pay as well as consumer loyalty. Among others, the findings of Stöckl, Rinke, and Eisingerich (2016), Gross and Brown (2006), Hammitt, Backlund, and Bixler (2006), Hidalgo and Hernandez (2001), and Orth, Limon, and Rose (2010) suggest that affective consumer experiences such as the ones visitors and/or holidaymakers live through have the power to influence their attachments to targets prominently featured during these experiences.

Furthermore, wine tourism includes various relationships between wineries, wine regions, and the visitor/consumer that have to be considered. While it is known that the wineries' offer and the appeal of the destination to be visited are the main motives for participating in wine tourism, further aspects are needed to understand the key elements of the wine tourism experiences and to develop comprehensive packages: The characteristics of the tourist (e.g. demographics, wine consumption, consumer behaviour), the characteristics of the wine region (e.g., geographic location, brand image, cellar door profiles, wine products, landscape, attractions, accessibility, proximity, infrastructure, climate, economic development), and the visit dynamic (first time or repeat visit) must be evaluated before developing wine tourism strategies. Therefore, it is necessary to determine the elements that motivate a tourist to participate in wine tourism as well as potential barriers or restrictions to take part in wine tourism activities (Marzo-Navarro and Pedraja-Iglesias 2012, p. 314).

Correia et al. (2019, p. 206) identified many aspects influencing wine tourism experiences, including wine quality, regional gastronomy, service provided to the tourists, variety of events and activities, accommodation, architecture and heritage, staff, setting/surroundings, and winery attributes. Furthermore, they specifically mention the aspect of the marketing of the wine region, as it influences positioning and differentiation of the wine tourism destination in the competitive business environment.

Koch, Martin, and Nash (2013) propose that close cooperation between tourism organizations and vintners as well as knowledge of wine tourism as well as marketing is needed to increase wine sales and brand loyalty. They claim that successful wine tourism marketing strategies need to include customer segmentation, e-business marketing as well as networking. For this, market research is needed to understand all aspects related to visitors' winery experiences (Koch, Martin, and Nash 2013).

Strategic alliances and collective action are therefore to be seen as further success factors in marketing a wine tourism destination. Wine routes, culinary events, and itineraries which connect wineries, agritourist farms, and restaurants can yield these synergistic effects (Wargenau and Che 2006, p. 47). For this, Sigala (2019a, p. 99) proposes the idea of coopetition which allows competing businesses to cooperate for generating competitive value at an individual as well as a collective level.

Marketing destinations based on their wine offer: three Austrian case studies

The success of wine tourism in Austria is based on initiatives coming from three different levels: (1) the national service body, (2) regional wine committees, and (3) at the micro-level of individual businesses.

Case Study 25.1 Austrian Wine Marketing Board (AWMB)

The Austrian Wine Marketing Board (Österreich Wein Marketing GmbH) is the national service body for the Austrian wine industry. The organization was founded in 1986 with the aim to strategically support, coordinate, and maintain quality and sales. The main goal for the domestic market is to strive for an increased market share of quality wine, along with the export of bottled quality wine, with a particular focus on the continued increase in value. Besides, the AWMB aims at promoting the origin and wine-growing regions in close cooperation with regional wine committees. The revenues of the AWMB amount annually to approximately €4 million in marketing contributions from the wine industry (1.1 cents per liter of harvested wine and 1.1 cents per liter of bottled wine placed on the market). Additionally, it receives about €3.5 million from the federal provinces of Lower Austria, Burgenland, Styria, and Vienna. Finally, the AWMB receives funding from the European Union (AWMB 2020c).

While explicit wine tourism is still a niche in Austria (approximately 5% of all tourists), a clear trend towards traveling to Austrian wine and culinary tourism destinations can be seen (AWMB 2020d). Approximately 30% of all tourists in Austria are interested in wine and culinary activities. This target group is especially attractive to marketers, as these culinary and wine tourists show significantly higher tourism spending compared to travelers in general. This value-added is generated directly in the wine-growing regions – at winemakers, restaurants, and accommodation providers – and thus creates considerable value for the respective regions (AWMB 2020a).

The AWMB, as well as the Austrian National Tourist Office (ANTO), therefore, set a clear focus on marketing Austria's wine tourism offer (ANTO 2020). In 2016, Austria was even awarded with the Travvy Award as the Best Wine Destination in Europe (AWMB 2020b). Approximately 5% of all travelers to Austria can be characterized as wine and culinary travelers, whereas the vast majority of travelers come from Austria itself (43%) and Germany (27%). On average, they spend €196 per day, which is slightly above the average of all travelers. The main decision factor for their holiday destination is the specific wine and culinary tourism offer at the chosen destination (ANTO 2019).

After a period of massive trade and travel restrictions in spring 2020, the aim of the AWMB was to support trade with a special focus on gastronomy on the one hand and wine tourism on the other. Because of the COVID-19 lockdown in spring 2020, the main distribution channels of Austrian vintners suddenly collapsed. This situation prompted the AWBM to initiate the wine tourism campaign called '*Auf zum Wein*' (literally translated as '*On to the wine*'). The AWMB assumed that summer tourism would not recover quickly and, therefore, a high need for short vacations and nature-based trips to destinations in the near proximity of guests would arise. The idea was to inspire these domestic travelers to visit the Austrian wine-growing regions. Consequently, the *Auf zum Wein* campaign aims to bundle and promote wine tourism events in all Austrian wine-growing regions. Some examples of advertisement motifs can be seen in Figure 25.1. The ultimate goal of this campaign was to increase direct ex cellar door sales and to increase value-added in the wine-growing destinations (AWMB 2020a).

The main sales channel via restaurants suffered the most; approximately 23 million liters of Austrian wine could not be sold because of the complete gastronomy industry lockdown. Therefore, the campaign tried to compensate for this loss and to encourage tourists to travel to wine regions in order to discover regional wines and culinary delights. The main communication of this campaign is the campaign's website (www.aufzumwein.at), which presents the wine tourism offer of more than 1,800 wineries coming from all 17 Austrian

Figure 25.1 Advertisement motifs.

Source: AWMB (2020a)

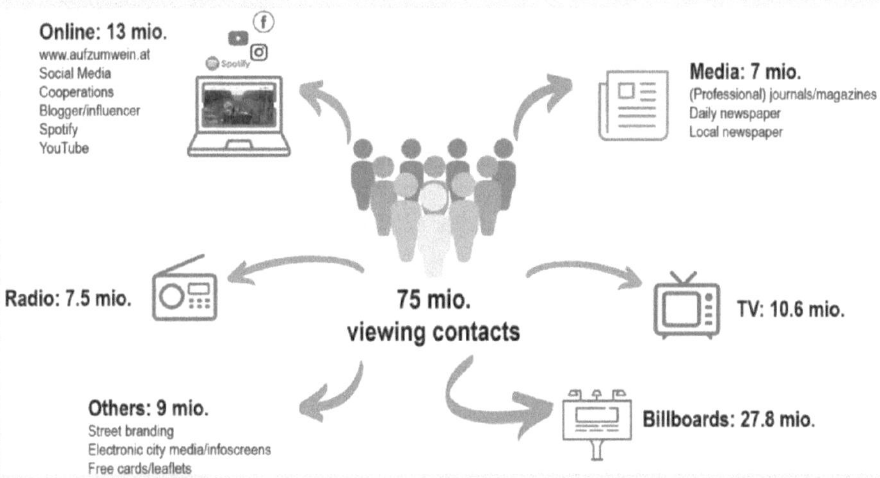

Figure 25.2 Communication channels used for the *Auf zum Wein* campaign and resulting viewing contacts.

Source: AWMB (2020b)

wine-growing regions. The website provides an overview of accommodation, wine-related events, wine taverns, and wine routes in the region, information related to ex cellar door sales, as well as other culinary activities. Besides the online platform, printed advertisements and billboard campaigns as well as radio and video spots were launched. In two neighboring countries of Austria (Germany and Switzerland), the campaign was launched on different online platforms as well. With more than 75 million viewing contacts, this campaign is the largest in the AWMB's history (AWMB 2020a). Figure 25.2 summarizes the communication channels used for this campaign.

Case Study 25.2 Wein Burgenland

Founded in 2004, "Wein Burgenland" is the official wine marketing organization of the Austrian federal province of Burgenland. The organization supports and coordinates the strategic efforts of the Burgenland wine trade and industry. Its board of directors is comprised of different representatives coming from the Regional Wine Committee of Burgenland, the Chamber of Agriculture, the Chamber of Commerce, and representatives of the agricultural policy institutions in the province. The owners are the Chamber of Agriculture, the Chamber of Commerce, the Province of Burgenland, and the Regional Wine Committee (each 25%) (Wein Burgenland 2020a).

The key objectives of the association are to raise awareness for Burgenland wines both domestically and abroad and to promote sales, as well as to consolidate all activities related to Burgenland wines, including the development of new potential and products. The ultimate goal for the domestic market is to be represented in all segments for quality wine. The main efforts on the international markets are concentrated on Germany, Switzerland, the United States, and Slovakia. In a few years, the Nordic countries as well as Russia, Poland, and the Asian market are to be considered as well. All other relevant markets are covered with the support of the Austrian Wine Marketing Board (Wein Burgenland 2020a).

Strategic marketing efforts attempt to tailor a specific marketing mix for each respective target market. A clear focus is thereby set on communication efforts, which include image campaigns, sales promotion, public relations, and supportive media advertising. The marketing activities of Wein Burgenland include the implementation of operational measures to increase the added value of Burgenland wines, e.g., by advertising in media, participating in trade fairs and exhibitions, and organizing special events. Furthermore, it is a clear objective to establish cooperation and networks as well as to find synergy effects when marketing Burgenland wines. Finally, the organization's activities cover market research as well as the implementation of marketing activities in cooperation with the Austrian Wine Marketing Board (Wein Burgenland 2020a).

Finally, the organization cooperates closely with other Burgenland institutions such as the official tourism destination management organization 'Burgenland Tourismus,' the agricultural marketing initiative 'Genuss Burgenland,' and other businesses and initiatives marketing culinary and cultural activities in the province (Wein Burgenland 2020a).

The initiative '*Martiniloben*' is an event that takes place on the weekends before and after November 11, which is the public holiday of Saint Martin, the patron of Burgenland. Vintners from numerous wine-growing regions open their cellars and invite their guests to wine tastings and try culinary delicacies. The event is accompanied by art exhibitions in press houses and cellars, guided tours and excursions, and culinary events (Wein Burgenland 2020b).

The idea of *Martiniloben* came up in 1987, where young wines were officially baptized for the first time in the local basilica of *Frauenkirchen*, a main wine spot in Burgenland. The first official *Martiniloben* took place in 1989. At first, the idea was to celebrate *Martiniloben* in a different town each year. However, the event received enormous interest from locals as well as guests in the first years. More and more towns took up this idea, which resulted in an event on a yearly basis taking place in several towns simultaneously. In 2019, 24 towns welcomed their guests to *Martiniloben*. The main concept of the event is centered on young wine tastings which are traditionally presented for the first time during *Martiniloben* (Neusiedler See Tourismus GmbH 2020).

Today, *Martiniloben* builds the core of all wine tourism marketing activities of the region as well as local wine-growing associations each autumn. In 2019, the event generated more than 43,000 visitors in the region. Meanwhile, ancillary services such as the so-called 'Martini' buses bringing the guests to the different towns are well established. Related marketing activities include media cooperation, PR, press trips, newsletters, and advertising media such as brochures and leaflets (Neusiedler See Tourismus GmbH 2020).

Case Study 25.3 Wine experience center 'SANDGRUBE 13 wein.sinn'

Winzer Krems eG is a winegrowers' cooperative based in Krems, Lower Austria. The valley of the small river Krems, which flows into the Danube, is one of Austria's viticultural centers (Winzer Krems 2021a). The Winzer Krems, meaning 'Vintners of Krems,' have the legendary address 'Sandgrube 13' – literally, 'sandpit,' which gives an idea of the soil type in that part of the valley – known to most Austrian and German wine enthusiasts through radio advertising and TV spots. The producer association has 90 employees and processes grapes from around 900 member winegrowers from a contract vineyard area of 1,200 hectares. Fifty percent of the wine is exported, and 50% is marketed domestically through the business segments of food trade, gastronomy in own distribution, wine and beverage trade, and private customers. The annual turnover is currently around €27 million (Stephan Nessl – Head of Marketing Winzer Krems, email communication, February 8, 2021).

At Winzer Krems, groups of customers had been guided through the winery for many years, but without a clear concept. This was no longer compatible with the upcoming IFS certification of the company in the early 2000s, which imposes strict rules for food (in this case, drink) security and safety. Until the end of the 1990s, guided tours with wine tastings were offered for bus groups at Kloster UND, a former monastery in Krems/Stein. Due to the sale of the ancient building of the Kloster UND, this way of presenting wines to locals and tourists no longer existed. A market niche became obvious, and the Winzer Krems management reacted. In the following years an entire concept was developed and €12.5 million were invested in adaptions of existing production areas as well as new cellar buildings, including a wine experience center (Stephan Nessl – Head of Marketing Winzer Krems, email communication, February 8, 2021).

With the aforementioned market gap in mind as well as with a new cellar building already in the planning stage, it was decided to invest in 're-inventing' cellar door sales. The main aim was to create an attractive experience for the visitor in order to strengthen brand awareness and brand equity. Consequently, the goal was to create conceptually planned tours, mainly for groups but also for individual visitors. The objectives were to create an authentic wine experience consistent with the quality and market 'philosophy' of the Winzer Krems, to enhance the wine tourism attractiveness of the entire region, to enhance consumer knowledge and understanding of wine by offering tours with experience and also fun character, to arouse enthusiasm for wine among 'average' consumers, to increase customer loyalty, to augment cellar door sales, to increase brand awareness and image, to create a spatial separation of production and visitor areas in the sense of IFS and, last but not least, to bring additional jobs into the rural area (Stephan Nessl – Head of Marketing Winzer Krems, email communication, February 8, 2021).

The creation of *SANDGRUBE 13 wein.sinn* started with the decision of Managing Director Franz Ehrenleitner and Marketing Manager Stephan Nessl to establish a wine experience center. Further milestones included study trips to the United States, Germany, France, Italy, Spain, Portugal, and other Austrian regions, consulting with various

Figure 25.3 Wine cellar lane 'Vinothekkeller Bründlgraben' for events, feasts, and weddings.

Source: Winzer Krems (2021c)

recognized experts in the field of tourism and architecture as well as universities. In 2002, flood waters of the Kamp River destroyed parts of the external grape takeover and processing station in Hadersdorf, which slowed down the process. On the other hand, it led to the idea to combine a new grape takeover facility and a new fermentation cellar with the planned wine experience and visitor center. Construction took place between May 2003 to July 2005. The grand opening was celebrated in August 2005. In 2014, another milestone was achieved: the opening of the entirely revitalized wine cellar lane – the Bründlgraben, depicted in Figure 25.3 – located right behind the wine experience center as an authentic event venue took place (Stephan Nessl – Head of Marketing Winzer Krems, email communication, February 8, 2021).

The guided *wein.sinn* tour through parts of the winery comprises eight stations with three wine tastings in different places and settings. To accompany the wine, the classic and very delicious bread roll of the area called *Wachauer Laberl* is offered. The eight stations of the tour with explanations of the guide include: (1) *wine.garden*, a wine tasting of Austria's signature grape varietal Grüner Veltliner in the vineyard with information on vineyard work and history; (2) *wine.region*, a 64 m^2 walk-in aerial view of the region, a winegrowers' board, and soil profiles (see Figure 25.4); (3) *wine.treasure*, a light-sound-fog-laser-staged historical bottle cellar; (4) *wine.cellar*, a wine tasting of Blauer Zweigelt in the barrique cellar with a view of the stainless steel fermentation and maturing cellar (see Figure 25.5); (5) *wine.film*, a 4D film (3D + scent and wind); (6) *wine.art*, a video installation; (7) *wine.gallery*, a wine tasting in the Riesling-brand world of WINZER KREMS – the 'living room

Figure 25.4 Station wine.region – 'walk-on map' of the Krems wine region.

Source: Winzer Krems (2021b)

Figure 25.5 Station wine.cellar – art and wine tasting in the barrique cellar wine-cathedral.

Source: Winzer Krems (2021b)

of the winemaker'; and (8) *wine.purchase*, a cellar door sale including the additional possibility of tasting (a lot of different) wines (Winzer Krems 2021b).

Altogether, the eight stations create an exciting mix of information and interpersonal interaction provided by the guide on the one hand, and funny or exciting multimedia elements on the other. The light, sound, and laser show, which takes place in the old oak-cask cellar, entertains visitors of all age groups and wine lovers of all knowledge and involvement levels; the 4D cinema is exciting, informative, and good fun; while the art installation in the huge modern wine hall, which resembles an under-earth 'wine cathedral,' is aesthetic and also impressive due to its size. Wine cellar lane revitalizations were also continued after the aforementioned 'milestones.' Today, the 'Vinothekkeller Bründlgraben' comprises a total of seven historic press houses. One after the other and bit by bit, they were acquired, renovated, and converted into a venue for celebrations, weddings, and tastings. Today, they form one of the most carefully renovated and beautiful wine cellar ensembles in Lower Austria. Unseen from the outside, all cellars are connected within. Five hundred wine compartments were built in the cellar vaults and rented out to private customers and companies for the storage of Winzer Krems wines under optimal conditions (Winzer Krems 2021c).

For groups, guided tours are organized upon appointment. Although the 4D cinema offers 48 seats, the management always seeks to maintain the personal contact between guides and visitors. Thus, the group size must not exceed 48 people. Bigger groups are divided into two or more smaller units to make sure questions can be asked and answered and a familiar atmosphere is established. For individual guests guided tours take place without any reservation required between 10 a.m. and 2 p.m. Sandgrube 13 is open daily from May to October; from November to April they are closed on Sundays and public holidays. The majority of visitors come with a group (approximately 60%), about 40% are individual visitors. A high proportion of individual guests possess a Niederösterreich-Card – a benefit card for the federal state of Lower Austria. Roughly 75% of visitors are from Austria, the rest are mostly Danube cruise ship passengers and mainly from the United States (Stephan Nessl – Head of Marketing Winzer Krems, email communication, February 8, 2021).

As of this writing, the success story of the wine experience world *wein.sinn* resulted in an increase of cellar door sales by 63% as well as 411,000 visitors since its opening (Stephan Nessl – Head of Marketing Winzer Krems, email communication, February 8, 2021).

Summary, discussion, and prospects

Marketing destinations through wine tourism have earned academic attention, achieved practical relevance, and shown impressive progress since the 1990s. Since then, especially the wine tourism destinations of the 'Old World' find themselves in a worldwide competition (Getz and Brown 2006; Williams 2001a). This global competitive situation was, above all, triggered by geopolitical incidents such as the fall of the iron curtain and the expansion of the EU, which enables Europe's citizens to access and visit any neighboring country with ease. Also, a general increase of leisure time throughout the 'Western world,' cheap travel by car or plane, the development of inexpensive ways to market destinations via the internet, a trend toward (more) short-haul trips as well as an overall rise in wine interest and 'culture' favored the development of wine tourism. Using the example of growing wine tourism in Austria allowed the authors to showcase initiatives on three levels:

- A nationwide strategy developed and launched by the national wine marketing board *AWMB* (Austrian Wine Marketing Board)
- Regional initiatives bundled by the wine marketing board of one specific area (in this example the wine region of Burgenland)
- An entrepreneurial venture from one of Austria's biggest cooperatives – a wine experience center called *wein.sinn*.

The insights gained by presenting and analyzing the three cases suggest that successful initiatives and campaigns in a wine destination marketing context can be found in all featured levels. It appears that managers can enhance the potential effects of positive tourism experiences by gaining high levels of brand strength and equity for their destination and its wine brands. Such a view is consistent with theoretical works by Keller (2003) and others who suggest that consumer attachment is multidimensional. These insights signify, among other aspects, that even small and internationally unknown wine regions have the chance to create additional value for visitors by 'enriching' holiday offers with wine experiences, and the other way round, create brand awareness and customer loyalty for locally produced wine brands when pleasureful holidays were spent in the region. Furthermore, the cases highlight the importance of close cooperation between tourism organizations and destinations as well as vintners and wine promoting bodies e.g., regional marketing boards.

The COVID-19 pandemic – which as of this writing (in 2020 and 2021) afflicts all countries in the world, but especially Europe and the United States – meant an abrupt end to the freedom of travel. The measures taken to reduce COVID's fast spread by governments across the world brought air travel to a halt. Vienna Airport reported a drop of between 60% and 98% in passenger volumes (Vienna Airport 2020), depending on which months and/or destinations are considered. Between March 2020 and January 2021, restaurants, inns, (wine) bars, taverns, and hotels in Austria were open for only a couple of weeks. Despite all these limitations, the Austrian Wine Marketing Board launched its campaign *On to the wine* (*Auf zum Wein*) in spring 2020 and ended up being right. Mostly unable to travel elsewhere, many Austrians stayed in the country. Therefore, the wine regions close to Vienna in Lower Austria, Burgenland, and Styria were not as severely affected by COVID limitations as others. In ordinary times, these areas are definitely not the tourism hot spots of the country, as the alpine regions Tyrol, Salzburg, and Vorarlberg in the west of Austria have far more incoming tourists and overnight stays than the rural areas in the east, where all wine regions are located (Tischler and Stöckl 2015).

In summer 2019, about 1.1 million beds in 65,000 accommodation establishments were available for guests in Austria. Usually, spring, summer, and autumn are not the peak season in Austria. Nevertheless, 78.93 million overnight stays in Austria were registered in the period from May to October 2019, more than half of which were in the alpine regions in the west (Statistik Austria 2020). COVID-19 and its side effects may change Austria's and other European countries' tourism landscape strongly. Where mass tourism generated most income – sea and beach, mountains, city trips – we saw a massive decline during COVID months. Restrictions and limitations might last longer than everybody in the industry is hoping.

We might see a strong and lasting shift towards holidays within consumers' own countries. Remote and internationally rather unknown regions could thus even see increasing numbers of visitors in the years to come. Citizens of densely populated areas in Europe – the countries' capitals and/or formerly intensively industrialized zones – are especially likely to seek alternatives for their former holiday destinations. Nature-based tourism in rural areas might therefore offer high potential in the future. Local/regional culinary delights and wine experiences as an additional attracting and motivating factor could rise further in importance. The regionality trend, which in the 2000s involved a countermovement to food globalization and industrialization, might additionally trigger this development. To conclude, marketing destinations through wine experiences may therefore become an 'accessory' of relevance in the future.

References

Austrian National Tourism Office (ANTO). (2019). *Wein- und Kulinarikurlauber in Österreich*. https://www.austriatourism.com/tourismusforschung/studien-und-berichte/wein-und-kulinarikurlauber-in-oesterreich-gesamtjahr-201819/

Austrian National Tourism Office (ANTO). (2020). *Weintourismus*. https://www.austriatourism.com/blog/bulletin-texte-2017/oktober-november-2017/weintourismus/

Austrian Wine Marketing Board (AWMB). (2020a). *'Auf zum Wein': Österreich bringt den Urlaub zu den Winzern*. https://www.oesterreichwein.at/presse-multimedia/pressetexte/news-1-1/article/auf-zum-wein-oewm-bringt-den-urlaub-zu-den-winzern

Austrian Wine Marketing Board (AWMB). (2020b). *'Auf zum Wein': Ein Rückblick*. https://www.oesterreichwein.at/presse-multimedia/pressetexte/news-1/article/auf-zum-wein-ein-rueckblick

Austrian Wine Marketing Board (AWMB). (2020c). *Austrian Wine Marketing Board*. https://www.austrianwine.com/awmb-partners/austrian-wine-marketing-board

Austrian Wine Marketing Board (AWMB). (2020d). *Weintourismus*. https://www.oesterreichwein.at/unser-wein/weintourismus

Austrian Wine Marketing Board (AWMB). (2022). *History*. http://www.austrianwine.com/our-wine/a-taste-of-culture-a-special-treat/history/

Bernetti, I., Casini, L., and Marinelli, N. (2006). Wine and globalisation: changes in the international market structure and the position of Italy. *British Food Journal*, 108(4): 306–315.

Byrd, E. T., Canziani, B., Hsieh, Y.-C., Debbage, K., and Sonmez, S. (2016). Wine tourism: Motivating visitors through core and supplementary services. *Tourism Management*, 52: 19–29.

Bruwer, J., and Alant, K. (2009). The hedonic nature of wine tourism consumption: an experiential view. *International Journal of Wine Business Research*, 21(3): 235–257.

Cohen, E., and Ben-Nun, L. (2008). The important dimensions of wine tourism experience from potential visitors' perception. *Tourism and Hospitality Research*, 9(1): 20–31.

Correia, A. I., Cunha, R., Matos, O., and Fernandes, C. (2019). Wine tourism experiences and marketing: the case of the Douro Valley in Portugal. In Sigala, M., and Robinson, R. N. S. (eds.), *Wine Tourism Destination Management and Marketing – Theory and Cases* (pp. 203–220). Cham: Palgrave/Macmillan.

Croce, E., and Perri, G. (2010). *Food and Wine Tourism: Integrating Food, Travel and Territory*. Wallingford: CABI.

Dixit, S. K. (ed.). (2020). *The Routledge Handbook of Tourism Experience Management and Marketing*. London: Routledge.

Everett, S. (2016). *Food & Drink Tourism*. London: Sage.

Fournier, S. (1998). Consumers and their brands: developing relationship theory in consumer research. *Journal of Consumer Research*, 24(4): 343–373.

Getz, D., and Brown, G. (2006). Critical success factors for wine tourism regions: a demand analysis. *Tourism Management*, 27: 146–158.

Gomez, M., Lopez, C., and Molina, A. (2015). A model of tourism destination brand equity: the case of wine tourism destinations in Spain. *Tourism Management*, 51: 210–222.

Gross, M. J., and Brown, G. (2006). Tourism experiences in a lifestyle destination setting: the roles of involvement and place attachment. *Journal of Business Research*, 59: 696–700.

Hammitt, W. E., Backlund, E. A., and Bixler, R. D. (2006). Place bonding for recreation places: conceptual and empirical development. *Leisure Studies*, 25(1): 17–41.

Hidalgo, C., and Hernandez, B. (2001). Place attachment: conceptual and empirical questions. *Journal of Environmental Psychology*, 21: 273–281.

Hudson, S., and Ritchie, J. R. (2009). Branding a memorable destination experience: the case of 'Brand Canada.' *International Journal of Tourism Research*, 11: 217–228.

Keller, K. L. (2003). Brand synthesis: the multidimensionality of brand knowledge. *Journal of Consumer Research*, 29(4): 595–600.

Knoll, R. (2000). *Die großen Weine Österreichs*. München: Heyne.

Koch, J., Martin, A., and Nash, R. (2013). Overview of perceptions of German wine tourism from the winery perspective. *International Journal of Wine Business Research*, 25(1): 50–74.

Marzo-Navarro, M., and Pedraja-Iglesias, M. (2012). Critical factors of wine tourism: incentives and barriers from the potential tourist's perspective. *International Journal of Contemporary Hospitality Management*, 24(2): 312–334.

Morgan, M. (2006). Making space for experiences. *Journal of Retail and Leisure Property*, 5(4): 305–313.

Neusiedler See Tourismus GmbH (2020). *Geschäftsbericht*. https://www.neusiedlersee.com/de/ueber-uns/geschaeftsbericht.html#:~:text=Die%20Neusiedler%20See%20Tourismus%20GmbH,den%20Sie%20hier%20einsehen%20k%C3%B6nnen.&text=Gesch%C3%A4ftsbericht%202020%20wird%20im%202,Quartal%202021%20ver%C3%B6ffentlicht (last accessed 13 November 2020).

Okech, R. N. (2016). Developing wine tourism: an exploratory study of wineries in Newfoundland. *Advances in Hospitality and Tourism Research (AHTR)*, 4(1): 70–82.

Orth, U. R., Limon, Y., and Rose, G. (2010). Store-evoked affect, personalities, and consumer emotional attachments to brands. *Journal of Business Research*, 63(11): 1202–1208.

Ritchie, J. R., and Crouch, G. I. (2003). *The Competitive Destination: A Sustainable Tourism Perspective*. Wallingford: CABI.

Robinson, J. (2015). *The Oxford Companion to Wine* (4th ed.). Oxford: Oxford University Press.

Saayman, M., and van der Merwe, A. (2014). Factors determining visitors' memorable wine-tasting experiences at wineries. *Anatolia: An International Journal of Tourism and Hospitality Research*, 26(3), 372–383.

Shaw, C., and Ivens, J. (2002). *Building Great Customer Experiences*. London: Palgrave/Macmillan.

Sigala, M. (2019a). Building a wine tourism destination through coopetition: the business model of ultimate winery experiences Australia. In Sigala, M., and Robinson, R. N. S. (eds.), *Wine Tourism Destination Management and Marketing – Theory and Cases* (pp. 99–112). Cham: Palgrave/Macmillan.

Sigala, M. (2019b). Designing experiences: developing and innovating wine destinations. In Sigala, M., and Robinson, R. N. S. (eds.), *Wine Tourism Destination Management and Marketing – Theory and Cases* (pp. 139–141). Cham: Palgrave/Macmillan.

Sigala, M., and Robinson, R. N. S. (2019). Introduction: the evolution of wine tourism business management. In Sigala, M., and Robinson, R. N. S. (eds.), *Management and Marketing of Wine Tourism Businesses* (pp. 1–21). Cham: Palgrave/Macmillan.

Soteriades, M. (2012). Tourism destination marketing: approaches improving effectiveness and efficiency. *Journal of Hospitality and Tourism Technology*, 3(2): 107–120.

Statistik Austria. (2020). *Tourism Summer Season 2019 Reached All-Time High with 79 Million Overnight Stays*. http://www.statistik.at/web_en/statistics/Economy/tourism/accommodation/arrivals_overnight_stays/123055.htmlon(last accessed 29 January 2021).

Steiner, D., and Seiler, C. (2005). *WeinArchitektur-WineArchitecture*. Ostfildern: Hatje Cantz.

Stöckl, A. F. (2006). *Austrian Wine: Developments after Wine Scandal of 1985 and Its Current Situations*. Presented at the 3rd International Wine Business Research Conference, July 6–8, Montpellier.

Stöckl, A. F., Rinke, W., and Eisingerich, A. B. (2016). Using ANNs to determine place evoked affective consumer reactions in wine tourism. In Kyuho, L. (ed.), *Strategic Winery Tourism and Management* (pp. 167–186). Oakville: CRC Press/Taylor & Francis Group.

Tischler, S., and Stöckl, A. (2015). *Kostbares Österreich – a project report publication for the Austrian Federal Ministry of Economics*. https://www.academia.edu/15008164/Kostbares_%C3%96sterreich

Vienna Airport. (2020). *News*. https://www.viennaairport.com/unternehmen/investor_relations/news?news_beitrag_id=1597719494556

Wargenau, A., and Che, D. (2006). Wine tourism development and marketing strategies in Southwest Michigan. *International Journal of Wine Marketing*, 18(1): 45–60.

Wein Burgenland. (2020a). *About Us*. http://weinburgenland.at/en/about-us/

Wein Burgenland. (2020b). *Martiniloben*. http://www.weinburgenland.at/wein-tourismus/martiniloben/

Williams, P. (2001a). Positioning wine tourism destinations: an image analysis. *International Journal of Wine Marketing*, 13(3): 42–58.

Williams, P. (2001b). The evolving images of wine tourism destinations. *Tourism Recreation Research*, 26(2): 3–10.

Winzer Krems. (2021a). *About Us*. https://www.winzerkrems.at/

Winzer Krems. (2021b). *Sandgube 13, wein.sinn*. https://www.sandgrube13.at/

Winzer Krems. (2021c). *Vinothekkeller Bründlgraben*. https://www.vinothekkeller-bruendlgraben.at/

26

THE ROLE OF WINE FESTIVALS IN SWEDISH WINE TOURISM MARKETING

Olga Rauhut Kompaniets

Introduction

Being a part of wine tourism experience, wine festivals play an important role in wine tourism marketing (Cagnina, Cicero, and Osti 2019; Dixit 2019; Yuan and Jang 2008). Wine festivals are the second-most valuable activity after vineyard visitation (Hall et al. 2000). These festivals stimulate wine tourism opportunities for destinations and vineyards, giving marketing and promotional benefits (Axelsen and Swan 2010). While these special events are popular in well-established wine countries and wine tourism destinations, new wine countries have just began to use festivals for their wine tourism marketing purpose.

The growing interest in different wine festivals and events taking place all over the world over the last two decades is generated by both vineyards and tourists, as well as researchers (Carlsen and Getz 2006). Research points out that it is even impossible to count how many wine festivals are being organized in Europe every year (Lee and Arcodia 2011; Santini and Cavicchi 2014). There are a few explanations for it. First, each organizer can decide how to promote the festival and how to inform target markets about it. Second, while some festivals have a hundred-year tradition, others are newbies trying to establish themselves in a wine tourism market. Wine festivals and events in established wine countries and destinations are well researched, while the research of wine festivals in new wine countries is very fragmented.

Sweden, the youngest and northernmost wine country, entered the EU wine market in 1999. Its wine sector has grown and commercialized since then. Currently, there are about 55 vineyards in Sweden's southernmost region, Scania. Several have been commercially successful in linking wine production with wine tourism and an emerging Swedish Wine Route (Rauhut Kompaniets and Nilson 2020). Factors such as grapes, climate, or technical issues regarding wine production actually cause few problems. The challenges to overcome lay in structural problems, e.g., a monopoly market due to the government alcohol retail monopoly and its legal constraints, but also a lack of marketing and promotion, as well as networking and cooperation with local enterprise (Rauhut Kompaniets and Nilson 2020; Säfwenberg 2019). The wine festivals, however, are considered to play an important role in Swedish wine tourism marketing (Skånes Matfestival 2020; Skånska drycker 2020; Vinfestival 2021).

This chapter aims at discussing the role of wine festivals in Swedish wine tourism marketing. Every year several regional wine festivals take place in Sweden; the most known is 'Taste

DOI: 10.4324/9781003143628-31

Swedish Wines' in Scania, organized by the local vineyards. A conceptual framework based upon tourism destination life cycle theory and Experience Economy theory explains the role wine festivals play in marketing wine tourism in southern Sweden. The empirical material consists of official statistics, existing research, policy documents, and other secondary data sources (e.g., reports, promotion materials, media coverage etc.). The material is analyzed by a content analysis methodology.

Conceptual framework

The *Tourism Destination Life Cycle* (TDLC) model correlates with the stages models of an organizational life cycle and marketing product life cycle. According to Butler (1980), tourism destination development goes through the following life cycle stages: exploration, involvement, development, consolidation, stagnation, and rejuvenation or decline (Butler 1980). The TDLC stages represent the life cycle of many contemporary festivals (Holmes and Ali-Knight 2017). However, research focusing on the correlation between festivals and TDLC is quite fragmented.

Wine events organizers have to recognize the importance of the marketing product life cycle in developing a wine event marketing strategy. While analyzing the evolution of wine festivals, researchers modified TDLC, creating the six stages of wine festival development such as conception, launch, growth, consolidation, decline, and revival (Beverland, Hoffman, and Rasmussen 2001). This wine festival life cycle model, following Butler's TDLC construct, proposes goals and focus for each stage of wine event development.

Based on the idea that every festival has its own goals and measurements of success, Holmes and Ali-Knight (2017) identified the key factors influencing a festival's life cycle, taking as a starting point the TDLC model. The authors then extended Butler's model by formulating seven trajectories of a festival development, such as accelerated development, continued growth, survival, redevelopment, hiatus, decline, and cancellation.

The *Experience Economy* model, introduced by Pine and Gilmore (1998), is successfully used for explaining event and festival marketing (Culha 2020). According to Pine and Gilmore (1998, p. 99) consumer experience is internal personal characteristics existing just in the mind of the individual consumer 'who have been engaged at an emotional, physical, intellectual, or even spiritual level.' Experience is an interaction between the event and the consumer's mind; it means that two persons will have a completely different experience from the same event.

Several studies conclude that the festival experience consists of four dimensions such as education, entertainment, escapism, and aesthetics. In other words, the experiential product stimulates consumers' 'multisensory connection to feel, learn, be, and do' (Mehmetoglu and Engen 2011, p. 243). These dimensions have to be considered while designing a festival program and activities (Culha 2020; Manthiou et al. 2014; Mehmetoglu and Engen 2011).

Wine festivals and wine tourism

Wine festivals are defined as special temporal events with a focus on locally produced wines with the main goal of marketing wine tourism in a specific wine destination (Carlsen and Getz 2006; Holmes and Ali-Knight 2017); and they are considered to attract visitors to local vineyards and build relations between vineyards, wineries, wine tourism actors, and destination stakeholders (Alonso and Bressan 2013). Using a cultural learning component – i.e., when during the festival wine tourists not only get acquainted with a wine but also learn more about the wine destination, traditions, and culture – wine festivals contribute to sustainable wine tourism development of the destination (Bruwer and Kelley 2015; Velikova, Slevitch, and Mathe-Soulek 2017).

Wine tourism destinations use wine festivals and events as a promotion tool, and research points out that successful wine festivals can positively influence vineyard visitations (Cagnina, Cicero, and Osti 2019; Galvez, Fernandez, and Lopez-Guzman 2015; Yuan and Jang 2008). Wine events are considered as an inspiring occasion for tourists and locals to visit a wine destination (Dawson et al. 2011). However, not just wine-related reasons motivate tourists and locals to attend wine festivals. Among other motives, previous research identifies entertainment (Carlsen and Getz 2006; Yuan et al. 2005), socialization component (Charters and Ali-Knight 2002), unique experience (Hubbard et al. 2012), and educational purpose (Alonso and Bressan 2013; Kruger, Rootenberg, and Ellis 2013). Moreover, interaction with wine professionals and learning opportunities became the main motivation for consumers to attend wine festivals (Velikova, Slevitch, and Mathe-Soulek 2017).

Growing consumer interest in attendance of wine festivals gives a marketing tool to vineyards and wine destinations in hosting festivals and leads to a rapid increase of the number of wine festivals organized all around the world as well as an amount of research in the wine festival field (Carlsen and Getz 2006). Wine festivals and wine tourism research show significant *geographical patterns* (Rauhut Kompaniets and Nilson 2019). New World wine countries, especially Australia and New Zealand, dominate in research (Axelsen and Swan 2010; Beverland, Hoffman, and Rasmussen 2001; Carlsen and Getz 2006; Getz 2000; Staggs and Brenner 2019). Also, among New World wine countries, the United States, with its wine festivals both in established and emerging wine regions, is presented (Hubbard et al. 2012; Velikova, Slevitch, and Mathe-Soulek 2017) as well as South African wine festival research (Marais and Saayman 2011). Wine festivals and similar wine events have also been surveyed in Old World wine countries like Italy (Alonso and Bressan 2013; Novelli 2003), Spain (Galvez, Fernandez, and Lopez-Guzman 2015; Gómez and Molina 2012), and France (Frochot 2000). In addition, wine events in Serbia (Jević, Jević, and Barović 2016), Hungary (Szabo, Komaromi-Gergely, and Szeles 2014), and Bulgaria (Dimitrova and Yoveva 2014) have been surveyed in research.

Besides geographical clustering, wine festival research can be defined into the following categories: (1) visitor attraction, experience, and satisfaction; (2) stakeholder perspective; (3) cooperation between wine tourists and vineyards, as well as between residents and local vineyards; and (4) wine festivals as a source for destination identity and branding.

Visitor attraction, satisfaction, and experience of the event are the most common purposes of wine festival research. Frochot (2000) argues that French wine festivals often target local and domestic markets. Carlsen and Getz (2006) conclude that wine festivals are mostly missing international focus, attracting just local visitors. On the other hand, research also shows a valuable contribution of wine festivals in attracting international tourists to the wine destination (Dimitrova and Yoveva 2014). Festival location divides wine festivals into urban and rural events with different target groups, goals, tactics and programs (Novelli 2003; Vitale et al. 2018). While urban festivals predominantly attract international wine tourists, rural events have locals as a target group utilizing wine as a differentiation tool for the destination (Staggs and Brenner 2019). Moreover, urban wine festivals, due to the venue, size, and economic impact, are of interest to wine tourism researchers (Taylor 2006).

Visitor satisfaction with an event is considered as a loyalty-building element leading to future consumer preferences of local wines as well as intention to revisit (Axelsen and Swan 2010; Yuan and Jang 2008). It also encourages positive word-of-mouth when attendees recommend and influence others' decision to visit wine destinations and to taste wines (Hubbard et al. 2012). Visitor satisfaction can be measured by wine festival attributes. While analyzing wine festivals in an emerging wine region in the United States, Velikova, Slevitch, and Mathe-Soulek (2017) concluded that festival attributes differ from event to event and differ between emerging wine regions and established wine countries. Moreover, the festival's atmosphere and different as well as new experiences play a role as sources of positive consumer satisfaction (Axelsen and Swan 2010), as do the organizational attributes of the festival such as location, facilities, and number of participated vineyards (Yuan and Jang 2008). Furthermore, facility-oriented attributes of visitor satisfaction, such as location, admission, parking, friendliness, music, food and tasting, education, and activities for children, matter as well to visitor satisfaction (McDowall 2011).

As wine is considered an experiential product (Storchmann 2011), visitor *experience* becomes the key factor of wine festivals (Culha 2020). Festival experience consists of four dimensions: education, entertainment, escapism, and aesthetics (Manthiou et al. 2014; Mehmetoglu and Engen 2011). Hence, the experiential product stimulates consumers' multisensory connection to feel, learn, be, and do (Mehmetoglu and Engen 2011).

Stakeholder perspective and benefits from wine festivals are widely discussed in research, combining different points of view on the importance of wine festival goals and strategy for varied groups of stakeholders (Hede 2008; Santini and Cavicchi 2014). Additional wine sales during the festivals are common and appreciated by both visitors and vineyards and helps festivals create profit and develop the festival experience (Beverland, Hoffman, and Rasmussen 2001; Carlsen and Getz 2006; Dixit 2020; Galvez, Fernandez, and Lopez-Guzman 2015). Cooperation between different destination stakeholders is the key to success (Santini and Cavicchi 2014).

Destination marketing organizations (DMOs) worldwide recognized the marketing opportunities wine festivals provide to destination development (Galvez, Fernandez, and Lopez-Guzman 2015). Building relations and cooperation between wine tourists and vineyards, as well as between residents and local vineyards, is another component of a successful wine festival (Getz 2019; Houghton 2008; Hubbard et al. 2012). The cooperation can have a different way from matching interests of tourists and vineyards through attraction, satisfaction, and experience while organizing a wine festival (Getz 2019; Houghton 2008), providing a partnership between locals and vineyards focusing on an educational experience of attendees (Hubbard et al. 2012).

As a worldwide practice, wine festivals are a common way for off-season tourist attractions, balancing high and low tourist seasons (Getz and Page 2016). Common in research is to use events as an off-season activity attracting locals as well as domestic and international tourists.

Wine festivals can also be seen as a source for destination image and identity. Wine events offer a destination image-making opportunity (Getz 2000; Gómez and Molina 2012), an identity creation based on a sense of belonging and community pride (Kim 2015; Hubbard et al. 2012), and a place identity based on positive emotions created by the unique event (Staggs and Brenner 2019). Wine and food events not only contribute to destination brand building but also serve as drivers of destination rebranding (Dimitrova and Yoveva 2014; Lee and Arcodia 2011).

Wine tourism in Sweden

Since 1999 when Sweden officially became a wine-producing country with a quota of 250,000 liters of wine per annum (Skjöldebrand 2010), the Swedish wine sector has grown and commercialized, transforming from a 'pensioners hobby occupation' to the professional award-winning sector (Henley 2020; Säfwenberg 2019). Most of the Swedish vineyards are located in Scania, the southernmost region of Sweden. Up to the autumn of 2020, 55 vineyards officially operated in the region, with 34 of them active in the emerging regional wine tourism sector (Rauhut Kompaniets and Nilson 2020). Additionally, 200 hobby vineyards are ready to commercialize (Hultgren Karell 2015; Livsmedelsföretagen 2019). Therefore, hosting most of the commercial Swedish vineyards, Scania is considered 'the Vineyard of Sweden' (Föreningen Svenskt Vin 2019; Säfwenberg 2019).

Scania, with its 1.3 million inhabitants, is the third-most visited destination in Sweden. It is a popular summer destination known for its cultural as well as nature tourism activities (Rauhut Kompaniets and Nilson 2020; Sjöberg 2019). The region is also famous for its local culinary culture; *The New York Times* and *Vogue* praise products from numerous organic farms in Scania (Abend 2016; Wergeland 2017), as well as Swedish wine (Henley 2020; Hultgren Karell 2015). Wine tourism plays an important role for Scania as a factor for regional wine sector development. As domestic wine production is still relatively unknown to locals, wine tourism marketing and promotion activities are needed for local and international tourist attractions.

Scanian vineyards offer several wine tourism activities (Rauhut Kompaniets and Nilson 2019), such as vineyard visitations, guided tours, wine tasting, wine festivals, and events, biking gourmet tours, and vineyard camping. Moreover, Scanian wine tourism combines historical, cultural, and sports activities with wine (Säfwenberg 2019).

The Scanian wine tourism sector faces structural challenges. Sweden is a wine country without any wine region (Säfwenberg 2019; Hultgren Karell 2015), i.e., all wines classify as *table wine*. Furthermore, the government alcohol retail monopoly remains in Sweden, with all bureaucratic restrictions and permits, limiting the wine tourism sector development (Malm, Gössling, and Hall 2013; Rytkönen 2012; Skjöldebrand 2010). Internationally, cellar door sales and festival wine sales complement the tourist experience (Carlsen and Getz 2006), but it is not permitted in Sweden. Moreover, vineyards have to get special permission to serve wine and even to offer wine tasting. To buy a bottle of local wine, wine tourists have to visit the closest to the vineyard *Systembolaget* (government alcohol retail store). Meanwhile, Swedish alcohol law allows serving local wine at hotels and restaurants, so the local hospitality and tourism sector actively helps to promote local wine tourism (Livsmedelsföretagen 2019; Näringsdepartementet 2017). The first wine bar specializing in Swedish wine was opened in Scania in April 2020 (Swedish Wine Center 2020). Scania also developed its Wine Route in 2017, with a list of 21 vineyards offering guided tours and wine tasting (Vinvägen 2020). Local wine festivals and events become a unique opportunity for marketing and promotion of the Swedish wine tourism sector.

Case Study 26.1 Wine festival *Taste Swedish Wines*

Wine festivals and events actively contribute to the Swedish wine tourism experience. Due to the growing interest in Swedish wine, some local festivals were created in Scania. At the wine festival *Taste Swedish Wines*, visitors can meet wine experts, learn about local wine production and taste various local wines (Säfwenberg 2019; Vinfestival 2021; Vinvägen 2020). Scania also hosts gastronomy and beverage events, with participation opportunities for Swedish wineries (Rauhut Kompaniets and Nilson 2020) such as the Scanian Beverages Festival in Malmö (Skånska drycker 2020) and the Scanian Gastronomy Festival in Landskrona (Skånes Matfestival 2020).

Taste Swedish Wines is a popular summer experience for local and international visitors. The festival was established in 2016 as an annual two-day event. It takes place at the Skepparps vineyard during the peak of tourism season, the last Friday and Saturday of July.

In 2019, almost 4,000 people visited the festival facilities. According to the organizers, the interest to local wines and Scania as a wine destination is growing; the festival has become a tradition, with attendees revisiting the event every year. Visitors also stay longer at the festival facilities, enjoying the program and socializing with others (Vinfestival 2021).

The wine festival is organized by local vineyards. In 2019 nine vineyards participated in the festival with 40 different wines presented, including white, red, rosé, orange, and sparkling wines. Unfortunately, in July 2020, the festival was canceled due to COVID-19. During the last weekend of July 2021, the fifth wine festival took place, with nine vineyards participating and presenting more than 40 wines. As of this writing, the sixth festival was expected to take place in July 2022, with 12 vineyards presenting their wines. Following the pandemic restrictions, visitors have to pre-book tickets for the festival and all the extra activities. The festival facilities could host just 600 persons simultaneously, thus significantly limiting the number of visitors. Still, the festival gives Scanian vineyards an opportunity to market and promote their wines. All the tickets for the 2021 festival were sold out within one week (Vinfestival 2021).

Taste Swedish Wines provides a unique opportunity to get acquainted with local wines at the vineyard location. The festival experience consists of a reach program of activities where visitors can communicate with local wine professionals; learn more about wine growing, tasting, and production; taste local wines; meet wine experts; and take part in a wine quiz. A few times every day, a *wine school* opened its doors, where visitors met a famous sommelier and learned more about wine. They were also welcomed to meet the international food expert Tareq Taylor, who prepared festival lunches using just local ingredients and discussed how to match local wine and food. Every festival's day finished off with an evening grill-wine-music activity where the participants learned how to match local wine and food.

Located in Österlen, a tourist destination in the southeast of Scania known for its natural, cultural, and historical environment, the wine festival provides a cultural and aesthetic experiences for visitors. Unforgettable views of the surroundings, nearest beach facilities, and the picturesque Scanian landscape all contribute to the wine experience. The participants will really have a taste of Scania when visiting *Taste Swedish Wines*.

Discussion and conclusion

The case study illustrates a rural wine festival in a new wine tourism country without any history of wine growing and wine production. Scania, as an emerging wine tourism destination which achieved a lot in the first two decades of the 21st century, increased in number of vineyards as well as in wine growing and production. Challenged by structural problems associated with the government alcohol retail monopoly and restrictions, local vineyards need marketing and promotion strategies for their wine products and their emerging wine tourism destinations. However, wine festivals and events play an important role in it.

In line with *TDLC* (Butler 1980), *Taste Swedish Wines* can be placed in the exploration stage of the event development, when tourists started to explore wine tourism activities and events related to it, but vineyards and destination are still unexplored by both domestic and international tourists. However, to overcome this stage, marketing efforts are needed with a massive information concentrated on wine actors as well as on wine tourism activities. One reason for this is that the wine sector is unprofitable due to a full occupation with wine production and sales, not applying any marketing concepts to wine activities (Spawton 1986). Moreover, rural wine festivals often lack the marketing capacity and cooperation needed to attract international tourists (Carlsen and Getz 2006). Although wine festivals can become a promotional vehicle for small vineyards, they generally have difficulties in wine product marketing (Galvez, Fernandez, and Lopez-Guzman 2015).

On the other hand, Axelsen and Swan (2010) conclude that quite limited research is done on how local wine festivals successfully can market and promote local wines and vineyards. Moreover, if this is the case even for well-established wine tourism destinations, it becomes a real problem in new wine countries like Sweden (Rauhut Kompaniets and Nilson 2019).

According to the wine festival life cycle by Beverland, Hoffman, and Rasmussen (2001), *Taste Swedish Wines* appears located somewhere in between conception and launch stages – when the destination is still relatively unknown, and it is important to build awareness among the target groups. On the other hand, gaining a critical mass of attendees is important. What is missing to consider *Taste Swedish Wines* as a launch-stage festival is the measure of the visitors' satisfaction level and marketing plan focus.

Regional wine festivals involve several local wineries cooperating to develop and market their wines to target groups. The involvement of Tareq Taylor, an international food expert, explicates the unique nature of the wine festival when it is likely to get attention from wine and food celebrities (Beverland, Hoffman, and Rasmussen 2001).

The program and activities of *Taste Swedish Wines* follow the Experience Economy model (Culha 2020; Pine and Gilmore 1998). During the festival, visitors can get an educational experience by attending varied learning activities, such as sessions at wine school about wine growing, production, and tasting local wines. An entertainment experience comes from communication with wine and food experts and professionals, and a wine quiz, not to mention evening grill-wine-music activities. The main goal for the local wine festival in an emerging wine tourism destination is mostly information providing. A region can produce the best wine in the world, but if nobody knows about it, the quality becomes insignificant, concludes the international wine expert Mattias Säfwenberg (2019). The main priority is to inform the local and domestic communities about a new wine festival and product experience. It is also of high interest to give local and domestic visitors an opportunity to taste local wines, to educate them about the new wine culture, showing wine growing facilities, and explaining that wine grapes can grow in Sweden and that it is possible to produce high-quality wines.

Taste Swedish Wines provides attendees with a novel and unique experience for Sweden – what is considered an escapism experience of the wine festival, when visitors can 'escape' their everyday life and routine and try something extraordinary (Manthiou et al. 2014). An aesthetic experience is provided by a festival's atmosphere and facilities at the vineyard. Located at the field with a view of the wine grapes, the festival gives an opportunity to experience real local rural life and culture.

As previous research has shown, additional wine sales during the festivals are common and appreciated by both visitors and vineyards, which helps festivals create profit and gives opportunities for development (Beverland, Hoffman, and Rasmussen 2001; Carlsen and Getz 2006; Galvez, Fernandez, and Lopez-Guzman 2015). This is not the case for Sweden, with its government alcohol retail monopoly. However, for the wine festival's development according to the festival life cycle model (Beverland, Hoffman, and Rasmussen 2001; Butler 1980), a marketing strategy is required.

An organizational paradox occurs for the wine festival *Taste Swedish Wines*. While wine festivals are a common solution for off-season tourist attractions worldwide and the way of balancing high and low tourist seasons (Getz and Page 2016), it is not the case for Scania. Here, the wine festival takes place during the peak of tourist season in July, when locals are usually leaving their home region, spending their summer vacation somewhere else. On the other hand, international tourists come to spend their summer vacation in southern Sweden. Considering that *Taste Swedish Wines* mostly focuses on local visitors, the marketing problem becomes more visible. Local vineyards need to change their marketing and promotion strategy by considering international visitors.

The information about the wine festival on the festival's official website is only in Swedish, which makes it difficult for an international audience to know more about the festival. Furthermore, the festival's name is *Taste Swedish Wines*, and this name is stated on the official festival website, in the program for the upcoming festival, and on the festival's page on Facebook (Vinfestival 2021). However, the website's name is Österlen Wine Festival, and the website of the Swedish Wine Route mentions 'Österlen Wine Festival at Skepparps vineyard' (Vinvägen 2020). To refer to the same wine festival using different names is very unfortunate from a marketing perspective. The official website of the Scanian Destination Marketing Organization (DMO), Visit Skåne, does not mention any wine festival opportunity in Scania at all (Visit Skåne 2020). The festival's name is very important as a marketing and promotion part, and if locals can realize that all the names mean the same festival, confusion occurs with an international audience.

The findings in this review on the role of wine festivals in Swedish wine tourism marketing lead to the following conclusions. Marketing and promotion efforts are crucial for the wine festival and wine destination development (Novelli 2003; Cagnina, Cicero, and Osti 2019). It is important to consider the characteristics of a wine country that organizes a wine festival. There is a big difference between the goals and plans of wine festivals in the traditional wine-growing countries and new wine countries without any wine-growing history. Well-established wine tourism destinations organize bigger wine festivals for international tourists to a region with a goal of economic destination development. It is even simpler to promote an emerging wine destination in the wine region of an established wine country (Velikova, Slevitch, and Mathe-Soulek 2017). In new wine countries, we are talking about curiosity and a novel, unexpected experience as a part of the wine festival experience. Sweden is famous not for its wines, but for its culinary and nature heritage attractions, and it explains that the wine festival visit is not a goal for a destination visit; it is just an extra valuable activity during the destination stay. It becomes even more unexpected knowing that Sweden has a government alcohol retail monopoly and wine-related restrictions.

From the managerial perspective, the study contributes to market Scania as a wine tourism destination. The case of the wine festival enhances both marketing goals of local vineyards and a new wine destination.

Wine festivals provide an important research opportunity to analyze consumer experience (Staggs and Brenner 2019). It is quite common for small vineyards, and *Taste Swedish Wines* is not exception, to put most of the efforts and resources into wine festival organization and to fail to learn about visitor behavior and experience (Velikova, Slevitch, and Mathe-Soulek 2017). Thus, it is of importance for *Taste Swedish Wines* to measure attendees' perceptions, satisfaction and loyalty during the next wine festival.

Cooperation with regional DMOs and Swedish Wine Route is also important from a marketing and promotion perspective for the wine festival as well as for the emerging wine tourism destination (Hede 2008; Santini and Cavicchi 2014). Local DMOs, wine route officials, and other local tourism actors must coordinate with vineyards to build a complex tourist experience together.

Scania is an emerging wine tourism destination, where the regional wine festivals primarily focus on the locals and domestic tourists. This is in line with the theoretical framework. The conclusion is that Scanian regional wine festivals are in the early stages of the event life cycle, while the wine festivals in most established wine countries are in the more mature stages. While most other studies on wine festivals focus on the consumer experience and satisfaction of the event, this chapter focuses on the vineyards' perspective on wine marketing through the wine festival in emerging wine destinations. This is the added value of this chapter.

Finally, two suggestions for future research can be identified. The first deals with a comparison of wine festivals in cool climate wine countries to fill up the knowledge gap regarding new wine tourism destinations. A second proposed research topic is based on the conclusion of Taylor (2006), mentioning that urban wine festivals are more attractive for researchers due to the venue, size, and economic impact. A descriptive and empirical analysis of rural and regional wine festivals in new wine countries is needed, with a comparison to Swedish wine festivals, if other destinations struggle with the same marketing problems as the Swedish wine tourism sector or other problems occur.

References

Abend, L. (2016). No. 9: Skane. Sweden. Nordic cuisine's next big thing: 52 places to go 2016. *The New York Times*, January 7. https://www.nytimes.com/interactive/2016/01/07/travel/places-to-visit.html

Alonso, A. D., and Bressan, A. (2013). Stakeholders' perspectives on the evolution and benefits of a traditional wine festival: the case of the Grape Festival ('Festa dell'Uva') in Impruneta, Italy. *Journal of Convention & Event Tourism*, 14(4): 309–330. https://doi.org/10.1080/15470148.2013.834806

Axelsen, M., and Swan, T. (2010). Designing festival experiences to influence visitor perceptions: the case of a wine and food festival. *Journal of Travel Research*, 49: 436–450. https://doi.org/10.1177/0047287509346796

Beverland, M., Hoffman, D., and Rasmussen, M. (2001). The evolution of events in the Australasian wine sector. *Tourism Recreation Research*, 26(2): 35–44. https://doi.org/10.1080/02508281.2001.11081341

Bruwer, J., and Kelley, K. (2015). Service performance quality evaluation and satisfaction in a USA wine festivalscape: buying behavioural effects. *International Journal of Event and Festival Management*, 6(1): 18–38.

Butler, R. W. (1980). The concept of a Tourist Area Cycle of Evolution: Implications for management of resources. *The Canadian Geographer*, 24(1): 5–12.

Cagnina, M. R., Cicero, L., and Osti, L. (2019). Knowledge and consumption of organic and biodynamic wines. In Sigala, M., and Robinson, R. N. S. (eds.), *Management and Marketing of Wine Tourism Business* (pp. 81–100). https://doi.org/10.1007/978-3-319-75462-8_5

Carlsen, J., and Getz, D. (2006). Strategic planning for a regional wine festival: the Margaret River Wine Region Festival. In Carlsen, J., and Charters, S. (eds.), *Global Wine Tourism: Research, Management and Marketing* (pp. 209–224). Wallingford: CAB International

Charters, S., and Ali-Knight, J. (2002). Who is the wine tourist? *Tourism Management*, 23(3): 311–319.

Culha, O. (2020). The effect of food festival quality on place attachment and destination recommendation intention through festival experience and festival satisfaction: the case of the Didim International Olive Festival. *Journal of Convention & Event Tourism*, 21(5): 387–416. https://doi.org/10.1080/15470148.2020.1775743

Dawson, H., Holmes, M., Jacobs, H., and Wade, R. I. (2011). Wine tourism: winery visitation in the wine appellations of Ontario. *Journal of Vacation Marketing*, 17(3): 237–246.

Dimitrova, M., and Yoveva, I. (2014). Food festivals as purpose-created tourism attractions and their impact on destination branding. In Cavicchi, A., and Santini, C. (eds.), *Food and Wine Events in Europe: A Stakeholder Approach* (pp. 172–181). London: Routledge.

Dixit, S. K. (ed.). (2019). *The Routledge Handbook of Gastronomic Tourism*. Oxon: Routledge.

Dixit, S. K. (ed.). (2020). *The Routledge Handbook of Tourism Experience Management and Marketing*. London: Routledge.

Föreningen Svenskt vin (2019). *Svenska Viner. Sverige - ett 'Vinland' som andra?* http://svensktvin.se/svenska-viner/

Frochot, I. (2000). Wine tourism in France: a paradox? In Hall, C. M., Sharples, L., Cambourne, B., and Macionis, N. (eds.), *Wine Tourism around the World: Development, Management and Markets* (pp. 67–80). Oxford: Butterworth-Heinemann.

Galvez, J. C. P., Fernandez, G. A. M., and Lopez-Guzman, T. (2015). Wine festivals as a vehicle of promotion of a tourist destination. *Mediterranean Journal of Social Sciences*, 6(3): 574–581.

Getz, D. (2000). *Explore Wine Tourism: Management, Development & Destinations*. New York: Cognizant Communication Corporation.

Getz, D. (2019). Wine and food events: experiences and impacts. In Sigala, M., and Robinson, R. N. S. (eds.), *Wine Tourism Destination Management and Marketing* (pp. 143–164). https://doi.org/10.1007/978-3-030-00437-8_11

Getz, D., and Page, S. (2016). Progress and prospects for event tourism research. *Tourism Management*, 52: 593–631. https://doi.org/10.1016/j.tourman.2015.03.007

Gómez, M., and Molina, A. (2012). Wine tourism in Spain: denomination of origin effects on brand equity. *International Journal of Tourism Research*, 14: 353–368.

Hall, C. M., Johnson, G., Cambourne, B., Macionis, N., Mitchell, R., and Sharples, L. (2000). Wine tourism: an introduction. In Hall, C. M., Sharples, L., Cambourne, B., and Macionis, N. (eds.), *Wine Tourism around the World: Development, Management and Markets* (pp. 1–23). Oxford: Butterworth-Heinemann.

Hede, A.-M. (2008). Food and wine festivals: stakeholders, long-term outcomes and strategies for success. In Hall, C. M., and Sharples, L. (eds.), *Food and Wine Festivals and Events around the World* (pp. 85–100). Oxford: Butterworth-Heinemann.

Henley, J. (2020, March 20). Wine cooler: global heating helps Sweden's vineyards to success. *The Guardian*. https://www.theguardian.com/world/2020/mar/20/wine-cooler-global-heating-helps-swedens-vineyards-to-success

Holmes, K., and Ali-Knight, J. (2017). The event and festival life cycle – developing a new model for a new context. *International Journal of Contemporary Hospitality Management*, 29(3): 986–1004. https://doi.org/10.1108/IJCHM-10-2015-0581

Houghton, M. (2008). Wine connoisseurs or curious tourists? The participants in wine festivals. In Hall, C. M., and Sharples, L. (eds.), *Food and Wine Festivals and Events around the World: Development, Management and Markets* (pp. 201–217). Oxford: Butterworth-Heinemann.

Hubbard, K. W., Mandabach, K. H., McDowall, S., and VanLeeuwen, D. M. (2012). Perceptions of quality, satisfaction, loyalty, and approximate spending at an American wine festival. *Journal of Culinary Science & Technology*, 10(4): 337–351. https://doi.org/10.1080/15428052.2012.735962

Hultgren Karell, G. (2015). Scandinavian vineyards. *Scandinavian Traveler*, March 17. https://scandinaviantraveler.com/en/lifestyle/scandinavian-vineyards

Jević, G., Jević, J., and Barović, S. (2016). The role of Belgrade wine fair BeoWine on development of wine tourism in Serbia. *Economics of Agriculture*, 63(4): 1347–1363. https://doi.org/10.5937/ekoPolj1604347J

Kim, S. (2015). Understanding the historical and geographical contexts of food festival tourism development: the case of the Tatebayashi noodle grand prix in Japan. *Tourism Planning and Development*, 12(4): 433–446.

Kruger, S., Rootenberg, C., and Ellis, S. (2013). Examining the influence of the wine festival experience on tourists' quality of life. *Social Indicators Research*, 111(2): 435–452.

Lee, I., and Arcodia, C. (2011). The role of regional food festivals for destination branding. *International Journal of Tourism Research*, 13: 355–367. https://doi.org/10.1002/jtr.852

Livsmedelsföretagen (2019). *Dryckesbranschrapporten 2019*. https://www.livsmedelsforetagen.se/app/uploads/2019/04/dryckesbranschrapporten-2019.pdf

Malm, K., Gössling, S., and Hall, C. M. (2013). Regulatory and institutional barriers to new business development: the case of Swedish wine tourism. In Hall, C. M., and Gössling, S. (eds.), *Sustainable Culinary Systems. Local Foods, Innovation, Tourism & Hospitality* (pp. 241–255). London: Routledge.

Manthiou, A., Lee, S. A., Tang, L. R., and Chiang, L. (2014). The experience economy approach to festival marketing: vivid memory and attendee loyalty. *Journal of Services Marketing*, 28(1): 22–35. https://doi.org/10.1108/JSM-06-2012-0105

Marais, M., and Saayman, M. (2011). Key success factors of managing the Robertson Wine Festival. *Acta Academica*, 43(1): 146–166.

McDowall, S. (2011). The festival in my hometown: the relationships among performance quality, satisfaction, and behavioral intentions. *International Journal of Hospitality & Tourism Administration*, 12(4): 269–288.

Mehmetoglu, M., and Engen, M. (2011). Pine and Gilmore's concept of experience economy and its dimensions: an empirical examination in tourism. *Journal of Quality Assurance in Hospitality & Tourism*, 12(4): 237–255. https://doi.org/10.1080/1528008X.2011.541847

Näringsdepartementet (2017). *Ett land att besöka. En samlad politik för hållbar turism och växande besöksnäring*, SOU 2017:95, Stockholm: Wolters Kluwers.

Novelli, M. (2003). Wine tourism events: Apulia, Italy. In Yeoman, I., Robertson, M., Ali-Knight, J., Drummond, S., and McMahon-Beattie, U. (eds.), *Festival and Events Management: An International Arts and Culture Perspective* (pp. 329–345). New York: Routledge.

Pine, B. J., and Gilmore, J. H. (1998). Welcome to the experience economy. *Harvard Business Review*, 76(4): 97–105.

Rauhut Kompaniets, O., and Nilson, H. (2019). Wine tourism and family enterprises in southern Sweden: problems, challenges and potentials. In Deffner, A., and Psatha, E. (eds.), *Proceedings of the 4th Annual Conference of the International Place Branding Association (IPBA)* (pp. 243–262). Volos, Greece: Department of Planning and Regional Development, University of Thessaly.

Rauhut Kompaniets, O., and Nilson, H. (2020). Wine tourism in Southern Sweden: opportunities and challenges. In Haller, C., and Sigala, M. (eds.), *Proceedings of the 2nd International Research Workshop on Wine Tourism: Challenges, Innovation and Futures* (pp. 67–77), September 17–18. https://winetourism2020.sciencesconf.org/

Rytkönen, P. (2012). *Sweden – an Emerging Wine Country – Valorising Marginal Lands for New Types of Production*. Paper presented at *10th European IFSA Symposium*, Aarhus, Denmark. http://ifsa.boku.ac.at/cms/index.php?id=131&L=0.#c368

Säfwenberg, M. (2019). *Svenskt vin: En vinvärld som växer*. Sverige: Idus förlag.

Santini, C., and Cavicchi, A. (2014). The importance of a stakeholder approach to investigate (and plan) food and wine events: research issues from a European perspective. In Cavicchi, A., and Santini, C. (eds.), *Food and Wine Events in Europe: A Stakeholder Approach* (pp. 3–14). London: Routledge.

Sjöberg, S. (2019). Sommarturismen i Sverige 2019. *Tillväxtverket*. https://tillvaxtverket.se/statistik/vara-undersokningar/resultat-fran-turismundersokningar/2019-10-07-sommarturismen-2019.html

Skånes Matfestival (2020). *Skånes Matfestival i Landskrona*. http://skanesmatfestival.se/

Skånska drycker (2020). *Skånska drycker. Superlokal dryckesfestival*. https://skanskadrycker.se/

Skjöldebrand, C. (2010). *Svensk vinakademi – innovativ metod att utveckla Matlandet*. Dnr 19-12802/09, Rapport 20101130

Spawton, T. (1986). Marketing planning for small wineries. *Australian and New Zealand Wine Industry Journal*, 1(3):54–57.

Staggs, J., and Brenner, M. (2019). Pouring new wines into old wineskins? Sub-regional identity and the case of the Basket Range Festival. In Sigala, M., and Robinson, R. N. S. (eds.), *Wine Tourism Destination Management and Marketing: Theory and Cases* (pp. 165–183). https://doi.org/10.1007/978-3-030-00437-8_12

Storchmann, K. (2011). Wine economics: emergence, developments, topics. *Agrekon*, 50(3): 1–28.

Swedish Wine Center (2020). https://swedishwinecenter.se/

Szabo, Z., Komaromi-Gergely, A., and Szeles, Z. (2014). *The Role of Event Marketing in Case of the Buda Castle Wine Festival*. Paper presented at the 8th Academy for Wine Business Research Conference, Geisenheim, Germany, June 28–30.

Taylor, R. (2006). Wine festivals and tourism: developing a longitudinal approach to festival evaluation. In Carlsen, J., and Charters, S. (eds.), *Global Wine Tourism* (pp. 179–195). Perth: CAB International.

Velikova, N., Slevitch, L., and Mathe-Soulek, K. (2017). Application of Kano model to identification of wine festival satisfaction drivers. *International Journal of Contemporary Hospitality Management*, 29(10): 2708–2726, https://doi.org/10.1108/IJCHM-03-2016-0177

Vinfestival (2021). Vinfestival på Österlen 'Smaka Svenska Viner [Taste Swedish wines]. http://www.vinfestivalosterlen.se/

Vinvägen (2020). *Upplev vinvägen i Skåne*. http://www.vinvagen.se/

Visit Skåne (2020). *Wine Tastings and Winery Tours*. https://visitskane.com/food-drinks/wine-tastings-and-winery-tours-skane

Vitale, L., López-Guzmán, T., Galvez, J. C. P., and Di Clemente, E. (2018). Profiling wine tourists: a comparison between rural and city Wine Festivals in Campania, Italy. *African Journal of Hospitality, Tourism and Leisure*, 7(6): 1–14.

Wergeland, M. (2017). A guide to Sweden's charming, relaxed southernmost county: Skåne. *Vogue*, August 4. https://www.vogue.com/article/skane-county-scania-sweden-travel-guide?mbid=social_facebook

Yuan, J., Cai, L. A., Alastair, M. M., and Linton, S. (2005). An analysis of wine festival attendees' motivations: a synergy of wine, travel and special events? *Journal of Vacation Marketing*, 11(1): 41–58.

Yuan, J., and Jang, S. (2008). The effects of quality and satisfaction on awareness and behavioral intentions: exploring the role of a wine festival. *Journal of Travel Research*, 46(3): 279–288. https://doi.org/10.1177/0047287507308322

27

THE EMERGENCE OF A NEW WINE TOURISM FIELD

Entrepreneurial struggles

Jon Sundbo

Introduction

This chapter deals with wine tourism in the making. It describes and analyzes how entrepreneurs in the new field of modern fruit wine production in Denmark struggle to develop new types of fruit wine and establish a fruit wine interest regime (Sundbo and Dixit 2020; Sundbo and Sundbo 2018), and how tourism is a means to achieving that goal. The chapter provides an understanding of this early phase of wine tourism creation and the difficulties in creating wine tourism from the beginning within a framework in which tourism is a means of creating an interest regime rather than the main goal. An 'interest regime' is a loosely coupled group of people who share the same interest. The group can be widespread, and the members do not necessarily have to know each other; their common interest is the basis for identification and interaction, both in person and via the media. There are common agendas and norms within an interest regime. For example, membership is a social signal of the status group to which one belongs and also a signal of individual choice (Sundbo 2021; Sundbo and Sundbo 2018).

Wine production and wine tourism are established ventures (e.g., Croce and Perri 2010; Mitchell and Hall 2006). However, new innovative phenomena do emerge from time to time. This chapter presents and discusses the innovation and diffusion processes of modern fruit wine, a new wine product introduced in Denmark from about 2005. Fruit wine is not new, but Danish entrepreneurs have developed new production methods for fruit wine types that have elevated their quality, and they are imbibed to new heights. The focus is on how this new fruit wine is marketed and the role that tourism plays in this marketing.

The emphasis is on entrepreneurship. The market-creation process is a process of innovation diffusion, with a few entrepreneurs introducing an innovation (in this case, modern fruit wine) and followers subsequently adopting this idea and developing their own fruit wines (Rogers 1995). The market-creation process is also an attempt to raise consumer interest in new types of fruit wine by emphasizing the experiential aspects of this new product (Sundbo and Sørensen 2013). Raising consumers' interest in new fruit wines has included public relations campaigns in the media, the promotion and sale of fruit wines in bars and restaurants, and the publication

DOI: 10.4324/9781003143628-32

of fruit wine recipes in cookbooks and mixology books. The market-creation process also includes wine tourism, which will be emphasized in this chapter.

This chapter also analyzes the entrepreneurial process and the challenges of establishing a new type of wine and wine-related tourism. Entrepreneurial processes are challenging, and things do not always go as intended (Gartner et al. 2004; Allen 2006). This chapter aims to discuss the challenges that entrepreneurs face and the factors that determine their success. It tells the story of an innovation in the winemaking industry and the path from innovation to diffusion into the wine tourism sector. The chapter also recounts entrepreneurs' struggles in developing and marketing their fruit wine and the challenges they encounter in their journey. Hereby can a story about one of the rare innovations within winemaking, and how that may lead to wine tourism, be told, and we can learn from the challenges and struggling that the entrepreneurs have met. The latter can enrich both knowledge within tourism theory and entrepreneurship theory.

The content in this chapter is drawn from a research project whose methodology is presented in the Method section. The subsequent sections will (a) describe the development of fruit wine and the tourism elements related to it; (b) discuss entrepreneurship and experiential marketing theories; (c) describe tourism-related experiences, fruit wine festivals, and visits to fruit wine plants; and (d) present a fruit wine winery case study that illustrates all of these factors. The chapter ends with a conclusion.

Method

The research project dealt with the investigation of the development of modern fruit wines in Denmark (Sundbo and Sundbo 2018, 2019). The theoretical approaches used were entrepreneur theory (Nielsen et al. 2012), diffusion theory (Rogers 1995), and Experience Economy theory (Pine and Gilmore 1999).

The project was an experimental research project (Sørensen, Mattsson, and Sundbo 2010) where researchers procured new knowledge but also helped a group of fruit wine producers in developing their fruit wines and businesses. Two researchers from the business and Experience Economy field (including the author of this chapter) participated in the research project together with a group of biologists and chemists. Four fruit wine producers and the Fruit Wine Producers Association were involved. The task of the biologists and chemists was to collaborate with the fruit wine producers in experimenting with the fruit wine production process and thereby learn how the process works. The task of the business and Experience Economy researchers was to study how the entrepreneurs developed their businesses and how they used experiences and tourism for that purpose. These researchers also suggested improvements concerning the producers' and the association's tourism activities and marketing strategies.

We – the business and Experience Economy researchers – followed the fruit wine producers and the association for four years, from 2014 to 2017. The collected data were observations from meetings, visits to the producers with observation and regular interviews with the fruit wine producers, documentary materials, and photos from the fruit wine plants and fruit wine festivals. The association organized fruit wine festivals, which we observed. We used surveys to question guests on their attitudes toward the fruit wine and their experiences at the festivals. From 2017 to 2020, we collected follow-up data, primarily from documents, using netnographic methods (Mkono 2012). The case we present in this chapter is that of a fruit wine producer who developed and added experiential and tourism elements to fruit wine production.

The new fruit wine and how it was developed in Denmark

Fruit wine refers to wine made from different kinds of fruits and berries, including cider. Making fruit wine is an old tradition in Denmark, where cherry wine has been produced for the last 100 years. Whereas cider is an internationally well-known product, there is no Danish tradition for producing and drinking cider; instead, Danes produce and drink apple juice. Danish cherry wine has been considered a very sweet wine of low and uninteresting gastronomic quality. However, some Danish entrepreneurs have begun to use new methods for producing fruit wine and cider. They use scientific knowledge and experiment with new production methods. This has resulted in new types of fruit wine that have a much more refined taste and gastronomic value than the old cherry wine. Thus, a new tradition of Danish viticulture has emerged.

The basis for developing new types of fruit wine is the climate in Denmark, which is optimal for fruit and berries. They grow slowly and have an intense taste and a good balance between acidity and sweetness. Examples are Danish apples that are unsurpassed in taste, and an old variety of Danish cherries, the 'Stevns' cherry, which are suitable for juice and wine production. Thus, apples and cherries are the fruits most used for wine production. However, producers also experiment with all sorts of fruit and berries, including pears, rhubarbs, redcurrants, and quince. Producers have also developed new types of cider, but there is an internal discussion among the Danish fruit wine producers on whether to call this wine or cider.

A few individuals started independently to experiment with producing fruit wine. These are the entrepreneurs who have developed modern fruit winemaking. They all started as amateurs for whom wine production was a hobby. Only one producer, the owner of a manor house, began by producing cherry wine professionally in collaboration with a chef and a journalist with an interest in wine. Very soon, the entrepreneurs connected with each other and with university biologists with expertise in fruit growing. Together they created a community that exchanged technical and scientific knowledge and shared the results of their experiments using different methods for producing fruit wine. Later, the network established itself as a formal association.

Several of the entrepreneurs developed their hobbies into real businesses. They started to make a living from fruit wine production, invested in production equipment, sales, and marketing channels, and employed staff. They then arrived at the stage where they needed to become business managers, a new and unusual challenge. Some succeeded at this stage of the business creation, but many did not. None of them developed into large businesses, but a few now run small/medium-sized businesses.

The marketing of fruit wine was the biggest challenge, not only because the entrepreneurs were not trained as marketers, but also because people in the wine-drinking market were not used to the new, modern fruit wine. The producers had to create a new market by creating interest in a completely new product. Those producers who succeeded in converting their hobby into a business used experiences to create an interest in the market for fruit wine. Specifically, the producers used the Danish and international media to promote their fruit wines and persuaded famous Danish chefs and international bartenders to use the fruit wine to make their dishes and cocktails. The entrepreneurs also used tourism elements in their business and market-creation efforts – they organized fruit wine festivals and visits to their production facilities and requested and received publicity for fruit wine production from official Danish tourism promotion and marketing organizations.

To meet the marketing and tourism outreach challenges, the fruit wine producers decided, in 2017, to merge their association with the grape wine producers association. This new association attempts to develop both fruit and grape wine tourism products by telling stories about

Danish wine production and entrepreneurs, organizing visits to wine producers, creating wine routes, and including the routes and visits in existing organized tourism tours. However, there is more to be done to optimize the tourism and marketing potential of Danish wine and its production and to tell the story of its entrepreneurs.

Entrepreneurship and the diffusion of innovation theory as an explanation for this process

Fruit wine tourism development as described here is a two-step entrepreneurial innovation process: first, the innovation of modern fruit wine and the establishment of a fruit wine business; second, the addition of tourism and other marketing aspects to fruit wine production by some of the entrepreneurs.

No one person can be identified as the original inventor and innovator of modern fruit wines in Denmark. Modern fruit wine enterprises were developed by a few independent entrepreneurs, and the process is in line with classic entrepreneurship theory (Schumpeter 1934/1961; Shane 2003). However, it was a situation with several simultaneous sources of this innovation, which has been referred earlier in innovation theory (Tidd and Bessant 2018). Soon after beginning their fruit wine businesses, the first entrepreneurs created a network, a phenomenon that has been written about in the entrepreneurship literature (Hoang and Antoncic 2003; Johannisson 1988). This network was not the kind of network that a sole entrepreneur creates around himself with investors, researchers, and consultants to procure resources for his business (Johannisson 1988); instead, it was a homogeneous network of peer entrepreneurs (Aldrich and Zimmer 1986) that was created to exchange technical knowledge. However, as often happens in social networks (Burt 1992), some entrepreneurs became more dominant than others. The dominant entrepreneurs in the network were those who had developed their fruit wine production businesses into professional firms with a higher turnover and employment of people – the business growers (Allen 2006).

A particular element in the Danish fruit wine growers' network was the involvement of researchers – biologists and chemists from universities, who researched and experimented with the biological and chemical processes in producing fruit wine. Their effort was a crucial condition for the development of modern fruit wine techniques. Later, social scientists and business researchers also became involved in developing marketing strategies for fruit wines and wine tourism. The researchers were not entrepreneurs, as they did not want to establish fruit wine firms. Their interest was in procuring new scientific knowledge about the wine-growing and production process and the process of developing marketing strategies for fruit wines and wine tourism. The researchers' role was to function as catalysts for the developments and innovations around the fruit wine business (cf. Sørensen, Mattsson, and Sundbo 2010).

In time, new entrepreneurs emerged. Several of them acted independently and did not involve themselves in the established network. They developed their fruit wine production into real businesses and saw the established producers as competitors. Simultaneously, the dominant entrepreneurs in the network began to grow their businesses with investments in large-scale production technology. They also increased their production scale and stepped up their professional marketing efforts. In this second phase, the entrepreneur network weakened, and the industry entered a new phase.

The whole development process can be understood as a diffusion of the innovation process (Rogers 1995). The diffusion process starts with a few early innovators, followed by a communication process whereby awareness of the innovation is diffused, followed by early adopters who imitate the original idea. Diffusion theory says that as more adopters arrive, the innovation diffuses widely into society. However, this has not been the case with modern fruit wine production, at least not yet. In fact, the diffusion process and the growth of the industry have slowed. Innovation theory (Sundbo 1998) states that innovations often fail because they are dropped by the market or the entrepreneurs or because the diffusion process stops at an early stage. Sometimes a new phase of increased diffusion starts up again, but that requires new innovative elements added to the original innovation. This renewal could be a new version of the original innovation, new production processes, or new marketing strategies. Some fruit wine growers in Denmark followed this pattern by initiating new marketing elements, including the introduction of new wine tourism approaches, innovations that may revitalize the diffusion process in the near future.

Experiential marketing: development of fruit wine tourism

As described, most of the fruit wine growers have not left the hobby or early business stage. This implies that they produce a limited amount of fruit wine and do not utilize professional marketing strategies or a well-developed distribution system. Only a few producers managed to develop and grow their business. To help other small producers, the fruit wine producers' association is attempting to promote modern fruit wine – to drink and to add to cocktails and food dishes.

Fruit wine entrepreneurs soon realized that developing fruit wine into a daily consumption commodity was not practicable because fruit wine is quite expensive (the price can easily be €20 to €40 for a half-liter bottle). Only one producer attempted to sell fruit wine in supermarkets but was not very successful. The producers used other sales channels – direct sales to customers from the production plants or via the internet, to bars and restaurants in Denmark and abroad, and to specialized wine merchants. The producers also realized that modern fruit wine is best sold as an extraordinary experience (Mossberg 2007), so experiential marketing would be the best way to market fruit wine (Schmitt 2003; Smilanski 2009). Consumption of fruit wine should be seen within the framework of the Experience Economy (Sundbo and Dixit 2020; Sundbo and Sørensen 2013), whereby people experience fruit wine during leisure time and as tourists (Dixit 2020). The extraordinary experiences of fruit wine can be experienced in cocktails and aperitifs or extraordinary meals in gastronomic restaurants and consumers' homes.

Experiential marketing emphasizes giving customers' extraordinary experiences. This cannot be done by traditional marketing means such as advertising or media promotions. It requires that people experience fruit wine on the spot (Smilanski 2009). To provide an extraordinary experience, fruit wine producers need to present the new fruit wines to a broader market. This market could include famous bars that launch new cocktails with fruit wine, and chefs who develop dishes with fruit wine and present them on TV. In fact, the fruit wine producers and researchers collaborated with students from the Danish School for Chefs. The students experimented with dishes that included fruit wine and developed a fruit wine cookbook. Journalists wrote about the modern fruit wine in Danish and international newspapers and food magazines. Other initiatives included storytelling, a core element in experiential marketing (Mossberg 2008). Entrepreneurs told stories about their new fruit wine product and their struggles developing the fruit wine innovation and establishing their fruit wine firms. A fruit wine entrepreneur with a good story to tell can do much to promote his fruit wine company. In the cases we studied, storytelling was a challenge because many of the fruit wine entrepreneurs were introverts and possessed technically oriented personalities.

One way to create extraordinary, on-the-spot experiences is through tourism. In the first diffusion phase, fruit wine entrepreneurs did not think about tourism. They were technologically oriented and occupied with biological and chemical growth and production issues. However, entrepreneurs who had the drive to develop their fruit wine hobby into a real business began to contemplate wine tourism as a way to promote consumers' interest in and the sales of modern fruit wine. They saw that they could not sell fruit wine as just another new commodity because people either were not used to drinking fruit wine or thought of fruit wine as a low-quality drink. The entrepreneurs realized that to create a market for modern fruit wine, they needed to create a new interest regime (i.e., a wide interest in fruit wine and how to use it; Sundbo and Sundbo 2018). They also realized that tourism and experiential events would help to create a new market and fruit wine interest regime. Thus, entrepreneurs took on two initiatives. The first initiative was to launch fruit wine festivals (Sundbo and Sundbo 2018). The second initiative was for sole entrepreneurs to introduce wine tourism elements into their businesses. These two initiatives will be described in the next two sections.

Fruit wine festivals

The Fruit Wine Producers Association begun in 2013 to organize annual fruit wine festivals. The first fruit wine festival took place in a fruit wine production plant located within the estate of a manor house in a peripheral region of Denmark (Sundbo and Sundbo 2018). Producers presented their fruit wines at the festival and talked about the wines and their businesses. Awards for the best fruit wines were presented, and the guests could taste the fruit wines. However, no other experiences were provided. For example, there were only a few simple food items for guests to purchase. The total and extraordinary experience was geared toward guests who already had an extraordinary interest in wine – frontrunners in the emerging fruit wine interest regime who could promote the fruit wine market and interest regime. The attendance of local inhabitants who came to the festival out of curiosity (primarily to see inside the manor house) did little to promote the development of a fruit wine market and interest regime.

After the initial festivals, the association understood that it needed to create a more holistic and extensive tourist event if it wanted to attract more people from outside the local area as well as people who were not especially interested in wine. To attract this audience, it would be necessary to create a wider market and interest regime. The social science researchers involved with the association were catalysts in this cognizance. The association introduced more experiential foods at the festivals; demonstrated fruit, wine, and meal pairings; held seminars; made speeches; and gave guided tours of the winery and the manor's fields and gardens. These improvements boosted the number of guests and attracted new types of people from outside the local community, most of whom were from Denmark. A few foreigners also began to attend the festival.

In 2015, the festival moved to Copenhagen, Denmark's capital city, to attract more people of the kind that would buy and promote modern fruit wine, together with more foreign tourists. The 2015 festival attracted the target audience from Denmark, but the number of foreign tourists remained limited. Even though the festival had relocated to a geographic center inhabited by the type of people who could boost the fruit wine interest regime, there were many other competing experiences and tourism activities in the capital. Thus, in 2017, the producers discontinued the fruit wine festivals. This decision was related to the merger of the fruit wine association with the Danish grape wine producers' association. Although Denmark has never been a wine-producing country, some Danish entrepreneurs began in 2000 to grow and produce grape wine. The intent of the merger of the two associations was to promote wine tourism in Denmark that included both grape and fruit wine production and experiences.

Visits to the fruit wineries

Improvements in promoting fruit wine led to new kinds of wine tourism in Denmark, including traditional tourism elements such as visits to wineries. Tourists visited several entrepreneurs' fruit wine wineries, but only a few of the entrepreneurs had developed experiential programs in their wineries to give visitors a comprehensive, holistic experience. The industry still struggles to create a wider market and an international fruit wine interest regime. Although the sales of fruit wine have continuously increased, both nationally and internationally, the growth of an international market for fruit wine and a fruit wine interest regime has slowed down. Several attempts to overcome this barrier have been made, including visits to the fruit wineries and other tourism activities, e.g., promotion in food and wine fairs abroad and participation in Danish food festivals. One of the cases we studied, which is this chapter's case study, shows how one entrepreneur has been successful in adding tourism elements to his winery.

Case study 27.1 A case of experiential wine tourism development: Cold Hand Winery (https://www.coldhandwinery.dk/)

Jens Skovgaard, owner of Cold Hand Winery, is a fruit wine entrepreneur who has developed his fruit winery into a business. He has also developed the winery's tourism aspects. Like other fruit wine entrepreneurs, Jens started as an amateur – he was director of a continuing education school and developed an interest in fruit wine. In 2010 he left his job and bought a farm where, in alliance with an apple grower, he started experimenting with producing apple wine. At first, he focused on developing new techniques to produce new fruit wine and on acquiring science-based knowledge about fruit wine from universities and other sources around the world. He also joined and became active in the Fruit Wine Producers Association. He developed several new types of apple wine, including ice wine, which implies that the wine has got frost (from which the winery's name is derived), and wines from a variety of fruits and berries – cherries, rhubarbs, blackcurrants, redcurrants, plums, and blackberries. After experimenting with the winemaking process (he used different production methods to improve his wines, dropping those that did not make the grade), he invested in more advanced production equipment and gained more production experience. As a result of his efforts, the quality of the wines improved. Some of the wines were not good, thus, he had to improve the production method or skip it; some were really good; and the best wines became a permanent part of the wine selection offered at Cold Hand Winery.

Jens developed Cold Hand Winery as a business because he wanted to make a living from it. He established several sales channels, which was a challenging because fruit wines are expensive due to their long production time and small quantities. For example, he contacted Danish wine dealers and started to sell his wines to them. He also started selling his wines to consumers via a website. He made promotional trips to other countries where he presented his wines at wine and food fairs and contacted well-established bars and restaurants in these countries to sell his fruit wines. Promotions of his fruit wines in the media also became part of the sales strategy. His business model was to market to the upper end of the gastronomic market – bars, restaurants, and people who were willing to pay a high price for the special experience of tasting and serving fruit wine. His wines are not sold in the supermarkets as that would signal a lower-quality, less exclusive wine and force the wines into price competition. He was successful in selling his fruit wines to the higher end of the market, including several fashionable bars and restaurants around the world. Consequently, he was able to expand the winery by adding new buildings and production capacity and employing more people. A shop selling fruit wines and other products was established on the premises. He procured investment capital for the expansion via crowd-funding (Ordanini et al. 2011), a process in which entrepreneurs raise capital through the collective effort of friends, family, customers, and individual investors who become share-owners and are invited to experiences in the winery.

Experiences have become a crucial part of the fruit wine business. To create an interest in the new fruit wine, owners of such wineries, including Cold Hand Winery, share their stories with the media and with visitors. The stories are about the taste and gastronomic quality of the fruit wines and the production process. In the case of Cold Hand Winery, Jens tells stories about his struggle with developing new and innovative production methods and his personal life history. Jens is a brilliant storyteller, and listening to him is one of the most exciting experiences in winery tourism. He has even published a book about his entrepreneurial journey, with photos of the winery and recipes that utilize his fruit wines.

After the winery was extended with new buildings (Figure 27.1) and capital, wine tourism in Cold Hand Winery was off to a running start. Visitors had always come to the winery, but to buy wine rather than for an experience. With the launch of wine tourism in Cold Hand Winery, Jens set out to attract more people, including journalists. He serves meals and offers wine tasting while simultaneously telling stories about the wine, himself, and his business. Companies began to use the winery for meetings and seminars. More visitors came to the winery, both local people and tourists. The winery expanded its tourism and visitor-oriented activities. Cottages for overnight stays were built, and guests were offered menus listing fruit wine-infused meals (Figure 27.2). Meeting activities became more organized and better marketed; stories became more elaborate; guidance on wines, food, and activities was further developed. More tour companies began to include Cold Hand Winery in their tours. The shop began to attract more people.

Figure 27.1 New buildings at Cold Hand Winery.

Figure 27.2 Summer grill meal at Cold Hand Winery.

The visitors were mostly Danish, but an increasing number of foreign tourists also begin to visit Cold Hand Winery.

These initiatives have contributed to modern Danish fruit wine becoming a part of people's tourist interest regimes and have increased interest in Cold Hand Winery and its wines. Cold Hand Winery has created a wine tourism market for itself, and the turnover from wine tourism has become increasingly important for the business. However, fruit wine tourism remains a modest part of the overall tourism activity in Denmark. In fact, wine tourism is only a small part of Denmark's official tourism marketing plan. This is evident from the ongoing lack of wine routes in Denmark, which is surprising, as wine is such an important part of European culture. However, the merging of the fruit wine growers association and grape wine growers association could increase wine tourism in Denmark, and Cold Hand Winery could profit from this in the future.

Conclusion

The Cold Hand Winery case study showcases the development of new fruit wine in Denmark. It also demonstrates that wine tourism can emerge from areas outside traditional tourism tracks. The development of new types of fruit wine flowed from innovation based on the biological and chemical processes of producing quality fruit wine. This technologically produced fruit wine was the creation of entrepreneurs who wanted to develop this innovation into a business. To do so, they needed to market the new fruit wine and to create a new market and a new interest regime for it (Sundbo and Sundbo 2018). Tourism such as festivals and visits to producers became means in the market and interest regime creation at a second stage just after the first entrepreneurial initiatives. Tourism and other experiential elements created an interest in fruit wine and boosted its sales, thereby actualizing the diffusion of the new fruit wine innovation. The evolution of the process from innovation to diffusion took place in three phases:

Phase 1. Entrepreneurial initiatives drive the development of new fruit wines and fruit wine businesses.
Phase 2. Market outreach takes place, with entrepreneurs participating in tourism events such as festivals and hosting visitors to their wineries.
Phase 3. Tourism becomes a more prominent element for some winemaking businesses such as Cold Hand Winery.

The drivers of this process have been:

1 *The emergence of entrepreneurs with a particular interest in an innovative new idea.* In this case, several entrepreneurs emerged simultaneously, and the new idea was about new technical-biological production methods.
2 *The early involvement of researchers in the process of developing the new fruit wine.* This has contributed to overcoming many technical barriers.
3 *Diffusion of this new idea and an interest in developing it into businesses.* The diffusion process was, in this case, enhanced by the entrepreneurs in the beginning who were seeking collaboration with each other to exchange knowledge about the production methods and business problems.
4 *The necessity of creating a new market for the fruit wine and an interest regime.* This has led to tourism becoming an element in this creation.

The Cold Hand Winery case study also shows that the diffusion process is difficult and may run into impediments at a later stage. One impediment is that diffusion and growth can stop. Impediments, in this case of developing of fruit wines in Denmark, were as follows:

1 There were difficulties in creating a new interest regime. Few people developed an interest in fruit wine.
2 Tourist organizations had insufficient interest in fruit wine. Even though innovative tourism and other experience elements may have helped consumers develop a greater interest in fruit wine, producers had difficulties finding the right tourism and other types of experiences that can engage people in an interest in fruit wine and to increase tourist organizations' involvement in marketing fruit wine tourism.
3 Many entrepreneurs do not have sufficient interest or competencies in developing and marketing their businesses.

Lack of interest and competencies in marketing and business development are well-known impediments for technically oriented, idea-based entrepreneurs seeking to implement their idea as a business (Gartner et al. 2004). However, if new innovations and initiatives, like a novel type of tourism, are introduced at this later stage, the diffusion and growth process can be revitalized.

Implications

From this example of the development of Danish fruit wine entrepreneurship, we learned how new fruit wine and wine tourism innovations can be launched, as well as the conditions for the growth and diffusion of these innovations. These conditions include both the drivers and barriers for the diffusion of the innovations. The learnings are useful because they help to identify where support is necessary when an innovation such as new fruit wine needs to be developed as a substantial new tourism attraction. This learning may also be applied to other types of tourism areas.

Tourism entrepreneurs such as fruit wine ones can be supported by three actors to develop their business and create innovative tourism development. The first is private advisors, who can advise and help the entrepreneurs to innovate, market their products, create new interest regimes, and develop tourism products on the top of the core product (such as fruit wine). Such advisors might be the entrepreneurs' own associations (such as the Danish wine growers association) or consultancy firms. The second actor is the public sector, which in its industrial policy must be aware of supporting such entrepreneurs in research, training programs, and advisory systems. The third actor is tourism promotion organizations, which could be aware of promoting such new interest regimes as tourist attractions.

References

Aldrich, H., and Zimmer, C. (1986). Entrepreneurship through social networks. In Sexton, D., and Smilor, R. (eds.), *The Art and Science of Entrepreneurship* (pp. 3–23). New York: Ballinger.

Allen, K. (2006). *Launching New Ventures: An Entrepreneurial Approach*. Boston: Houghton Mifflin.

Burt, R. (1992). *Structural Holes – The Structure of Competition*. London: Harvard University Press.

Croce, E., and Perri, G. (2010). *Food & Wine Tourism: Integrating Food, Travel and Tourism*. Wallingford, UK: CABI.

Dixit, S. K. (ed.). (2020). *The Routledge Handbook of Tourism Experience. Management and Marketing*. London: Routledge.

Gartner, W. B., Shaver, G. K., Carter, N. M., and Reynolds, P. D. (eds.) (2004). *Handbook of Entrepreneurial Dynamics – The Process of Business Creation*. Thousand Oaks, CA: Sage.

Hoang, H., and Antoncic, B. (2003). Network-based research in entrepreneurship – a critical review. *Journal of Business Venturing*, 18: 165–187.

Johannisson, B. (1988). Business formation – a network approach. *Scandinavian Journal of Management*, 4(3–4): 83–99.

Mitchell, R., and Hall, C. M. (2006). Wine tourism research: the state of play. *Tourism Review International*, 9: 307–332.

Mkono, M. (2012). Investigating hospitality consumers through netnographic methods. *Advances in Hospitality and Leisure*, 8: 235–246.

Mossberg, L. (2007). A marketing approach to the tourist. *Scandinavian Journal of Hospitality and Tourism*, 7(1): 59–74.

Mossberg, L. (2008). Extraordinary experiences through storytelling. *Scandinavian Journal of Hospitality and Tourism*, 8(3): 195–210.

Nielsen, S. L., Klyver, K., Evald, M. R., and Bager, T. (2012). *Entrepreneurship in Theory and Practice*. Cheltenham: Edward Elgar.

Ordanini, A., Miceli, L., Pizzetti, M., and Parasuraman, A. (2011). Crowd-funding: transforming customers into investors through innovative service platforms. *Journal of Service Management*, 22(4): 443–470.

Pine, J., and Gilmore, J. (1999). *The Experience Economy*. Boston: Harvard Business School Press.
Rogers, E. M. (1995). *Diffusion of Innovations* (4th ed.). New York: Free Press.
Schmitt, B. (2003). *Customer Experience Management: A Revolutionary Approach to Connecting with Your Customers*. Hoboken, NJ: Wiley.
Schumpeter, J. (1934/1961). *The Theory of Economic Development*. New York: Oxford University Press.
Shane, S. (2003). *A General Theory of Entrepreneurship*. Cheltenham: Edward Elgar.
Smilanski, S. (2009). *Experiential Marketing. A Practical Guide to Interactive Brand Experiences*. London: Kogan Page.
Sørensen, F., Mattsson, J., and Sundbo, J. (2010). Experimental methods in innovation research. *Research Policy*, 39(3): 313–322.
Sundbo, D., and Sundbo, J. (2018). Interest regimes in fruit wine tourism. *Journal of Gastronomy and Tourism*, 3(3): 163–176.
Sundbo, J. (1998). *The Theory of Innovation*. Cheltenham: Edward Elgar.
Sundbo, J. (2021). *Advanced Introduction to the Experience Economy*. Cheltenham: Edward Elgar.
Sundbo, J., and Dixit, S. K. (2020). Conceptualizations of tourism experience. In Dixit, S. K. (ed.), *The Routledge Handbook of Tourism Experience Management and Marketing* (pp. 15–26). London: Routledge.
Sundbo, J., and Sørensen, F. (eds.) (2013). *Handbook on the Experience Economy*. Cheltenham: Edward Elgar.
Sundbo, J., and Sundbo, D. (2019). Gastronomy as a means to sustainable tourism development in peripheral destinations: triple helix as a driver. In D. Jalincic and Y. Mansfeld (eds.), *Creating and Managing Experiences in Cultural Tourism* (pp. 255–271). Singapore: World Scientific.
Tidd, J., and Bessant, J. (2018). *Managing Innovation*. London: Wiley.

28

CREATIVE STAR WINE DISTRICTS

An exploratory research

Pasquale Sasso, Anne Gombault, Fabrice Thuriot, and Ludovico Solima

Introduction

Wine is more than just a fundamental local economic resource or an important part of gastronomy. Wine's features exalt it as a 'cultural and creative product.' If wine is a creative product, the industry of wine rightfully belongs to the category of creative industries (Sasso 2016). The global wine competition (Carlsen 2004) has developed various wine regions, among them renowned attractive wine regions known as star wine districts (Rosen 1981). Their success stories require not only the talent of individuals but many assets and involvement of strategic partnerships among local players. In fact, in recent years, the contrast between the old and the new world of wine has stimulated what we can define as a 'creative turning point' in the sector, similar to other creative industries (Caves 2000). Star wine districts have put in place an 'offensive' strategy based on creative tourism (Richards and Wilson 2007). In this context, Italy and France have always represented the world reference points in the world in the field of quality wine production and tourism. It is within these two countries that the research has been conducted, through a multicase comparative analysis, on the 'ingredients' of the success of three wine regions in particular. Resulting from a PhD work (Sasso 2016), this chapter tries to describe and understand the main features of a creative star wine district by showing how the wine industry can cooperate with other creative industries and local institutions in order to develop this ambitious project. This work is composed as follows: The next section presents the theoretical framework, followed by a description of the methodology. The cases are then presented, and the final section discusses the main results and conclusions.

Theoretical framework

Wine in economics of beauty

Wine has always represented a fundamental element in humans' table, and its symbolic and aesthetic features have encouraged some cultures to associate it with various meanings of cultural and creative nature (Gombault 2010). Wine is a cultural heritage. It is possible to classify wine's cultural value in two typologies of heritage: *tangible and intangible*. In the first category, there is the natural heritage (geology, landscapes) and its conformation (terraces, small

DOI: 10.4324/9781003143628-33

walls), wine cultural heritage (terroir, types of wine, savoir-faire, production instruments), and built heritage (vineyard villages, chateaux, and wineries). The second category, represented by intangible heritage, sees the presence of artistic and literary testimonies on wine, folk heritage (rituals, traditions, festive moments), and the art of gastronomy (Deyrieux 2011). Nowadays, wine culture and vineyard landscapes are celebrated by important awards, like a chance to be entered in UNESCO's lists. Finally, wine is 'a localized cultural good,' which, in the era of glocalization, becomes the spokesperson of a territorialization process, considering its status of 'heritage, founder of identity' and its ability to 'produce authenticity' (Crenn and Techoueyres 2007). Great brands have asserted their leadership on the market, heavily promoting their history, which has conferred authenticity on them (Beverland and Luxton 2005).

The idea that wine is a creative product is the key to the competitive advantage of new world wine. Robert Mondavi defined wine as 'the liquid art' (Roberto 2002), because its production process is a cultural combination of emotions, creativity, savoir-faire, wine branding, bottle design, wine label, and wine cork. Beyond the aesthetic and emotional features, it is possible to affirm that the wine industry is a creative industry since it is possible to apply the seven properties that, according to Caves (2000), characterize industries operating in creative fields. In the field of creative industries, wine can specifically make part of the style industry in so far as it produces style goods with predominant sensory and aesthetic features (Barrère and Chossat 2004). Besides the physical features of wine, some scholars were engaged in underlying the aesthetic nature of wine, and it is easier to understand why adjectives such as 'wonderful' are associated to wine (Charters and Pettigrew 2005).

Culture and creativity as drivers of territorial development

The definition of cultural district refers to urban areas and represents a concentration of activities connected to intangible elements. For this reason, we talk about 'cultural driven' districts (Sacco, Ferilli 2006). Valentino's work considers the topic of cultural districts in a wide manner, since according to him, the cultural sector is presented as a whole made up of cultural goods, performing arts, contemporary arts, music, design, fashion, and taste industry. According to Valentino (2003), the cultural district should organize itself around its 'most valuable cultural good.' It has also the duty to bring an advantage to the cultural section and to the territory on which it exercises its effects. The importance of Valentino's work resides in the fact that he integrated the cultural district with a net of relations, whose central node is represented by the process of valorization of the main asset, made up of cultural goods. The other nodes represent other resources of the territory (environmental and cultural manifestations), human resources (professionalism and talents), social (identity and values), and infrastructures (from transport to leisure time). Santagata (2001) emphasizes the importance given to the cultural capital of a specific place, namely local culture, traditions and sediment social capital, explicit and implicit knowledge, intellectual, aside from its tangible forms (Sacco and Ferilli 2006). Last, when talking about the topic of cultural districts, it is important to mention the definition of 'evolved cultural district.' According to the definition, this district has the ability to govern knowledge's production processes created within it (Sacco et al. 2013).

The concept of creative region, instead, is well represented by territories where the innovative climate and cultural field attracts artists and researchers from outside. A creative region is characterized by the presence of economic and social infrastructures, where artists and creativity play a central role. But what makes a place creative? The relation between creative industries and a place is the result of a balance among factors. According to Comunian, Chapain, and Clifton (2010), four dimensions make possible the birth of a place of the creative economy:

(a) infrastructures; (b) governance; (c) soft infrastructures (all those activities that contribute to the creation of a place's image and identity); and (d) market. However, according to Hagoort (2009), the idea of the creative region demands:

- A strong cultural identity, supported by the presence of personal social and economic infrastructures
- The presence of a city with strong relations in the scenery of small urban and extra-urban centers surrounding it
- The ability of cultural local institutions to establish solid connections with public and private cultural initiatives which aim at the enhancement of a territory in a double way: the creation of new audiences and the professionalization of cultural initiatives
- The presence of a deep connection between the professional and amateur art
- A system of relation among companies and scientific and vocational institutions
- The central role of artists and creators in every sector
- The strategic role of local institutions in order to promote cultural policies

In this context, and according to the new economy of culture, the competitive advantage of territory not only lies in the richness of raw materials (in their traditional connotation), manufacturing industries, or specific geographic conditions, but also in the creative and innovative power which exists in a place or a territory (Throsby 2008).

Finally, thanks to the media, the tourism sector has been facing increasing awareness, information, and prepared demand in recent years. In this context, tourist destinations need to tame how they can stand out from the crowd and attract the attention of tourist demand. According to Richards (2009), there is a shift from the demand for tangible to intangible resources, including tourism. In fact, while previously the tourist offer was dominated by museums and monuments to which it was almost 'obligatory' to visit, today the tourist is looking for participative experiences, to stop playing the role of spectator and become more the protagonist of his or her trip.

Methodology

In a global market, territories compete with each other, trying to become more attractive in order to accelerate social and economic development. The context where local meets global is called 'glocalization' (Robertson 1995). When territories are exposed to global competition, a glocal strategy based on regional resources shows its effectiveness as to the development of a creative wine district becoming a valid solution.

Research goal and question

The exploratory goal of this research is to understand and describe what a 'creative star wine district' is, which its characteristics are, which creative movements and processes are generated within it, and which topics are involved in this mechanism (Sasso 2016). The research question is formulated as: 'What are the main features of a creative star wine district?'

Research strategy

In order to answer this question, this phenomenon was observed through the story of some particular territories: the vineyard landscapes. Through a qualitative research approach, three case studies of wine districts were compared, all of which characterize and distinguish themselves in a different way from one another, due to the characterizing weight of a resource of extraordinary value: wine. The choice of the case studies sample is the result of the union of some different tools (Della Porta 2010). The first tool is the Expanded model for local development, with its competitive factors: *system territorial attractions* (social and cultural, heritage, habitat and landscape, leisure, craft and food), *system of touristic structure* (touristic infrastructure, transport, accommodations, information services), and *system of culture and creativity* (culture industries and creative industries) (Solima and Minguzzi 2014). Moreover, the choice of triangulating these three wine districts in the sample came from their potentialities, as: old wine-growing tradition; the extension of lands under vineyards; presence of 'quality marks' (DOCG, AOC, for example); and notoriety as touristic destinations. Consequently, it chosen three 'stars' (Rosen 1981) among world wine regions through the interaction of these tools: the Langhe-Roero Monferrato district and Chianti district for Italy, and the Saint-Emilion district for France.

Data collection and analysis methods

The choice of who to involve in the interviews was preceded by a series of reflections on 'where,' 'when,' 'who,' and 'what' to observe. Ultimately, respondents were selected based on their representativeness (Merriam 2001). Research reports, newspaper and magazine articles, and specialized websites were used to analyze the three case studies. Finally, in order to obtain information that had the same starting point of view, in addition to consulting secondary sources, 33 semi-structured interviews were carried out. These interviews were organized within them to wine producers and their consortia, cultural and creative bodies, public institutions, between tourist bodies, municipalities, and provinces.

Case studies

The observations collected during the study cases, together with the analysis of the interviews to some of the actors who work on the territory, have shown the presence of three dominant topics (Sasso 2016).

Case Study 28.1 The Saint-Emilion district

This region is a territory within the southwest of France. This district, one of the wine districts of the Bordeaux region, a few kilometers north of the city of Bordeaux, with special attention to the grapevine. The whole district is composed of eight municipalities, and Saint-Emilion (Figure 28.1) is a small 'capital' that every year welcomes more than one million tourists. The quality and variety of the wines of Saint-Emilion are linked to the diversity of terroirs and the skillful blending of grape varieties, mainly Merlot, Cabernet Franc, Cabernet Sauvignon, or Malbec. This area is known as 'the hill of 1000 Châteaux' where, more than museums, producers are ambassadors of the territory. This cultural landscape, as the first vineyard landscape in the world, was listed as a UNESCO World Heritage Site in 1999.

Figure 28.1 The Saint-Emilion district.

Source: Pasquale Sasso

Cooperation among actors and resources for a creative dynamism

In Saint-Emilion, various activities in which producers are involved are organized. However, the free or guided visit to the wine cellars and the wine tasting are not enough to fight off the fierce competition. Their winemakers take advantage of the local architecture and gastronomy. Architecture is a fundamental element that differentiates this amazing landscape, dotted with these classic medieval châteaux, from others. Nowadays, modern structures, perfectly integrated with the landscape and designed by famous architects such as Mario Botta, are pulled alongside the châteaux. Over the last several years, winemakers decided to break the rule of classicism and embrace modern paths when planning their wine cellars; representative examples are Château Cheval Blanc. In these places, modern architecture and advanced technology represent an element of attractiveness. The architecture of the châteaux has become a real brand that identifies this region, making it different from others in the rest of the world. Concerning gastronomy, many châteaux such as Château Ambe Tour Pourret have organized local and gourmet cooking classes or dinners in wine cellars such as Franc Mayne. Some others have decided to open restaurants inside their châteaux. Music, especially jazz and classical music, seems to be the second great passion felt by winemakers. In fact, concerts and other music events are often held in the châteaux, Châteaux Angelus being an example. Some winemakers have realized perfumes with wine aroma, as in the case of Franc Mayne. Others, such as Château Fombrauge, offers the visitor the once-in-a-lifetime chance to produce wine for a day from the flavor until the label. Finally, others participate in cultural programs in which châteaux host one or more artists who realize works of art for it or promote cultural events. Among all the events, some have the incredible ability to attract visitors and local actors; they are the Jazz Festival and Philosophia Festival.

Governance and collective action

The cultural and tourism offer in Saint-Emilion shares a strict relation with the Tourism Office. The latter, created in 1930, is one of the oldest in France and enjoys an exclusive autonomy and independence in comparison to others. The Tourism Office has an associative nature, and for this reason, it is not bound to territorial institutions' ties. Its budget amounts to €2 million, and just 5% come from municipalities. According to its director, this independence is fundamental to the office, necessary to the creation of an efficient and quality tourism offer. Consequently, people are deeply satisfied with the actions taken by the office, defining it 'absolutely essential.' The Tourism Office promotes a series of events that combine wine with other creative sectors. This organization collaborates with more than 400 structures on the territory, including restaurants, artisans, wine shops, and more than 150 châteaux. The link with local producers is so strong and witnessed by the fact that the president of the Office is a winemaker.

For instance, the City Hall is especially focused on the recovery of historical buildings, but this is also due to insufficient financial resources. Another important actor of this territory is the Wine Council, which has the duty to promote wine and its denomination. Saint-Emilion is the most concrete example of a cultural and tourism offer, in which the role of coordinator and intermediary between the private and public sector and between châteaux is performed by the Tourism Office.

The perception of UNESCO recognition

Since the district entered the World Heritage List in 1999, the Tourism has done a first-rate job and takes enormous pride in its territory; in fact, they use the brand UNESCO when making marketing decisions. Additionally, local people derive great satisfaction from being part of the World Heritage. Reading the Tourism Office's annual report from 1999 until today, it is possible to notice a massive increase in the number of tourists visiting the region. This figure has also been confirmed by the director of the Office, who estimated a substantial increase of 20% in the number of tourists from 1999 until today. In fact, according to him, this award represents the key to success because: 'people that come from Bordeaux could make choice Saint-Emilion or Médoc, the UNESCO represents an element of differentiation for us in comparison with other wine regions.' In the end, it seems that the award has favored the development of events and festivals focused on the relationship between wine and other creative industries in the region.

Case Study 28.2 The Langhe-Roero Monferrato district

This region is a recent tourist destination in southern Piedmont, and its 'capital' is Barolo (Figure 28.2). The process of transformation of the territory from a territory with a predominantly rural vocation to a tourist destination was started in the late 1990s, and the morphological characteristics of the area have determined its tourist identity, with a clear food-and-wine matrix. The main factors that have allowed the Langhe to become a globally known tourist destination today are: the hilly and rural landscape of viticulture; the culinary culture, strongly rooted in the area and an element wisely translated also in a tourist key; the consolidation of the local culinary tradition and its combination with haute cuisine and high-level gastronomic experimentation (the 2021 edition of the *Michelin Guide* lists 17 starred restaurants from this area); and the entrepreneurial attitude, as from here come highly successful entrepreneurial experiences such as Ferrero.

Cooperation among actors and resources for a creative dynamism

In the Langhe, the wine industry and those connected to it have been able to cooperate synergistically with other industries in the creative sector. The existence of a close relationship between wine and architecture is clearly evident from the new cellars designed by internationally renowned architects, in order to create artistic structures that are more innovative and sustainable. Many wineries have created as many private museums related

Figure 28.2 Barolo town.

Source: Mario Ferrero

to wine. The first and most famous of these, the Ceretto family, really kicked off the positioning of the Langhe as a place of culture particularly suited to contemporary art. The birth of the wine museum in Barolo is certainly an important piece of the coherent overall strategy of interrelationships between wine and cultural heritage. The WiMu is based in the Falletti castle in Barolo and is managed by a specially established private structure, the Barolo & Castles Foundation, which has networked the Barolo castle with other important local manors directly linked to the world of wine. Finally, the relationship between wine and gastronomy is perhaps the most immediate. Whether it is to keep a tradition alive or to innovate, the relationship between wine companies and cuisine is close and dense. The cooperation between private economic players, such as the local wineries, becomes evident if one thinks of the network projects between the wineries themselves. Among the many, it is worth mentioning that of the Strada del Barolo and Grandi Vini di Langa and the Consorzio Vini del Piemonte. The wineries find different ways of collaborating, giving life to shared projects at the regional level which, not surprisingly, find their origin and headquarters here, such as, for example, Piedmont Land of Perfection, which was born in 2011 to offer consortia for the protection of Piedmont's wine a meeting point to discuss, identify common operations and strategies useful for the promotion of Piedmont's wine. Piemonte Land, which today represents all 14 wine consortia, promotes the over 44,000 hectares of Piedmont's vineyards. Also, the Collisioni festival and Barolo Fashion Show can be seen as virtuous examples of public–private collaboration, receiving funding from local authorities and gathering numerous private sponsors.

Governance and collective action

The systemic strength cannot fail to be traced back to the presence, since the early 2000s, of a Local Tourist Agency particularly suited to carry out a destination management service and to make the Langhe brand known well beyond local borders and nationally, with a promotion at a European and international level that was ahead of the geographically close or similar destinations from the point of view of positioning and specificity. Today, the local tourism body is also at the forefront in the form of governance, as demonstrated in 2019 by the merger with the closer Asti ATL to give life to a common tourist company, which has competence over a large part of the region.

The perception of UNESCO recognition

The candidacy of this region as a UNESCO World Heritage site was long and complex. The process that led to the application dossier involved the various components of the civil society concerned, which has evolved around a double principle:

- The visibility and prestige that this recognition would have generated for the territory, especially in terms of tourism.
- The adoption of territorial development policies aimed at sustainability and the virtuous management of the environment and context.

Six years after obtaining this recognition, it is possible to state that the territory has benefited as a whole, especially, in terms of communication and notoriety of the area as a tourist destination. According to the Langhe-Roero and Monferrato Wine Landscapes Association, in the first six years on the List, the inclusion has contributed to the increase in tourist flows in recent years in the hilly areas with a 20% increase in presence and an estimated economic impact on the entire area of about €425 million.

Case Study 28.3 The Chianti district

The actual notoriety of the Chianti region is clear: wine, the agricultural vocation, and its location in the heart of Tuscany. Nowadays, the Chianti region is more extensive than the area of Chianti Classico production, which is more delimitated, including only wine-related activities (Radda in Chianti appears in Figure 28.3). In this region, typical productions continue to represent a source of economic and social development for residents. Chianti represents an evolutionary cultural landscape, based on an evolved agriculture made up by many small farmers as well as landowners, that, with their farms, castles, and residences, make this region a landscape that leaves one breathless. In fact, after the end

Figure 28.3 The Chianti district.

Source: Claudia D'Aliasi

of the Second World War, residents abandoned their farms in order to find better conditions in cities. Later, many foreigners decided to live in this area, buying old abandoned farms and transferring their residence in these places. This phenomenon is known as 'Chiantishire' (Sasso 2016).

Cooperation among actors and resources for a creative dynamism

The peripheral offer of Chianti cellars seems to be oriented towards two different directions: gastronomy and relaxation. Regarding relaxation, many companies have decided to build agritourism inside their structures, giving visitors the chance to spend a night there. Regarding gastronomy, some companies sustain wine production with the cultivation of olive oil, saffron, cheese, cured meats, and other products of local cuisine. In this region, it is easy to find wonderful medieval castles, abbeys, and rural houses but also innovative wine cellars, in perfect harmony with Tuscan hills dotted with vineyards, olive trees, and poplars. Chianti is also a synonym for luxury. Some wine cellars have decided to become luxury resorts and venues for unforgettable weddings. Even though gastronomy has achieved unprecedented success, many other farms, for financial, historical, and family reasons and for their owner's sensitivity, have decided to use art and culture as an element of competition. Some have drawn upon the historical heritage of their family, opening museums, while others have aimed at recovering ancient methods of cultivation and dry-stone walling in collaboration with universities. On the contrary, some companies have decided to use creativity in order to stand out from the crowd. Chianti is also the land of star chefs who search for a creative and quality cuisine. Moreover, music establishes fruitful collaborations with wine, examples are the festival Blues Divino and the music festival Chianti in Musica. Furthermore, visitors also have the chance to embark on tours focused on the discovery of the new architecture. An example is the project New Art Project and Wine in Tuscany. The project aims at entertaining visitors with a new tour in order to get to know wine cellars designed by great architects.

Governance and collective action

To discuss critically a region whose territorial borders are not well defined and still challenged with arms and marketing strategies between Florence and Siena means going beyond the brand famous all around the world. It especially means looking at the singularities and difficulties of the territory. Chianti is a tiny plot of land at the center of Tuscany, challenged between the provinces of Florence and Siena; this element has its own consequences in terms of organization and territorial identity. Even though this place seems to be the birthplace of the 'campanilismo,' a sort of local pride, remarkable efforts have been made at an institutional level to reduce these differences, making Chianti a single territory. However, the cultural events are still different from one place to another and independent among them.

In the Chianti region, in 2011, the regional agency Toscana Promozione was created. Moreover, in Chianti, various municipalities and hamlets independently organize events to promote the whole territory. In this region, as well as in other wine regions, 'Pro loco' and consortiums perform a fundamental role. In Chianti, a sense of general confusion about 'who has to do that' has emerged from the interviews. As a result, institutions are distant from people and often are judged inefficient. This situation does not favor collaboration between private and public institutions. Moreover, the local community has a strong sense of participation, and the social capital seems structured; but a sense of coordination, which may be carried out by a mid-level institution exclusively engaged in the promotion of the territory, is missing. In conclusion, Chianti cannot be considered the result of a 'top-down' model for its territory, its history, and its administration but a varied system in the competences and in the responsibilities that need to be greatly improved.

The perception of UNESCO recognition

How is it possible that one of the most celebrated wine landscapes in the world does not nurture the ambition to be recognized as a World Heritage site? Acceptable explanations may justify the absence of this territory in the WHL. Some of those are the vastness of the territory, the absence of clear territorial borders, and belonging to different provinces. These explanations may impede people from thinking about this territory as unique. The second and more important explanation, emphasized by interviewed subjects, is represented by the well-founded fear of turning Chianti into 'an open sky museum.' According to the sample, the award of Chianti in the WHL would be a synonym of quality and a way to appreciate the admirable job done during these years. Inside the sample, while provincial institutions strongly stress the importance of Chianti as one of the most famous vineyard landscapes in the world, emphasizing greater efforts than benefits coming from to stay on the WHL, but the prospective changes completely in the small towns. Finally, according to winemakers, the award instead may constitute a serious impediment to the process of evolution that has characterized the area during these years, turning Chianti into the great place everyone knows.

Discussion and conclusion

The comparison of the three case studies allows us to formulate proposals on the characteristics of a creative star wine district.

Case analysis: different approaches to the same territorial issues

From the comparison of these cases, a composite picture of three similar regions has been built up. In all cases, the adopted competitive strategy focuses on three elements: wine, gastronomy, and architecture. In all regions, wine producers have turned the severe competition into an element of creative dynamism. Anyway, if the majority of them have concentrated on the quality of products as a tool to stand out from the crowd, some have chosen a peripheral creative vision.

As also analyzed by Giacosa et al. (2019), Italian wine producers have paid close attention to activities engaged in the promotion of wine and other local products such as oil, saffron, cheese, cured meats, and food service through agritourism and accommodation. As claimed by Faugère et al. (2013), on the other hand, in Saint-Emilion, wine producers have widened their peripheral offer focusing on those activities concentrated on wines, such as visits to vineyards, châteaux, and tastings. Concerning the way producers start collaborations with local actors, in all cases, producers are not very keen on cooperating with each other. From the point of view of a total and intersectoral competition, all regions encourage a deep awareness that the visitor's quality of experience should depend not only on the features of the main attraction (wine), but also on services' quality and efficiency such as hotels, restaurants, transport, and cultural events. In all the regions examined, practical initiatives, in which wine producers work with local actors and institutions collaboratively, are widely developed in order to lay the basis for a complete cultural offer.

The topic of territorial governance shows a different picture between Chianti and the other two cases. In Chianti, people feel a great sense of confusion about 'who has to do what.' Based on their opinions, it is not yet clear how the regional office will organize, coordinate, and promote the territory. Consequently, institutions are totally detached from people, and they are often judged inefficient by those working in the territory. Thus, this unstable situation does not favor the collaboration between the private and public sectors. Associative institutions such as consortiums and cultural associations, which join local communities with the different institutional and organizational bodies present in the territories, belong to this context. Moreover, in Chianti, the sense of belonging to the local community is strong and the social capital is well structured; but the presence of a coordinating body, such as a park or a foundation engaged in the promotion of Chianti, is missing. Centralizing the promotion of a territory in the hands of an agency detached from people does not help integration either among private citizens or among the public and private sectors. Moreover, local and national institutions should overcome some practical problems that arose from a detailed reading of the interviews. In line with Santoro, Venturi, and Agnoletti (2020), one problem is the lack of commitment by institutions to promoting the territory through the aid of territorial marketing. In fact, since the cultural offer is truly varied and disorganized, investing money in public relations to inform visitors about events in Chianti, without any administrative or border distinction, would be of paramount importance. The second problem concerns the scarcity of infrastructural resources in the territory. In Chianti, the absence of the railway transport system does not facilitate the discovery of the beauties and the excellence of this place. As a matter of fact, this place can be just visited by car, with tourists driving through scenic but winding hill roads. In conclusion, Chianti, for its historical, territorial, and administrative features, cannot be considered the result of a 'top-down' development but a varied system for competences and responsibilities. This system needs to be substantially improved.

In the other two cases, the situation radically changes. Here, tasks have been clearly assigned to local actors. All of them have recognized the territorial leadership of the Tourism Offices, which represents the focal point of the cultural and tourism offer. In line with Giacosa et al. (2019), these offices are able to coordinate local actors, recommending a varied cultural offer, and it could easily represent the example of a typical 'destination marketing organization.' Therefore, Saint-Emilion and Langhe are representative examples of what is known as 'community destination.' This type of destination is characterized by a strong integration among local actors and is launched on the market as a brand.

Another topic of this work concerns the recognition of cultural landscape by UNESCO of wine territories around the world. The answers have outlined a complex situation. In Saint-Emilion, Langhe–Roero, and Monferrato, all the interviewed expressed great appreciation towards the UNESCO award. The number of tourists has enormously increased, encouraging the development of creative projects on the territory. As claimed by Peira et al. (2018) and Gullino and Larcher (2013), therefore, the award has transformed the territory for good, not only as a wine landscape but also especially as a region rich of history, art and culture.

In line with Brunori and Rossi (2007) in Chianti, there are contrasting opinions. First, there are various reasons why Chianti does not make part of the World Heritage List. They are: (a) the vastness of the territory; (b) the administrative responsibility (it is important to remember that Chianti is collocated in two provinces); (c) the scarcity of financial and human resources to promote and enhance a project like this. Concerning the potential advantages deriving from the award, people aired conflicting opinions. People agree on the fact that the award would result in only a small increase of Chianti's notoriety, given that it is already famous. However, it is important to point out the differences among opinions within the sample. On one hand, public bodies and cultural and creative institutions think that the brand UNESCO would represent a 'proof of quality.' On the other hand, vine growers are completely indifferent to the brand. They are also contrary to it because they fear the brand may introduce further limits and Chianti would be transformed in an 'open sky museum.'

Contribution: common features of creative star wine districts

The three cases have common features. All of them have organized their social, productive, cultural, and touristic offers around their most prestigious resource, wine. All of them are open to a collaboration among local actors but less among producers, there is an ongoing professionalization of cultural offers and creativity is also expressed by the lively social fabric.

Moreover, the element that does not allow these stars wine districts to become a 'wine Disneyland,' a sort of playground for wine lovers, is cultural heritage. Unlike other wine regions that do not have such a widespread cultural heritage and thus have to resort to the spectacularization of the tourist offer, these districts can focus on an offer based on the heritage, creativity, and creative tourism trilogy. Aiming at the authenticity of the localized cultural heritage and promoting the cultural heritage creatively, and pursuing a strategy based on heritage, creativity, and creative tourism is the key to success for these wine regions.

Finally, in line with Richards (2009), who argued that a creative approach to tourism was the key to the competitive advantage of small local communities, as well as those that make up the wine districts examined in this chapter, perhaps the thing that most unites these districts of the wine and which allows them to be considered stars on the world scene, is that of understanding long before others that it was necessary to start building a tourist offer that would actively involve tourists and put them in close relationship with the daily culture of local communities. In fact, in a different way, in the three cases examined, activities are carried out that immerse the tourist in an all-encompassing experience, such as active participation in harvesting activities or the possibility of being a winemaker for a day making one's own blend, or even take the art of cooking and the use of local raw materials and matching with local wines through cooking classes. Finally, another element that unites these regions is the use of gaming as a tool to actively involve tourists, in laboratory activities, treasure hunts, or escape rooms, making the narration of tangible and intangible cultural heritage more accessible and involving a slice audience far from this type of destination, such as the youngest.

This work has tried to describe and understand the main features of a creative star wine district, but some limitations emerge. One observation that can be raised in this work is represented by the choice to focus on the role of a single node of an ambitious project, which necessarily requires the involvement of many other actors. The wine industry is an essential component in developing a creative wine region, but they will never realize this project with their work only. So, observing and describing the role of other creative industries and other economic and institutional actors involved in this process could be an important starting point for future research.

References

Barrère, C., and Chossat, V. (2004). *Intellectual Property Rights and Creative Heritages*. Presented at the Annual Meeting of the Society for Economic Research on Copyright Issues (SERCI).

Beverland, M., and Luxton, L. (2005). Managing integrated marketing communication (IMC) through strategic decuopling. *Journal of Advertising*, 34(4): 103–106.

Brunori, G., and Rossi, A. (2007). Differentiating countryside: social representations and governance patterns in rural areas with high social density: the case of Chianti, Italy. *Journal of Rural Studies*, 23(2): 183–205.

Carlsen, P. J. (2004). A review of global wine tourism research. *Journal of Wine Research*, 15(1): 5–13.

Caves, R. E. (2000). Creative industries: contracts between art and commerce. *Journal of Economics Perspectives*, 17(2): 73–83.

Charters, S., and Pettigrew, S. (2005). Is wine consumption an aesthetic experience? *Journal of Wine Research*, 16(2): 121–136.

Comunian, R., Chapain, C., and Clifton, N. (2010). Location, location, location: exploring the complex relationship between creative industries and place. *Creative Industries Journal*, 3(1): 5–10.

Crenn, C., and Techoueyres, I. (2007). Local heritage to singularize a wine terroir: the example of pays Foyen (Gironde, France). *Anthropology of Food*, S2(March).

Della Porta, D. (2010). *L'intervista Qualitative*. Laterza: Editori.

Deyrieux, A. (2011). Le vin, un patrimoine à valoriser, Vin, vignoble et tourisme, Espaces. *Tourisme et Loisirs*, 111: 15–19.

Faugère, C., Bouzdine-Chameeva, T., Pesme, J. O., and Durrieu, F. (2013). The impact of tourism strategies and regional factors on wine tourism performance: Bordeaux vs. Mendoza, Mainz, Florence, Porto and Cape Town. Mendoza, Mainz, Florence, Porto and Cape Town. *SSRN*, January 16. https://doi.org/10.2139/ssrn.2201563

Giacosa, E., Rossi, M., Festa, G., and Ferraris, A. (2019). Wine and the 'spirit' of the territory: the Langhe case as a successful wine tourism destination 'system.' *Tourism Analysis*, 24(3): 291–304.

Gombault, A. (2010). Patrimoine, création et tourisme. Pour développer des 'régions créatives.' *Espaces, Tourisme et Loisirs*, 283: 8–9.

Gullino, P., and Larcher, F. (2013). Integrity in UNESCO World Heritage Sites: a comparative study for rural landscapes. *Journal of Cultural Heritage*, 14(5): 389–395.

Hagoort, G. (2009). Région créative. In Gombault, A., Livat-Pècheux, F., and Durrieu, F., *L'AlphaBEM des Industries Créatives* (pp. 249–252). Bordeaux: BEM Bordeaux Management School.

Merriam, S. (2001). *Qualitative Research and Case Study Application in Education* (2nd ed.). San Francisco: Jossey-Bass.

Peira, G., Beltramo, R., Pairotti, M. B., and Bonadonna, A. (2018). Foodservice in a UNESCO site: the restaurateurs' perception on communication and promotion tools. *Sustainability*, 10(8): 2911.

Richards, G. (2009). Creative tourism and local development. In Wurzburger, R., Pattakos, A., and Pratt, S. (eds.), *Creative Tourism: A Global Conversation* (pp. 78–90). Santa Fe, NM: Sunstone Press.

Richards, G., and Wilson, J. (2007). *Tourism, Creativity and Development*. London: Routledge.

Roberto, M. A. (2002). *Robert Mondavi and the Wine Industry*, Boston: Harvard Business School.

Robertson, R. (1995). Glocalization: time-space and homogeneity-heterogeneity. In Featherstone, M., Lash, S., and Robertson, R., *Global Modernities* (pp. 25–44). London: Sage.

Rosen, S. (1981). The economics of superstars. *The American Economic Review*, 71(5): 845–858.

Sacco, P., and Ferilli, G. (2006). Il distretto culturale evoluto nell'economia post industriale. *IUAV Università di Venezia, W.P*, 4: 1–28.

Sacco, P., Ferilli, G., Tavano Blessi, G., and Nuccio, M. (2013). Culture as an engine of local development processes: system-wide cultural districts. *Theory, Growth and Change*, 44(4): 555–570.

Santagata, W. (2001). Economia creativa e distretti culturali. *Economia Della Cultura*, 2: 167–174.

Santoro, A., Venturi, M., and Agnoletti, M. (2020). Agricultural heritage systems and landscape perception among tourists: the Case of Lamole, Chianti (Italy). *Sustainability*, 12(9): 3509.

Sasso, P. (2016). *Chianti vs Saint-Emilion. A double-case comparative study approach on creative wine regions*, Ph.D. thesis, Economics Department, Seconda Università degli Studi di Napoli.

Solima, L., and Minguzzi, A. (2014). Territorial development through cultural tourism and creative activities. *Mondes du Tourisme*, 10: 6–16.

Throsby, D. (2008). The concentric circles model of the cultural industries. *Cultural Trends*, 17: 147–164.

Valentino, P. (2003). *Le trame del territorio. Politiche di sviluppo dei sistemi territoriali e distretti culturali*, Milano: Sperling & Kupfer.

29

FOLK WINE FESTIVALS IN ITALY

A grassroots strategy for developing wine tourism

Michele F. Fontefrancesco

Introduction

From the North of the Alps and the Po Valley to the South of the Mediterranean Sea and its islands, folk wine festivals are a common recurrence in Italy. They are popular wine and food touristic events that take place mostly early in autumn, during September and October. Generally, they are not wine industry events, such as Vinitaly in Verona (www.vinitaly.com). Rather, they are touristic events held mostly in the rural centers over one or more days, during which the communities celebrate wine and grapes produced locally through kermesses that combine gastronomic events with recreational and cultural occurrences (such as concerts, art exhibitions, and dance nights), commercial initiatives (such as open markets and exclusive auctions), and other moments dedicated to viticulture (through presentations, tastings, and conferences).

Folk wine festivals are one aspect of the broader phenomenon of the Italian folk food festival, often called *sagra* (plural *sagre*, pronounced: /ˈsagra/ and / ˈsagre/) or, a more generic *festa* (plural *feste*, pronounced: /ˈfɛsta/ and /ˈfɛste/). To grasp the relevance of the phenomenon, in 2019, it was calculated that over 42,000 *sagre* were celebrated in the country with an estimated turnover of €900 million (Pascale 2019). Moreover, in the same summer, four out of five Italians considered these food festivals as one of the most appreciated attractions during the summer and one of the main drivers to visit rural areas because they enjoy buying and tasting the products they offer (Coldiretti 2019). Many hundreds among these festivals have names directly connected to wine and grapes, but in all the events, despite their name and gastronomic focus, wine is always present, being part of the Italian foodscape (Capatti and Montanari 2003; Montanari 2013; Sassatelli 2019) and wine production a key part of the Italian primary sector (Bravo 2013; Farolfi and Fornasari 2011; Guaschino and Martinotti 1984; Zangheri 1977).

This chapter presents the characteristics and history of these events and how they answer the needs of tourism and the communities. While the focus is specifically on folk wine festivals, considering their peculiarity and link with the broader phenomenon of the *sagre*, the chapter will also often refer to them and, where scholarship did not specifically investigate the folk wine festivals, will consider the data gathered about *sagre* in order to highlight the role of the wine festivals.

DOI: 10.4324/9781003143628-34

The chapter opens by presenting the most common characteristics of folk wine festivals. Then, it describes the specific history of these events and the role played by the Fascist regime in shaping the contemporary aspects of these celebrations. The chapter then investigates how folk wine festivals answer the needs of contemporary tourism and concludes by highlighting the main contributions of these events in the local touristic and economic development of rural communities.

The characteristics and history of folk wine festivals

Folk wine festivals, as most of the folk food festivals in Italy (Fontefrancesco 2020b) and abroad (Adema 2009; Blichfeldt and Halkier 2014; Kruger, Rootenberg, and Ellis 2013; Laing, Frost, and Kennedy 2019; Sabanpan-Yu 2007), are hallmark events, 'of limited duration, developed primarily to enhance the awareness, appeal and profitability of a tourism destination in the short and/or long term' (Ritchie 1984, p. 2). They come with straightforward names that identify the event, the products that are promoted, the edition, and the place where the festival is celebrated. An example is the 'LXXXIV Edizione della Festa del Vino di Caluso' (84th Wine Festival of Caluso) celebrated in 2019.

As most of the folk food festivals in Italy, they are not linked to the main urban centers (Laing, Frost, and Kennedy 2019). They are mostly organized by rural communities in order to promote local wine production. Their presence across the country can be explained by considering the characteristics of the Italian agricultural landscape. According to ISMEA (www.ismea.it), 666,400 hectares were planted with vines in 2020. Viticulture is conducted across the country, making wine, together with grain, the two most common cultivations of Italy. However, differently from grain, which is one of the most representative food commodities (Fontefrancesco 2020a), the wine market in the past two centuries transformed, making the idea of terroir and localization of production key in the perception of the product itself (Shapin 2012). Thus, through the use of these concepts, as well as the peculiarity of the vintage, the link between community, landscape, and product is established and the tourists reassured of the authenticity of the experience (Dixit 2020; Richards 2015). In this respect, thus, the wide presence of viticulture across the country is not a limit, but a resource for the spreading of wine festivals, intended as tourist attractions as well as a moment of celebration of the identity and history of the feasting community.

The deep history of the festivals stems from the late medieval ages (i.e., from the 12th to the 16th century CE). In fact, there is a red thread that connects modern events with the festivals organized by the rural communities to celebrate the end of the harvest season (Grimaldi 1993). They were important moments in which rural households opened their mainly subsistence economy to the outside through the organization of markets; an exceptional event in a context in which markets were rare in the countryside, being organized few times a year (Cipolla 1974; Di Francesco 2013; Fontefrancesco 2018; Grimaldi 1993; Le Goff 1988). These feasts had a crucial role for the communities but were not aimed at tourism. The touristification (del Romero Renau 2018, p. 1) of the wine festival started at the beginning of the 20th century and in particular with the invention of the National Grape Festival.

The 'First National Day of the Grape Harvest Festival' was celebrated on September 28, 1930 (Cavazza 1997, pp. 122–125), turning the initiative of Arturo Marescalchi, the then under-secretary of agriculture, into a reality. In fact, he wanted to find an effective way to boost grape and wine sales nationwide through an extensive schedule of festivals centered on grapes and wine. Before this celebration was made official, Fascism had already promoted and advertised single-themed festivals centered on specific crop production, such as strawberries, wheat, and grapes. These proved to be effective tools to relaunch local economies (Cavazza 1997).

From 1930 onwards, based on these first experiences, the Fascist regime enforced a vast national program of grape harvest festivals in order to revive the wine-growing and producing sector that suffered from the recession of the entire national viniculture market. These festivals had to be organized by local committees and had to conform to certain criteria: provide a contest for 'the best offer of grape sales' and set up a parade with people dressed up as traditional peasants and decorated festival floats, which represented themes linked to wine growing and agricultural life. The archive of the Luce Institute possesses several video accounts of some of the first grape harvest festivals (https://www.archivioluce.com/). The model of the Grape Harvest Festival became very popular and widespread across the whole of Italy, triggering a deep transformation in the way local, rural communities perceived their wine production, including its economic and tourism potential. It also radically affected the way in which local communities communicated and promoted their production, creating a standard that was continued throughout Italy after World War II and the fall of Fascism in the postwar period and beyond to the present day. Indeed, after 80-plus years, some of the grape harvest festivals created during the Fascist period continue to take place every year (Di Francesco 2016; Fontefrancesco 2014).

Contemporary folk wine festivals have different names, from the more generic name of *Festa del Vino*, wine festival, or *Festa dell'Uva*, grape festival, such as the one of Lu (see Case Study 29.1), to the one of local wine, such as *Festa dell'Erbaluce* in Caluso (see Case Study 29.2). They present programs centered on the promotion of grape and wine served in temporary restaurants and bars managed by local grassroots associations and organized for the event and other activities that range from religious services (e.g., Mass, benedictions, processions), to leisure and cultural activities (e.g., exhibitions, shows, theatrical performances), to official events (e.g., public speeches by local dignitaries, prize-giving ceremonies, parades).

Common elements that connect these modern initiatives to the model invented during the Fascist period is the presence of float parades as well as wine markets. Moreover, differently from most of the contemporary *sagre*, which date back at most to the 1960s–1970s, it is common to find folk wine festivals that count over 80 editions, having been originally launched during the Fascist period, although their contemporary structure may differ from the original ones (encompassing more initiatives aimed at a better exploration of the local wine production, starting from guided tours to the wineries, tasting session, thematic conferences, etc.), and the very origins of the festivals may be unknown to the local population that links it to a remote past (Giovannini 2009).

Matching the demand of the tourist

Folk wine festivals, as well as *sagre* in general, are highly regarded by national and international tourists (Di Francesco 2013; Guigoni 2019; Zannini, Cesarini Mason, and Ciani 2020). Their touristic success should be considered within the broader rise of food tourism that marked the beginning of the 21st century (Albala 2013).

Food tourism is a recent cultural and economic phenomenon that in less than 20 years has become one of the key areas of contemporary tourism (Dixit 2019; Garibaldi 2018b; Getz and Robinson 2014; Guigoni 2019; Hall and Sharples 2008). It is based on traveling, cultural encounters, and gastronomic experience (Hall and Gossling 2013; Kivela and Crotts 2006; Mkono 2011; Wolf 2006), making food the object through which tourists satisfy their demand for cultural and geographical otherness (Aime and Papotti 2012).

Italy has been one of the key international destinations for food tourism since the 1990s (Croce and Perri 2015; Garibaldi 2018a). The complexities of Italian gastronomy, as well as the wide variety of opportunities in terms of restaurants and other culinary attractions, have been important assets in matching the shifting expectations of tourists. In particular, while food tourists are interested in discovering new, unique products, tasting the flavors of different places as well as exploring new localities which can provide the experience of authenticity (Hall and Sharples 2008), the *sagre* have proved particularly capable of answering to these expectations (Skinner and Theodossopoulos 2011) by promoting a specific experience that draws from a specific bundle of selected foods and landscapes, both elements ostensibly unique to the event. In so doing, *sagre* addresses both the tourists' desire for a unique culinary experience and their search for new exotic and beautiful locations to be explored in the moment of their maximum splendor, turning food into an accessible platform for the cultural encounter. It is a way in which the tourist can access and explore the 'authenticity' of the countryside, appreciate the local heritage, and become part of that heritage through the 'genuineness' of their products, providing a memorable experience (Bessière 2013; Bessière and Tibère 2013; Brulotte and Di Giovine 2016; Timothy and Ron 2013).

Similarly to what is happening for the *sagre*, in general, the folk wine festivals appear to answer to the specific needs of wine tourists, in particular those who are interested in knowing more about the communities and landscapes that are at the basis of a particular wine and terroir, completing and boosting the wine experience (Perullo 2020).

The effect for the communities

In terms of touristic and economic development, folk wine festivals play a crucial role in the communities. First, in a context in which rural communities are experiencing a steep cultural and economic marginalization (De Rossi 2018), the organization of a festival answers to 'the desire of communities to put themselves on the map, creating positive images and symbols for themselves [providing the opportunity to tourists] to feel a part of social groups (even if they are contrived and last only for a day or two)' (Lewis 1997, p. 76). In particular, in a context of limited resources to invest in destination marketing, the festivals play a central role in reinforcing local recognizability, becoming the fulcrum for strategies of place branding (Hospers 2011). As in the case of Caluso (Case Study 29.2), thanks to its characteristics, the festival becomes the object through which the place is communicated to the public and the image of the locality is conveyed in the market, making the festival the actual destination brand (Morgan et al. 2011). In this respect, the festival is not the touristic destination, but rather the device through which the community is able to affirm its presence in the market, establish its reputation (Bell 2016), spur interest in the history and peculiarities of the place, and generate an inbound flow of tourists during the entire year.

The festivals, moreover, have positive economic effects. On the one hand, in the days of the events, and potentially also during the rest of the year, the festival is a moment of visibility for local producers. The event creates a privileged opportunity for enhancing exposure to final consumers who come to the festival and try the products. This first contact, which commonly ends up in a purchase, provides the opportunity of creating a more stable relationship with a consumer that may be interested in coming back to the place to buy its products or in ordering products from home.

On the other hand, the celebration of a festival generates market possibilities for all the other local economic activities: local shops, restaurants, and hotels. This actual effect of folk wine festivals can be inferred by considering the turnout of other *sagre*. In the case of the 2017 edition of the Toma cheese festival in Usseglio, a five-day event that attracted over 27,000 visitors to a community of 200 inhabitants, the direct economic impact (Attanasi and Poli 2017) was €86,670 (which includes the revenues of the restaurant and the events of the festival) while the indirect impact was €69,336 (which includes the purchases made from local shops and producers as well as the amount spent on accommodation and meals outside the festival restaurant). Even bigger was the impact of the Bell Pepper Festival of Carmagnola in 2016 (Attanasi and Rotondi 2016), a ten-day event that attracted 250,000 visitors to the city of Carmagnola. The direct impact was €1,333,718, while the indirect impact was €2,159,804.

In light of these figures, it is evident that the folk wine festivals can be a fundamental device for development and potentially able to create stable assets for their economies. However, the very success of the events may raise issues concerning, first, sustainability (e.g., Ding and Lee 2017; Hu 2010; Kruger, Rootenberg, and Ellis 2013; Meretse, Mykletun, and Einarsen 2016; Park, Reisinger, and Kang 2008; Wu, Wong, and Cheng 2014). In fact, it may lead to forms of local commodification (Gyimóthy and Mykletun 2009; Hall and Gossling 2013) and overintensification of production triggered by the intent of satisfying an always-increasing number of visitors.

Case Study 29.1 The Grape Harvest Festival of Lugano (AL)

This festival was held on the second week of September from 1969 to 2013 in Lu (nowadays Lu e Cuccaro Monferrato), an agricultural community on the border of the UNESCO site of Langhe, Roero, and Monferrato (Fontefrancesco 2014).

The festival aimed at promoting local grape and wine production. It was organized by a grassroots nonprofit organization, the Pro Loco Luese, with the twofold objective of involving the local population and relaunching tourism. In order to achieve these objectives, in 1967, the organizers decided to design the festival according to an example seen on television: the Grape Harvest Festival of Lugano, Switzerland (Busto 1992).

The Grape Harvest Festival of Lugano was first organized in 1933, and since the very first event, festival floats have been its main attraction (Chiesa 1949). Over time, whilst remaining anchored in the traditional theme of the grape harvest, the floats in the Lugano festival began to explore new combinations of themes, linking the agricultural ideal with elements drawn from cinema, fashion, and current affairs. Like that of Lugano, the organizers of the festival in Lu decided to organize a parade themed on wine. The parade managed to adapt a custom that was already known to the town, the procession of symbolic floats, and exploit the proven ability of at least some of the organizing committee to design and build floats. These elements combined to create a festival that was wholly unique to the Alessandria region of the 1970s. In addition to the float parade, every year, the Pro Loco Luese organized other activities, such as dance nights, art exhibitions, and wine tasting. Since the mid-1970s, the role of food became more and more prominent in the festival. The prominence of catering can be explained from an economic perspective: On the one hand, the restaurant was a profitable activity for the Pro Loco, which financed its annual activities through the profit generated by the meals served during the festival. On the other hand, it was a distinctive feature of the festival that stood out from the others, if not for the menu, for the quality of the food and its reasonable prices.

Apart from the expansion of the gastronomic offerings, the arrangement of the festival remained mostly unchanged from the 1980s onwards: a program taking place on one or two weekends in which the core attractions were the restaurant and dance, and then, on the second Sunday of September, in the afternoon, the float parade.

Thanks to its originality and the enthusiasm it was able to generate for over 30 years, the Grape Harvest Festival remained the key summer celebration for the village until the mid-2000s. It then started a decline that can be linked to the difficulties for the local Pro Loco to motivate local participation, as well as to the inability of the festival as it remained to continue attracting tourists for the event, in a landscape that over 40 years grew crowded with new food festivals and other touristic initiatives. In 2014, the unavailability of some facilities used for festival activities was the last straw, and the *sagra* was suspended.

Case Study 29.2 The Erbaluce Festival in Caluso (TO)

The Erbaluce grape variety is one of the indigenous white grapes of Italy and is officially registered in the National Vine Variety Catalogue since 1970. It is a versatile grape, thanks to the strong acidity and sweetness of its grapes. It is used to produce different types of wine, from dry to sparkling and sweet wines. Erbaluce production is, however, quite limited. In 2019 the esteemed area of production was restricted to 319 hectares, concentrated between Piedmont and Lombardy in the northern part of the country. Its main area of production is the Canavese, an area in the northwestern part of Piedmont on the border with Valle d'Aosta. The city of Caluso is the main center of production, and, after this city in 2010, it was established as a specific denomination that guarantees the quality and origin of this wine: the *Denominazione di Origine Controllata e Garantita Caluso*.

The institution of the denomination is only the last public acknowledgement of the deep bond that links this small city of 7,000 people with viticulture, a bond that the inhabitants of Caluso proudly date back to antiquity. Another important element is the *Festa dell'Erbaluce*, the Erbaluce festival, which is held in the third week of September. This successful event is able to attract tourists from across the Italian Northwest. Having turned its 87th edition in 2019, it represents a direct continuation of the innovations introduced to the celebration of folk wine festival by the Fascist regime during the 1930s.

In fact, the first edition of the Erbaluce Festival was organized in 1934 by the newspaper *La Stampa*. Following the success of the national *Giornata dell'Uva*, the Grape Day, celebrated since 1930 by the Fascist regime (Cavazza 1997, pp. 122–125) the newspaper sponsored the organization of a local *Festa dell'Uva*. The Erbaluce festival was organized as a one-day event that had as a key event a float parade and a beauty competition during which the organizing committee proclaimed the 'Albaluce Nymph,' the official testimonial for a year of Caluso wines.

The festival was suspended during the Second World War and resumed in 1948. Over the years, it became a multiday event, with various initiatives aimed at promoting the local wine and Caluso historical and cultural heritage, which include theatrical and musical performances, sports, and gastronomic events as well as wine exhibitions. At the center of the program was the nomination of the Albaluce Nymph and the float parade (Zampicinini 2007).

The festival begins on Friday evening with an opening parade in which the Nymph elected the previous year strolls for the last time in the cities of Caluso accompanied by a municipal band and representative teams of each neighborhood of the town dressed in folkloric costumes.

On Saturday, the festival continues with the food and wine initiatives. In particular, in the streets of the historic center, they set up gastronomic stands, the so-called *Veje Piole*, old taverns in the local dialect, where the tourists and locals can taste local wine and food.

On Sunday, in the morning, a solemn Mass is celebrated in the presence of the local authorities, as well as the Nymph of the previous year. At the end of the celebration, a new Nymph is proclaimed in the town council room. After the proclamation, a parade starts. Only the representative teams participate in this procession. At noon, the *Veje Piole* opened. In the afternoon, there is the float parade. The procession is open by the Nymph who parades sat on a horse-drawn chariot. Then the allegorical floats and folk groups follow, marching along the city streets. After the parade, representative teams compete in the Palio, which is not a horse race but rather a grape-pressing competition.

On Monday, the festivities end at night with a closing torchlight procession with the new Nymph, the representative teams, and the band. It follows the award ceremony of the best float participating in the parade and a firework show.

Conclusions

Overall, the folk wine festivals are an interesting sociocultural phenomenon and a potential model for harnessing the trends in wine tourism, however, with some potential shortcomings. The chapter has presented the main features of these events, focusing on the case of Italy, one of the most important countries in terms of wine production, in a context in which the 'exportation of the dolce vita,' the Italian lifestyle and foodways, appears to be one of the most promising directions for the national tourism industry (Confindustria and Prometeia 2016). The chapter has provided elements to understand how these events answer to some of the main trends in international wine tourism, both from the perspective of the tourists and the hosting communities.

While the past years were characterized by the spreading of new wine festivals and new *sagre*, it is open to questions concerning the impact of the COVID-19 crisis on the efficaciousness of these events in the post-pandemic period. However, at the moment, they appear an interesting model to be considered in order to promote wine tourism in marginal and rural wine territories.

Implications

The folk wine festivals show the potentialities of a format of touristic event based on local participation, voluntary work, and deep collaboration among the local economic actors (e.g., public institutions, local enterprises, and voluntary associations). This format appears particularly promising in reinforcing the link between community, place, and local wine. In so doing, the organization of a festival, such as the Italian ones, can be a useful initiative to reinforce marketing strategies aimed at creating a strong identification between specific wine productions and specific places/communities. However, the chapter shows the risks in terms of the commodification of products and places, and, heavily relying on voluntary and popular participation, the format may not be easily implemented without reaching a strong collaboration among the local economic actors and a strong engagement with the local community.

The folk wine festivals answer to the rising demand for food tourism and rural tourism. Thus they represent a fundamental opportunity to intercept this economic trajectory for those local rural communities. This is because the festivals propose a message coherent with the socioeconomic characteristics of the place and create an appealing interface through which tourists can access both the rural landscape and local food heritage. To reach their full potentiality, however, the festivals should be developed in communities where wine production plays a relevant role in the local agriculture or at least where consistent wine production development policies and initiatives are in place. This is to preserve and convey that sense of authenticity of an experience that tourists research in these events.

Finally, the folk wine festivals differ from other wine events in terms of being able to address a larger public, thanks to the combination of initiatives and experiences they provide. In this respect, they proved to be a powerful tool in order to introduce new people to the wine world and to wine consumption, expanding the market to new consumers as well as creating a strong, affective bond between them and specific products and production areas. A promising perspective, in particular, for those small wine producers that have otherwise difficult access to the market and limited media exposure.

References

Adema, P. (2009). *Garlic Capital of the World: Not Only Is It Good to Eat.* Jackson: University Press of Mississippi.

Aime, M., and Papotti, D. (2012). *L'altro e l'altrove.* Turin: Einaudi.

Albala, K. (ed.). (2013). *Routledge International Handbook of Food Studies.* London: Routledge.

Attanasi, G., and Poli, S. (2017). *Effetti economico-sociologo-turistici della valorizzazione del patrimonio culturale Piemontese: Il ruolo della 'Mostra regionale della Toma di Lanzo' sul turismo culturale nel territorio di Usseglio.* Usseglio: Usseglio Eventi.

Attanasi, G., and Rotondi, V. (2016). *Effetti economico-sociologo-turistici della valorizzazione del patrimonio culturale Piemontese: il ruolo della Sagra del Peperone.* Carmagnola: Città di Carmagnola.

Bell, F. (2016). *Beyond Place Branding? The Emergence of Place Reputation: A Comparative Study of Newcastle, Gateshead, Hull and Bristol* [PhD Thesis]. Newcastle University.

Bessière, J. (2013). 'Heritagisation': a challenge for tourism promotion and regional development: an example of food heritage. *Journal of Heritage Tourism*, 8(4): 275–291. https://doi.org/10.1080/1743873X.2013.770861

Bessière, J., and Tibère, L. (2013). Traditional food and tourism: French tourist experience and food heritage in rural spaces. *Journal of the Science of Food and Agriculture*, 93(14): 3420–3425. https://doi.org/10.1002/jsfa.6284

Blichfeldt, B. S., and Halkier, H. (2014). Mussels, tourism and community development: a case study of place branding through food festivals in rural North Jutland, Denmark. *European Planning Studies*, 22(8): 1587–1603. https://doi.org/10.1080/09654313.2013.784594

Bravo, G. L. (2013). *Italiani all'alba del nuovo millennio.* Milano: Franco Angeli.

Brulotte, R. L., and Di Giovine, M. A. (eds.) (2016). *Edible Identities: Food as Cultural Heritage.* London: Routledge.

Busto, G. (1992). La Sagra d'argento. *Al Païs d'Lü*, 16(9): 8.

Capatti, A., and Montanari, M. (2003). *Italian Cuisine: A Cultural History.* New York and Chichester: Columbia University Press.

Cavazza, S. (1997). *Piccole Patrie. Feste Popolari tra Regione e Nazione durante il Fascismo.* Bologna: Il Mulino.

Chiesa, V. (1949). *L'Opera della Pro Lugano.* Lugano: Arti grafiche già Veladini.

Cipolla, C. M. (1974). *Storia economica dell'Europa pre-industriale*. Bologna: Il Mulino.
Coldiretti. (2019). Vacanze, sagre e feste di paese per 8 italiani su 10. www.coldiretti.it.
Confindustria, C. S., and Prometeia (eds.). (2016). *Exporting La dolce vita. Italian BBF Products in New Markets: The Forces Tranforming Consumption*. Rome: Sipi.
Croce, E., and Perri, G. (2015). *Il Turismo Enogastronomico*. III ed. Milan: Franco Angeli.
De Rossi, A. (ed.). (2018). *Riabitare l'Italia. Le aree interne tra abbandoni e riconquiste*. Rome: Donzelli.
del Romero Renau, L. (2018). Touristification, sharing economies and the new geography of urban conflicts. *Urban Science*, 2(4): 104. https://doi.org/10.3390/urbansci2040104
Di Francesco, G. (2013). Sagre enogastronomiche e sviluppo locale. I correlati socio-antropologici. In Cipolla, C., and Di Francesco, G. (eds.), *La Ragione Gastronomica* (pp. 261–275). Milan: Franco Angeli.
Di Francesco, G. (2016). Food and wine tourism and urban local development. *Transnational Marketing Journal*, 4(2): 133–142. https://doi.org/10.33182/tmj.v4i2.396
Ding, D., and Lee, H. M. (2017). A measurement scale for food festival visitor experience. *International Journal of Tourism Sciences*, 17(3): 180–197. https://doi.org/10.1080/15980634.2017.1349069
Dixit, S. K. (2019). Gestronomic tourism: a theoretical construct. In Dixit, S. K. (ed.), *The Routledge Handbook of Gastronomic Tourism* (pp. 13–23). London: Routledge.
Dixit, S. K. (ed.). (2020). *The Routledge Handbook of Tourism Experience Management and Marketing*. London: Routledge.
Farolfi, B., and Fornasari, M. (2011). Agricoltura e sviluppo economico: il caso italiano (secoli XVIII-XX). *Quaderni-Working Paper DSE*, 756: 1–63.
Fontefrancesco, M. F. (2014). Of grape, feast and community: an ethnographic note on the making of the grape harvest festival in an Italian town in Piedmont. *Journal of Ethnology and Folkloristics*, 8(1): 75–90.
Fontefrancesco, M. F. (2018). Food festivals and expectations of local development in Northern Italy. *Ethnologia Actualis*, 18(2): 118–134. https://doi.org/10.2478/eas-2019-0006
Fontefrancesco, M. F. (2020a). Food commodity market: history and impact of food trading toward SDG2. In Leal Filho, W., Azul, A., Brandli, L., Özuyar, P., and Wall, T. (eds.), *Zero Hunger. Encyclopedia of the UN Sustainable Development Goals* (pp. 304–312). Cham: Springer.
Fontefrancesco, M. F. (2020b). *Food Festivals and Local Development in Italy: A Viewpoint from Economic Anthropology*. London: Palgrave.
Garibaldi, R. (2018a). L'enogastronomia in viaggio: da elemento accessorio a fattore determinante nelle scelte di viaggio degli italiani. In Becheri, E., Micera, R., and Morvillo, A. (eds.), *Rapporto sul Turismo Italiano. XXII Edizione 2017/2018*. Naples: Rogiosi Editore.
Garibaldi, R. (2018b). *Primo rapporto sul turismo enogastronomico italiano 2018*. Bergamo: CELSB.
Getz, D., and Robinson, R. N. S. (2014). Foodies and food events. *Scandinavian Journal of Hospitality and Tourism*, 14(3): 315–330. https://doi.org/10.1080/15022250.2014.946227
Giovannini, F. (2009). *La festa dell'uva di Firenze e dell'Impruneta: Storia di una tradizione inventata*. Florence: Florence Art Edizioni.
Grimaldi, P. (1993). *Il calendario rituale contadino: il tempo della festa e del lavoro fra tradizione e complessità sociale*. Milan: FrancoAngeli.
Guaschino, M., and Martinotti, M. (eds.) (1984). *Contadini di collina: viticoltura e condizioni materiali nella cultura orale del Basso Monferrato casalese*. Turin: Regione Piemonte Assessorato all'Agricoltura e Foreste.
Guigoni, A. (ed.). (2019). *Foodie con la valigia: Il turismo enogastronomico in Italia*. Rome: Aracne.
Gyimóthy, S., and Mykletun, R. J. (2009). Scary food: commodifying culinary heritage as meal adventures in tourism. *Journal of Vacation Marketing*, 15(3): 259–273. https://doi.org/10.1177/1356766709104271
Hall, C. M., and Gossling, S. (2013). *Sustainable Culinary Systems: Local Foods, Innovation, Tourism And Hospitality*. London and New York: Routledge.
Hall, C. M., and Sharples, L. (eds.) (2008). *Food and Wine Festivals and Events around the World: Development, Management and Markets*. Amsterdam: Butterworth-Heinemann.
Hospers, G.-J. (2011). City branding and the tourist gaze. In Dinnie, K. (ed.), *City Branding: Theory and Cases* (pp. 27–35). London: Palgrave.
Hu, Y. (2010). *An Exploration of the Relationships between Festival Expenditures, Motivations, and Food Involvement among Food Festival Visitors* [PhD thesis]. University of Waterloo.
Kivela, J., and Crotts, J. C. (2006). Tourism and gastronomy: gastronomy's influence on how tourists experience a destination. *Journal of Hospitality & Tourism Research*, 30(3): 354–377. https://doi.org/10.1177/1096348006286797
Kruger, S., Rootenberg, C., and Ellis, S. (2013). Examining the influence of the wine festival experience on tourists' quality of life. *Social Indicators Research*, 111(2): 435–452. https://doi.org/10.1007/s11205-012-0013-0

Laing, J., Frost, W., and Kennedy, M. (2019). Food and wine festivals and rural hallmark events. In Mair, J. (ed.), *The Routledge Handbook of Festivals* (pp. 285–294). London: Routledge.

Le Goff, J. (1988). *Your Money or Your Life: Economy and Religion in the Middle Ages*. New York: Zone Books.

Lewis, G. H. (1997). Celebrating asparagus: community and the rationally constructed food festival. *Journal of American Culture*, 20(4): 73–78. https://doi.org/10.1111/j.1542-734X.1997.00073.x

Meretse, A. R., Mykletun, R. J., and Einarsen, K. (2016). Participants' benefits from visiting a food festival – the case of the Stavanger food festival (Gladmatfestivalen). *Scandinavian Journal of Hospitality and Tourism*, 16(2): 208–224. https://doi.org/10.1080/15022250.2015.1108865

Mkono, M. (2011). The othering of food in touristic eatertainment: a netnography. *Tourist Studies*, 11(3): 253–270. https://doi.org/10.1177/1468797611431502

Montanari, M. (2013). *Italian Identity in the Kitchen, or, Food and the Nation*. New York: Columbia University Press.

Morgan, N., Pritchard, A., Pride, R., and Morgan, N. D. B. (2011). *Destination Brands: Managing Place Reputation* (3rd ed.). Oxford: Butterworth-Heinemann.

Park, K. S., Reisinger, Y., and Kang, H. J. (2008). Visitors' motivation for attending the South Beach Wine and Food Festival, Miami Beach, Florida. *Journal of Travel & Tourism Marketing*, 25(2): 161–181. https://doi.org/10.1080/10548400802402883

Pascale, A. (2019). *Sagre: ad Agosto il picco di eventi*. Rome: Federazione Italiana Pubblici Esercizi.

Perullo, N. (2020). *Epistenology: Wine as Experience. Arts and Traditions of the Table: Perspectives on Culinary History*. New York: Columbia University Press,.

Richards, G. (2015). Evolving gastronomic experiences: from food to foodies to foodscapes. *Journal of Gastronomy and Tourism*, 1(1): 5–17. https://doi.org/10.3727/216929715X14298190828796

Ritchie, B. (1984). Assessing the impact of hallmark events: conceptual and research issues. *Journal of Travel Research*, 23(1): 2–11. https://doi.org/10.1177/004728758402300101

Sabanpan-Yu, H. (2007). Cebuano food festivals: a matter of taste. *Philippine Quarterly of Culture and Society*, 35(4): 384–392. https://www.jstor.org/stable/29792629

Sassatelli, R. (ed.). (2019). *Italians and Food*. London: Palgrave.

Shapin, S. (2012). The tastes of wine: towards a cultural history. *Rivista di Estetica*, 51(3): 49–94. https://doi.org/10.4000/estetica.1395

Skinner, J., and Theodossopoulos, D. (eds.) (2011). *Great Expections: Imagination and Anticipation in Tourism*. New York: Berghahn.

Timothy, D. J., and Ron, A. S. (2013). Understanding heritage cuisines and tourism: identity, image, authenticity, and change. *Journal of Heritage Tourism*, 8(2–3): 99–104. https://doi.org/10.1080/1743873X.2013.767818

Wolf, E. (2006). *Culinary Tourism: The Hidden Harvest*. Portland: International Culinary Tourism Task Force.

Wu, H.-C., Wong, J. W.-C., and Cheng, C.-C. (2014). An empirical study of behavioral intentions in the food festival: the case of Macau. *Asia Pacific Journal of Tourism Research*, 19(11): 1278–1305. https://doi.org/10.1080/10941665.2013.844182

Zampicinini, F. (2007). La Festa dell'Uva. *Atlante delle feste popola*. http://www.atlantefestepiemonte.it/

Zangheri, R. (1977). *Agricoltura e contadini nella storia d'Italia*. Turin: Einaudi.

Zannini, A., Cesarini Mason, M., and Ciani, S. (2020). Gli Eventi. In Garibaldi, R. (ed.), *Rapporto sul Turismo Enogastronomico Italiano* (pp. 1–4). Bergamo: CLBS.

30

MATCHING BUSINESS TRAVEL WITH WINE TOURISM

The case of the canton of Grisons, Switzerland

Philippa Golding and Thorsten Merkle

Introduction

Wine tourism offers regions an opportunity to blend cultural and culinary elements, creating unique regional experiences attractive to a broad demographic and flexible enough to integrate with other tourism offers. Current trends in MICE (Meetings, Incentives, Conferences, and Events) tourism call for regional products, unique venues, and authentic experiences. We, therefore, suggest that culinary and wine tourism offers would be of interest to this sector. Wine tourism has the potential to offer meaningful connections to local culture, natural settings, and sensory elements to provide unique and sustainable experiences. Business tourism is often thought of as attracting international markets, while local businesses contributing to this sector are overlooked. However, local businesses represent an important market that can create less seasonal revenue and drive innovation in developing tourism offers. In this chapter, we examine key themes of wine tourism and business tourism and explore the case of wine tourism in the Bündner Herrschaft in Switzerland. We demonstrate the potential for combining wine and MICE tourism to support local value creation in tourism and associated sectors. The findings not only demonstrate potential for local companies to participate in local wine tourism, but also outline practical considerations for tourism suppliers creating offers for this sector.

Literature review

In this section, we describe key themes in wine and business tourism necessary to demonstrate their potential to interact. Along with an introduction to the Swiss wine region, the Bündner Herrschaft, this will give context for the case study presented.

Wine tourism

Wine has always played a central role in cultural, culinary, spiritual, and entertainment events (Anđelić, Garabinović, and Šormaz 2019). Wine tourism takes advantage of these interactions and blends them to create a diverse range of offers. As such, wine tourism can attract a range of demographics and easily integrate with other types of tourism to add value to various activity programs (Anđelić, Garabinović, and Šormaz 2019; Croce and Perri 2010).

DOI: 10.4324/9781003143628-35

Despite this potential, historically, little thought has been given to promoting wine tourism beyond the context of vineyard or cellar door tasting experiences (Macionis 1998). Only recently has the academic interest in wine tourism grown, with Anđelić, Garabinović, and Šormaz (2019) finding over half the papers on wine and wine tourism being published in the last decade.

Niche market

Wine tourism is part of the niche segment of culinary tourism (Poitras and Donald 2006). Niche markets have been one of the biggest recent trends as tourists seek authentic experiences and want to connect with the environment they visit (Morrison 2013). While the tourism experience is often seen as a product or service, experiential tourism offers, such as culinary tourism, should be seen as a consumption process, relying on the interaction between consumer and service. In that sense, the senses are being triggered to create memorable experiences (Meacci and Liberatore 2018). Tourists want experiences that enhance their senses, something wine tourism provides the opportunity to deliver, as culinary tourism offers are the only tourism products that can be experienced using all the senses (Dixit 2019; Everett 2016; López-Guzmán, Vieira-Rodríguez, and Rodríguez-García 2014). Regional variations in food, culture, and wine provide opportunities to create unique experiences, a source of competitive advantage for destinations. Consumption of food and drink, besides serving physiological needs, has also been shown to have a cultural dimension (Merkle et al. 2020). When connected to the local environment, it can serve as a means to enhance guest experience and build connections with nature and past times (Dixit 2020; Everett 2016). Visiting the locations where food is grown and produced gives visitors the opportunity to see where, how, and by whom it is produced, thus allowing them to connect with the people and the stories of the products. Visiting working farms or wineries to buy products, enjoy entertainment, eat a meal, or participate in activities are especially popular with domestic tourists (Morrison 2013).

Wine tourism is an important tool for regional development (Gómez, Pratt, and Molina 2019). In 2018, the UNWTO concluded that wine tourism offers opportunities for rural development and can be used within a cultural tourism approach to contribute to local economies and to create jobs (UNWTO 2018; Xu et al. 2016). Culinary tourism has the potential to stimulate and regenerate rural and peripheral areas and thus help to sustain and preserve cultural traditions (Everett 2016). Wine tourism allows guests to learn about the culture and traditions of winemaking and provide an opportunity to sample other regional products and learn about the stories behind them (Wine in Moderation 2018).

Winemakers

While wine tourism offers clear benefits to the tourism sector, it can also help vineyards and wine producers to establish customer relationships, promote brand loyalty (Koch, Martin, and Nash 2013), and be leveraged to improve the performance of wineries (Montella 2017). Additionally, participating in wine events and activities can help consumers overcome the reluctance felt by some who are intimidated by wine or visiting a winery for a tasting, thus attracting new segments (Getz 2000). Wine tourism has become popular in many parts of the world as its potential for regional and economic development is recognized (Gómez, Pratt, and Molina 2019; Morrison 2013). But while wine tourism may be able to balance unfavorable trends in wine sales, the seasonality of wine tourism has been noted as difficult to overcome (Poitras and Donald 2006).

Wine tourism destinations

The availability of unique food and wine in a destination has given rise to festivals and other events which appeal to tourists and locals and create opportunities to stimulate the reinvention of historical food traditions (Hjalager and Corigliano 2000). Wine tourism can thus help a region to develop a unique and positive destination image and overcome slow demand periods, as it can be used as a tool for the development of distinctive tourism activities (Getz 2000; Salvat and Boque 2009). Strengthening linkages between wine and other sectors can facilitate the development and marketing of a regional culinary image, encourage collaborative marketing, and increase the appeal of a destination (Macionis 1998; Winemakers' Federation of Australia 2011). To ensure the successful positioning of a destination, it is important to promote both a consistent image and a range of activities to satisfy the desires of the consumer (Everett 2016).

Stakeholder collaboration

Integrating stakeholders is a key requirement when professionally developing tourism offers (Knaus and Merkle 2020). Although wine tourism can exist with little or no coordination or planning, some form of marketing is needed to stimulate growth (Getz 2000). It needs to be noted, however, that wine tourism and winemakers have different priorities. Winemakers want to produce and sell their wine, and the tourism industry wants to attract visitors. As a result, any marketing and development strategies need to ensure there is a balance between them (Koch, Martin, and Nash 2013). This way, stakeholders are acknowledged in the partnership and can add value to one another rather than work against each other (Bruwer et al. 2017). Care should be taken to reduce competitiveness between wineries in a destination, as this could threaten the sustainability of individual companies (Poitras and Donald 2006).

Marketing

Unlike many products, in most countries, wine is marketed using the region of origin. Careful analysis of the image is necessary to inform marketing strategies as the perception of the region as a destination, and a wine region is therefore very important (Bruwer et al. 2017). Marketing wine tourism is a role that can be filled by destination management organizations (DMOs) but should be made in partnership with the suppliers of wine tourism products and offers (Getz 2000).

Wine tourism is a niche form of tourism; therefore, it cannot be marketed as mass tourism. When selecting a target market, wine destinations should aim to identify markets that fit the region and deliver a high yield, such as corporate events (Getz 2000). The corporate events sector is less seasonal than leisure tourism. Having pre-planned events or tours would allow the suppliers from the wine industry to better predict and manage the flow of visitors. Business tourism has been noted to contribute to mitigating seasonality and reducing environmental impact (Nicula and Elena 2014). While much research focuses on recognizing the needs of the wine tourist as someone having the primary motivation to visit a wine region, wine tourism can make an excellent leisure addition to a business trip and offers an opportunity to experience the destination and its culture and to interact with locals (Gómez, Pratt, and Molina 2019).

Questions about wine tourism's sustainability have been posed at the community level with regard to the impact on the natural environment, residents' lifestyles, and economic sustainability due to market shifts (Poitras and Donald 2006). Business tourists, on the other hand are generally considered high yield and have a minimal detrimental impact on the area (Abulibdeh and Zaidan 2017). Xu et al. (2016) suggest that wine tourism can, in fact, offer local residents sociocultural benefits and lead to an increase in quality of life; however, it is important to educate locals about the positive impacts of wine tourism.

Business tourism

Multiple definitions exist for business tourism. Generally, it may be considered as travel for the purpose of work-related activities that encompasses all aspects of the experience of the business traveler, being applicable to individuals or groups (Nicula and Elena 2014; Swarbrooke and Horner 2001). Business tourism often contains a leisure component (Abulibdeh and Zaidan 2017; Millán, Fanjul, and Moital 2016), although this is secondary to the primary purpose, which may be educational, governmental, or commercial (Nicula and Elena 2014). Business tourism represents the largest non-leisure type of tourism and is widely recognized as being the highest spending and least cost-sensitive category (Abulibdeh and Zaidan 2017; Holloway and Humphreys 2012; International Trade Centre 2001). Positive and memorable experiences can encourage business tourists to return to the destination for personal travel.(Chung et al. 2020; Mureşan, Chiripuci, and Nistoreanu 2017).

Segmentation

As with leisure tourism, business tourism may be segmented into domestic, inbound, and outbound travel (Mureşan, Chiripuci, and Nistoreanu 2017; Swarbrooke and Horner 2001) and contributes at both the B2C (business-to-consumer) and B2B (business-to-business) customer level.

B2C refers to business travelers using the opportunity to explore local areas in what has been called bleisure travel (Lichy and McLeay 2018). A global survey found 60% of business travelers find time to get out and explore while on business trips (The Economist 2019). From a B2C perspective, tourism organizations should try to appeal to individual business travelers to engage in activities during their free time; however, it may not be possible to distinguish these guests from other tourist groups. Bleisure travelers spend less time researching than leisure travelers will, so marketers have a smaller window to reach and influence them, although they also tend not to book activities in advance. This leaves opportunities to target them during travel by providing information on local offers or discount coupons (Chung et al. 2020; Expedia 2018).

B2B business tourism is generally thought of as attracting international markets to host events in a destination, and research into this field appears to focus more on attracting event attendees to participate in leisure activities. When considering local business partnerships, these event organizers are generally targeted to offer ancillary activities for event attendees. While this type of program stimulates additional spending and can improve attendee satisfaction, other sources of B2B customers are often disregarded (Chung et al. 2020). Local businesses contributing to business tourism in the region through their normal activities and internal events should be considered as an important B2B relationship.

International tourism often overshadows domestic tourism in literature and policy development (Bayih and Singh 2020). According to Canh and Thanh (2020), however, domestic tourism not only contributes to economic development but also creates jobs and helps reduce vulnerability to external influences on the economy when not overdeveloped into mass tourism. Local B2B customers can be a source of innovation in developing tourism offers or packages and should be engaged with, as it is their needs that facilitate the necessity for business tourism, and offers should create value for all involved (Tuan et al. 2019). When considering the interests of companies, a number of motivational factors have been recognized, including cost, location, accessibility, the standard of facilities, and an attractive image that appeals to the expectations of the business traveler (Mureşan, Chiripuci, and Nistoreanu 2017).

Types

MICE, short for variations of Meetings, Incentives, Conferences, and Exhibitions, is a sector of business tourism that includes a range of activities. Conferences and exhibitions are large-scale events that require specific infrastructure and are usually organized by specialized event services (Swarbrooke and Horner 2001). The large scale of these events suggests these may not be suitable from a B2B perspective for the wine tourism sector. Although wine tourism offers may appeal to participants as a leisure activity, it is by definition a niche sector that caters to smaller segments.

Meetings and incentives, however, provide a number of opportunities for wine tourism. Incentive travel and events use leisure tourism activities to reward productivity (Cook, Hsu, and Taylor 2018); however, the destination is not necessarily decided by the tourist (Swarbrooke and Horner 2001). Corporate events are the most common form of meetings and may include team-building and training events (Nicula and Elena 2014; Pike 2008). They are events funded by an organization for their staff, stakeholders, or customers, and they are planned for a specific purpose occurring outside of the normal activities of the organization (Getz 2012). These types of events often include experiences such as workshops, tours, or tasting sessions in the program to set them apart and provide a wow factor (Eventbrite 2019; Merkle and Lewis 2016). As with other forms of business tourism, this market is generally more resilient than leisure segments and can offer destinations the opportunity to generate additional business during off-peak periods (Pike 2008).

Both employers and employees benefit from holding corporate events outside the normal workplace setting. Social events can be used to help integrate new employees, reduce inter-departmental barriers, help teams recharge, bring management and employees closer, celebrate company and employee milestones, and add value to customer and stakeholder events (Thornley-Brown 2014). Good corporate events can boost morale, bond teams, and improve communication, all of which can lead to increased productivity (Leonard 2019).

While there is potential in the partnerships between DMOs, meeting organizers, and local businesses to improve the quality of business travelers' experiences and stimulate extra spending when attending conferences, DMOs do not typically invest resources into the development of team-building activities or pay attention to economic impacts. Rather, they tend to rely on event organizers to develop such programs (Chung et al. 2020). As many companies do not have internal event departments, corporate event organizers may be difficult to identify. While some companies may employ the services of professional corporate event planners, the task is often left to staff who are given the job of organizing the event in addition to their normal functions (Davidson and Rogers 2006). As a result, while wanting to plan the best corporate event possible, they have limited time to research and contact venues (Lavin 2015).

Trends

Business tourism is evolving, and organizers and suppliers should be aware of how trends and changes affect their events. The demand for experiential events has grown in the business tourism sector (CWT Meetings & Events 2020). Trends in business tourism are about engagement; as Millennials join the MICE industry, they are demanding experiences that enrich them and go beyond technology (Alderton 2017; Karantzavelou 2020). Today's attendees want to have multisensory and participative destination experiences and to return having learned and experienced something unique about the destination they have visited (Karantzavelou 2020; Campos et al. 2018; CWT Meetings & Events 2020).

Business tourism trends are for venues that tell a story and experiences which build a connection to the destination and give behind-the-scenes experiences with local experts (CWT Meetings & Events 2020). Travelers want to experience local culture, visit cultural sites, discover remote places, or experience destinations in new ways, such as discovering Scotland via whisky trails (Alderton 2017; Karantzavelou 2020). When planning events, organizers should accommodate dietary restrictions and preferences, as well as trends for plant-based or sustainable, locally sourced and produced food which is 'instagrammable' (CWT Meetings & Events 2020; Karantzavelou 2020). These trends indicate potential in destinations that offer niche experiences linked with local traditions, culture, and gastronomy. Wine tourism has the potential to offer the connection to local culture, the natural setting, and sensory elements needed to provide business tourists with unique and meaningful experiences.

Case Study 30.1 Bündner Herrschaft, Switzerland

Justification for a case approach

Exploring the benefits of combining wine tourism and MICE tourism allows the use of a case study approach as an appropriate framework (Bell, Bryman, and Harley 2018). As Yin (2009) points out, a case study is preferred in situations when 'a researcher has little or no control over behavioral events; and ... the focus of the study is a contemporary phenomenon.' Case research allows the study of the multiple features of cases that occur naturally (Hartley 2004). It does not claim to control variables; however, it allows the use of a multitude of methods and sources of data (Thomas 2015).

According to Flick (2018), case research is often not very strongly linked to generalization issues. Nevertheless, any research assumes that its conclusions or findings will have some relevance beyond its immediate context. Based on the nature of one case being examined, statistical generalizability can obviously not be reached (Bell, Bryman, and Harley 2018; Thomas 2015). The issue of generalizability, however, comes up in a different way, asking whether findings are valid in contexts with similar features as the one they were discovered in (Yardley 2008). We thus consider generalizability in the sense of 'a transferability of knowledge from one situation to another, taking into account the contextuality and heterogeneity of social knowledge' (Kvale 2008, p. 7). As such, we claim that learnings derived from the case in question may be transferred to other, similar contexts. In this sense, the rationale followed when selecting the Bündner Herrschaft as the case was in line with Stake (1995), so as to maximize what can be learned from them.

Materials and methods

Trends in MICE tourism suggest that regional tourism products such as culinary and wine offers would be of interest to this sector; however, little research was found on the needs of B2B customers when organizing corporate events locally. In order to bridge this gap, semi-structured interviews were conducted. The semi-structured approach is suitable when a topic is being explored, as interviewers have greater flexibility and are enabled to ask follow-up questions to significant responses while maintaining focus on the topic (Bryman and Bell 2007; Veal 2011). Interviews were conducted with 12 event organizers from nine companies in Switzerland and Liechtenstein of various sizes and sectors. While responsible for organizing corporate events, this was not the interviewees' primary function.

Location and environment

Switzerland has rich cultural, agricultural, and culinary traditions and wine production dating back to the Roman Empire. Alpine cheese, dried meats, beers, herbs, teas, honey, and spirits are just some of the products produced in the Swiss canton of Grisons, also home to one of Switzerland's top wine-producing regions; the Bündner Herrschaft. The Bündner Herrschaft covers a 74.2-km^2 area. Part of the Rhine Valley and bordering Austria, Liechtenstein, and the canton of St. Gallen, it encompasses four smaller regions – Fläsch, Maienfeld, Jenins, and Malans. About 90 minutes from Zürich airport, the region is easily reached by road or rail. It is the largest wine-growing region of the canton, with more than 70 wineries (Graubünden Ferien 2016).

Multiple organizations are involved in the management and promotion of tourism in the Bündner Herrschaft, which has had its own tourism association only since 2010. While part of the canton of Grisons, the mandate to operate the tourism information office and promote the region was given to Heidiland Tourismus (AWT 2013), which is based in St. Gallen, rather than DMO Graubünden Ferien, which is responsible for tourism promotion for Grisons. While tourism is a key industry for Grisons, it is also home to many other industries with almost 21,000 individual businesses (AWT 2020). Additionally, there are over 4,500 companies nearby in Liechtenstein (Amt für Statistik, 2020), where several large multinational companies have head offices.

Tourism management

Despite the large number of nearby companies, neither DMO actively targets them as B2B customers for business tourism. Although business tourism has been recognized as a potential growth market for the Grisons, this appears related to attracting visitors to the region as a destination for conferences and congresses, with congress infrastructure and accommodation being considered the most important assets (Plaz and Schmid 2015). The goal of DMOs is to generate overnight stays, meaning local markets for business or leisure tourism are not considered target markets (Golding 2018). Statistics for overnight stays in Switzerland are not differentiated by the purpose of travel, making it impossible to know the business-induced share of overnight stays in the Grisons.

Wine tourism in the Bündner Herrschaft

The Bündner Herrschaft is known for Pinot Noirs and for a rare indigenous grape variety called Completer, making it home to some unique wines. Viticulture in the Grisons was first mentioned in 765 AD (Graubünden Wein 2020). As an Old World wine region, the Bündner Herrschaft is characterized by small wineries, making it an attractive tourism destination. In contrast to their New World counterparts, Old World wineries tend to have limited space and facilities to host tastings and cellar door experiences. Still, a number of wineries offer tastings, and several wine events also occur each year from April to October.

MICE offers

Information about wine tourism in the Bündner Herrschaft is mainly available online. Golding (2018) identified 18 websites in addition to the 70+ websites of individual wineries, only the Wine Tours Switzerland website provided offers specifically for corporate clients. While having many sources of information could be considered positive, nearly 100 websites with information for a region of only 74 km^2 could also be confusing, especially for customers unfamiliar with the region. DMO websites contained limited information on wine-based activities, opting to link to additional websites for details about the offers. The multitude of websites signals the lack of a clear image as a greater wine destination. No single information source presenting the region as a wine destination with information about the wine and tourism offers could be found.

Wine Tours Switzerland

Wine Tours Switzerland (wine-tours.ch) is the main supplier of wine events and activities in the Bündner Herrschaft. Founded in 2013 by Gian Carlo Casparis in Maienfeld, Wine Tours Switzerland (WTS) offer wine-themed activities to experience Grisons's culinary and cultural specialties. Gian Carlo's local roots have helped him cultivate close relationships with local producers and knowledge of their products. The range of offers includes wine and culinary activities partnered with the walking, bike, and chauffeured tours, as well as complete event solutions for private and corporate customers. Offers are by booking only, available in five languages, and include non-alcoholic tasting options.

The WTS website includes a section dedicated to corporate customers with tours, unusual meeting and seminar spaces, and organized team-building events. Each offer includes information about the duration, capacity, price-per-person, contact details, and a booking request form. It also communicates the option to customize offers or create tailored solutions to corporate needs. Cooperation with a range of local partners ensures offers can be adapted to the individual requirements of corporate clients. Through partnerships with local wineries, venues, restaurants, hotels, transportation companies, and regional producers, WTS is able to deliver complete event solutions.

According to Gian Carlo, working with local partners adds value to the region in a number of ways. Cooperation can build social and business relationships around shared culture and values. Wine tourism offers can preserve and even create local jobs and provide local producers with new opportunities to showcase their products. Gian Carlo recognizes the importance of not overburdening the region with overtourism, so corporate groups are of particular interest as they tend to be smaller and higher-paying, which leaves a positive impact on the region's hotels, restaurants, and producers. His corporate customers demand high-quality, customizable offers to provide increasingly authentic and sustainable experiences for their events. Wine tourism offers opportunities to discover the region and gives them access to unique locations they may otherwise not find. Local producers are provided a platform to share the authentic stories of their products. Corporate customers tend to be higher paying, as they are willing to pay for the added value of local, sustainable, authentic experiences and high-quality service. Additionally, Gian Carlo finds corporate guests have the budget and affinity to appreciate and purchase local products while participating in corporate events.

By offering a range of flexible and customizable experiences and through building trust and personal relationships with corporate customers, Gian Carlo has created a loyal clientele. He now earns around 65% of his revenue from corporate event customers, the majority of whom are repeat customers.

Findings

The Bündner Herrschaft is surrounded by companies engaging with local tourism providers for corporate events. However, this group is not targeted by any of the DMOs. Despite a strong wine industry with existing wine tourism offers and the growing popularity of wine and culinary experiences, the DMOs responsible for the Bündner Herrschaft do not appear to have leveraged this potential and marketed the region as a culinary and wine destination. Wine tourism offers of the Bündner Herrschaft thus represent an underutilized tourism potential for the region. In that sense, they could be an ideal way to connect with the business tourism sector by offering regionally unique experiences through local food, wine, and culture. Although cooperation could be mutually beneficial, very little appears to have been done to bridge this gap.

Identification of needs (of local companies)

Local companies were found to be organizing a large number of events throughout the year, in some cases even weekly, from small business lunches to events of over 1,000 people. Organizers of large groups are limited by the capacity of many restaurants and subsequently make use of city halls, outdoor locations, or their own facilities. As a result, they tend to reuse the same locations requiring activities to come onsite to provide variety from year to year. Most groups generally have between 20 and 50 participants. This makes it particularly difficult for organizers, as there are few tourism activities marketed to groups this size, even if they can actually accommodate them.

Budget is an important aspect of event organization and varies between companies, groups, and events, although many receive a fixed budget from working with. Two main budget ranges were identified with means at 70 and 125 USD, respectively. In most cases, meal, transport, and the activity cost had to be covered.

Travel time is another important factor in deciding the location of events. Most companies like to organize events close to their premises. The average acceptable travel time was

45 minutes (one-way). This finding shows a limited radius of companies for tourism suppliers who want to attract them. Additionally, as they are local, making contact with these companies should not be as difficult as with other market segments.

Many of the companies have international staff working at or visiting their offices. Since many of those do not speak German, offers catering for English speakers are required. When hosting international guests, the events and activities are an opportunity to show off their country and region. Although respondents were generally satisfied with the variety of offers, they would appreciate a greater range of activities. Some companies felt there were many sport-based offers (which were not always suitable for all members of their group) but found a lack of other activities on offer. While larger companies have the resources to have and arrange their own ideas, many would prefer ready-made solutions.

When organizing events for local staff, it is more important to have region-specific specialties and experiences; however, for international guests, regional differentiation is less important than the overall 'Swissness' of the experience. Wine is an attractive topic, and many would be interested in tastings, tours, or associated cultural activities, especially for smaller team events. However, concerns were raised about alcohol consumption and the need to drive home.

Interviewees also felt there were too many websites and that they did not provide the information needed, particularly capacity, price, and contact information. The DMO websites were not considered helpful, as there was little or no information for their needs. Many wished for a single platform with an overview of event spaces, restaurants, and activities that could be searched using various filters.

Findings and implications

Whilst Old World wine regions are increasingly recognizing wine tourism's potential, collaboration is needed when developing wine tourism experiences (Gómez, Pratt, and Molina 2019). Wineries cannot stand alone in the wine tourism experience (Getz and Brown 2006). Locally owned and operated, Wine Tours Switzerland benefits the Bündner Herrschaft by facilitating the creation of tourism experiences in partnership with local stakeholders suitable for the needs of local businesses' tourism activities. The development of such offers helps the region have a consistent visitor flow. Business customers are less price-sensitive than leisure travelers. The smaller, higher-paying corporate groups benefit the region, and pre-booking ensures they are not burdened with overtourism. Combining wine and MICE tourism thus supports local value creation and contributes to both job creation and economic development.

Organizers of MICE events require unique locations and engaging experiences surrounding their workplace. These activities benefit the companies as well as the regions where they take place. They can introduce guests to local culture, new products, and locations and thus help preserve them. Additionally, these activities may contribute to the distribution of tourism income to less popular areas of a destination (Chung et al. 2020).

Based on the case of the Bündner Herrschaft, the following factors have been found to contribute to creating a barrier to cooperation:

1 A lack of focus on corporate group offers and local customers by the tourism organizations
2 An oversupply of online information sources, and
3 An unclear regional image

While DMOs may not be involved in the development of corporate offers, they could act as a centralized information source. Serving as a hub, they could identify local tourism offers that meet corporate customers' needs. When considering platform design, the possibility to search based on budget, location, activity type, and group size has shown to be of great importance.

Tourism suppliers wishing to develop events and activities targeted at local companies should approach them to assess their requirements. Attention should be given to maintaining relationships and their changing needs. Wine tourism suppliers should develop local partnerships in order to create customizable packages to provide variety for corporate customers. The focus should lie on creating unique experiences, and thought should be given to offering alcohol-free alternatives.

Our case approach has shown that wine tourism can support regional development for tourism and other sectors, especially when a clear destination image is created. The development of wine tourism offers must be balanced with the interests of the wine producers, as it is important to create a destination image that is consistent with the one needed for the marketing of wine products.

References

Abulibdeh, A., and Zaidan, E. (2017). Empirical analysis of the cross-cultural information searching and travel behavior of business travelers: a case study of MICE travelers to Qatar in the Middle East. *Applied Geography*, 85: 152–162. https://doi.org/10.1016/j.apgeog.2017.06.001

Alderton, M. (2017, October 13). Pacific world predicts top MICE trends of 2018, says Engagement Trumps Technology. *Successful Meetings*. https://www.successfulmeetings.com/News/Research-Whitepapers/Pacific-World-Report-Top-Meeting-Incentive-Trends 2018

Amt für Statistik. (2020). *Liechtenstein in Zahlen 2020* (p. 48). Amt für Statistik Fürstentum Liechtenstein. https://www.llv.li/files/as/liechtenstein_in_zahlen_2020.pdf

Anđelić, S., Garabinović, D., and Šormaz, G. (2019). A review of wine and wine tourism presence in the scientific papers in journals in the field of tourism. *Ekonomika Poljoprivrede*, 66(4): 1055–1090. https://doi.org/10.5937/ekoPolj1904055A

AWT. (2013). *Wettbewerbsfähige Strukturen und Aufgabenteilung im Bündner Tourismus: Tourismusreform 2006–2013 (Kurzfassung)*. Amt für Wirtschaft und Tourismus Graubünden. https://www.gr.ch/DE/institutionen/verwaltung/dvs/awt/Dokumente/Kurzfassung_D_311013.pdf

AWT. (2020). *Unternehmen – 06 Industrie, Dienstleistungen*. Graubünden Amt für Wirtschaft und Tourismus. https://www.gr.ch/DE/institutionen/verwaltung/dvs/awt/statistik/Industrie,%20Dienstleistungen/Seiten/Unternehmen.aspx

Bayih, B. E., and Singh, A. (2020). Modeling domestic tourism: Motivations, satisfaction and tourist behavioral intentions. *Heliyon*, 6(9): e04839. https://doi.org/10.1016/j.heliyon.2020.e04839

Bell, E., Bryman, A., and Harley, B. (2018). *Business Research Methods*. Oxford: Oxford University Press.

Bruwer, J., Pratt, M. A., Saliba, A., and Hirche, M. (2017). Regional destination image perception of tourists within a winescape context. *Current Issues in Tourism*, 20(2): 157–177. https://doi.org/10.1080/13683500.2014.904846

Bryman, A., and Bell, E. (2007). *Business Research Methods* (2nd ed). Oxford: Oxford University Press.

Campos, A. C., Mendes, J., do Valle, P. O., and Scott, N. (2018). Co-creation of tourist experiences: a literature review. *Current Issues in Tourism*, 21(4): 369–400. https://doi.org/10.1080/13683500.2015.1081158

Canh, N. P., and Thanh, S. D. (2020). Domestic tourism spending and economic vulnerability. *Annals of Tourism Research*, *85*, 103063. https://doi.org/10.1016/j.annals.2020.103063

Chung, J. Y., Choi, Y.-K., Yoo, B.-K., and Kim, S.-H. (2020). Bleisure tourism experience chain: implications for destination marketing. *Asia Pacific Journal of Tourism Research*, 25(3): 300–310. https://doi.org/10.1080/10941665.2019.1708760

Cook, R. A., Hsu, C. H. C., and Taylor, L. L. (2018). *Tourism: The Business of Hospitality and Travel* (6th ed.). Harlow: Pearson.

Croce, E., and Perri, G. (2010). *Food and Wine Tourism: Integrating Food, Travel, and Territory*. Wallingford: CAB International.

CWT Meetings & Events. (2020). *2020 Future Trends in Meetings and Events*. CWT Meetings & Events. https://www.cwt-meetings-events.com/futuretrends/

Davidson, R., and Rogers, T. (2006). *Marketing Destinations and Venues for Conferences, Conventions and Business Events*. New York: Elsevier/Butterworth-Heinemann.

Dixit, S. K. (2019). Gastronomic tourism: a theoretical construct, In Dixit, S. K. (ed.), *The Routledge Handbook of Gastronomic Tourism* (pp. 13–23), Oxon: Routledge.

Dixit, S. K. (ed.). (2020). *The Routledge Handbook of Tourism Experience Management and Marketing*. London: Routledge.

Eventbrite. (2019, June 22). *80+ Event Ideas to Inspire Your Next Event*. Eventbrite UK. https://www.eventbrite.co.uk/blog/70-event-ideas-and-formats-to-inspire-your-next-great-event-ds00/

Everett, S. (2016). *Food and Drink Tourism: Principles and Practice*. London: Sage.

Expedia. (2018). *Unpacking Bleisure Traveller Trends*. Expedia Media Solutions. https://info.advertising.expedia.com/bleisure-travel-trends

Flick, U. (2018). *Designing Qualitative Research*. London: Sage.

Getz, D. (2000). *Explore Wine Tourism: Management, Development and Destinations*. New York: Cognizant Communication.

Getz, D. (2012). *Event Studies: Theory, Research and Policy for Planned Events* (2nd ed.). Oxon: Routledge.

Getz, D., and Brown, G. (2006). Critical success factors for wine tourism regions: a demand analysis. *Tourism Management*, 27(1): 146–158. https://doi.org/10.1016/j.tourman.2004.08.002

Golding, P. (2018). *Connecting regional tourism offers with companies generating business tourism: a conceptual approach and a practical example of wine tourism in the Bündner Herrschaft* [Unpublished Bachelor Thesis]. University of Applied Sciences HTW Chur.

Gómez, M., Pratt, M. A., and Molina, A. (2019). Wine tourism research: a systematic review of 20 vintages from 1995 to 2014. *Current Issues in Tourism*, 22(18): 2211–2249. https://doi.org/10.1080/13683500.2018.1441267

Graubünden Ferien. (2016, January 21). *Bündner Herrschaft – a small region of great wines*. Graubünden Tourism. https://www.graubuenden.ch/en/experience-graubunden/culinary-arts/bundner-herrschaft-small-region-great-wines

Graubünden Wein. (2020). *Geschichte*. Graubuenden Wein. https://www.graubuendenwein.ch/de/geschichte

Hartley, J. (2004). Case study research. In Cassell, C., and Symon, G. (eds.). *Essential Guide to Qualitative Methods in Organizational Research* (pp. 323–333). London: Sage.

Hjalager, A.-M., and Corigliano, M. A. (2000). Food for tourists – determinants of an image. *International Journal of Tourism Research*, 2: 281–293.

Holloway, J. C., and Humphreys, C. (2012). *The Business of Tourism* (9th ed.). New York: Pearson.

International Trade Centre. (2001). *Business Tourism*. International Trade Forum Magazine. http://www.tradeforum.org/Business-Tourism/

Karantzavelou, V. (2020, February 5). *Five Trends Transforming Meetings in 2020 Announced by Benchmark* [News]. TravelDailyNews International. https://www.traveldailynews.com/post/five-trends-transforming-meetings-in-2020-announced-by-benchmark

Knaus, D. L., and Merkle, T. (2020). Open innovation in sports management – the case of the FIFA Museum in Zurich, Switzerland. In *Sport Entrepreneurship* (pp. 43–53). Emerald Publishing Limited. https://doi.org/10.1108/978-1-83982-836-220201008

Koch, J., Martin, A., and Nash, R. (2013). Overview of perceptions of German wine tourism from the winery perspective. *International Journal of Wine Business Research*, 25(1), 50–74. https://doi.org/10.1108/17511061311317309

Kvale, S. (2008). *Doing Interviews*. Thousand Oaks, CA: Sage.

Lavin, N. (2015, November 4). *Four Ways to Generate More Corporate Event Bookings*. Restaurant Hospitality. https://www.restaurant-hospitality.com/marketing/4-ways-generate-more-corporate-event-bookings

Leonard, K. (2019, February 4). *What Are the Benefits of Team Building?* Chron.Com. https://smallbusiness.chron.com/benefits-team-building-1979.html

Lichy, J., and McLeay, F. (2018). Bleisure: motivations and typologies. *Journal of Travel & Tourism Marketing*, 35(4): 517–530. https://doi.org/10.1080/10548408.2017.1364206

López-Guzmán, T., Vieira-Rodríguez, A., and Rodríguez-García, J. (2014). Profile and motivations of European tourists on the Sherry wine route of Spain. *Tourism Management Perspectives*, 11: 63–68. https://doi.org/10.1016/j.tmp.2014.04.003

Macionis, N. (1998). Wine and food tourism in the Australian capital territory: exploring the links. *International Journal of Wine Marketing*, 10(3): 5–22. https://doi.org/10.1108/eb008683

Meacci, L., and Liberatore, G. (2018). A senses-based model for experiential tourism. *Tourism & Management Studies*, 14(4): 7–14. https://doi.org/10.18089/tms.2018.14401

Merkle, T., and Lewis, R. (2016). *RB Leipzig – A Novel Approach to Sport Stadium Hospitality*. https://doi.org/10.4135/9781529713404

Merkle, T., Tajeddini, K., Vlachos, I., and Keane, J. (2020). Entrepreneurship within airside food and beverage outlet patronage: the creation of ecosystems using outlet context and passengers' emotions. In Ratten, V. (ed.), *Entrepreneurship as Empowerment: Knowledge Spillovers and Entrepreneurial Ecosystems* (pp. 127–150). Bingley: Emerald Publishing Limited.

Millán, A., Fanjul, M. L., and Moital, M. (2016). Segmenting the business traveler based on emotions, satisfaction, and behavioral intention. *Psychology & Marketing*, 33(2): 82–93. https://doi.org/10.1002/mar.20856

Montella, M. M. (2017). Wine tourism and sustainability: a review. *Sustainability*, 9(1): 113. https://doi.org/10.3390/su9010113

Morrison, A. M. (2013). *Marketing and Managing Tourism Destinations*. Oxon: Routledge.

Mureşan, M. L., Chiripuci, B. C., and Nistoreanu, P. (2017). Business tourism – a sustainable form to relaunch Romanian tourism. *Proceedings of the International Conference on Business Excellence*, 11(1): 1101–1110. https://doi.org/10.1515/picbe-2017-0113

Nicula, V., and Elena, P. R. (2014). Business tourism market developments. *Procedia Economics and Finance*, 16: 703–712. https://doi.org/10.1016/S2212-5671(14)00858-2

Pike, S. (2008). *Destination Marketing: An Integrated Marketing Communication Approach*. Oxford: Elsevier/Butterworth-Heinemann.

Plaz, P., and Schmid, S. (eds.). (2015). *Langfristige Entwicklung des Bündner Tourismus: Grundlagenberichtim Rahmen des Projekts 'Strategien für Bündner Tourismusorte'*. Wirtschaftsforum Graubünden Unterstützt vom Staatssekretariat für Wirtschaft (SECO). https://www.wirtschaftsforum gr.ch/uploads/files/a1_analyse.pdf

Poitras, L., and Donald, G. (2006). Sustainable wine tourism: The host community perspective. *Journal of Sustainable Tourism*, 14(5): 425–448. https://doi.org/10.2167/jost587.0

Salvat, J., and Boque, J. B. (2009). *New Opportunities and Challenges in Wine Tourism* [Grup de Recerca d'Anàlisi Territorial i Estudis Turístics, Rovira i Virgili University, Tarragona]. http://www.pct-turisme.cat/intranet/sites/default/files/URV_project_salvat_blay_seminar_pctto.pdf

Stake, R. E. (1995). *The Art of Case Study Research*. Thousand Oaks, CA: Sage.

Swarbrooke, J., and Horner, S. (2001). *Business Travel and Tourism*. Oxford: Butterworth-Heinemann.

The Economist. (2019). *The 2019 Bleisure Barometer: Asia's Best Cities for Work and Recreation*. London: The Economist Intelligence Unit.

Thomas, G. (2015). *How to Do Your Case Study*. Thousand Oaks, CA: Sage.

Thornley-Brown, A. (2014). *Corporate Events: Hidden Benefits of Recreational Events for Corporate Teams* [Blog]. LinkedIn. https://www.linkedin.com/pulse/20140913163215-2498009-corporate-events-hidden-benefits-of-recreational-events-for-corporate-teams

Tuan, L. T., Rajendran, D., Rowley, C., and Khai, D. C. (2019). Customer value co-creation in the business-to-business tourism context: The roles of corporate social responsibility and customer empowering behaviors. *Journal of Hospitality and Tourism Management*, 39: 137–149. https://doi.org/10.1016/j.jhtm.2019.04.002

UNWTO. (2018, September 10). *UNWTO Hails Wine Tourism as an Approach to Sustainable Rural Development*. UNWTO. https://www.unwto.org/global/press-release/2018-09-10/unwto-hails-wine-tourism-approach-sustainable-rural-development

Veal, A. J. (2011). *Research Methods for Leisure and Tourism: A Practical Guide* (4th ed.). New Jersey: Prentice-Hall.

Wine in Moderation. (2018). *Wine Tourism*. Wine in Moderation. https://www.wineinmoderation.eu/en/content/Wine-Tourism.7/

Winemakers' Federation of Australia. (2011). *2020 Harnessing the Tourism Potential of Wine and Food in Australia: A Strategic Framework for the Future of Wine and Food in Tourism to 2020*. https://www.agw.org.au/assets/strategies-plans/pdfs/Wine_food_tourism_strategy.pdf

Xu, S., Barbieri, C., Anderson, D., Leung, Y.-F., and Rozier-Rich, S. (2016). Residents' perceptions of wine tourism development. *Tourism Management*, 55: 276–286. https://doi.org/10.1016/j.tourman.2016.02.016

Yardley, L. (2008). Demonstrating validity in qualitative psychology. *Qualitative Psychology: A Practical Guide to Research Methods*, 2: 235–251.

Yin, R. K. (2009). *Case Study Research: Design and Methods*. Thousand Oaks, CA: Sage.

31

MEMORABILITY, SATISFACTION, AND INTENTION TO RECOMMEND TO OTHERS

A segmentation-based study of winery visitors in Australia

Giacomo Del Chiappa, Ester Napolitano, and Marcello Atzeni

Introduction

The experiential paradigm originated in the early 1980s (Holbrook and Hirshman 1982), when scholars started to analyze tourist behavior focusing on emotional, hedonic, aesthetic, and symbolic aspects of consumption, rather than only functional/utilitarian aspects (Holbrook 1999). In current tourism markets, and according to more recent studies, customers do not 'just' desire high-quality experiences and satisfaction; rather, they seek outstanding and memorable experiences (Kim, Ritchie, and McCormick 2012). This explains why, in recent years, both researchers and practitioners have been trying to understand how to provide memorable experiences (Dixit 2020; Sthapit et al. 2019) as a main strategy to differentiate themselves from competitors and to gain a sustainable competitive advantage (Pine and Gilmore 1998).

In this vein, the variables traditionally considered as positive marketing outcomes (i.e., satisfaction and loyalty) are currently replaced by, or blended with, an innovative concept highly linked with customers' emotions and thoughts, namely the memorability of the experience. Specifically, memorability refers to 'the subjective feeling that one will remember and relive the emotional event in the future vividly, confidently and accurately' (Sthapit et al. 2019, p. 6).

In tourism literature, most of the existing experiential-related studies have investigated the tourist experience on site, with less attention given to experiences in supporting services, such as wineries (Del Chiappa and Gallarza 2019). According to Hall et al. (2000), wine tourism can be defined as the 'visitation to vineyards, wineries, wine festivals and wine shows for which grape wine tasting and/or experiencing the attributes of a grape wine region are the prime motivating factors for visitors' (p. 3).

Nowadays, wine regions and wineries located around the world are in fierce competition as a result of the significant economic growth of the worldwide wine industry (Cvijanovi, Novakovi, and Vojinovi 2017). Since wine producers are struggling to differentiate themselves from competitors, they are looking for other ways to survive and to increase their profits, such

DOI: 10.4324/9781003143628-36

as attracting visitors willing to undertake wine-related tourism experiences (Charters and Ali-Knight 2002). In this scenario, obtaining knowledge on wine tourism consumers' profiles and behavior is certainly a key success factor (e.g., Marzo-Navarro and Pedraja-Iglesias 2009). In this vein, numerous researchers have shed light on the motivations that drive consumers to visit wine regions and wineries (Bruwer and Rueger-Muck 2018), others have applied different market segmentation methods to profile wine tourists according to several sociodemographic characteristics and travel-related variables (e.g., Bruwer, Prayag, and Disegna 2018; Gu et al. 2018). However, even more important for the evolution of wine regions and the long-term competitiveness of wineries would be to be able to design and market memorable wine tourism experiences (Gunasekar et al. 2021; Quadri-Felitti and Fiore 2013). Despite this, there is still limited research aimed to analyze memorability in a winery-based experience setting, with no published academic paper profiling winery visitors based on their perceived memorability of the visit experience.

Hence, this study was carried out in order to fill this gap and to serve as guidance to wine managers and producers attempting to understand the extent to which their offer is able to render the overall visit experience vividly, confidently, and accurately memorable in the future, as well as satisfy their visitors and encourage them to recommend the visit to others (both online and offline).

More specifically, this research aims to investigate whether perceived memorability can be appropriately employed to profile different types of wine visitors. Furthermore, it seeks to examine whether subgroups differ in terms of their sociodemographics, visit-related aspects, overall satisfaction, intention to recommend to others (both offline and online), and purchasing behavior. To reach these objectives, a cluster analysis, different statistical chi-square tests, and *t*-tests were performed on data from a convenience sample of 394 visitors of Australian wineries obtained in the period January–February 2019.

Literature review

Memorability and memorable tourism experiences

Marketing literature suggests that experiences have value as long as they are stored in the memory of the customers (Kim 2010). Indeed, when the holiday ends, memories are all that remains in tourists' minds and, because of this, memories have the power to direct their behavior now and in the future (Tung and Ritchie 2011). Personal memories are crucial for individuals since they are the most reliable source of information during their decision-making process (Kim, Ritchie, and McCormick 2012).

Several researchers have provided different definitions of memorable experiences and tourism memorability. Larsen (2007) argued that a personal tourism experience could become memorable when it is strong and able to enter into long-term memory. Others, similarly, suggest that a memorable experience is something that is positively remembered and recalled after the event has occurred (Kim, Ritchie, and McCormick 2012). In view of this, a memorable experience is a meaningful occasion that can be repeatedly accessed and remembered through recollection (Kim, Ritchie and McCormick 2012) which arises out of the personal evaluation of the place's attributes.

Memorability results from unforgettable and extraordinary experiences, and it can be considered as a predictor of later memory (Oh, Fiore, and Jeoung 2007). This latter can be qualified as ordinary (Carù and Cova 2003); instead, memorability pertains to the cognitive ability of people to remember positive past experiences in an easy and emotional way (Kim, Ritchie, and McCormick 2012). As a result, memorability of tourism experiences is vital to ensure that tourists gain memorable experiences during their travels and holiday (Quadri-Felitti and Fiore 2013).

In tourism studies, several approaches and models have been proposed to ensure a real understanding of the essence and process of memorable tourism experience formation.

The so-called expansive perspective (Coelho, Gosling, and Almeida 2018) includes theoretical studies aimed to conceptualize the construct and shed light on what makes certain experiences special, spectacular, and memorable (Knobloch, Robertson, and Aitken 2014). For example, through findings from in-depth interviews, Tung and Ritchie (2011) proposed four factors as critical components of a memorable tourist experience, namely affect, expectations, consequentiality, and recollection.

Conversely, other studies have adopted a so-called modeling perspective to investigate the dimensions of Memorable Tourism Experience (hereafter MTE) (Kim, Ritchie, and Tung 2010). In this vein, Chandralal, Rindfleish, and Valenzuela (2015) identified seven MTE dimensions, i.e., authentic local experiences, personally significant experiences, shared experiences, perceived novelty, perceived serendipity, professional guides and tour operator services, and affective emotions. However, to date, the seven-dimension view of MTE proposed by Kim, Ritchie, and McCormick (2012) is the most acknowledged and used measurement scale and includes the following dimensions: hedonism, refreshment, local culture, meaningfulness, knowledge, involvement, and novelty.

Finally, the influential perspective has recently gained ground, owing to its particular focus on exploring the antecedents and consequences of tourism memorability (Kim 2014; Oh, Fiore, and Jeoung 2007). In this stream of research, Oh, Fiore, and Jeoung (2007) and Hosany and Witham (2009) have investigated the predictive power of the four experience realms (i.e., education, aesthetic, escapism, entertainment) of the Experience Economy model (Pine and Gilmore 1999) in shaping tourism memorability and, in turn, tourism satisfaction and loyalty. Yet, Coelho, Gosling, and Almeida (2018) have proposed three influential factors of memorability (i.e., personal, relational, and environmental), which highlight the critical importance of the ambiance, socialization, emotions, and reflections in shaping memorable tourism experiences and positive future behavioral intentions.

Kim, Ritchie, and McCormick (2012) suggest that not all experiences are memorable, and several variables (e.g., sociodemographics, motivations, experiential components of the destination) might be responsible for changes in the level of memorability of tourists (Chandralal and Valenzuela 2013). Furthermore, based on some studies (e.g., Kim, Ritchie, and McCormick 2012; Oh, Fiore, and Jeoung 2007), the experiential factors are believed to be able to positively influence tourism satisfaction and loyalty only if the experience with a destination/attraction trip left a mark on the consumers' memory. In this vein, memorable experiences – in the form of recollections of outcomes of previous experiences – are considered to be able to predict future behavior better than any other traditional marketing variables (Sthapit, Björk, and Coudounaris 2017; Zhang, Wu, and Buhalis 2018). For instance, Kim, Ritchie, and Tung (2010), Hung, Lee, and Huang (2014), and Gohary et al. (2020) found that individuals who indulged in more memorable activities were found to express a higher level of satisfaction and loyalty (intention of revisit and recommendation). Thus, delivering memorable experiences promotes competitive advantages (Pine and Gilmore 1998), encourages positive word-of-mouth (Chandralal and Valenzuela 2013), drives sales, and builds loyalty (Tung and Ritchie 2011).

Wine tourism experience

A wine tourist is a person who decides to visit a certain region with a special interest in wine and wine-related activities such as tasting and purchasing wine. However, wine tourists are also motivated by other elements (Carmichael 2005), including culture and heritage, recreation and education, and local gastronomy (Cohen and Ben-Nun 2009).

Accordingly, Byrd et al. (2016) note that tourists who visit wineries demand other supplementary services, encompassing wine education, guided cellar visits and walking through the vineyards, enjoying stunning landscapes, events and entertainment, tourism, and hospitality services. All these experience components stimulate the five senses of visitors and their imagination and creativity (Galloway et al. 2008), while providing opportunities for them to socialize with other tourists and with winery personnel (Joy et al. 2018; Marzo-Navarro and Pedraja-Iglesias 2009). Since wine tourists seek more for pleasant and authentic experiences (Quadri-Felitti and Fiore 2013) than ever before, a mere experience appears neither adequate to meet wine tourists' expectations and desires (Saayman and Van der Merwe 2015) nor sufficient to inspire future behavioral intentions of the wine tourists (Wirtz et al. 2003). Therefore, Quadri-Felitti and Fiore (2013) and Bruwer and Rueger-Muck (2018), among many others, maintain that wineries and wine tourism destinations must invest in creating conditions that offer visitors memorable tourism experiences.

In this direction, previous research in wine regions has examined the influence of some experiential factors (i.e., aesthetic, education, escapism, and entertainment) on tourists' memories by integrating Pine and Gilmore's (1998) experience model (Quadri-Felitti and Fiore 2013). Other authors attempted to understand the determining factors affecting wine tourists' memorable experiences (Bruwer and Rueger-Muck 2018; Thanh and Kirova 2018). From their findings, it appears that the hedonic and physical attributes, as well as other attributes (e.g., education, socialization and novelty, themes and activities) have a considerable role in shaping the memorability of a wine tourism experience.

Market segmentation in wine tourism research

Tourists carry out different backgrounds (e.g., motivations, demographics, travel-related variables, and personality traits), and various subjective feelings might arise when engaging with the physical and social environment. No winery and wine tourism destination can guarantee that the same experience is considered to be memorable to the same extent by different individuals (Oh, Fiore, and Jeoung 2007). In this vein, Horvat et al. (2018) recently called for further research aimed at exploring individual perceptions of memorability to segment the tourism demand in order to provide winemakers, destination marketers, and policymakers with new and useful information regarding the design elements and processes necessary to deliver memorable experiences, by customizing offers according to the characteristics of the different types of travelers.

Existing marketing theory widely recognizes that market segmentation is a critical technique for efficiency and cost-effectiveness and agree that different variables can be applied to profile consumers (e.g., demographic, geographic, benefit, psychographics, and behavioral traits, etc.) (Kotler 1980). Accordingly, Molina et al. (2015) consider market segmentation as a valuable method to profile the wine tourism demand in different subgroups of actual and prospective wine tourists and to target these with a tailored offer. In the existing literature, wine tourism segmentation has been employed in many studies, the most carried out in New World countries such as Australia, Canada, and South Africa, and with data collection carried out mostly in a winery setting. Considering that there is no unique, typical wine tourist (Charters and Ali-Knight 2002), several variables have been used to profile wine tourists. Among these, we can refer to sociodemographic traits (e.g., age, gender, income, home country) (Dodd and Bigotte 1997) and travel-related variables (e.g., travel party), psychographic variables (e.g., wine interest and involvement, knowledge, lifestyle) (Nella and Christou 2014), personality and values (Galloway et al. 2008), wine tourism experiential aspects (Del Chiappa, Napolitano, and Atzeni 2019), motivations and barriers (Kim, Del Chiappa, and Napolitano 2019), and visitation rate to wineries (e.g., Molina et al. 2015). Although many studies have been published about this topic, Molina et al. (2015) argue that more research is needed to further substantiate and deepen our knowledge about wine visitors, particularly considering new and neglected variables pertaining more closely to experiential aspects of consumption, among which memorability can be surely considered. This study aims to answer this call for future research by investigating whether memorability can contribute to developing distinct segments of wine visitors with similar characteristics based on their sociodemographics, satisfaction, revisiting and recommendation intentions (online and offline), and purchase behavior.

Methodology

Research setting: a brief overview of the wine industry in South Australia

Wine is a key attractor of international and domestic visitors in Australia, representing the strong natural synergies between the wine industry and the tourism sector. According to Tourism Research Australia's National and International Visitor Surveys (Wine Australia 2019), in 2018–2019, 1 million international and 7.4 million domestic visitors spent their holidays in Australian wine regions, with an overall wine-related expenditure of $9.6 billion. South Australia is the largest wine producer in Australia (50% of all bottled wine and almost 80% of premium wine), with more than 350 cellar doors disseminated in 18 wine regions (South Australian Tourism Commission, December 2017).

From 2016, South Australia's capital city, Adelaide, distinguished itself by joining the Great Wine Capital Global Network, a system of worldwide cities whose wine regions are awarded for their excellent grape and wine production, for the effort made in research, and for the development of educational and wine tourism services. South Australia currently attracts a greater number of international wine visitors when compared to any other Australian state (i.e., 34% of the total number of international visitors to Australian wine regions). The most important source markets are the United Kingdom, New Zealand, China, the United States, and Germany.

Barossa Valley and McLaren Vale have been acknowledged as two of South Australia's top five wine regions for grape production and tourism experiences, along with Clare Valley, Coonawarra, and Adelaide Hills. Located at a 40-minute drive south and a 50-minute drive northeast from Adelaide, respectively, McLaren Vale and Barossa offer a wide range of opportunities to taste and enjoy international and Mediterranean wine varieties, as well as various tourist attractions spreading out in both regions. Cellar door businesses (of which there are 70 in McLaren Vale and 80 in Barossa) provide facilities such as tasting rooms, restaurants, regional produce platters, and accommodations, and they are members of wine routes. In conjunction with traditional wine tasting offerings, wineries are now benefitting from the inclusion of more innovative and creative activities such as wine blending sessions, winemaking and label creation workshops, tasting and cookery masterclasses, food and wine pairing activities, winery tours, and art museums which personally engage wine tourists in education and entertainment experiences while visiting the winery.

Survey design

Following the purposes of the research, a survey questionnaire was developed in accordance with previous literature and was made up of two sections. The first section asked respondents to answer some sociodemographic (i.e., age, gender, marital status, education, occupation, home country, and annual family income) and visit-related questions (e.g., previous visit to the winery, time spent at the winery, etc.). The second section asked respondents to evaluate their degree of agreement/disagreement with a series of items aimed to measure their perceived memorability of the visit experience, their level of satisfaction, and their behavioral intentions. Measurement scales were taken from previous research and adapted to fit the study setting. 'Experience memorability' applied seven items adopted from Oh, Fiore, and Jeoung (2007), and Kim (2010); 'Satisfaction' (described by three items) was based on Oliver (1999) and Zhong, Busser, and Baloglu (2017); 'Storytelling behavior' (i.e., willingness to share both online and offline stories and photos of their own experience in the winery) was measured with three items from Zhong, Busser, and Baloglu (2017); and 'Intention to recommend' (i.e., willingness to recommend via positive word of mouth) was measured with two items from Zeithaml, Berry, and Parasuraman (1996). A 7-point Likert scale was used (1 = strongly agree; 7 = strongly disagree) to indicate their answers.

Prior to starting the data collection, the measurement scales were validated by a panel of highly respected international scholars in tourism and wine marketing research; no concerns were reported in this panel assessment. Hence, the questionnaire was pilot-tested by 20 tourists intercepted at South Australian wineries in order to check content validity and comprehensibility of the question and the measurement scales.

Data collection

According to data provided by the South Australian Tourism Commission and the Fleurieu Peninsula Tourism Regional Office, 150 wineries and 80 cellar doors exist in Barossa, whilst 100 exist in McLaren Vale. To prevent self-selection attribution bias, the selection of wineries and cellar doors where data were collected was based on the fact that they all satisfy certain criteria, namely, being members of the Barossa Grape and Wine Association or of the McLaren Vale Grape Wine Tourism Association, offering wine tasting experiences and a wide array of services (e.g., wine-related events and tourism activities). Hence, 48 wineries were identified. Each of these was contacted by phone to explain the purpose of the study and to verify their willingness to collaborate. Four wineries, located in both Barossa and McLaren Vale, agreed to participate in the research. The survey was administered face-to-face by one researcher who intercepted visitors at the end of their visit. Between January and February 2019, a total of 800 visitors (aged 18 and over) was approached and asked to complete the survey. On average, in this period, the interviewer had access to the wineries twice a week during specific time frames, in accordance with the winery managers so as not to interfere with the visit experience (which resulted in 16 days of data collection). At the end of the data collection period, a convenience sample of 394 completed questionnaires was collected (response rate: 49.25%) and used to run statistics.

Results

Overall profile of the sample

Respondents were mostly male (64.8%), middle-aged (35–54 years old: 43.1%) or Millennial (18–34 years old: 38.8%), highly educated (university degree: 43.2%; master's/PhD: 29.1%), married (53.2%) or engaged (26.1%), employed (49.2%) and with an annual household income higher than AUD 100,000 (44.4%). Most visitors were Australians (83.8%) (Table 31.1).

The majority of the respondents were revealed to be first-time visitors at the winery where they were intercepted (63.4%). A reasonable number of visitors were excursionists spending part of the day (35.3%) or the full day (26.6%) in the wine region. Around two-thirds of the sample stayed no more than two hours at the winery (69.8%). Quite interestingly, 57.6% of the sample reported buying at least one bottle of wine, with 77% of respondents being interested in seeking the winery's products once they returned to their place of residence (Table 31.2). These findings confirm the important role of wine tourism as a direct-to-consumer channel and to support the export activity (Gurau and Duquesnois 2008).

In general, respondents reported that they had many positive (M = 5.76) and wonderful (M = 5.71) memories about their winery experience and will not forget it (M = 5.62) (Figure 31.1). Further, most of the respondents were highly satisfied (M = 6.18) and pleased (M = 6.17) with their experience, and they enjoyed the activities offered by wineries (M = 5.67) (Figure 31.2). Moreover, visitors reported to be prone to revisit the wine region (M = 6.11), to visit more South Australian wineries in the future (M = 6.30), and to recommend the experience to others (M = 6.18). However, they were moderately eager to spread traditional word-of-mouth (WOM) by sharing comments and narratives with friends and relatives (M = 4.81), as well as e-WOM by posting photos (M = 4.84) or comment on social media channels (M = 4.51) (Figure 31.2).

Table 31.1 The sample

Variables	%	*Variables*	%
Gender		Master/PhD	29.1
Male	64.8	**Employment status**	
Female	35.2	Employee	49.2
Age		Manager/executive	16.8
18–24	10.9	Freelance	13.1
25–34	27.9	Retired	8.3
35–44	21.5	Occasional worker	1.3
44–54	21.6	Unemployed	1.1
55–64	10.7	Student	7.0
≥ 65	7.4	Other	3.2
Marital status		**Annual Household income**	
Single	17.6	Less than AUD 20,000	8.3
Engaged	16.0	AUD 20,001–30,000	7.1
De facto	10.1	AUD 30,001–40,000	6.8
Married	53.2	AUD 40,001–50,000	3.6
Divorced	2.7	AUD 50,001–70,000	10.7
Widow	0.4	AUD 70,001–100,000	19.1
Education		Greater than AUD 100,000	44.4
Primary school	0.3	**Place of Residence**	
Secondary school	17.4	Australia	83.8
High school	10.0	Overseas	16.2
University degree	43.2		

Table 31.2 Travel-related characteristics of the sample

Variables	%
Prior visitation experience in the wine region	
Yes	89.1
No	10.9
Time spent in the region	
Half day	35.3
All day	26.6
Two days	13.4
More than two days	12.8
Local resident	11.9
Time spent at the winery	
< 2 hours	69.8
Half day	19.8
All day	6.6
2 days	0.5
> 2 days	3.3
Frequency of visitation for the winery	
First visit	63.4
Second visit	23.9
Once every 1–6 months	5.3
Once every 7–12 months	3.3
Less than once every 12 months	4.1
Purchasing of wine bottles	
Yes	57.6
No	18.0
I prefer not to answer	24.4
Seeking wine bottles once returned to the place of residence	
Yes	77.0
No	23.0

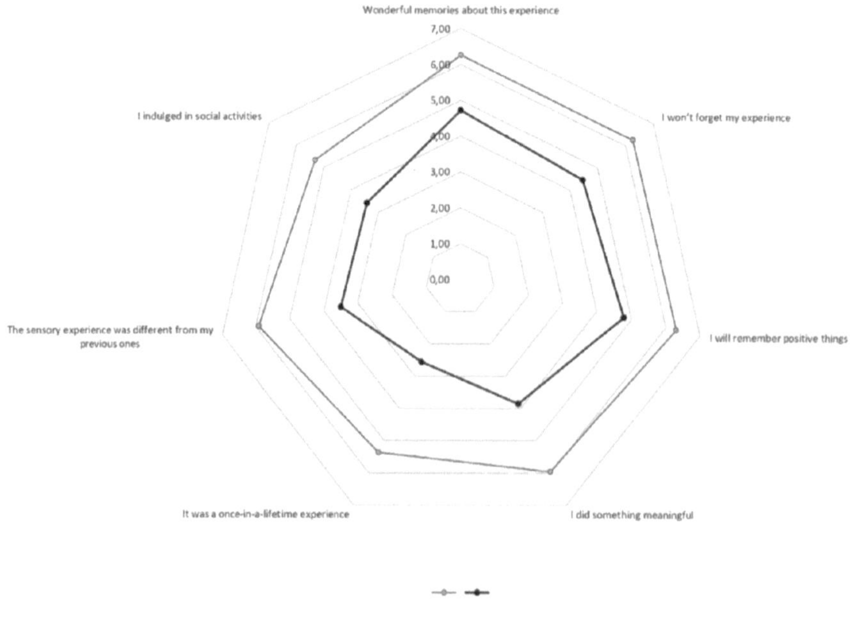

Figure 31.1 Results of cluster analysis (by memorability).

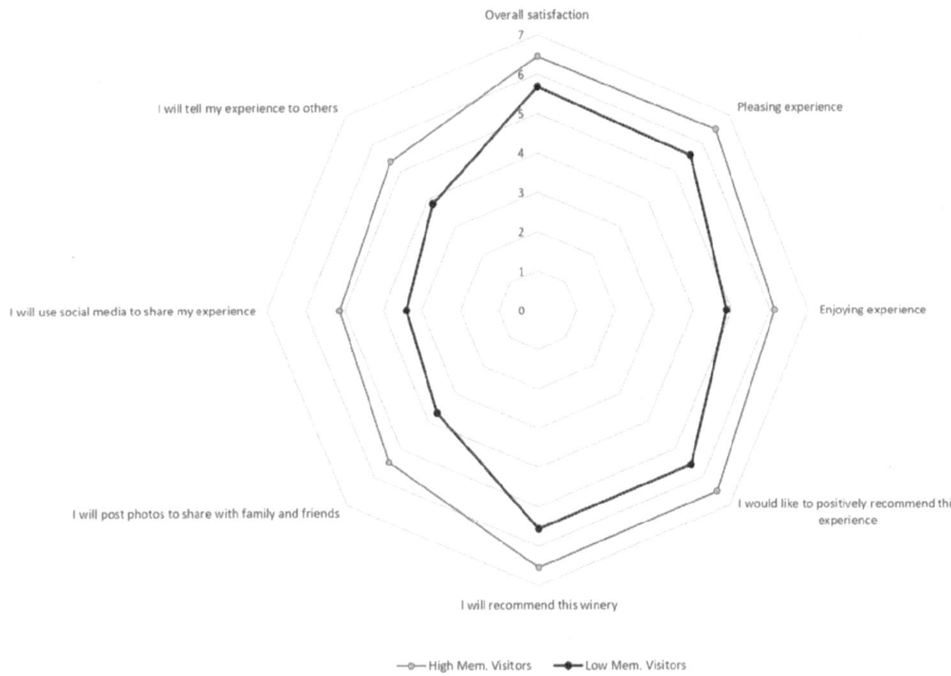

Figure 31.2 Satisfaction and intention to recommend by cluster: results of *t*-test analysis.

In order to reach the purposes of the study, a cluster analysis was performed on memorability items to profile wine visitors. In accordance with Hair et al.'s (2010) recommendations, two clusters emerged, named 'High Memorability Visitors' and 'Low Memorability Visitors' (Figures 31.1 and 31.2). A series of *t*-tests confirmed the existence of significant differences among clusters, corroborating the adequate group pattern ($p < 0.000$).

Results of cluster analysis (by memorability)

High Memorability Visitors (N = 254) made up the larger segment and included individuals scoring high or particularly high in all the memorability items. For example, they perceived the overall visit experience to be pleasing ($M = 6.26$) and memorable ($M = 6.26$), while also offering meaningful moments ($M = 5.96$) and social opportunities ($M = 5.32$), as well as novelty and uniqueness ($M = 5.91$). This cluster mostly included males (59.8%), belonging to the age bracket 25–34 (29.6%) or 44–54 (21.4%), married (52.7%), with a university degree (40.0%) or master's degree (30.2%), employed (47.7%), and with an annual household income higher than AUD 100,000 (42.5%). Most respondents in this cluster reported being first-timers at the winery (68.50%), where they spent a relatively limited amount of time (< 2 hours: 65.70%). They made purchases at the winery (60.20%) and resulted in being prone to keep seeking local wine once they returned to their place of residence (80.80%).

Low Memorability Visitors (N = 140) scored lower than the previous group in all memorability items. For example, they did not perceive the visit as a 'once-in-a-lifetime experience' ($M = 2.56$), or a place where they indulged in social activities ($M = 3.41$) and/or able to offer something they can consider to be meaningful ($M = 3.85$) or unique and different from prior visitations at other wineries ($M = 3.50$). The cluster consisted mostly of males (68.6%), aged

25–34 (24.8%) or 35–44 (24.7%), married (54.1%), with a university degree (48.9%) or master's degree (27.4%), employees (51.9%) or executives (19.3%) and with an annual household income higher than AUD 100,000 (47.9%).

In order to verify whether significant differences exist between the clusters, a series of *t*-tests were completed based on items used to measure their satisfaction and future behavioral intentions (i.e., these variables were considered as segment descriptors). Significant differences existed among clusters for all the items (*p*-value < 0.001). Overall, High Memorability Visitors are more satisfied ($M = 6.45$) and pleased ($M = 6.49$) when compared to their counterparts ($M = 5.68$; $M = 5.57$). Furthermore, High Memorability Visitors are more prone to recommend the winery to others both offline ($M = 6.52$) or online by posting a photo ($M = 5.46$) or a comment on social media ($M = 5.13$), when compared to Low Memorability Visitors (offline WOM: $M = 5.57$; e-posting: $M = 3.69$; e-review: $M = 3.39$) (Figure 31.2).

Finally, results from a series of chi-square tests (χ^2) (Table 31.3) indicate that significant differences between the two clusters emerged in regard to their annual household income ($\chi^2 = 20.675$; $p < 0.001$) and marital status ($\chi^2 = 20.675$; $p < 0.001$). As far as the latter is concerned, the majority of interviewees belonging to the Low Memorability Visitors segment were married (54.1%) or in a de facto relationship (18.5%). This finding seems to further confirm previous exploratory study studies' results (Horvat et al. 2018).

No significant differences emerged based on gender, age, level of education, and employment status. When considering the moderator effect of travel-related variables, significant differences among clusters were reported to correlate with the period of time spent in the winery ($\chi^2 = 17.723$; $p < 0.001$), the purchasing of wine bottles ($\chi^2 = 7.481$; $p < 0.05$), and the willingness to seek winery products once returned to the place of residence ($\chi^2 = 6.652$; $p < 0.05$). In particular, High Memorability Visitors include a higher percentage of visitors staying at the winery half a day (24.40%) or all day (7.10%), when compared to Low Memorability Visitors (half day: 11.40%; all day: 5.8%). Our results suggest that memorability increases as visitors stay longer so that they have a greater possibility to fully enjoy/experience the winery offers. Furthermore, High Memorability Visitors are more prone to buy wines after their visit (62.60%) and to look for them once returned in their country of residency (80.8%) compared to their counterparts (48.57% and 70.1%, respectively). Overall, findings suggest that higher perceived memorability tends to generate a higher level of satisfaction, a higher willingness to recommend the winery experience to others (both online and offline), and a higher likelihood of buying behavior. No significant differences emerged based on the frequency of visitation of

Table 31.3 Clusters by sociodemographics and travel-related variables: chi-square test

	Chi-square	*Sig.*
Gender	3.298	0.071
Age	55.795	0.52
Marital status	20.675	0.001**
Education	6.420	0.17
Employment status	5.516	0.597
Annual household income	20.675	0.001**
Time spent at the winery	17.723	0.007**
Frequency of visitation for the winery	7.258	0.298
Purchasing of wine bottles	7.481	0.024*
Seeking wine products once returned to the place of residence	6.652	0.036*

* Significant at the .05 level.

** Significant at the .01 level.

the winery ($\chi^2 = 7.258$; $p = 0.298$), thus somehow suggesting that the wineries where visitors were surveyed are over time able to innovate their offer to keep 'surprising' their guests and to render their experience always new and memorable.

Discussion and conclusion

Existing tourism literature concurs that wine tourists have become more educated, knowledgeable, and experienced (Carmichael 2005) and has stressed the need to develop a greater understanding of who these visitors are, what the reasons are behind their decision to visit a wine tourism destination/winery, and which beneficial outcomes they are looking for before, during, and after the experience (Charters and Ali-Knight 2002; Kim, Del Chiappa, and Napolitano 2019; Molina et al. 2015).

In this scenario, several researchers have suggested that considering marketing outcomes such as consumer satisfaction is no longer sufficient to stimulate visitors' actual and future behavior towards the destination and wine attraction (Wirtz et al. 2003). In fact, wine tourists now desire appealing and unique experiences (Kastenholz et al. 2018) that can inspire and engage them emotionally, physically, and intellectually (Fernandes and Cruz 2016; Lee and Chang 2012; Schmitt 1999), until eventually flowing into positive and long-lasting memories of their winery visit, a phenomenon that is usually identified by the term 'memorability' (Joy et al. 2018). Hence, memorable tourism experiences have become the cutting edge of tourism experience as they enable consumers to vividly and accurately recollect and relive moments of the trip long after the event took place.

Against this background, and to the best of our knowledge, this is the first empirical study whose aims are profiling wine visitors based on their perceived level of memorability towards the visit experience and ascertaining whether they significantly differ based on sociodemographics and travel-related variables of visitors.

Findings revealed that visitors at Australian wineries do not have homogeneous levels of memorability and that two main clusters can be identified, namely High Memorability Visitors (N = 254) and Low Memorability Visitors (N = 140), with the former group being the greater in size. Hence, this study proves that perceived memorability is adequate for consumer segmentation when it comes to the wine tourism demand. Moreover, this research contributes to the growth of the scientific knowledge on wine tourists' behavior (Hall and Prayag 2017) by particularly considering new and neglected experiential-related variables, among which memorability should be contemplated (Molina et al., 2015). Our findings also reveal that visitors with higher levels of memorability of their experience reported to express significantly higher level of satisfaction and behavioral intentions (i.e., intention to recommend the visit experience to others both offline and online), corroborating previous tourism research (e.g. Kim, Ritchie, and McCormick 2012; Quadri-Felitti and Fiore 2013; Zhong, Busser, and Baloglu 2017). Meanwhile, our findings also add to the current body of knowledge, suggesting that a higher level of memorability is more likely to lead visitors to buy wine bottles at the winery and to search for the visited winery's products once returned to their place of residence, thus confirming the relevant role wine tourism development can have in stimulating direct sales and in supporting exports (e.g., Chandralal and Valenzuela 2015). Additionally, inferential statistical analysis indicates the fact that significant differences exist among the clusters based just on certain sociodemographics (i.e., marital status and annual household income) and visit-related variables (i.e., time spent at the winery).

From a managerial perspective, findings of this study also offer insights devoted to support the efforts of winemakers, policymakers, and destination marketers in attracting tourists to visit wineries, in providing visitors with satisfaction, and in making them ambassadors of their business abroad, while increasing profit through the direct sales channel. Our findings indicate that tourism managers should plan and implement their experience design (e.g., multisensory cues, aesthetic attributes, social servicescapes, etc.) to be able to deliver memorable, rather than 'merely' satisfactory, tourism experiences. Furthermore, segmenting visitors based on the perceived memorability of their visit experience, and tracking this aspect in the long run, would contribute to better understanding of the overall visitors' experience and measuring of the adequate level of investment in enhancing the design and marketing of memorable experiences. On one hand, this would assist wine tourism experts in the development and promotion of always-innovative services and activities without forgoing visitors' expectations. On the other hand, it will make wine tourists more likely to remember and relive the experience in the future vividly, confidently, and accurately and, thus, to be more prone to buy winery products and to continue to seek them once they return to their place of residence. While doing this, as suggested by our findings, winemakers should also recognize nuances in the way memorability is differently perceived by visitors according to their sociodemographic and visit experience. In this vein, for example, our findings show that those who tend to stay longer at the winery are the High Memorability Visitors when compared to counterparts. Hence, winemakers should adopt marketing strategies aiming to incentivize their visitors to stay longer at the winery and/or to experience a wider variety of activities, so that they can fully enjoy/experience the winery offers. This could be done, for example, by using promotion and/or bundle-pricing strategies.

This study has some limitations. Firstly, it uses a convenience sample and considers visitors of wineries located in two specific Australian subregions; thus, findings cannot be generalized. Future studies might repeat the investigation in other tourism destinations in an attempt to further generalize our findings. Secondly, our study did not discern the extent to which the likelihood to buy wine bottles while at the winery is affected by visitors' level of perceived memorability rather than their available economic budget during their holiday (e.g., Chang, Chen, and Meyer 2013). Thirdly, our study did not examine clusters based on their travel-related characteristics and/or motivations for their traveling/winery visitation; these aspects would also merit future investigations. Finally, this chapter did not inspect any antecedents that might influence the perceived memorability of the visit experience; this aspect should be explicitly investigated in future research so that further insights can be offered to both theory and practice.

References

Bruwer, J., Prayag, G., and Disegna, M. (2018). Why wine tourists visit cellar doors: segmenting motivation and destination image. *International Journal of Tourism Research*, 20(3): 355–366. https://doi.org/10.1002/jtr.2187

Bruwer, J., and Rueger-Muck, E. (2018). Wine tourism and hedonic experience: a motivation-based experiential view. *Tourism and Hospitality Research*, 19(4): 488–502. https://doi.org/10.1177/1467358418781444

Byrd, E. T., Canziani, B., Hsieh, Y. C. J., Debbage, K., and Sonmez, S. (2016). Wine tourism: motivating visitors through core and supplementary services. *Tourism Management*, 52: 19–29. https://doi.org/10.1016/j.tourman.2015.06.009

Carmichael, B. A. (2005). Understanding the wine tourism experience for winery visitors in the Niagara region, Ontario, Canada. *Tourism Geographies*, 7(2): 185–204. https://doi.org/10.1080/14616680500072414

Carù, A., and Cova, B. (2003). Revisiting consumption experience: a more humble but complete view of the concept. *Marketing Theory*, 3(2): 267–286. https://doi.org/10.1177/14705931030032004

Chandralal, L., Rindfleish, J., and Valenzuela, F. (2015). An application of travel blog narratives to explore memorable tourism experiences. *Asia Pacific Journal of Tourism Research*, 20(6): 680–693. https://doi.org/10.1080/10941665.2014.925944

Chandralal, L., and Valenzuela, F. R. (2013). Exploring memorable tourism experiences: antecedents and behavioural outcomes. *Journal of Economics, Business and Management*, 1(2): 177–181. https://doi.org/10.7763/JOEBM.2013.V1.38

Chandralal, L., and Valenzuela, F. R. (2015). Memorable tourism experiences: scale development. *Contemporary Management Research*, 11(3): 291–310. https://doi.org/10.7903/cmr.13822

Chang, K. L., Chen, C. M., and Meyer, T. J. (2013). A comparison study of travel expenditure and consumption choices between first-time and repeat visitors. *Tourism Management*, 35: 275–277. https://doi.org/10.1016/j.tourman.2012.06.006

Charters, S., and Ali-Knight, J. (2002). Who is the wine tourist? *Tourism management*, 23(3): 311–319. https://doi.org/10.1016/S0261-5177(01)00079-6

Coelho, M. D. F., Gosling, M. D. S., and Almeida, A. S. A. D. (2018). Tourism experiences: core processes of memorable trips. *Journal of Hospitality and Tourism Management*, 37: 11–22. https://doi.org/10.1016/j.jhtm.2018.08.004

Cohen, E., and Ben-Nun, L. (2009). The important dimensions of wine tourism experience from potential visitors' perception. *Tourism and Hospitality Research*, 9(1): 20–31. https://doi.org/10.1057/thr.2008.42

Cvijanovi, D., Novakovi, D. J., and Vojinovi, Z. (2017). Wine industry as a source of rural growth and development. In *Proceeding of the International Scientific Conference, 'Strategies for the Agri-Food Sector and Rural Areas–Dilemmas of Development'* (pp. 19–21), Licheń Stary, Poland.

Del Chiappa, G., and Gallarza, M. G. (2019). Past, present and future challenges on customer experience: digging knowledge across sectors. *Mercati & Competitività*, 3: 7–15. https://doi.org/10.3280/mc3-2019oa8499

Del Chiappa, G., Napolitano, E., and Atzeni, M. (2019). Perceived authenticity, satisfaction and behavioural intentions at wineries. *Micro & Macro Marketing*, 28(1): 117–138. https://doi.org/10.1431/93019

Dixit, S. K. (ed.). (2020). *The Routledge Handbook of Tourism Experience Management and Marketing*. London: Routledge.

Dodd, T., and Bigotte, V. (1997). Perceptual differences among visitor groups to wineries. *Journal of Travel Research*, 35(3): 46–51. https://doi.org/10.1177/004728759703500307

Fernandes, T., and Cruz, M. (2016). Dimensions and outcomes of experience quality in tourism: the case of Port wine cellars. *Journal of Retailing and Consumer Services*, 31: 371–379. https://doi.org/10.1016/j.jretconser.2016.05.002

Galloway, G., Mitchell, R., Getz, D., Crouch, G., and Ong, B. (2008). Sensation seeking and the prediction of attitudes and behaviours of wine tourists. *Tourism Management*, 29(5): 950–966. https://doi.org/10.1016/j.tourman.2007.11.006

Gohary, A., Pourazizi, L., Madani, F. F., and Chan, E. Y. (2020). Examining Iranian tourists' memorable experiences on destination satisfaction and behavioral intentions. *Current Issues in Tourism*, 23(2): 131–136. https://doi.org/10.1080/13683500.2018.1560397

Gu, Q., Qiu Zhang, H., King, B., and Huang, S. (2018). Wine tourism involvement: a segmentation of Chinese tourists. *Journal of Travel & Tourism Marketing*, 35(5): 633–648. https://doi.org/10.1080/10548408.2017.1401031

Gunasekar, S., Das, P., Dixit, S. K., Mandal, S., and Mehta, S. R. (2021). Wine-experienscape and tourist satisfaction: through the lens of online reviews. *Journal of Foodservice Business Research*. https://doi.org/10.1080/15378020.2021.2006039

Gurau, C., and Duquesnois, F. (2008). Direct marketing channels in the French wine industry. *International Journal of Wine Business Research*, 20(1): 38–52. https://doi.org/10.1108/17511060810864606

Hair, J. F., Black, W. C., Babin, B. J., and Anderson, R. E. (2010). *Multivariate Data Analysis: A Global Perspective*. New Jersey: Pearson Prentice Hall.

Hall, C. M., Johnson, G., Cambourne, B., Macionis, N., Mitchell, R., and Sharples, E. (2000). Wine tourism: an introduction. In Hall, C. M., Sharples, L., Cambourne, B., and Macionis, N. (eds.), *Wine Tourism around the World: Development, Management, and Markets* (pp. 1–23). Oxford: Butterworth-Heinemann.

Hall, C. M., and Prayag, G. (2017). Guest editorial. *International Journal of Wine Business Research*, 29(4): 338–345. https://doi.org/10.1108/IJWBR-09-2017-0053

Holbrook, M. B. (1999). *Consumer Value. A Framework for Analysis and Research*. London: Routledge.

Holbrook, M. B., and Hirshman, E. C. (1982). The experiential aspects of consumption: consumer fantasies, feelings, and fun. *The Journal of Consumer Research*, 9(2): 132–140. https://doi.org/10.1086/208906

Horvat, J., Mijoč, J., Matković, T., and Barač Miftarević, S. (2018). Tourist segmentation in memorable tourism experience survey. *Congress Proceedings: Tourism & Hospitality Industry 2018* (pp. 110–119). https://thi.fthm.hr/congress-proceedings/send/5-2018/105-tourist-segmentation-in-memorable-tourism-experience-survey

Hosany, S., and Witham, M. (2009). Dimensions of cruisers' experiences, satisfaction, and intention to recommend. *Journal of Travel Research*, 49(3): 351–364. https://doi.org/10.1177/0047287509346859

Hung, W.-L., Lee, Y.-J., and Huang, P.-H. (2014). Creative experiences, memorability and revisit intention in creative tourism. *Current Issues in Tourism*, 19(8): 763–770. https://doi.org/10.1080/13683500.2013.877422

Joy, A., Belk, R. W., Charters, S., Wang, J. J. F., and Peña, C. (2018). Performance theory and consumer engagement: wine-tourism experiences in South Africa and India. In Cross, S. N. N., Ruvalcaba, C., Venkatesh, A., and Belk, R. W. (eds.), *Consumer Culture Theory* (pp. 163–187). Bingley, UK: Emerald Group Publishing Limited.

Kastenholz, E., Carneiro, M. J., Marques, C. P., and Loureiro, S. M. C. (2018). The dimensions of rural tourism experience: impacts on arousal, memory, and satisfaction. *Journal of Travel & Tourism Marketing*, 35(2): 189–201. https://doi.org/10.1080/10548408.2017.1350617

Kim, A. K., Del Chiappa, G., and Napolitano, E. (2019). Understanding the wine tourist markets' motivations, travel constraints, and perceptions of destination attributes: a case study of winery visitors in Sardinia, Italy. In Sigala, M., and Robinson, R. N. S. (eds.), *Wine Tourism Destination Management and Marketing: Theory and Cases* (pp. 9–27). Switzerland: Palgrave Macmillan. https://doi.org/10.1007/978-3-030-00437-8

Kim, J. H. (2010). Determining the factors affecting the memorable nature of travel experiences. *Journal of Travel & Tourism Marketing*, 27(8): 780–796. https://doi.org/10.1080/10548408.2010.526897

Kim, J. H. (2014). The antecedents of memorable tourism experiences: the development of a scale to measure the destination attributes associated with memorable experiences. *Tourism Management*, 44: 34–45. https://doi.org/10.1016/j.tourman.2014.02.007

Kim, J. H., Ritchie, J. B., and McCormick, B. (2012). Development of a scale to measure memorable tourism experiences. *Journal of Travel Research*, 51(1): 12–25. https://doi.org/10.1177/0047287510385467

Kim, J. H., Ritchie, J. R., and Tung, V. W. S. (2010). The effect of memorable experience on behavioral intentions in tourism: a structural equation modeling approach. *Tourism Analysis*, 15(6): 637–648. https://doi.org/10.3727/108354210X12904412049776

Knobloch, U., Robertson, K., and Aitken, R. (2014). (Mis)Understanding the nature of tourist experiences. *Tourism Analysis*, 19(5): 599–608. https://doi.org/10.3727/108354214X14116690097891

Kotler, P. (1980). *Principles of Marketing*. Englewood Cliffs, NJ: Prentice-Hall.

Larsen, S. (2007). Aspects of a psychology of the tourist experience. *Scandinavian Journal of Hospitality and Tourism*, 7(1): 7–18. https://doi.org/10.1080/15022250701226014

Lee, T. H., and Chang, Y. S. (2012). The influence of experiential marketing and activity involvement on the loyalty intentions of wine tourists in Taiwan. *Leisure Studies*, 31(1): 103–121. https://doi.org/10.1080/02614367.2011.568067

Marzo-Navarro, M., and Pedraja-Iglesias, M. (2009). Wine tourism development from the perspective of the potential tourist in Spain. *International Journal of Contemporary Hospitality Management*, 21(7): 816–835. https://doi.org/10.1108/09596110910985304

Molina, A., Gómez, M., González-Díaz, B., and Esteban, Á. (2015). Market segmentation in wine tourism: strategies for wineries and destinations in Spain. *Journal of Wine Research*, 26(3): 192–224. https://doi.org/10.1080/09571264.2015.1051218

Nella, A., and Christou, E. (2014). Segmenting wine tourists on the basis of involvement with wine. *Journal of Travel & Tourism Marketing*, 31(7): 783–798. https://doi.org/10.1080/10548408.2014.889639

Oh, H., Fiore, A. M., and Jeoung, M. (2007). Measuring experience economy concepts: tourism applications. *Journal of Travel Research*, 46(2): 119–132. https://doi.org/10.1177/0047287507304039

Oliver, R. L. (1999). Whence consumer loyalty. *Journal of Marketing*, 63: 33–34. https://doi.org/10.2307/1252099

Pine, B. J., and Gilmore, J. H. (1998). Welcome to the experience economy. *Harvard Business Review*, 76: 97–105. https://hbr.org/1998/07/welcome-to-the-experience-economy

Pine, B. J., and Gilmore, J. H. (1999). *The Experience Economy: Work Is Theatre & Every Business a Stage*. Boston: Harvard Business Press.

Quadri-Felitti, D. L., and Fiore, A. M. (2013). Destination loyalty: effects of wine tourists' experiences, memories, and satisfaction on intentions. *Tourism and Hospitality Research*, 13(1): 47–62. https://doi.org/10.1177/1467358413510017

Saayman, M., and Van Der Merwe, A. (2015). Factors determining visitors' memorable wine-tasting experience at wineries. *Anatolia*, 26(3): 372–383. https://doi.org/10.1080/13032917.2014.968793

Schmitt, B. (1999). Experiential marketing. *Journal of Marketing Management*, 15 (1–3): 53–67. https://doi.org/10.1362/026725799784870496

South Australian Tourism Commission (2017, December). South Australia regional profile. *Adelaide*. https://tourism.sa.gov.au/media/3tpbzxgf/regional-profile.pdf

Sthapit, E., Björk, P., and Coudounaris, D. N. (2017). Emotions elicited by local food consumption, memories, place attachment and behavioural intentions. *Anatolia*, 28(3): 363–380. https://doi.org/10.1080/13032917.2017.1322111

Sthapit, E., Del Chiappa, G., Coudounaris, D. N., and Björk, P. (2019). Tourism experiences, memorability and behavioural intentions: a study of tourists in Sardinia, Italy. *Tourism Review*, 75(3): 533–558. https://doi.org/10.1108/TR-03-2019-0102

Thanh, T. V., and Kirova, V. (2018). Wine tourism experience: a netnography study. *Journal of Business Research*, 83: 30–37. https://doi.org/10.1016/j.jbusres.2017.10.008

Tung, V. W. S., and Ritchie, J. B. (2011). Exploring the essence of memorable tourism experiences. *Annals of Tourism Research*, 38(4): 1367–1386. https://doi.org/10.1016/J.ANNALS.2011.03.009

Wine Australia (November 2019). Wine Australia providing insights on Australian wine. Wine tourism snapshot 2018–19. https://www.wineaustralia.com/market-insights/australian-wine-tourism-snapshot#:~:text=The%20latest%20figures%20on%20wine,an%20average%20of%207%20nights.&text=The%20largest%20international%20market%20that,China%2C%20followed%20by%20the%20UK

Wirtz, D., Kruger, J., Scollon, C. N., and Diener, E. (2003). What to do on spring break? The role of predicted, online, and remembered experience in future choice. *Psychological Science*, 14(5): 520–524. http://www.jstor.org/stable/40064178

Zeithaml, V. A., Berry, L. L., and Parasuraman, A. (1996). The behavioral consequences of service quality. *Journal of Marketing*, 60(2): 31–46. https://doi.org/10.2307/1251929

Zhang, H., Wu, Y., and Buhalis, D. (2018). A model of perceived image, memorable tourism experiences and revisit intention. *Journal of Destination Marketing & Management*, 8: 326–336. https://doi.org/10.1016/j.jdmm.2017.06.004

Zhong, Y. Y. S., Busser, J., and Baloglu, S. (2017). A model of memorable tourism experience: the effects on satisfaction, affective commitment, and storytelling. *Tourism Analysis*, 22(2): 201–217. https://doi.org/10.3727/108354217X14888192562366

PART V

The wine tourism experience: management and success stories

32

INTRODUCING THE WINE TOURISM EXPERIENCE (WTE) FRAMEWORK TO DESIGN THE ULTIMATE TOURIST EXPERIENCE

Wided Batat

In different marketplaces and sectors such as food and tourism, companies are turning to customer experience offers to retain actual customers and attract potential buyers.

(Batat 2019a)

Introduction

The development of wine tourism is nowadays considered to be linked to three essential factors, including the growing interest in wine, international travel, and the aesthetics of the vineyard experience that delights visitors (Hall, Johnson, and Mitchell 2000). Despite the established existence of wine tourism, the concept of wine tourism continues to evolve alongside the joint relationship between wine and the tourism industry (e.g., Beames 2003; Cho, Bonn, and Brymer 2017; Gázquez-Abad et al. 2015; Humphreys and Carpenter 2018). In recent decades, wine tourism has experienced a significant growth rate around the world; many new wine tourism destinations have emerged, increasing competition between wine tourism regions and established tourism destinations (Getz and Brown 2006). Hall, Johnson, and Mitchell (2000) state that in addition to wine, wine tourism is about experiencing space through its various natural, social, and cultural elements. Thus, besides wine, at the center of tourism demand is recreation. Thus, the wine tourism experience includes various components such as visits to vineyards, wineries, festivals, aesthetics, sensory, and wine exhibitions organized for recreation, among others (e.g., Ye, Zhang, and Yuan 2017). Without all these components, there is a lack of quality wine tourism experience valuations in winegrowing areas. Although wine tourism emerged during the 1990s, the related literature is relatively recent, and research is still in a development phase and predominately examines wine tourism from a product perspective instead of being experiential and consumer/tourist-oriented.

This chapter adopts a consumer-centric perspective to introduce the wine tourism experience (WTE) framework and explain its drivers, forms, and outcomes. The proposed framework can help both tourism scholars and practitioners capture the mechanism of designing a comprehensive and engaging wine tourism experience from a holistic perspective. This perspective

DOI: 10.4324/9781003143628-38

adopts a bottom-up approach, focusing on the consumer/tourist and his/her functional, emotional, and experiential needs. The chapter first starts with the definition of wine tourism and examines the components of wine tourism experiences through a multidisciplinary review of the literature in marketing, consumer behavior, and tourism studies. This perspective provides an understanding of wine tourism experience production and consumption, with guidance on how to design the ultimate wine tourism experience. Then, an integrative framework identifying different dimensions, stages, and outcomes will be introduced and explained, followed by presenting avenues for research in the tourism field.

Wine tourism: from product to experience

The term *wine tourism* refers to the characteristics of two fields, namely tourism and wine. On the one hand, to define wine as a tourism activity, there is a need to create and develop accommodation, restaurants, and multiple activities related to wine (Cohen and Ben-Nun 2009). On the other hand, tourism is defined by activities undertaken by individuals during their travels and stays in places outside their usual environment for leisure and any other reason not related to a long-term paid activity in the place visited (Bruwer and Thach 2013). Considering the two words, tourism and wine, we can offer a generic definition of wine tourism as a set of activities implemented by individuals during their trips and stays in places and wine-producing regions located outside their usual setting for a limited period, not exceeding one year. This definition, therefore, emphasizes two critical aspects of wine tourism: space and time. Besides, wine tourism is not limited to the spacio-temporal dimension. It is also a social, cultural, and economic phenomenon that involves the movement of people to countries or places outside their usual environment, to wine-producing countries in order to experience wine-related activities, some of which involve an economic dimension such as purchasing wine as a souvenir of the lived experience (e.g., Carlsen and Boksberger 2015; Nisco et al. 2015; Sampaio 2012).

Hence, wine tourism aims at discovering wine-producing regions through wine and vines, but also through the local culture by transforming the place of production into a cultural heritage to be shared through wine production and consumption, which certainly transform the role of winegrowers from producers to key actors involved in several activities related in the wine tourism ecosystem. Some authors offered a holistic definition of wine tourism as the visitation to vineyards, wineries, wine festivals, and wine shows for which grape wine tasting and/or experiencing the attributes of a grape wine region are the prime motivating factors for visitors (e.g., Back, Tasci, and Milman 2020; Charters and Ménival 2011). This definition emphasizes the crucial role of the tourist's experience in interaction with different elements and actors of the local wine environment. Consequently, wine tourism is more about the wine as a hedonic and emotional experience than merely about the wine as a product.

Wine consumption is an intense sensory experience that combines several aspects and is subjective and personal; its tasting differs from one person to another. Besides, wine consumption practices have unique features that are intrinsically linked to the individuals' mental representations as well as symbolic and ritual dimensions of this product (e.g., Gu et al. 2020; Lunardo and Ponsignon 2020). Mental representations refer to the cognitive dimensions and the individuals' knowledge about wine accumulated through interactions with their environment (Ravenscroft and Westering 2001). Thus, consumers' perception of wine is related to their perception of the world, and wine, as a product, defines the worlds in which wine consumers locate themselves according to their representations and preferences (Brown and Getz 2005). Besides, authors have underlined that tasting wine will stimulate consumers' sensory systems and allow drinkers to experience a pleasure. In this sense, we can note that drinking wine is not an ordinary consumption practice; it also nourishes consumers' imagination, emotions, and feelings. Besides, wine also has a solid symbolic dimension that leads consumers to assign different meanings to their wine consumption. Therefore, wine consumption activities encompass several representations. One of these dimensions is hedonism, which is related to the notion of pleasure related to wine consumption (e.g., Bruwer and Rueger-Muck 2019). Indeed, the organoleptic aspects of wine generate happiness, well-being, relaxation, and a cheerful feeling. Another dimension is related to the social aspect of wine consumption. Indeed, wine tasting is considered an act of unity and an initial rite, which is regularly repeated within different groups throughout a socialization process that allows individuals to access the wine's universe. Therefore, wine consumption allows individuals to gather and connect where activities such as sharing rituals are the heart of the social representation of wine consumption.

Then, the third dimension of wine consumption emphasizes its cultural dimension thanks to its connection with the past, heritage, and traditions (Ravenscroft and Westering 2001). Wine consumption often gives rise to both collective and individual narratives charged with symbolic meanings ranging from religious representations to ancient myths and stories anchored in a particular culture and region where the wine is produced. For instance, in France, the cultural perception of wine consumption and production includes origins and labels that are powerfully grounded in meaning embedded within a particular cultural setting. Besides, in the French case, wine is perceived as part of the individual's identity and food consumption culture (Batat 2020). Finally, authenticity is another dimension that is an integral part of wine consumption, which refers to elements such as label, know-how, the variety of grape used to produce wine, the story and history around the wine producer family, the region, and the field (Batat 2021).

To sum up, wine consumption is an experience that is defined in terms of its holistic and multidimensional aspects than merely the product (wine) and its tangible attributes. Therefore, wine consumption encompasses dimensions, including culture, socialization, hedonism, and authenticity, that together will create a unique wine experience charged by emotions and meanings. Thus, the wine experience cannot be dissociated from hedonism, sensory, emotions, storytelling, and magic that constitute an integral part of the production and consumption process while guaranteeing the heist quality of wine that provides consumers with an authentic and pleasurable drinking experience. That is why it is essential for tourism scholars and professionals to shift the focus from wine tourism, principally focused on the product's features, to a more experiential approach (Batat 2021). This shift leads to identify both tangible (e.g., quality) and intangible (e.g., emotions) factors that are part of the wine tourism experience (WTE) framework and examine how they can affect positively or negatively the tourist's wine experience.

The wine tourism experience (WTE) framework

More than in any other tourism sector, wine tourism responds to a search for experiences that offer quality, emotions, pleasure, originality, consideration, and additional services (Batat 2019a; Batat et al. 2019). Tourists want to live experiences – not just buy wine and other products or services (Batat 2019b) – when they visit a destination. Although it is one of the most traditional industries, the wine tourism sector is very dynamic and is changing rapidly under the influence of new tourist trends, the advent of technology, and the rise of young Millennials and Postmillennials (Batat 2019a). However, while wine tourism is a field of research that has been the subject of several studies, most of this research focuses almost exclusively on wine products and brands, ignoring the newly booming economy of experiential wine tourism.

To our knowledge, no research has examined wine tourism from an experiential theory approach, and most of the empirical studies remain focused on a product-oriented approach.

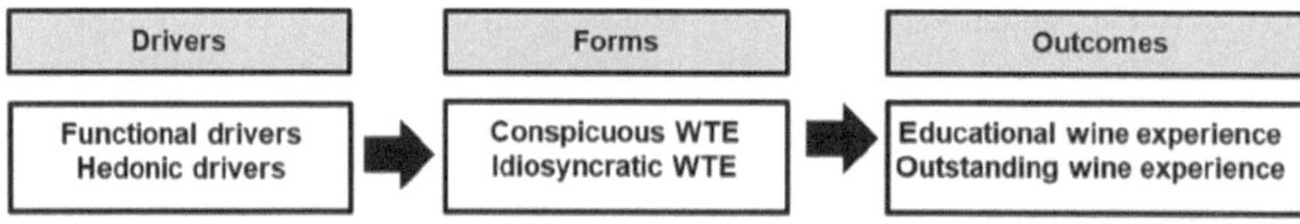

Figure 32.1 The wine tourism experience (WTE) framework.

Therefore, this chapter aims to fill this gap in the wine tourism literature by introducing a conceptual framework of the wine tourism experience (WTE), as shown in Figure 32.1. The WTE framework introduces the drivers, forms, and outcomes of the wine tourism experience that can help tourism scholars and professionals rethink their approaches by considering a holistic and comprehensive analysis (Batat 2019a). In the proposed framework, derived from recent research in several disciplines ranging from tourism and marketing to sociology and cultural anthropology, I discuss the drivers, forms, and outcomes of the wine tourism experience. It is defined as *an evolving and multidimensional tourism experience that brings together hedonic and functional meanings that tourists assign to their wine experience according to forms that might be either conspicuous or idiosyncratic, leading to educational or outstanding wine experience outcomes.*

WTE drivers: Functional vs. hedonic

Following an extensive review of recent research on the works that focused on the experiential theory in consumption domains such as food, tourism, and luxury (Batat 2019a, 2019b), I identified two key driving factors – functional and hedonic – that motivate consumers to experience the consumption of wine while visiting a tourism destination.

Functional drivers refer to tourists' motivations related to wine consumption attributes such as taste, quality, and sensory that constitute the key drivers enhancing the wine tourism experience. Indeed, wine is embedded within the region where it has been generated. Its taste is rooted more in this soil than in the wine producer's know-how. Thus, tourists are motivated by the region and the place of production that gives its prestige to the uniqueness of the wine experience that they expect. In this sense, the origin is a theme used as part of the positioning strategies, because it enhances the wine tourism experience by providing the destination a superior and unique meaning, and it informs about the good taste and quality of the wine being produced in the region.

On the other hand, the sensory factor suggests integrating the five senses (sight, smell, taste, hearing, and touch) into the wine tourism experience to enhance the tourists' immersion. Unlike the traditional wine tourism perspective, which views tourists as rational decision-makers interested in the quality of the wine, the experiential logic looks at tourists as rational and emotional actors who want pleasurable, sensory, and immersive experiences (Batat and De Kerviler 2020). Research has highlighted that the sensory factor of experiential tourism engages tourists' senses and affects their perception, judgment, and behavior (Batat 2019a). Although many studies explored the impact of the five senses on individuals' behaviors, little research has been conducted in the area of wine experience in the tourism field. Indeed, tourists' perception of the wine senses plays a fundamental role in the appreciation of the wine experience. Moreover, focusing on tourists' perception of sensory can lead wine producers to create emotional connections with tourists and visitors. Furthermore, sensory marketing studies highlight the idea that a consumer who feels good in a consumption or purchase context (e.g., store, restaurant, hotel, museum, gym) tends to spend more time there, consume, and generate a positive word-of-mouth effect as a result of their experience (Skandalis, Byrom, and Banister 2019).

Hedonic drivers refer to meanings related to the symbolic or social aspects of the wine consumption tourists assign to their definition and wine tourism experiences. Although experiences, as Holt (1995) argues, may have a shared and collective dimension, wine tourism experiences are inherently personal, and their meanings belong to the individuals who have lived them. Consumers give meaning to their consumption experiences according to the symbolic aspects (Batat 2019a) that they attribute to them, and how wine tourism experiences contribute to the construction of their own personal and social identity. Indeed, it is a matter of going beyond thinking about the tangible aspects of wine and considering more the symbolic and subjective facets from the tourist's perspective (Batat 2019a). Thus, the wine tourism experience is not only related to a tangible object (wine) defined by its taste, shape, design, color, brand, or composition, but it can also symbolize love, commitment, or other values according to the different perceptions of tourists. As Batat 2019a states: 'only by giving meaning to the consumption experience can consumers remember it … and repeat it if they are satisfied' (p. 69).

Regarding the social meaning, socialization through the creation of the social bond is integral to improving the quality of the wine tourism experience. Individuals develop a social connection to live desired experiences (Belk and Costa 1998). They can participate in tourism and consumption activities centered on shared interests and lifestyles related to wine consumption to create union and harmony among them. This act of bonding helps to improve the perception of the wine tourism experience. While the human dimension is essential in the tourism experience, tourist engagement can be affected by the quality of the tourist's interactions with the wine producer as well. Thus, a good wine alone is not enough, and since it does not guarantee a positive wine tourism experience, wine producers should be trained to develop their empathy capital (Batat 2019a) to ensure an enchanting wine tourism experience for their visitors.

WTE forms: conspicuous vs. idiosyncratic

Recent studies classify consumer experiences into two primary forms, namely conspicuous vs. idiosyncratic. Following a conspicuous perspective, Veblen's (1953) theory has amply contributed to understanding consumer motivations and attitudes towards highly experiential products such as luxury goods. Thus, a conspicuous wine tourism experience emphasizes that tourists' motivation in terms of wine consumption is to define a social status recognized by all actors belonging to the same social sphere. Based on the exploration of consumption within the old aristocracy to the adoption of luxury and its codes by new emerging social classes, Veblen identified the salient features (e.g., power, strength, level of influence, ostentation) that characterize luxury consumption in these different social classes. Like Veblen's insights, wine consumption follows the same codes and norms that underline tourists' motivations. Thus, the wine experience can take many forms, among them 'ostentatious leisure,' which consists of dedicating one's time to unproductive tasks such as tourism activities around wine. Following the theory of capital (Liu 2018), different types of conspicuous wine experiences that integrate several motivations, including economic, social, cultural, symbolic, and emotional. In the conspicuous perspective, the wine experience is not only related to the utility of the objects and the types of wine products consumed. It is also considered according to its perception anchored in a defined cultural context as well as in tourists' motivations (Holt 1995). Indeed, the individual does not consume or purchase wine to satisfy a tangible need and with regard to the usefulness of the good, but to defend his/her social status in society and preserve his/her honor (Batat 2019a). Thus, a conspicuous wine tourism experience serves to send social signs and tacit and explicit messages, visible or invisible and directly or indirectly, to social actors to decode by referring to consumer objects and lifestyles. In the conspicuous wine tourism experience, tourists' motivations and behaviors are subject to the dominant values and meanings in the social group they belong to.

The second form, as indicated in the WTE framework, refers to an idiosyncratic tourism experience that is personal. It can be explained by Bourdieu's (1980) theory of distinction. Drawing on Bourdieu's works, the idiosyncratic tourism experience is seen as a means by which tourists can distinguish themselves from the massive tourism trend prevalent in society. In the theory of distinction, Bourdieu explains that the upper classes, attempting to distinguish themselves, construct and assert their own identity. The contribution of distinctive theory to the wine consumption experience in the field of tourism highlights the importance of status and a social life structured around tourism activities and wine consumption codes, symbols, and negotiation strategies. Each individual plays a social role determined by his or her consumption of certain types of wine tourism experiences or certain wine consumption practices to assert social status while differentiating identity. By living certain wine tourism experiences, tourists make a representation of themselves and express their identity and values. Wine as a tourism experience becomes a symbol of distinction that conveys the tourist's specific culture via communicating a high social status. Consequently, wine tourism experiences allow the establishment of links and relationships with other social actors belonging to reference groups (Mason 2006). In other words, to be distinctive, the wine tourism experience must be inaccessible, exclusive, rare, and new. If it becomes accessible and widely spread, it loses its distinctive value and becomes common and commoditized.

In summary, the two forms of the wine tourism experience guide the preferences and choice of wine products and destinations visited by tourists. Nevertheless, the two forms, although different, are complementary and allow tourists to visit destinations and live wine experiences where activities are defined by the social, emotional, and symbolic dimensions linked to wine consumption and production.

WTE outcomes: educational vs. outstanding

The WTE framework suggests different outcomes depending on the profiles of individuals and their motivations (functional or hedonic) related to their wine tourism experiences. Thus, the outcomes related to these experiences can be classified into two main categories: educational and outstanding.

Regarding *educational* expectations related to wine tourism experiences, tourists are more willing to engage in wine tourism to develop their knowledge about wine products and production processes and thus refine their tastes. Education as an outcome of the wine tourism experience will transform tourists into experts who are more knowledgeable about the wine world and can analyze the offers to satisfy their needs according to their values and lifestyle. Indeed, wine should not be reduced to alcohol; consumers need to acquire and develop skills and knowledge through their tourism experience to understand the complexity of what makes wine cultures. By visiting wine destinations, tourists can then capture the tacit meanings of a glass of wine, which can refer to the culture of local people and the whole civilization. Thus, tourists expect their wine tourism experiences to help them to develop cognitive and cultural skills to decode the wine language and thus describe and appreciate wine. It is not only a simple drink, but an object of civilization. Therefore, tourists build and develop their wine culture through education and learning by embracing wine tourism activities to learn more about their own wine culture and others. In this case, the wine tourism experience is considered an influential factor of personal development.

On the other hand, an outstanding development of competence as an outcome of the wine tourism experience is related to novelty and the 'Wow' effect aiming to consolidate the cognitive dimension and build consumption skills in the individual. Although the idea of a tourists' competence and knowledge as a consequence of their lived wine consumption experience is a relevant approach to explain how designing wine experiences in the tourism field can be a vector of elitism in consumer society, very little research has explicitly used consumer competence as an outcome of the consumption and/or tourism experience. The majority of work, such as Carlson et al. (2009) and Park, Mothersbaugh, and Feick (1994), focuses on the knowledge developed by consumers using the concept of 'consumer expertise,' which refers to cognitive knowledge and the process by which consumers process and remember relevant information to optimize their purchase. Thus, while the cognitive dimension has been emphasized in marketing research, the competence derived from an experiential journey and its creative potential as an outcome of the consumption experience has rarely been examined, except for a few works. In marketing, authors define consumer competence that emerges in the past and present consumption experiences along three main dimensions: cognitive skills related to the ability to interpret marketing messages, instrumental skills that translate into the ability to use products before, during, and after the consumption process, and behavioral skills related to the ability of individuals to appropriate the products and services provided by the firm (Batat 2019a).

Beyond learning via past or present wine tourism experiences, tourists can also develop creative potential and social skills through their socialization experiences. Thus, the knowledge developed in the wine tourism field evolves throughout the life cycle and experiences, allowing tourists to build an elite identity by developing fine expertise in the wine sector. In the experiential approach to wine tourism, an outstanding outcome of the experience can be defined by its intellectual and creative dimensions resulting from tourist experiences that also include perception (through the five senses), emotion, intellect, and action. Besides, the intellectual dimension is also linked to reflection, the curiosity of tourists regarding their wine tourism experiences. Studies have highlighted that consumers value intellectual aspects and appreciate the brand's ability to arouse their curiosity and reflection and thus nourish and stimulate their intellect.

Conclusions and implications

This review of recent advances in experiential marketing and the wine tourism literature highlights the central idea that experiential wine tourism is a growing field of research that scholars can examine from various angles and perspectives. Thus, understanding the wine tourism experience has become a primary goal for both academics and tourism scholars. Thus, experiential thinking is considered an effective strategy that allows wine producers to differentiate themselves, strengthen tourist loyalty, and improve tourist satisfaction while developing a positive attitude towards wine consumption and tourism. On a practical level, this synthetic literature review identifies several drivers, forms, and outcomes of the wine tourism experience by proposing the WTE framework to develop our understanding of the tacit and explicit aspects related to wine consumption. Therefore, developing a better understanding of these three elements of experiential wine tourism is a necessity that will help tourism professionals create wine experiences tailored to the tangible and intangible needs of different profiles of tourists worldwide. By introducing new perspectives on the experiential approach to wine tourism, I hope to pave the way for further theoretical and empirical research on this topic.

Regarding the managerial implications, the WTE framework can help wine tourism professionals better understand tourists' functional, social, and emotional needs by designing wine experiences that respond to both tangible and intangible expectations and thus differentiate them from their competitors. In order to distinguish wine tourism offers and transform them into highly attractive wine experiences, winegrowers can create additional activities related to wine and the vine to augment tourists' experience and enhance their positive memories, which will entice them to come back because they lived an enchanting experience. Thus, the wine experience can be associated with several activities, including proposing accommodation in a new themed place to help visitors escape their daily routine, creating social activities with local people, and offering gastronomic experiences linking wine with food. These activities can also integrate high-end services, such as offering a spa space that generates positive emotions and makes visitors feel unique and thus responsive to the outstanding wine experience clients seek, as demonstrated in the WTE framework introduced in this chapter. At the cognitive level, tourists can also discover the secrets of the production process and broaden their knowledge of oenology. The value of exemplarity is essential to confirm the customer's convictions about the vineyard, which aligns with the educational expectation highlighted in the proposed framework.

References

Back, R. M., Tasci, A., and Milman, A. (2020). Experiential consumption of a South African wine farm destination as an agritourism attraction. *Journal of Vacation Marketing*, 26: 57–72.

Batat, W. (2019a). *Experiential Marketing: Consumer Behavior, Customer Experience, and the 7Es*. London: Routledge.

Batat, W. (2019b). *Food and Experiential Marketing*. New York: Routledge.

Batat, W. (2020). Pillars of sustainable food experiences in the luxury gastronomy sector: a qualitative exploration of Michelin-starred chefs' motivations. *Journal of Retailing and Consumer Services*, 57: 102255. https://doi.org/10.1016/j.jretconser.2020.102255

Batat, W. (2021). The role of luxury gastronomy in culinary tourism: an ethnographic study of Michelin-starred restaurants in France. *International Journal of Tourism Research*, 23(2): 150–163. https://doi.org/10.1002/jtr.2372

Batat, W., and De Kerviler, G. (2020). How can the art of living (art de vivre) make the French luxury industry unique and competitive? *Marché & Organisations*, 1(37): 15–32.

Batat, W., Peter, P. C., Moscato, E. M., Castro, I. A., Chan, S., Chugani, S., and Muldrow, A. (2019). The experiential pleasure of food: a savoring journey to food well-being. *Journal of Business Research*, 100: 392–399. https://doi.org/10.1016/j.jbusres.2018.12.024

Beames, G. (2003). The rock, the reef and the grape: the challenges of developing wine tourism in regional Australia. *Journal of Vacation Marketing*, 9: 205–212.

Belk, R., and Costa, J. A. (1998). The mountain man myth: a contemporary consuming fantasy. *Journal of Consumer Research*, 25: 218–240.

Bourdieu, P. (1980). La distinction: critique sociale du jugement. *Revue Française De Sociologie*, 21: 439–444.

Brown, G., and Getz, D. (2005). Linking wine preferences to the choice of wine tourism destinations. *Journal of Travel Research*, 43: 266–276.

Bruwer, J., and Rueger-Muck, E. (2019). Wine tourism and hedonic experience: a motivation-based experiential view. *Tourism and Hospitality Research*, 19: 488–502.

Bruwer, J., and Thach, L. (2013). Wine tourists' use of sources of information when visiting a USA wine region. *Journal of Vacation Marketing*, 19: 221–237.

Carlsen, J., and Boksberger, P. (2015). Enhancing consumer value in wine tourism. *Journal of Hospitality & Tourism Research*, 39: 132–144.

Carlson, J. P., Vincent, L., Hardesty, D., and Bearden, W. (2009). Objective and subjective knowledge relationships: a quantitative analysis of consumer research findings. *Journal of Consumer Research*, 35: 864–876.

Charters, S., and Ménival, D. (2011). Wine tourism in champagne. *Journal of Hospitality & Tourism Research*, 35: 102–118.
Cho, M., Bonn, M. A., and Brymer, R. A. (2017). A constraint-based approach to wine tourism market segmentation. *Journal of Hospitality & Tourism Research*, 41: 415–444.
Cohen, E., and Ben-Nun, L. (2009). The important dimensions of wine tourism experience from potential visitors' perception. *Tourism and Hospitality Research*, 9: 20–31.
Gázquez-Abad, J., Huertas-Garcia, R., Vázquez-Gómez, M., and Romeo, A. C. (2015). Drivers of sustainability strategies in Spain's wine tourism industry. *Cornell Hospitality Quarterly*, 56, 106–117.
Getz, D., and Brown, G. (2006). Critical success factors for wine tourism regions: a demand analysis. *Tourism Management*, 27(1): 146–158.
Gu, Q., Qiu, H., King, B., and Huang, S. (2020). Understanding the wine tourism experience: the roles of facilitators, constraints, and involvement. *Journal of Vacation Marketing*, 26: 211–229.
Hall, C. M., Johnson, G., and Mitchell, R. (2000). Wine tourism and regional development. In Hall, C. M., Sharpies, L., Cambourne, B., and Macionis, N. (eds.), *Wine Tourism around the World, Development, Management and Markets* (pp. 196–225). Oxford: Butterworth-Heinemann.
Holt, D. (1995). How consumers consume: a typology of consumption practices. *Journal of Consumer Research*, 22: 1–16.
Humphreys, A., and Carpenter, G.S. (2018). Status games: market driving through social influence in the U.S. wine industry. *Journal of Marketing*, 82: 141–159.
Liu, C. (2018). Examining social capital, organizational learning and knowledge transfer in cultural and creative industries of practice. *Tourism Management*, 64: 258–270.
Lunardo, R., and Ponsignon, F. (2020). Achieving immersion in the tourism experience: the role of autonomy, temporal dissociation, and reactance. *Journal of Travel Research*, 59: 1151–1167.
Mason, C. (2006). Exploring the processes underlying within-group homogeneity. *Small Group Research*, 37: 233–270.
Nisco, A. D., Mainolfi, G., Marino, V., and Napolitano, M. (2015). Tourism satisfaction effect on general country image, destination image, and post-visit intentions. *Journal of Vacation Marketing*, 21: 305–317.
Park, C. W., Mothersbaugh, D. L., and Feick, L. (1994). Consumer knowledge assessment. *Journal of Consumer Research*, 21: 71–82.
Ravenscroft, N., and Westering, J. V. (2001). Wine tourism, culture and the everyday: a theoretical note. *Tourism and Hospitality Research*, 3, 149–162.
Sampaio, A. (2012). Wine tourism and visitors' perceptions: a structural equation modelling approach. *Tourism Economics*, 18, 533–553.
Skandalis, A., Byrom, J., and Banister, E. (2019). Experiential marketing and the changing nature of extraordinary experiences in post-postmodern consumer culture. *Journal of Business Research*, 97: 43–50.
Veblen, T. (1953). *The Theory of the Leisure Class: An Economic Study of Institutions*. New York: Macmillan.
Ye, B., Zhang, H., and Yuan, J. (2017). Intentions to participate in wine tourism in an emerging market: theorization and implications. *Journal of Hospitality & Tourism Research*, 41: 1007–1031.

33

THE STAGING OF WINE TOURISM EXPERIENCES IN AUSTRALIA

Paul Strickland and Kim M. Williams

Introduction

Wine tourism is favored by government bodies within numerous New World wine countries, including Australia, Canada, South Africa, and New Zealand. The popularity of partaking in wine tourism has expanded considerably in recent decades and remains critical to entice visitors to wine regions to purchase local products and services (Sigala and Robinson 2018). This component of the tourism sector has grown into a multibillion-dollar industry and is particularly crucial to regional producers where small to medium-sized wineries rely on travelers for sales and the viability of their tasting room (also known as cellar door) (Pratt 2019). This is especially true for regional and rural wine areas that are located vast distances from major cities and large populations, as witnessed in Australia. In order to attract visitors, wine regions need to develop inventive strategies to stage enticing and exciting wine tourism experiences which suit different market segments, such as Baby Boomers and Millennials. These encounters also need to fulfill the desire for a variety of sensorial, hedonic, and experiential consumption experiences – for instance, offering entertaining and interactive attractions as well as utilizing state-of-the-art technological applications (Roberts and Sparks 2006; Strickland et al. 2020).

In 1998, the Wine Makers' Federation of Australia, via their National Wine Tourism Strategy, put forward the following definition of wine tourism:

> Visitation to wineries and wine regions to experience the unique qualities of contemporary Australian lifestyle associated with enjoyment of wine at its source, including wine and food, landscape, and cultural activities.
>
> *(Winemakers' Federation of Australia 1998)*

Furthermore, wine tourism attracts individuals that are principally inspired by a desire to engage in wine and food activities who are known as wine tourists. However, these individuals may also be subsequently interested in encountering local gastronomy, culture, arts, education, and entertainment activities when staying in these wine destinations (Cohen and Ben-Nun 2009).

This chapter investigates current and contemporary staging opportunities for wine tourism experiences that would attract visitors for day trips and short stays. Wine regions provide countless opportunities to attract the tourist, which also include strategically staged eventualities.

DOI: 10.4324/9781003143628-39

The provision of relaxed wine tasting rooms and cellar doors, informative and interactive wine notes and wine rating apps on smartphones, degustation and wine matching classes, tours of the winery's production facilities, and meeting the vintner can all be considered as staged occurrences. In addition, complementary experiences which value-add can include shopping for local produce, hot air ballooning, Segway tours, wine tours by cycling or equestrian, art instillations, music festivals/events, grape stomping, and website/social media activities (Prahalad and Ramaswamy 2004). Thus, it is important to recognize and understand that select consumers have expectations for services beyond a rudimentary wine tasting, and it is vital for management of wineries to take advantage of ways to address these more sophisticated service expectations to motivate the visitor to engage and return on numerous occasions (Beames 2003; Canziani 2019; Marzo-Navarro and Pedraja-Iglesias 2009). To maximize the return on investment from these encounters, wineries are advised to consider a range of encounters that may appeal to different target markets and price points.

Wine segmentation: target markets for staged winery experiences

The Australian wine industry uses segmentation to categorize wine consumers. These distinct market segments have different attributes and are defined by the era in which these consumers were born (Gardiner, Grace, and King 2013), namely:

- Traditionalists, or the Silent Generation: born 1945 and prior
- Baby Boomers: born 1946–1964
- Generation X: born 1965–1976
- Millennials or Gen Y: born 1977–1995
- Gen Z, iGen, or Centennials: born 1996–2015

Baby Boomers were previously the largest wine consumer segment who preferred high-quality wine and older grape varieties such as Chardonnay, Shiraz, and Cabernet Sauvignon (Strickland et al. 2020). Generation X is considered middle-aged and is strongly influenced by wine critiques and wine awards (Barber, Dodd, and Ghiselli 2008). Generation X members are well educated and prefer medium-bodied wines and Sauvignon Blanc to Chardonnay, for example. Higgins and McGarry Wolf (2016) have suggested Millennials make up a new, fastest-growing wine consumer group and a significant contributor to the wine sector. They prefer social, hedonic wine marketing campaigns and imagery rather than critiques of wine. Millennials have a willingness to learn and prefer lighter, sweeter grape varieties such as Riesling, Rosé, or Pinot Noir. Generation Z is an emerging market segment, as people born in this era may not have reached legal alcohol consumption age in Australia. Although it is anticipated they will also be a significant wine-consuming segment, due to a growing understanding of alcohol and health issues, this generation may consider wine consumption in moderation.

Staging experiences

Staging is a concept that can be transported from the most central aspects associated with theatrical productions to the application of successful event design. Staging can be associated with both the layout of the environment and what is occurring on the stage to enhance the audience's satisfaction with the experience (Getz and Page 2016). Staging experiences in wine tourism entails that winemakers in a locality 'intentionally use services as the stage and goods as props, to engage individual customers in a way that creates a memorable event' (Pine and Gilmore 1999, p. 11).

Staging elements can be agreed upon, planned, and then delivered. A myriad of elements must come together holistically and portray synergy for an event to be effective and successful; these include a specific audience, a complementary theme, suitable timing and program, efficient and friendly service, a wow-factor setting, entertainment, use of technology, and the tying together of logistics. Aligned with this, Mueller and Scheurer (2004) developed seven mechanisms to assist in effectively staging experiences: theming, target market (discussed earlier), activities/attractions, aesthetics, visitor guidance, well-being, and evaluation.

Theming delivers consistency, should be genuine, and could possibly connect with the culture and history of the region or specific location. The target audience or visitor should be able to easily understand, interpret, and relate to all-inclusive aspects being presented to elevate the staged experience. Customer needs, expectations, and behaviors should be determined to assist with each staging element, including appropriate and achievable theming. Activities or attractions can be passive or active and absorptive or immersive, providing a range of participant experiences known as the four realms of the Experience Economy (Pine and Gilmore 1999). These activities/attractions require alignment with the chosen theme as well as with visitors' expectations and contribute to the overall experience setting (Peters and Weiermair 2000). The aesthetic of a staged experience is of great importance, including the countryside, weather conditions, and architecture of the human-made environment, all of which impact the customers' overall encounter in a positive or negative way.

Similarly, the management of traffic flow, both vehicles and people, are key to early satisfaction for visitors to an experience. Parking, queuing times, and signage should be readily found and of an acceptable standard. Providing ease of facility access, reception/information desks, and maps, for instance, should be available on arrival so visitors can rapidly feel welcome and relaxed. Information may also be provided in advance on websites or specific apps where visitors can obtain information in real time.

Well-being is a supporting tool for visitor's positive evaluation of the experience setting and is strongly linked with visitor guidance. Only visitors who feel comfortable will perceive positive experiences. To provide well-being, a setting may offer family-friendly facilities, appropriate toilets, and food and beverage outlets and allow for all conditions of peoples' physicality. Furthermore, utilizing an evaluation instrument to gauge the satisfaction levels of the visitor is especially important for facilitating improvements to staged experiences in the future. In addition, visitors can be co-creators of an experience and be part of the attraction and contribute to the overall experience themselves (Dixit 2020; Pikkemaat et al. 2009).

Engaging the five senses

Staged wine encounters require the winery management to be progressive and start to design multisensorial wine experiences that can both arouse tourists' senses and construct passionate impressions about their winery engagement. This type of engagement can have the capacity to prompt inquisitiveness, intellect, and cognitive processes whilst satisfying facets of conventional enjoyment and fun. As Sigala (2019a) commented, 'it has become imperative that effective wine tourism experiences are those that can engage wine tourists behaviourally, emotionally, mentality/cognitively but also spiritually' (p. 139) and offer a positive and satisfying visitor experience.

Pivac (2012) indicated that the 'total wine experience' includes ten features: (1) wine tasting and buying, (2) socializing with friends, (3) enjoying the day outdoors, (4) enjoying the rural environment and vineyards, (5) becoming familiar with wine production, (6) learning about wine, (7) wine tasting in restaurants, (8) tour of the winery, (9) visitor experience of other attractions and activities, and (10) relaxation. Each of these encounters facilitates the incorporation of a sensory engagement which can heighten the visitors' pleasure and satisfaction. Humans have five senses: sight, touch, taste, hearing, and smell (Brochado, Stoleriu, and Lupu 2021). Winery experiences provide a possibility of managing the staging of memorable and holistic experiences, which include as many aspects of our senses as is practical. Sensory experiences can highlight memories and reinforce future relationships with a product such as wine. This may assist with repeat visitation or additional wine purchases.

Memorable encounters

Sigala (2019a) pointed out that not only has wine tourism demand grown and flourished, but wine tourists, and consumers in general, have become more well-informed, savvy, and educated and demand superior customer service. Thus, winery management strategies could include innovative memorable experiential encounters at the winery to satisfy discerning visitors. Wine regions now need to create novel and imaginative wine experiences to stand out from the crowded wine tourism arena. To attract and satisfy the current wine tourists, the staging of these encounters could occur in distinguished and specific wineries. Conversely, staged experiences can occur on a much larger scale, such as wine regions, which differentiate themselves from other competing locations and tourism attractions based on wine.

In general, tourists tend to revisit a location that has provided positive memories for them. Significant memorable experiences influence tourist retention and destination loyalty (Marschall 2012; Seyfi, Hall, and Rasoolimanesh 2020). Memorable and unique encounters assist in creating positive engagement and loyalty (Chen and Chen 2010; Dixit 2017). De Freitas Coelho, de Sevilha Gosling, and de Almaida (2018) proposed a theoretical framework by identifying the core processes that interfere with memorability and make sense and meaning in a memorable tourism experience (MTE). The MTE has three dimensions: (1) the personal dimension, incorporating the degree of perceived novelty, dreams and desires of fulfillment, live emotions, and also travel motivations; (2) the relational dimension, including travel planning, travel companionship, and interpersonal interaction; and (3) the environmental dimension, involving tourism attractions and cultural exchange (Santos et al. 2019). From these dimensions, three key processes emerged: ambiance, socialization, and emotion with reflection. The presence of all three dimensions strongly assist in achieving an encouraging MTE (Santos et al. 2019). Wine tourism experiences that employ the findings framework from this research have the capacity to benefit from developing a staged experience, which focuses on all three dimensions to obtain an advantageous outcome for their business.

Hedonism and hedonistic well-being

As discussed previously, Santos et al. (2019, p. 721) suggested that 'wine is the central product of wine tourism, which on its own, has an experimental dimension based on the wine tourism experience, essentially related to the hedonic perspective, as it involves sensations, through the five senses' (Candi, Beltagui, and Riedel 2013). Hedonistic pleasures are often associated with self-indulgences and decadence, which in turn can be allied with the overconsumption of alcohol and food. However, moderate wine consumption can be linked with hedonistic well-being and favorable health benefits, especially if connected to stress release and social activities (Fiore, Alaimo, and Chkhartishvil 2019).

Generating a staged and controlled wine consumption environment which assists with hedonistic well-being can be fruitful to the wine tourism sector. Getz (2019) suggested that placing a spotlight on the lifestyles of food and wine lovers is appropriate for purposes of target marketing and understanding consumer motivations. Hence, creating a total wine tourism experience is suggested. Charters and Mitchell (2014) stated that there is a favorable synergy between wine, music, and the arts, and that all could be considered hedonistic pursuits. Wang, Tang, and Cheng (2018) discussed how art is utilized by various hotels for creating an artistic, sensory, and intellectually enhancing encounter. By staging a sophisticated, luxury experience via the hosting of music festivals or art installations, wineries have been able to increase visitations of day trippers or those that may be motivated to stay for a short period of time in the wine area. Sigala (2019b) provided an example of a scarecrow art installation that took place in the winescape of Domaine Sigalas on the island of Santorini, Greece, during the month of September 2016. The winery successfully used art to bring together local artisans, wine tourists, and the local community to co-create additional value for the region.

Another innovative way to generate a hedonistic or personalized staged experience is creating opportunities for the wine tourist to be exposed to tuition by specialists, especially with celebrity chefs hosting cooking classes and winemakers divulging their winemaking techniques. Personal narratives and sharing of family history and heritage, picking grapes, wild-food scavenging (for example, mushrooming), and other novel events are examples of combining active participation, consumption, and education (Getz 2019).

The 2020 wineries predicament

At the end of 2019 and the start of 2020, the devastating Australian summer bushfires took hold and made news around the world. Australia was burning, and some of the areas affected were in renowned wine regions. Wittwer (2020, p. 6), from the grape and wine grower's perspective, wrote:

> First, in the Adelaide Hills, one third of vineyards were destroyed by the catastrophic fire which swept through the Adelaide Hills shortly before Christmas 2019. Second, in the Hunter Valley and Yass-Young, there may have been no vineyards destroyed by fire, but weeks of bushfire smoke resulted in around 80% of the 2020 vintage being unusable for wine. Third, in regions heavily reliant on international tourism, turnover may be disrupted for several years into the future due to an altered perception of Australia as a tourism destination, beyond the collapse in international tourism arising from COVID-19. That is, Australia's image as relatively unspoilt with ample natural assets has been damaged by the bushfires and may take years to restore.

Many vineyards and wineries were destroyed by fire. Others that escaped the direct flames were subjected to smoke, tainting the grapes and rendering them unusable. That vintage literally went up in smoke. The King Valley experienced the threat of fire in 2020, and Dal Zotto could retain only an extremely limited amount of their 2020 vintage. Pizzini winery was able to produce some white wine products; however, they did not manage to produce any 2020 red wines and only a very limited 2021 red wine selection. The bushfires had devastated their sales expectation for the next three years.

To compound this tragic circumstance for Australia, the world was to face a one in one hundred years' pandemic. The impacts of COVID-19 on the tourism industry were to be felt around the globe. When federal and state governments of Australia closed their borders in March 2020, wineries were instructed to shut their doors to visitors for an extended period of time, with some being closed until November 2020. The celebrated Australian wine critic James Halliday summed it up when he stated that COVID-19 had 'been brutal in its impact on small to medium-sized wineries, the vast majority relying on tourism for the bulk of their sales income' (Halliday 2020, p. 4). The immediate closure of wineries meant mass cancellations of cafés and restaurant reservations, cellar door walk-ins, wine tours, and staged face-to-face events and/or festivals such as weddings. Wineries had to quickly reinvent their business models to online sales and take-away food services and focus on other revenue streams. Some wine regions were able to host virtual wine events and festivals where the wine visitor and/or consumer purchased wine from certain wineries in advance and gained exclusive access to tutored tasting classes via protected online links. Although Australia's wineries are starting to re-open, the legacy of the bushfires and COVID-19 will impact the wine tourism industry for a decade or longer. In mid-November 2020, Victorians were able to resume statewide travel, and wineries began to welcome visitors back at the cellar door. It is now more important than ever for these businesses to develop a range of staged offerings to attract all possible wine tourism markets back to their regions.

To further complicate these challenging times, in November 2020, China placed tariffs of up to 212% on all Australian wine, severely reducing their competitiveness. If the Australian export market to China dried up, huge pressure would be placed on the industry to find other viable distribution outlets. The domestic market must now rally to support our winemakers, from day-to-day consumers to restaurants and retailers, to strive to make up the shortfall of revenues (Black 2020). This may force wineries to create innovative and staged experiences focused on selling more wines to domestic markets or finding other international markets such as countries in Southeast Asia as well as in Great Britain.

The use of technology to facilitate staged experiences

Wollan (2011) identified seven main social media platforms: blogs, ratings and reviews, referrals and sharing, forums, user-generated content, member profiles, and social networking websites. From these, the wine industry in Australia adopted four of them: (1) referrals and sharing, (2) user-generated content, (3) member profiles, and (4) social networking (Strickland et al. 2013). Furthermore, the most common social media used in the Australian wine industry are Facebook, Twitter, Instagram, TripAdvisor, and YouTube, all of which fall into one or more of the four categories. The wine industry selected these social media platforms because they are the most regularly accessed by potential consumers in Australia.

Technology has provided a channel for interactive applications to increase availability and another avenue to access product information, whereas social media enhances consumer engagement with wine tourism products, including promotion of staged encounters. For instance, virtual reality tours and Zoom wine tasting encounters allow for wider accessibility and educational opportunities, which may convert into greater purchasing preference and brand loyalty. This may be of great interest to wine tourists who do not want to, or cannot, travel to wine regions, although most wine tourists do, as demonstrated by the case study that explores the La Dolce Vita wine and food festival.

Case Study 33.1 La Dolce Vita wine and food festival

The La Dolce Vita wine and food festival is located within the King Valley in northeast Victoria, Australia. This region is approximately a three-hour drive north of Melbourne and four hours south of Canberra (the national capital city of Australia). The location can be accessed by road, rail, and cycling and also provides equestrian trails. 'The King Valley wine region has one of the greatest concentrations of varietals of any wine region in Australia whilst [having] an Italian bias reflecting [the] region's history' (Wine of the King Valley 2020a). The La Dolce Vita wine and food festival, or 'the sweet life,' first appeared in 2001 to pay homage to the region's Italian heritage and culture, whilst also enticing increased tourist visitations (Laing and Frost 2013, p. 323). 'The Italian Spirit' of nine local wineries – Brown Brothers, Dal Zotto, Darling Estate, King River Estate, La Cantina King Valley, Pizzini, Politini, Sam Miranda, and Symphnia – participate annually in the festival (Wine of the King Valley 2020b).

The La Dolce Vita festival attracted over 4,000 visitors to the region in 2010 (Laing and Frost 2013) and has increased since. The staging of the festival is based on nonlocal visitors for short stays and weekend breaks, which have become popular in Victoria. This trend will continue into the foreseeable future; however, due to extended COVID-19 travel restrictions, which predominantly curbed intrastate and eventually interstate, there has been a lack of movement in many wine regions, especially in Victoria and for a longer period of time compared to other Australian states and territories.

King Valley has a strong connection to family and traditions alongside Italian heritage:

> The King Valley lifestyle is influenced by Italian roots but grounded in its Australian bush setting. Our winemaking families explain how an Italian style influences their own lives and work within the King Valley … People say Italians don't take life seriously, but there is something we take very seriously: food and wine, and what that brings to our lives. When growing up, food was always a big part of our life. Mum and Dad were working long hours in tobacco fields and in the vineyard, but we always had to have dinner together, no matter what. At the end of each day there was a call of, 'Tutti a tavola!' which means, 'Everyone at the table!' And it's there that I learnt the importance of getting together with my family, eating and talking about our day and having some fun. That's something I do with my own family now.
>
> *(Wine of the King Valley 2020c)*

It is the wine families' attitudes reflected by these quotes that inspired the La Dolce Vita festival. The staging of the festival has been well coordinated. Tickets in the past have only been AUD $25.00 for adults and free entry for children. The money raised is reinvested back into the region. Ticket holders gain access to participating wineries, free wine tasting, and entertainment such as gnocchi making and Italian music, for example (Bodycomb and O'Neill 2014). The festival is unique in Victoria, as it focuses specifically on Italian food and wine. In 2019, the program offered a variety of Italian music, pasta cooking classes, picnics, cheese platters, circus workshops, face painting, children's playgrounds, Italian sausage sizzle, lawn games, free porchetta rolls, and a unique wine bobbing game of prosecco fishing, among others. Each winery had its own particular Italian atmosphere with a strong focus on sensorial, hedonic, and experiential consumption, which facilitated fun and memorable encounters, with the view of tourists visiting many wineries and sharing the benefits through inclusive experiences. The aim of the festival is to positively impact the wine region, giving wineries and complementary businesses (local farms, shops, cafes, restaurants, accommodations, and natural and historic attractions) increased social and economic benefits, essentially through wine tourism.

With the COVID-19 pandemic, the festival organizers had to determine if La Dolce Vita would proceed in 2020. The largest cohort of visitors come from Melbourne, and at the time of writing and the event decision-making deadline, winery visitation was forbidden, and no winery experience was possible. Only takeaway food and wine or online sales were sanctioned by the government for wineries to sell. An option that could be considered was offering a virtual festival, similar to the successfully delivered Winery Walkabout on June 6, 2020, which is located in the Rutherglen region in Victoria, or canceling the festival altogether. Owing to the continual changing and severely harsh COVID-19 restrictions in Victoria, the decision was made to completely cancel the 2020 festival, the 2021 festival was scaled back to an online experience, and in June 2022, the festival returned to the usual format. It is noted on the Wine of the King Valley (2020b) website that individual wineries would provide other alternative experiences for the wine tourist who might be able to visit the region in November and into the summer season. Pizzini Wines, for example, developed a fabulously fresh tasting experience that allowed the wine tourist to really understand the wines they make from Italian varietals. Tasting can be booked in advance, are a duration of approximately 45 minutes, and are now a seated tutored experience. There are two levels of tasting, and both incur a reasonable tasting fee. Tasting charges are new for this area, although this procedure is common for tastings across the globe.

Spending a weekend in the King Valley in late 2020 provided an opportunity to experience and observe several new staged winery cellar door wine-tasting and meal encounters. Seated wine tasting is required, and this has provided a more personalized educational encounter which also has increased expectations from the consumer. However, the success of this experience is predicated on the engagement between the customer and the winery representative (Carlsen and Boksberger 2015). It has become even more essential to have a winery employee, someone who understands and knows the heritage of the winery and their products to be able to engage the wine tourist at this time and build brand loyalty into the future (Williams 2013).

Many of the King Valley wineries visited at this time are considering a variety of new staged cellar door engagement strategy: pay or not, or a combination of both, and inclusion of a wine tasting which provides regional food samples. Dal Zotto is not charging, and Pizzini is, but there is a huge complexity behind any of these cellar door marketing strategies. Does the consumer want to make an advance booking and stay for an hour or more and be satisfied by paying for an educational encounter? Or do they want a quick couple of tastings and move on to see the range of wineries in the region? In addition, where is the designated driver considered in this mix? Other wine regions, such as the Barossa Valley in South Australia, offer a complimentary bus service from designated pick-up points. It may be time for all wineries across Australia to develop a range of price points to engage the wine tourist to enjoy a range of satisfying experiences.

Chrismont, Dal Zotto, and Pizzini, which the authors visited in late November 2020, provided a quality customer service and engaging wine encounters in a semi COVID-19 normal environment. All these wineries mentioned that a shortage of staff due to travel restrictions provided challenges, although this has allowed the consumer to have the potential to meet and discuss the wines with the owner, viticulture, or blender, providing a richer staged encounter.

Future of staging wine events

With the impacts of COVID-19, future spikes of the virus being transmitted to the general public may influence government policy to limit the number of visitors to wineries and large gathering such as festivals without a vaccination. For the La Dolce Vita wine and food festival to be offered in the future, it may have to have a reduced number of visitors to the King Valley or move to a virtual model. Virtual festivals have already been successfully delivered in Australia (i.e., Winery Walkabout) and may be a solution to continue to offer staged wine experiences in the future. If a virtual solution were to occur, wineries would need to adapt and innovate to remain relevant. This means creating online and remote wine experiences that suit families, individuals, and a mix of target markets. Virtual wine festivals have streamed live music, conducted online wine and cheese tasting with ingredients being delivered to homes prior, wine trivia, costume competitions, and virtual treasure hunts. Due to the reliance of technology, tickets to attend were still sold, with revenues going directly back to the wine region. Virtual festivals have the potential to increase visitor numbers without the need for travel which reduces the traveler's carbon footprint. The wine regions triple bottom line is still considered as the impacts of the local environment are reduced, the social aspect of a festival is still offered albeit in a different form, and the economic benefits still reach the intended wine stakeholders. It should be noted that negative impacts will continue to exist, such as traditional visitors not able to support complementary local businesses like restaurants and accommodation facilities; however, it encourages further innovations in the wine sector, and staged experiences will be paramount.

Conclusion

In conclusion, wine regions are extending beyond the winery tourism experience and reinvigorating repeat patronage through social media exposure and personal messaging (Thach, Lease, and Barton 2016). It also demonstrates the need for the adaption of business models such as offering virtual wine events and tastings when unforeseen circumstances arise, such as bushfires and global pandemics that severely restrict the movement and travel of people.

The staging of wine experiences is something most wineries in Victoria strive for. From offering only wine tastings at the cellar door to providing genuine and authentic family connections, heritage, and cultural links. The La Dolce Vita wine and food festival is a good example of a wine region that has tapped into Italian heritage and culture. However, it is geographically located in Victoria, Australia, which presents challenges. To attract visitors to the King Valley, the La Dolce Vita festival has used the shared economy philosophy and brought together nine wineries, all offering different wine experiences for both adults and children targeting short stays from visitors coming mainly from Melbourne. Using the underpinning theme of 'the Italian spirit,' the winery stakeholders have developed a two-day festival that includes wine tasting, pasta-making, Italian music, 'prosecco fishing,' Italian food, and lawn games to showcase their Italian heritage and culture. It is also evident that the staging of wine experiences must be authentic, hedonic, fun, and interactive and offer a range of experiences that appeal to both adults and children, as family and friends' interactions is at the heart of the festival's conception. All wine experiences (except wine tasting for under 18 years of age) are designed for families to participate, and this is key to the success of encouraging short breaks for the entire family.

Implications

The implications for the wine tourism sector are to strategically respond to challenges presented by changing, innovating, and adapting in this dynamic environment. Creating quality wine products are just not enough to survive financially in the long term. Therefore, individual wineries and wine regions need to 'value add' by offering a variety of co-created and collaborative products, services, and experiences which complement the surrounding wine region. This means creating inventive wine tourism marketing strategies which focus on the strengths of the wine region, with realistic investments in infrastructure, and also obtaining positive government and community support, whilst increasing employee skill sets in sales and wine education to meet and preferably exceed customer service expectations. Creating and staging wine experiences and festivals also need to occur for the wine regions to have a competitive advantage and provide desirable experiences that act as a tourist draw card for family-oriented short stays and repeat visitation.

References

Barber, N., Dodd, T., and Ghiselli, R. (2008). Capturing the younger wine consumer. *Journal of Wine Research*, 19(2): 123–141. https://doi.org/10.1080/09571260802622225

Beames, G. (2003). The rock, the reef and the grape: the challenges of developing wine tourism in regional Australia. *Journal of Vacation Marketing*, 9(3): 205–212. https://doi.org/10.1177/135676670300900301

Black, Z. (2020). James Halliday tells Australians: drink up and drink local to save our industry. *The New Daily* (Australia). https://thenewdaily.com.au/life/eat-drink/2020/11/22/james-halliday-wines-australia/

Bodycomb, T., and O'Neill, T. (2014). Festivals and feasts. *Green: Sustainable Architecture and Landscape Design*, 39: 86–91. https://search.informit.org/doi/10.3316/informit.724254775975180

Brochado, A., Stoleriu, O., and Lupu, C. (2021). Wine tourism: a multisensory experience. *Current Issues in Tourism*, 24(5): 597–615. https://doi.org/10.1080/13683500.2019.1649373

Candi, M., Beltagui, A., and Riedel, J. C. (2013). Innovation through experience staging: motives and outcomes. *Journal of Product Innovation Management*, 30(2): 279–297. https://doi.org/10.1111/j.1540-5885.2012.00999.x

Canziani, B. F. (2019). Wine tourism: balancing core product and service-dominant strategies. In Sigala, M., and Robertson, R. N. S. (eds.), *Wine Tourism Destination Management and Marketing* (pp. 185–201). Cham: Palgrave Macmillan.

Carlsen, J., and Boksberger, P. (2015). Enhancing consumer value in wine tourism. *Journal of Hospitality and Tourism Research*, 39(1): 132–144. https://doi.org/10.1177/1096348012471379

Charters, S., and Mitchell, R. (2014). Food and wine events in Europe and the New World. In Cavicchi, A., and Santini, C. (eds.), *Food and Wine Events in Europe: A Stakeholder Approach* (pp. 41–53). London: Routledge.

Chen, C. F., and Chen, F. S. (2010). Experience quality, perceived value, satisfaction and behavioral intentions for heritage tourists. *Tourism Management*, 31(1): 29–35. https://doi.org/10.1016/j.tourman.2009.02.008

Cohen, E., and Ben-Nun, L. (2009). The important dimensions of wine tourism experience from potential visitors' perception. *Tourism and Hospitality Research*, 9(1): 20–31. https://doi.org/10.1057/thr.2008.42

De Freitas Coelho, M., de Sevilha Gosling, M., and de Almaida, A. S. A. (2018). Tourism experiences: core processes of memorable trips. *Journal of Hospitality and Tourism Management*, 37: 11–22. https://doi.org/10.1016/j.jhtm.2018.08.004

Dixit S. K. (ed.). (2017). *The Routledge Handbook of Consumer Behaviour in Hospitality and Tourism*. London: Routledge.

Dixit, S. K. (ed.). (2020). *The Routledge Handbook of Tourism Experience Management and Marketing*. London: Routledge.

Fiore, M., Alaimo, L. S., and Chkhartishvil, N. (2019). The amazing bond among wine consumption, health and hedonistic well-being. *British Food Journal*, 22(8): 2707–3323. https://doi.org/10.1108/BFJ-05-2019-0344

Gardiner, S., Grace, D., and King, C. (2013). Challenging the use of generational segmentation through understanding self-identity. *Marketing Intelligence and Planning*, 26(1): 39–55. https://doi.org/10.1108/MIP-06-2012-0062

Getz, D. (2019). Wine and food events: experiences and impacts. In Sigala, M., and Robertson, R. N. S. (eds.), *Wine Tourism Destination Management and Marketing* (pp. 143–164). Cham: Palgrave Macmillan.

Getz, D., and Page, S. J. (2016). *Event Studies: Theory, Research, and Policy for Planned Events*. London: Routledge.

Halliday, J. (2020). *Halliday Wine Companion 2021: The Bestselling and Definitive Guide to Australian Wine*. E-book: Hardie Grant Publishing.

Higgins, L., and McGarry Wolf, M. M. (2016). Millennials as luxury wine buyers in the United States? *International Journal of Wine Business Research*, 28(3): 190–205. https://doi.org/10.1108/IJWBR-08-2015-0029

Laing, J., and Frost, W. (2013). Food, wine … heritage, identity? Two case studies of Italian diaspora festivals in regional Victoria. *Tourism Analysis*, 18(3): 323–334. https://doi.org/10.3727/108354213X13673398610817

Marschall, S. (2012). Tourism and memory. *Annals of Tourism Research*, 39(4): 2216–2219. https://doi.org/10.1016/j.annals.2012.07.001

Marzo-Navarro, M., and Pedraja-Iglesias, M. (2009). Wine tourism development from the perspective of the potential tourist in Spain. *International Journal of Contemporary Hospitality Management*, 21(7): 816–835. https://doi.org/10.1108/09596110910985304

Mueller, H. R., and Scheurer, R. (2004). *Tourismus-Destination als Erlebniswelt: Ein Leitfaden zur Angebotsinszenierung* [*Tourism Destination as an Experience World: A Guideline to Stage Experiences*]. Bern: Forschungsinstitut für Freizeit und Tourismus (FIF) der Universität Bern [Research Institute for Leisure and Tourism, University of Berne].

Peters, M., and Weiermair, K. (2000). Tourist attractions and attracted tourists: how to satisfy today's fickle tourist client. *The Journal of Tourism Studies*, 11(1): 22–29.

Pikkemaat, B., Peters, M., Boksberger, P., and Secco, M. (2009). The staging of experiences in wine tourism. *Journal of Hospitality Marketing and Management*, 18(2–3): 237–253. https://doi.org/10.1108/WHATT-09-2019-0053

Pine, B. D., and Gilmore, J. J. (1999). *The Experience Economy*. Boston: Harvard Business Press.

Pivac, T. (2012). *Vinski Turizam Vojvodine, Monografija*, Novi Sad: Departmanza geografiju, turizam I hotelijerstvo.

Prahalad, C. K., and Ramaswamy, V. (2004). Co-creation experiences: the next practice in value creation. *Journal of Interactive Marketing*, 18(3): 5–14. https://doi.org/10.1002/dir.20015

Pratt, M. (2019). The image of a wine tourist and impact on self-image congruity. In Sigala, M., and Robinson, R. N. S. (eds.), *Wine Tourism Destination Management and Marketing* (pp. 45–65). Cham: Palgrave Macmillan.

Roberts, L., and Sparks, B. (2006). Enhancing the wine tourism experience: the customers' viewpoint. In Carlsen, J., and Charters, S. (eds.), *Global Wine Tourism: Research Management and Marketing* (pp. 47–55). London: CABI.

Santos, V. R., Ramos, P., Almeida, N., and Santos-Pavón, E. (2019). Wine and wine tourism experience: a theoretical and conceptual review. *Worldwide Hospitality and Tourism Themes*, 11(6): 718–730. https://doi.org/10.1108/WHATT-09-2019-0053

Seyfi, S., Hall, C. M., and Rasoolimanesh, S. M. (2020). Exploring memorable cultural tourism experiences. *Journal of Heritage Tourism*, 15(3): 341–357. https://doi.org/10.1080/1743873X.2019.1639717

Sigala, M. (2019a). Managing and marketing wine destinations with and through art: a framework for designing wine experiences. In Sigala, M., and Robertson, R. N. S. (eds.), *Wine Tourism Destination Management and Marketing* (pp. 221–243). Cham: Palgrave Macmillan.

Sigala, M. (2019b). Scarecrows: an art exhibition at Domaine Sigalas inspiring transformational wine tourism experiences. In Sigala, M., and Robertson, R. N. S. (eds.), *Wine Tourism Destination Management and Marketing* (pp. 313–343). Cham: Palgrave Macmillan.

Sigala, M., and Robinson, R. N. S. (2018). Introduction: the evolution of wine business tourism management. In Sigala, M. and Robertson, R. N. S. (eds.), *Wine Tourism Destination Management and Marketing* (pp. 1–14). Cham: Palgrave Macmillan.

Strickland, P., Frost, W., Williams, K. M., and Laing, J. (2013). The acceptance or rejection of social media: a case study of Rochford Winery Estate in Victoria, Australia. *Tourism, Culture and Communication*, 13: 19–27. https://doi.org/10.3727/109830413X13769180530602

Strickland, P., Laing, J., Frost, W., and Williams, K. M. (2020). Trends in experience design: Strategies for attracting Millennials to wineries in Victoria, Australia, In Dixit, S. K. (ed.), *The Routledge Handbook of Tourism Experience Management and Marketing* (pp. 214–225). London: Routledge.

Thach, L., Lease, T., and Barton, M. (2016). Exploring and special wine club offers the impact of social media practices on wine sales in US wineries. *Journal of Direct, Data and Digital Marketing Practice*, 17(4): 272–283. https://doi.org/10.1057/dddmp.2016.5

Wang, T. C. Tang, T. W., and Cheng, J. S. (2018). Art-oriented model of hotel service innovation. *International Journal of Contemporary Hospitality Management*, 30(1): 160–177. https://doi.org/10.1108/IJCHM-02-2016-0059

Williams, K. M. (2013). A profile of cellar door personnel: an exploratory case study analysis of two Australian wine regions. *Journal of Human Resources in Hospitality and Tourism*, 12(1), 91–108. https://doi.org/10.1080/15332845.2013.723268

Wine of the King Valley. (2020a). Australia's home of Italian wines. *Australia*. http://www.winesofthekingvalley.com.au/king-valley-wine-region/

Wine of the King Valley. (2020b). Italian style. *Australia*. http://www.winesofthekingvalley.com.au/italian-style-in-the-king-valley/

Wine of the King Valley. (2020c). La Dolce Vita Festival 21 + 22 November 2019, Victoria. *Australia*. http://www.winesofthekingvalley.com.au/la-dolce-vita-festival-2019-whats-on/

Winemakers' Federation of Australia. (1998). *National Wine Tourism Strategy*. Adelaide: WFA.

Wittwer, G. (2020). *The 2019–20 Australian Economic Crisis Induced by Bushfires and COVID-19 from the Perspective of Grape and Wine Sectors*. Centre of Policy Studies, Victoria University, CoPS Working Paper No. G-299, March. *Melbourne*. http://vuir.vu.edu.au/40403/1/g-299.pdf

Wollan, R. (2011). *The Social Media Management Handbook: Everything You Need to Know to Get Social Media Working in Your Business*. Hoboken, NJ: Wiley.

34

THE TASTE OF EXPERIENCE

The sensitivities of wine tourists

Wagner Otávio Gabardo, Vander Valduga, and Marcos Alberto Torres

Introduction

Wine as a food strongly characterized in cultural terms can be a mediator of new eating experiences and a way of drawing closer to other places' cultures during a trip. It constitutes an agri-food product with high added value (Elías 2008), in addition to being a basic ingredient of the Mediterranean diet. Drinking is a cultural achievement that synthesizes the harmonic relationship of farmers with the nature they inhabit and through which they think and build landscapes. The wine landscapes act as a territory's identity element, bring forth geographical indications for food, and constitute an asset to be preserved, in addition to being a primary resource for wine tourism.

Winescapes are the essence of the *terroir*, an expression of French origin that seeks to define the junction between nature and culture that makes each wine location unique and unrepeatable. For Jacquet (2010), the terroir is an association of environmental factors and cultural, sociability, and political practices of those who live there, and it impacts how grape and wine production is disseminated in a given region. For the French, the wine terroir has a taste – translated into the term '*goût du terroir*' (Parker 2017) – that results from environmental aspects of the wine-growing landscape. Among winegrowers, for example, there is a strong belief that the taste of certain wines stems from the type of soil from which the vine feeds, which would reinforce the idea that the experience of a wine landscape is also understood through taste.

For Luginbühl (2005), there is not only one, but several possible wine landscapes that demonstrate human inventiveness in adapting the cultivation of a plant to different soil and climate conditions. Such landscapes bring together traditional know-how, beliefs, and knowledge of nature. The author recognizes that the wine landscape goes beyond the visual domain and emphasizes the relevance of taste in order to understand it:

> [T]he name refers to a flavor first, even before inspiring a spectacle. This gustatory quality of the vineyard's name underlines the multi-sensorial dimension of the landscape that is not only founded on sight and appearance. Perhaps this is also the reason why painters have rarely painted vineyards, since the brush was unable to express the wine aromas, more significant and evocative than the vines' shapes, textures and colors.
> *(Luginbühl 2005, p. 17)*

DOI: 10.4324/9781003143628-40

Peters (1997) is a pioneer in suggesting the concept *of winescape* – wine landscape – and defining it as an agricultural landscape where the growing of vines – the raw material for wine production – dominates among other crops. The author highlights the role of wine in the construction of a possible representation of the wine landscape, when quoting James Newman's argument about the identity of a wine region being also marked by the color, smell, taste, and quality appraisal of a wine.

Luginbühl and Peters reinforce the argument that the landscape is not only seen, but felt. The landscape has aromas and flavors, and they are referents in the imagery that we create of a wine place. A wine's taste can make us travel to its origins, acting as an identity and memorable element of the experience of a wine region.

Wine, in fact, is a reality where the visual should not play a hegemonic role. Although studies reinforce the role of wine labels in the perception of their qualities (Henley et al. 2011), the visual is not enough to satisfy a wine taster as his interest in the beverage increases, and the search for stimulating nuances of aromas and flavors gains relevance. Likewise, a wine tourism experience perhaps would not be better described in images but in memories of flavors and aromas of a wine tasted on a rich table. Wine tourism would, therefore, be a way of living the experience based on taste. Aesthetic pleasure thus becomes mediated by the senses of proximity which penetrate the individual: the senses of smell and taste. And taste, as a combination of these two senses, is considered primitive and not very rational by Westerners (Le Breton 2016). Senses, with strong evocative power of emotions and memories, arouse feelings in those who taste wine.

The experience that derives from taste, according to Perullo (2013), unfolds in three ways: that of *pleasure*, *knowledge*, and *indifference*. We will focus on the first two dimensions as they are relevant to the experience of most wine tourism. The pleasure path would be a kind of 'naked perception,' 'a fundamental way of feeling and perceiving the experience' (p. 47), the satisfaction of a desire, the voluptuousness. The *knowledge* path would be that of 'knowledge and culture, a conscious construction of one's own identity, biographical and cultural … as a progressive development' (p. 81). Through frequent wine tasting, horizons are broadened and a kind of taste wisdom is acquired, a 'flexible and elastic attitude' (p. 25) resulting from the expansion of the perceptive abilities of a wine taster that happen in the concrete context of experience.

Wine tourism as pleasure would be part of the hedonic instinct of the traveler, fueled by the appeal of wine and its landscapes, the satisfaction of tasting a wine, accompanied by a tasty meal, sharing the table with good company and enjoying a beautiful view of the landscape (Tonini and Lavandoski 2011). Wine tourism as knowledge, on the other hand, would be the practice of a lifestyle based on wine culture, seeking to know the history, the processes of a winery, and to improve the practice of wine tasting (Cambourne and Macionis 2004). Pleasure and knowledge tend to coexist in the experience of many wine tourists, intertwining with varying intensities and ranges, except for the indifferent. To visit a winery, paraphrasing Tuan (2013), would be 'to experience in the active sense … to venture into the unknown and experience the illusory and the uncertain' (p. 18). Not only the aromas and flavors of a wine are tasted, but also a series of activities that make up the experience of wine tourism: tasting grapes straight from the vine when they are ripening, sharing the table with the author of the wine tasted, and tasting a wine straight from the barrel are experiences that far exceed contemplation. The flavors and odors of these experiences make them unique and make up the perception of a visited place.

The reflections herein presented shed light on the relationship between enogastronomy and the signification of touristic places, based on human senses within wine tourism experiences. It discusses the results of a case study in which Brazilian tourists describe their experiences in different wine-producing places, where the memories evoked by the flavors of the wines blend with the recollections of the places where they were produced and tasted.

Enogastronomy and the senses

Wine tasting can be the consolidating experience for the wine tourist, as already expressed by researchers on the subject (Alant and Bruwer 2004; Charters and Ali-Knight 2002; Getz and Brown 2006; Gunasekar et al. 2021; Mitchell and Hall 2006; Tzimitra-Kalogianni et al. 1999). It is unlikely for people to get involved in wine tourism without tasting the wine or at least smelling its aromas, if for whatever reason, they do not want to drink it. When tasting a glass of wine, the most diverse senses are stimulated, especially taste, smell, and touch. For Tuan (2013), there is a tendency to neglect the cognitive power of these senses, which, however, can reach an extraordinary refinement. One can smell the wine when swirling the glass. Upon swallowing it, the mouth is flooded by salivation, liquid density, freshness, warmth, astringency, and aftertaste. One perceives its quality, which, according to Peynaud and Blouin (2010), is the meeting of the wine characteristics – reflections of the grapes' *terroir* and the producer's *savoir faire* – combined with the taster's experience, preference, and sensitivity: 'quality exists only through him, through his judgment, his taste, his pleasure' (Peynaud and Blouin 2010, p. 193). Tasting a wine requires perception and cognition from the taster; the attribution of a rational and emotional meaning to the liquid that is ingested.

For Peynaud and Blouin (2010), the tasting experience is the conversion of 'a stimulus – luminous photon or molecule – into a sensation (a color, an odor)' through a 'complicated physical, chemical, electrical and psychic path' (p. 13) that will always be subject to subjectivity and its hedonic field. The authors argue that the conscious exercise of tasting represents a cognitive challenge for the taster, since at that moment he or she must attribute a meaning to the perceived aroma, flavor, or texture. Such meanings would be constructed and expressed from each individual's cultural repertoire and sensitive memory. After the tasting, the taster's aptitude would be knowing how to express the objective qualities of the beverage. Among them are the aromas that volatilize from the glass, the level of unctuosity, astringency, acidity, and other liquid flavors that flood the palate. In the description and evaluation stage, the authors recognize the difficulty of detaching the hedonic dimension from the objective one, and assume that one can influence the other. Through the articulation of the hedonic, subjective, and cultural dimension of wine tasting with the influence of the environment where it takes place, it is understandable how tasting wine in a beautiful glass, in the surroundings where it was produced, and guided by a sommelier or oenologist can positively impact the impression created about the beverage. The synesthesia of wine tasting is impressive.

Gastronomy, in turn, is a *sine qua non* condition of wine tourism. For a start, for the European community, wine is considered food, an essential component of the Mediterranean diet (UNESCO 2013). That is, it is made to accompany regional gastronomy and, at the same time, to be an integral part of it. A gastronomy founded on popular, peasant cuisine, connected to the land and tradition, which seeks

> to explore the regional and seasonal products, in close agreement with nature, resting on ancestral knowledge transmitted through unconscious ways of imitation and habit, applying cooking procedures … utensils and vessels from the kitchen clearly fixed by tradition.
>
> *(Revel 1996, p. 38)*

Or in other cases, an 'erudite' and professional cuisine by chefs, which rests on the invention, renovation, and experimentation of a tasty, varied, and consolidated popular cuisine that serves as a foundation. In many wine territories, the beverage is an adjunct to the table and, together with the food, makes up the food experience. From a marriage of cultural practices at the table, where wine was usually drunk together with the typical local food, the concept of enogastronomy arises. For Silva (2015), enogastronomy is formed by a tripod that unites rural producers, wineries/restaurants, and the territory where they live, produce, and reproduce know-how:

> Through enogastronomy it is possible to identify customs and traditions that are not revealed in a tangible way, linked to the imagination of each person, in the awakening of the senses (sight, hearing, taste, touch and smell), in the expression of differences, similarities, beliefs and the feeling of belonging, as it is full of cultural marks and social expressions.
>
> *(p. 746)*

When visiting the Vale dos Vinhedos (Brazil), for example, one can enjoy homemade cappelletti soup, followed by a barbecue accompanied by table wines from the Serra Gaúcha. Or one can eat at a restaurant that serves dishes made with ingredients from small organic crops from the surrounding area, paired with fine wines presented by a sommelier. Whatever the meal choice, favoring the tradition of the typical settler lunch or the sophistication of the chef, in both cases we observe that wine tourism stimulates the valorization of regional gastronomy and enriches the travel experience, also mediated by the smells and flavors of the places experienced. Eating at a winery while drinking the wines it produces is one of wine tourists' favorite activities, as found in research carried out in wine regions in Spain (Gálvez, Fernández, and Guzmán 2015; Ricolfe, Barrio, and Vera 2010), Australia (Alant and Bruwer 2004; Charters and Ali-Knight 2002), and Vale dos Vinhedos and Vale do São Francisco, Brazil (Zanini and Rocha 2010), among others.

According to Le Breton (2016), food is a full sensory object as it stimulates all the senses. The gastronomic experience is consolidated on the palate. The gustatory sensation is, at the same time, knowledge and affectivity in action. Knowledge, rationally constructed, that guides individuals in their dietary practice and ethics. Affectivity as emotion spontaneously aroused by the taste of something eaten.

The sense of taste, just as the sense of smell, has a strong evocative power of what Proust (2003) defines as 'involuntary' memory, capable of bringing to awareness feelings of well-being and past memories which were asleep until then. For Brillat-Savarin (1995), the sense of taste is pleasure and synesthesia, existing at all times and conditions, and can mix with the other senses. In the context of enogastronomy, to eat is to enter the hedonic field where pleasure aroused by aromas, flavors, and textures from the mixture of wine and food is the beginning and the end of the experience. To eat is to internalize, physiologically and mentally, part of the place visited, of its culture expressed in its eating habits. It is to absorb sensations based on subjectivity, difficult to measure, volatile and invisible, but not ephemeral:

> But when nothing remains of an ancient past, after the death of beings, after the destruction of things, alone, more fragile but more vivacious, more immaterial, more persistent, more faithful, the aroma and the taste remain for a long time, like souls.
>
> *(Proust 2003, p. 50)*

Tasting the vineyard

The possibilities of tasting the winescape go beyond enogastronomy and include the vineyard, the beverage's source of raw material. Close to the harvest season, the wine grower usually tastes grape berries picked randomly from the vineyard in search of flavors and textures that characterize a grape ripe enough to be harvested, something well described in Potter's novel (2015), about the Romanée–Conti story:

> He tugged off a berry, placed it in his mouth. He bit down on it … to release the juice onto his palate, where he could savor it … the texture was good. The juice was good and sweet.
>
> *(p. 25)*

Le Breton (2016) points out the olfactory and taste skills that oenologists must cultivate in their professional careers, a knowledge as important as that acquired by hard science, by oenology and organic chemistry teachings. Trusting the senses has always been an indicator for a winegrower's decision making before the emergence of modern oenology in the mid-19th century (Phillips 2014). An example would be the sensory measuring of the grape berry's ripeness, compared to technological measuring. Traditionally, before the arrival of grape pulp sugar measuring devices such as the refractometer, winegrowers had the habit of tasting grape berries randomly picked from the vineyard, to taste the skin's astringency, the crunchy seed texture, and the pulp sweetness and acidity. Through these gustatory sensations, they were able – and many still are – to tell whether the grape was perfectly ripe for harvest. In the same way, wine tourists who visit the vineyard at harvest time have the opportunity to taste grapes. According to previous research by Gabardo (2019), this is one of the experiences frequently mentioned by travelers. Tasting the grape that will be used to make the wine of one's liking adds airs of authenticity to the wine tourism experience, making it memorable, as we will describe in the case study.

Case Study 34.1 Sensitivities of wine tourists

The following accounts are part of a broader research on the sensitive experience in wine tourism (Gabardo 2019). Experiences based on taste that made the trip memorable and of high added value were selected. They consist of accounts from Brazilian wine tourists with a wide repertoire of trips to wine territories, participants of a qualitative research of narrative interviews based on the generating question: 'What is your unforgettable wine tourism experience? Why?,' followed by the respondents' free expression and submitted to subsequent content analysis. The following inferences illustrate how memorable experiences around wine are steeped in flavor.

In almost all of the accounts, memories of the tasted wine set the tone: 'we discovered that inside there was a shop for small producers only … and we tasted a wonderful wine there, a chardonnay that even my husband, who does not like chardonnay, loved' (V.P., civil servant). 'There I tasted the best viognier I have ever tasted and to this day' (J.P., art curator). For both respondents, the taste of the Chardonnay and Viognier wines is remembered as an unforgettable aspect of the wine tourism experience. And they reinforce Luginbühl

(2005) and Peters (1997) assumptions in which the wine taste and quality are identity elements of a place and its landscape. To remember a wine place is to remember wines tasted there, to bring to the present the taste of a lived experience.

The wine tourism experience is synesthetic, and visual aesthetic satisfaction can impact the aesthetic perception of taste. Wine and food gain a remarkable flavor when consumed at their place of origin. The testimony of G.B. (sommelier) is a good illustration: 'the best scrambled egg of your life is the one you eat at breakfast, 12 km from the Cordilheira [Andes Mountains]. It is something fantastic.'

The taste of the experience can also be authentic, sublime, and transformative. Such is the case of wine tourists who find their occupation and passion in wine, like two sommeliers interviewed (G.B. and C.M.) when describing different trips they made to Mendoza, Argentina. For an aficionado, to participate in the process of winemaking or to taste a wine straight from the barrels means immersing oneself in the elaboration of the beverage, what requires a sensitive, perceptual, and cognitive effort from the subjects involved. The accounts are a good example:

> what delighted me the most was the oenologist's welcoming attitude, the more we talked about wines, the more he would get excited, we tasted the wines straight from the barrel, what delighted me the most were the blends [wines from a mixture of grapes] that he is yet to launch, he gave me from three barrels to taste, explaining what each wine will give to the blend ...it was a *surreal experience.*
>
> *(C.M. 2018, emphasis added)*

> I created a blend [wine from a mixture of grapes] that was then sold in our store, only 660 bottles, *something that is very special for me ... my hand, I mean, my nose and my taste buds are in this wine* [laughs].
>
> *(G.B. 2018, emphasis added)*

C.M., a sommelier, reveals the climax and transforming moment of the experience in his account. Participating in the process of creating a wine, tasting, and understanding through taste the role of each grape in the final beverage is considered 'a surreal experience.' It could be assumed that, for the interviewee, such an experience would become a kind of ritual process, as described by Turner (1974), where he moves from the role of mere visitor and spends a few hours in the role of 'oenologist,' tasting and discussing the possibilities of a wine composition, maturing as a taster, and actually expanding his understanding of the beverage's processes. The emotion in C.M.'s account is evident in the next stage of the interview, when he describes the feelings that a barrel room evokes in him, the venue in which this account takes place. Similarly for G.B., having participated in the composition of a wine using one's own 'nose and taste buds' is something unforgettable and 'very special.' This would be for him the experience's liminal moment, since he was involved in the ritual of making a wine by defining its qualities, instead of just tasting the final beverage, as he usually does in the exercise of his profession.

As an outcome of the same research, the interview with wine tourists continued with inquiries about sensitive memories, that is to say, if they were able to recall specific sensations (tastes, aromas, textures, temperatures) surrounding the wine experience. The mapping of these sensitivities was based on the *Sensory Experience Elicitation Protocol* proposed by Gretzel and Fesenmaier (2010). This protocol was developed by the authors based on two other qualitative techniques that seek to break communication barriers by allowing access to sensitive data, less spontaneously verbalized and, therefore, more difficult to abstract. One of them is Zaltman (2003) 'evocation by image' in which, from a visual metaphor, respondents are asked to describe and explain a certain situation related to the research scope. *Laddering* is the other evocation technique, and it departs from the narration of a hypothetical situation that the interviewees could face. It consists of a personal and in-depth interview, gradually analyzed in detail, a kind of survey that begins with more concrete situations in order to move on to more abstract situations. It allows unconscious emotions and manifestations to be captured.

The dynamics of the research started from a wine tasting, along the lines of Peynaud and Blouin (2010), in order to stimulate the senses of smell, taste, and touch. Then, while the wines were being tasted, images of different wine tourism experiences were projected – vineyards, winemaking room, barrel room, sparkling wines cellar, tasting room, enogastronomic meal, and harvest feast. At each experience suggested by the images, the participants were prompted to remember whether or not they went through a situation similar to that portrayed, before moving on to the sensory evocation itself. If so, they were asked about the recalled sensitivities: feelings and sensations, smells, flavors, textures, and temperatures, in order to compose the synesthetic experience of taste. The accounts reveal a wide menu of rescued memories.

The enogastronomy experiences depicted reinforce Le Breton (2016) argument that 'food is the total sensory object.' The diversity of descriptors, led by flavor, draw more clearly the possibilities of wine and food pairing, together with their aromas and textures. The flavors and aromas of roasted, smoked, and juicy meat often inhabit the memories of diners. But seafood, cheese, bread, and pasta also do. Particular wine flavors are evoked, such as of Chardonnay and Cabernet Franc grapes, orange or full-bodied wines. The intrinsic qualities of the flavor are mentioned: salty, sweet, acid, bitter. The dishes' aromas, both from ingredients and cooking processes, also come up: spices, herbs, sauce, olive oil, lemon, smoke. The textures and temperatures described refer to the external surroundings and to the internalized, ingested food: a pleasant weather warmth with pleasant human warmth. The texture of the food is diverse: soft, crunchy, fluffy, mousse, delicate, light, hot and fresh. The memory of the aroma also carries affection in the speech: 'mom's food,' 'good food,' and 'warm bread.' It was the experience with the greatest number of descriptors, even superior to that of wine tasting only.

Participating in a wine tasting is an exercise in cognition and sensory memory that includes a wide repertoire of descriptive vocabularies for a wine, as expressed by Peynaud and Blouin (2013). There are so many references that the use of generic words such as 'diverse/innumerable,' 'pleasant/good' flavors and aromas, and 'red/white wine' was frequent. Palate-specific descriptors are used again: acid, bitter, sweet, together with references to fruit in various stages – red, ripe, green, citrus – which might be recognized in the beverage. Textures and temperatures are mentioned: cold, cool, icy, astringent, dry. Flavors of food that usually accompany tasting are also remembered, such as bread and cheese.

To go through a winemaking room is to be inebriated by the aromas of grapes, wine, yeast, and a 'sweet smell' of the fermenting must. Smells that are confused in the respondents' imagination, since fermentation is the transformation of grape must into wine by the action of yeast. Wine flavors and its typical descriptions – fresh fruit, fruit in compote, tobacco, coffee, chocolate – begin to become apparent along with the accompanying foods – cheese, meat, breads, and sausages. Precise and specific flavor descriptors – acid, bitter – and some reference to flavors in the fermentation process are mentioned.

Celebrating a harvest feast proved to be a broad sensory experience, full of senses and Dionysian, as stated by Lacoste (2013). Evoked feelings of 'authenticity' and 'ancestry' are seasoned with aromas and flavors of grapes harvested and tasted at different stages – ripe, fresh, crushed, and fermented. Scents of bush, herbs, and flowers mix with those of the harvested fruit. The flavor is 'delicious,' a mixture of acid and sweet. The feast also tastes like wine and polenta, a typical Italian dish associated with immigrants who founded viticulture in Brazil.

Visiting the underground sparkling wine cellar where bottles rest is to register particular sensations in the memory – such as humidity, cold, and dust – and especially the textures of the effervescent beverage that matures there: bubbles, freshness, light, and delicate are descriptors mentioned by those surveyed. The flavor of the yeast used in the wine refermentation is often remembered, along with typical flavor descriptors of a sparkling wine: acid/fresh, citric, light, soft, refreshing, bread and hazelnut. Figure 34.1 illustrates the sensitive memories of experiences in sparkling wine cellars.

Figure 34.1 Sparkling wine cellar sensitivities.

Source: Adapted from Gabardo (2019)

Figure 34.2 Barrel room sensitivities.

Source: Adapted from Gabardo (2019)

To walk into a barrel room is to be surrounded by the emanations of the wood from the recipients that contain the wine. The aromas of oak and wood, wet or soaked in wine, are commonly remembered by the participants. The smell is the most mentioned sensation: aromas of mold, sulfur – used to sterilize barrels – and vinegar, which should indicate the early oxidation of a wine, are remembered. References to specific wine flavors emerge, with mentions of the grape or appellation of origin as: Cabernet, Sangiovese and Porto. The wine palate becomes woody and full-bodied, as well as its descriptive vocabulary: acid, bitter, balanced, and mineral. Once again, meats, cheeses, breads, and sausage are foods mentioned in flavors, drawing a possible enogastronomy associated with woody wines, disseminated in the imaginary. Figure 34.2 illustrates sensitive memories of experiences in barrel rooms:

From this description, many wine tourism experience possibilities become evident: from freshly harvested grapes in a vineyard to wine maturing in barrels or a sparkling wine resting in the cellar. Wine tasting brings with it several memorable taste records for the individual: sometimes an attentive taste in the pursuit of learning, sometimes drinking in a relaxed way, around a large table. Table 34.1 summarizes the taste of the landscape at different times:

Table 34.1 Sensitive memories of wine tourists

Landscape/Experience	*Aromas and Flavors*
Vineyard	The smell of flowering, the flavor of ripening grapes, their sweetness and acidity.
Winemaking room	The smell of fresh grapes being squeezed. The flavor of the freshly extracted must, its sweetness and acidity. The aromas and flavors of the fermenting must/freshly fermented wine.
Barrel room	The aroma of wood; the intense flavor and texture of a wine tasted straight from the barrel. Tastes of grapes and wines from specific regions.
Sparkling wines cellar	The aroma of moisture, mold, earth and wet stone; the flavor of a sparkling wine tasted with its yeasts; bubbles, freshness and acidity.
Wine tasting	The aromas, flavors, and textures of the wine tasted (sweetness, acidity, body, tannins, persistence, etc.). Tastes of grapes and wines from delimited regions.
Harvest feast	The tasting of the grapes during the harvest, the aromas released when treading the grapes with bare feet, simulating a traditional practice.
Enogastronomy	The aromas and flavors of the dishes, wines, and their combinations. The salty, sweet, spicy, acid, and bitter. The hot and the cold; soft and crunchy. The aromas emanating from the kitchen where they are prepared. Tastes of grapes and wines from delimited regions.

Source: Adapted from Gabardo (2019)

Final considerations

Once the visit to the wine region comes to an end, it is time to go back home, to return to the comfort and familiarity of everyday life. But to return transformed, full of memories and affections awakened by wine tourism. Such experiences feed our existence and strengthen our topophilia with the winescape. Wine is taken back home as a *souvenir*, to be opened on a special occasion. Once the occasion arises, the wine is tasted, and the tasters are flooded by its aromas, flavors, and textures. At this very moment when our senses are activated, we remember the original experience.

The emotional dimension can perhaps make one feel the wine not in the exact same way as when it was tasted in its productive surroundings. The memories are never repeated *ipsis litteris*, but they certainly bring pleasure and make one wonder about the experience. And the perception of the beverage will never be exactly the same, even between two people who went on the same trip. Neither will the memories of the place, since they inhabit the subjective terrain.

The important thing is to understand that wine has the ability to take us far, where memories and affections are triggered by the exercise of taste. The beverage of god Bacchus or Dionysus – whatever the mythology – has the power to make people travel inwardly. When tasting a wine, one is penetrated by its aromas, flavors, and textures. The same occurs with the food that occasionally accompanies it. Smell, taste, and touch are exercised in a more intense way than one is aware of.

The wine tourism experience is sensitive, is intangible, and bears sensitivities perceived in an individual and subjective way by the one who experiences it. The discovery of these places is a way to experience and understand the world based on taste, exploring other narratives that can describe an experience in its complexity. We must not neglect the cognitive and affective power of the other senses. To go on a wine tourism experience is to combine pleasure and knowledge, affections and sensitivities, capable of transforming the subjects who experience them.

Implications

Research that correlates the senses to wine tourism experiences can contribute to the understanding of the complex relationship between the two, and to the signification of places and products that derive from such a relationship. Wines, bearers of specific elements of their terroir and, as such, products that merge with the identity of the place, carry in their flavor, texture, color, and odor the marks from where they were produced. They enrich the experience of those who travel and interact with such places. Stimulating the tourists' perceptions during guided tours enhances the experience, making them unforgettable and of high added value.

References

Alant, K., and Bruwer, J. (2004). Wine tourism behaviour in the context of a motivational framework for wine regions and cellar doors. *Journal of Wine Research*, 15(1): 27–37. https://doi.org/10.1080/0957126042000300308

Brillat-Savarin, J. A. (1995). *Fisiologia do Gosto*. São Paulo: Companhias das Letras.

Cambourne, B., and Macionis, N. (2004). Meeting the wine-maker: wine tourism product development in an emerging wine region. In Hall, C. M., Sharples, L., Cambourne, B., and Macionis, N. (eds.), *Wine Tourism around the World: Development, Management and Markets* (pp. 81–101). Boston: Elsevier.

Charters, S., and Ali-Knight, J. (2002). Who is the wine tourist? *Tourism Management*, 23(3): 311–319. https://doi.org/10.1016/S0261-5177(01)00079-6

Elías, L.V. (2008). Paisaje del viñedo: Património y recurso. *PASOS. Revista de Turismo y Patrimonio Cultural*, 6(2): 137–158. https://doi.org/10.25145/j.pasos.2008.06.012

Falcade, I. (2011). *A paisagem como representação espacial: a paisagem vitícola como símbolo das indicações de procedência de vinhos das regiões Vale dos Vinhedos, Pinto Bandeira e Monte Belo (Brasil)* [PhD Thesis]. Instituto de Geociências, UFRGS, Porto Alegre.

Gabardo, W. O. (2019). *A paisagem sensível do enoturismo: uma abordagem fenomenológica* [Master's dissertation]. PPGTUR, UFPR, Curitiba.

Gálvez, J. C. P., Fernández, G. A. M., and Guzmán, T. L. G. (2015). Motivación y satisfacción turística en los festivales del vino: XXXI ed. cata del vino Montilla-Moriles, España. *Tourism & Management Studies*, 11(2): 7–13. https://doi.org/10.18089/tms.2015.11201

Getz, D., and Brown, G. (2006). Critical success factors for wine tourism regions: a demand analysis. *Tourism Management*, 27(1): 146–158. https://doi.org/10.1016/j.tourman.2004.08.002

Gretzel, U., and Fesenmaier, D. (2010). Capturing sensory experiences through semi-structured elicitation questions. In Morgan, M., Lugosi, P., and Brent Ritchie, J. R. (eds.), The Tourism and Leisure Experience (pp. 137–162). Bristol, UK: Channel View Publications.

Gunasekar, S., Das, P., Dixit, S. K., Mandal, S., and Mehta, S. R. (2021). Wine-experienscape and Tourist satisfaction: through the lens of online reviews. *Journal of Foodservice Business Research*. https://doi.org/10.1080/15378020.2021.2006039

Henley, C. D., Fowler, D. C., Yuan, J. J., Stout, B. L., and Goh, B. K. (2011). Label design: impact on millennials' perceptions of wine. *International Journal of Wine Business Research*, 23(1): 7–20. https://doi.org/10.1108/17511061111121371

Jacquet, O. (2010). Les appellations d'origine et le debát sur la tipicité dans la première moitié du XXe siècle: le rôle du syndicalisme vitivinicole bourguignon et la création des AOC. In *Faire vivre le terroir: AOC, terroirs et territoires du vin* (pp. 117–128). Pessac: Presses Universitaires de Bordeaux.

Lacoste, P. (2013). *Argentina: Vinos de Capa y Espada*. Mendoza: Antacura.

Le Breton, D. (2016). *Antropologia dos sentidos*. Petrópolis: Vozes Press.

Luginbühl, Y. (2005). Paysages viticoles. In: *Les paysages culturels viticoles* (pp. 25–30). Paris: Convention du patrimoine mondiale UNESCO, Icomos.

Mitchell, R., and Hall, C. M. (2006). Wine tourism research: the state of play. *Tourism Review International*, 9(4): 307–332. https://doi.org/10.3727/154427206776330535

Parker, T. (2017). *Le goût du terroir: histoire d'une idée française*. Rennes: Presses universitaires de Rennes.

Peynaud, E., and Blouin, J. (2010). *O gosto do vinho: o grande livro da degustação*. São Paulo: WMF Martins Fontes.

Perullo, N. (2013). *O gosto como experiência: ensaio sobre filosofia e estética do alimento*. São Paulo: Sesi Press-SP.

Peters, G. L. (1997). *American Winescapes: The Cultural Landscapes of America's Wine Country*. Boulder, CO: Westview Press.

Phillips, R. (2014). *Uma Breve História do Vinho*. Rio de Janeiro: Record.

Potter, M. (2015). *A história do Romanée-Conti: e a trama para destruir o melhor vinho do mundo*. Rio de Janeiro: Zahar.

Proust, M. (2003). *No caminho de Swan*. São Paulo: Folha de S. Paulo.

Revel, J. F. (1996). *Um banquete de palavras: uma história da sensibilidade gastronômica*. São Paulo: Companhia das Letras.

Ricolfe, J. S. C., Barrio, J. E. R., and Vera, J. M. B. (2010). Análisis del mercado potencial enoturístico en Valencia. *Papers de Turisme*, 47–48: 93–108.

Silva, R. d. N. (2015). *A enogastronomia no processo de formação de identidade territorial do Vale dos Vinhedos/RS (Tese de Doutorado)*. Brasil: Universidade de Santa Cruz do Sul, Santa Cruz, RS.

Tonini, H., and Lavandoski, J. (2011). Enoturismo: experiências e sensações no Vale dos Vinhedos (RS). *Revista Turismo em Análise*, 22(1): 25–43. https://doi.org/10.11606/issn.1984-4867.v22i1p25-43

Tuan, Y. F. (2013). *Espaço e Lugar: A perspectiva da Experiência*. Londrina: Eduel.

Turner, V. (1974). *O processo ritual*. Petrópolis: Vozes.

Tzimitra-Kalogianni, I., Papadaki-Klavdianou, A., Alexaki, A., and Tsakiridou, E. (1999). Wine routes in Northern Greece: consumer perceptions. *British Food Journal*, 101(11): 884–892. https://doi.org/10.1108/00070709910301391

UNESCO, United Nations Educational, Scientific and Cultural Organization. (2013). *Mediterranean Diet. Nomination file no. 00884 for inscription in 2013 on the Representative List of the Intangible Cultural Heritage of Humanity*. https://ich.unesco.org/en/RL/mediterranean-diet-00884

Valduga, V. (2011). *Raízes do Turismo no Território do Vinho: Bento Gonçalves e Garibaldi – 1870 A 1960 (RS/Brasil)* [Tese (PhD in Geography)]. Instituto de Geociências, UFRGS, Porto Alegre.

Zaltman, G. (2003). *How Customers Think: Essential Insights into the Mind of the Market*. Boston: Harvard Business Press.

Zanini, T. V., and Rocha, J. M. (2010). O Enoturismo no Brasil: um estudo comparativo entre as regiões vinícolas do Vale dos Vinhedos (RS) e do Vale do São Francisco (BA/PE). *Revista Turismo em Análise*, 21(1): 68–88. https://doi.org/10.11606/issn.1984-4867.v21i1p68-88

35

ANTECEDENTS, ELEMENTS, AND OUTCOMES OF THE WINE TOURISM EXPERIENCE AT THE DESTINATION

Gökhan Yılmaz and Ayşe Şahin

Introduction

Wine tourism is defined differently by various researchers. For example, Getz (2000) defined wine tourism as the 'travel related to the appeal of wineries and wine country, a form of niche marketing and destination development, and an opportunity for direct sales and marketing on the part of the wine industry.' Likewise, a widely accepted definition of wine tourism is proposed by Hall et al. (2000), which embodies visitation of wineries, attending wine festivals, and tasting/experiencing wines. These definitions mainly focus on tourist behaviors. Saayman and van der Merwe (2015) also defined wine tourism as visiting vineyards and wineries and attending wine festivals and wine shows, where tasting wine, and 'experiencing the attributes' of the grape-growing region are the basic reasons for the visit. As for this definition, it is considered that wine tourism is about traveling mainly to experience the attributes of a wine region (such as winescape). Therefore, the importance of the wine tourist's experience is emphasized with this definition (Saayman and van der Merwe 2015; Stone, Garibaldi, and Pozzi 2019).

Wine tourism is a holistic experience that includes many ways of visiting a wine-producing region, to feel its lifestyle and culture. This holistic experience occurs in a number of ways, primarily by events and festivals, cultural heritage, dining, hospitality, education, tasting and cellar door sales, and winery tours. On the other hand, wine tourism experience involves many specialties, including: a lifestyle experience; education; linkages to art, wine, and food; tasting and cellar door sales; winery tours; incorporation within the destination image; and a marketing opportunity (Charters and Ali-Knight 2002; Gunasekar et al. 2021; Madeira, Correia, and Filipe 2019). Moreover, the experience of wine tourism is an association between wine tourist motivations and the tourism and wine industries, with the region and the winescape as scenery (Hall et al. 2000).

Wine tourism experience activities are fundamentally based on wine tastings and visits to wine museums, wine houses, and cellars (Santos et al. 2019). Studies related to wine tourism experience generally focus on wine tourists (Asero and Patti 2011), wine tourist behaviors (Loureiro and da Cunha 2017), wine involvement (Rahman and Reynolds 2015), emotions (Leri and Theodoridis 2019), the Experience Economy and wine experience (Thanh and Kirova 2018), wine tourism destinations (Bruwer, Roediger, and Herbst 2017), and memorability (Quadri-Felitti and Fiore 2013). For example, according to Quadri-Felitti and Fiore (2012),

 DOI: 10.4324/9781003143628-41

there are three main groups of research in wine tourism experience: motivations and satisfaction of wine tourists; demographic or psychographic profiles of wine tourists; and marketing to tourists through wine routes, festivals, and events. Mason and O'Mahony (2007) identified six themes as important for a memorable wine route experience for the culinary tourist: cuisine, lifestyle, regionalism, environment, rurality, and health. Adding to this investigation, Charters and Ali-Knight (2000) found that different wine tourists seek different experiences, and that research concerning wine tourists indicated that wineries attract a diverse mix of consumers, which implies different needs and experiences. Quadri-Felitti and Fiore (2013) adapted the 4Es of experience to wine tourism. Moreover, Pivac (2012) attests that 'total wine experience' includes ten features: wine tasting and buying; socializing with friends; enjoying the day outdoors; enjoying the rural environment and vineyards; becoming familiar with wine production; learning about wine; wine tasting in restaurants; a tour of the winery; visitor experience of other attractions and activities; and relaxation. Most of these studies focus on understanding visitors as wine buyers and wine tourist behaviors (Dixit 2017; Saayman and van der Merwe 2015). In the current situation, most academic studies consider wine tourism experience to be related to the visitor (e.g., group size and visit frequency), destination (e.g., destination settings and culture), and different attributes (Charters, Fountain, and Fish 2009; Quadri-Felitti and Fiore 2012).

Destinations can offer a wide range of wine-related activities to tourists, such as wine tasting and purchasing at wineries, attending wine seminars, participating in organized tours to wine-related areas, participation in the grape harvesting and festivals in vineyards, and learning how to make wine. Destinations can provide many restaurants where they can consume local wines and souvenir shops where they can buy and taste local wines. Furthermore, purchasing local wine can be considered as a way of embodying the tourism experience, which is intangible in nature, bringing the experience home, and ensuring the memorability of the experience at a later time. To this end, the tourists visiting the destination can engage in wine-related activities, consume wine during their travel, and buy wines as souvenirs. In light of these arguments, the aim of the study is to theoretically discuss the wine experience of the tourists at the destination in greater depth as a holistic model. More clearly, the following questions are sought regarding the experiences of tourists at the destination: (1) What are the antecedents of the wine experience (such as motivation, information search behavior) at the destination? (2) What are the elements of the wine experience (such as elements, characteristics) at the destination? (3) What are the outcomes of the wine experience (such as satisfaction, revisiting, recommendation, memorability, well-being, emotions) at the destination?

Antecedents

Each experience, such as a tourism experience or a museum experience, has a quintessentially wide range of antecedents. Similarly, it is seen that the wine experience has also fundamentally consisted of motivations, lifestyles, involvement, information sources, and tourist typologies as antecedents. To this end, it is necessary to explain in greater detail the antecedents of wine tourism experiences.

Motivations

Motivation is described as a 'need-induced tension' (Schiffman et al. 2008, p. 94) that refers to a consumer to do something to relieve the tension (Bruwer and Alant 2009). Effective marketing is impossible without an understanding of consumers' motivations. Wine and winery products are designed and marketed to meet consumers' needs, as well. As for wine experience motivations, it is not merely a tool to understand the wine destination; it also explains the needs, wants, expectations, and behaviors of wine tourists (Grybovych, Lankford, and Lankford 2013, p. 287). Wine tourists' motivations are an essential factor in the overall understanding of wine regions and wineries/cellar doors of the needs and expectations of their customers (Alant and Bruwer 2004). Hall et al. (2000) also indicate that wine tourism is based on wine tourists' motivations. Although wine tourists' motivation is one of the main antecedents, few studies are focused on wine tourist motivations. For example, Alant and Bruwer (2004) stated that the primary motivations of wine tourists are considered to be tasting and buying wine. Moreover, previous studies (Charters and Ali-Knight 2002; Getz and Brown 2006) reveal that wine tourists also seek shopping, dining, and cultural and recreational outlets. On the other hand, wine and wine-related products may not be a primary motivation for each tourist. Tourists can enrich their experience by visiting wineries, purchasing wine, and attending wine-related festivals and events. In this sense, secondary motivations such as socializing, hedonic experience, uniqueness, having a good time with friends and family, supporting local wine producers, tasting locally produced foods, learning about wine, being entertained, traveling in a rural setting, scenery, escape, relaxation, health or diet, and just having a day out also gain importance in the wine tourism experience (Charters and Ali-Knight, 2000; Grybovych, Lankford, and Lankford 2013; Hall et al. 2000; Koksal 2020; López-Guzmán, Vieira-Rodríguez, and Rodríguez-García 2014).

Involvement

Involvement is 'a person's perceived relevance of the consumption object-based in inherent needs, values, and interests' (Zaichkowsky 1985, p. 342). Wine involvement is an important element in understanding tourists' wine experiences and consumption. Similarly, as wine involvement is a driver of tourists' purchase intention and experience, wine-related tourist behavior depends dramatically on wine involvement (Lockshin et al. 2006; Rahman and Reynolds 2015). Prior research has generally investigated tourists' interests in wine (Crespi-Vallbona and Mascarilla-Miró 2020), the impact of wine tourism involvement on winery owners' identity (Canovi, Mordue, and Lyon 2020), and the relationship between drinking frequency and wine involvement (Rahman and Reynolds 2015). Moreover, according to Canovi, Mordue, and Lyon (2020), wine involvement is an integral part of the identity process for destinations. Rahman and Reynolds (2015) also state that wine drinking frequency significantly and positively influenced wine involvement.

Wine-related lifestyle

Wine-related lifestyle refers to a general condition of consumers' interest in wine. Previous studies (Santos et al. 2019; Sharma, Singh, and Misra 2020) have usually focused on market segmentation and emphasized that wine-related lifestyle is a strong antecedent for tourists' wine experience and purchase intentions. For example, Sharma, Singh, and Misra (2020) indicate that wine-related lifestyle is a determinant for wine consumer behavior. In a similar vein, Corsi et al. (2020) state that personal values also are a vital factor for tourists' wine behavior, like wine-related lifestyles. Moreover, Spielmann, Babin, and Verghote (2016) emphasize the relationship between tourists' personality and wine experiences. Taking all of these into account, it is clear that wine-related lifestyle, wine personality, and values are indispensable parts of the wine tourism experience.

Information sources

Information searching plays an important role in consumers' or tourists' wine selection process (Yılmaz and Gültekin 2016). The decision to visit wine regions is generally referred to as the decision to visit a tourist destination, and it contains an information search before the visit. Information search for wine experience points out to information sources used in the planning of a visit to a winery or wine regions and destinations. Wine tourism information sources are investigated as potential factors in identifying tourists' wine experiences. To this end, wine tourists' information sources may include recommendations from family and friends, brochures, previous exposure to a wine regions label, various internet sites, visitor information centers, print articles, festivals, billboards, television programs, and so on (Byrd et al. 2017; Hyde 2006). Moreover, Bruwer and Thach (2013) indicate that there are more than 16 different information sources in wine tourism, but they found that only a few of these sources are majorly used by the wine tourists, such as recommendations from family and friends, brochures, and previous exposure to a regions winery label.

Tourist typology

The wine tourist is 'a person with needs to relate to both wine and the location (wine region) where wine is produced' (Bruwer and Alant 2009, p. 3). In addition to this, wine tourist is used for someone who has visited a winery, wine trail, or wine festival while traveling in the past two years (Stone, Garibaldi, and Pozzi 2019, p. 90). The wine tourism experience is affected by many characteristics, such as tourist typology. The wine tourist typology refers to the profile of wine tourists and is a critical factor to better understand tourist behavior in general (Charters and Ali-Knight 2002). Lifestyle, behavior, sex, age, culture, nationality, and involvement are the most extensive kinds of elements to indicate wine tourists' typologies (Rodríguez-Donate et al. 2020; Santos et al. 2020). To this end, some researchers have studied the profile and typology of the wine tourist. According to Hall and Macionis (1998), wine tourists consisted of three categories: the 'wine lovers,' the 'wine interested,' and 'curious tourists.' The typology of the wine lover refers to 'highly interested'; the wine-interested typology indicates 'interested,' and the curious tourist typology shows 'limited interest.' In the same vein, Bruwer, Li, and Reid (2002) found a typology of five wine tourists: enjoyment-oriented social wine drinkers; fashion/image-oriented wine drinkers; ritual-oriented, conspicuous wine enthusiasts; purposeful, inconspicuous premium wine drinkers; and basic wine drinkers. Santos et al. (2020) also classified tourists into four wine typologies as high, medium, and low wine product involvement levels. In this context, while tourists with a low number of participation in wine-related activities (such as buying wine, visiting wineries and wineries, attending wine courses, and so more) are in the low group, while tourists with more participation are in the high group. As a result, all of these studies show that wine tourist typology is an important antecedent to investigate wine tourists and their behaviors.

Elements

Pine and Gilmore's (1998, 1999) Experience Economy model stands out through operational design such as tourism experiences. This Experience Economy model has four dimensions: educational, aesthetic, entertainment, and escapist experiences. The wine tourism experience is also classified as a form of tourism experience in wine tourism destinations. Moreover, the wine tourism experience is provided in several ways, including 'festivals, cultural heritage, dining, hospitality, education, tasting and cellar door sales, and winery tours' (Charters and Ali-Knight 2000). In this sense, it is seen that educational, aesthetic, entertainment, and escapist dimensions are the major components of the wine tourism experience.

Aesthetic

The aesthetic experience involves immersion in a sensual environment. Places are selected for the pleasure and uniqueness they offer that differs from everyday life. Driving rural roads lined with vineyards and orchards is a basic tourism activity for a day in wine destinations. Likewise, well-appointed country inns and hotels populate wine regions to accommodate the wine tourist. Moreover, the aesthetics applied to the wine tourism experience involve a core of different features that derive from the design of the routes, the architecture of the wineries, and the environment where the wine and food tasting experience takes place (Madeira, Correia, and Filipe 2019). Having all these in mind, it is seen that the winescape provides these opportunities and has indicated to be fundamental for a wine tourism experience (Quadri-Felitti and Fiore 2012; Thanh and Kirova 2018). In this sense, the aesthetic experience is an essential element for the wine experience. For instance, Quadri-Felitti and Fiore (2012) stated that activities such as vineyard hiking, cycling tours, hot air ballooning over vineyards, vineyard tour by horse and carriage, harvesting grapes, and riding a grape picker are indicated aesthetic dimensions of the wine experience. Besides the beauty of the winescape, signage and information about wine attractions also support tourists' aesthetic experience in terms of the wine tourism experience (Pikkemaat et al. 2009).

Entertainment

According to Pine and Gilmore (1999), entertainment is described as the passive/absorption dimension of the model wherein performers engage a customer's attention. As expressed by Quadri-Felitti and Fiore (2013), whereas agricultural and viticultural activities taking place at a winery, farm, or vineyard may be absorbed, cultural attractions and events that contribute to wine destinations may also be a considerable draw. Visiting the wine regions is included by a wide range of entertainment activities that vary from region to region (Carlsen and Boksberger 2013). Madeira, Correia, and Filipe (2019) indicate that

> the entertainment experience may occur through a number of cultural and educational activities, when tourists observe other people's activities and/or performances, including listening to music and reading or in an active way which includes wine tours in vineyards with tastings, visits to museums, exhibitions, and local cultural sites, participating in harvesting, hiking, cycling and photo-taking.

Within this context, it is clear that cultural attractions and events within wine destinations may intensify the entertainment dimension of tourists' wine experience (Quadri-Felitti and Fiore 2013). Wine festivals, shows, and specialty shops (e.g., wine shops) are also important elements to entertain tourists in wine destinations (Thanh and Kirova 2018). Moreover, Quadri-Felitti and Fiore (2013) indicated that cellar concerts, music in vineyards, wine blending demonstrations, farm and food demonstrations, and museum and heritage site visits are generally activities for the entertainment dimension of the wine tourism experience.

Escape

The escapist dimension emphasizes seeking experiences that differ from tourists' daily experiences (Pine and Gilmore 1999). Tourists visit wine destinations to escape daily reality during their holidays (Thanh and Kirova 2018). Research about wine tourists' motivations has supported the escape dimension within the model and confirmed tourists want more to do than just visit wineries and taste wine. For instance, tourists may also immerse themselves by participating in wine country recreation, abundant in the natural, rural setting of wine appellations. Thus, the escape dimension is supported by this wine tourism experience (Quadri-Felitti and Fiore 2013). On the other hand, it is considered that the greater the number of activities offered within the wine destination, the greater the potential for tourists to wholly realize the escapist dimension (Ali-Knight and Carlsen 2003). Considering the experience model of Pine and Gilmore (1999), some activities enrich the escape dimension of the wine tourism experience such as vineyard hiking, cycling tours, hot air ballooning over vineyards, vineyard tour by horse and carriage, harvesting grapes, and riding a grape picker (Quadri-Felitti and Fiore 2013).

Education

The education dimension of experiences ensures that the tourist absorbs the events and involves the active participation of the individual. Tourists increase the knowledge or skill with the educational activities of experience as well (Pine and Gilmore 1999). Education is one of the main motivations in wine tourism research literature. Therefore, personal development is identified as a critical factor desired by wine tourists. Wine tourism includes accommodation, culinary, cultural, and recreational activities – all of which may include an educational component. Wine tasting and wine-and-food pairing events are the most common educational activities in terms of the wine tourism experience. Home winemaking seminars and partnering with chefs and farmers to offer culinary classes also enhance the education dimension of wine experiences (Pikkemaat et al. 2009; Sparks 2007; Thanh and Kirova 2018). Moreover, the wine tourism experience provides tourists with opportunities to learn about wine characteristics, as well as the making process, and thus developing the wine industry (Madeira, Correia, and Filipe 2019). Based on Pine and Gilmore's (1999) experience model, Quadri-Felitti and Fiore (2013) classified the education dimension of the wine tourism experience with some activities such as wine tastings and seminars, culinary-wine pairing events, home winemaking seminars, cooking, and craft-making classes.

Outcomes

The wine tourism experience has similar outcomes to those of other touristic experiences. Some studies show that wine tourism experiences have contributed to building a strong brand image for wine destinations, eliciting different emotions for tourists, involving many behaviors, and cultivating various marketing outcomes such as satisfaction, loyalty, trust, and memorability (Quadri-Felitti and Fiore 2013; Pelegrín-Borondo, Olarte-Pascual, and Oruezabala 2020; Thanh and Kirova 2018). To this end, the outcomes of the wine tourism experience are fundamentally considered as behavioral intentions, satisfaction, and emotions in this study.

Behavioral intentions

Intention is indicated as the motivation necessary to engage in a particular behavior (Quadri-Felitti and Fiore 2013). According to Yuan et al. (2008), behavioral intention (hereinafter BI) is the 'relationship of the tourist's general level of interest in wine to their immediate purpose of visiting a specific winery and the reasons for their overall presence in the region' in terms of the wine tourism experience. Understanding the reasons that trigger the tourist to purchase wine and learning how the decision-making process of wine choice has occurred are essential factors for the wine destinations (Lanfranchi et al. 2020). Previous studies have supported that BI positively affects the wine tourism experience, in general. For instance, Santos et al. (2019) state that involvement, emotions, and place attachment can significantly predict BI as antecedents. Pikkemaat et al. (2009) associated well-being as behavioral outcomes for the wine tourism experience. Moreover, several outcomes of wine tourism experiences involve behavioral intentions such as revisit intention (Yen 2020), word-of-mouth intention (Loureiro and da Cunha 2017), quality of life (Kruger, Rootenberg, and Ellis 2013), memories (Quadri-Felitti and Fiore 2013), trust or loyalty (Fernandes and Cruz 2016), and wine buying intention (Pivac 2012). Consequently, it is clearly seen that behavioral intentions are vital for the wine tourism experience as an outcome.

Satisfaction

Important marketing outcomes, such as satisfaction, are important in the wine experience as well as in tourist experience. Satisfaction shows the customer's likes or dislikes of the service. Moreover, satisfaction is usually indicated as a function of whether visitors' expectations or their psychological needs are met. In the winery setting, several visit motives have been determined to explain tourist satisfaction with a wine destination, such as purchasing wine, viewing local landscapes, expanding wine knowledge, socializing with others, experiencing entertainment, and enjoying ancillary hospitality services (i.e., dining and accommodations) (Canziani 2019). Moreover, wine is an important factor of a destination's attractiveness and can be considered as more than just a product; it is associated with quality tourism and travelers seeking authenticity and high levels of satisfaction as a multisensory experience (Bonarou, Tsartas, and Sarantakou 2019). In addition to this, it is expected that positive wine experiences may result in positive satisfaction (Fernandes and Cruz 2016). Some studies indicate that the natural beauty, geographical setting, and rurality of the wine regions and the cultural product of the winescapes influence wine tourism demand and tourists' satisfaction (Leri and Theodoridis 2019; Quadri-Felitti and Fiore 2013). Therefore, satisfaction is a key factor in revisiting the wine destination. For example, Hubbard et al. (2012) associated perceptions of quality and satisfaction about a wine festival to loyalty (i.e., future intention to return). Likewise, Yuan et al. (2008) indicate that satisfaction with a wine tourism experience creates intentions to buy the wines and visit the wine regions.

Emotions

Motivations, perceptions, previous experiences, preferences, information, and expectations are involved in wine tourists' demand. Specifically, destination choices and visits are affected by previous experiences and the degree of enjoyment with these experiences (Yuan et al. 2008). Tourists have emotional and experiential expectations from visiting a wine destination. Wine regions cannot offer only wine tasting to encompass the wine experience; focus should also be placed on emotional aspects of wine experiences. The emotions in wine tourism are very strong with wine tourists, and this is a result of its experiential nature (Pratt 2019). Hedonic services, such as tourism and wineries, are good examples of experiences that are able to elicit emotional reactions (Fernandes and Cruz 2016). Wine experiences are a multisensory tourism experience by nature (Sigala 2019). In this sense, Pratt and Sparks (2014) indicate that a winescape as an important component of the wine experience has often elicited an emotional response and feelings towards the setting. Dann (1996) also emphasizes that the affective component of destination image reflects the individual's feelings or emotional response towards the destination, such as winescape or sensorial factors of the wine experience. Previous research shows emotions as a determinant in deciding to visit a wine region (Pratt 2019).

Emotions are also one of the outcomes of the wine experience. For this reason, it is expected that the wine tourism experience elicits tourists' emotions. In this respect, emotions are a key factor for both tourists and destinations. For instance, Calvo-Porral, Lévy-Mangin, and Ruiz-Vega (2020) state that four clusters emerge experiencing different emotions in wine consumption: 'emotionally unattached,' 'negatives,' 'contented circumspect,' and 'wine lovers.' According to these results, 'wine lovers' are the most attractive segment due to their strong wine emotional bond, and emotions have a strong relationship to satisfaction, loyalty, and involvement. In a similar vein, Pelegrín-Borondo, Olarte-Pascual, and Oruezabala (2020) emphasize that positive emotions are associated with wine purchase intention.

Case Study 35.1 Wine experience in Cappadocia, Turkey, as a wine destination

Cappadocia, meaning the 'Land of Beautiful Horses' in the Persian language, comprises the provinces of Aksaray, Nevşehir, Niğde, Kayseri, and Kırşehir in Turkey. Major towns include Uçhisar, Avanos, Ürgüp, Derinkuyu, Kaymaklı, and Ihlara. Millions of years ago, the volcanoes of Erciyes, Hasadağ, and Melendiz mountains erupted and covered Cappadocia with a layer of tuff. This tuff layer has eroded and produced various earth formations since that time. Moreover, wind and rainwater flowing down the sides of valleys eroded the tuff structure and sculpted the formations known as 'fairy chimneys.' The earliest human settlements in Cappadocia date to the Paleolithic period, and the written history of the region goes back to the Hittites. The Hittite settlers were followed by the empires of Phrygian, Persian, Roman, Byzantine, Seljuk, and Ottoman (Republic of Turkey Ministry of Culture and Tourism General Directorate of Information 2014, pp. 1–2). To this end, fairy chimneys, ancient cities, mosques, churches and monasteries, underground cities, caravanserais, rock carvings, and valleys are among the attractions of the Cappadocia region, which has hosted various civilizations. In Cappadocia, activities such as Turkish night, hot air balloon tours, bicycle and safari tours, horse riding, paragliding, pottery making, local food, and wine are also organized. Furthermore, the Cappadocia region was cited on the World Heritage List by UNESCO in 1985.

Cappadocia is one of the oldest wine regions in the world. Considering the availability of grape chewing and wine production areas in the underground cities of Cappadocia, it is known that the history of grape cultivation and wine goes back to the 3000s BCE, the Hittite period in the region (Özgül-Katlav, Yönet-Eren, and Tuna 2019). To this end, the favorable climatic conditions and soil structure of the region have also led to the emergence of viticulture with thousands of years of history. Red wine grapes such as Cabernet Sauvignon, Merlot, and Syrah and white wine grapes such as Chardonnay and Sauvignon Blanc are the best-known examples of grape varieties grown in Cappadocia. It is also known that many native Turkish grapes, such as Dimrit, Emir, Kalecik Karası, etc., are grown in the region. A total of 13% of wine grapes in Turkey are grown in Cappadocia.

Wine tourism products or experiences have become one of the important products for the Cappadocia region. To this end, many products and experiences of wine tourism – such as wine tasting, seminars on wine and viticulture, wine competitions, factory tours of vineyards, wineries, and wine factories, wine festivals, and wine routes –are offered to tourists in the destination (Yılmaz and Özdemir 2017). Local wines and wine-related

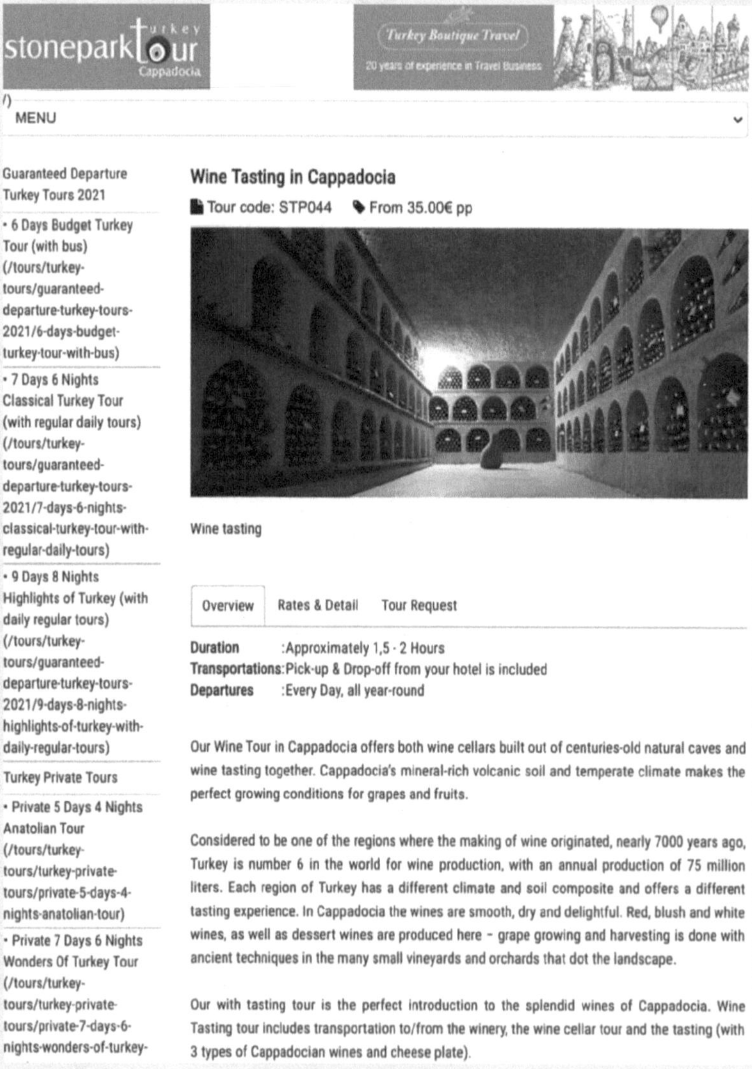

Figure 35.1 An example of a wine tourism experience in Cappadocia.

activities are an important motivation for tourists to visit Cappadocia. Previous studies (Kılıçhan and Birdir 2014; Akdağ 2015) clearly stated that local wine is a key motivation for visiting Cappadocia. Tourists, especially wine tourists, can obtain information from various sources such as promotional materials and websites before traveling to the destination. Destination marketing and management organizations provide information on local wines and wine-related activities on their websites and promotional materials to meet the information needs of tourists. In this sense, Yılmaz and Özdemir (2015, p. 771) stated that information was provided on the presentation, characteristics, and production process of local wines, the historical development of local wine and viticulture, grape varieties, events and festivals related to grapes, wine, and vintage.

At the famous wine houses of Cappadocia, local wines are served to tourists in authentic terracotta decanters and ceramic goblets. In a conducted study by Kılıçhan and Birdir (2014), it is stated that local wine presentation is perceived as an important component affecting consumption for tourists. Moreover, tourists emphasize the qualities of wine-related activities and experiences such as learning, relaxation, escape, memorability, and so forth (Yılmaz 2017). In the light of these, details on the wine tourism experience organized by a travel agency in Cappadocia are shown Figure 35.1.

Conclusion

Considering the studies in the literature, it is thought that the concept of wine experience should be addressed in a holistic perspective consisting of antecedents, elements, and outcomes, as shown in Figure 35.2. To this end, tourists' wine experience in a particular destination can be explained in greater detail. Wine experience has different antecedents such as motivations, wine involvement, wine-related lifestyle, information sources, and tourist profile and typologies. At the same time, tourists can be attentive to various components and qualities such as aesthetic, entertainment, escape, and education to evaluate the wine experience in the destination. Lastly, the wine experience has several outcomes such as satisfaction, behavioral intention such as revisit and recommend, destination attachment, well-being, and memorability.

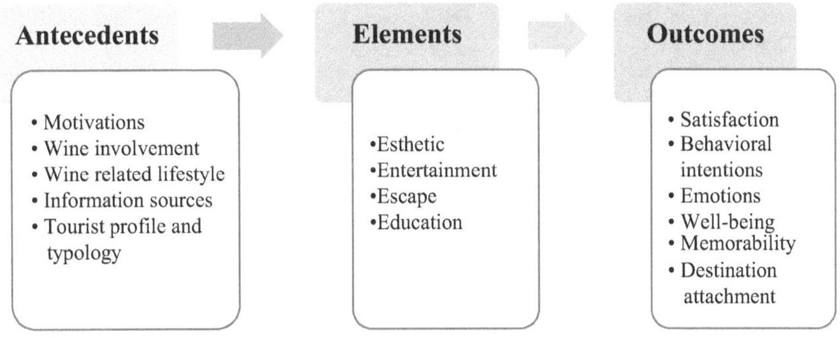

Figure 35.2 A holistic model of wine tourism experience.

Implications

It is necessary to enrich wine tourism experiences in order to increase the attractiveness of wine destinations. When tourists gain value from their wine experience, their satisfaction and positive feelings increase, and they tend to revisit the destination. In this context, destination marketers can highlight not only wine tasting but also all wine-related activities. For example, tourists can be offered a combination of activities such as participating in the grape harvest, walking around the vineyards, attending homemade wine classes, and wine tasting.

References

Akdağ, G. (2015). Tourism experiences of local tourists following Thrace Vineyard Route. *Journal of Tourism and Gastronomy Studies*, 3(4): 3–11.

Alant, K., and Bruwer, J. (2004). Wine tourism behaviour in the context of a motivational framework for wine regions and cellar doors. *Journal of Wine Research*, 15(1): 27–37. https://doi.org/10.1080/0957126042000300308

Ali-Knight, J., and Carlsen, J. (2003). An exploration of the use of 'extraordinary' experiences in wine tourism. In Lockshin, L., and Rungie, C. (eds.), *Proceedings of the International Colloquium in Wine Marketing* (pp. 1–7). Adelaide: University of South Australia.

Asero, V., and Patti, S. (2011). Wine tourism experience and consumer behavior: the case of Sicily. *Tourism Analysis*, 16(4): 431–442. https://doi.org/10.3727/108354211X13149079788936

Bonarou, C., Tsartas, P., and Sarantakou, E. (2019). E-storytelling and wine tourism branding: Insights from the wine roads of northern Greece. In Sigala, M., and Robinson, R. N. S. (eds.), *Wine Tourism Destination Management and Marketing* (pp. 77–98). Switzerland: Palgrave Macmillan.

Bruwer, J., and Alant, K. (2009). The hedonic nature of wine tourism consumption: an experiential view. *International Journal of Wine Business Research*, 21(3): 235–257. https://doi.org/10.1108/17511060910985962

Bruwer, J., Li, E., and Reid, M. (2002). Segmentation of the Australian wine market using a wine-related lifestyle approach. *Journal of Wine Research*, 13(3): 217–242. https://doi.org/10.1080/0957126022000046510

Bruwer, J., Roediger, B., and Herbst, F. (2017). Domain-specific market segmentation: a wine-related lifestyle (WRL) approach. *Asia Pacific Journal of Marketing and Logistics*, 29(1): 4–26. https://doi.org/10.1108/APJML-10-2015-0161

Bruwer, J., and Thach, L. (2013). Wine tourists' use of sources of information when visiting a USA wine region. *Journal of Vacation Marketing*, 19(3): 221–237. https://doi.org/10.1177/1356766712464768

Byrd, E. T., Canziani, B., Boles, J. S., Williamson, N. C., and Sonmez, S. (2017). Wine tourist valuation of information sources: the role of prior travel. *International Journal of Wine Business Research*, 29(4): 416–433. https://doi.org/10.1108/IJWBR-04-2017-0020

Calvo-Porral, C., Lévy-Mangin, J.-P., and Ruiz-Vega, A. (2020). An emotion-based typology of wine consumers. *Food Quality and Preference*, 79: 103777. https://doi.org/10.1016/j.foodqual.2019.103777

Canovi, M., Mordue, T., and Lyon, A. (2020). The impact of wine tourism involvement on winery owners' identity processes. *Tourism Planning & Development*, 17(5): 573–590. https://doi.org/10.1080/21568316.2020.1730945

Canziani, B. (2019). Wine tourism: Balancing core product and service-dominant strategies. In Sigala, M., and Robinson, R. N. S. (eds.), *Wine Tourism Destination Management and Marketing: Theory and Cases* (pp. 185–201). Switzerland: Palgrave Macmillan.

Carlsen, J., and Boksberger, P. (2013). Enhancing consumer value in wine tourism. *Journal of Hospitality & Tourism Research*, 39(1): 132–144. https://doi.org/10.1177/1096348012471379

Charters, S., and Ali-Knight, J. (2000). Wine tourism – a thirst for knowledge? *International Journal of Wine Marketing*, 12(3): 70–80. https://doi.org/10.1108/eb008715

Charters, S., and Ali-Knight, J. (2002). Who is the wine tourist? *Tourism Management*, 23(3): 311–319. https://doi.org/10.1016/S0261-5177(01)00079-6

Charters, S., Fountain, J., and Fish, N. (2009). You felt like lingering ...': experiencing 'real' service at the winery tasting room. *Journal of Travel Research*, 48(1): 122–134. https://doi.org/10.1177/0047287508326508

Corsi, M. A., Modroño, J. I., Mariel, P., Cohen, J., and Lockshin, L. (2020). How are personal values related to choice drivers? An application with Chinese wine consumers. *Food Quality and Preference*, 86: 103989. https://doi.org/10.1016/j.foodqual.2020.103989

Crespi-Vallbona, M., and Mascarilla-Miró, O. (2020). Wine lovers: their interests in tourist experiences. *International Journal of Culture, Tourism, and Hospitality Research*, 14(2): 239–258. https://doi.org/10.1108/IJCTHR-05-2019-0095

Dann, G. M. S. (1996). Tourists' images of a destination – an alternative analysis. *Journal of Travel & Tourism Marketing*, 5(1–2): 41–55. https://doi.org/10.1300/J073v05n01_04

Dixit, S. K. (ed.). (2017). *The Routledge Handbook of Consumer Behaviour in Hospitality and Tourism*. London: Routledge.

Fernandes, T., and Cruz, M. (2016). Dimensions and outcomes of experience quality in tourism: the case of Port wine cellars. *Journal of Retailing and Consumer Services*, 31: 371–379. https://doi.org/10.1016/j.jretconser.2016.05.002

Getz, D. (2000). *Explore Wine Tourism: Management, Development & Destinations*. New York: Cognizant Communication Corporation.

Getz, D., and Brown, G. (2006). Benchmarking wine tourism development: the case of the Okanagan Valley, British Columbia. *International Journal of Wine Marketing*, 18(2): 78–97. https://doi.org/10.1108/09547540610681077

Grybovych, O., Lankford, J., and Lankford, S. (2013). Motivations of wine travelers in rural Northeast Iowa. *International Journal of Wine Business Research*, 25(4): 285–309. https://doi.org/10.1108/IJWBR-07-2012-0023

Gunasekar, S., Das, P., Dixit, S. K., Mandal, S., and Mehta, S. R. (2021). Wine-experienscape and tourist satisfaction: through the lens of online reviews. *Journal of Foodservice Business Research*. https://doi.org/10.1080/15378020.2021.2006039

Hall, C. M., Johnson, G., Cambourne, B., Macionis, N., Mitchell, R., and Sharples, L. (2000). Wine tourism: an introduction. In Hall, C. M., Sharples, L., Cambourne, B., and Macionis, N. (eds.), *Wine Tourism around the World: Development, Management and Markets* (pp. 1–23). Oxford: Elsevier.

Hall, C. M., and Macionis, N. (1998). Wine tourism in Australia and New Zealand. In Butler, R. W., Hall, C. M., and Jenkins, J. M. (eds.), *Tourism and Recreation in Rural Areas* (pp. 267–298). Sydney: John Wiley & Sons.

Hubbard, K. W., Mandabach, K. H., McDowall, S., and Vanleeuwen, D. M. (2012). Perceptions of quality, satisfaction, loyalty, and approximate spending at an American wine festival. *Journal of Culinary Science & Technology*, 10(4): 337–351. https://doi.org/10.1080/15428052.2012.735962

Hyde, K. F. (2006). Contemporary information search strategies of destination-naïve international vacationers. *Journal of Travel and Tourism Marketing*, 21(2–3): 63–76. https://doi.org/10.1300/J073v21n02_05

Kılıçhan, R., and Birdir, K. (2014). Lezzetin peşinde şarap turizmi: Kapadokya'nın eko-gastronomik potansiyeli. *Eko-Gastronomi Dergisi*, 1(1): 103–122.

Koksal, H. M. (2020). Segmentation of wine consumers based on level of involvement: A case of Lebanon. *British Food Journal*, 123(3): 926–942. https://doi.org/10.1108/BFJ-03-2020-0183

Kruger, S., Rootenberg, C., and Ellis, S. (2013). Examining the influence of the wine festival experience on tourists' quality of life. *Social Indicators Research*, 111(2): 435–452. https://doi.org/10.1007/s11205-012-0013-0

Lanfranchi, M., Zirilli, A., Alibrandi, A., and Giannetto, C. (2020). The behaviour of wine consumers towards organic wine: a statistical analysis through the non-parametric combination test. *International Journal of Wine Business Research*. https://doi.org/10.1108/IJWBR-04-2020-0015

Leri, I., and Theodoridis, P. (2019). How do personality traits affect visitor's experience, emotional stimulation and behaviour? The case of wine tourism. *Tourism Review*, 74: 480–502. https://doi.org/10.1108/TR-05-2019-0148

Lockshin, L., Jarvis, W., d'Hauteville, F., and Perrouty, J. P. (2006). Using simulations from discrete choice experiments to measure consumer sensitivity to brand, region, price, and awards in wine choice. *Food Quality and Preference*, 17 (3–4): 166–178. https://doi.org/10.1016/j.foodqual.2005.03.009

López-Guzmán, T., Vieira-Rodríguez, A., and Rodríguez-García, J. (2014). Profile and motivations of European tourists on the sherry wine route of Spain. *Tourism Management Perspectives*, 11: 63–68. https://doi.org/10.1016/j.tmp.2014.04.003

Loureiro, S. M. C., and da Cunha, N. P. (2017). Wine prestige and experience in enhancing relationship quality and outcomes: wine tourism in Douro. *International Journal of Wine Business Research*, 29(4): 434–455. https://doi.org/10.1108/IJWBR-04-2017-0033

Madeira, A., Correia, A., and Filipe, J. A. (2019). Modelling wine tourism experiences. *Anatolia: An International Journal of Tourism and Hospitality Research*, 30(4): 513–529. https://doi.org/10.1080/13032917.2019.1642922

Mason, R., and O'Mahony, B. (2007). On the trail of food and wine: the tourist search for meaningful experience. *Annals of Leisure Research*, 10(3–4): 498–517. https://doi.org/10.1080/11745398.2007.9686778

Özgül-Katlav, E., Yönet-Eren, F., and Tuna, M. (2019). Evaluation of viticulture in Cappadocia in the scope of gastronomy tourism. *Journal of Tourism and Gastronomy Studies*, 7(3): 2167–2186. https://doi.org/10.21325/jotags.2019.466

Pelegrín-Borondo, J., Olarte-Pascual, C., and Oruezabala, G. (2020). Wine tourism and purchase intention: a measure of emotions according to the PANAS scale. *Journal of Wine Research*, 31(2): 101–123. https://doi.org/10.1080/09571264.2020.1780573

Pikkemaat, B., Peters, M., Boksberger, P., and Secco, M. (2009). The staging of experiences in wine tourism. *Journal of Hospitality Marketing & Management*, 18(2–3): 237–253. https://doi.org/10.1080/19368620802594110

Pine, B. J., and Gilmore, J. H. (1998). Welcome to the experience economy. *Harvard Business Review*, 76(4): 97–105.

Pine, B. J., and Gilmore, J. H. (1999). *The Experience Economy: Work Is Theatre and Every Business a Stage*. Boston: Harvard Business School Press.

Pivac, T. (2012). *Wine Tourism of Vojvodina*. Novi Sad: University of Novi Sad, Faculty of Sciences.

Pratt, M. (2019). The image of a wine tourist and impact on self-image congruity. In Sigala, M., and Robinson, R. N. S. (eds.), *Wine Tourism Destination Management and Marketing* (pp. 45–66). Switzerland: Palgrave Macmillan.

Pratt, M. A., and Sparks, B. (2014). Predicting wine tourism intention: destination image and self-congruity. *Journal of Travel and Tourism Marketing*, 31(4): 443–460. https://doi.org/10.1080/10548408.2014.883953

Quadri-Felitti, D., and Fiore, A.M. (2012). Experience economy constructs as a framework for understanding wine tourism. *Journal of Vacation Marketing*, 18(1): 3–15. https://doi.org/10.1177/1356766711432222

Quadri-Felitti, D. L., and Fiore, A. M. (2013). Destination loyalty: effects of wine tourists' experiences, memories, and satisfaction on intentions. *Tourism and Hospitality Research*, 13(1): 47–62. https://doi.org/10.1177/1467358413510017

Rahman, I., and Reynolds, D. (2015). Wine: intrinsic attributes and consumer's drinking frequency, experience, and involvement. *International Journal of Hospitality Management*, 44(1): 1–11. https://doi.org/10.1016/j.ijhm.2014.09.004

Republic of Turkey Ministry of Culture and Tourism General Directorate of Information. (2014). Cappadocia. *Ankara*. https://nevsehir.ktb.gov.tr/Eklenti/60483,kapadokyapdf.pdf

Rodríguez-Donate, M. C., Romero-Rodríguez, M. E., Cano-Fernández, V. J., and Guirao-Pérez, G. (2020). Gender and wine consumption: sociodemographic profiles. *British Food Journal*, 122(1): 242–257. https://doi.org/10.1108/BFJ-02-2019-0128

Saayman, M., and van der Merwe, A. (2015). Factors determining visitors' memorable wine-tasting experience at wineries. *Anatolia: An International Journal of Tourism and Hospitality Research*, 26(3): 372–383. https://doi.org/10.1080/13032917.2014.968793

Santos, V., Ramos, P., Almeida, N., and Santos-Pavon, E. (2020). Developing a wine experience scale: a new strategy to measure holistic behaviour of wine tourists. *Sustainability*, 12: 8055. https://doi.org/10.3390/su12198055

Santos, V. R., Ramos, P., Almeida, N., and Santos-Pavon, E. (2019). Wine and wine tourism experience: a theoretical and conceptual review. *Worldwide Hospitality and Tourism Themes*, 11(6): 718–730. https://doi.org/10.1108/WHATT-09-2019-0053

Schiffman, L. G., Bednall, D., O'Cass, A., Paladino, A., Ward, S., and Kanuk, L. (2008). *Consumer Behaviour*. 4th ed. Australia: Pearson Education.

Sharma, P., Singh, S., and Misra, R. (2020). Wine-related lifestyle segmentation in the context of urban Indian consumers. *International Journal of Wine Business Research*, 32(4): 503–522. https://doi.org/10.1108/IJWBR-05-2019-0036

Sigala, M. (2019). Supporting tourists' mobility in wine destinations: the hop-on hop-off bus in Swan Valley, Western Australia. In Sigala, M., and Robinson, R. N. S. (eds.), *Wine Tourism Destination Management and Marketing* (pp. 357–372). Switzerland: Palgrave Macmillan.

Sparks, B. (2007). Planning a wine tourism vacation: factors that help to predict tourist behavioral intentions. *Tourism Management*, 28(5): 1180–1192. https://doi.org/10.1016/j.tourman.2006.11.003

Spielmann, N., Babin, B. J., and Verghote, C. (2016). A personality-based measure of the wine consumption experience for millennial consumers. *International Journal of Wine Business Research*, 28(3): 228–245. https://doi.org/10.1108/IJWBR-09-2015-0035

Stone, M. J., Garibaldi, R., and Pozzi, A. (2019). Exploring additional food and beverage activities of wine travelers. In Dixit, S. K. (ed.), *The Routledge Handbook of Gastronomic Tourism* (pp. 89–96). New York: Routledge.

Thanh, T. V., and Kirova, V. (2018). Wine tourism experience: a netnography study. *Journal of Business Research*, 83: 30–37. https://doi.org/10.1016/j.jbusres.2017.10.008

Yen, T. (2020). The determinates of revisit intentions in wine cultural event: the moderating role of experience authenticity. *Asian Journal of Education and Social Studies*, 6(4): 27–40. https://doi.org/10.9734/ajess/2020/v6i430183

Yılmaz, G. (2017). An evaluation of the relationship between gastronomy and tourism. *Journal of Travel and Hospitality Management*, 14(2): 171–191. https://doi.org/10.24010/soid.335117

Yılmaz, G., and Gültekin, S. (2016). Consumers and tourists' restaurant selections. In Avcikurt, C., Dinu, M. S., Hacioglu, N., Efe, R., Soykan, A., and Tetik, N. (eds.), *Global Issues and Trends in Tourism* (pp. 217–230). Sofia: St. Kliment Ohridski University Press.

Yılmaz, G., and Özdemir, B. (2015). The role of local gastronomy in promotional materials: The case of Cappadocia region. *Proceedings of the I. Eurosia International Tourism Congress* (Vol. II, pp. 760–780). Konya: Selçuk University Tourism Faculty.

Yılmaz, G., and Özdemir, B. (2017). Restaurant configurations in tourism destinations: a qualitative study about Cappadocia. *Anatolia: Turizm Araştırmaları Dergisi*, 28(1): 81–95. https://doi.org/10.17123/atad.316075

Yuan, J., Morrison, A. M., Cai, L. A., and Linton, S. (2008). A model of wine tourist behavior: a festival approach. *International Journal of Tourism Research*, 10(3): 207–219. https://doi.org/10.1002/jtr.651

Zaichkowsky, J. (1985). Measuring the involvement construct. *Journal of Consumer Research*, 12(3): 341–352.

36

MULTISENSORY EXPERIENCE OF WINE TOURISM

Anthony J. Saliba and Genevieve d'Ament

Theoretical underpinnings of the chapter

Consumers participate in wine tourism out of an interest in new experiences, not solely based on wine. In fact, the sensory experience associated with wine tourism is so broad that a single chapter could not scratch the surface. The scope of this chapter is specifically the cellar door experience, with reference to how multisensory factors influence individual responses. The chapter will adopt hedonism to interpret how multisensory experiences influence emotion, memory, and buying behavior both at the cellar door and thereafter. Hedonism has historical antecedents to at least Aristotle (384–322 BCE), who mused that sensory experiences were part of a fulfilling life. In more modern times, psychology has focused on how sensory experiences influence human behavior; although an oversimplification, the basic tenant of hedonism in psychology is that humans are motivated by the desire to increase pleasure and decrease pain.

To interpret human processing of sensory information, this chapter will use the VACP model (Visual, Auditory, Cognitive, and Psychomotor; McCracken and Aldrich 1984). The model presumes that the loading of sensory channels is relatively independent. For instance, even when the visual channel is fully utilized, more information can be processed if that information is delivered via another channel (e.g., a complex visual environment while driving would still allow a driver to hear and successfully encode auditory information from the radio). A logical conclusion of the model is that where multiple sensory channels are highly loaded, the overall mental workload can be excessive. Most theories of cognitive processing presume that humans have a limited capacity to process information, and that ability declines slowly throughout the lifespan (Salthouse 1988). Where information presented to humans exceeds processing resources, that information is not encoded and therefore cannot be recalled. Complex multisensory environments may lead to poor encoding and therefore recall of key information. Early flight simulator research, for instance, found that highly complex simulators with rich visual information did not lead to the transfer of training as successfully as simple stick-figure simulators, provided that the latter presented all of the key information pilots need to learn to take off and land. What was happening was that the complex, highly attractive simulators were presenting superfluous details such as blades of grass that were essentially distracting attention from pilots learning key facts such as aircraft orientation, horizon, and airspeed. While

DOI: 10.4324/9781003143628-42

the cellar door is not as complex as an aircraft, it can still present much complex multisensory information that is superfluous to buying wine.

A comparative case study approach is adopted (Ragin 1987) where two cellar door case studies are described and then compared to elucidate interactions between individual differences, multisensory experience, and the wine tourism experience. The authors have over 30 years of combined experience working in and researching the cellar door process within a wine tourism context. This allowed for the creation of two archetypes that were used for the case studies. That is, two case studies were created not from one cellar door experience, but from multiple examples of highly similar cellar door experiences. Although case studies are in themselves not considered to be generalizable, this approach ensured that the two case studies were a reasonable representation of a moderate number of cellar doors found throughout the world. The archetype approach was adapted from Cooper's (1999) archetype method, which he called personas. A pragmatic focus on cellar door sales is adopted to ensure that this chapter is of benefit not only to the academic community but also to the wine business owner.

Multisensory research

Sensory research has traditionally looked at individual senses (e.g., taste) in a reductionism paradigm, usually within a laboratory. This approach has led to a vast knowledge of sensory experience and, within a hedonism framework, has been valuable to the food and beverage industry (for a review of early discoveries, see Amerine, Pangborn, and Roessler 1965). More recently, sensory research has started to examine sensory experiences within the context of multiple senses, as well as other relevant factors. For instance, it has been demonstrated that taste preference in wine is influenced by cultural experience and health expectation (Yoo et al. 2012; Yoo et al. 2013). The term 'multisensory' research has become suggestive of the rich interactions that occur when different senses are experienced together. Moving away from purely laboratory-based research has enriched our understanding of cellar door sales, as it has embraced truly ecologically valid scenarios. Since consumers do not taste wine in a vacuum, the findings from multisensory research have helped to explain some of the phenomena that have previously been understood only anecdotally. For instance, wine tasted at the cellar door can often taste vastly different when consumed at home due to the environment and context being substantially different. So, what happens when two cellar doors present vastly different multisensory experiences?

Case Study 36.1 The exciting and busy cellar door

As the wine tourist, you have planned your wine trail of the region pre-visit. Some winery visits are planned, but you're prone to impulsive decisions and decide to visit this winery due to the glossy brochure you picked up at the last winery. As you approach down the winding driveway, a building resplendent in both size and color looms large among the vines. A lack of clear signage leaves you searching for the cellar door. You have no doubt that this impressive facility is multipurpose, with signage indicating the inclusion of a restaurant, casual dining cafe, and gift shop. After visually scanning the many options, you eventually find the entrance to the cellar door. Already you are feeling visually overwhelmed and slightly stressed, but relieved that you are now in what should be familiar surroundings, and excitement builds for the new wines to be discovered.

On entering what you perceive to be the cellar door, your senses are pervaded by a myriad of lights, smells, and noises, and you feel even more overwhelmed. However, clear grandeur is conveyed by your surroundings, and you are filled with confidence that wine is taken seriously. The range of shelves, sections, artwork, and people make it difficult to know exactly where to enjoy the wine tasting experience. A staff member makes eye contact, motions for you to join them, asking how they can help, as in such a multipurpose space, it is not clear to them that you are there to taste wine. Soon you're comfortably settled and tasting an impressively large range of wines. You are, however, distracted by the multiple conversations of other patrons given their proximity, and by the smells and jarring sounds from the nearby restaurant and family-oriented café, which give the effect of surrounding the cellar door. The staff member expertly focuses your attention on tasting each wine. Encased in the familiarity of wine tasting, you notice the taste, mouthfeel, look, and aroma of each wine. You're engaged in conversation with the staff member who provides the rich history of each label, the winemaker's ethos, and the location of the grapes. They also invite you to attend the winemaker's lunch if you're still in the region next week. At the moment, you feel thoroughly impressed and convey this through your body language, which provides positive feedback to staff.

Soon, you have tasted 14 wines in total. When asked if you would like to purchase any of the wines tasted, you recall liking several of the wines but also not liking others. At this point, you regret not taking notes as you're feeling a little overwhelmed and having difficulty recalling which of the wines you liked best. Your emotional connection to the cellar door experience is high; you feel aroused by the multisensory experience. In the multisensory environment, you feel so overwhelmed, you do not feel comfortable interrupting the busy cellar door staff to ask for clarification. You decide to purchase two of the wines you recall liking but were concerned that you had recalled them incorrectly. You decide to mitigate your risk by buying 6-bottle cases rather than full 12-bottle cases. Upon taking the wines home, you noticed that one of the wines you purchased was indeed the 'wrong' wine, it was one that you did not like relative to others, and you felt disappointed. The whole experience was impressive at the moment; you had a terrific hedonic experience, so your revisit intentions remain high. However, it was difficult for you to recommend that winery's wines to your friends, as you had a poor recollection of tangible aspects of your experience.

Case Study 36.2 The simple, clear cellar door

As a wine tourist, you were attracted to this winery by the clear signage on the road. You approach a building that is small with clear signage and enter after noticing that there was a winery, vineyard, and cellar door on site. The cellar door is well-lit, emphasizing bare walls with an open design. Your attention is drawn to the one scene in the room – a back-lit shelf containing seven wine bottles, with a bench afront. You approach one of the staff standing behind the bench and ask to taste some wines. The staff member introduces you to the winery by offering a brief history, and you begin your tasting flight.

There are virtually no distracting smells, and your visual attention is continually drawn to the back-lit shelf containing the seven wines. You taste each wine, noticing the look, taste, aroma, and, finally, the mouthfeel. There is a good separation between you and other consumers. The lighting and open design affords an excellent visual experience without extraneous distractions. As you experience each wine, you feel that they are good, clean examples, but you do not feel a deep emotional connection to the wines. Upon completing the tasting experience, you decide to purchase two of the wines. You are confident that you recall which two you liked best, so you agree to buy a 12-bottle case of each wine.

When you return home and try the wines, you are pleased with your purchases and feel positive about your experience. You have a good recollection of the details of your experience and tell several friends about your experience. You do not have a strong emotional connection to the winery and feel that while you may buy that wine brand again if available, you would not revisit the cellar door.

Method and analysis of case studies

The common themes found in the case studies were used to compare each experience and outcome. Table 36.1 contains a summary of the differences, which will be discussed in greater detail in the following section. A comparative case study approach was used (Ragin 1987); the entity studied was the cellar door experience, and cases were selected based on the experience of the researchers. The case studies were ostensibly archetypes of two common cellar door experiences. The archetypal method was adapted from Cooper's (1999) method for creating personas. The concept is that the personas, or case studies in this chapter, are based on real experiences and, as such, represent segments of the cellar door market. That is, the two cellar doors can be taken to represent a moderate portion of the cellar door market. In this way, Cooper (1999) attempted to bring an aspect of generalizability to the findings. Case study

Table 36.1 Comparison of the two case studies

Theme	*Case Study Study 36.1* *Large, complex*	*Case Study Study 36.2* *Small, simple*
	Experience	
Venue	Overwhelmed, impressed	Comfortable
Décor	Emotionally connected	No emotional connection
Staff	Helpful, necessary	Helpful, necessary
Mental workload	High	Low
Encoding/recall	Poor encoding and recall	High encoding and recall
Revisit	Likely	Unlikely
	Wine Purchase	
Confidence	Moderate	High
Cellar door	Moderate purchase	Moderate to high purchase
Post-cellar door	Moderate multiplier	Low multiplier
Outcome summary	Better long-term sales	Better sales on the day of visit

methods do not intend to report results that are generalizable, and we acknowledge that there are limits to how generalizable our results can be since they are based on more than 30 years of combined experience of two people. However, we do want to assert that the breadth of experience that led to the creation of the case study archetypes does allow our results to represent two different categories of cellar door – large and complex, compared with small and simple. There is no suggestion that our results can reasonably be extrapolated beyond those two categories.

Discussion

This chapter presents two case studies that represent different categories of cellar door – the large and complex cellar door and the small and simple cellar door. These cellar doors present vastly different multisensory experiences, and these interact with individual differences of the customers to influence outcomes.

Emotional connection with the wine business

One intention of wine tourism is to create an emotional connection with a region or winery. That emotional connection can lead to repeat visits, a good impression of wine from that region, and increased trust around wine brands. From a marketing perspective, it is important and developed to a greater extent in Case Study Study 36.1. While overwhelming, the cellar door generated a greater emotional connection through size, depth of experience, and breadth of experience. The cellar door in Case Study Study 36.2 failed to generate an emotional connection. This could have been mitigated in several ways. For instance, cellar door staff can play a substantial role in generating an emotional connection, although this does rely on developed recruitment and training practices. Retention of well-trained staff in the wine industry has been an issue in some countries.

Multisensory factors

Providing an environment rich in sensorial stimulation is a key way to create an emotional connection with the consumer. When designing the multisensory cellar door, several factors need to be considered. Sensory information delivered via multiple channels can interfere with the encoding of memories, and that will be discussed in the next section on mental workload. Some sensory information can interfere with the experience in other ways. For instance, the color of the mug can influence taste perception of coffee – thereby influencing liking (Van Doorn, Wuillemin, and Spence 2014). Images can evoke the perception of smells, so various objectionable flavors can be primed with the wrong label or wall pictures. Cheese served during wine tasting may reduce perceived astringency and does not appear to negatively influence wine liking (Bastian et al. 2009). However, the wine may dominate preferred wine and cheese matching, which may not be ideal if the winery is also selling cheese or partnering with a cheesemaker.

Food manufacturers take advantage of well-established food-color relationships to elicit hunger, enhance taste and freshness. Sonic seasoning is a term adopted by Spence and Wang (2015a, 2015b, 2015c) where wine styles were matched with classical music. Their work was possibly inspired by wine writer and sound artist Stanier Black-Five, whose multisensory installation work Oenosthesia incorporated vineyard and winery sounds into a wine tasting experience

(Stanier Black-Five (Burzynska, J.) 2018). Expanding on Stanier Black-Five's work, Spence and Wang found the addition of sonic seasoning has shown a small but significant increase in reported enjoyment of the whole experience. Further collaborations with Stanier Black-Five have found bass significantly increases the aromatic intensity of New Zealand Pinot Noir and Spanish Garnacha (Spence et al. 2019). Caution is urged as even the authors admit the complex multidimensional attributes of wine makes concluding definitive causal links challenging if not impossible (for an approachable review of cognitive and sensory wine research, see Spence 2020). Music is often a component of the cellar door experience, with many wineries offering live music as an added incentive to visit their winery. However, wine is a complex product, and more research is needed before we can predict how wine preference is influenced by extraneous multisensory information. For now, the best practice is to minimize multisensory interactions during the cellar door experience and consider locating live music performances a discreet distance from the cellar door experience.

There is an interaction between multisensory experience and individual differences. Personality acts as a processing moderator of the abundant stimuli present in our environment and the comparison of our current experience against developed schemas from past experience. There are personality types that crave greater sensory stimulation, and there are also those who prefer lower levels of stimulation. Take, for example, the authors of this chapter: both contributed to the case study creations from their many experiences. One, who is quite extroverted, prefers the sensory stimulation from Case Study Study 36.1; the other, a more introverted soul, craves the peace of Case Study Study 36.2. The cellar door staff can play an important role in identifying visitors who are not enjoying the noise and other sensory stimulation during the tasting experience.

The personality traits of impulsiveness and openness have been linked to sweet taste preference in wine (Saliba, Wragg, and Richardson 2009). Those higher in impulsiveness and lower in openness prefer sweeter wines. Low openness is related to not wanting to try new things, and the wine tourist's first experience of wine is often with sweeter wines. Where visitors report liking sweet wines, it may simply be because they are reluctant to try something new. The cellar door is the ideal environment to carefully expose visitors to drier wine, whereas in a retail setting, they would be unlikely to change to a new experience, even with some prompting from sales staff. Understanding these subtle behavioral markers of personality and their influence on wine tourism experiences allows cellar door and winery staff to integrate equally subtle changes to their presentation style in what can be a confronting multisensory setting.

Galloway et al. (2008) found that the personality trait of sensation seeking was influential in wine tourism behavior. Wine tourists with higher sensation seeking were more likely to explore further than tourism-focused wine trails, utilize the internet to research a region, drink more, and buy more wine. Tourists high in sensation seeking would most likely enjoy the multisensory energy afforded them during the first case study experience. In addition to engaging the senses, connecting on a personal level engages the wine tourist beyond their first visit. Case Study Study 36.1 included staff who provided a rich history of the winery and their winemaker's ethos, and they invited their customers to meet the winemaker. Conveying the history and story of the winery and region during the cellar door experience is important for development of brand loyalty (Frost et al. 2020). Finding the right balance of information to be conveyed to the consumer during the cellar door experience is important. The highly wine-involved consumer might enjoy learning about the winemaking processes or be happier holding the floor and delight in showcasing their wine knowledge. However, there are limits to the amount of information we can encode for later recollection.

Mental workload

The cellar door is a complex multisensory environment. Humans can encode complex information across multiple sensory channels, though there are limits. Wine is a unique beverage that requires moderately high processing resources to experience. The consumer needs to attend to taste, smell, look, mouthfeel, and extrinsic factors such as price. With mental workload already moderately high, the tasting room environment can play an important role in encoding and recall of key information. The overwhelming environment in Case Study Study 36.1 interfered with the encoding and recall of wine information. This can lead to lower confidence and lower purchases. Case Study Study 36.2 described a scenario that allowed for an optimal recall. Given that the cellar door is principally aimed to sell wine, careful consideration should be given to the impact of mental workload in the tasting setting.

Attempts have been made to characterize wine consumer preferences in terms of levels of experience. There is evidence that macro-categories such as novice, expert, and winemaker have within-group similarities and between-group differences for taste preferences. The relationship is not linear, with novices and experts similar for sweet taste preference and winemakers being almost an outlier (Blackman, Saliba, and Schmidtke 2010). This can be explained by their role; an expert might research wine and consume wine regularly, but they cannot be as *involved* with wine as a winemaker, who is surrounded by wine the entire day.

Research has demonstrated that wine involvement is a useful construct that accounts for the full continuum of levels of experience. Moreover, for extrinsic factors such as price and how information about wine is used, wine involvement has been shown to be a measure that better differentiates wine consumers (Bruwer et al. 2014). At the cellar door, wine experience can be more easily assessed than wine involvement. Individual differences in the level of experience are an important consideration with respect to mental workload. A novice can easily exceed processing resources just by tasting wine. A novice is therefore likely to benefit more from Case Study Study 36.2, where processing resources were focused on the wine tasting experience, with some brand promotion. An expert would recall more than a novice in Case Study Study 36.1.

Wine involvement and past taste preference work show us that those working in the wine industry often characterize wine differently to everyday consumers. Often, the wine cellar door and tasting room are designed by those who work in the wine industry – by definition, those who are highly involved with wine. To those of us so involved in the wine industry, it is easy to forget what it was like to first drink wine. Processing resources are easily filled with the concentration required to attend to so many variables. A more experienced taster is able to use template matching and other cognitive 'shortcuts' which require lower levels of processing resources.

The optimal cellar door is one that considers multiple levels of experience. Some cellar doors have a traditional tasting room for more advanced wine consumers, along with break-out rooms that host regular guided tasting sessions. These sessions provide step-by-step guidance for novices, allowing them to become more comfortable within the limits of their wine tasting skills and cellar door experiences. These sessions are not purely to educate but are also an acceptance that forcing novices into a traditional cellar door tasting experience could easily overwhelm and discourage a future enduring customer. Consider, for instance, if Case Study Study 36.1 had break-out rooms aimed for novices. Provided staff successfully filtered consumers to the correct situation, optimal emotional connection and recall could have been achieved. Perhaps only the largest wineries could develop such a sophisticated process.

The future – virtual reality?

Virtual reality (VR) equipment is becoming more accessible to consumers. There has been hastened interest in VR and tourism during COVID-19 lockdowns around the world. While isolated to the home environment, people have been able to participate in tourism through video and more immersive technologies. Wen and Leung (2021) compared traditional and VR video of winery tours and found that VR wine tours elicited higher intentions to purchase and willingness to pay. The elicitation was strongest for experienced consumers, suggesting that a frame of reference assisted the emersion of the VR cues. Immersive technologies such as VR could also be used at cellar door, to expose interested consumers to the winemaking process. However, it is in-home where VR offers the greatest promise. Individuals who prefer not to tour or find it difficult to tour through disability, for instance, can experience immersive wine tourism through VR. Wine samples can be sent to virtual tourists as part of the activity. Future research needs to establish how traditional cellar door sales research relates to in-home VR experiences. As the availability of various forms of VR become more commonplace, this is an area that will certainly experience growth in opportunities for the wine tourist and winery owner.

Post-cellar door experience

The principal role of the cellar door is usually to sell wine and make money. The amount of money made at the cellar door far exceeds that made in a retail setting. This chapter has outlined several things that wineries can implement to enhance cellar door sales. A secondary role of the cellar door is to increase brand awareness and create a cellar door multiplier – to generate sales after the cellar door visit. This can be achieved in two ways: (1) the visitor can enjoy the wine experience and recall enough detail to purchase wine, and/or (2) the visitor can make a recommendation to others that leads to a purchase. Both types of wine sales are valuable, but the latter is especially important, as recommendations to others can lead to further recommendations. The cellar door experience can become a powerful marketing tool, not just a point of sale. Moreover, research has shown that wine tourists can change their ongoing wine purchasing after the cellar door experience (e.g., increase price point, buy more wine from the region visited; Bruwer et al. 2015). The same researchers have shown that visitors who buy more wine at the cellar door are more likely to buy more three months post visit. Wine purchases during the cellar door experience are maximized when visitors have a clear recollection of the wines they liked. Maximizing cellar door sales seems to also increase the potential for sales months later. There are gender and age differences in wine purchasing behavior during wine tourism, though this is out of the scope of the current chapter (for a review, see Bruwer, Saliba, and Miller 2011).

Conclusion

As wine tourists search for new experiences, they embark on a complex multisensory journey, generally motivated by the pursuit of pleasurable experiences. In order to understand the influence of sensory overload on the behavior of wine tourists, this chapter focused on the cellar door experience. Understanding the impact of multisensory load during the cellar door experience provides insight into buying behavior, post-visit recollection, future revisit intentions, and pre-visit expectations of future wine tourism escapades.

According to the VACP model, senses load relatively independently. Separate loading allows us to watch wine being poured into a glass, noting its sound and color, while listening to the accompanying explanation of winemaking techniques and flavor profile. However, increased loading of multiple sensory channels can overload our limited processing resources, in turn affecting our ability to encode and therefore recall the input with clarity. The two case studies presented in this chapter demonstrated how different multisensory environments, external to the sensory involvement of wine tasting, in a cellar door experience can impact on the ability to correctly recall judgments, and also how that uncertainty impacts buying behavior.

While taste preference of wine can be influenced by experience and expectations, the context of wine consumption is a powerful influencer. The contextual characteristics in the comparative case studies in this chapter demonstrated the extremes of cellar door experience: one overwhelming and large in every sense, the other small, simple, and sparce. The first scenario presented an overwhelming sensory experience and provided opportunity for greater emotional connection at every turn. The second simpler presentation failed to create an emotionally engaging ambiance. However, such a void can be filled through the co-creation of emotional connection between staff and customers. The first scenario risked losing that connection amongst sensory clutter. The second provided an instant sales connection which faded with time. It is a fine balancing act that, when perfected, creates an easy space where staff and customers can enjoy their co-created hedonic experience and encode relevant information for future retrieval and reference.

During the COVID-19 lockdowns, some wineries re-created the experience online, couriering wine samples to enjoy while watching a video or when joining others in a zoomed experience from a lonely cellar door. Recent advances in VR experience will open wine tourism to those unable to venture to their favorite wine regions due to restrictions imposed by budget, travel, or disability. As the main role of a cellar door is to deliver sales, providing a pleasurable hedonic experience to those unable to attend the cellar door in person opens up new sales opportunities. Future research could explore the connection of visual attention and the co-creation of emotional attachments, which build brand loyalty and foster the relationships required with enduring customers. Incorporating the findings of current and future research into training for cellar door and winery staff will assist in finding the multisensory balance which provides customer engagement, emotional connection, clear recollection of judgment, and stories of an enjoyable experience. Understanding the influence of the multisensory environment enhances the experience, which increases sales, heightens revisit intentions, and boosts word-of-mouth recommendations.

Implications

There is a paucity of research on multisensory experience in wine tourism, despite being a major form of tourism and revenue base for some wine businesses. In the absence of applied research into the cellar door context, this chapter used a theoretical model to understand the implications of the cellar door environment. It suggests that cellar doors need to be carefully planned, and that the consequence of getting it wrong is that the experience is not remembered. Memorability is a major factor in the cellar door multiplier – where visitors buy and promote wine that they experience at the cellar door. Cellar door sales can lead to profits for the wine business, but the cellar door multiplier offers a greater benefit in both total sales and the number of consumers generated.

More research is needed on the multisensory experience in wine tourism, in particular how cellar door staff mediate environment and sales. The case studies used in this chapter highlight how good staff can facilitate encoding of key information, thereby increasing the likelihood of post-cellar door sales and promotion. In fact, cellar door staff are likely to be one of the most important factors in sales at the cellar door. Paradoxically, there is no formalized training, job title, or career path in most countries for cellar door staff.

References

Amerine, M. A., Pangborn, R. M., and Roessler, E. B. (1965). *Principles of Sensory Evaluation of Food.* New York: Academic Press.

Bastian, S. E. P., Payne, C. M., Perrenoud, B., and Joscelyne, V. L. (2009). Comparisons between Australian consumers' and industry experts' perceptions of ideal wine and cheese combinations. *Australian Journal of Grape and Wine Research*, 15(2): 175–184. https://doi.org/10.1111/j.1755-0238.2008.00043.x

Blackman, J. B., Saliba, A. J., and Schmidtke, L. M. (2010). Sweetness acceptance of novices, experienced consumers and winemakers in Hunter Valley Semillon wines. *Food Quality and Preference*, 21(7): 679–683. https://doi.org/10.1016/j.foodqual.2010.05.001

Bruwer, J., Burrows, N., Chaumont, S., Li, E., and Saliba, A. J. (2014). Consumer involvement and associated behaviour in the UK high-end retail off-trade wine market. *The International Review of Retail, Distribution and Consumer Research*, 24(2): 145–165. https://doi.org/10.1080/09593969.2013.839464

Bruwer, J., Lockshin, L., Saliba, A., and Hirche, M. (2015). Cellar door, trial purchase-repurchase of the brand: how does a cellar door visit impact future sales? *Wine and Viticulture Journal*, 30(1): 56–59. https://doi.org/10.3316/informit.921130750289321

Bruwer, J., Saliba, A. J., and Miller, B. (2011). Consumer behaviour and sensory preference differences: implications for wine product marketing. *Journal of Consumer Marketing*, 28(1): 5–18. https://doi.org/10.1108/07363761111101903

Cooper, A. (1999). *The Inmates Are Running the Asylum.* Indianapolis, IN: Macmillan Publishing.

Frost, W., Frost, J., Strickland, P., and Maguire, J. S. (2020). Seeking a competitive advantage in wine tourism: heritage and storytelling at the cellar-door. *International Journal of Hospitality Management* 87: 102460. https://doi.org/10.1016/j.ijhm.2020.102460

Galloway, G., Mitchell, R., Getz, D., Crouch, G., and Ong, B. (2008). Sensation seeking and the prediction of attitudes and behaviours of wine tourists. *Tourism Management*, 29(5): 950–966. https://doi.org/10.1016/j.tourman.2007.11.006

McCracken, J. H., and Aldrich, T. B. (1984). *Analysis of Selected LHX Mission Functions Implications for Operator Workload and System Automation Goals.* Alabama: Anacapa Sciences, INC. Military Programs.

Ragin, C. (1987). *The Comparative Method.* Berkeley: University of California Press.

Saliba, A. J., Wragg, K., and Richardson, P. (2009). Sweet taste preference and personality traits using a white wine. *Food Quality and Preference*, 20(8): 572–575. https://doi.org/10.1016/j.foodqual.2009.05.009

Salthouse T. A. (1988). The role of processing resources in cognitive aging. In Howe M. L., and Brainerd, C. J. (eds.), *Cognitive Development in Adulthood. Springer Series in Cognitive Development* (pp. 185–239). New York: Springer. https://doi.org/10.1007/978-1-4612-3852-2_7

Spence, C. (2020). Wine psychology: basic and applied. *Cognitive Research: Principles and Implications*, 5: 22. https://doi.org/10.1186/s41235-020-00225-6

Spence, C., Burzynska, J., Wang, Q., and Bastian, S. E. P. (2019). Taste the bass: low frequencies increase the perception of body and aromatic intensity in red wine. *Multisensory Research*, 32(4–5): 429–454. https://doi.org/10.1163/22134808-20191406

Spence, C., and Wang, Q. J. (2015a). Wine & music (I): on the crossmodal matching of wine & music. *Flavour*, 4(1): 1–14. https://doi.org/10.1186/s13411-015-0045-x

Spence, C., and Wang, Q. J. (2015b). Wine & music (II): can you taste the music? Modulating the experience of wine through music and sound. *Flavour*, 4(1): 1–14. https://doi.org/10.1186/s13411-015-0043-z

Spence, C., and Wang, Q. J. (2015c). Wine & music (III): so what if music influences taste? *Flavour*, 4, 36. https://doi.org/10.1186/s13411-015-0046-9

Stanier Black-Five (Burzynska, J.). (2018). *Assessing Oenosthesia: Sound and Wine.* March 27. http://joburzynska.com/assessing-oenosthesia-blending-sound-and-wine/

Van Doorn, G. H., Wuillemin, D., and Spence, C. (2014). Does the colour of the mug influence the taste of the coffee? *Flavour*, 3(1): 1. https://doi.org/10.1186/2044-7248-3-10

Wen, H., and Leung, X. Y. (2021). Virtual wine tours and wine tasting: the influence of offline and online embodiment integration on wine purchase decisions. *Tourism Management*, 83(2021): 104250. https://doi.org/10.1016/j.tourman.2020.104250

Yoo, Y. J., Saliba, A. J., MacDonald, J. B., and Prenzler, P. D. (2013). A cross-cultural study of wine consumers with respect to health benefits of wine. *Food Quality and Preference*, 28(2): 531–538. https://doi.org/10.1016/j.foodqual.2013.01.001

Yoo, Y. J., Saliba, A. J., Prenzler, P. D., and Ryan, D. (2012). Total phenolic content, antioxidant activity, and cross-cultural consumer rejection threshold in white and red wines functionally enhanced with catechin-rich extracts. *Journal of Agriculture and Food Chemistry*, 60(1): 388–393. https://doi.org/10.1021/jf203216z

37

THE INFLUENCE OF WINE INVOLVEMENT ON WINE TOURISM MOTIVATIONS AND EXPERIENCES

Joanna Fountain

Introduction

To date, there has been relatively limited research on the impact of involvement on motivations to engage in wine tourism, and specifically wine festivals, or the benefits sought from such experiences. This information is important for providers, as understanding tourists' motivations can better ensure that products and experiences are crafted to meet their needs.

This chapter begins with a brief overview of the wine tourism experience, the concept of involvement, and its role in understanding engagement with wine and wine tourism. Previous research that has explored involvement and other forms of psychographic segmentation in the context of wine tourism is reviewed to form a context for analysis of the role of wine involvement in motivations to engage in wine tourism via a wine and food festival. Results confirm previous research that there are significantly different motives for attending festivals based on levels of wine involvement. In particular, low-involvement participants are primarily interested in socializing with friends and family, relaxing, and enjoying the setting and the atmosphere. By contrast, the wine product is central for high wine involvement– trying it, buying it, talking about it with winemakers – but this does not mean the festive and social atmosphere is unimportant for this group. The chapter concludes by discussing the finding's implications for wine tourism product development and future research.

The wine tourism experience

The wine tourism experience is often treated as synonymous with cellar-door (also known as tasting-room) visitation; however, the most cited definition of wine tourism acknowledges that the phenomenon includes 'visitation to vineyards, wineries, wine festivals and wine shows for which grape wine tasting and/or experiencing the attributes of a grape wine region,' where these experiences are 'the prime motivating factors for visitors' (Hall et al. 2000, p. 3). This chapter assumes this broader view of the scope of wine tourism experiences, whilst also questioning the necessary centrality of 'wine' as a key motivator in wine tourism experiences (e.g., Fountain 2018). Wine tourism is primarily a hedonic experience, with a focus on personal enjoyment and engagement (e.g., Gill, Byslma, and Ouschan 2007; Pikkemaat et al. 2009;

DOI: 10.4324/9781003143628-43

Roberts and Sparks 2006), and the social and emotional benefits of these experiences may be as important – or even more important – than the experiences with wine (e.g., Bruwer and Alant 2009; Carmichael 2005; Carlsen and Boksberger 2015; Charters, Fountain, and Fish, 2009; Chen et al. 2016; Getz and Brown 2006). For example, the opportunity to relax and socialize with friends and family are often cited as key benefits sought from wine tourism (e.g., Byrd et al. 2016; Lee and McCole 2016; Roberts and Sparks 2006), and the winescape, incorporating the built setting of the winery and the regional surroundings, also significantly contributes to enjoyable wine tourism experiences (e.g., Bruwer and Lesschaeve 2012; Carlsen and Boksberger 2015; Carmichael 2005; Fountain and Thompson 2019; Gunasekar et al. 2021; Pikkemaat et al. 2009; Quadri-Felitti and Fiore 2016). Encounters with hospitality and service staff at winery cellar doors and wine festivals are also critical, providing both formal and informal opportunities to learn about wine and wine culture (Byrd et al. 2016; Carlsen and Boksberger 2015; Charters et al. 2009; Dodd 1995; Fountain and Charters 2010; O'Neill, Palmer, and Charters 2002; Quadri-Felitti and Fiore 2016). Research on wine festival motivation generally reveals a similar broad range of motivations, including opportunities to relax and escape, specific focus or product of the event (in this case wine), socialization with friends and family, and entertainment (Park, Reisinger, and Kang 2008; Tanford, Montgomery, and Hertzman 2012; Yuan et al. 2008). The findings echo those of more general festival studies (e.g., Crompton and McKay 1997; Li and Petrick 2006).

A focus on the experiential nature of wine tourism means acknowledging that it is constituted out of an amalgam of various elements. The aesthetic surroundings (natural and built), the cellar door or other service personnel, companions and other visitors, and the visitor's relationship with the product (wine) all contribute to the overall experience (Charters, Fountain, and Fish 2009; see also Bruwer and Alant 2009; Fountain and Charters 2010; Gill, Byslma, and Ouschan 2007; Pikkemaat et al. 2009). The nature and quality of a wine tourism experience is central to the satisfaction and emotional attachment that visitors may develop for a brand or an entire wine region (e.g., Bruwer and Alant 2009; Charters, Fountain, and Fish 2009; Chen et al. 2016; Nowak and Newton 2006). Therefore understanding what wine tourists seek from their wine tourism experience is critical.

Wine involvement and the wine tourism experience

While there are multiple elements and types of wine tourism experiences, wine tourists also differ on the basis of sociodemographic, cultural, psychological, and behavioral characteristics. Consumers' varying wine interests, cultural or educational backgrounds, age, and previous wine tourism experiences mean that while participating in the same activity, at the same time, their expectations, experience, and satisfaction with an encounter will differ (Kim, Ritchie, and McCormick 2012; Ooi 2005). For example, Bruwer and Alant (2009) report that first-time winery visitors seek novel experiences and to learn more about wine, while repeat visitors are primarily there to purchase wine. First-time visitors are also likely to be generalists, adding a winery visit as a component of an overall holiday experience, while for repeat visitors, the wine is a central motivation. This finding supports that of Charters and Ali-Knight (2002), who found novice wine consumers are interested in the whole experience of relaxing, socializing with friends, and enjoying the setting, rather than wine-specific experiences (see also Brown, Havitz, and Getz 2007).

As is the case at winery cellar doors, wine festival attendees are not homogeneous (Yuan et al. 2008) and come to events with varied characteristics and motivations (Houghton 2001; Park, Reisinger, and Kang 2008). There is some evidence that wine festivals attract younger, less experienced wine consumers who seek to learn about wine in a more social and less intimidating environment than at a wine class (Dodd et al. 2006; Fountain 2014; Houghton 2001). Some researchers suggest that the focus on fun and entertainment and 'party atmosphere' at wine festivals deters serious wine consumers (Houghton 2008). Yet even in these contexts, there is great variation in wine festival attendees, as is apparent in Yuan et al.'s (2005) observations about the motives of festival participants, who attend 'for the satisfaction of their interest in wine and/or for the entertainment made available by other leisure activities' (p. 39; see also Fountain 2014).

Wine tourists have been described primarily in demographic and behavioral terms (Brown, Havitz, and Getz 2007); however, research is increasingly utilizing elements of psychographic segmentation, recognizing that it may have greater ability to profile market segments than segmenting on sociodemographic characteristics alone. Wine consumers, and by extension wine tourists, differ in their attitudes to wine and the role of wine in their lifestyle and consumption patterns (Bruwer, Li, and Reid 2002). To date, wine segmentation research has explored the influence of lifestyles (Bruwer and Li 2007; Thach and Olsen 2006), personality characteristics (Galloway et al. 2008), self-congruity or self-image (Pratt and Sparks, 2014), and wine interest and knowledge on wine consumption and wine tourism behavior (Charters and Ali-Knight, 2002; Marzo-Navarro and Pedraja-Iglesias, 2010). Two of the first researchers to explore this approach, Charters and Ali-Knight (2002), argue that visitor's interest in wine, *combined with* wine knowledge, is a better predictor of the purpose of a winery visit than knowledge alone. Thus, they conclude that visitors who are 'wine lovers' seek new wine experiences and to improve their knowledge at a winery; wine is central (see also Brown, Havitz, and Getz 2007). Fountain and Charters (2010) assert that younger consumers are more interested in wine education experiences than Baby Boomers, suggesting that it may be interest in wine, rather than knowledge itself, that determines the benefits sought.

A concept that has received substantial attention in wine segmentation research in recent years is wine involvement. Product involvement has been defined as 'the degree of personal relevance to a product' (Zaichkowsky 1985, p. 342), and relates to the idea that 'certain product classes may be more or less central to an individual's life, their attributes about themselves, their sense of identity and their relationship with the rest of the world' (Bruwer et al. 2014, p. 147). Being highly involved with a product, in this case, wine, is also about interest in, and engagement with the product on a regular basis and relates to one's propensity to enjoy activities related to wine and wine consumption. This interest may manifest in spending time reading about wine, and talking about wine with friends or wine professionals (Lockshin and Spawton 2001; Pratt and Sparks 2014; Yuan et al. 2008). High involvement in wine relates also to a sense of excitement about the product. In other words, wine is an important part of the consumer's lifestyle and sense of identity (Yuan et al. 2008) and affects the way they engage with wine and wine tourism activities (Charters and Ali-Knight 2002; Gross and Brown 2008; Sparks 2007).

Wine involvement has been studied extensively in relation to wine consumer behavior, particularly around wine-buying behavior, with great variations in involvement found amongst wine consumers, suggesting that wine is the 'type of product with which consumers may become totally immersed or which may leave them relatively ambivalent' (Barber, Almanza, and Dodd 2008, p. 34; see also Hollebeek et al. 2007; Ogbeide and Bruwer 2013; Quester and Smart 1996; Yuan et al. 2005). Lockshin and colleagues were some of the first to apply the concept of involvement specifically to wine segmentation. They identified two broad groups of wine consumers. The highly involved are wine enthusiasts who view wine as an important part of their lifestyle and spend leisure time reading about and conversing with others about wine. By comparison, the low-involvement wine consumers view wine as more about fun and socializing in a relaxed setting (e.g., Lockshin, Quester, and Spawton 2001; Lockshin and Spawton 2001; Lockshin, Spawton, and Macintosh 1997).

Involvement has been one of the most widely used psychographic variables in tourism market segmentation and is increasingly used in wine tourism research. The assumption behind it is that one's level of wine involvement may influence the wine tourism motivations, visit intentions, and behavior at winery cellar doors, in wine regions (Brown, Havitz, and Getz 2007; Lockshin and Spawton 2001; Nella and Christou 2014; Pratt and Sparks 2014) and at wine festivals (Fountain and Ryan 2016; Gu et al. 2020; Yuan et al. 2008). One of the key findings of this body of research is that wine tourists differ considerably in their wine involvement levels. For example, Brown, Havitz, and Getz (2007) identified five types of wine consumers based on involvement, with each cohort having a different foci and level of interest in wine tourism experiences (see also Gu et al. 2018). As the level of wine involvement increased, so too did nearly all wine-related purchase and consumption behaviors, but this was less predictable of wine-related travel contexts compared with daily wine consumption contexts (see also Pratt and Sparks 2014). This research also highlights the range of motives for wine tourism participation. For example, Nella and Christou (2014) report that the most highly wine-involved segment had the highest proportion of people visiting wineries for wine tasting and wine buying, and to acquire wine information and to meet the producer. By comparison, the low-involvement segment had the highest percentage of respondents visiting to experience the atmosphere of a winery and to participate in a group visit (see also Gu et al. 2020). In contrast, in a study of wine festival attendees, Yuan et al. (2008) report that low- and medium-involvement groups were significantly *more* interested in the motivational factors 'festival and escape,' 'socialization,' and, surprisingly, 'wine' to explain wine festival attendance than high-involvement respondents.

Not only do the more highly involved wine consumers have different motivations, but they also have different emotional or affective connections to the wine tourism experience. Pratt and Sparks (2014) report that high-involvement wine consumers reported high levels of self-congruity with wine tourism. High-involvement wine consumers also had a stronger affective destination image of wine regions. Gu et al.'s (2018) findings support this; high-involvement participants generally express more positive emotional attitudes towards wine tourism. There is also some evidence amongst highly wine-involved Chinese tourists that they will go to considerable effort to experience wine tourism, negotiating a range of constraints brought by language and transport difficulties (Gu et al. 2020).

As has been reported in leisure research elsewhere (Havitz and Dimanche 1997), there is an inconsistent relationship between sociodemographic variables and involvement (Fountain and Ryan 2016; Nella and Christou 2014), meaning there is plenty of scope for further research. Furthermore, to date, most of the wine involvement research has focused on either its relationship to motivation for winery experiences or intention to participate in regional wine tourism, with limited research focused on the influence of involvement on wine festival motivations and benefits sought. The remainder of this chapter reports on research that has explored the impact of wine involvement on motivations in a wine festival context.

Case Study 37.1 Case study, materials, and methods

The Christchurch/South Island Wine and Food Festival is a relatively recent addition to the growing number of wine festivals and events in New Zealand. This festival was first held in December 2011 and occurs on an annual basis. Unlike the many wine festivals which are located in rural regions, this festival is based in the center of the city of Christchurch and features wines from all South Island wine regions. While wine is undoubtedly central to the festival, this is supplemented by a strong program of musical entertainment and a range of food options. There is a significant program of free educational events also, which are generally oversubscribed (Christchurch/South Island Wine and Food Festival 2020). The price of tickets to the event range from NZ$30 to NZ$150, with general admission tickets costing around the NZ$40 mark. This entry fee entitles ticket holders to entry and access to the musical performances and seminars, but wine tasting incurs an extra cost. Children attending the event with an adult received free entry. With a total capacity of 8,000, all tickets generally sell out before the day of the event.

This research is based on an intercept survey conducted at the South Island Wine and Food Festival. The survey was conducted using convenience sampling, using a 'next to pass' procedure of festival visitors aged 18 and over. The final sample included 249 wine festival participants. The survey addressed a range of questions about wine consumption behavior, information sources and decision making, travel behavior, and sociodemographic characteristics, but the focus of this chapter specifically explores the influence of wine involvement on wine festival motivations.

To date, there is no universally accepted measurement of wine involvement, but most researchers have used a modification of either Zaichkowsky's (1985) Product Involvement Index (PII; e.g., Gu et al. 2018; Yuan et al. 2008) or Laurent and Kapferer's (1985) Consumer Involvement Profile (CIP). This latter measurement identifies five facets of involvement: *interest*, which refers to the centrality of a product (in this case, wine) in one's life; *pleasure*, or the extent of hedonic involvement or rewards; *sign*, the perceived symbolic value of the product class; *risk importance*, referring to the perceived negative consequences of a poor purchase decision; and *risk probability*, the subjective assessment of the probability of making a poor purchase. Brown, Havitz, and Getz (2007) modified the CIP to create a Wine Involvement Profile (WIP), which itself has been adapted by other researchers (e.g., Gu et al. 2020; Pratt and Sparks 2014) and was the basis for the current scale.

The foundation of the scale used by Brown, Havitz, and Getz (2007) are seven items from Laurent and Kapferer's scale, to which they added wine-specific items after focus group discussion. Their final scale incorporated three factors, which they labeled expertise (nine items), enjoyment (eight items), and symbolic centrality (five items), which were measured using a 7-point Likert scale. A number of these items loaded onto more than one factor. The current research incorporated 11 of the 15 items used by Brown, Havitz, and Getz (2007), reducing the number of items to avoid respondent fatigue. The wording of a few of these statements was modified slightly. For example, the item relating to interest in leisure travel to wine regions was altered from 'My interest in wine makes me want to visit wine regions' to 'I make an effort to visit wine regions when traveling.' One statement was added to explore the behavioral element of high wine involvement in daily life: 'I am a member of a wine club or wine-related organization.'

In order to assess the dimensionality of the involvement scale applied in this study, an exploratory factor analysis was performed on the data. A single factor solution was reached, explaining 55.25% of the total variance. The reliability test gave a Cronbach's alpha score of 0.922. The Kaiser-Meyer-Olkin measure of sampling adequacy was 0.937. It was therefore considered appropriate to combine all items to form a total involvement score for current purposes. Involvement levels for respondents were determined by adding the scores for the 12 involvement statements for a sum involvement score. From this total, the most involved quartile was identified and labeled 'high involvement,' and the least involved quartile labeled 'low involvement.' The remainder of the sample was labeled 'moderate involvement.' Cross tabulations were used to describe the basic characteristics of the three segments, and analysis of variance (ANOVA) tests were run as appropriate to detect possible differences between the involvement groups in regard to their motivations to attend this wine festival.

Findings

Sociodemographic characteristics

The sociodemographic characteristics of the sample are presented in Table 37.1. Females outnumbered males at a ratio of two to one amongst festival respondents. This high proportion of females is similar to that found in previous studies in New Zealand (Fountain 2014; Nicholson and Pearce 2000) and globally (Park, Reisinger, and Kang 2008; Weiler, Truong, and Griffiths 2004; Yuan et al. 2008), and reflects the fact that females represent a higher proportion of wine consumers than males (Kolyesnikova, Dodd, and Duhan 2008; Pettigrew 2003). The age distribution was bimodal, peaking in the 26–35 year and 46–55 year groups. As reported by Fountain (2014), this festival seems to attract an older age profile than many wine festivals (Houghton 2008; Nicholson and Pearce 2000; Park, Reisinger, and Kang 2008), which may reflect the family-friendly environment (children gain free entry). The educational food and wine seminars may attract an older audience than other wine festivals. Furthermore, the price

Table 37.1 Demographic profile of wine festival visitors

Demographic Variable	%
Gender	
Female	67.5
Male	32.5
Age	
18 to 25 years	9.4
26 to 35 years	31.1
36 to 45 years	18.9
46 to 55 years	29.1
56 to 65 years	10.7
Over 65 years	0.8
Highest qualification	
No formal qualifications	3.7
High school qualification	23.1
Trade certificate	26.4
Degree	21.1
Postgraduate degree	16.5
Other tertiary qualification	3.6

of entry (excluding wine tasting) is relatively high, perhaps deterring younger people from attending. In relation to educational qualifications, 37.6% of the respondents had a university degree or postgraduate degree, while a quarter of respondents had a trade qualification (26.4%) and a quarter had a high school qualification or no formal qualifications, reflecting previous research (Houghton 2008; Park, Reisinger, and Kang 2008; Yuan et al. 2008). This spread of educational qualifications is reflected in the personal incomes reported.

Wine involvement of respondents

As outlined above, involvement levels for respondents were determined by adding the scores for the 12 involvement statements for a sum involvement score. The minimum score possible (that is, least wine involved) was 12, with a maximum score possible of 84. The low-involvement segment was defined as those scoring in the first quartile of the distribution score range (12–37), while the high-involvement segment consisted of those from the fourth (upper) quartile of distribution (58–84). The middle 50% formed the moderate-involvement group. The mean score for the sample was 47.73 (SD = 15.1699). The reason for using the top and bottom quartile of wine involvement samples for high- and low-involvement groupings was to ensure there were two groups that were bimodal and separate.

Analysis of variance (ANOVA) revealed that education [$F(2, 233) = 3.955$, $p = 0.020$] and age [$F(2, 235) = 5.370$, $p = 0.005$] had a significant correlation with involvement, but gender did not. There were significant differences in frequency of wine consumption by wine consumption [$F(2,234) = 53.995$, $p = 0.000$]. The high-wine-involvement segment was much more likely to state that wine was their most frequently consumed alcohol [$F(2, 202) = 27.299$, $p = 0.000$].

Wine involvement and motivations for wine festival attendance

Respondents were presented with a number of statements regarding motivation to attend this wine festival on a 7-point Likert scale (1 = not at all important; 7 = extremely important). These statements were drawn from previous festival research, incorporating generic motives (e.g., opportunities to relax and escape, socialize with friends and family) and more specific wine-related motives (Park, Reisinger, and Kang 2008; Tanford, Montgomery, and Hertzman 2012; Yuan et al. 2008). In the sample overall (see Table 37.2), the highest motivations for attendance were generic, reflecting a desire 'to have fun with my friends' (M = 6.23), 'to enjoy the festival atmosphere' (M = 6.09), and 'to relax and unwind from the stress of everyday life' (M = 5.98). It is interesting to note that enjoying the food at the event was a more important motivation than trying new, or a range of, wines; increasing knowledge about wine (M = 4.42) and buying wine (M = 3.40) were relatively unimportant motives for the sample as a whole. It is also noticeable that the supplementary attractions – the music, the cooking demonstrations – have little appeal for the sample overall.

An analysis of variance (ANOVA) of wine motivation reveals significant differences on many of these statements by wine involvement, particularly those related to specific, wine-related motives (see Table 37.3). A post-hoc Tukey's *b* test shows that each of the three involvement groups differed significantly at the $p < .05$ level for the motives: 'to increase my knowledge about wines,' 'to try new wines,' to buy wine,' 'to experience new wines,' and 'to exchange ideas with winemakers.' The high-involvement segment differed significantly from the other two groups in terms of seeking a novel experience ('to do and see new things') and 'to meet people with similar interests to me.' There were no significant differences on the majority of generic

Table 37.2 Motivations for attending the wine festival

Motivation	*Mean (n = 248)*
Have fun with my friends	6.23
Enjoy a festive atmosphere	6.09
Relax and unwind from the stress of everyday life	5.98
Enjoy interesting and tasty food	5.91
Try new wines	5.67
Experience a range of different wines	5.54
Do and see new things	5.47
Have fun with my family	5.20
Be entertained by musical performances	5.13
Enjoy a change of scenery	5.05
Increase my knowledge about wine	4.42
Have new things to tell my friends about after the event	4.32
Listen to some of my favorite bands	3.96
Meet people with similar interests to me	3.55
Learn about cooking through cooking demonstrations	3.50
Buy wine	3.40
Exchange ideas with winemakers	2.88

Table 37.3 One-Way Analysis of Variance of motivation to attend the wine festival: influence of wine involvement

Source	*df*	*SS*	*MS*	*F*	*p*
To do and see new things					
Between groups	2	15.761	7.880	5.207	.006
Within groups	235	355.634	1.513		
Total	237	371.395			
To increase my knowledge about wine					
Between groups	2	155.733	77.866	35.737	.000
Within groups	239	520.751	2.179		
Total	241	676.483			
To try new wine					
Between groups	2	97.135	48.568	31.916	.000
Within groups	239	363.691	1.522		
Total	241	460.826			
To buy wine					
Between groups	2	109.812	54.906	18.958	.000
Within groups	239	692.188	2.896		
Total	241	802.000			
To experience a range of wines					
Between groups	2	93.806	46.903	31.397	.000
Within groups	239	357.037	1.494		
Total	241	450.843			
To meet people with similar interests to me					
Between groups	2	51.837	25.918	9.478	.000
Within groups	238	650.835	2.735		
Total	240	702.762			
To have new things to tell my friends about after the event					
Between groups	2	36.461	18.230	5.339	.005
Within groups	238	812.643	3.414		
Total	240	849.104			
To exchange ideas with winemakers					
Between groups	2	147.924	73.962	28.527	.000
Within groups	236	611.867	2.593		
Total	238	759.791			

Table 37.4 Motivations to attend the wine festival: high- and low-wine-involvement *t*-test results

	High Involvement (n=67)		*Low Involvement (n=60)*				
	M	*SD*	*M*	*SD*	*df*	*T*	*Sig.*
To try new wines	6.42	0.907	4.76	1.729	128	6.895	.000
To have fun with my friends	6.39	1.014	6.32	1.162	128	.370	ns
Experience a range of different wines	6.28	1.012	4.65	1.667	128	6.699	.000
Enjoy a festive atmosphere	6.24	0.854	6.00	1.016	128	1.454	ns
Enjoy interesting and tasty food	6.18	1.167	5.84	1.345	127	1.530	ns
Relax and unwind from the stress of everyday life	6.08	1.328	5.78	1.054	127	1.407	ns
Do and see new things	5.88	0.993	5.23	1.511	125	2.881	.006
Increase my knowledge about wine	5.54	1.470	3.35	1.567	128	8.213	.000
Be entertained by musical performances	5.40	1.558	4.90	1.593	128	1.801	ns
Enjoy a change of scenery	5.02	1.705	5.00	1.699	126	0.050	ns

motives around socialization (e.g., 'to have fun with my friends,' 'to have fun with my family'), relaxation and escape (e.g. 'to relax and unwind from the stress of everyday life,' 'to enjoy a festive atmosphere'), or entertainment (e.g., 'to be entertained by musical performances').

A *t*-test comparison of the means of the top ten motivations of the high-wine-involvement group compared to the low-wine-involvement group reveals some substantial differences in motivations (Table 37.4). In particular, the high-wine-involvement group ranked two wine-specific motivations in the highest three motives for attendance, whereas socialization and escape-and-relax motivations ranked much more highly for the low-wine-involvement groups. While the most significant differences between the high- and low-involvement groups relate to the focus on extending the former's wine interest, it should be acknowledged that socializing with friends and family, relaxing, and enjoying a festive atmosphere were still very important motives for the high-involvement group. In fact, on all of these measures, the high-involvement group ranked these motives more highly than the low-involvement group.

Discussion and conclusion

This chapter has outlined the concept of involvement as it relates to wine consumption and wine tourism, specifically in the form of a wine and food festival. Supporting previous research, these findings highlight the fact that wine festival attendees differ in their level of engagement with wine and the centrality of wine in their lives and lifestyles. While wine involvement levels was influenced by age and education, there was no significant influence by gender. This latter finding is somewhat surprising, given the fact that females constituted 67.5% of the sample and represent a higher proportion of wine consumers than males (Kolyesnikova, Dodd, and Duhan 2008; Pettigrew 2003). Not surprisingly, high-involvement wine consumers were significantly more likely than low-involvement consumers to report that wine was their alcoholic beverage of choice and consumed wine more frequently.

Varying levels of wine involvement are reflected in differing motivations to attend this wine and food festival. While across the sample as a whole, generic motivations dominated, a finding reported in studies of many other wine festivals (Park, Reisinger, and Kang 2008; Tanford, Montgomery, and Hertzman 2012; Yuan et al. 2008), the high-wine-involved consumers were significantly more likely than respondents with low or moderate involvement to report wine-specific motives as important in their festival attendance motivations.

For those with low wine involvement, this event primarily provided the opportunity to relax, socialize with friends and family, and enjoy the festive atmosphere. While enjoying the food on offer was the third-highest motivation for this group, trying new wines and experiencing a range of different wines were of moderate importance at most. Learning about wine is clearly not a priority at all for those with low wine involvement. This study is not able to fully test the proposition by Dodd et al. (2006) that wine festivals provide less experienced wine consumers with a good opportunity to learn about wine, as levels of wine experience or subjective and objective wine knowledge was not investigated in this research. However, what is apparent is that for low-wine-involvement consumers, learning about wine is not an important benefit sought from a festival of this type.

By comparison, the high-involvement group had a significantly higher motivation to learn about wines, although this ranked as only as the eighth-highest motivation for attending the festival, suggesting that this formal learning may not have been a goal. However, experiential learning – trying new wines and experiencing a range of different wines – represented two of the three highest motivations for this group. Other avenues to increase knowledge are also ranked significantly higher than for the low-involvement group, including the opportunity to exchange ideas with winemakers and meet people with similar interests. Having new things to tell friends about is ranked significantly higher for this group than low-involvement respondents. This may reflect the role of wine as part of a wider lifestyle, which involves talking to friends about wine, but those with moderate involvement were also motivated to attend to have new things to talk to friends about. As this statement didn't specify *what* these new 'things' might be, we do not know if they might be specifically wine-related.

One feature of motivation that is consistent across wine festival attendees of every involvement level is the centrality of the festive atmosphere and opportunities to socialize; and in fact, the high-involvement group ranked these motives more highly than the low-involvement group. This may suggest that the role of wine in their life implicitly makes these types of events more exciting and more of a celebration than for the low-wine-involved, and perhaps contradicts the suggestion that the focus on a festive atmosphere is less likely to attract serious wine consumers (Houghton 2008). However, for the high wine involved, the festive atmosphere and opportunity to relax provides a context in which to extend their wine knowledge and support a wine lifestyle; for low-involvement consumers, this festivity is an end in itself.

There are a number of implications of these findings for wine festival organizers and wine producers considering attending wine festivals as suppliers. First, it would seem that the festive and social atmosphere of a festival is appealing to wine consumers of differing involvement levels. It is clear also that the motives to attend a wine festival differ significantly between segments; therefore, it is important to be attentive to the warning of Nella and Christou (2014, p. 793) that the 'co-existence of wine tourists with totally different motives could destroy the overall wine tourism experience, as these non-compatible visitors interact with each other' (see also Carlsen 2004). Recognizing the diversity of motives might lend itself to thinking about the layout and design of the eventscape of an event. For example, it may be possible to introduce greater demarcation between areas for tasting and drinking wine and spaces for learning about wine and engaging with winemakers. This is managed to some extent with the allocation of tents and marques for workshops, but ensuring that these are further away from the musical performances could help establish 'learning' and 'playing' zones. It may be possible also to demarcate festival attendees through pricing and ticketing strategies, with different types of tickets for those with distinct motives for attendance.

While for event organizers it seems that there is something for all involvement levels in a wine festival of this type, the finding that a substantial proportion of attendees at wine festivals have no interest in wine – in tasting new wines or learning about wine – may be more problematic for wine producers, and the assumption that the festive atmosphere provides a relaxing environment for those with low involvement in wine to get to know the product category – or a winery's specific wine – may be somewhat misplaced. What has not been studied in this project is the relationship between wine knowledge (objective and subjective), wine involvement, and the motivations to experience new wine and learn about wine at wine and food festivals, and there is further scope to investigate these relationships.

References

Barber, N., Almanza, B., and Dodd, T. (2008). Relationship of wine consumers' self-confidence, product involvement, and packaging cues. *Journal of Foodservice Business Research*, 11(1): 45–64, https://doi.org/10.1080/15378020801926692

Brown, G. P., Havitz, M. E., and Getz, D. (2007). Relationship between wine involvement and wine-related travel. *Journal of Travel & Tourism Marketing*, 21(1): 31–46.

Bruwer, J., and Alant, K. (2009). The hedonic nature of wine tourism consumption: an experiential view. *The Australian and New Zealand Grapegrower and Winemaker*, 503: 50–55.

Bruwer, J., Burrows, N., Chaumont, S., Li, E., and Saliba, A. (2014). Consumer involvement and associated behaviour in the UK high-end retail off-trade wine market. *The International Review of Retail, Distribution and Consumer Research*, 24(2): 145–165.

Bruwer, J., and Lesschaeve, I. (2012). Wine tourists' destination region brand image perception and antecedents: conceptualization of a winescape framework. *Journal of Travel & Tourism Marketing*, 29(7): 611–628.

Bruwer, J., and Li, E. (2007). Wine-Related Lifestyle (WRL) market segmentation: demographic and behavioural factors. *Journal of Wine Research*, 18(1): 19–34.

Bruwer, J., Li, E., and Reid, M. (2002). Segmentation of the Australian wine market using a wine-related lifestyle approach. *Journal of Wine Research*, 13(3): 217–242.

Byrd, E. T., Canziani, B., Hsieh, Y., Debbage, K., and Sonmez, S. (2016). Wine tourism: motivating visitors through core and supplementary services. *Tourism Management*, 52: 19–29. https://doi.org/10.1016/j.tourman.2015.06.009

Carlsen, J. (2004). A review of global wine tourism research. *Journal of Wine Research*, 15(1): 5–13.

Carlsen, J., and Boksberger, P. (2015). Enhancing consumer value in wine tourism. *Journal of Hospitality and Tourism Research*, 39(1): 132–144. https://doi.org/10.1177/1096348012471379

Carmichael, B. (2005). Understanding the wine tourism experience for winery visitors in the Niagara region, Ontario, Canada. *Tourism Geographies*, 7(2): 185–204.

Charters, S., and Ali-Knight, J. (2002). Who is the wine tourist? *Tourism Management*, 23(3): 311–319.

Charters, S., Fountain, J., and Fish, N. (2009). 'You felt like lingering …' Experiencing 'real' service at the winery tasting room. *Journal of Travel Research*, 48(1): 122–134.

Chen, X., Goodman, S., Bruwer J., and Cohen, J. (2016). Beyond better wine: the impact of experiential and monetary value on wine tourists' loyalty intentions. *Asia Pacific Journal of Tourism Research*, 21(2): 172–192. https://doi.org/10.1080/10941665.2015.1029955

Crompton, J. L., and McKay, S. L. (1997). Motives of visitors attending festival events. *Annals of Tourism Research*, 24(2), 425–439.

Dodd, T. (1995). Opportunities and pitfalls of tourism in a developing wine industry. *International Journal of Wine Marketing*, 7(1): 5–16.

Dodd, T., Yuan, J., Adams, C., and Kolyesnikova, N. (2006). Motivations of young people for visiting wine festivals. *Event Management*, 10(1): 23–33.

Fountain, J. (2014). *Who's Here? An Exploratory Study of the Characteristics and Wine Consumption Behaviours of Visitors at a New Zealand Wine Festival.* Paper presented at the *8th Academy of Wine Business Research Conference*. June 27–30, Geisenheim, Germany.

Fountain, J. (2018). The wine tourism experience in New Zealand: an investigation of Chinese visitors' interest and engagement. *Tourism Review International*, 22(1): 67–79. https://doi.org/10.3727/154427218X15202734130459

Fountain, J., and Charters, S. (2010). Generation Y as wine tourists: their expectations and experiences at the winery cellar door. In Benckendorff, P., Moscardo, P. G., and Pendergast, D. (eds.), *Tourism and Generation Y* (pp. 47–57). Oxford: CABI.

Fountain, J., and Thompson, C. (2019). Wine tourists' perception of winescape in Central Otago, New Zealand. In Sigala, M., and Robinson, R. N. S. (eds.), *Wine Tourism Destination Management and Marketing: Theory and Cases* (pp. 29–44). London: Palgrave Macmillan.

Fountain, J. M., and Ryan, G. J. (2016). Motivations to attend a New Zealand wine and food festival: the role of involvement. In Bruwer, J., Lockshin, L., Corsi, A., Cohen, J., and Hirche, M. (eds.), *9th Academy of Wine Business Research: Conference Proceedings* (pp. 563–571). Adelaide, Australia. http://academyofwinebusiness.com/?page_id=1035

Galloway, G., Mitchell, R., Getz, D., Crouch, G., and Ong, B. (2008). Sensation seeking and the prediction of attitudes and behaviours of wine tourism. *Tourism Management*, 29(5): 950–966.

Getz, D., and Brown, G. (2006). Critical success factors for wine tourism regions: a demand analysis. *Tourism Management*, 27(1): 146–158.

Gill, D., Byslma, B., and Ouschan, R. (2007). Customer perceived value in a cellar door visit: the impact on behavioural intentions. *International Journal of Wine Business Research*, 19(4), 257–275.

Gross, M. J., and Brown, G. (2008). An empirical structural model of tourists and places: progressing involvement and place attachment into tourism. *Tourism Management*, 29(6): 1141–1151.

Gu, Q., Qiu, H., King, B. E. M., and Hua, S. (2020). Understanding the wine tourism experience: the roles of facilitators, constraints, and involvement. *Journal of Vacation Marketing*, 26(2): 211–229.

Gu, Q., Zhang, H. Q., King, B., and Huang, S. (2018). Wine tourism involvement: a segmentation of Chinese tourists. *Journal of Travel and Tourism Marketing*, 35(5): 633–648. https://doi.org/10.1080/10548408.2017.1401031

Gunasekar S., Das, P., Dixit, S. K., Mandal, S., and Mehta, S. R. (2021). Wine-experienscape and tourist satisfaction: through the lens of online reviews. *Journal of Foodservice Business Research*. https://doi.org/10.1080/15378020.2021.2006039

Hall, C. M., Johnson, G., Cambourne, B., Macionis, N., Mitchell, R., and Sharples, L. (2000). Wine tourism: an introduction. In Hall, C. M., Sharples, L., Cambourne, B., and Macionis, N. (eds.), *Wine Tourism around the World: Development, Management and Markets* (pp. 1–23). Oxford: Elsevier.

Havitz, M. E., and Dimanche, F. (1997). Leisure involvement revisited: conceptual conundrums and measurement advances. *Journal of Leisure Research*, 29(3): 245–278.

Hollebeek, L. D., Jaeger, S. R., Brodie, R. J., and Balemi, A. (2007). The influence of involvement on purchase intention for new world wine. *Food Quality and Preference*, 18: 1033–1049.

Houghton, M. (2001). The propensity of wine festivals to encourage subsequent winery visitation. *International Journal of Wine Marketing*, 13(3): 32–41.

Houghton, M. (2008). Wine connoisseurs or curious tourists? The participants in wine festivals. In Hall, C. M., and Sharples, L. (eds.), *Food and Wine Festivals and Events around the World: Development, Management and Markets* (pp. 179–194). Oxford: Butterworth Heinemann.

Kim, J.-H., Ritchie, J. R. B., and McCormick, B. (2012). Development of a scale to measure memorable tourism experiences. *Journal of Travel Research*, 51(1): 12–25.

Kolyesnikova, N., Dodd, T. H., and Duhan, D. F. (2008). Consumer attitudes towards local wines in an emerging region: a segmentation approach. *International Journal of Wine Business Research*, 20(4): 321–334.

Laurent, G., and Kapferer, J. (1985). Measuring consumer involvement profiles. *Journal of Marketing Research*, *22*, 1–53.

Lee, J., and McCole, D. (2016). *Understanding Winery Visitors: The Relationship among Wine Involvement, Perceived Value, and Winery Loyalty*. Paper presented at Tourism Travel and Research Association: Advancing Tourism Research Globally, June 3. http://scholarworks.umass.edu/cgi/viewcontent.cgi?article=1250&context=ttra

Li, R., and Petrick, J. (2006). A review of festival and event motivation studies. *Event Management*, 9(4): 239–245.

Lockshin, L., Spawton, A. L., and Macintosh, G. (1997). Using product, brand and purchasing involvement for retail segmentation. *Journal of Retailing and Consumer Services*, 4(3): 171–183.

Lockshin, L., and Spawton, T. (2001). Using involvement and brand equity to develop a wine tourism strategy. *International Journal of Wine Marketing*, 13(1): 72–81.

Lockshin, L. S., Quester, P., and Spawton, T. (2001). Segmentaton by involvement or nationality for global retailing: a cross-national comparative study of wine shopping behaviours. *Journal of Wine Research*, 12(3): 223–236.

Marzo-Navarro, M., and Pedraja-Iglesias, M. (2010). Wine tourism development from the perspective of the potential tourist in Spain. *International Journal of Contemporary Hospitality Management*, 21(7): 816–835.

Nella, A., and Christou, E. (2014). Segmenting wine tourists on the basis of involvement with wine. *Journal of Travel & Tourism Marketing*, 31(7): 783–798.

Nicholson, R., and Pearce, D. G. (2000). Who goes to events: a comparative analysis of the profile characteristics of visitors to four South Island events in New Zealand. *Journal of Vacation Marketing*, 6(3): 236–253.

Nowak, L. I., and Newton, S. K. (2006). Using the tasting room experience to create loyal customers. *International Journal of Wine Marketing*, 18(3): 157–165.

O'Neill, M., Palmer, A., and Charters, S. (2002). Wine production as a service experience – the effect of service quality on sales. *Journal of Services Marketing*, 16(4): 342–362.

Ogbeide, O. A., and Bruwer, J. (2013). Enduring involvement with wine: predictive model and measurement. *Journal of Wine Research*, 24(3): 210–226.

Ooi, C. S. (2005). A theory of tourism experiences: the management of attention. In O'Dell, T., and Billing, P. (Eds.). *Experiencescapes: Tourism Culture and Economy* (pp. 51–58). Copenhagen: Copenhagen Business School Press.

Park, K. S., Reisinger, Y., and Kang, H. J. (2008). Visitors' motivation for attending the South Beach Wine and Food Festival, Miami Beach, Florida. *Journal of Travel & Tourism Marketing*, 25(2): 161–181.

Pettigrew, S. (2003). Wine consumption contexts. *International Journal of Wine Marketing*, 15(2): 37–45.

Pikkemaat, B., Peters, M., Boksberger, P., and Secco, M. (2009). The staging of experiences in wine tourism. *Journal of Hospitality Marketing and Management*, 18: 237–253.

Pratt, M. A., and Sparks, B. (2014). Predicting wine tourism intention: destination image and self-congruity. *Journal of Travel & Tourism Marketing*, 31(4): 443–460.

Quadri-Felitti, D., and Fiore, A. M. (2016). Wine tourism suppliers' and visitors' experiential priorities. *International Journal of Contemporary Hospitality Management*, 28(2), 397–417. https://doi.org/10.1108/IJCHM-05-2014-0224

Quester, P., and Smart, J. (1996). Product involvement in consumer wine purchases: its demographic determinants and influence on choice attributes. *International Journal of Wine Marketing*, 8, 37–56.

Roberts, L., and Sparks, B. (2006). Enhancing the wine tourism experience: the customer's viewpoint. In Carlsen, J., & Charters, S. (eds.). *Global Wine Tourism: Research, Management and Marketing* (pp. 141–152). Wallingford, UK: CABI.

Sparks, B. (2007). Planning a wine tourism vacation? Factors that help to predict tourist behavioural intentions. *Tourism Management*, 28(5): 1180–1192.

Tanford, S., Montgomery, R., and Hertzman, J. (2012). Towards a model of wine event loyalty. *Journal of Convention & Event Tourism*, 13(2): 77–99.

Thach, E. C., and Olsen, J. E. (2006). Market segment analysis to target young adult wine drinkers. *Agribusiness*, 22(2): 307–322.

Weiler, B., Truong, M., and Griffiths, M. (2004). *Visitor Profiles and Motivations for Visiting an Australian Wine Festival.* Working paper [19/04], Monash University, Faculty of Business and Economics, April 2004.

Yuan, J., Morrison, A., Cai, L. A., Dodd, T., and Linton, S. (2008). Segmenting wine festival visitors using their personal involvement with wine. In Hall, C. M., and Sharples, L. (eds.), *Food and Wine Festivals and Events around the World: Development, Management and Markets* (pp. 146–165). Oxford: Butterworth Heinemann.

Yuan, J. J., Cai, L. A., Morrison, A. M., and Linton, S. (2005). Segmenting wine festival attendees: a factor cluster approach. *Tourism Review International*, 8(4), 297–309.

Zaichkowsky, J. L. (1985). Measuring the involvement construct. *Journal of Consumer Research*, 12(3): 341–352.

38

WINE TASTING AND APPRECIATION EXPERIENCES AT WINERIES

Ana Brochado, Maite Sena, and Fernando Oliveira

Introduction

Wineries seek to balance wine tourism and sustainable tourism in order to contribute to the preservation of vineyards and surrounding rural areas and the valorization of cellar doors and ancient buildings (Brochado et al. 2019b). Wine tasting rooms are vital to wineries' success (Shapiro and Gómez 2014), as visits to these facilities are a crucial component of wine tourism experiences, playing an important role in developing wineries' competitiveness and long-term sustainability. Gómez, Pratt, and Molina (2019) report that one of the main topics in wine tourism research is winery and wine cellar service quality. Tasting rooms and/or cellar doors are a key component of wine producers' value chain, and providing service quality is an important way to enhance visitors' experiences of wineries' tasting offers as it affects brand awareness, loyalty, and sales (Carlsen 2011).

Wine-cellar studies first focused on wine tourism's most important concrete attributes, such as wine tasting, physical environment, and information. Various authors have highlighted the role of both tangible and intangible aspects of tasting rooms (O'Neill, Palmer, and Charters 2002). More recently, researchers have confirmed that winery visits can be included in the Experience Economy (Charters, Fountain, and Fish 2009). Wine quality and service encounters' efficiency in tasting rooms are thus significant components of winery experiences. In addition, scholars have suggested that visitors might want more than just core tangible benefits (i.e., wine tasting) and look for added value from other elements of winery experiences.

Wineries need to make their mark in varied ways in order to differentiate their offer and provide added value to visitors. Recent studies (Brochado, Stoleriu, and Lupu 2019a; 2019b; Gunasekar et al. 2021; Thanh and Kirova 2018) have underlined the role of web reviews shared online by wine tourists staying in wine hotels and used this content to identify overall wine tourism experiences' main dimensions. The present chapter sought to contribute to the existing knowledge by focusing on the winery and/or wine cellar visits. A case study approach was selected to offer an in-depth analysis of the most common dimensions of winery experiences shared by tourists online and the themes associated with different approaches, namely, each winery's added value.

The research outlined in this chapter, therefore, focused on identifying the key dimensions of the overall winery visit experience. Two research questions were addressed:

 DOI: 10.4324/9781003143628-44

- What are the main winery experience dimensions in narratives shared online by the winery and/or wine cellar visitors?
- How do wineries differentiate their wine tourism products in the relevant markets?

The methodology applied was a dual case study approach (Yin 2018). The present study concentrated on the village of Azeitão, which is part of a broader wine region of Portugal – an Old World wine country – called the Península de Setúbal. This region is famous, among other things, for its Moscatel (Muscat) sweet wine. The peninsula is located on the Atlantic Ocean and between two rivers (i.e., the Tagus and Sado), which creates a diverse climate across the region and has a positive impact on the diversity of wines produced.

José Maria da Fonseca's House Museum offers wine tourism experiences that include a presentation on the company's history, visits to its local *adegas* (wine cellars), Adega da Mata and Adega dos Teares Novos, and wine tastings of a variety of wines. This winery also offers opportunities to pair wines with regional gastronomy and a wine shop in which tourists can buy wine and gourmet regional products. The experiences offered by the Bacalhôa winery in Azeitão include the Wine Cellar Museum, which includes art exhibitions (i.e., African art, Art Nouveau, and Portuguese tiles), wine tastings led by professionals, and the Bacalhôa Palace. The latter component allows visitors to enjoy landscapes incorporating gardens and vineyards and a tour focused on the winery's history, specific architecture, and art. This part of the visit also offers a wine tasting, and all experiences end with the company's wine shop.

This chapter is structured as follows. The next section describes wine tourism experiences and attributes and wine tasting and appreciation at wineries as dimensions of overall experiences. The third section presents the dual case study, which facilitated the identification of Azeitão winery experiences' main dimensions. Content analysis identified not only the main themes in online reviews shared by tourists who visited the two wineries, but also the approaches these companies use to differentiate their products. The final section discusses the findings based on the case study.

Winery experiences

According to Pine and Gilmore (1998, p. 98), '[a]n experience occurs when a company intentionally uses services as the stage, and goods as props, to engage individual customers in a way that creates a memorable event. Commodities are fungible, goods tangible, services intangible, and experiences memorable.' Andresson (2007), in turn, provides a broad definition of a tourism experience as the moment in which both production and consumption come together. Getz and Brown (2006) maintain that wine tourism comprises core wine products, wine regions, and cultural experiences. Getz (2000) also asserts that wine tourism exists only when tourists travel to wine tourism destinations for two main reasons: (1) an interest in the relevant regions' wines, or (2) these regions' specific appeal generated by tourists' imagination.

Wine tourism experiences include, among other elements, tastings, cellar-door sales, and winery tours (Charters and Ali-Knight 2002). Williams (2001) argues that these experiences are not restricted to wine purchases or winery visits, and that wine tourism must be understood as more complete experiences in which tourists can engage. Getz and Brown (2006) suggest that wine tourists want a holistic experience of wine regions that the cited authors call a 'bundle of benefits,' which the wine tourism sector still considers to be its most important feature. Gu et al. (2018, p. 644) state, 'few tourists focus exclusively on the winery experience, with more regarding winery visits as part of a wider mix of attractions.'

Bruwer and Alant (2009) further define wine tourism as a hedonic experience that is based on wine tasting and consumption. This experience is defined as hedonic because '[i]n essence the hedonism value … involves pleasure and sensuous gratification for oneself' (Bruwer and Alant 2009, p. 240). Wine tourism is thus still perceived as experiences that involve hedonic activities rather than just the required daily wine consumption (Gu et al., 2018). Brochado, Stoleriu, and Lupu (2019a) concluded that wine tourism experiences can be viewed as sensorial activities that appeal to visitors' smell, taste, sight, hearing, and touch.

These experiences are a complex research topic 'shaped by complex interactions of visitors and the service staff, the wine product, the cellar door setting, winery management systems, and other winery attributes' (Carlsen and Boksberger 2013, p. 132). Roberts and Sparks (2006) identified eight enhancement factors in wine tourism experiences: authenticity, value for money, service interactions, settings and surroundings, products offered, opportunities to learn, and indulgence and/or lifestyle. Griffin and Loersch (2006) also classified wine tourists' main expectations regarding winery experience attributes as exterior, interior, service, staff, wine, and convenience. Getz and Brown (2006) found that wine tourism's critical success factors are the core wine products and the destinations' appeal, as well as cultural products.

In addition, Bruwer and Alant (2009) report that the primary reasons visitors engage in wine tourism activities are usually directly related to products (i.e., tasting or purchasing opportunities). More specifically, the main underlying motivations for visiting winery cellars are to taste, buy, and learn about wine, experience wineries' atmosphere; and drink unique wines. Martin, O'Neill, and Palmer (2004) emphasize that cellar door service quality involves multiple service interactions (e.g., tours and tastings) critical to consumers' satisfaction with their overall experience. The latter cited authors suggest that cellar doors are a vital link in the service chain because these facilities are often the first contact that visitors have with wineries and their wines. Mauracher, Procidano, and Giovanna (2016), in turn, concluded that the most important attributes of winery visit experiences are wine quality, wine prices, staff friendliness and competence, and wine tasting.

The four characteristics of experiences defined by Pine and Gilmore (1998) – educational, esthetic, entertainment, and escapist – were applied by Oh, Fiore, and Jeong (2007) to the tourism industry and, in later studies, specifically to wine tourism (Quadri-Felitti and Fiore 2012; Thanh and Kirova 2018). The education components are new knowledge or skill acquired during visits and wine tasting, that is, learning about the wine industry and products (e.g., courses or guided tours), wine culture (e.g., the relationship between wines and regional gastronomy), and wine history (e.g., museum visits). The aesthetic aspects are related to an appreciation of the regional landscape's beauty. The entertainment elements of wine tourism are tours with tastings, demonstrations, and events directly linked with culture and wine. Finally, escapism is linked with tourists' desire for out-of-the-ordinary experiences regarding time and space, including walks around vineyards or other activities, as long as these are combined with wine. More recently, Thanh and Kirova's (2018) netnographic study found that the education and aesthetic dimensions are the most important in terms of describing overall wine tourism experiences.

Duan et al. (2018) combined the four-realms Experience Economy model with theories of service and product levels. The latter include core (i.e., wine tastings, education, and wine shops), basic (i.e., transportation, buildings, and landscapes), expected (i.e., services), augmented (i.e., accommodations, events, and festivals), and potential (i.e., social interactions) products. The Experience Economy concept and four realms of experience defined by Pine and Gilmore (1998) have been applied in other research on winery experiences. For instance, Charters, Fountain, and Fish (2009) suggest that visitors might not just want core tangible benefits (i.e., wine tasting) but also added value from other components of overall winery experiences. Wineries can, for example, add value to and differentiate their offer in the relevant markets through the overall experiences these companies provide.

Gómez, Pratt, and Molina's (2019) results further highlight that the main theme of wine tourism studies is winery and wine cellar service quality. Previous research has found that quality is of utmost importance to winery visits and tasting rooms because it affects positive post-visit behaviors (Charters, Fountain, and Fish 2009). Winery service quality can be understood as a combination of different attributes such as cellar doors' interior, service, staff, wineries' exterior, and convenience (Carlsen 2011). Carlsen adopted a service mapping approach to describe overall winery experiences' steps: approaching the winery (e.g., signs), entering the winery (e.g., physical evidence), parking (e.g., convenience), experiencing the cellar door, and departing. Wine cellar visits are described as sensory experiences that need to be pleasing to all five senses. For instance, sight can be associated with wine labels and merchandise, smell with aromas in wine tasting, hearing with the staff, taste with wine quality and temperature, and touch with contact with surfaces.

In addition, O'Neill and Charters (2000) used the SERVQUAL instrument's five dimensions to evaluate service quality in wine cellars and concluded that responsiveness is the most important in this context. O'Neill, Palmer, and Charters's (2002) study revealed that winery tourism experiences combine tangible components such as wine tasting and physical environments with the intangible elements of staff responsiveness and interactions. Charters, Fountain, and Fish (2009) also applied a modified mystery shopping approach to examine winery tasting-room experiences, confirming that wineries should be aesthetically appealing, manage service encounters through their staff's engagement with visitors, and consider pricing issues carefully. Finally, Santos et al. (2020) developed a scale to assess wine cellar experiences, combining four dimensions: wine storytelling (i.e., wine tour guides), wine tasting excitement (i.e., delightful sensory experiences), wine involvement (i.e., during visits), and winescapes (i.e., wine-related scenery).

Case Study 38.1 Wine tourism experiences in Azeitão

The research context was Azeitão, a village in Portugal's Península de Setúbal wine region. This peninsula is bordered by the Atlantic Ocean and the Tagus and Sado Rivers, generating a varied regional climate and ensuring a diverse wine production. Bacalhôa and José Maria da Fonseca are Azeitão's two main wineries.

Methodology

Research Design

A case study was selected as the best approach. The research design relied on the netnography method.

Data collection

The data included 552 reviews shared online on TripAdvisor. The reviewers were visitors to the two Azeitão wineries: José Maria da Fonseca (235) and Bacalhôa (317).

Data treatment

The content analysis of the reviews was conducted with Leximancer software. This program's main output is concept maps of the main themes extracted automatically from texts with minimal intervention from researchers (Brochado et al. 2019b).

Sample

The sample thus consisted of TripAdvisor reviews of the Bacalhôa and José Maria da Fonseca wineries. The reviewers were 60% males and from 21 different countries. Three-quarters (77%) of the visitors were between 35 and 64 years old. A majority (86%) described their experiences as 'excellent' or 'very good.'

Results

The analysis focused on dimensions of overall winery experiences. The results reveal a total of nine themes: 'wine,' 'visit,' 'guided tour,' 'art,' 'garden,' 'price,' 'recommend(ation),' 'Moscatel,' and 'friendly staff' (see Figure 38.1).

Wine theme

Wine is the most important theme for tourists who participate in wineries' activities. This theme includes the concepts of wine, tasting, winery, production, store, and shop. Typical reviews include information about how wine can be purchased. Tastings were prominent in these visitors' experiences and explanations of production processes.

One visitor said:

> One of the most traditional wine producers in Portugal[,] … José Maria da Fonseca shows … some secrets of the production of its award-winning wines [during the visit]… At the end of the tour, there is a tasting of two wines in the wine store (usually a purple Moscatel and another award-winning wine).
>
> *(male, Brazilian nationality, José Maria da Fonseca winery)*

Another visitor shared:

> After the tour, we did sample some of the wines in a private wine tasting with the guide who drove us to the winery. We got explanations on the types of wines and regions in Portugal. The Fonseca brand produces wines from all the regions in Portugal.
>
> *(male, from the United States [US], José Maria da Fonseca)*

Visit theme

This theme comprises five concepts: visit, place, beautiful (place), worth, and museum. Tourists generally describe their visits in positive terms (e.g., a 'beautiful place' that is 'worth a visit'), especially the museum component. A visitor wrote that this was a 'very well-spent afternoon. [It is a] very beautiful place … [w]ell worth … visit[ing] and explor[ing]' (Female, Portuguese, Bacalhôa).

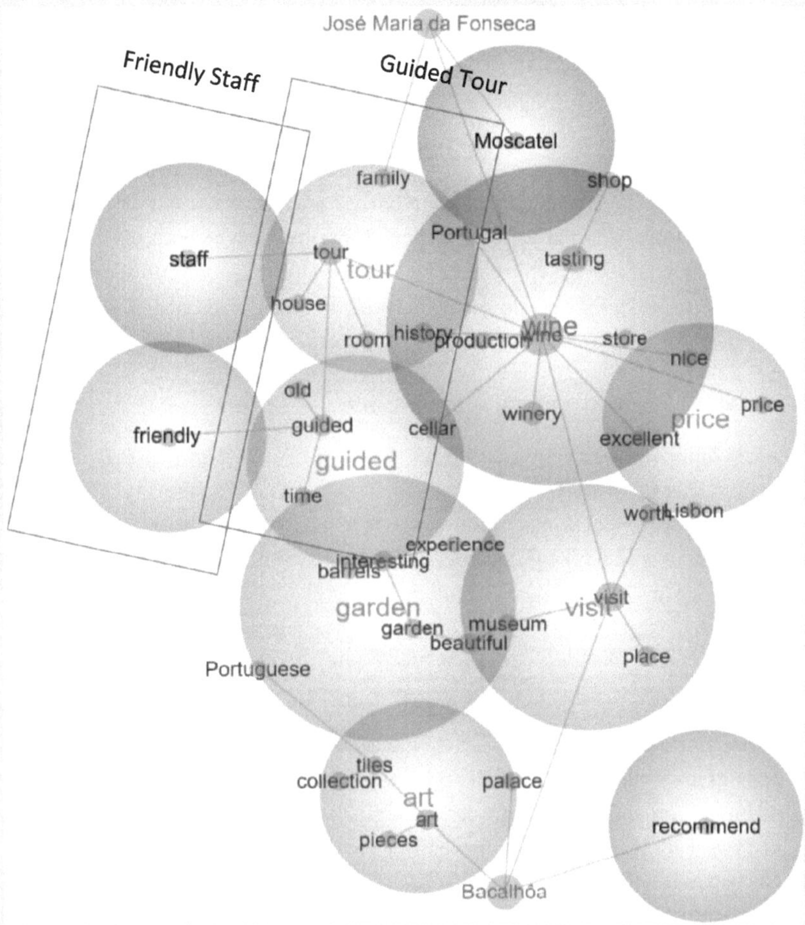

Figure 38.1 Concept map of themes in tourists' reviews of Azeitão wine tourism experiences.

Guided tour theme

This theme covers 11 concepts: tour, history, family, Portugal, room, house, guided (tour), interesting (experience), cellar, time, and old (winery). For instance, Bacalhôa's tour includes the wine cellar, the palace, and a vineyard. This dimension is linked with opportunities to learn about wine history, the two wineries, and their families, as well as Portugal's history and wine production. A reviewer reported, '[w]e were driven by our excellent tour guide' (female, from the United Kingdom [UK], Bacalhôa).

Art theme

This theme incorporates five concepts: art, palace, collection, tiles, and pieces. These features are associated with the Bacalhôa winery, which presents wine tourism experiences based on a combination of art- and wine-related activities. One visitor described this as 'a very interesting experience full of unexpected surprises between history, wine, and art,

[and] a very powerful cocktail' (female, UK, Bacalhôa). Another tourist noted, '[w]e then visited the wine cellar … [and] some fascinating art collections (African art and [A]rt Nouveaux)' (male, Portuguese, Bacalhôa).

Garden theme

This theme includes the concepts of garden, barrels, experience, and Portuguese. A close connection exists between the art and garden components, which are part of Bacalhôa's package tour. A review reads, '[t]he palace is a beautiful example of Portuguese Renaissance architecture with a magnificent courtyard as well as a splendid garden with a beautiful maze thrown in for good measure' (female, UK, Bacalhôa).

Price theme

This theme comprises four concepts – price, nice (price), excellent (price), and Lisbon – linked to two other themes – wine and visit. Tourists include feedback on their own pricing of experiences or prices of wines in the wineries' shops. A visitor said, '(t)hey make [*sic*] tastings of their wines at very good prices…. They have their own wines at a very good price' (male, Spanish, Bacalhôa).

Moscatel theme

This theme is related to the wine region's identity, which is widely referred to by reviews of Azeitão wine tourism experiences. This theme is used in particular by visitors to the José Maria da Fonseca winery, which offers regional products with wine and Moscatel tastings. This pairing of food and wine reinforces the region's distinctive identity through not only tastings of sweet Moscatel wine but also pairings with typical products.

A reviewer wrote:

> In the heart of the homeland of Moscatel de Setúbal, … [w]e made a reservation for a specialized wine tasting… The tasting paired these [wines] with a rich, creamy version of the local Azeitão cheese, bread, and the special torta, or rolled cake of the region.
>
> *(female, US, José Maria da Fonseca)*

Friendly staff theme

This theme is directly related to the wineries' personnel and their overall performance. One visitor said, '[t]he staff is friendly and try [*sic*] to answer … questions from customers and visitors' (male, Portuguese, Bacalhôa).

Recommend(ation) theme

This final theme is directly related to post-purchase feedback, which, in the case of these two wineries, is quite positive. A typical review asserts, this winery is '[r]ecommended for anyone' (male, Brazilian, Bacalhôa).

Conclusion

This research sought to answer two research questions by adopting a case study approach. The two largest wineries located in the village of Azeitão, Portugal, were selected as the focus. Regarding the first research question (i.e., 'What are the main winery experience dimensions in narratives shared online by the winery and/or wine cellar visitors?'), content analysis of winery visitors' reviews showed that Azeitão winery experiences can be organized around nine dimensions: wine, visit, guided tour, art, garden, price, recommend(ation), Moscatel, and friendly staff. Wine is the core product and thus the central theme.

The theme of wine is mainly linked with wine tasting and appreciation and with visitors' experiences in the wineries' tasting rooms and shops. Wine tasting and appreciation are the main product offered during winery visits (Duan et al. 2018), providing delightful sensory experiences to tourists (Santos et al. 2020). The guided tour theme is in accordance with the storytelling dimension identified by Santos et al. (2020). The price of visits and tastings is also a frequent theme in online winery reviews (Charters, Fountain, and Fish 2009). The friendly staff theme highlights hospitality's role in winery visits. Previous studies have also confirmed the importance of staff responsiveness (O'Neill and Charters 2000; O'Neill, Palmer, and Charters 2002; Carlsen 2011). The garden theme is associated with wineries' surrounding scenery (Santos et al. 2020).

Notably, the themes of art, museum visit, and Moscatel are unique, as they have not been mentioned previously by researchers, so these novel components reflect differentiated experiences in Azeitão wineries (Charters, Fountain, and Fish 2009). The final theme of recommendation describes positive post-visit behaviors among these wineries' visitors (Carlsen 2011). The present study's findings, therefore, confirm that various attributes are needed to ensure complete wine tourism experiences (Quadri-Felitti and Fiore 2012).

The second research question addressed was as follows: How do wineries differentiate their wine tourism offers in the relevant markets? Winery experiences in Azeitão are rich and diversified because they involve both traditions and art in different forms. José Maria da Fonseca offers a quite traditional experience in which the focus is on learning about the winery's family, Portugal, and wine history and production, as well as paired tastings of wines and traditional products from the region. In contrast, the experiences provided by the Bacalhôa winery concentrate on combining wine (i.e., tasting wines and appreciating beautiful vineyards) with visiting the owners' gardens, maze, and art collection. The case study analysis thus highlighted two examples of differentiation in winery visits (Charters, Fountain, and Fish 2009).

To summarize, the two wineries in question simultaneously share some dimensions of winery experiences and ensure product differentiation during visits. Both Azeitão wineries include a 'bundle of benefits' in their own unique way. These companies offer different packages for consumption in which tourists have to decide among various options, namely, guided tours that can include the wineries, art exhibitions, gardens, and/or tastings at the end. Several reviewers in this study's sample wrote that they prefer a complete package that permits them to have a more holistic experience.

The main themes found in reviews of Azeitão wineries can be linked with the 4Es theory of the Experience Economy (Pine and Gilmore 1998; Quadri-Felitti and Fiore 2012; Dixit 2020), which classifies attributes into four dimensions: entertaining, educational, (a)esthetic, and escapist. These characteristics are essential to finding the sweet point of a perfect wine tourism experience, and Azeitão's wineries have most of these attributes. From this model's perspective, the Azeitão winery experiences offer visitors the four key experience dimensions (Pine and Gilmore 1998).

Visits begin with the educational dimension, and the current results underline tourists' interest in learning as part of their experiences as reviewers often focus on the wineries' history, especially regarding José Maria da Fonseca. In addition, Bacalhôa visits include the owners' art collection as a central point in the tour, which allows tourists to learn about something more than just tastings and the winemaking process. As for entertaining tourists, art exhibits are a part of experiences due to Bacalhôa's art collection and José Maria da Fonseca's House Museum. By opening their gardens and vineyards to the public, both wineries enhance visitors' appreciation of the surrounding landscape, which contributes to the aesthetic component of winery experiences. Bacalhôa's tourists also report that the entire experience is connected to the winery owners' beautiful gardens, art, and palace architecture.

Managerial implications

These results have clear managerial implications. For instance, the analysis revealed that wine tasting and appreciation are the most important components of winery experiences. Tasting rooms and shops should, therefore, be designed to offer visitors delightful sensory experiences.

Other key elements to include are guided tours and friendly staff. These tours' itineraries need to provide opportunities for visitors to taste and buy wines, and learn about the wine production process, surrounding region, and winery's history. In addition, managers should be aware that their winery's staff plays an important part in creating stimulating learning opportunities and entertaining experiences during guided tours. Thus, recruiting and training staff members is of the utmost importance. As visitors also value the surrounding scenery, tour itineraries need to include breaks that allow guests to appreciate the landscape in order to enhance their experience's aesthetic dimension.

One of the most interesting results is that reviewers make comments about Moscatel wine that highlight the Setubal wine region's special identity. The narratives shared online further emphasize that the two adjacent wineries in question actively seek to make their mark in different ways. To enhance these differences, winery managers could combine visits' core components and provide added value by including other unique features of the overall winery experience (e.g., art, wine, and food pairing). This differentiation contributes to strengthening experiences' escapism dimension.

Many winery visitors use social media platforms (e.g., TripAdvisor) to share their experiences in the post-purchase phase. The case study presented in this chapter offers managers an example of how to analyze this user-generated content to gain a fuller understanding of the key dimensions of visitors' wine tourism experience and to assess their own winery's performance.

References

Andresson, T. D. (2007). The tourist in the experience economy. *Scandinavian Journal of Hospitality and Tourism*, 7(1): 46–58. https://doi.org/10.1080/15022250701224035

Brochado, A., Stoleriu, O., and Lupu, C. (2019a). Wine tourism: a multisensory experience. *Current Issues in Tourism*, 24(5): 597–615. https://doi.org/10.1080/13683500.2019.1649373

Brochado, A., Troilo, M., Rodrigues, H., and Oliveira-Brochado, F. (2019b). Dimensions of wine hotel experiences shared online. *International Journal of Wine Business Research*, 32(1): 59–77. https://doi.org/10.1108/IJWBR-12-2018-0072

Bruwer, J., and Alant, K. (2009). The hedonic nature of wine tourism consumption: an experiential view. *Journal of Wine Business Research*, 21(3): 235–257. https://doi.org/10.1108/17511060910985962

Carlsen, J. (2011). Assessing service quality at wineries and cellar doors through service mapping. *International Journal of Wine Business Research*, 23(3): 271–290. https://doi.org/10.1108/17511061111163087

Carlsen, J., and Boksberger, P. (2013). Enhancing consumer value in wine tourism. *Journal of Hospitality & Tourism Research*, 39(1): 132–144. https://doi.org/10.1177/1096348012471379

Charters, S., and Ali-Knight, J. (2002). Who is the wine tourist? *Tourism Management*, 23: 311–319. https://doi.org/10.1016/S0261-5177(01)00079-6

Charters, S., Fountain, J., and Fish, N. (2009). 'You felt like lingering …': experiencing 'real' service at the winery tasting room. *Journal of Travel Research*, 48(1): 122–134. https://doi.org/10.1177/0047287508326508

Dixit, S. K. (ed.). (2020). *The Routledge Handbook of Tourism Experience Management and Marketing*. London: Routledge.

Duan, B., Arcodia, C., Ma, E., and Hsiao, A. (2018). Understanding wine tourism in China using an integrated product-level and experience economy framework. *Asia Pacific Journal of Tourism Research*, 23(10): 949–960. https://doi.org/10.1080/10941665.2018.1512506

Getz, D. (2000). *Explore Wine Tourism: Management, Development, Destinations*. New York: Cognizant.

Getz, D., and Brown, G. (2006). Critical success factors for wine tourism regions: a demand analysis. *Tourism Management*, 27(1): 146–158. https://doi.org/10.1016/j.tourman.2004.08.002

Gómez, M., Pratt, M. A., and Molina, A. (2019). Wine tourism research: a systematic review of 20 vintages from 1995 to 2014. *Current Issues in Tourism*, 22(18): 2211–2249. https://doi.org/10.1080/13683500.2018.1441267

Griffin, T., and Loersch, A. (2006). Determinants of quality experience. In Carlsen, J., and Charters, S. (eds.), *Global Wine Tourism: Research, Management and Marketing* (pp. 80–90). Wallingford, UK: Cabi.

Gu, Q., Zhang, H. Q., King, B., and Huang, S. (2018). Wine tourism involvement: a segmentation of Chinese tourists. *Journal of Travel & Tourism Marketing*, 35(5): 633–648. https://doi.org/10.1080/10548408.2017.1401031

Gunasekar, S., Das, P., Dixit, S. K., Mandal, S., and Mehta, S. R. (2021). Wine-experienscape and tourist satisfaction: through the lens of online reviews. *Journal of Foodservice Business Research*. https://doi.org/10.1080/15378020.2021.2006039

Martin, A., O'Neill, M. A., and Palmer, A. (2004). Wine production and tourism: adding service to a perfect partnership. *Cornell Hotel and Restaurant Administration Quarterly*, 45(3): 269–284. https://doi.org/10.1177/0010880404263075

Mauracher, C., Procidano, I., and Giovanna, S. (2016). Wine tourism quality perception and customer satisfaction reliability: the Italian Prosecco District. *Journal of Wine Research*, 27(4): 284–299. https://doi.org/10.1080/09571264.2016.1211514

O'Neill, M., and Charters, S. (2000). Service quality at the cellar door: implications for Western Australia's developing wine tourism industry. *Managing Service Quality: An International Journal*, 10(2): 112–122. https://doi.org/10.1108/09604520010318308

O'Neill, M., Palmer, A., and Charters, S. (2002). Wine production as a service experience –the effects of service quality on wine sales. *Journal of Services Marketing*, 16(4): 342–362. https://doi.org/10.1108/08876040210433239

Oh, H., Fiore, A. M., and Jeong, M. (2007). Measuring experience economy concepts: tourism applications. *Journal of Travel Research*, 46: 119–132. https://doi.org/10.1177/0047287507304039

Pine, B. J., and Gilmore, J. H. (1998). Welcome to the experience economy. *Harvard Business Review*, July–August: 97–105.

Quadri-Felitti, D., and Fiore, A. M. (2012). Experience economy constructs as a framework for understanding wine tourism. *Journal of Vacation Marketing*, 18(1): 3–15. https://doi.org/10.1177/1356766711432222

Roberts, L. and Sparks, B. (2006). Enhancing the wine tourism experience: the customers' viewpoint. In Carlsen, J., and Charters, S. (eds.), *Global Wine Tourism: Research, Management and Marketing* (pp. 47–55). Wallingford, UK: Cabi.

Santos, V., Ramos, P., Almeida, N., and Santos-Pavón, E. (2020). Developing a wine experience scale: a new strategy to measure holistic behaviour of wine tourists. *Sustainability*, 12: 8055. https://doi.org/10.3390/su12198055

Shapiro, M., and Gómez, M. (2014). Customer satisfaction and sales performance in wine tasting rooms. *International Journal of Wine Business Research*, 26: 45–60. https://doi.org/10.1108/IJWBR-09-2012-0026

Thanh, T. V., and Kirova, V. (2018). Wine tourism experience: a netnography study. *Journal of Business Research*, 83: 30–37. https://doi.org/10.1177/1356766711432222

Williams, P. (2001). The evolving images of wine tourism destinations. *Tourism Recreation Research*, 26(2): 3–10. https://doi.org/10.1080/02508281.2001.11081338

Yin, R. K. (2018). *Case Study Research and Applications: Design and Methods* (6th ed.). London: Sage.

39
THE LOUISE
A luxury wine hotel experience in the vineyards

Sylvie Hertrich, Coralie Haller, and Jodie Conduit

Introduction

International travelers consider Australia as an aspirational, once-in-a-lifetime destination, which involves extensive planning, long-haul travel, and higher-than-average costs. Though international visitors have driven the Australian tourism industry, locals' desire to travel domestically has contributed to the sector's double-digit growth in 2018–2019. Domestic spending exceeds AU$100 billion, with nearly 350 million domestic and international visits to and within Australia (Australian Government Austrade 2020). As Australia is worldwide recognized for its wine regions, it welcomes around 8.4 million visitors to wineries from both domestic (day and overnight) and international travelers (Wine Australia 2019). However, recent bushfires, together with the COVID-19 pandemic, are expected to have a significant impact on the wine tourism business for wineries. Though it would take no less than two years for wine tourism to return to the pre-COVID level, the Australian wineries seemed to be confident considering their wine tourism turnover improvement in 2020 compared to other wine regions in the world (WineTourism.com 2020).

In this context, it might be interesting to investigate what are the best wine tourism practices developed in Australia. More specifically, a particular interest should be given to South Australia (SA), as it is the wine region with a strong reputation for quality wine and which has the highest instance of wine region visits across Australia, especially in three key regions; 62% visited the Barossa Valley, 27% visited McLaren Vale, and 22% visited the Adelaide Hills, with some visiting multiple regions (Tourism South Australia 2016, p. 6). Thus, this chapter provides a case study on the wine tourism experience developed by 'The Louise,' a world-class wine destination for fine dining situated within the Barossa Valley wine region. The objective is to explore the role of exceptional service design in the context of a luxury wine experience and how it can contribute to building a reputation of a world-class destination, aligned with trends of future travelers in 2030.

Wine tourism: a multifaceted activity

There is an increasing interest in wine tourism within the broader context of tourism development. Wine tourism is flourishing around the world, with more than 70 countries

DOI: 10.4324/9781003143628-45

having documented wine regions (MacNeil 2015) and thousands of unique wineries offering distinctive portfolio of wines (Haller, Hess-Misslin, and Méreaux 2021). It is a multifaceted activity inspired and driven by the appeal that goes beyond simply the tasting and purchasing of wine. Wine tourism includes a combination of food, accommodation, recreational activities (wine events, festivals, shows), cultural attractions, and enjoyable rural and natural landscapes (Brown, Havitz, and Getz 2006; Getz et al. 2008; Dawson et al. 2011; Park, Bufquin, and Back 2018; Haller, Thach, and Olsen 2020). Wine tourism is commonly categorized as both rural tourism and agritourism, with wine tourists staying with local people in rural areas and spending their time on farms, experiencing different activities with a recreational purpose (Pezzi, Faggian, and Reid 2020). Wine tourism can also be linked to gastronomy or culinary tourism, with the United Nations World Tourism Organization (UNWTO) defining it as defining as a sub-type of gastronomy tourism, which 'refers to tourism whose purpose is visiting vineyards, wineries, tasting, consuming and/or purchasing wine' (UNWTO 2019, p. 44).

Wine tourism is a major source of revenue for both the wine and tourism industries. It enables the marketing of a destination and provides opportunities for direct sales for large and small wineries (Getz 2000) and implies marketing opportunities for distributors as well as retailers (Taylor, Barber, and Deale 2010). Wine producers work together with tourism industry networkers and government entities (e.g., destination marketing organizations [DMOs]) to leverage the synergies that arise from coordinated efforts among wine consumers, tourists, and wine tourism destinations (Hall et al. 2009). Wine tourism has also become a strategic priority for national governments, as shown by the recent creation of a Global Wine Tourism Organization (GWTO), aligned with the United Nations 2030 Agenda for Sustainable Development Goals (SDGs) framework (GWTO 2020).

Wine tourism experience: an emerging research topic

Wine tourism emerged as a subfield of tourism research in the late 1990s, with mostly descriptive studies. It is now recognized by the international academic community as a notable research area (Bonn, Cho, and Um 2018) and is among the most predominant and arguably the fastest-growing field of academic research in the wine business (Haller, Hess-Misslin, and Méreaux 2021), resulting in approximately 250.000 publications in this field according to Google Scholar in 2020. Wine tourism appears as one of the top ten most commonly occurring keywords in *Journal of Wine Research* publications between 1990 and 2020 (Lam et al. 2020). Several studies conducted a wide-ranging review of wine tourism literature (Sánchez, Del Río, and García 2017) or even provided a 30-year review on wine tourism research (Durán-Sánchez et al. 2016), offering scholars a number of potential avenues for research to explore on wine tourism. Indeed, some of the latest research explores different perspectives of wine tourism (Vo Thanh and Kirova 2018; Sigala and Robinson 2019), including wine destination strategy (Ben Tahar et al. 2018; Bruwer and Lesschaeve 2012) and wine stakeholders in the wine tourism industry (Hojman and Hunter-Jones 2012). Researchers have also identified key factors that motivate travelers to visit wineries and wine regions, including wine products and experiences, entertainment and events, and educational and social-related factors (Bruwer and Alant 2009; Getz and Brown 2006; Haller, Thach, and Olsen 2020; Thach and Charters 2016). Wine tourism is strongly associated with the destination image, and there is a growing interest in the concept 'winescape' – that is, those attributes that are complementary to the core wine product (Thomas, Quintal, and Phau 2018).

Wine tourism is known for its hedonic nature (Bruwer and Alant 2009), and despite this – or perhaps because of it – an academic focus on the wine tourism experience has just begun to emerge. Building on the concept of the 'Experience Economy' (Pine and Gilmore 1998) and presenting a new consumption model based on experiential marketing, wine tourism experiences have been recognized as embodied in the natural, physical, human, and cultural heritage of the winescape (Quadri-Felitti and Fiore 2012; Thach and Charters 2016; Vo Thanh and Kirova 2018; Haller, Hess-Misslin, and Méreaux 2021). There is a constant need for a better understanding of the Experience Economy approach in order to inform the debate about wine tourism (Vo Thanh and Kirova 2018; Haller, Hess-Misslin, and Méreaux 2021). Considering the increasing sophistication of wine tourism demand based on experiences, there is a need to better identify wine tourists' expectations and the customer journey related to wine tourism.

Wine tourism in Australia

Across domestic (day and overnight) and international travelers, there were 8.4 million visits to wineries in Australia in 2019. Travelers that included a visit to a winery spent a total of $9.6 billion during their trip and are typically higher spenders with an average spend per person of $1,136 (compared with an average of $380 across all visitors) (Wine Australia 2019).

It is estimated that one in eight international travelers to Australia visited a winery during their trip, while only 2% of Australians visited a winery. However, as illustrated in Figure 39.1, in aggregate domestic visitors made up the largest total number of visits to a winery, with 7.4 million visits in 2018–2019 versus 1 million international visits. The majority of all these domestic winery visitors (93%) were visiting friends and family and/or on holiday (Wine Australia 2019). The international visitors come mainly from China, the United Kingdom, the United States, Singapore, Korea, Malaysia, New Zealand, Hong Kong, and Japan. Food and wine experiences are a strong driver of choice for visiting Australia, and thus, the potential of the wine tourism market in Australia is clear (Wine Australia 2020a). The total surface under vines in Australia approximates 147,000 hectares in 2019. The top three wine regions, by

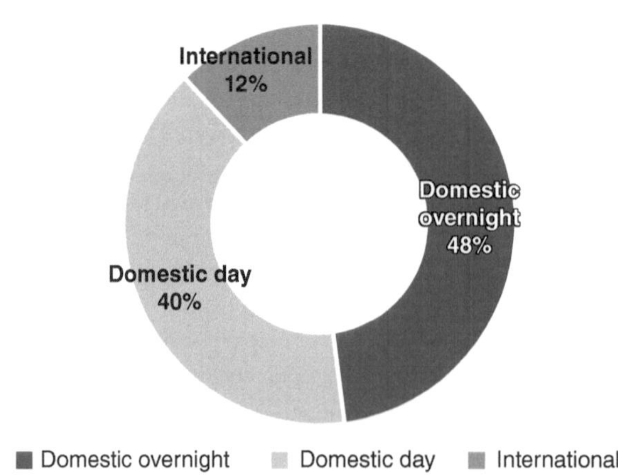

Figure 39.1 Split between the usual place of residence.

total vineyard area, are South Australia (52%), New South Wales (24%), and Victoria (15%), respectively (Wine Australia 2020b).

Wine tourism in South Australia (SA)

SA has a reputation for delivering a premium food and drink experience. According to the SA Tourism Plan 2030, 'SA leads the nation for "good food and wine" as well as "credible wineries and wine regions"' and notes that this is 'critically important as research shows that these themes are the strongest driver of destination selection for the interstate audience' (South Australian Visitor Economy Sector Plan 2030, 2019, p. 8). SA produces 50% of all bottled wine in Australia and almost 80% of premium wine (Tourism South Australia 2016). It includes 18 wine regions, with Coonawarra in the South East, Kangaroo Island, the Southern Flinders, and the Riverland, among some of the lesser-known regions (South Australian Visitor Economy Sector Plan 2030, 2019, p. 6.). Those regions have a diverse variation in rainfall, climate, topography, and terroir, resulting in a variety of wines and styles and providing a diverse range of experiences for visitors.

International tourists are mostly visiting the Barossa Valley (62%), McLaren Vale (24%), and Adelaide Hills (22%), with some visiting multiple regions. All these wine regions are close to Adelaide, which is the local entry point for the majority of international visitors (South Australian Visitor Economy Sector Plan 2030, 2019, p. 6). As part of the global network of the *Great Wine Capitals* of the world, Adelaide has been recognized as a top wine tourism destination, providing more than 200 cellar doors within an hour's drive, offering diverse wine tourism experiences (Great Wine Capitals Global Network 2020).

SA produces some of the oldest and finest wines in the world. In 2020, SA was the largest contributor, with 47%, of the national crush with 700,000 tonnes (Wine Australia 2020b). SA wine exports achieved the best global market result for ten years, with $1.56 billion earned in 2017. International expenditure in SA has grown by 51% since 2013 and now reaches $1.1 billion. When it comes to international wine tourism, demand has largely been driven by the Chinese market, while traditional markets such as the United States and United Kingdom also continue to grow. Wine has been a key driver for the SA region, with 34% of all international visitors visiting one or many of the 350 cellar doors in the region (South Australian Visitor Economy Sector Plan 2030, 2019, p. 2).

Table 39.1 recalls four different types of visitors to SA, which have relatively comparable size but with different behaviors, itineraries, booking patterns, and expenditure. Interstate overnight visitors represent one-third (31%) of the total visitors coming to South Australia, following closely by the intrastate overnight visitors (28%) and domestic day-trip visitors (24%).

For visitors coming to SA, the most important purpose of the visit is 'Holiday,' accounting for 42% of total expenditure by the purpose of visit; visits for education, for business, and to visit friends and relatives respectively representing 22%, 18%, and 17%. Decisions concerning infrastructure, accommodation, and budget are common to all purposes of visit (South Australia Visitor Economy Sector Plan 2030, 2019, p. 15). The state has the highest level of visitor satisfaction for wine experiences compared with other regions in Australia (Tourism South Australia 2016, p. 7). Even if food and tour options represent significant volumes of visitors and participate in the increase in overall yield per visitor, wine tasting is by far the most common activity at a winery (Tourism South Australia 2016, p. 8). A focus on product and

Table 39.1 Typology of visitors to South Australia (SA)

(2018)	*Traveler behaviors (socio-styles)*
International Visitors: (17%) - \$2.435 average spend per visitor - 22 nights average length of stay - 452,200 visitors	- Long-staying, high spending visitors. - Diverse market comprising visitors from Eastern and Western Hemispheres) - Likely to have a low awareness of the experiences (education) SA has to offer.
Interstate Overnight Visitors (31%) - \$851 average spend per visitor - 5 nights average length of stay - 2,598,000 visitors	- Average spending visitors - Key markets from Victoria and New South Wales - Moderate understanding of the SA experiences
Intrastate Overnight Visitors (28%) - \$478 average spend per visitor - 3 nights average length of stay - 4,035,000 visitors	- South Australians traveling within their own state. - Lower spending visitors on shorter itineraries (familiar destinations) - Feel they have a solid understanding of SA and its attractions - Booking lead times tend to be short
Domestic Day Trip Visitors (24%) - \$133 average spend per visitor - N/A average length of stay - 13,456,000 visitors	- Australians who are away from their homes for more than 4 hours, but do not stay overnight. - Interstate travel (largely business purpose), day trips are predominantly leisure trips to nearby locations. - Over 13 million of these trips/year every - Overall contribution is significant, especially to tourism regions bordering Adelaide.

Source: Adapted from South Australian Visitor Economy Sector Plan 2030 (2019)

service development in this area is worth strong consideration and offers a relatively untapped opportunity for increased turnover and business diversification.

The Barossa Valley is Australia's best-known wine region and SA's most visited tourist destination. As it built its success on European cultural traditions of food and wine, it is known as an 'Old New World' region (The South Australia Wine Story 2019, p. 13). It finds its origin in 1842 and today accounts for more than 550 grape-growing families, with some of them having the sixth generation of winemakers working in the vineyards. This wine region is comprised of both the Barossa and Eden Valleys, and is one of the only wine regions in Australia to have neighboring warm and cool climate growing conditions (Barossa Grape and Wine Association 2020).

The South Australian Visitor Economy Sector Plan 2030 (2019) forecasts significant changes in consumer behavior, including travel inspiration, planning, booking, the actual experience, and the eventual sharing of the trip. The key trends for the future traveler in the Barossa Valley have been analyzed considering the following key trends: shorter stays, growth from Asia, digital connectivity, more likely to fly, and high expectations of service. While conditions of 2020 (including the COVID-19 pandemic) have seen a short-term variation, the longer-term vision of the future traveler remains. Table 39.2 outlines some of the key trends expected of future travelers to South Australia.

Table 39.2 The future traveler 2030

Key trends	*Description*
Growth in Eastern markets	- Expecting growth from Western markets - Expansion continue to be the strongest from the Eastern Hemisphere, including China. - Continued growth in the demand for international education - Eastern Hemisphere continuing to be the largest source market.
Shorter stays	- The trend towards a shorter average length of stay is likely to continue across all markets - International tourism will reflect a changing profile towards the Eastern markets - Shorter stays, but with a higher spend.
More spontaneous	- The expectation for 'On Demand services' will only increase through to 2030. - Traveler's expectations for personalized, high-quality experiences with short lead times will represent both an opportunity and a challenge for tourism businesses, particularly regarding opening hours, seasonal operations, and the availability of skilled labor.
Older	- Baby Boomer generation will continue its insatiable thirst for travel experiences. - The typical visitor in 2030 is likely to be, on average, older than those of today.
More diverse	- The visitor mix, both domestically and internationally, will continue to diversify. - Defining a 'typical visitor' will become increasingly challenging with a more ethnically diverse visitor mix demanding a range of appropriate visitor experiences
More connected	- It is impossible to predict the direction that technology will take us - It will be disruptive and present new opportunities and challenges.

Source: Adapted from the South Australian Visitor Economy Sector Plan 2030 (2019)

Case Study 39.1 The Louise Hotel

The story

Located in the heart of the Barossa wine region, The Louise houses 15 suites of pure comfort. A founding member of Luxury Lodges of Australia and proud affiliate of the prestigious association, *Relais et Châteaux*, this bespoke vineyard accommodation is an ideal location from which to enjoy the renowned wine region and a peaceful sanctuary to return to at the end of each day. At The Louise, customers find premium luxury and inspired design with private spaces, generous suites, and spectacular vineyard views. No stay at The Louise is complete without experiencing the onsite restaurant Appellation, a world-class destination for dining with food that creatively reflects a sense of place. Executive Chef Daniel Murphy and his team pride themselves on being dedicated storytellers of the Barossa Valley region, its produce, and its people. For a more casual culinary delight, The Three75Bar offers mouth-watering comfort food and cocktails daily.

Jim and Helen Carreker, current owners of The Louise, spent more than ten years determining the ideal location for the business and ultimately selecting the Barossa Valley. After travels through France, Spain, Portugal, and their then local Napa and Sonoma Valley areas of California in the 1980s and 1990s, the couple concluded that luxurious accommodation is best paired with a truly memorable food and wine experience. As with many highly regarded hospitality venues, the luxurious accommodation is supportive of the original wine and food experiences. This concept is relatively rare in Australia and unique within the Barossa region. The Louise has built a reputation for this unique combination of seasonal, regional food, world-class wine, and private secluded accommodation (The Louise 2020).

Box 39.1 The Genesis of the project

From personal travels with my wife Helen, we had observed that the great wine regions of the world often had outstanding small restaurants, and many had complementary luxury hotels. We decided to become hoteliers and find a wine region where we could develop a business together that offered guests the types of experiences that we enjoyed. After a search that included Spain, Portugal, Argentina, Chile, New Zealand, and all the regions of Australia, we relocated to South Australia, having decided to launch a restaurant and accommodation in the Barossa Valley.

The challenge was to develop a destination restaurant with co-located guest rooms at a time when virtually all regional Australian properties were focused on filling rooms. Back then, the food-and-beverage cost centre was simply adequate to feed overnight guests.

We recognized that the Barossa Valley had a fine international reputation for its wine exports, but there was really no appreciation that the local food sources were of equal quality to the wines. By establishing the restaurant Appellation, sourcing outstanding local produce, and operating the business to celebrate the dining experience, we set out in a new direction.

Source: Jim Carreker, *Restaurant and Catering* magazine (2015)

Key success factors of a luxury wine hotel experience

Natural, exclusive signature and 'extraordinary' experiences

Experiences are a key mechanism for developing and enhancing the appeal of wine tourism destinations (Charters, Fountain, and Fish 2009; Quadri-Felitti and Fiore 2012; Vo Thanh and Kirova 2018). Wine tourists seek an aesthetic experience in the form of an attractive and charming environment (Haller, Thach, and Olsen 2020). Thus, when developing an immersive offer, it is important to build on the surrounding environment like The Louise plays on its ideal location in the Barossa Valley. The region 'is full of unique natural experiences from holding a koala, hand feeding kangaroos, to enjoying the outback' according to Jim Carreker, who is convinced that 'experiences such as these are the strongest thematic driver for international visitors.'

The Louise goes beyond its natural appeal and develops 'exclusive signature experiences' including: breakfast with the kangaroos, ultimate legend of Seppeltsfield tour, Harvest with Izway Wines, or regional experiences such as traditional hands-on pasta making, and the Hutton Vale Farm visit. The hotel also builds a distinctive competitive advantage for its national guests by developing 'extraordinary experiences' which allow them to 'do something they cannot get in their daily busy life.' Jim identified that they

> will come to (The Louise) – whether they stay with us or not – for a special occasion like a birthday, an anniversary, an engagement.... The interstate overnight visitors represent 25% of our guests. They usually come for the weekend: they arrive on Friday afternoon and leave on Sunday afternoon.

These interstate overnight visitors mainly come from Victoria and New South Wales, have moderate understanding of South Australian experiences, and are considered as average spending visitors (see Table 39.1). The Louise also 'welcome 25% of intrastate overnight visitors. They arrive any day of the week and stay one night to celebrate a special occasion.' The intrastate overnight visitors are South Australian traveling in their own state and are familiar with the destination (see Table 39.1). This is the reason why Jim wants to 'create "magic" and deliver it to each individual guest.'

The Louise's objective is to 'bring a WOW experience for the guest to come back, to get that experience again. Loyalty is important for us: half of our Australian guests are repeaters.' Thus, providing customers with a combination of authenticity, service and product quality, and environmental embedded offers, The Louise develops a 'total experience' (Roberts and Sparks 2006) to the customer. 'That feeling can be inspired by all sorts of things: the taste of a drink or a meal, its presentation, the ambiance, the décor, the people,' stated Jim. As such, the experience can be considered as a 'sweet-spot' (Pine and Gilmore 1998; Vo Thanh and Kirova 2018), encompassing entertainment, education, aesthetic, and escapism dimensions which facilitates the success of an experience.

Inter-industry collaborations

The research found a distinct lack of inter-industry cooperation, which challenged the capacity to develop a sustainable wine tourism experience in rural Australia (Beames 2003). However, the ability to develop and implement a collective strategy is a determining factor in the successful promotion of a given wine tourism destination (Baker and Cameron 2008). Thus, to achieve this, stakeholders from the wine, tourism, food, and hospitality industries need to go beyond their scope of expertise and work in a transversal way across industries' value chains and across regional professional committees (Maaninou, Boland, and Haller 2021), especially when promoting a region on the international market. 'International visitors represented only 10% of (The Louise's) guests a decade ago; nowadays they represent 50%. They stay on average for 3 nights, arriving any day of the week,' as these international visitors spend three times more than the interstate overnight visitors (see Table 39.1), The Louise understood the value of inter-industry collaboration in order to attract them. The company has been working with 'Tourism Australia and took part in seminars, travel fairs, workshops, [and] site inspections' and has been meeting with 'travel specialists selling trips to Australia from the UK or the US.' Jim recognized that

> people visiting Australia for the first time did not intend to visit South Australia, but were only planning to visit the 'classic' destinations (Sydney, Melbourne, Barrier Reef, Ayers Rocks …). He decided to work '… with travel agents to target customers coming to Australia for a second time…, this type of traveler has a strong interest in nature and wildlife.

Jim's strategy seems to be aligned with what the South Australian Visitor Economy Sector Plan 2030 (see Table 39.1), which has been pinpointing concerning international travelers likely to have a low awareness of the experiences South Australia has to offer.

Building a collective brand reputation

The company even recognized the power of international professional networks like 'Relais et Chateaux.' As 'when we started 15 years ago, the hotel did not meet the "**Relais et Châteaux**" standards,' remembers Jim. The Louise had, thus first, to build its brand reputation to meet the requirements of excellence in hospitality and culinary arts. 'Our first step was to develop a reputation around the fine dining experience. It was a real challenge due to the facilities that existed at that time' The Louise had to 'rethink the design of all (its) rooms' which took some time, 'as our aim was to satisfy demanding customers,' recalls Jim. He took his 'inspiration from some of the best places (he) had visited all over the world, while also adding a personal touch.' Jim also invited '10 owners of properties similar to The Louise to create a marketing network: **Luxury Lodges of Australia**, a collection of nineteen independent luxury lodges and camps offering unforgettable experiences in Australia's most inspiring and extraordinary locations.' The Louise could then rely on a collective website which 'provides valuable information about each property and useful tools to optimize transfers between lodges.' Jim put forward the 'new distinctive style of Australian elegance and experiential luxury' of the 'experience offered by these lodges.' As the company's first market was 'Adelaide locals coming for special occasions,' the challenge was to 'attract customers from Sydney, Melbourne or Perth.' To help build up its national reputation The Louise decided to 'invite media that specializes in lifestyle and travel magazines.'

Customer centric approach

For The Louise to stand out in a crowded tourism market, such as in the Barossa, Jim recognized they needed to adopt a customer-centric approach and create a truly extraordinary customer experience.

> I have learnt that to understand the customers, you have to be a good listener. If you're not a good listener and if you're not receptive to the feedback that you receive from your clients, you will not be able to refine your approach and adjust your intended solution.

Extraordinary consumption experiences, such as those that Jim has created at The Louise, are memorable and extend beyond the realm of everyday life; they are depicted by high levels of emotional response, and are most often triggered through interpersonal interaction (Arnould and Price 1993). Recent academic literature defines customer experiences as the non-deliberate and spontaneous (cognitive, emotional, behavioral, sensorial, and social) responses and reactions to particular stimuli (Becker and Jaakkola 2020). The Louise offers a suite of engaging customer experiences that stimulate all of these responses, including meeting with up-and-coming and boutique winemakers to learn about their wines (i.e., cognitive engagement), being directly involved in the harvest during vintage (i.e., behavioral engagement), enjoying a sunrise picnic with kangaroos (i.e., affective engagement) and plenty of opportunities for tastings of premier wines and foods with other visitors and the staff (i.e., sensorial and social engagement). This suite of signature and regional experiences is thus central to the extraordinary customer experiences that Jim has created at The Louise. It also answers the future traveler expectations for personalized and high-quality experiences (see Table 39.2). Engaging customers in this way leads to increased customer satisfaction, brand attachment, and loyalty, and ultimately firm performance (Hollebeek, Srivastava, and Chen 2019).

Internal customer management

Customers establish an impression of a tourism business from their interactions with frontline service employees (Hartline, Maxham, and McKee 2000). The employees have personal contact with the customers, and thus are the face and voice of the tourism business. Jim is cognizant of this important role and noted,

> Employees convey the hotel's unique culture ... its DNA and spirit. It is not just about qualifications and diplomas, although they help of course. It is about knowing your client and having the right attitude. It is about understanding the psychology of the guest. This isn't something you can accomplish through training. It's about being innovative, caring, passionate and committed. It's more than a job. Hiring people is a challenge.

As Jim is aware, in a customer-centric business, it is important to nurture and develop a customer orientation among employees. This is achieved through a careful mix of recruitment strategies, management support, training, internal communication, and an internal reward and recognition program (Conduit and Mavondo 2001).

Frontline employees are also important contributors to the innovation process, as they have extensive knowledge about customers, which can be used to enhance the value proposition (Karlsson and Skålen 2015). Jim recognizes this and encourages his staff. 'So, one of the most important messages that I share with my staff is to be a good listener and have a conversation with our customers and then share the feedback with the entire team.' By having a focus on the inherent value of employee knowledge of customers, Jim will be able to leverage this knowledge and keep the experiences offered by The Louise relevant for customers having high expectations for 'on demand services' (see Table 39.2).

Conclusion

Luxury wine hotel experiences are considered 'extraordinary' consumer experiences, and therefore the memories formed are long-lasting. However, customers are discerning and have high expectations and expect premier quality food and wine experiences that stimulate cognitive, emotional, behavioral, social, as well as sensorial responses. Thus, it is important that luxury wine hotel accommodation provides a range of unique and authentic experiences to stimulate these responses and appeal to a customer base (i.e., international and domestic travelers) that is diversifying. The quality of the wine hotel experience is often assessed through personal interactions; therefore, internal customer management of frontline service employees is crucial to achieving consistent, premium experiences, and the development of a truly customer-centric approach throughout the organization will allow for ongoing innovation and improvements. Further, inter-industry collaborations, particularly those that help to build a collective brand reputation such as with the *Relais et Châteaux*, are critical for success.

Planning for the future

The customer-centric approach adopted by Jim will allow The Louise to continue to evolve to meet the needs of future visitors. It is predicted that by 2030, visitors will be more demographically diverse, stay for shorter periods, and be more spontaneous and less planned with their experiences. Jim notes this: 'Our visitor mix, both domestic and international, will continue to diversify. Defining a typical visitor will become increasingly challenging, with a more ethnically diverse visitor mix growing, demanding a range of suitable visitor experiences.'

Recent studies indicate that future travel behavior will also differ among the generational cohorts of Baby Boomers, Generation X, and Generation Y, which will have a significant impact on the future of domestic travel in Australia (Gardiner, Grace, and King 2014). In the near future, Baby Boomers will dominate the domestic travel market, as they have more discretionary money and time. This segment is ideal for The Louise, as they seek high-quality, hedonic experiences that promote emotional reactions. While Generation X and Y prefer to travel internationally, factors such as COVID-19 are reminding them of the value of domestic travel. These segments are becoming more prominent in the domestic travel market, and they typically seek unique experiences that allow them to stand out and differentiate themselves from others. They require experiences that offer social value and align with the individual's self-concept and how they wish to be perceived by others. As such, The Louise will need to allow these tourists to customize their experiences to reflect their individuality. This need aligns well with the core value proposition of The Louise (i.e., high quality, extraordinary wine, and food experiences), and with only minor adjustments, such as allowing more customization, the extraordinary experiences are likely to appeal to these segments as well as to Baby Boomers.

Perhaps the biggest differences expected in 2030 will emerge from the role of technology and how tourists seek to use it in their decision making. While the digital economy has forever shaped the tourism industry over the last decade, it is impossible to predict the effect that this may have by 2030. Generations X and Y already gather information and use technology to inform their decision making and facilitate their purchase (Gardiner, Grace, and King 2014). However, despite the increased role of technology, 'travelers will still be seeking out high quality immersive or relaxing experiences with friendly customer service. Core draw-cards are likely to remain the established strengths such as food & wine, nature-based tourism, and cultural experiences' (Tourism South Australia 2019), all of which are offered at The Louise. Thus, the future looks bright for a destination such as The Louise, that has built a reputation for its high-quality, extraordinary experiences. However, there is still a need to adapt, be creative, and remain current, as Jim highlighted:

> We need to play to our strengths. We need a clear and consistent vision, which is to keep moving in order to give the very best to each guest in the hotel and the destination. I call it the WOW effect. The main thrust is to surprise and delight the guests at all the points of contact, with a need for a subtle balance of perfection, discretion and recognition. This starts with his arrival. Creativity is essential to the guest the sense of being expected and desired. But do not do too much. A high trend in luxury today is the 'Less is more.'
>
> It seems like the more money the guests have, the more excited they are by simple and authentic experiences. What touches the guest is also to live an 'extra-ordinary' experience, something he cannot get in his daily busy life.

As Jim states, maintaining its core value proposition and offering high-quality, 'extraordinary' experiences to both international and domestic travelers into the future is likely to be an effective marketing strategy for The Louise. Utilizing their deep knowledge of their customers and making small adaptations for the next generation of domestic travelers, should enable The Louise to continue to thrive. Through industry collaborations and a customer-centric approach, The Louise should retain its reputation as a destination for extraordinary food and wine experiences in Australia.

References

Arnould, E. J., and Price, L. L. (1993). River magic: extraordinary experience and the extended service encounter. *Journal of Consumer Research*, 20(1): 24–45. https://doi.org/10.1086/209331

Australian Government Austrade. (2020). *Tourism Research Australia, State of the Industry 2018–2019* [Report]. March 2020. https://www.coursehero.com/file/60192264/TRA-State-of-the-Industry-2018-1pdf/

Baker, M. J., and Cameron, E. (2008). Critical success factors in destination marketing. *Tourism and Hospitality Research*, 8(2): 79–97. https://doi.org/10.1057/thr.2008.9

Barossa Grape and Wine Association. (2020). *Barossa*. https://www.barossawine.com/barossa/

Beames, G. (2003). The rock, the reef and the grape: the challenges of developing wine tourism in regional Australia. *Journal of Vacation Marketing*, 9(3): 205–212. https://doi.org/10.1177/135676670300900301

Becker, L., and Jaakkola, E. (2020). Customer experience: fundamental premises and implications for research. *Journal of the Academy of Marketing Science*, 48: 630–648. https://doi.org/10.1007/s11747-019-00718-x

Ben Tahar, Y., Haller, C., Massa, C., and Bédé, S. (2018). Designing and creating tourism experiences: adding value for tourists. In Sotiriadis, M. (ed.), *The Emerald Handbook of Entrepreneurship in Tourism, Travel and Hospitality* (pp. 313–328). Bingley: Emerald Publishing Limited. https://doi.org/10.1108/978-1-78743-529-220181024

Bonn, M. A., Cho, M., and Um, H. (2018). The evolution of wine research: a 26 year historical examination of topics, trends and future direction. *International Journal of Contemporary Hospitality Management*, 30(1): 286–312.

Brown, G. P., Havitz, M. E., and Getz, D. (2006). Relationship between wine involvement and wine-related travel. *Journal of Travel and Tourism Marketing*, 21(1): 31–46. https://doi.org/10.1300/J073v21n01_03

Bruwer, J., and Alant, K. (2009). The hedonic nature of wine tourism consumption: an experiential view. *International Journal of Wine Business Research*, 21(3), 235–257.

Bruwer, J., and Lesschaeve, I. (2012). Wine tourists' destination region brand image perception and antecedents: conceptualization of a winescape framework. *Journal of Travel and Tourism Marketing*, 29(7): 611–628. https://doi.org/10.1080/10548408.2012.719819

Carreker, J. (2015). Biography. *Restaurant and Catering*. https://magazine.rca.asn.au/jim-carreker/

Charters, S., Fountain, J., and Fish, N. (2009). You felt like lingering ... Experiencing 'real' service at the winery tasting room. *Journal of Travel Research*, 48(1): 122–134. https://doi.org/10.1177/0047287508326508

Conduit, J., and Mavondo, F. T. (2001). How critical is internal customer orientation to market orientation? *Journal of Business Research*, 51(1): 11–24. https://doi.org/10.1016/S0148-2963(99)00044-2

Dawson, H., Holmes, M., Jacobs, H., and Wade, R. I. (2011). Wine tourism: winery visitation in the wine appellations of Ontario. *Journal of Vacation Marketing*, 17(3): 237–246. https://doi.org/10.1177/1356766711409185

Durán-Sánchez, A., Álvarez-García, J., del Río, M., and González-Vázquez, E. (2016). Literature review of wine tourism research: Bibliometric analysis (1984–2014). In Peris-Ortiz, M., de la Cruz Del Río Rama, M., and Rueda-Armengot, C. (eds.), *Level of Implementation of Quality in the Designation of Origin and Monterrei Wine Route* (pp. 257–273). Cham: Springer International Publishing.

Gardiner, S., Grace, D., and King, C. (2014). The generation effect: the future of domestic tourism in Australia. *Journal of Travel Research*, 53(6): 705–720. https://doi.org/10.1177/0047287514530810

Getz, D. (2000). *Wine Tourism: Management, Development and Destinations*. New Jersey: Cognizant Communications.

Getz, D., and Brown, G. (2006). Critical success factors for wine tourism regions: a demand analysis. *Tourism Management*, 27(1): 146–158.

Getz, D., Carlsen, J., Brown, G., and Havitz, M. (2008). Wine tourism and consumers. In Woodside, A. G., and Martin, D. (eds.), *Tourism Management: Analysis, Behaviour and Strategy*. (chap. 15, pp. 245–268) Wallingford: Cab eBook. https://www.cabi.org/cabebooks/ebook/20083075462

Global Wine Tourism Organisation (GWTO). (2020). https://gwto.org/

Great Wine Capital Global Network. (2020). https://www.greatwinecapitals.com/capitals/adelaide-south-australia

Hall, C. M., Sharples, L., Cambourne, B., and Macionis, N. (2009). *Wine Tourism around the World*. New York: Routledge.

Haller, C., Hess-Misslin, I., and Méreaux, J.-P. (2021). Aesthetics and conviviality as key factors in a successful wine tourism experience. *International Journal of Wine Business Research*, 33(2): 176–196. https://doi.org/10.1108/IJWBR-12-2019-0063

Haller, C., Thach, L., and Olsen, J. (2020). Understanding eWineTourism practices of European and North America wineries. *Journal of Gastronomy and Tourism*, 4(3): 141–156.

Hartline, M. D., Maxham, J. G., III, and McKee, D. O. (2000). Corridors of influence in the dissemination of customer-oriented strategy to customer contact service employees. *Journal of Marketing*, 64(2), 35–50. https://doi.org/10.1509/jmkg.64.2.35.18001

Hojman, D. E., and Hunter-Jones, P. (2012). Wine tourism: Chilean wine regions and routes. *Journal of Business Research*, 65(1): 13–21. https://doi.org/10.1016/j.jbusres.2011.07.009

Hollebeek, L. D., Srivastava, R. K., and Chen, T. (2019). SD logic–informed customer engagement: integrative framework, revised fundamental propositions, and application to CRM. *Journal of the Academy of Marketing Science*, 47(1): 161–185. https://doi.org/10.1007/s11747-016-0494-5

Karlsson, J., and Skålen, P. (2015). Exploring front-line employee contributions to service innovation. *European Journal of Marketing*, 49(9/10): 1346–1365. https://doi.org10.1108/EJM-10-2012-0568

Lam, J., Feng, M. C., Treen, E., and Ferreira, C. (2020). The *Journal of Wine Research*: a 30-year bibliographic analysis. *Journal of Wine Research*, 31(3): 176–193, https://doi.org/10.1080/09571264.2020.1816535

Maaninou, N., Boland, G., and Haller, C. (2021). Regional wine branding for US consumers: effects of wine Consumer Expertise, Regional Brand Attitude and Regional Wine Brand Image on Regional Wine Brand Equity – aerial mediation. *7ème Journée de Recherche en Marketing du Grand Est (JRMGE7)*, 2nd April.

MacNeil, K. (2015). *The Wine Bible*. New York: Workman Publishing.

Park, J.-Y., Bufquin, D., and Back, R. M. (2018). When do they become satiated? An examination of the relationships among winery tourists' satisfaction, repeat visits and revisit intentions. *Journal of Destination Marketing and Management*, 11: 231–239. https://doi.org/10.1016/j.jdmm.2018.04.004

Pezzi, M. G., Faggian, A., and Reid, N. (2020). *Agritourism, Wine Tourism, and Craft Beer Tourism: Local Responses to Peripheral Through Tourism Niches*. London: Routledge.

Pine, B. J., II, and Gilmore, J. H. (1998). Welcome to the experience economy. *Harvard Business Review*, 76(4): 97–105.

Quadri-Felitti, D., and Fiore, A. M. (2012). Experience economy constructs as a framework for understanding wine tourism. *Journal of Vacation Marketing*, 18(1): 3–15. https://doi.org/10.1177/1356766711432222

Roberts, L., and Sparks, B. (2006). Enhancing the wine tourism experience: the customers' viewpoint. In Carlsen, J., and Charters, S. (eds.), *Global Wine Tourism: Research, Management and Marketing* (pp. 47–56). Wallingford, UK: CABI.

Sánchez, A., Del Río, M., and García, J. (2017). Bibliometric analysis of publications on wine tourism in the databases Scopus and WoS. *European Research on Management and Business Economics*, 23(1): 8–15. https://doi.org/10.1016/j.iedeen.2016.02.001

Sigala, M., and Robinson, R. N. S. (2019). *Management and Marketing of Wine Tourism Businesses: Theory, Practice and Cases*. Cham: Palgrave Macmillan.

South Australian Visitor Economy Sector Plan 2030. (2019). https://tourism.sa.gov.au/media/s3ynh5hk/satc_tourism-plan-2030_final_aug2019.pdf

Taylor, C., Barber, N., and Deale, C. (2010). Environmental attitudes towards wine tourism. *International Journal of Wine Research*, 2(1): 13–16. https://doi.org/10.2147/IJWR.S6685

Thach, L., and Charters, S. (2016). *Best Practices in Global Wine Tourism: 15 Case Studies from Around the World*. New York: Miranda Press.

The Louise. (2020). https://thelouise.com.au/

The South Australia Wine Story. (2019). https://www.pir.sa.gov.au/data/assets/pdf_file/0006/274065/South_Australian_Wine_Story.pdf

Thomas, B., Quintal, V. A., and Phau, I. (2018). Wine tourist engagement with the winescape: scale development and validation. *Journal of Hospitality and Tourism Research*, 42(5): 793–828. https://doi.org/10.1177/1096348016640583

Tourism South Australia. (2016). *South Australian International Wine Tourism Strategy*. https://tourism.sa.gov.au/media/bg5ljup2/south-australian-international-wine-tourism-strategy.pdf

United Nations World Tourism Organization (UNWTO). (2019). *UNWTO Tourism Definitions*, UNTWO, Madrid. https://doi.org/10.18111/9789284420858

Vo Thanh, T., and Kirova, V. (2018). Wine tourism experience: a netnographic study. *Journal of Business Research*, 83: 30–37. https://doi.org/10.1016/j.jbusres.2017.10.008

Wine Australia. (2019). *Australian Wine Tourism Snapshot*. https://www.wineaustralia.com/market-insights/australian-wine-tourism-snapshot#:~:text=The%20latest%20figures%20on%20wine,an%20average%20of%207%20nights.andtext=It%20is%20also%20a%20strong%20driver%20of%20choice%20for%20visiting%20Australia

Wine Australia. (2020a). *Australian Wine Sector at a Glance*. https://www.wineaustralia.com/getmedia/dbcfd666-748a-4464-9ed7-b7b6e03b7251/MI_Sector_at_a_glance_Aug2020_v2.pdf
Wine Australia. (2020b). *National Vintage Report 2020*. https://www.wineaustralia.com/getmedia/7dcd66a7-86b5-4606-9ab7-416086077099/MI_VintageReport2020.pdf
WineTourism.com. (2020). *Impact of COVID-19 on Global Wine Tourism*. https://www.winetourism.com/wp-content/uploads/2020/12/WineTourism-Covid-19-impact-report.pdf.pdf

40

MARKERS, BENEFITS, AND PITFALLS OF AUTHENTICITY IN WINE TOURISM

A case study in Melnik, Bulgaria

Ilinka Terziyska

Introduction

Ever since it first appeared in the scientific tourism literature in the 1970s, the concept of authenticity has been contradictory. On the one hand, the question of whether tourists seek authenticity during their travels provoked a broad discussion (Boorstin 1962; MacCannell 1976), on the other hand, the very idea of authenticity evolved through several stages, from objectivism, through the theory of constructivism, and finally to the existential authenticity of the postmodern tourist (Wang 1999). Authenticity is claimed to be one of the central topics in the sociology of tourism, and although the debates have been going on for years, it continues to be one of the most controversial constructs in the scientific literature (Kim and Bonn 2016). One of the most accurate comparisons likens authenticity to jelly – its form and nature change depending on the way one looks at it (Heitmann 2011).

Although most studies are related to cultural and heritage tourism (Chhabra, Healy, and Sills 2003; Hede et al. 2014), there is one form of tourism where authenticity is of particular importance – wine tourism. Broadly defined as 'visitation to vineyards, wineries, wine festivals and wine shows for which grape wine tasting and/or experiencing the attributes of a grape wine region are the prime motivating factors for visitors' (Hall 1996, p. 109), this type of travel is intrinsically connected with the identity and culture of the destination (Corigliano and Mottironi 2020). Although this definition limits wine tourism to trips motivated primarily by an interest in wine, in recent years, the scope has been significantly expanded to cover not only the wine connoisseur or wine lover but also any tourist who experiences the wine culture of the destination, without necessarily having this as a primary travel motivation (Sigala and Robinson 2019, p. 1).

Despite the need to better understand authenticity in the context of the wine-related tourist experience, the scientific literature on this topic is still scarce. This chapter aims to contribute to the body of knowledge by offering a conceptual model of authenticity markers in wine tourism based on the winescape concept. Two dimensions (wine and people) are further explored through a wine destination case study to identify potential opportunities and threats in offering an authentic wine tourism product. The destination chosen for the case study is Melnik – the smallest heritage town in Bulgaria and a landmark in terms of local wine.

 DOI: 10.4324/9781003143628-46

Authenticity and tourism

The topic of authenticity has long been present in tourism research and is mainly explored through the prism of sociology. The three main approaches are objectivism, constructivism, and postmodernism.

The *objectivist theory* sees reality as mind-independent; hence, authenticity is inherent to the toured object and is not linked with tourists' perceptions. Thus, the object and the tourist are detached, and authenticity is seen as an absolute objective quality (Terziyska 2012).

The authenticity debate between Boorstin (1962) and MacCannell (1976), which started in the 1970s, is an example of the objectivist view of authenticity – it did not focus on the nature of authenticity, but whether it was important as a travel motivation or not. The discussion was triggered by the question of what tourists really seek. Boorstin argues that modern tourists do not seek authenticity – they are fully aware that what the tourism industry offers them is identical to the purposefully created pseudo-events of their usual environment, i.e., it is not authentic, and they willfully accept this inauthenticity. MacCannell, for his part, insists that authenticity is a central motivation for tourist travel. In his view, tourism can be seen as a quest for authenticity, caused by the alienation of modern people and their belief that the 'real' can be experienced through 'the pristine, the primitive, the natural, that which is as yet untouched by modernity' (MacCannell 1976, p. 384). However, if there is disagreement about the importance of authenticity as a driving force for tourist travel, the opinions on the nature of the tourist product in this debate do not differ. Like Boorstin, MacCannell (1976) argues that what tourists actually receive is inauthentic and introduces the notion of staged authenticity to denote a space developed especially for tourists, which resembles but is not truthful to the original. Based on Goffman's (1956) dramaturgic sociology, MacCannell introduced the concepts of 'frontstage' and 'backstage,' denoting, respectively, the areas developed especially for tourists, and the 'true' local areas where the hosts live and to which tourists usually have no access. It is the backstage where authenticity resides, while frontstage areas are considered inauthentic. Although this division is based on the physical environment, the author emphasizes that social roles are even more important.

Authenticity through the prism of objectivism is a largely museum-linked type of construct (Wang 1999), in which the authenticity of the object is strictly defined and attached to the original and is not influenced by the perceptions of the tourist. This approach is sometimes considered oversimplistic because 'the tourist's recognition of the original as original is deemed an authentic experience' (Ouston 2015, p. 260).

On the other hand, the *constructivist approach* sees authenticity as socially constructed, something that goes beyond the objectivist understanding as original and true (Jones 2010; Mkono 2012). The main representative of this approach (and the one who introduced it in tourism studies) is Erik Cohen. In his seminal work 'Authenticity and Commoditization in Tourism' (Cohen 1988), he turns the notion of authenticity by presenting it as socially negotiated rather than a permanent quality of the toured object. Thus, an object or event that is not authentic in the strict sense of the word can gain the 'patina of authenticity' through cultural change (Cohen 1988). Cohen links the quest for authenticity with the five modes of tourist experience previously proposed by him, claiming that different types of tourists would have differing criteria for the judgment of authenticity. Thus, existential tourists, who seek the most in-depth experience, would be the ones with the strictest criteria, while recreational tourists would be prone to accept a playful, make-believe attitude (Cohen 1988, p. 377). Apart from introducing the notion of the tourist experience in the authenticity debate and overcoming the limitations of the objectivist approach, Cohen's work is important in one more aspect: it draws the attention to the criteria for judging authenticity – or, in other words, the markers of authenticity – which would vary for different types of tourists. A preceding study also examines this issue by identifying the elements that make tourists perceive a tourism product as authentic (Pearce and Moscardo 1986). Most studies on tourism authenticity performed after the 1990s adopt this approach and would use the term *perceived authenticity* to denote the phenomenon's negotiable character (Hede et al. 2014; Ramkissoon and Uysal 2011).

The third approach to conceptualizing authenticity is the *existential* one. The objectivist approach was entirely focused on the toured object, and the constructivist linked the toured and touree. Existential authenticity refers entirely to the tourist and their experience, where tourism experiences are the medium to achieving a state of Being true to oneself (Wang 1999). Thus, it is not only the authenticity of the Other that tourists are after, but authenticity within themselves, and their relations with others (Wang 1999). In this sense, existential authenticity can be both intra- and interpersonal dimensions. This approach has its roots in the postmodernism movement and the notion of the post-tourist, previously described by Urry (1990, p. 11) as someone who not only does not seek (objective) authenticity but enjoys the standard inauthentic tourist product.

Apart from the continuing debate on authenticity as a driving force for tourism travel, many studies have been dedicated to its impact on tourist behavior. Although most of them are related to cultural tourism (Hede et al. 2014; Kolar and Zabkar 2010), a positive relationship has also been established between authenticity and perceived value among Airbnb consumers (Liang, Choi, and Joppe 2018), and event visitors (Wong, Ji, and Liu 2018).

Authenticity has been identified as a driver for customer satisfaction (Chhabra, Healy, and Sills 2003) and contributes to memorable gastronomy tourism experiences (Antón et al. 2019; Dixit 2020). In food tourism, both subjective and object-based authenticity was found to positively affect purchase intention (Sidali and Hemmerling 2014).

As a concept, authenticity in tourism is closely related to sustainable development, mainly in terms of preserving local cultural and social identity (Cohen 2002). Winemaking is often considered a part of a region's or community's heritage; there are even instances of winemaking practices inscribed in the UNESCO Representative List of World Intangible Heritage (UNESCO 2013). A recent study highlights the importance of storytelling in wineries as a powerful tool for heritage branding and a source of competitive advantage (Frost, et al., 2020). Authenticity was identified as the thread that went through the various perceptions expressed by interviewees.

The tourists' quest for authenticity can be used for boosting the sustainable development of both the tourism and the agricultural sector; on the other hand, the authenticity of local food contributes to sustainable experiences by facilitating a deeper connection with the culture and heritage of the toured region (Sims 2009).

A conceptual model of authenticity in wine tourism

As seen from the literature review just presented, authenticity is considered to be among the main motivations for tourism travel and directly impacts customer satisfaction, perceived quality, loyalty, and intention to repurchase/revisit. Due to the connection of authenticity with cultural heritage, most studies have been in the field of cultural tourism, while in the field of wine tourism, these are still very limited. As early as 2004, Cohen and Avieli noted how unusual it was to neglect 'the most bodily of the senses, taste, and more specifically eating and drinking' (Cohen and Avieli 2004) in the sociology and anthropology of tourism, especially given their key role in the tourist experience.

Research on authenticity in wine tourism usually uses previous work to operationalize authenticity elements (Chiappa, Napolitano, and Atzeni 2019). The models are most commonly based on the popular division between objective, constructive, and existential authenticity (Chiappa, Napolitano, and Atzeni 2019; Kim and Bonn 2016). An example of such a model is the one of Kim and Bonn (2016). In it, objective authenticity includes items such as winery location and atmosphere, the possibility to observe the wine production process, wine bottle labeling, information on the wine bottle labels, and wine tasting. Constructive authenticity refers to communication with winery staff (and the wine-related information they provide), visitors' expectations, and experiences at the winery. Existential authenticity is marked by two items: freedom from daily routine, and participation in wine-related activities. Similarly, Chiappa, Napolitano, and Atzeni (2019) use factor analysis to distribute authenticity elements into two groups – objective and existential authenticity. Such a division, however, has some drawbacks. First, the distinction between objectivist, constructed, and existential authenticity should be based not on different markers/elements but on the different foundation upon which the concept is constructed. As explained in the previous paragraph, objective authenticity presupposes independence/detachment of the object from the tourist (the object is either authentic or inauthentic based on certain qualities it possesses). In contrast, in the constructivist approach, we speak of perceived authenticity, i.e., it is the tourist who decides on the elements of the toured object that make it authentic. Thus, a model of authenticity markers should be based on the objectivist approach (by listing the features that constitute authenticity) and then tested to see which of these markers are valid for different types of tourists. Previous research suggests that in the general case, tourists do not seek total authenticity; their judgments are often

> based on some aspects or dimensions which metonymically stand for the authenticity of the object as a whole: the ethnic origins of the producer, the production techniques, the materials used, the designs or colors of the object, or even the locality in which it has been purchased.
>
> *(Littrell, Anderson, and Brown 1993)*

The study of existential authenticity, on the other hand, has a totally different logic – the question it needs to answer is: In what way do wine tourism activities contribute to an authentic state of Being, and could any of these markers contribute to such an experience?

This study proposes a conceptual model rooted in the objectivist understanding of authenticity (by listing possible objective markers) but is then intended to be implemented in a constructivist manner (by testing the validity of these markers for different types of wine tourists. For the study, authenticity is operationalized as 'a judgement or value placed on the setting or product by the observers' (Chhabra 2005, p. 65), and authenticity markers are the product/setting elements that may independently or in total lead to the observer perceiving the object as authentic. Structurally, the model is organized around the concept of winescape, as it is believed to most comprehensively describe the environment that shapes tourist experiences (Terziyska and Damyanova 2020). Several models present winescape dimensions, and these can take either a micro (winery) or a macro (destination perspective). The proposed model is meant to refer to the broader wine tourism experience, so we will use Bruwer and Alant's (2009) model because it portrays the macro perspective. In this model, the winescape has the following dimensions: scenery setting, wine product, people and hospitality, ambience of the region, wineries and wine estates. Table 40.1 presents a slightly adapted version of the winescape dimensions and the respective authenticity markers for each of these dimensions, which are derived from existing literature.

Table 40.1 Markers of authenticity in wine tourism

Winescape dimension	*Marker*	*Study*
Setting/ destination features	Nature-related	Kolar and Zabkar (2010)
	Heritage-related towns	
	Cultural and/or traditional presence of winemaking over time	Groves (2001, p. 252)
	Local food	Chiappa, Napolitano, and Atzeni (2019)
Wineries and vineyards	Small, family-run business	Skinner, Chatzopoulou, and Gorton (2020), Frost et al. (2020)
	Winery blends harmoniously with the environment	Kolar and Zabkar (2010)
	Information and storytelling (family-based, ethnic, place-based)	Frost et al. (2020), Groves (2001), Chiappa, Napolitano, and Atzeni (2019)
	Active participation in wine-related activities	Chiappa, Napolitano, and Atzeni (2019), Kim and Bonn (2016)
Wine product	Local varieties	Littrell, Anderson, and Brown (1993)
	The presence of an authority (assurance of authenticity from a trusted body) – geographic appellations	Groves (2001, p. 252), Kim and Bonn (2016), Downing and Parrish (2019)
	Unique	(Groves, 2001, p. 252), (Littrell, Anderson, and Brown 1993)
	Traditional methods of production	Ivanova, Terziyska, and Trifonova (2014)
	Bio/natural wines	Ivanova, Terziyska, and Trifonova (2014)
Ambient factors	Unique atmosphere, special ambience	Skinner, Chatzopoulou, and Gorton (2020)
People and hospitality	Local staff, opportunity to meet the winemaker	Skinner, Chatzopoulou, and Gorton (2020), Frost, et al., (2020)
	Sincerity of both service staff and local people	Taylor (2001)

Source: Author

The first winescape dimension (setting/destination features) comprises three authenticity markers. Nature was included because the very essence of the quest for authenticity is related to the search for the 'pristine, the primitive, the natural, that which is as yet untouched by modernity' (Cohen 1988, p. 374). Unique natural features/attractions could enhance perceived authenticity because uniqueness is often linked with authentic products (Sharpley 1994). Heritage towns could contribute to perceived authenticity because they provide an environment where cultural heritage can be experienced. Groves (2001) suggested cultural presence over time as an important constituent of the authenticity of food products, and in this model, it refers to winemaking.

Wineries and vineyards are the core attractions of wine tourism. Wine cellars can be perceived as authentic if they are family-run, the architecture blends well with the environment (e.g., traditional/vernacular architecture), and they offer participative activities. Storytelling is also an important aspect (Frost et al. 2020).

Several features may induce perceived authenticity of the core wine product: if it is produced from local varieties (which is enhanced if they are unique for the region), if traditional winemaking methods are employed, or if an authority has verified its authenticity (geographic appellations). Bio and natural wines have been included because they are thought to be clean (close to nature) products, and – in the case of natural wines – as close as possible to traditional methods of home-making with maximum preservation of the character of the grapes.

The ambient factors include a more general perception of the ambiance as special or unique.

People and hospitality are an essential dimension as far as authenticity is concerned, given that research has identified authentic encounter is desired by many tourists (Timothy and Boyd 2006, p. 4).

Case Study 40.1 Positive and negative effects of authenticity

Despite its thousand-year history of wine production, Bulgaria appeared on the stage of world wine production and wine tourism only relatively recently. In the 1980s, the country was among the top five wine exporters (Anderson and Nelgen 2009, pp. 176,196). However, with the end of the socialist era, this changed, and the sector suffered major defeats; recovery did not begin until after 2000. The last decade has been marked by an increased focus on quality resulting in numerous awards at world competitions (Concours Mondial de Bruxelles 2022) and the emergence of many new and promising wineries, as well as specialized wine tour operators. The country is perceived as an emerging wine tourism destination, attractive with its fascinating local varieties and award-winning wines, as well as the fact that it is relatively unknown and, therefore, interesting (Terziyska 2020). This study will consider a specific destination for wine tourism in Bulgaria - Melnik - as it has characteristics that make it a particularly suitable example of the possible consequences of focusing on authenticity in the wine tourism offering.

With a population of no more than 250 people, Melnik is the smallest town in Bulgaria and is emblematic in at least four areas - history, architecture, nature, and wine. Thanks to this combination, it is one of the most popular destinations in Bulgaria. One of the symbols of the city is the wine of the same name - Melnik wine. The autochthonous grape variety from which it is made is called Shiroka Melnishka vine/Broad-leaved Melnik vine, but Melnik is also an umbrella term for several wines that are later modifications of the original, such as Melnik 55 vine and Melnik 1300. Winemaking is a major part of the local population's traditional livelihood and an essential part of the local identity. What is specific to the region and the country as a whole is that much of the production is homemade - in 2017, it accounted for one-fifth of the total (Boshnakova 2019, p. 4).

Regarding the development of wine tourism, this feature has a twofold effect. On the one hand, it gives a special character to the region. It distinguishes the region from other destinations, where wine production is mainly industrial (in wine cellars). On the other hand, it adversely affects wine's domestic consumption and the attitudes towards commercially produced wine in general. The latter is expressed in idealizing homemade wines and the belief that industrially produced wines are of lower quality and not 'real.' Fortunately, the trend is towards a gradual change, especially in younger generations (Boshnakova 2019).

Wine tourism in the region of Melnik can take several forms. There is the conventional winery visit (about seven wine tourism–hosting wineries are in close proximity) and a town visit. The town of Melnik can be called a wine destination in its own right, with a wine museum, wine shops, traditional-style restaurants serving house wine, and some family-owned micro wineries situated in the town, which feature a restaurant and wine tastings. The two options (micro wineries within the town and commercial wineries in the larger region) stand at the two extremes of a continuum, ranging from a strong focus on conservatism in wine production to modern winemaking and keeping in line with the latest trends. The micro wineries are owned by local people, who are proud of their family history as wine producers and adhere to the local variety (Broad-leaved Melnik) and the traditional winemaking process. In order to identify the possible benefits and pitfalls of authentic wine tourism products, this study will focus on these micro producers. The choice is grounded on the following reasons:

1 The wine they offer is made of the autochthonous variety – Broadleaved Melnik vine.
2 The winemaking process is as close to the traditional home-making techniques as possible.
3 The owners are local people and winemaking is a traditional activity in their family history.
4 The owners/winemakers communicate directly/face-to-face with tourists on a regular basis.
5 The wine made from a Broad-leaved Melnik vine can be considered unique to the region, given that all previous attempts to grow the variety in other regions have failed.

This list corresponds to at least six authenticity markers from the proposed conceptual model: it offers a local variety, local staff, history in winemaking, unique grape variety, family-run business, and traditional methods of production.

The case was found suitable for identifying the effects of authenticity because a previous study found an unusual discrepancy in consumer assessments of the perceived wine and service quality in two of these micro wineries (Terziyska 2020, p. 113). Based on user-generated content on TripAdvisor, it became evident that the perceptions were distributed in the two extremes: one-fifth of reviewers thought the quality of the wine was bad, or even terrible, while almost the same share (28% of all reviews in TripAdvisor) perceived it as good or excellent. We performed a qualitative analysis of reviews using a passive Netnography approach to see if it can be attributed to perceived authenticity to explain this anomaly. A total of 61 reviews were included. All reviews are publicly available at the following links: https://www.tripadvisor.com/Attraction_Review-g608698-d2371699-Reviews-Shestaka_Wine_Cellar-Melnik_Blagoevgrad_Province.html and https://www.tripadvisor.com/Restaurant_Review-g608698-d4543937-Reviews-Mitko_Manolev_Winery-Melnik_Blagoevgrad_Province.html.

The latest review as of the writing of this chapter was posted in February 2020, and although it does not refer to wine or service quality, the review is worth mentioning because it validates, on the one hand, the choice of the research object, and on the other, some of the authenticity markers included in the proposed model:

> The place is super authentic. The owner tells about traditions and winemaking. The inside of the cave is very 'game-of-thrones' -ish and it's a worthwhile experience to visit.

> Also, sitting outside to taste the wine with the view of the town and surroundings is really pleasant.

Moving on to older reviews, we read that a visitor received 'not the friendliest of welcomes,' with the owner 'warming up' after chatting about his wines. The winery owner is frequently mentioned, and the visitors' impressions range from 'very nice' and 'welcoming' to 'very aggressive' and 'unpleasant.' One review incorporates the two extremes by describing him as 'Fantastic guy, has his own style, might seem rude to some people but actually, that who he is' (the original spelling of reviews has been kept throughout the chapter). This is an obvious reference to sincerity, which is an element of authenticity emphasized by several authors (Cohen 1988; Taylor 2001). In this case, rude behavior is justified because it is perceived as a sign of sincerity. The 'pride and passion for winemaking' are also pointed out as a feature that makes rude behavior more acceptable.

The other feature that receives conflicting assessments is wine. Some visitors describe it as 'great,' while others think it is 'awful.' Sometimes the good quality is mentioned alongside some authenticity markers, such as storytelling and the attractiveness of nature (as a symbol of the pristine and authentic): 'Their wine is rich and dense and can tell long stories … especially when you try it if you are sitting in front of the cellar … and enjoying the view of the nature around the town of Melnik.' At the same time, a significant share of consumers perceive wine as terrible, but nonetheless, this does not affect their general positive rating. The most common explanations for this are that it is a traditional wine and 'it's not for everyone, but it is worth the experience and expanding your horizon,' 'a matter of taste,' and we should 'keep in mind it's a rustic experience and so is the taste.' The most striking comment describes the low quality of wine as fun: 'Some of the wines are ok, some are terrible, but hey that is the fun!'

From these comments, we can conclude that it is quite possible that for some visitors, the perceived authenticity will have a strong influence on the overall level of satisfaction and mitigate the effect of the low quality of service or of the wine. This is beneficial for the individual business but has a negative impact, too, which relates mostly to the destination. The biggest threat is to foster the belief that authentic wine is associated with low quality. This is particularly important for destinations such as Melnik, where traditional winemaking methods are still preserved. On the one hand, this nourishes perceived authenticity, but on the other hand, when quality standards are not met, this will have a negative effect in the long run, especially on the destination image. It should be borne in mind that only a small share of visitors was ready to justify the low quality of the wine with its authenticity; there are many cases where the perceived low quality had a negative impact on the overall assessment.

Conclusion

Authenticity in tourism has been the subject of debate for more than half a century. Taking a constructivist perspective, we offer a conceptual model for markers of authenticity in wine tourism, which is based on the dimensions of the winescape and covers elements related to both the wider region and the micro level (wineries, the core wine product, and the people who act as mediators of the tourist encounter with the wine culture – the winery staff and owners). The model can serve as a basis for future research on perceived authenticity in wine tourism.

The case study in the town of Melnik confirmed the results of previous research on the positive effect of perceived authenticity on consumer satisfaction in wine tourism (Chiappa, Napolitano, and Atzeni 2019). However, the study is entirely qualitative, and no definite conclusions can be drawn on the scale of this impact. Based on visitor comments, we can only conclude that, for some of them, authenticity was an important factor for a memorable experience and resulted in a positive assessment. One of the interesting findings is that perceived authenticity can, in some cases, not only lead to higher satisfaction but also significantly mitigate the effect of low wine quality or service, especially if the latter is perceived as an authentic encounter. However, this can be seen as both a benefit and a potential danger, as it can lead to low quality being considered a marker of authenticity (some visitors commented it was normal for a traditional wine to taste bad or strange) and building an unfavorable image in the long run. This is especially dangerous in wine destinations like Melnik, which are the 'face' of the region and, because of their attractiveness as a cultural destination, are visited by many tourists – both domestic and foreign. Thus, a large part of the incidental wine tourists form their first and sometimes only impressions from the meeting with the town's wine culture. Those who would suffer most severely from this are the wineries in the area that have been working for years to build a good image of the local wine sector, including local wines.

The study's main limitation is that the proposed model is entirely based on a literature review and will therefore need to be validated through empirical evidence.

Implications

Future research should focus on identifying perceived markers of wine tourism authenticity by including the viewpoints of various stakeholders – tourists, winery owners, tour operators, and DMOs. In practical terms, the findings can help businesses operating in the sphere of wine tourism to meet the quest for authenticity by focusing on the aforementioned markers. DMOs and local authorities can use the case study to ensure sustainable destination development, accounting for the pitfalls of increased local community participation while making use of the benefits it brings.

References

Anderson, K., and Nelgen, S. (2009). *Global Wine Markets, 1961–2009: A Statistical Compendium*. Adelaide: Adelaide University Press.

Antón, C., Camarero, C., Laguna, M., and Buhalis, D. (2019). Impacts of authenticity, degree of adaptation and cultural contrast on travellers' memorable gastronomy experiences. *Journal of Hospitality Marketing & Management*, 28(7): 743–764. https://doi.org/10.1080/19368623.2019.1564106

Boorstin, D. J. (1962). *The Image: A Guide to Pseudo-Events in America*. New York: Vintage Books.

Boshnakova, M. (2019). *Wine Annual Report (Bulgaria)*, Sofia: USDA Foreign Agricultural Service.

Bruwer, J., and Alant, K. (2009). The hedonic nature of wine tourism consumption: an experiential view. *International Journal of Wine Business Research*, 21(3): 235–257. https://doi.org/10.1108/17511060910985962

Chhabra, D. (2005). Defining authenticity and its determinants: toward an authenticity flow model. *Journal of Travel Research*, 44(1): 64–73. https://doi.org/10.1177/0047287505276592

Chhabra, D., Healy, R., and Sills, E. (2003). Staged authenticity and heritage tourism. *Annals of Tourism Research*, 30(3): 702–719. https://doi.org/10.1016/S0160-7383(03)00044-6

Chiappa, G. D., Napolitano, E., and Atzeni, M. (2019). Perceived authenticity, satisfaction and behavioural intentions at wineries. Micro & Macro Marketing. *Rivista Quadrimestrale*, 1: 117–138. https://doi.org/10.1431/93019

Cohen, E. (1988). Authenticity and commoditization in tourism. *Annals of Tourism Research*, 15: 371–386. https://doi.org/10.1016/0160-7383(88)90028-X

Cohen, E. (2002). Authenticity, equity and sustainability in tourism. *Journal of Sustainable Tourism*, 10(4): 267–276. https://doi.org/10.1080/09669580208667167

Cohen, E., and Avieli, N. (2004). Food in tourism: attraction and impediment. *Annals of Tourism Research*, 31(4): 755–778. https://doi.org/10.1016/j.annals.2004.02.003

Concours Mondial de Bruxelles. (2022). Results 2022. https://resultats.concoursmondial.com/en/results

Corigliano, M. A., and Mottironi, C. (2020). Wine and food tourism and place identity. In Foroudi, P., Mauri, C., and Melewar, C. D. T. C. (eds.), *Place Branding: Connecting Tourist Experiences to Places* (pp. 285–301). London: Routledge.

Dixit, S. K. (ed.). (2020). *The Routledge Handbook of Tourism Experience Management and Marketing*. London: Routledge.

Downing, J. A., and Parrish, D. C. (2019). Welcome to my house, do you like the neighborhood? Authenticity differentiation within strategic groups of wineries. In Sigala, M., and Robinson, R. N. S. (eds.), *Management and Marketing of Wine Tourism Business: Theory, Practice and Cases* (pp. 277–294). Cham: Springer International Publishing. https://doi.org/10.1007/978-3-319-75462-8_14

Frost, W., Frost, J., Strickland, P., and Smith Maguire, J. (2020). Seeking a competitive advantage in wine tourism: heritage and storytelling at the cellar-door. *International Journal of Hospitality Management*, 87: 102460. https://doi.org/10.1016/j.ijhm.2020.102460

Goffman, E. (1956). *The Presentation of Self in Everyday Life*. Edinburgh: University of Edinburgh Social Sciences Research Centre.

Groves, A. (2001). Authentic British food products: a review of consumer perceptions.. *International Journal of Consumer Studies*, 25(3): 245–254. https://doi.org/10.1046/j.1470-6431.2001.00179.x

Hall, C. M. (1996). *Wine tourism in New Zealand. Proceedings of Tourism Down Under II: A Research Conference* (pp. 109–119). Dunedin: University of Otago.

Hede, A.-M., Garma, R., Josiassen, A., and Thyne, M. (2014). Perceived authenticity of the visitor experience in museums: conceptualization and initial empirical findings. *European Journal of Marketing*, 48(7/8): 1395–1412. https://doi.org/10.1108/EJM-12-2011-0771

Heitmann, S. (2011). Authenticity in tourism. In Peter Robinson, S. H. P. D. (ed.), *Research Themes for Tourism* (pp. 45–58). Wallingford: CABI.

Ivanova, L., Terziyska, I., and Trifonova, J. (2014). Characteristics of traditional food – the viewpoint of the tourism business. *Service Management*, 14(3): 123–130.

Jones, S. (2010). Negotiating authentic objects and authentic selves: beyond the deconstruction of authenticity. *Journal of Material Culture*, 15(2): 181–203. https://doi.org/10.1177/1359183510364074

Kim, H., and Bonn, M. (2016). Authenticity: do tourist perceptions of winery experiences affect behavioral intentions? *International Journal of Contemporary Hospitality Management*, 28(4): 839–859. https://doi.org/10.1108/IJCHM-05-2014-0212

Kolar, T., and Zabkar, V. (2010). A consumer-based model of authenticity: an oxymoron or the foundation of cultural heritage marketing? *Tourism Management*, 31: 652–664. https://doi.org/10.1016/j.tourman.2009.07.010

Liang, L. J., Choi, H. C., and Joppe, M. (2018). Understanding repurchase intention of Airbnb consumers: perceived authenticity, electronic word-of-mouth, and price sensitivity. *Journal of Travel & Tourism Marketing*, 35(1): 73–89. https://doi.org/10.1080/10548408.2016.1224750

Littrell, M., Anderson, L. F., and Brown, P. J. (1993). What makes a craft souvenir authentic? *Annals of Tourism Research*, 20: 197–215. https://doi.org/10.1016/0160-7383(93)90118-M

MacCannell, D. (1976). *The Tourist: A New Theory of the Leisure Class*. New York: Schocken Books.

Mkono, M. (2012). A netnographic examination of constructive authenticity in Victoria Falls tourist (restaurant) experiences. *International Journal of Hospitality Management*, 31(2): 387–394. https://doi.org/10.1016/j.ijhm.2011.06.013

Ouston, A. (2015). The absence of authenticity and the authenticity of absence in Robert Dessaix's *Twilight of Love*. *Studies in Travel Writing*, 19(3): 259–273. https://doi.org/10.1080/13645145.2015.1074810

Pearce, P., and Moscardo, D. (1986). The concept of authenticity in tourist experiences. *Journal of Sociology*, 22(1): 121–132. https://doi.org/10.1177/144078338602200107

Ramkissoon, H., and Uysal, M. S. (2011). The effects of perceived authenticity, information search behaviour, motivation and destination imagery on cultural behavioural intentions of tourists. *Current Issues in Tourism*, 14(6): 537–562. https://doi.org/10.1080/13683500.2010.493607

Sharpley, R. (1994). *Tourism, Tourists and the Society*. Huntingdon: ELM.

Sidali, L. K., and Hemmerling, S. (2014). Developing an authenticity model of traditional food specialties: does the self-concept of consumers matter? *British Food Journal*, 116(11): 1692–1709. https://doi.org/10.1108/BFJ-02-2014-0056

Sigala, M., and Robinson, R. N. S. (2019). *Management and Marketing of Wine Tourism Business*. Cham, Switzerland: Palgrave Macmillan.

Sims, R. (2009). Food, place and authenticity: local food and the sustainable tourism experience. *Journal of Sustainable Tourism*, 17(3): 321–336. https://doi.org/10.1080/09669580802359293

Skinner, H., Chatzopoulou, E., and Gorton, M. (2020). Perceptions of localness and authenticity regarding restaurant choice in tourism settings. *Journal of Travel & Tourism Marketing*, 37(2): 155–168.

Taylor, J. (2001). Authenticity and sincerity in tourism. *Annals of Tourism Research*, 28(1): 7–26. https://doi.org/10.1016/S0160-7383(00)00004-9

Terziyska, I. (2012). Interpretations of authenticity in tourism. *Science & Research*, 4: 85–94.

Terziyska, I. (2020). *Wine Tourism: Critical Success Factors for an Emerging Destination*. Sofia: Gea Libris.

Terziyska, I., and Damyanova, R. (2020). Winescape through the lens of organized travel: a netnography study. *International Journal of Wine Business Research*, 32(4): 477–492. https://doi.org/10.1108/IJWBR-09-2019-0050

Timothy, D. J., and Boyd, S. W. (2006). Heritage tourism in the 21st century: valued traditions and new perspectives. *Journal of Heritage Tourism*, 1(1): 1–16. https://doi.org/10.1080/17438730608668462

UNESCO (2013). Ancient Georgian traditional Qvevri winemaking method. https://ich.unesco.org/en/RL/ancient-georgian-traditional-qvevri-wine-making-method-00870

Urry, J. (1990). *The Tourist Gaze: Leisure and Travel in Contemporary Societies*. London: Sage.

Wang, N. (1999). Rethinking authenticity in tourism experience. *Annals of Tourism Research*, 26(2): 349–370. https://doi.org/10.1016/S0160-7383(98)00103-0

Wong, I. A., Ji, M., and Liu, M. (2018). The effect of event supportive service environment and authenticity in the quality–value–satisfaction framework. *Journal of Hospitality & Tourism Research*, 42(4): 563–586. https://doi.org/10.1177/1096348015614957

41

CAPTURING CORE EXPERIENTIAL ASPECTS IN WINERY VISITORS' TRIPADVISOR REVIEWS

Netnographic insights from Santorini and Crete

Georgia Papadopoulou and Maria Alebaki

Introduction

In the Experience Economy era, wine tourists – as consumers of destination resources – are increasingly seeking a rich set of diverse elements that expand beyond on-site tasting (Quadri-Felitti and Fiore 2012; Martins et al. 2017; Yeoman, McMahon-Beattie, and Wheatley 2015). The so-called 'wine tourism terroir' incorporates a unique combination of physical, cultural, and natural attributes of the wine region as a whole (Alebaki and Iakovidou 2011; Hall and Mitchell 2002). Thus, the experience of wine tourism is formed by 'a bundle of activities, services, and benefits' (Medlik and Middleton 1973, quoted in Carmichael 2005) that include, inter alia: rural landscape, history and heritage, gastronomy, arts, lifestyle, and education (Charters and Ali-Knight 2002; Dixit 2019; Kirova 2020; Madeira, Correia, and Filipe 2019).

Wine tourism service providers strive to create and deliver valuable and memorable experiences (Sotiriadis and Gursoy 2016). Since visitors' expectations are not always fully met (Quadri-Felitti and Fiore 2016), obtaining information into what satisfies or dissatisfies customers is crucial for companies (Hsieh, Lee, and Yin 2019). To this end, the present chapter aims to improve our understanding of the particular structural components that constitute the wine tourism experience.

The material contained in the chapter is organized as follows. An overview of the relevant literature on the concept of wine tourism experience is provided in the next section. This is followed by a description of the aim and methodology of empirical research conducted in Greece, exploring visitors' comments on social media with respect to their wine tourism experiences on two Greek islands. The results of this netnographic study are then presented, followed by some concluding remarks.

DOI: 10.4324/9781003143628-47

Theoretical background and research questions related to the wine tourism experience

The concept of tourism experience has been approached from many different perspectives, with no widely accepted definition available for the term at present (Dixit 2020; Ritchie, Tung and Ritchie 2011). This lack of consensus can be attributed to the multidimensionality of the construct and the heterogeneity of the individual structures embraced in the tourism experience. However, most of the relevant research efforts have focused either on the travel journey per se or on tourists' activities and expectations, while limited studies have sought to explore the particular components that constitute the tourism experience (Dixit 2020; Dodds 2020). Regarding the latter, Pine and Gilmore's (1998, 1999) 'Four Experience Economy Realms Framework' (i.e., education, entertainment, escapism, and [a]esthetics) stands as one of the most influential models in this area.

Several authors have attempted to build a theory for the wine tourism experience or broaden the existing theoretical perspectives (Gómez, Pratt, and Molina 2019). Table 41.1 presents a summary of the literature review.

Amongst the studies included, a stream of research has focused on testing Pine and Gilmore's Experience Economy model (e.g., Haller, Hess-Misslin, and Mereaux 2021; Quadri-Felitti and Fiore 2013; Thanh and Kirova 2018) and its adaptation to wine tourism. Addressing core elements (Cohen and Ben-Nun 2009; Fernandes and Cruz 2016) and enhancement factors (Roberts and Sparks 2006) of the wine tourism experience have also attracted academic interest. Yet, as Table 41.1 indicates, most efforts focus on specific features of the experience (e.g., sensory impressions), rather than adopting a holistic approach to what wine tourism encapsulates.

Quoting O'Mahony et al. (2006, p. 132), 'it is significant to note that wine experience is a shared experience.' The rapid development of the World Wide Web brought up 'a less personal but more ubiquitous' (Jalilvand and Samiei 2012, p. 592) form of word-of-mouth, the so-called electronic word-of-mouth (eWOM), with online consumer reviews being one of its most accessible and prevalent forms. Therefore, a number of authors have stressed the advantages of utilizing information shared on online platforms in order to explore wine tourists' experiences and satisfaction (Brochado et al. 2020; Simeon and Sayeed 2011). In this respect, content analysis of web reviews is receiving growing attention among wine tourism scholars (Terziyska and Damyanova 2020).

Despite the important progress attained, the majority of researchers in this area have adopted a micro-level approach, focusing on a specific wine route, a winery, or a wine tourism event (Thanh and Kirova 2018). A widening of this perspective was first proposed by Getz and Brown (2006), who called for more empirical evidence that takes into account comparative considerations. Gómez, Pratt, and Molina (2019) further underlined the need for researchers to 'expand the notion of mapping the visitor experience' and conduct a comparative analysis between different regions. Similarly, Quadri-Felitti and Fiore (2012) call for more qualitative research that triangulates the experiential view of wine tourism. It is upon this perspective that the current chapter focuses.

Aim and methodology of the empirical study

The study presented in this chapter was carried out with the objective to provide empirical data on the wine tourism experience and its particular structural components, as perceived by visitors in Greek wineries. Relying on Pine and Gilmore's (1999) 4Es theoretical framework,

Table 41.1 A review of published research on the wine tourism experience

Author(s)	*Contribution*
Hall and Mitchell (2002)	Stressed the significance of region in the experience of wine tourism
Carmichael (2005)	Conceptualized attributes that affect the regional wine tourism experience
Roberts and Sparks (2006)	Identified factors that enhance the wine tourism experience
Cohen and Ben-Nun (2009)	Addressed 'winery atmosphere,' 'cultural activities,' and 'family activities' as core features of the wine tourism experience
Simeon and Sayeed (2011)	Examined the wine tourism experience as offered online by California wineries
Quadri-Felitti and Fiore (2012)	Adjusted Pine and Gilmore's (1999) 4Es model in the case of wine tourism
Quadri-Felitti and Fiore (2016)	Used Pine and Gilmore's framework to compare wine tourism suppliers' design priorities and perceptions with tourists' perceived experiences
Bruwer et al. (2013)	Tested the effects of the experience in tasting rooms on wine brand loyalty
Fernandes and Cruz (2016)	Developed a model to capture the quality of the wine tourism experience
Martins et al. (2017)	Proposed a multisensory virtual-reality wine tourism model
Duan et al. (2018)	Used Pine and Gilmore's (1999) framework to explore China's wine tourism
Thanh and Kirova (2018)	Applied the netnography method and Pine and Gilmore's (1999) 4Es model to address the dimensions of the wine tourism experience
Brochado, Stoleriu, and Lupu (2021)	Identified aspects of sensory experiences in wine tourists' TripAdvisor reviews
Bruwer and Rueger-Muck (2019)	Examined the nature of wine tourist motivations from an experiential-hedonic point of view
Correia et al. (2019)	Evaluated visitors' wine tourism experiences based on online reviews
Madeira, Correia, and Filipe (2019)	Proposed a conceptual model for the wine tourism experience, incorporating motivations, brand equity, satisfaction, and image
Massa and Bédé (2018)	Applied netnography and the consumption value approach to look into the winery experience
Napolitano, Kim, and Del Chiappa (2019)	Proposed a conceptual framework for the wine tourism experience, which incorporates external attributes and tourists' psychological characteristics
Santos et al. (2019)	Examined the influence of the wine tourism experience on visitors' memorable sensory impressions
Sigala (2019)	Stressed the importance of art-based initiatives in designing wine tourism experiences
Haller, Hess-Misslin, and Mereaux (2021)	Used Pine and Gilmore's (1998) and Quadri-Felitti and Fiore's (2012) frameworks to explore discrepancies between wine tourism experiences offered and winery visitors' expectations
Gu et al. (2020)	Investigated the impact of facilitators, constraints, and involvement on wine tourism experience
Santos et al. (2020)	Developed a scale to measure wine tourism experiences

Source: Authors' compilation

we employed a comparative case-study methodology, which was guided by the following questions:

- In which ways do wine tourists express their experiences through social media?
- Which specific aspects of the Experience Economy realms (E's) are most valued by visitors?
- Does wine tourism terroir have an impact on shared experiences?

Following Thanh and Kirova (2018), a qualitative research design was developed for this exploratory study in order to 'provide a more meaningful analysis of all components of the wine tourism experience' (p. 33). In particular, the netnography method was applied to collect and analyze winery visitors' comments shared on social media. Netnography, or *ethnography on the internet*, is a stand-alone tool that is more flexible, unobtrusive, convenient, and economical than focus-groups or interviews (Kozinets 2015). It assists in utilizing digitally contextualized data and is gradually gaining wide acceptance in tourism research (Kozinets 2015; Tavakoli and Wijesinghe 2019).

TripAdvisor was selected as the main data source because it is the largest travel platform globally, having 463 million monthly average unique visitors and more than 878 million reviews and opinions of 8.8 million tourism businesses and experiences (TripAdvisor 2020). Moreover, this platform is regarded as most relevant to examine the experience of wine tourism (Gunasekar et al. 2021; Thanh and Kirova 2018). For the purpose of the comparative analysis, two island wine tourism destinations of the Mediterranean, i.e., Santorini and Crete, were selected on the basis of the following criteria:

- Both islands are wine-centric parts of the South Aegean region with a long tradition in viticulture and winemaking. Santorini is known for its world-class Assyrtiko, and, to a lesser extent, for Athiri and Aidani. Nevertheless, Crete has 11 unique indigenous varieties with a good export potential (Jefford 2016; Wines of Crete 2020).
- Both regions are established as popular travel destinations for national and international tourists.
- Santorini attracts the vast majority of Greece's winery visitors (Triantafyllou-Pitsaki 2005). Wine tourism in Crete is still in its infancy; Yet, its recent development is largely attributed to the collective actions of the local association of winemakers, which, since 2006, are jointly promoting the island's terroir under the trademark 'Wines of Crete.'
- Apart from wine, the 'wine tourism terroir' of both islands is made up of unique elements, including local food and gastronomy, the remarkable natural environment, cultural richness, as well as supporting businesses.
- Both islands have plenty of wineries open to visitors and a high number of reviews on TripAdvisor (In December 2020, there were 8,145 reviews for Santorini and 2.367 for Crete).

In line with Kozinets's (2015) recommendations and Thanh and Kirova's (2018) approach, only comments associated with Hall et al.'s (2000) definition[1] for wine tourism were included in the analysis. This resulted in the selection of 800 comments, following a stratified sampling which was based on the total number of reviews for each winery. Of them, 400 were related to the wine tourism experience in Santorini, while the remaining 400 concerned the Cretan case. Although there was no limitation in terms of length (comments counted from 5 to 300 words), only reviews written in English were analyzed since the majority of them (6,455 or 63.7%) have been written in that language. Moreover, despite that both authors are native speakers, reviews expressed in Greek were not taken into account since they represented only 6.4% of the total (655 reviews). Finally, 800 original reviews – which were posted on TripAdvisor between March 2020 and July 2020 – were extracted.

Pine and Gilmore's (1999) Experience Economy framework 4Es (i.e., the entertainment, educational, escapist, and the aesthetic experiences), adapted by previous studies (Quadri-Felitti and Fiore 2013; Terziyska and Damyanova 2020; Thanh and Kirova 2018), informed the development of a dictionary of themes, which was further enriched by the empirical data. QSR NVivo 10 software was used to support the thematic analysis.

Empirical results

As shown in Figures 41.1 and 41.2, the educational component of the wine tourism experience is the most salient amongst all four Es, as – on both islands – the largest majority of references concern this particular aspect (986 out of 1,792 in Crete, 744 out of 1,411 in Santorini). In line with Thanh and Kirova's (2018) findings, analysis shows that the escapist dimension is underrepresented in both cases.

The large majority of coded items express positive references (93.25% in Crete, 91.21% for Santorini, 92.35% of the total number). Figure 41.3 presents their distribution across the four Es, with education receiving most of the positive evaluations. According to the results, Crete seems to generate more flattering comments with respect to education, compared to Santorini.

It is important to underline that winery visitors particularly appreciate the opportunity to meet in person the wine producer:

> We were honored to meet Mr. … himself, who was an extremely gracious host and treated us as if we were the most important items on his busy agenda.
>
> *(dangK8940XP, 06/08/2018)*

In both regions (23.65% in Crete, 23.53% in Santorini), negative references were mostly related to 'signage and information about wineries,' which is a subdimension of the aesthetic component:

> It is really not easy to find the winery in (region in Crete).
>
> *(joern76870, 11/07/2019)*

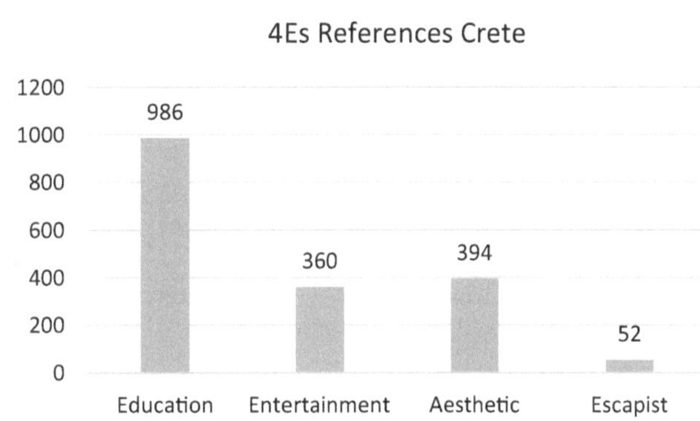

Figure 41.1 Total coded items (references) per dimension in Crete.

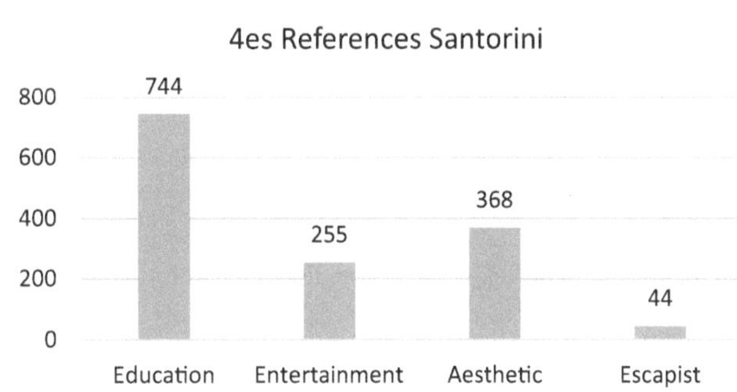

Figure 41.2 Total coded items (references) per dimension in Santorini.

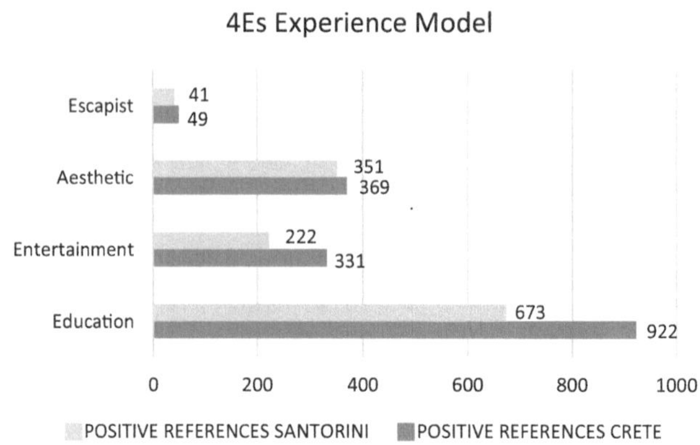

Figure 41.3 Positive references per dimension.

According to Pine and Gilmore's (1998) initial definition, the richest experiences incorporate aspects of all 4Es, forming a 'sweet spot' around the area where the spectra meet. In the case of wine tourists' reviews on TripAdvisor, we adopted Thanh and Kirova' s (2018) assertion that 'a sweet spot comprises at least one item representing each of the four dimensions.' The former study found only one such review. Notably, we identified 24 reviews which incorporated a combination of references related to all 4Es (16 in the case of Crete and 8 in the case of Santorini):

> This is a must for any visitor to Crete. Outstanding winery tour just outside of Chania in the town. Easy to get to and so worth the short drive. The meal is outstanding, the wine incredible and the staff is top notch. It's a peaceful place that really makes you feel like you are one of the natives of Crete. I highly recommend this excursion and make sure you buy the wine. I love it all!
>
> *(Hillsborough, CA. 14/07/2019)*

> We visited this wine Estate on our tour of Santorini. It was such a delight. Very beautiful place. The interior is virtually appealing. Our wine tasting session was very good. The wine expert was very knowledgeable, taking us through each of the wines produced. We also went outside to see the vines. I learnt so much about how vines are grown! I did a similar tour in Cape Town but this was much better. We also saw the production unit. The tasting was good and we bought some wine as take away. This is a must do if you visit Santorini. I strongly recommend.
>
> *(MsMot, 16/10/2019)*

Discussion and conclusions

The concept of tourism experience has attracted research interest from both academia and industry, as the provision of memorable experiences to visitors is a key factor for the successful management of destinations (Agapito, Mendes, and Valle 2013; Kim and Fesenmaier 2017). This study was carried out with the objective to identify the key features that affect wine tourists' post-visit impressions. Pine and Gilmore's (1999) 4Es theoretical model was employed to assist in data interpretation. Netnography was used to analyze 800 winery visitors' online reviews on TripAdvisor. Following a comparative approach, we attempted to assess the experience offered in two major wine tourism destinations of Greece, i.e., the islands of Santorini and Crete.

Empirical findings highlighted the complexity and multidimensionality of the wine tourism experience (Brochado, Stoleriu, and Lupu 2021) as an amalgam of four distinct components, namely, entertainment, education, escapism, and aesthetics. The aforementioned 4Es encompass both wine-related aspects, i.e., tasting and purchase along with supplementary characteristics, i.e., the landscape; the atmosphere; services; human interactions; and local culture (Bruwer and Alant 2009; Cohen and Ben-Nun 2009). In terms of the 'sweet spot,' although it is rare to be achieved (Oh et al. 2007), it was reached in both cases of the present study, confirming Thanh and Kirova's work (2018) as well as the validity of Pine and Gilmore (1998) conceptual model. In total, comments incorporating elements of all four dimensions were found in 24 reviews.

Amongst the four realms, results reveal the critical role of education as a core element of the wine tourism experience, since this particular dimension is prominent in winery visitors' positive references. Findings confirm previous work stressing the importance of education as the main motivation for visiting the winery (Alonso et al. 2014; Brochado, Stoleriu, and Lupu 2021; Bruwer and Alant 2009; Getz and Carlsen 2008), inextricably linked to the degree of satisfaction of visitors (Quadri-Felitti and Fiore 2013). Visitors' online references to education are prominent in both cases (Crete and Santorini) and mainly concern wine tasting, meeting the wine producer, having a tour at the winery, and delving into the history of the wine region.

Results also support other studies suggesting a strong link between wine consumption and the aesthetic component of wine tourism (Brochado, Stoleriu, and Lupu 2021; Charters and Pettigrew 2005; Lin and Mao 2015). In both regions, there was a particular reference to the 'winescape' along with the local buildings and architecture, reinforcing the view that the overall experience in a wine region is significantly affected by the 'aesthetics' (Brochado, Stoleriu, and Lupu 2021; Thanh and Kirova 2018). For instance, several visitors of Santorini made positive comments about the overall experience in the winery, mainly referring to the aesthetic component of the visit.

With respect to the dimension of entertainment, the large majority of reviews concerned wine purchase, product and service prices, special blends, and local produce. In this context, the role of local food products is deemed crucial and could be further promoted through wine tourism. For example, results showed that visitors to Crete made positive references to olives as part of the dining experience. Therefore, synergies toward the organization of wine festivals and exhibitions aimed at promoting local wines and food could make a positive contribution towards this direction. Notably, in both regions, few references concerned music in vineyards, concerts, etc., despite the fact that such activities are often associated with the entertaining component of wine tourism.

Remarkably, comments related to the escapist component of the wine tourism experience were rather limited. As Bosangit (2012) has also stated, reviews to escape stories on travel blogs are minimal. References to this dimension were found in only 96 reviews, 52 of which concerned the island of Crete. Almost all of them were positive and related to the experiences of 'participating in a different time or place' and 'walking in the vineyards.' In general, results agree with Thanh and Kirova's (2018) findings, where the majority of tourists' positive comments revolved around the experience of escape.

It is important to address another key finding: Several reviews reflected visitors' desire for interaction, not only with the winery staff but also with other individuals, indicating a need for active participation in the wine tourism experience. Thus, a new subdimension, 'interaction among visitors,' was added to the existing theoretical framework. Winery visitors on both Greek islands made positive comments regarding this aspect, leading to the conclusion that opportunities for socialization would enhance wine tourism experiences. In the case of Santorini, for instance, analysis of TripAdvisor ratings shows that wine tourism providers could benefit from pursuing a more personalized, tailor-made approach, focusing on providing specialized knowledge, rather than being based on more standardized services.

Implications

This chapter generates both theoretical and practical implications. Firstly, it offers insights into the particular aspects of the wine tourism experience that are most appreciated by visitors. Building on previous work on the 'digital winescape' (Terziyska and Damyanova 2020), the study reports new empirical data that allow for comparisons between two of the most popular wine tourism destinations in Greece. Of course, results presented in the current chapter are not sufficient to provide a comprehensive and detailed standardization framework. Nevertheless, the specific indicators influencing the wine tourism experience – as captured in this study – could be used as a reference point for further investigation. In this context, this piece of work could assist in developing an integrated tool for comparisons between wine tourism destinations in terms of the experience offered.

In particular, recommended directions for future research include: (1) extending the theoretical framework of wine tourism experience by including more items in each dimension, based on empirical findings from both Old World and New World regions; and (2) creating an analytical benchmarking guide for wine tourism experiences. Future research could also focus on exploring whether specific dimensions are related to stronger recommendations (e-WOM) and intentions to revisit a wine tourism destination in the future.

From a managerial point of view, this chapter offers useful insights into the particular mix of attributes that shape the demand for wine tourism in the Experience Economy era. The proposed framework could be utilized by both wineries and wine tourism associations in improving the visitor experience offered. Thus, this chapter could serve as a useful interdisciplinary resource for students, researchers, and industry practitioners who are interested in online marketing as a tool of enhancing the competitiveness of wine tourism destinations.

Note

1 'Visitations to vineyards, wineries, wine festivals, and wine shows for which grape wine tasting and/or experiencing the attributes of a grape wine region are the primary motivating factors for visitors' (Hall et al. 2000, p. 4).

References

Agapito, D., Mendes, J., and Valle, P. (2013). Exploring the conceptualization of the sensory dimension of tourist experiences. *Journal of Destination Marketing and Management*, 2(2): 62–73. https://doi.org/10.1016/j.jdmm.2013.03.001

Alebaki, M., and Iakovidou, O. (2011). Market segmentation in wine tourism: a comparison of approaches. *Tourismos*, 6(1): 123–140.

Alonso, D. A., Bressan, A., O'Shea, M., and Krajsic, V. (2014). Educating winery visitors and consumers: an international perspective. *Current Issues in Tourism*, 17(6): 539–556. https://doi.org/10.1080/13683500.2012.746650

Bosangit, C. (2012). Understanding consumption experiences: a discourse analysis of travel blogs [PhD thesis]. University of Nottingham. http://eprints.nottingham.ac.uk/12763/1/Bosangit_thesis_2012.pdf

Brochado, A., Stoleriu, O., and Lupu, C. (2021). Wine tourism: a multisensory experience. *Current Issues in Tourism*, 24(5): 597–615. https://doi.org/10.1080/13683500.2019.1649373

Brochado, A., Troilo, M., Rodrigues, H., and Oliveira-Brochado, F. (2020). Dimensions of wine hotel experiences shared online. *International Journal of Wine Business Research*, 32(1): 59–77.

Bruwer, J., and Alant, K. (2009). The hedonic nature of wine tourism consumption: an experiential view. *International Journal of Wine Business Research*, 21(3): 235–257.

Bruwer, J., Coode, M., Saliba, A., and Herbst, F. (2013). Wine tourism experience effects of the tasting room on consumer brand loyalty. *Tourism Analysis*, 18(4): 399–414.

Bruwer, J., and Rueger-Muck, E. (2019). Wine tourism and hedonic experience: a motivation-based experiential view. *Tourism and Hospitality Research*, 19(4): 488–502.

Carmichael, B. (2005). Understanding the wine tourism experience for winery visitors in the Niagara region, Ontario, Canada. *Tourism Geographies*, 7(2): 185–204.

Charters, S., and Ali-Knight, J. (2002). Who is the wine tourist? *Tourism Management*, 23(3): 311–319.

Charters, S., and Pettigrew, S. (2005). Is wine consumption an aesthetic experience? *Journal of Wine Research*, 16(2): 121–136.

Cohen, E., and Ben-Nun, L. (2009). The important dimensions of wine tourism experience from potential visitors' perception. *Tourism and Hospitality Research*, 9(1): 20–31.

Correia, A. I., Cunha, R., Matos, O., and Fernandes, C. (2019). Wine tourism experiences and marketing: The case of the Douro Valley in Portugal. In Sigala, M., and Robinson, R. N. S. (eds.) *Wine Tourism Destination Management and Marketing* (pp. 203–220). Cham: Palgrave Macmillan.

Dixit, S. K. (2019), Gastronomic tourism: A theoretical construct, In Dixit, S. K. (ed.), *The Routledge Handbook of Gastronomic Tourism* (pp. 13–23), Oxon: Routledge.

Dixit, S. K. (ed.). (2020). *The Routledge Handbook of Tourism Experience Management and Marketing*. London: Routledge.

Dodds, R. (2020). The tourist experience life cycle: a perspective article. *Tourism Review*, 75(1): 216–220.

Duan, B., Arcodia, C., Ma, E., and Hsiao, A. (2018). Understanding wine tourism in China using an integrated product-level and experience economy framework. *Asia Pacific Journal of Tourism Research*, 23(10): 949–960.

Fernandes, T., and Cruz, M. (2016). Dimensions and outcomes of experience quality in tourism: the case of Port wine cellars. *Journal of Retailing and Consumer Services*, 31: 371–379.

Getz, D., and Brown, G. (2006). Benchmarking wine tourism development. *International Journal of Wine Marketing*, 18(2): 78–97.

Getz, D., and Carlsen, J. (2008). Wine tourism among Generations X and Y. *Turizam: Međunarodni znanstveno-stručni časopis*, 56(3): 257–269.

Gómez, M., Pratt, M. A., and Molina, A. (2019). Wine tourism research: a systematic review of 20 vintages from 1995 to 2014. *Current Issues in Tourism*, 22(18): 2211–2249.

Gu, Q., Qiu, H., King, B. E., and Huang, S. (2020). Understanding the wine tourism experience: The roles of facilitators, constraints, and involvement. *Journal of Vacation Marketing*, 26(2): 211–229.

Gunasekar, S., Das, P., Dixit, S. K., Mandal, S., and Mehta, S. R. (2021). Wine-experienscape and tourist satisfaction: Through the lens of online reviews. *Journal of Foodservice Business Research*. https://doi.org/10.1080/15378020.2021.2006039

Hall, C. M., and Mitchell, R. D. (2002). The tourist terroir of New Zealand wine: the importance of region in the wine tourism experience. In Montanari, A. (ed.), *Food and Environment: Geographies of Taste* (pp. 69–91), Rome: Società Geografica Italiana.

Hall, C. M., Sharples, L., Cambourne, B., and Macionis, N. (eds.). (2000). *Wine Tourism Around the World*. Oxon: Routledge.

Haller, C., Hess-Misslin, I., and Mereaux, J. P. (2021). Aesthetics and conviviality as key factors in a successful wine tourism experience. *International Journal of Wine Business Research*, 33(2): 176–196. https://doi.org/10.1108/IJWBR-12-2019-0063

Hsieh, Y. C. J., Lee, Z. C., and Yin, P. (2019). Wine attractions at hotels: study of online reviews. *International Journal of Wine Business Research*, 31(1): 89–108. https://doi.org/10.1108/IJWBR-05-2018-0020

Jalilvand, M. R., and Samiei, N. (2012). The impact of electronic word of mouth on a tourism destination choice. *Internet Research*, 22(5): 591–612.

Jefford, A. (2016). Jefford on Monday: the call of Crete. *Decanter*, March 14. https://www.decanter.com/wine-news/opinion/jefford-on-monday/jefford-on-monday-the-call-of-crete-295571/

Kim, J., and Fesenmaier, D. R. (2017). Sharing tourism experiences: The posttrip experience. *Journal of Travel Research*, 56(1): 28–40.

Kirova, V. (2020). Value co-creation and value co-destruction through interactive technology in tourism: the case of 'La Cité du Vin' wine museum, Bordeaux, France. *Current Issues in Tourism*, 24(5): 637–650. https://doi.org/10.1080/13683500.2020.1732883

Kozinets, R. V. (2015). *Netnography: Redefined*. London: Sage.
Lin, L., and Mao, P. C. (2015). Food for memories and culture – a content analysis study of food specialties and souvenirs. *Journal of Hospitality and Tourism Management*, 22(1): 19–29.
Madeira, A., Correia, A., and Filipe, J. A. (2019). Modelling wine tourism experiences. *Anatolia*, 30(4): 513–529.
Martins, J., Gonçalves, R., Branco, F., Barbosa, L., Melo, M., and Bessa, M. (2017). A multisensory virtual experience model for thematic tourism: a Port wine tourism application proposal. *Journal of Destination Marketing and Management*, 6(2): 103–109.
Massa, C., and Bédé, S. (2018). A consumer value approach to a holistic understanding of the winery experience. *Qualitative Market Research: An International Journal*, 21(4): 530–548. https://doi.org/10.1108/QMR-01-2017-0031
Medlik, S., and Middleton, V. T. C. (1973). *Product Formulation in Tourism and Marketing*. Berne: AIEST.
Napolitano, E., Kim, A., and Del Chiappa, G. (2019). Re-conceptualising the memorable wine tourism experience: a conceptual model and implications for wineries. In *CAUTHE 2019: Sustainability of Tourism, Hospitality and Events in a Disruptive Digital Age: Proceedings of the 29th Annual Conference* (p. 466). Australia: Central Queensland University.
O'Mahony, B., Hall, J., Lockshin, L., Jago, L., and Brown, G. (2006). Understanding the impact of wine tourism on post-tour purchasing behaviour. In Carlsen, J., and Charters, S. (eds.), *Global Wine Tourism: Research, Management and Marketing* (pp. 123–137), Wallingford: CABI.
Oh, H., Fiore, A. M., and Jeoung, M. (2007). Measuring experience economy concepts: Tourism applications. *Journal of Travel Research*, 46(2): 119–132.
Pine, B. J., and Gilmore, J. H. (1998). Welcome to the experience economy. *Harvard Business Review*, 76: 97–105.
Pine, B. J., and Gilmore, J. H. (1999). *The Experience Economy: Work Is Theatre and Every Business a Stage*. Boston: Harvard Business Press.
Quadri-Felitti, D., and Fiore, A. M. (2012). Experience economy constructs as a framework for understanding wine tourism. *Journal of Vacation Marketing*, 18(1): 3–15.
Quadri-Felitti, D., and Fiore, A. M. (2016). Wine tourism suppliers' and visitors' experiential priorities. *International Journal of Contemporary Hospitality Management*, 28(2): 397–417.
Quadri-Felitti, D. L., and Fiore, A. M. (2013). Destination loyalty: effects of wine tourists' experiences, memories, and satisfaction on intentions. *Tourism and Hospitality Research*, 13(1): 47–62.
Ritchie, J. B., Tung, V. W. S., and Ritchie, R. J. (2011). Tourism experience management research. *International Journal of Contemporary Hospitality Management*, 23(4): 419–438.
Roberts, L., and Sparks, B. (2006). Enhancing the wine tourism experience: the customers' viewpoint. In: Carlsen, J., and Charters, S. (eds.), *Global Wine Tourism: Research, Management and Marketing* (pp. 47–55). Wallingford: CABI.
Santos, V., Caldeira, A., Santos, E., Oliveira, S., and Ramos, P. (2019). Wine tourism experience in the Tejo region: the influence of sensory impressions on post-visit behaviour intentions. *International Journal of Marketing, Communication and New Media*, 5: 54–75.
Santos, V., Ramos, P., Almeida, N., and Santos-Pavón, E. (2020). Developing a wine experience scale: a new strategy to measure holistic behaviour of wine tourists. *Sustainability*, 12(19): 8055.
Sigala, M. (2019). Managing and marketing wine destinations with and through art: a framework for designing wine experiences. In Sigala, M., and Robinson, R. N. S. (eds.), *Wine Tourism Destination Management and Marketing: Theory and Cases* (pp. 221–243). Cham: Palgrave Macmillan.
Simeon, R., and Sayeed, L. (2011). Examining the online wine tourism experience of California wineries. *International Journal of Online Marketing (IJOM)*, 1(1): 24–40.
Sotiriadis, M., and Gursoy, D. (2016). *The Handbook of Managing and Marketing Tourism Experiences*. Bingley: Emerald Group Publishing.
Tavakoli, R., and Wijesinghe, S. N. (2019). The evolution of the web and netnography in tourism: a systematic review. *Tourism Management Perspectives*, 29: 48–55.
Terziyska, I., and Damyanova, R. (2020). Winescape through the lens of organized travel – a netnography study. *International Journal of Wine Business Research*, 32(4): 477–492. https://doi.org/10.1108/IJWBR-09-2019-0050
Thanh, T. V., and Kirova, V. (2018). Wine tourism experience: a netnography study. *Journal of Business Research*, 83: 30–37.
Triantafyllou-Pitsaki, M. E. (2005). Promoting Greek wine through foreign visitors in Greece. In *1st Conference of Greek Wine*. Diaselo Conference Center, Metsovo, Greece (pp. 1–3).

TripAdvisor. (2020). https://www.tripadvisor.com.gr

Wines of Crete. (2020). http://www.winesofcrete.gr

Yeoman, I., McMahon-Beattie, U., and Wheatley, C. (2015). The future of food tourism: a cognitive map(s) perspective. In Yeoman, I., McMahon-Beattie, U., Fields, K., Albrecht, J. N., and Meethan, K. (eds.), *The Future of Food Tourism: Foodies, Experiences, Exclusivity, Visions and Political Capital* (pp. 237–278). Bristol: Channel View Publications.

42

WINE ROUTES AND TRAILS VIS-À-VIS WINE TOURISM

Experiences and examples from Slovenia

Janez Bogataj

Introduction

Viticulture and winemaking have a history dating back more than 2,300 years in the territory that makes up present-day Slovenia. Over the course of centuries, its geographical situation has facilitated the development of three different wine regions – Podravje (in the Drava Basin), Posavje (in the Sava Basin), and Primorska (the coastal region) – resulting in a diversity of terroirs and a wide range of varieties and styles, including many original (autochthonous) and regional varieties. Around 16,000 hectares of land in the country's three wine regions are officially registered as vineyards, with an annual output of around 70 million liters of wine (Nemanič and Bogataj 2004, p. 9). Numerous experts, both Slovene and foreign, rate Slovenia's wine production as being of the highest quality and at the same time as 'boutique.' The latter characteristic is a great advantage when it comes to connecting viticulture and winemaking to tourism, or rather, to that special segment known as wine tourism.

Wine tourism began to develop in Slovenia in an extremely modest and unsystematic fashion around two decades after the end of the Second World War, for the most part in connection with the development of village tourism (or farm tourism) in the late 1960s. This was also the period in which the subject of wine routes was raised for the first time. The response, however, was underwhelming, and the matter remained more or less at the level of amateur initiatives. For the most part, these revolved around paving roads through vineyards, which was seen at the time as one of the principal preconditions for the development of this sector of tourism. The consequences of these unsystematic first steps are still visible today, since every wine route in Slovenia is marked and signposted in a completely different way. The amateur approach was also reflected in the arrangement of the vineyard landscapes along the wine routes. Individual owners began placing items of cultural heritage, in particular various kinds of tools and equipment (e.g., old winepresses), outside their wineries, along with wine barrels converted into furniture. Meanwhile, old barrel staves were used to make sculptures and other unlikely creations.

It was not until 1993, two years after Slovenia gained independence from the former socialist Yugoslavia, that the first Wine Routes of Slovenia project was drawn up. This envisaged 20 wine routes with a total length of around 900 km. Over time, wine and wine-producing

DOI: 10.4324/9781003143628-48

regions increasingly began to be included in tourism promotion. The term 'wine tourism' did not truly establish itself in its own right until after the year 2000, in conjunction with efforts to give gastronomy a more prominent role in tourism.

This chapter discusses the development of wine tourism in Slovenia as an important part of gastronomic tourism, its structure, and the typology of wine (gastronomic) tourists. The Slovenian experience can be usefully used in countries where they want to develop wine tourism.

Evolution of wine tourism in Slovenia

Slovenia did not have a Gastronomic Tourism Strategy until 2006. The gastronomic strategy is a fundamental document for the development and promotion of gastronomy and cuisine and includes the cultural heritage of nutrition and modern development in this field. The role in education is also important. For international recognition, the Strategy established the culinary pyramid and gastronomic regions of Slovenia.

This document was the first tourism strategy to consistently include wine (or wine culture). Wines and the matching of wines with specific dishes were incorporated into Slovenia's gastronomic pyramid, which today consists of 24 gastronomic regions and more than 430 typical dishes and beverages. Naturally, the importance of wines is much broader than merely matching them with food. It is precisely because of wine's role in tourism that it is linked to numerous aspects of contemporary life in Slovenia and to the country's natural and cultural heritage.

In recent years Slovenia has seen the introduction of a new tourism product called vineyard cottage tourism. Known in Slovene as *zidanice* (Figure 42.1), vineyard cottages in the Posavje

Figure 42.1 Old vineyard cottage or *zidanica* in Posavje Region.

Source: Author's private photo documentation

wine region are traditional structures found in vineyards consisting of a wine cellar and a ground floor with temporary living accommodations, separated from the permanent dwelling of the winegrower or winemaker. The Podravje wine region is also led by stand-alone wine stores or cellars known by a variety of names. Wine cellars in the Primorska wine region are usually part of the winegrower or winemaker's home. *Zidanice* throughout the Posavje region experienced a remarkable boom during the period of socialist Yugoslavia. Owners began remodeling old wooden or masonry structures, in some cases demolishing them and building new ones in an unregulated fashion. These frequently self-built structures began to significantly change the identity of the vineyard landscape. Today vineyard cottage tourism is being introduced as a new and attractive tourism product. Unfortunately, the connection with tourism has not been seen to a significant extent as an opportunity for the conservation of older buildings, or even for the construction of modern buildings that could nevertheless play a part in shaping local and regional landscape values. It is evident from this that wine tourism has undergone interesting development, from the initial impulse to pave wine routes with asphalt to today's conception based on the discovery of Slovenia's diverse natural and cultural heritage and aspects of modern life through elements of the country's viticultural and winemaking traditions.

The relationship between gastronomic tourism and wine tourism

Gastronomic tourism is today a relatively well-established type of tourism around the world. This form of tourism builds its content on numerous components of the food culture of a specific locality or region, or wider geographical area. Gastronomic tourism is not only about individual dishes and their preparation and service; it involves the entire range of food culture and its connections with the material, social, and spiritual aspects of the life of an individual, family, or group, of the habitants of rural and urban environments, regions, countries, and even continents. The concept of gastronomy, therefore, also includes the role of beverages in the lives of individuals or groups, so the use of the term 'wine tourism' is actually nonsensical. The frequently used formulation *food and drink* is, in reality, erroneous, since all beverages, including wine, are merely one of the components of food in the sense of nourishment. The use of the term 'wine tourism' is therefore acceptable only in cases when we are talking about 'focused' tourism products and tourism programs, and only in wine-growing areas and involving the tasting of various varieties of wine or the attendance of specifically wine-related events. Even in these cases, however, we cannot ignore the dishes that traditionally accompany these special activities. Just as there can be no gastronomic tourism without beverages, so we cannot discover and understand beverages without the food that goes with them. Since the turn of the 21st century, there has also been an emphasis on tourism products that highlight the matching of wines with food and vice versa. In Slovenia, this tendency has made a significant contribution to raising the level of wine-drinking culture, particularly in catering establishments. A great deal of consideration is given to the selection of appropriate dishes and the most suitable wines to accompany them, something that is also increasingly reflected in the composition of menus. The same applies to the many local and regional tastings organized both in wine regions and elsewhere, where attempts are made to offer a complete presentation of a specific winery, location, or landscape with the help of selected wines and accompanying snacks. Understanding wine within broader gastronomic contexts is gradually becoming a tourism product in itself, and one that also has considerable importance and a notable effect on wine routes, above all by playing an important part in forming a characteristic typology of wine/gastronomic tourists, and through them, shaping the structure of wine/gastronomic tourism.

Structure of wine tourism and typology of wine tourists

The structure of wine tourism on wine routes consists of the following targeted or thematic contents:

Vineyard landscape, wine-growing area: a typical type of agrarian landscape consisting of different types of vineyards, wine cellars, winegrowers' settlements (villages) and the connections between them.

Wine route: a marked tourist or cultural route or road in a wine-growing area (Figure 42.2).

Vineyard and wine estate, winery: different types of vineyards. Wine cellars can be part of the farm (vineyard) residential farmhouses or independent buildings in vineyards.

Catering establishment(s): taverns and restaurants, wine shops, wine cellars with catering offer, other occasional forms of sale of wines, foodstuffs, and dishes.

Museum of viticulture and winemaking; viticultural and winemaking heritage: professionally arranged private collections and museums of viticultural and wine cultural heritage. The vast majority of these are dominated by amateurly presented objects of this heritage, which is often also dysfunctionally included in the offer.

Wine event, festival: a whole series of occasional or annual events, including calendar customs and habits associated with the traditions of viticulture and winemaking.

Wine shop: different types of wine sales directly in wine-growing areas or on wine-growing estates to professional buyers and tourists.

Professional wine event/meeting: various professional and other meetings in wine-growing areas for experts and other target groups.

Figure 42.2 Wine route in Podravje Region.

Source: Ponikvar, from Tourism Maribor, Slovenia

Wine societies and associations: the operation of societies and associations in wine-growing areas or the possibility of their occasional visits and meetings.

All of the these contents also include wine tasting on wine routes. Within the context of the latter, there are numerous smaller elements that also play a part in making up this typology. The list that follows contains those contents that are the most widespread and at the same time have the most positive impact on comprehensive wine route planning. We have divided them into two groups: first, those contents that are developed as part of viticultural and winemaking heritage, and then more recent contents.

First place in the first group undoubtedly belongs to the *harvesting of the grapes*, which is not merely an agricultural activity but a social event, and an opportunity to experience the conclusion of work in the vineyard and the start of the winemaking process. Involving tourists in this process fills more jobs by wine-growing families, and at the same time enables tourists to experience one of the most important milestones of the wine year. By tradition, of course, everyone who attends the harvest is also festively entertained with local and regional dishes. This process, introduced by tourism, often perfectly fills in the issues of human power required for manual grape picking, which is often lacking.

Next, there are a number of very important holidays or festivities, of which the most significant is *Saint Martin's Day or Martinmas* (November 11). In Slovenia, Saint Martin's Day has for centuries been the biggest celebration of the new vintage (Bogataj, Brejc, and Bratovž 2010, pp. 5–38). Particularly since the Second World War, it has spread from family contexts into wider society as a special type of public event. Every year there are smaller Martinmas celebrations that take place in a circle of family and/or friends, as well as big public Martinmas events in towns, cities, and individual wine regions (Figure 42.3). The number of these large-scale Martinmas events has been growing steadily since 1991, the year Slovenia became independent, as shown by Table 42.1.

In all wine-growing regions, great changes have taken place since the end of the Second World War in the field of the central celebration represented by Saint Martin's Day. These substantive changes in holiday rituals were both the result of a new ideology as well as the transfer of foreign ritual practices in the characteristics of wine-growing regions. Many inventions of traditions belong to this time, which in no way represent the continuation and identity of centuries-old ritual heritage. Saint Martin's Day becomes an opportunity for the often banal role-playing and costume of Bishop Martin with his entourage. It is a kind of drunken folk theater that travels from one wine cellar to another. Of course, these contents cannot be described as positive efforts in the context of modern martinis, as they have nothing to do with continuity from different bases in cultural heritage. This establishes a kind of global holiday that does not contribute to the range of cultural – i.e., creative – differences that are characteristic of individual wine-growing areas. For the needs of tourist products, of course, all such forms are acceptable, as they co-create uncritical content products that do not contribute in any way to the development of noble content for visitors to wine roads in the days before and after Saint Martin's Day. However, these holidays do not have as mass manifestations as is typical of Saint Martin's Day. This also applies to developments on wine roads.

Besides Martinmas, which may be characterized as the biggest secular holiday in Slovenia, there are celebrations on *St. Urban's Day* (May 25) and the *Feast of St. John the Evangelist* (27 December). St. Urban's Day coincides with the flowering of the grapevine, while the Feast of St. John the Evangelist is the festivity of the wine of the new vintage (Bogataj 2011, pp. 293–311).

Another special event deriving from the time of the emperor Charlemagne is an eight-day period of serving wine to seated or standing customers in the context of what is known as an

Figure 42.3 Saint Martin's Day at the wine cellar on wine route.

Source: Bogataj (2017)

Table 42.1 Number of large-scale Martinmas celebrations

Year	*Number of celebrations*
1994	7
1995	12
1997	35
2006	44
2009	55

The number of small family-and-friends celebrations is even greater and impossible to put a number to.

osmica (from the word *osem*, meaning eight) or, in some parts, *pušelšank* (*pušenšank*, *bušenšank*) all deriving from the German *Buschenschank*, meaning 'wine shop under the bundle of twigs.' The emperor allowed winegrowers and winemakers to sell their wine to customers tax-free for a period of eight or more days in order to boost the sale of old stocks of wine. The wineries that sold their wine in this way were indicated by a bundle of juniper, ivy, or pine twigs, hence the name *pušelšank*. The name *osmica* derives from the length of the period for which they were allowed to serve and sell wine. In Slovenia, this custom survives in the Primorska wine region, but not in the other two. This interesting aspect of wine heritage is undoubtedly an excellent opportunity to introduce original content to the wine route experience. The tradition of marking wineries with bundles of twigs could also serve as a welcome piece of visual communication.

All three of Slovenia's wine regions organize a wide variety of *wine days*, which are usually linked to the local and regional dishes characteristic of the food culture along a specific wine route. These events can be dedicated to a single variety (such as the Ribolla Festival in Višnjevik or the Teran and Prosciutto Festival in Dutovlje) or to multiple varieties (e.g., Brda and Wine in Šmartno, in the Brda wine district, or the Wine Spring event in Metlika). Wine and food days are also relatively frequent outside areas with wine routes, such as urban environments (e.g., the Orange Wine Festival in Izola).

A wide variety of *hikes* and cycling tours (such as the Vipava Valley Winery Tour or the Hike from Litija to Čatež) are tied in with experiencing the natural and cultural environment of wine routes. In these cases, recreation is complemented by educational and social contact.

Although opportunities for the organization of wine *fairs* and *exhibitions* are somewhat limited along wine routes (given the lack of suitable premises), numerous smaller locations nevertheless organize presentations of the wines (and typical dishes) produced along individual wine routes. Events of this type are particularly characteristic of autumn and spring. These events frequently involve *wine competitions*, which make an important contribution not only to raising awareness of the wines of specific wine routes but also to the education of winegrowers and winemakers.

Slovenia's wine routes and wine regions also play a part in shaping the image of cultural events, particularly in the *literary* and *artistic* fields. One of the bigger literary events is Days of Poetry and Wine in Ptuj. Some wineries organize art colonies, where as well as painting pictures, the participants decorate the heads of wine barrels, in a special interpretation of the symbiosis of wine and art (e.g., at the Ščurek winery in Plešivo, Brda wine district) (Figure 42.4).

Safety is an important consideration when it comes to wine routes, particularly in Slovenia, where alcohol is one of the leading contributing factors to road accidents. In recent years various new ways of following wine routes have been introduced (alongside the hiking and cycling options already mentioned). These include group transport in the form of *hop-on, hop-off services* using a variety of (electric) vehicles that take visitors from one winery to the next.

If wine tourism is considerably varied in terms of content, there is also a varied typology of wine/tourists who follow wine routes either individually or in groups. They can be divided into the following groups:

Friends of wine and wine connoisseurs: This group is numerically most strongly represented on wine roads. It is considered the most welcome. There is also an increasing number of international tourists who want to learn about the stories of wines and viticulture and its role in the life of the local population and their culture on the wine roads.

Insatiable drinkers: The main goal of this group of tourists is to drink large quantities of wine. They appreciate the pleasant ambiance of wine cellars, but they are not interested in the local environment and the culture of the local population.

False connoisseurs: Visiting wine-growing areas and tasting wines is not a question of (tourist) fashion but of systematically getting to know and acquiring knowledge. This orientation is often tried to be surpassed by individuals who claim to be great connoisseurs, but their fundamental tendency is to stand out socially with some knowledge that may also be fictional.

Figure 42.4 Wine region Goriška Brda in Slovenia.

Source: Bogataj (2019)

Wine snobs: Snobbery is also present in the field of wine and wine culture. This group of tourists is about exposing individuals who are consolidating their worth with apparent knowledge.

Experts (oenologists, sommeliers, wine judges, wine tasters, winegrowers, and winemakers): This group of tourists is the least problematic and is based on professionalism. Their visit to the wine roads is often also associated with various professional meetings.

Members of wine/gastronomic societies and associations: These are very desirable groups of tourists who have extensive knowledge about wine in general and gastronomy. They supplement the acquired knowledge with frequent visits to various wine roads, make comparisons of wines, and are interested in the life and culture along the wine roads.

Wine merchants: This group consists of only indirect tourists, as their main purpose is the wine trade. The residence time on the wine road is also relatively short.

Novices: As the name suggests, newcomers encounter wine and wine culture for the first time. Therefore, it should be given as much attention as possible, as they can become big fans who will often return to the wine roads.

Youngsters: This group of tourists is perhaps the most challenging. Due to various negative impacts (e.g., alcoholism), it is necessary to pay more attention to teaching. Bringing the proper understanding regarding consuming wine and the role of wine in the natural and cultural environment of wine roads is one of the basic tasks and content offerings on the wine roads.

One of the principal reasons for visiting wine routes in Slovenia is to buy wine directly from producers, which always involves a tasting, frequently with accompanying snacks. Many tourists buy their wine directly from wineries and combine their visit with a tour of one of Slovenia's three wine regions. A fundamental aspect of the successful functioning of a wine route and the products and services it offers is ensuring that wineries are open for people to visit them, which in Slovenia is quite a frequent problem, particularly during the week. For many vineyard owners and winemakers, wine is not their principal activity or occupation. Some wineries address the problem of opening hours by requiring customers to book in advance, which, however, is more of an obstacle than a boost to sales.

Conclusions

A wine route is a marked tourist/thematic/cultural route through wine country. It connects wineries, farms, wine shops, and other providers involved in viticulture, winemaking, catering, and tourism. The contents of a wine route include all the natural and cultural sites of interest along the way as well as surviving examples of immovable, movable, and (in some cases, depending on content) intangible cultural heritage. A wine route is, therefore, a synthesis of the entire gamut of activities connected to the cultivation of grapevines and the production of wine – in other words, to everything offered by the population of a given locality or region. Wine routes lead us to people and are therefore an opportunity to learn about cultural differences, something that numerous wines also offer. Wine routes offer individuals and groups a chance to leave the busy motorway of everyday life and experience the vineyard landscape and culture in all seasons of the year. As part of efforts to promote modern and sustainable tourism, wine routes should promote sustainable mobility (walking, hiking, cycling, riding, electric vehicles, etc.) in environmentally friendly ways. Through the work and expertise of vineyard owners and winemakers, through vineyards and grapevines, through wine cellars and wine, we learn about everyday life and festivities, heritage and tradition, contemporary commercial endeavors, and forms of spiritual creativity, in all of which wine occupies its own special place. Wine roads allow tourists to get to know the life of the vine and vineyard throughout the year, i.e., in all seasons. Thus, they face the life of nature, its changes and try to understand the role of man in this process and the positive components of his knowledge, which are necessary for the development of viticultural and wine culture. Only wine routes that are understood and conceived in this way can represent a positive form of tourism in the modern world.

Implications

We should not overlook the question of wine marketing, a sector that is constantly inventing new products connected with wine. Wine has been the subject of intensive marketing ever since it first began to be considered an essential part of the notion of human abundance and prosperity. We might therefore ask whether wine tourism is simply one more of the numerous plans for marketing wine that we have encountered in the recent past, including at the international level. These include the fashion for barrique wines and also the currently popular orange wines, the inventing of traditions such as the coronation of the wine queen, experiments with biodynamic methods of winemaking, the creation of a special type of wine tourist or wine lover, the founding of various wine societies, and so on. Winemaking in Slovenia has followed and continues to follow these generally global tendencies and wine 'fashions,' which frequently tend more in the direction of uniformity of tastes than of maintaining local and regional specificities and creating new ideas with local or regional character. It is, however, the latter that makes it possible to incorporate long-term, 'friendly' content linked to wine culture. It is, therefore, necessary to ask ourselves how well we actually know these contents and whether we are making sufficient efforts to discover them. With the support of wine marketing, we can anticipate an increase in the development of wine tourism in the future. However, we must take into account the growing efforts to limit drinking, which is associated with awareness of the negative effects on health and the rise of alcoholism. The presented strategy and typology indicate the correct direction of the future development of wine tourism. This means learning about natural environments with vines, discovering knowledge about wines, their connection with proper and healthy nutrition, and understanding viticulture and winemaking as a complete cycle of human economic endeavor, social relations, and forms of spiritual creativity.

References

Bogataj, J. (2011). *Slovenija praznuje/Celebrate Slovenia/*. Ljubljana: Mladinska knjiga.

Bogataj, J., Brejc, D., and Bratovž, J. (2010). *Martinov praznik/St Martin's Day*. Ljubljana: Založba Kmečki glas.

Nemanič, J., and Bogataj, J. (2004). *Wines of Slovenia*. Ljubljana: Rokus Gifts.

PART VI

Innovations and technological advancements in wine tourism

43

THE ROLE OF TECHNOLOGY IN WINE TOURISM

Roberta Garibaldi

Introduction

Since the late 1990s, the wine industry has constantly expanded from simply growing grapes and selling wine to providing experiences to visitors. Wine tourism has boomed as a consequence, with novel and more sophisticated proposals that have been raised for the amusement of travelers. New technologies have gradually become an integral part of this industry, from simply facilitating travelers in the information-seeking process to revolutionizing the way the wine experience is created and consumed (Duarte Alonso et al. 2013; Kirova 2020). Novel applications include wine booking platforms and virtual marketplaces as well as technology-based winery experiences (i.e., exploited through multitouch tables and screens, virtual and augmented reality) (Garibaldi and Sfodera 2019). The global health emergency in 2020, while severely affecting wine tourism businesses, stimulated the development of innovative concepts such as digital tastings and virtual tours, which are now being increasingly included as a part of the wine tourism offering (Garibaldi and Pozzi 2020).

This chapter focuses on the application of technology in the wine tourism experience to provide information to customers, allow the purchase of wine and experiences, and encourage future onsite visits. In doing so, it also outlines the impacts of the three main elements that constitute the wine tourism experience: learning, hedonism, and the 'winescape.' Recent data coming from international surveys (Garibaldi 2020a, 2020b) and examples are also included to highlight the digital transformation of the experience due to the health emergency.

The wine tourism experience

Beginning in the mid-1990s, wine tourism research began to be published in academic journals. Due to the experiential nature of wine tourism, academics have increasingly paid attention to the experience concept drawing from the model proposed by Pine and Gilmore (1998) (Quadri-Felitti and Fiore 2012; Thanh and Kirova 2018). By reviewing the literature, three main dimensions constituting the wine tourism experience can be identified: learning, hedonism, and the 'winescape.'

The educational aspect of the wine tourism experience deals with the fact that learning about the wine and the winemaking process is among the primary motivations to visit a winery

DOI: 10.4324/9781003143628-50

(Bruwer 2003). Tastings and guided tours are among the most practiced activity by visitors and trigger the educational aspects, allowing them to gain knowledge about and test the product (Alant and Bruwer 2010; Jago and Issaverdis 2001; O'Neil and Charters 2000). Learning is developing through meeting with the winemaker and the winery staff as well as devices (both nontechnological and technological, i.e., touchpoints, multitouch tables). The wine tourism experience has an inherent element of hedonism because it involves pleasure, socializing, and relaxation, which points to indulgent activity (Alant and Bruwer 2004). The special nature of products and activities, such as wines, wine tastings, tours in the cellar or the vineyard, wine demonstrations, events, and festivals, arise feelings of delight, enjoyment, joy, and relief. By meeting the pleasure (hedonic) motivation of visitors, they drive them to consumption (Bruwer and Rueger-Muck 2018). The 'winescape' refers to the 'combination of vineyards, wineries and supporting activities necessary for modern production' (Peters 1997, p. 124). This concept, drawing from Bitner's (1992) 'servicescape' is a multidimensional construct that includes physical, social, and cultural components of a wine region or of a winery (Quintal, Thomas, and Phau 2015). Each attribute (i.e., the natural landscape, the winery environment, its people, and staff) can positively or negatively influence travelers' attitudes and behaviors as well as their satisfaction and their visit or revisit intentions (Byrd et al. 2016; Dixit 2017; Thomas, Quintal, and Phau 2010, 2018). Despite that the theoretical framework is still under development, there is empirical evidence that the adoption of technologies can exploit the wine tourism experience, influencing its dimension of learning, hedonism, and the 'winescape.'

The strategic adoption of technologies in the wine tourism experience

Technology certainly provides facilitation and new opportunities to tourism businesses. Today there exist a great number of electronic tools that actually or potentially influence and enhance the tourism experience. Drawing from Neuhofer, Buhalis, and Ladkin (2014), it can be argued that their role mainly depends on the relative intensity of the adoption:

- Technology can simply provide a facilitation role of the tourism experience in assisting travelers in information-seeking and buying processes. Websites, blogs, social media (e.g., Facebook, Instagram, Twitter, Twitch), booking and reservation systems, virtual marketplaces, and e-commerce platforms are useful for tourism businesses to stay closer to their customers, establish a direct and customized relationship, and drive and facilitate consumption.
- From being merely supportive, technology can enhance, empower, and exploit the tourism experience. At this level, technology needs to exist for the experience to happen. Touchpoints, multitouch tables, 360°, and 180° projections, and augmented and virtual reality have more sophisticated and relevant functions, becoming both enhancers and enablers of experiences.
- Technology can also partially replace the onsite experience. The health emergency in 2020 obliged tourism businesses to rethink their traditional business model, developing innovative and technology-based concepts to engage with their customers, inspire travelers for future visits, and also grant revenue. Remote social dining and partying, online classes and courses, and augmented and virtual reality experiences have increased, significantly changing the traditional tourism experience.

The following section presents the applications of technologies in wine tourism by distinguishing their role in the wine tourism experience.

Technology supporting information-seeking and buying processes

Within the field of communication and selling, the adoption of technology in the wine tourism sector is an ongoing process. Web 2.0 – which refers to applications based on content predominantly generated by users who are often anonymous (Constantinides and Fountain 2008) – and its growing popularity among consumers have helped to create new links and relationships via online communications. Through the years, the wine tourism businesses have gradually created websites, blogs, social media accounts (e.g., on Facebook, Instagram, Twitter, and Twitch) with the purpose of staying closer to their customers, learning about their opinions and needs, and even providing a direct and personalized interaction.

Given the thousands of existing winery websites and social media accounts, that this technology is today widely adopted and considered among winery entrepreneurs and regional/national wine associations is undoubtful. The cases of Napa Valley in the United States (https://www.visitnapavalley.com), Hunter Valley in Australia (http://www.winecountry.com.au), and the wine regions in France (https://www.visitfrenchwine.com/), to name a few, illustrate the ways in which regional and national tourism agencies are using websites and social media. Despite that, the shift from their use as traditional and one-way communication tools to effective instruments that enable interaction with actual and potential customers, especially for tourism purposes, remains partly achieved (Duarte Alonso et al. 2013; Haller, Thach, and Olsen 2020).

Apart from promoting the winery and its products, Web 2.0 has also helped in establishing commercial relationships. Virtual marketplaces and online shops, from being viewed as merely interesting niches, have gradually become important retail channels for wine. The global health emergency accelerated their adoption as to partially replace economic damages due to the temporary closure of both direct sales and the HoReCa channel. In 2020, the global value of the online alcohol market had increased by 42% in 2020, reaching US$24 billion, with wine that is the dominant e-commerce category in most countries (IWSR 2020).

Its commercial potential remains partly underdeveloped in wine tourism. Today, a number of virtual travel marketplaces such as Viator (https://www.viator.com), GetYourGuide (https://www.getyourguide.com), Airbnb (https://www.airbnb.com) and Musement (https://www.musement.com/uk/) enlist wine tourism experiences. These platforms facilitate tourists in their choices and provide them the opportunity to buy a wide range of activities, from tastings to guided tours, while increasing the visibility of the offering for wineries. Even wineries' websites often allow booking of tourism experiences apart from buying wines. Castello di Amorosa in the United States (https://castellodiamorosa.com), Herederos del Marques de Riscal in Spain (https://www.marquesderiscal.com), and Château Smith Haut Lafitte in France (https://www.smith-haut-lafitte.com), to name a few, illustrate how wineries can sell tourism experiences along with their wines. Despite that, online channels still represent a relatively small fraction of sales: 26% according to the International Wine Tourism Think Tank survey,[1] of which 17% through the company website and 9% through virtual travel marketplaces. The majority are booked through direct contacts or onsite (Garibaldi 2020a). It can be argued that the adoption of technology for commercial purposes largely varies among countries and wineries, with small and medium-sized businesses that are less likely to use these tools due to their lack of capital, time, human resources, and/or experience.

Technology enabled wine tourism experience

Technology to enrich the tourism experience has been recently adopted in wineries. The need to offer new and distinctive experiences that go beyond traditional visits and tastings (Garibaldi and

Pozzi 2018; Getz 2000; Jaffe and Pasternak 2004) has stimulated wine entrepreneurs and associations to invest in technological tools and innovate their proposals. There exists a great number of electronic tools, i.e., interactive touch screens and multitouch tables that allow visitors to access a variety of videos, images, and sounds that make them more informed about the wines, their history, and manufacturing methods; interactive PODs where visitors can, for instance, have a person-to-person virtual interview with wine experts; 180° and 360° projections that get visitors immersed in a journey across the wines, the winery, and so on (Garibaldi and Sfodera 2019).

Among these tools, virtual and augmented reality are recognized as having a huge potential in the various subsegments of tourism. Regardless of the context, the VR and, especially, AR technology offer novel and interactive avenues to disseminate information, improve learning and interaction, create emotional environments, and provide public access to sites and attractions so that those for whom visitation is limited due to physical barriers can potentially benefit (Yung and Khoo-Lattimore 2019).

Virtual reality is 'the use of computer-generated 3D environment, that the user can navigate and interact with, resulting in real-time simulation of one or more of the user's five senses' (Yung and Khoo-Lattimore 2019, p. 2057). In the field of wine tourism, its application is less recent and has mainly served to encourage future visits by engaging customers with learning about the winery, its history, and its products. VR has also been used to exploit onsite visits, but its application is limited.

One of the earliest examples comes from the Spanish winery *Bodegas Torres* (https://www.torres.es/en). In 2014, it launched an audiovisual experience that allows users to view and discover the company's main wineries and vineyards. Conceived as a tool for the trade team to be used in events and at fairs and summits, it has gradually become available also for end-consumers to encourage future visits. More sophisticated VR-based experiences were later developed by the Spanish winery Bodegas Ramòn Bilbao (https://www.bodegasramonbilbao.es) and the New Zealand-based winery Brancott Estate (https://www.brancottestate.com/en). Since 2017, Bodegas Ramòn has been offering to its visitors the opportunity to enjoy the 'Experiencia Oculus' during the guided tour in the Haro cellar. This VR-based experience allows visitors to discover the entire winemaking process, from 'the birth of grape to the glass of wine,' and to be introduced to the range of products that can be appreciated in the tasting room. By using this technology, physical barriers can overcome, and contents about the history, the values, and the products of the winery can be provided in a more engaging way. Brancott Estate created the 'Red Shed Experience' in 2018: through the use of VR headsets, users at wine fairs and events could view the virtual replica of its vineyards and cellar, switch between different periods to discover the maturation of the wines, and see the entire winemaking process. To offer a more realistic taste of the winery, specific devices were developed to release scents of the grapes and fruits, contributing to immerse them in the experience of what it would be like to be hanging out in New Zealand, enjoying a glass of wine.

More traditional web-based virtual tours were recently developed by Castello di Amorosa in the United States (https://castellodiamorosa.com) and the Seppeltsfield winery in Australia (https://seppeltsfield.com.au). The former has a digital tour on the winery website that allows web users to take a virtual walkthrough its courtyard, chapel, and barrel room, while the latter has created a VR smartphone tour that recreates panoramic 'spheres' of its wine, food, and art precincts. The health emergency in 2020 boosted VR applications due to the restrictions to movements and temporary closure of the wineries. Glandore Estate in Australia (https://glandorewines.com), Perrier-Jouët in France (https://www.perrier-jouet.com), and Familiares Matarromera in Spain (https://matarromera.es), to name a few, have developed 360-degree virtual tours inside their cellars and through vineyards.

Augmented reality differs from the previous technology, and it is defined as the 'enhancement of a real-world environment using layers of computer-generated images through a device' (Yung and Khoo-Lattimore 2019, p. 2058). With AR, what the user sees is still the real world but implemented with digital content. Today the technology is recognized to help with engaging customers while increasing their sense of ownership.

In the field of wine tourism, it can be argued that AR is gradually replacing VR. Its greater potential in the context of marketing makes it a more suitable tool to provide information via a vicarious experience and to persuade actual and future travelers to visit the winery. One of the most renowned examples comes from the Treasury Wine Estates (https://www.tweglobal.com), an Australian wine company that created AR labels for its 19 Crimes wine brand. Customers could download an app that allows figures in the label to come to life, move, and talk. The 19 criminals on the label told their own individual stories about what crime they had committed and how their new life in Australia turned out. Along with astonishing their customers, the winery was committed to narrating an aspect of Australian history that is less known to people. Similarly, the US-based winery Chateau St. Jean (https://www.chateaustjean.com) has created a living wine label that tells the personal story of the winemaker when meeting her husband at the winery. In doing so, it creates a closer and more personal contact with the customers.

The American winery Siduri (https://www.siduri.com) has opted for something different from a living label, developing a holographic web augmented reality experience. This disruptive technology allows us to easily engage and educate actual and potential customers, encouraging their future onsite visits. By scanning a QR code (currently available only on the winery website), the hologram of the founder appears and tells the history of the 25-year-old winery in a lighthearted, immersive, but educational way (Wine Industry Advisor 2020). Visual and textual contents have been prepared with the purpose of capturing the attention of the users, mixing entertainment and learning. From this perspective, Siduri winery's holographic web AR experience represents a marketing tool that can encourage a more conscious visitation.

Along with providing content in a more astonishing way, AR can also serve to educate and entertain customers. The 'Augmented Glinvet Tasting Experience' well illustrates how the technology can be combined with the gamification technique. Glenlivet (https://www.theglenlivet.com/en-EN) is a distillery in the United Kingdom that has developed an augmented tasting experience in collaboration with the music app Shazam. This experience can be launched through the Shazam app and bring the master distiller to life, who explains the 'mysterious' story behind Glenlivet's distinctive taste. Users are challenged to decode the whisky's ingredients in a series of mini-games and are given a score that they can share to social media to see how they measured up against their peers.

Technology partly replacing the onsite wine tourism experience

The health emergency in 2020 severely affected wineries and their tourism businesses. The temporary restrictions to movements and the closure of nonessential activities (e.g., tourism) posed new challenges: How to maintain relationships with current customers in this new and uncertain scenario? How to stimulate future onsite visits while granting revenues during the closure? Wineries resorted to innovative tools, accelerating the adoption of technology.

Digital wine experiences were developed purposely, moving wine tourism to online channels. These home-based experiences consisted of live events via online platforms where participants can enjoy a guided tasting with the winery's entrepreneurs and/or the winemakers. There is large evidence of their adoption: an International Wine Tourism Think Tank survey found that 32% adopted digital wine tasting as a way to compensate for the loss of tourism services; other strategies included gift cards for future onsite visits (27%) and virtual tours (11%) (Garibaldi 2020a).

Earliest proposals were not sophisticated and simply offered an online guided tasting. For instance, Maeli (https://maeliwine.com) was among the first Italian wineries to adopt this new concept: at the beginning of March 2020 (when the first national lockdown began), the owner decided to create digital tastings to sell together with the bottles of their wines. Participants could choose among three different proposals according to their tastes; on a fixed date, they connected to Skype for the virtual tasting, during which they were explained about the organoleptic characteristics of each wine while enjoying the vintages at their home.

Their success stimulated both wine entrepreneurs and regional associations to enrich digital wine tastings, which gradually become more and more diversified and sophisticated. For instance, Tolosa Winery in the United States (https://www.tolosawinery.com) combined traditional tasting with sports activities. During the lockdown period, it developed the 'WOGA Wednesday experience' in partnership with 'The Yoga Standard' centre. The online yoga classes were streamed from the vineyards by a specialized instructor and were followed by a taste of Tolosa wines. The South African winery Creation Wines (https://www.creationwines.com) proposed two innovative concepts that creatively combined wine, shopping, and music: along with wine selection, the 'Creation Virtual Tasting pack' allowed customers to choose their personal shopper that would have conducted their tastings via the Zoom platform; the 'Creation Alchemy Tasting kit' included short videos introducing the six wines in the kit as well as a dedicated Spotify playlist to fully immerse in the world of the winery.

Moving to the regional association, the Paso Robles Wine Country Alliance in the United States (https://pasowine.com) created the so-called 'Home with Paso Wine' program, which included a variety of remote opportunities. Along with listing purchase specials for over 60 wineries in the Paso Robles wine region, the 'Home with Paso Wine' program also featured new and amusing ways to connect customers with local wineries. Online users could choose among a variety of proposals, including independent virtual experiences, Zoom hangouts, inspiring recipes, Spotify playlists, and binge-worthy pairings (e.g., the 22 'Netflix and Chill-ed Wine Pairings,' to be enjoyed with a glass of Paso Robles wines in hand). The 'Digital Wine Show' developed by the Italian start-up Vinhood (https://www.vinhood.com/en) provides another remarkable example of the 'new breed' of digital tasting. This concept successfully combined education and entertainment: participants at the event pre-ordered a pack of wines, connected with Zoom on a fixed date, and then they were guided in tasting games and quizzes on wine by a professional sommelier.

That wineries resorted to new and more sophisticated digital tasting experiences is also recognized by a recent study carried out by Garibaldi (2020b). Among the 298 proposals from Italian, French, and American wineries that were analyzed,[2] 55% were live wine tastings, 20% were live talks with the winery entrepreneur or the winemaker, 5% were food-and-wine tastings, and 2% were live wine tastings with virtual tours. More than half of the experiences (53%) required a tasting kit that one could pre-order through the website.

Perspectives on technology in wine tourism

Technology has been increasingly important in the wine tourism sector, from the first function of supporting information-seeking processes to the subsequent act of buying products, services, and experiences. In recent years, it has also taken on more sophisticated and relevant functions, becoming both enhancers and enablers of onsite experiences.

Despite its potential uses and benefits, the application of technology in wine tourism still remains underinvestigated in the academic literature (Garibaldi and Sfodera 2019). This chapter's illustrated examples highlight that the use of technology can facilitate the interaction with actual and potential customers; allow the development of pre-experiences that encourage future onsite visits by giving customers a 'first taste' of the winery (such in the case of virtual tours); help in creating enhanced experiences where customers actively participate and interact with virtual contents and places, while gaining knowledge on the winery, its history, and products. Despite the technology used (virtual and augmented reality, touchpoints, multitouch tables, 360° and 180° projections, etc.), there is increasing evidence that these tools can have a direct impact on the main three elements constituting the wine experience: learning, hedonism, and the 'winescape.'

This chapter also provided practical examples – although limitative in number – of the response from the wine tourism industry to the health emergency in 2020. Technology-based concepts such as digital tasting emerged, granting extra revenue and allowing the creation or maintenance of relationships with customers. This obliged shift from offline to online channels definitively stimulated or boosted innovation within the industry. Several recommendations for the future can also be drawn from the analysis of the past and current situation.

As the market is becoming increasingly overcrowded, there emerged the need of acquiring distinctiveness. Digital tastings cannot be longer considered amateur activity but should be carried out with the utmost professionalism and seriousness. This means an improvement in the equipment used –high-definition cameras and professional microphones, with quality videos and images that can also be used to space out the live sessions. In designing the experience, narrative methods to deeply engage participants should be taken into account, e.g., by including personal storytelling that introduces human touch. Actually, clearly defining who speaks, who tells the story, and building all those hooks would eventually bring to the final purchase or onsite visit to the wine cellar. Online platforms (e.g., Zoom, Microsoft Teams, Facebook, Instagram, YouTube) should be used to improve interaction and active engagement from the participants – for example, by giving them the possibility of giving an impression or evaluation of the various parameters of the wine they are tasting. Data deriving from this evaluation process can be analyzed to create an experience that is more personalized as well as to improve the service.

Digital tasting should also be considered to further exploit the onsite wine tourism experience. Integrating digital (e.g., virtual tours), 'phygital' (e.g., digital tastings), and physical (e.g., onsite visits) experiences may actually provide a 24-hour real and virtual access to the winery, overcoming physical barriers. It may also facilitate visitors in gaining knowledge about its history and its wines, the winemaking process, and the 'winescape' during all seasons. Lastly, the integration of different experiences may reinforce the hedonic dimension that characterizes wine tourism, creating new options for the entertainment and pleasure of the visitors.

Implications

From a theoretical point of view, the present chapter identifies and explains the three main scopes of technology in the wine tourism experience – supporting/assisting, enabling, and partly replacing. It, therefore, clarifies to what extent technology can reinforce the dimensions of learning, hedonism, and the 'winescape' that characterize the wine tourism experiences.

Several practical implications can be drawn. The use of technology can further exploit the wine tourism experience, making it a never-ending journey. First, it stimulates travelers before the departure, giving them a 'taste' of the onsite visit. Second, it can facilitate the discovery of the wine production by giving access to new and attractive contents. Finally, it can create a long-lasting relationship, improving loyalty on the part of travelers.

Notes

1 Data were collected among wineries from nine countries – Italy, France, Spain, Portugal, Mexico, Chile, Brazil, Argentina and South Africa – from July to September 2020. A total of 200 valid responses were collected.

2 The study was performed between May 12 and June 15, 2020. Proposals were collected from virtual travel marketplaces (Viator, GetYourGuide, Airbnb and Divinea) and the official website of wine tourism destinations in the considered country.

References

Alant, K., and Bruwer, J. (2004). Wine tourism behaviour in the context of a motivational framework for wine regions and cellar doors. *Journal of Wine Research*, 15(1): 27–37. https://doi.org/10.1080/0957126042000300308

Alant, K., and Bruwer, J. (2010). Winery visitation sets: intra-regional spatial movements of wine tourists in branded regions. *International Journal of Wine Business Research*, 22(2): 191–210. https://doi.org/10.1108/17511061011061748

Bitner, M. J. (1992). Servicescapes: the impact of physical surroundings on customers and employees. *Journal of Marketing*, 56(2): 57–71.

Bruwer, J. (2003). South African wine routes: some perspectives on the wine tourism industry's structural dimensions and wine tourism product. *Tourism Management*, 24(4): 423–435. https://doi.org/10.1016/S0261-5177(02)00105-X

Bruwer, J., and Rueger-Muck, E. (2018). Wine tourism and hedonic experience: a motivation-based experiential view. *Tourism and Hospitality Research*, 19(4): 488–502. https://doi.org/10.1177/1467358418781444

Byrd, E. T., Canziani, B., Hsieh, Y. C. J., Debbage, K., and Sonmez, S. (2016). Wine tourism: motivating visitors through core and supplementary services. *Tourism Management*, 52: 19–29. https://doi.org/10.1016/j.tourman.2015.06.009

Constantinides, E., and Fountain, S. J. (2008). Web 2.0: conceptual foundations and marketing issues. *Journal of Direct, Data and Digital Marketing Practice*, 9(3): 231–244. https://doi.org/10.1057/palgrave.dddmp.4350098

Dixit, S. K. (ed.). (2017). *The Routledge Handbook of Consumer Behaviour in Hospitality and Tourism*. London: Routledge.

Duarte Alonso, A., Bressan, A., O'Shea, M., and Krajsic, V. (2013). Website and social media usage: Implications for the further development of wine tourism, hospitality, and the wine sector. *Tourism Planning and Development*, 10(3): 229–248. https://doi.org/10.1080/21568316.2012.747989

Garibaldi, R. (2020a). The impact of COVID 19 on wine tourism. Intervention at the OIV-UNWTO webinar *Innovating in wine tourism in the context of COVID-19*, September 22.

Garibaldi, R. (2020b). Enoturismo e COVID-19. Intervention at the webinar *L'enoturismo in epoca COVID-19: quali soluzioni?* organized by Associazione Italiana Turismo Enogastronomico and DOC-COM, November 5.

Garibaldi, R., and Pozzi, A. (2018). Creating tourism experiences combining food and culture: an analysis among Italian producers. *Tourism Review*, 73(2): 230–241. https://doi.org/10.1108/TR-06-2017-0097

Garibaldi, R., and Pozzi, A. (2020). Gastronomy tourism and Covid-19: technologies for overcoming current and future restrictions. In Burini, F. (ed.), *Tourism Facing a Pandemic: From Crisis to Recovery* (pp. 35–40). Bergamo: Università degli studi di Bergamo.

Garibaldi, R., and Sfodera F. (2019). Technologies for enhancing wine tourism experiences. In Dixit, S. K. (ed.), *The Routledge Handbook of Tourism Experience Management and Marketing* (pp. 409–417). London: Routledge.

Getz, D. (2000). *Explore Wine Tourism: Management, Development, Destinations*. New York: Cognizant.

Haller, C., Thach, L., and Olsen, J. (2020). Understanding eWinetourism practices of European and North America Wineries. *Journal of Gastronomy and Tourism*, 4(3): 141–156. https://doi.org/10.3727/216929720X15846938923987

IWSR – Drink and Market Research (2020). Beverage alcohol ecommerce value grows by 42% in 2020, to reach US$24 billion. *IWSR*. https://www.theiwsr.com/beverage-alcohol-ecommerce-value-grows-by-42-in-2020-to-reach-us24-billion/

Jaffe, E., and Pasternak, H. (2004). Developing wine trails as a tourist attraction in Israel. *International Journal of Tourism Research*, 6(4): 237–249. https://doi.org/10.1002/jtr.485

Jago, L. K., and Issaverdis, J. P. (2001). Building cellar door business through accreditation – Victoria's perspective. *International Journal of Wine Marketing*, 13(3): 18–31. https://doi.org/10.1108/eb008724

Kirova, V. (2020). Value co-creation and value co-destruction through interactive technology in tourism: the case of 'La Cité du Vin' wine museum, Bordeaux, France. *Current Issues in Tourism*, 24(5): 637–650. https://doi.org/10.1080/13683500.2020.1732883

Neuhofer, B., Buhalis, D., and Ladkin, A. (2014). A typology of technology-enhanced tourism experiences. *International Journal of Tourism Research*, 16(4): 340–350. https://doi.org/10.1002/jtr.1958

O'Neill, M., and Charters, S. (2000). Service quality at the cellar door: implications for Western Australia's developing wine tourism industry. *Managing Service Quality*, 10(2): 112–122. https://doi.org/10.1108/09604520010318308

Peters, G. L. (1997). *American Winescapes: The Cultural Landscapes of America's Wine Country*. Boulder, CO: Westview Press.

Pine, B. J., and Gilmore, J. H. (1998). Welcome to the experience economy. *Harvard Business Review*, 76: 97–105.

Quadri-Felitti, D. L., and Fiore, A. M. (2012). Experience economy constructs as a framework for understanding wine tourism. *Journal of Vacation Marketing*, 18(1): 3–15. https://doi.org/10.1177/1356766711432222

Quintal, V. A., Thomas, B., and Phau, I. (2015). Incorporating the winescape into the theory of planned behaviour: Examining 'new world' wineries. *Tourism Management*, 46: 596–609. https://doi.org/10.1016/j.tourman.2014.08.013

Thanh, T. V., and Kirova, V. (2018). Wine tourism experience: a netnography study. *Journal of Business Research*, 83: 30–37. https://doi.org/10.1016/j.jbusres.2017.10.008

Thomas, B., Quintal, V., and Phau, I. (2010). Predictors of attitudes and intention to revisit a winescape. In *Proceedings of Australian and New Zealand Marketing Academy Conference*. Australian and New Zealand Marketing Academy.

Thomas, B., Quintal, V. A., and Phau, I. (2018). Wine tourist engagement with the winescape: scale development and validation. *Journal of Hospitality and Tourism Research*, 42(5): 793–828. https://doi.org/10.1177/1096348016640583

Wine Industry Advisor (2020). Siduri Winery serves up holographic augmented reality experience. *WineIndustryAdvisor.com*, July 28. https://wineindustryadvisor.com/2020/07/28/siduri-winery-holographic-augmented-reality-experience

Yung, R., and Khoo-Lattimore, C. (2019). New realities: a systematic literature review on virtual reality and augmented reality in tourism research. *Current Issues in Tourism*, 22(17): 2056–2081. https://doi.org/10.1080/13683500.2017.1417359

Sitography

Airbnb: https://www.airbnb.com
Bodegas Ramòn Bilbao: https://www.bodegasramonbilbao.es
Bodegas Torres: https://www.torres.es/en
Brancott Estate: https://www.brancottestate.com/en
Castello di Amorosa: https://castellodiamorosa.com
Chateau St. Jean: https://www.chateaustjean.com
Château Smith Haut Lafitte: https://www.smith-haut-lafitte.com
Creation Wines: https://www.creationwines.com
Familiares Matarromera: https://matarromera.es
GetYourGuide: https://www.getyourguide.com
Glandore Estate: https://glandorewines.com
Glinvet distillery: https://www.theglenlivet.com/en-EN
Herederos del Marques de Riscal: https://www.marquesderiscal.com
Hunter Valley in Australia: http://www.winecountry.com.au
Maeli: https://maeliwine.com

Musement: https://www.musement.com/uk/
Napa Valley: https://www.visitnapavalley.com
Paso Robles Wine Country Alliance: https://pasowine.com
Perrier-Jouët: https://www.perrier-jouet.com
Seppeltsfield winery: https://seppeltsfield.com.au
Siduri: https://www.siduri.com
Tolosa Winery: https://www.tolosawinery.com
Treasury Wine Estates: https://www.tweglobal.com
Viator: https://www.viator.com
Vinhood: https://www.vinhood.com/en
Visit French Wine: https://www.visitfrenchwine.com/

44
VIRTUAL WINE TOURISM EXPERIENCES

Sandra Maria Correia Loureiro

Introduction

The wine industry, born in Europe, is one of the world's oldest commodities. Since the turn of the 21st century, the European wine market has been confronted with major changes at economic and regulatory levels. The emergence of new wine-producing countries has significantly increased competition, reinforcing the global dimension of the wine market (Cafaggi and Lamiceli 2011). In 2020, the COVID-19 pandemic had also posed several challenges to producers with the limitations of drinking wine due to the reduced number of social events.

The wine sector has unique characteristics regarding, for instance, color and cultural issues (Drennan et al. 2015; Cunha and Loureiro 2017). These characteristics have not yet been deeply analyzed and could explain management specificities (Cunha, Loureiro, and Rego 2015). Thus, communicating a wine brand is an exercise that requires more than perfect control of the visible aspects of the product/service operations. Other aspects should not be discharged, such as: the location where the grapes were produced; the expertise of the wine-maker and his methods; the history of the company owner, and the culture of the brand (Zhao et al. 2017; Zhou 2017; Cunha, Loureiro, and Sarmento 2018). In a market with thousands of brands, this requires the wine producer to be truly passionate about the sector, but at the same time, this will help to develop a perfect relationship with the main players of the distribution channel (Anderson and Narus 1990; Bobot 2011; Claro and Claro 2010). Therefore, wine can enhance tourism experiences by generating an emotional connection with tourists (Dixit 2020; Loureiro and Cunha 2016; Loureiro and da Cunha 2017; Loureiro and Kaufmann 2012). Virtual experiences also contribute to promoting wine and have been created to mimic the reality of tourism destinations and attractions without requiring the tourists to leave their present location (Loureiro, Guerreiro, and Ali 2020). This is quite important in the pandemic situation due to COVID-19.

In this vein, the current chapter provides (1) an overview of virtual technologies such as virtual reality (VR) and augmented reality (AR), (2) an overview of scientific publication on tourism and VR and AR technologies, and (3) examples of applications of virtual technologies in wine tourism, where Port wine virtual simulation is highlighted. This chapter is of interest

DOI: 10.4324/9781003143628-51

to both practitioners and academics because it gives an overview of the meaning of VR and AR technologies as well as examples of applications of such technologies in wine tourism, and it points out critical areas for future research.

Overview of virtual technologies

Virtual technologies are changing the way tourists live their experiences (Guttentag 2010; Loureiro, Guerreiro, and Ali 2020). A VR environment represents a completely synthetic world that could be a completely new, unknown environment or a real-life mimicry. AR is an interactive experience where tourists can see the objects of the real-world 'augmented' by computer-generated perceptual information, sometimes across multiple sensory modalities, such as visual or hearing (Loureiro et al. 2019). Thus, in AR, tourists experience both the real world and virtual objects, while in VR the environment should be completely virtual (e.g., Bimber and Raskar 2005; Bishop and Gimblett 2000; Boes, Buhalis, and Inversini 2016; Buhalis and Law 2008; Milgram and Kishino 1994).

The reality–virtuality continuum mentioned by Milgram and Kishino (1994) ranges from the real environment (RE) to the virtual environment (VE). The real environment and the VR are at the opposite ends of the reality–virtuality continuum, where other types of combinations of virtuality and reality exist. In AR, the observer sees both the real and the virtual world, usually by wearing see-through displays (Bimber and Raskar 2005). Spatial augmented reality (SAR) is a type of AR that can be used, for instance, on large surfaces such as spatially aligned wall projections (named 3D video mapping). Augmented virtuality (AV) occurs when some real objects are interposed in a totally virtual environment (for example, when a projection of the real person is overlaid on a virtual environment). As Milgram and Kishino (1994) claim, between the two ends of the spectrum are the mixed reality environments (MR), where virtual and virtual objects coexist in different types of combination.

AR appears in the 1960s with head-mounted display (HMD) developed by Ivan Sutherland (Carmigniani, Carmigniani, and Furht 2011; Virtual-reality 2020), but the commercial development happened only in the 1990s (see, for instance, platforms such as Pokémon GO) (Javornik 2016). Smartphones allowed AR to become used in developing applications in many different sectors such as medicine, architecture, and retailing (e.g., Grewal, Roggeveen, and Nordfalt 2017; Fortin and Dholakia 2005; Freina and Ott 2015). AR has been very widely employed in the tourism sector. For instance, tourists can learn more about iconic landmarks without a tourist guide due to augmented information based on tracking tourists' location called ViewRanger Skyline (Skyline 2020). The Seattle Art Museum used the Layar AR app (2020) for the 'Kehinde Wiley: A New Republic' exhibition. ArtLens 2.0, developed by the Cleveland Museum of Art, uses image-recognition software to recognize a selection of two-dimensional pieces of art.

AR is based on the physical environment and does not transport the users to a new setting where they can experiment with what it would be like to be there enjoying a specific destination, service, or product (Loureiro, Guerreiro, and Ali 2020). VR is suitable in these cases and has played an important role in engaging tourists in new experiences. VR has been used successfully in various fields of study, such as architectural planning (Krasonikolakis, Vrechopoulos, and Pouloudi 2014; Krasonikolakis et al. 2018; van Herpen et al. 2016). SAR is more holistic technology than VR, as it allows tourists to see 3D virtual environments that are in sync with their physical environment. This way, tourists can feel as if the new stimuli are in fact, real (Roo and Hachet 2017). SAR creates augmented scene elements in 3D, which, combined with the real environment, allows the user to see virtual objects embedded in the physical world and interact with them (Benckendorff, Sheldon, and Fesenmaier 2014).

Overview of scientific publications

From the scientific perspective, we can highlight the first study in the field by Musil and Pigel (1994), who discussed whether tourism can be replaced by VR technology at the International Conference Information and Communications Technologies on Tourism in Innsbruck, Austria. The authors argued that public panic of polluted nature and social isolation can drive them to virtual worlds. Thus, in the Musil and Pigel (1994) example, tourists can practice sports indoors (e.g., golf or bicycle in the mountains) without having to be concerned with the weather and travel to the destination to play. A pandemic situation may also lead to tourists preferring to stay at home instead of traveling to distant places.

During the first decade of the 21st century, most of the articles on the topic were associated with computer science and technologies for multimedia applications or information visualization and were published in conference proceedings. In the second decade of the century, a proliferation of articles were published in scientific journals.

The first two articles published in top-tier tourism journals – regarding this connection between tourism and VR – are research notes giving the first ideas of VR and related issues in the tourism context. Williams and Hobson (1995) consider VR as a tourism tool, where tourists can tailor their experiences as they wish. Three key characteristics can be elected: visualization (e.g., stereoscopic vision in three dimensions), immersion into the experience, and interactivity involved in the experience (meaning the degree of control tourists have over the experience). Cheong (1995) draws attention to VR as a marketing tool, where tourists can preview the experience of destinations, attractions, or other events.

Guttentag (2010) was the first to give an overview of the conceptualization of VR and its application within the tourism sector. Then, several other papers extended well-known frameworks to the virtual technologies, such as the Technology Acceptance Model (TAM) or the SOR (Stimulus-Organism-Response) framework (Loureiro and Guerreiro 2018). For instance, Yoon and Bendle (2012) used VR to analyze approach/avoidance responses of crowding in a restaurant. Huang, Backman, Backman, and Moore (2013) extended TAM and hedonic theory to the virtual world of Second Life. Other studies integrated TAM and VR to understand customer experience (Huang et al. 2016) in wine tourism (Martins et al. 2017; Zhao et al. 2017), and in cultural and historic visitor attractions (Jung and tom Dieck 2017; Lagiewski and Kesgin 2017; Lee et al. 2016). Yeh et al. (2017) extended the SOR framework to incorporate tourists' responses, namely attention, interest, desire, and action (AIDA).

Virtual technologies in wine tourism

Experiences are associated with multiple stimuli and dimensionalities (Brakus, Schmitt, and Zarantonello 2009; Loureiro 2014; Pine and Gilmore 1999; Schmitt 1999; Verhoef et al. 2009). Tourists pay 'to spend time enjoying a series of memorable events that a company stages – as in a theatrical play – to engage him in a personal way' (Pine and Gilmore 1999, p. 2). Brakus, Schmitt, and Zarantonello (2009, p. 53) argue that experiences are subjective, internal tourist responses (sensations, feelings, and cognitions) and behavioral responses evoked by brand/product-related stimuli that are part of a brand/product's design: sensory (the impression that the experience of the brand makes on consumers visual or other senses), affective (emotional brand that induces feelings and sentiments), intellectual (stimulates tourist's curiosity and problem solving), and behavioral (the brand/product is action oriented and tourists could engage in physical actions and behaviors). Thus, experiences involve the interaction between the firm and its stakeholders, particularly tourists. These interactions occur before purchase, during the use of product/service, and after that.

In the case of virtual wine tourism, tourists can interiorize the stimuli through the five senses. The visual sense is the most used and studied in VR and AR. Visual stimulation can be delivered through the Cave Automatic Virtual Environment (CAVE), head mounted displays (HMDs), or panoramic displays. Audition sound is another extensively used sense that can be appraisal in 3D virtual environments through headphones (for instance, using oculus Rift or the mobile). Olfactory smell is not often employed, but some studies have already used this sense for training (Barfield and Danas 1996; Chen 2006). Tactile sense needs haptic devices, and currently, those devices have limited feedback capabilities (for such an oculus, Rift can give some support). Finally, only a few prototypes explore the taste sense. From those, the 'food simulator' is one of the most recognized (Iwata et al. 2004). One rudimentary option is to use real samples of the stimulus and give them to participants.

However, how have these stimuli been used in virtual wine tourism experiences in practice? In wine regions from Australia to Napa Valley, winemakers are experimenting with innovative uses for VR technology. There are cases that VR is used to up-sell tourists to more expensive bottles of wine. Thus, VR supports virtual tours and tastings by allowing tourists to visit a winery by putting on a pair of VR goggles or VR headsets. One example is Brancott Estate Wines in New Zealand, where tourists can pick up a bottle of Brancott Estate wine at the local retail wine shop and complement the experience virtually. Brancott created an multisensory VR experience called 'The Red Shed,' which includes a virtual tour of the vineyard and wine cellar and a tasting experience featuring wind machines and scented sprays that immersed tourists in the experience of what it would be like to be hanging out in New Zealand. Another example is in Spain, where VR is used to explain winemaking and horticultural practices at specific wineries.

Winemakers in Australia are using VR to boost local wine tourism initiatives in attracting foreign wine lovers to come to Australia. The VR trial of Seppeltsfield in the Barossa Valley is generally considered to be the first-ever for the promotion of local wine tourism. VR allows visualizing the vineyard, the wine cellar, the winery, the art gallery, and the on-site restaurant. In addition, QR links are included on the wine bottle labels.

Wine festivals in different parts of the world (from California to France) are using VR to provide users with a feeling about how it is like to be hanging out at a wine festival (Cafaggi and Lamiceli 2011; Loureiro and Kaufmann 2012). The Palatinate region of Germany uses VR to bring a wine festival to virtual festival-goers.

VR can even be used to conduct wine conferences, where participants can find information about a certain wine region, winery, or grape varietal and be used at wine retail to have a glimpse into a special VR cellar before purchase. AR wine apps tell stories about the winemaker and wines available. As an example, we can point out the SippClub startup, a very well-known AR wine club, which can recognize brands and recommend meals to a particular wine. This app gets the attention of Millennials to have more knowledge about wine and its stories.

In sum, AR is a good technology to explain to tourists in a wineyard what they are looking at using smartphones – the names of wines, age, soil, or the kind of wine that the vineyard produces (Monteiro, Guerreiro, and Loureiro 2019). VR is better to be employed for tourists to know the place where the vineyard is, showing the work of winegrowers in the present and past, in a very immersive way.

Case Study 44.1 Virtual wine tourism in Porto

Portugal is an old wine country where several brands from regions such as Douro, Vinho Verde, or Alentejo are internationally known. At the top of Portuguese wines is Port wine from the Douro Valley. The Douro Valley was the first officially demarcated wine appellation in the world (18th century) created under the leadership of the Marquis of Pombal (Loureiro and Kaufmann 2012; Drennan et al. 2015) to assure the authenticity of its wines. Although many port-style wines are made around the world (including South Africa, Australia, and United States), the strict usage of the name 'Porto' or 'Port' is reserved only for the wines produced in Portugal.

Wine consumption is associated to multisensory experiences through the bouquet, the color, the shape of the bottle, the brand and, above all, the taste. A particular wine can evoke memories of special events or stimulate fantasy related to the expectation of an enjoyable consumption experience. Wine consumers often seek cues to help them to decide what to purchase, such as price, variety, style, region, and recommendations made by other brand name and symbol (Drennan et al. 2015).

VR as an immersive technology has the potential to transport users to a virtual environment, with multisensory experiences in such a way as if they are experiencing a real environment. VR can be employed to promote wine destinations and wine brands (Cho, Wang, and Fesenmaier 2002; Hyun, Lee, and Hu 2009). The case of Portoalities (2020) can contribute to illustrating the relevance of VR for the wine tourism industry. Portoalities (2020) allows wine lovers to have a virtual experience through Zoom meetings, where they can learn the history of the Douro Valley region where Port wine is produced, discover the production process of Port wine and the different aging processes, learn about the three types of Port wine – Ruby's, Tawny's and White Ports –and how to serve, store, and age each bottle of Port wine according to its style and age. The tasting of wine is possible if the wine tourist purchases the wine in advance so it can be shipped to their house in time for the session. When interacting through the Zoom platform, tourists can follow the instructions of the guide. Although this is an interactive process, it does not allow the immersion in the environment. Alternatively, tourists can assist with a VR video and taste the wine they purchased previously. This second process is more immersive but less interactive.

In the future, virtual technologies need to evolve and provide an immersive and interactive system, where the stimulation of the five senses should be the key. Virtual sense is possible to experience through VR or other platforms as Zoom, leading to a high level of realism. The simulation of VR can be developing using the Cave Automatic Virtual Environment (CAVE), head-mounted displays (HMDs), or panoramic displays. Audition sense is better achieved when using VR equipment than the Zoom platform, where the sound can be delivered by using headphones or calibrated sound setups that ensure a 360-degree, 3D sound experience. Smell sense can increase the user's sense of presence (Pair et al. 2006). The tactile sense is possible in VR, but currently, the devices that allow haptic touch face diverse constraints, such as their limited feedback capabilities when compared to the human tactile. Finally, the gustative sense is the most complex due to the technology constraints. Hashimoto, Inami, and Kajimoto (2008) simulate food by using a virtual straw to allow users to suck the virtual food and from which they get sound and vibrations. Yet currently, this gustative sense is experienced by sending the wine product to users in advance.

In sum, although visual, sound, and smell are already possible to simulate in a closer way to the real environment, touch and taste are more complex senses that need to be enhanced in new devices.

Conclusions and implications

AR and VR technologies are extending and complementing the real environments, by introducing virtual objects and places and adding information about wine brands and wineries. The immersive VR experience allows for simulating of real places that wine lovers may visit without creating problems with overcrowding and damaging the actual place. As mentioned previously, VR can also be the first experience during the wine tourist decision process and create expectations before the real experience. Tourists with some disabilities can also have a VR experience at home. AR can enhance the wine experience by providing valuable information about the wine brands and wine stories. Yet, these new technologies also have negative aspects; one of the most discussed is related to the work organization and the reduction of the number of employees. Thus, new business models are needed, and citizens will assist in a reconfiguration of the meaning of work.

As for a research agenda, one can point out eight major lines: (1) how VR and AR can develop a relevant role after the winery visit; (2) how artificial intelligence can change the whole flow of the tourist experience process; (3) how VR can be a true alternative of a real wine visit, and how technology should evolve; (4) how the whole process of select, purchase, experience, and after experience can occur using VR and AR, and how these two technologies can be integrated; (5) how the five senses can effectively be incorporated in the VR environment; (6) how background music style in VR influences the whole wine experience; (7) development of frameworks devoted to VR and AR experiences (and not use the adaptations from other contexts); and (8) a full theory on the VR and AR experiences is needed. Regarding the eighth major line, we cannot keep using extensions and adaptations of theories that come from other fields of research.

References

Anderson, J. C., and Narus, J. A. (1990). A model of distributor firm and manufacturer firm working partnerships. *Journal of Marketing*, 54(January): 42–58. https://doi.org/10.2307/1252172

Barfield, W., and Danas, E. (1996). Comments on the use of olfactory displays for virtual environments. *Presence: Teleoperators and Virtual Environments*, 5(1): 109–121. https://doi.org/10.1162/pres.1996.5.1.109

Benckendorff, P., Sheldon, P., and Fesenmaier, D. R. (2014). *Tourism Information Technology*. Oxford: CAB International.

Bimber, O., and Raskar, R. (2005). *Spatial Augmented Rality Merging Real and Virtual Worlds*. Wellesley: A K Peters CRP Press.

Bishop, I. D., and Gimblett, H. R. (2000). Management of recreational areas: GIS, autonomous agents, and virtual reality. *Environment and Planning B: Planning and Design*, 27(3): 423–435. https://doi.org/10.1068/b2637

Bobot, L. (2011). Functional and dysfunctional conflicts in retailer-supplier relationships. *International Journal of Retail and Distribution Management*, 39(1): 25–50. https://doi.org/10.1108/09590551111104468

Boes, K., Buhalis, D., and Inversini, A. (2016). Smart tourism destinations: ecosystems for tourism destination competitiveness. *International Journal of Tourism Cities*, 2(2): 108–124. https://doi.org/10.1108/IJTC-12-2015-0032

Brakus, J. J., Schmitt, B. H., and Zarantonello, L. (2009). Brand experience: What is it? How is it measured? Does it affect loyalty? *Journal of Marketing*, 73(3): 52–68. https://doi.org/10.1509/jmkg.73.3.52

Buhalis, D., and Law, R. (2008). Progress in information technology and tourism management: 20 years on and 10 years after the internet-the state of eTourism research. *Tourism Management*, 29(4): 609–623. https://doi.org/10.1016/j.tourman.2008.01.005

Cafaggi, F., and Lamiceli, P. (2011). Inter-Firm Networks in the European Wine Industry. *American Association of Wine Economists*. http://cadmus.eui.eu/handle/1814/15654

Carmigniani, J., Carmigniani, J., and Furht, B. (2011). *Handbook of Augmented Reality*. Berlin: Springer.

Chen, Y. (2006). Olfactory display: development and application in virtual reality therapy. In Collins, K. (ed.), *Proceedings of the 16th International Conference on Artificial Reality and Telexistence* (pp. 580–584). Hangzhou, China.

Cheong, R. (1995). The virtual threat to travel and tourism. *Tourism Management*, 16(6): 417–422. https://doi.org/10.1016/0261-5177(95)00049-T

Cho, Y., Wang, Y., and Fesenmaier, D. (2002). Searching for experiences: the web based virtual tour in tourism marketing. *Journal of Travel and Tourism Marketing*, 12(4): 1–17. https://doi.org/10.1300/J073v12n04_01

Claro, D. P., and Claro, P. B. O. (2010). Collaborative buyer–supplier relationships and downstream information in marketing channels. *Industrial Marketing Management*, 39(2): 221–228. https://doi.org/10.1016/j.indmarman.2009.03.009

Cunha, N., and Loureiro, S. (2017). Wine tourism activities as facilitators of relationship quality in wine sector. *Journal of Tourism and Development*, 27(28): 1023–1029. http://revistas.ua.pt/index.php/rtd/issue/view/408/showToc

Cunha, N., Loureiro, S. M. C., and Sarmento, E. M. (2018). Social partner characteristics as facilitators of the wine co-creation experiences. In Paulo Nunes e Antonio González Molina (eds.) *International Congress CITADNS'2018 – Active Tourism, Sport, Nature and Sustainability*, 12–14 April, Setúbal, Portugal: Instituto Politécnico de Setúbal

Cunha, N. P., Loureiro, S. M. C., and Rego, A. (2015). Exploring the attitudes of bottled wine distributors toward wine producers in the Portuguese wine sector. *Journal of International Food and Agribusiness Marketing*, 27(1): 15–32. https://doi.org/10.1080/08974438.2013.805455

Dixit, S. K. (2020). Marketing gastronomic tourism experiences. In Dixit, S. K. (ed.), *The Routledge Handbook of Tourism Experience Management and Marketing* (pp. 322–335). London: Routledge.

Drennan, J., Bianchi, C., Cacho-Elizondo, S, Loureiro, S., Guibert, N., and Proud, B. (2015). Examining the role of wine brand love on brand loyalty: a multi-country comparison. *International Journal of Hospitality Management*, 49: 47–55. https://doi.org/10.1016/j.ijhm.2015.04.012

Guttentag, D. A. (2010). Virtual reality: applications and implications for tourism. *Tourism Management*, 31(5): 637–651. https://doi.org/10.1016/j.tourman.2009.07.003

Fortin, D. R., and Dholakia, R. R. (2005). Interactivity and vividness effects on social presence and involvement with a web-based advertisement. *Journal of Business Research*, 58(3): 387–396. https://doi.org/10.1016/S0148-2963(03)00106-1

Freina, L., and Ott, M. (2015, January). A literature review on immersive virtual reality in education: state of the art and perspectives. *The International Scientific Conference eLearning and Software for Education*. Vol. 1 (p. 133). Bucharest, Romania: 'Carol I' National Defence University. https://doi.org/10.1.1.725.5493

Grewal, D., Roggeveen, A. L., and Nordfalt, J. (2017). The future of retailing. *Journal of Retailing*, 93(1): 1–6. https://doi.org/10.1016/j.jretai.2016.12.008

Hashimoto, Y., Inami, M., and Kajimoto, H. (2008). Straw-like user interface (II): a new method of presenting auditory sensations for a more natural experience. In Ferre, M. (ed.), *Haptics: Perception, Devices and Scenarios* (pp. 484–493). Cham: Springer.

Huang, Y. C., Backman, K. F., Backman, S. J., and Chang, L. L. (2016). Exploring the implications of virtual reality technology in tourism marketing: an integrated research framework. *International Journal of Tourism Research*, 18(2): 116–128. https://doi.org/10.1002/jtr.2038

Huang, Y. C., Backman, S. J., Backman, K. F., and Moore, D. (2013). Exploring user acceptance of 3D virtual worlds in travel and tourism marketing. *Tourism Management*, 3: 490–501. https://doi.org/10.1016/j.tourman.2012.09.009

Hyun, M., Lee, S., and Hu, C. (2009). Mobile-mediated virtual experience in tourism: concept, typology and applications. *Journal of Vacation Marketing*, 15(2): 149–164. https://doi.org/10.1177/1356766708100904

Iwata, H., Yano, H., Uemura, T., and Moriya, T. (2004). *Food Simulator: A Haptic Interface for Biting*. Paper presented at the Virtual Reality, 2004. Proceedings. IEEE.

Javornik, A. (2016). 'It's an illusion, but it looks real!' Consumer affective, cognitive and behavioral responses to augmented reality applications. *Journal of Marketing Management*, 32(9–10): 987–1011. https://doi.org/10.1080/0267257X.2016.1174726

Jung, T. H., and tom Dieck, M. C. (2017). Augmented reality, virtual reality and 3D printing for the co-creation of value for the visitor experience at cultural heritage places. *Journal of Place Management and Development*, 10(2): 140–151. https://doi.org/10.1108/JPMD-07-2016-0045

Krasonikolakis, I., Vrechopoulos, A., and Pouloudi, A. (2014). Store selection criteria and sales prediction in virtual worlds. *Information and Management*, 51(6): 641–652. https://doi.org/10.1016/j.im.2014.05.017

Krasonikolakis, I., Vrechopoulos, A., Pouloudi, A., and Dimitriadis, S. (2018). Store layout effects on consumer behavior in 3D online stores. *European Journal of Marketing*, 52(5/6): 1223–1256. https://doi.org/10.1108/EJM-03-2015-0183

Lagiewski, R., and Kesgin, M. (2017). Designing and implementing digital visitor experiences in New York State: the case of the Finger Lakes Interactive Play (FLIP) project. *Journal of Destination Marketing and Management*, 6(2): 118–126.

Layar (2020). https://www.layar.com/

Lee, Y. Y., Ahmed, B., Lee, J. H., An, H., and Lee, K. H. (2016, September). Augmenting three-dimensional effects in digital exhibition of a cultural artifact using 3D pseudo hologram. In *2016 IEEE International Symposium on Mixed and Augmented Reality*.

Loureiro, S. M. C. (2014). The role of the rural tourism experience economy in place attachment and behavioral intentions. *International Journal of Hospitality Management*, 40(5 July): 1–9. https://doi.org/10.1016/j.ijhm.2014.02.010

Loureiro, S. M. C., and Cunha, N. (2016). Relationship quality in wine business to Business Portugal and China. In Capitello, R., Charters, S., Menival, D., and Yuan, J. (eds.), *The Wine Value Chain in China: Global Dynamics, Marketing and Communication in the Contemporary Chinese Wine Market* (pp. 163–186). Cambridge, MA: Elsevier – Chandos Publishing. ISBN 978-0-08-100754-9

Loureiro, S. M. C., and da Cunha, N. P. (2017). Wine prestige and experience in enhancing relationship quality and outcomes: wine tourism in Douro. *International Journal of Wine Business Research*, 29(4): 434–456. https://doi.org/10.1108/IJWBR-04-2017-0033

Loureiro, S. M. C., and Guerreiro, J. (2018). Psychological behavior of generation Y: living between real and virtual reality. In Gerhardt, M., and Peluchette, J. V. E. (eds.), *Millennials: Characteristics, Trends and Perspectives* (pp. 67–90). New York: NOVA Science Publishers, Inc.

Loureiro, S. M. C., Guerreiro, J., and Ali, F. (2020). Twenty years of research on virtual reality and augmented reality in tourism context: a text-mining approach. Tourism Management, 77: 104028. https://doi.org/10.1016/j.tourman.2019.104028

Loureiro, S. M. C., Guerreiro, J., Eloy, S., Langaro, D., and Panchapakesan, P. (2019). Understanding the use of Virtual Reality in Marketing: a text mining-based review. *Journal of Business Research*, 100: 514–530. https://doi.org/10.1016/j.jbusres.2018.10.055

Loureiro, S. M. C., and Kaufmann, H. R. (2012). Explaining love of wine brands. *Journal of Promotion Management*. 18(3): 329–343. https://doi.org/10.1080/10496491.2012.696460

Martins, J., Gonçalves, R., Branco, F., Barbosa, L., Melo, M., and Bessa, M. (2017). A multisensory virtual experience model for thematic tourism: a port wine tourism application proposal. *Journal of Destination Marketing and Management*, 6(2): 103–109. https://doi.org/10.1016/j.jdmm.2017.02.002

Milgram, P., and Kishino, F. (1994). A taxonomy of mixed reality visual displays. *IEICE Transactions on Information and Systems*, 77: 1321–1329.

Monteiro, P., Guerreiro, J., and Loureiro, S. M. C. (2019). Understanding the role of visual attention on wines' purchase intention: an eye-tracking study. *International Journal of Wine Business Research*, 32(2): 161–179. https://doi.org/10.1108/IJWBR-03-2019-0017

Musil, S., and Pigel, G. (1994). Can tourism be replaced by virtual reality technology? In Schertler, W., Beat Schmid, Tjoa, A. M., and Werthner, H. (eds.) *Information and Communications Technologies in Tourism* (pp. 87–94). Vienna: Springer.

Pair, J., Allen, B., Dautricourt, M., Treskunov, A., Liewer, M., Graap, K., and Reger, G. (2006). *A Virtual Reality Exposure Therapy Application for Iraq war Post Traumatic Stress Disorder.* Paper presented at the *Virtual Reality Conference*; 2006.

Pine, B. J., II, and Gilmore, J. H. (1999). *The Experience Economy: Work Is Theater and Every Business a Stage.* Cambridge, MA: Harvard Business School Press.

Portoalities. (2020). https://portoalities.com/en/virtual-port-wine-tasting/

Roo, J. S., and Hachet, M. (2017). Towards a hybrid space combining spatial augmented reality and virtual reality. In Marchal, M., Teather, R., and Thomas, B. H. (eds.). *IEEE Symposium on 3D User Interfaces (3DUI)* (pp. 195–198). Los Angeles, California. https://doi.org/10.1109/3DUI.2017.7893339

Schmitt, B. H. (1999). *Experiential Marketing.* New York: Free Press.

Skyline. (2020). https://www.viewranger.com/en-US/premium/skyline

van Herpen, E., van den Broek, E., van Trijp, H. C. M., and Yu, T. (2016). Can a virtual supermarket bring realism into the lab? Comparing shopping behavior using virtual and pictorial store representations to behavior in a physical store. *Appetite*, 107: 196–207. https://doi.org/10.1016/j.appet.2016.07.033

Verhoef, P. C., Lemon, K. A., Parasuraman, A., Roggeveen, A, Tsiros, M., and Schlesinger, L. A. (2009). Customer experience creation: determinants, dynamics and management strategies. *Journal of Retailing*, 85(1): 31–41. https://doi.org/10.1016/j.jretai.2008.11.001

Virtual-reality (2020). *History of Virtual Reality.* https://www.vrs.org.uk/virtual-reality/history.html

Williams, P., and Hobson, J. P. (1995). Virtual reality and tourism: fact or fantasy? *Tourism Management*, 16(6): 423–427. https://doi.org/10.1016/0261-5177(95)00050-X

Yeh, C. H., Wang, Y. S., Li, H. T., and Lin, S. Y. (2017). The effect of information presentation modes on tourists' responses in Internet marketing: the moderating role of emotions. *Journal of Travel and Tourism Marketing*, 34(8): 1018–1032. https://doi.org/10.1080/10548408.2016.1276509

Yoon, S. Y., Laffey, J., and Oh, H. (2008). Understanding usability and user experience of web-based 3D graphics technology. *International Journal of Human–Computer Interaction*, 24(3): 288–306. https://doi.org/10.1080/10447310801920516

Zhao, H., Huang, F. X., Spence, C., and Wan, X. A. (2017). Visual search for wines with a triangle on the label in a virtual store. *Frontiers in Psychology*, 8: 1–11. https://doi.org/10.3389/fpsyg.2017.02173

Zhou, Y. Y. (2017). The application and development of virtual reality technology in the advertisement design. *Agrofood Industry Hi-Tech*, 28(3): 3184–3187.

45

SOCIAL MEDIA MARKETING IN WINE TOURISM

Ige Pirnar and Aylin Caliskan

Introduction

Today, all over the world, people tend to ignore mass communication and are highly affected by reviews and word-of-mouth (WOM) when researching, deciding, purchasing, and after-sales activities; therefore, social media gains utmost importance in all industries. One of these industries includes wine suppliers and wineries as wine tourism suppliers. Social media movement is enabled by Web 2.0 and Web 3.0 technologies, and it engages visitors, customers, and wine suppliers to collaborate in transparent wine tourism marketing activities since social media creates new communication pathways among all parties.

Social media in wine tourism marketing is about the usage of trustworthy and popular social networks like Facebook, Twitter, Instagram, WhatsApp, WeChat, Snapchat, LinkedIn, Pinterest, Foursquare, YouTube, MySpace, WAYN, Flicker, Vero, Triller, TikTok, and personal or organizational web blogs where online activities take place among wine-interested people. Consumers and suppliers gather to produce and transfer user-generated content like wine-related information, suggestions, and opinions for marketing purposes on these social media platforms. Social media has recently become very popular; thus, wine marketers and wineries that effectively use it are gaining a unique differential competitive advantage in their wine tourism marketing activities.

Thus, this chapter reviews social media platforms that are available for wine tourism marketing. It details advantages of social media usage in wine tourism marketing, social media wine tourism marketing applications and suitable strategies, social media communication strategies for wine tourism, wineries' adoption of social media, and trends in social media wine tourism. Wine industry-specific social media strategies and success criteria in social media usage in wine tourism marketing are especially emphasized with social media terminology in wine tourism marketing, including examples.

Social media marketing and social media platforms in the context of wine tourism

Social media marketing is briefly defined as the usage of trustworthy and popular social networks like Facebook, Twitter, Instagram, WhatsApp, WeChat, Snapchat, LinkedIn, Pinterest,

DOI: 10.4324/9781003143628-52

Foursquare, TripAdvisor, YouTube, MySpace, WAYN, Flicker, Vero, Triller, TikTok, and personal or company web blogs. Social media marketing involves online activities among a group of people who gather to create and exchange user-generated content, including information, photos and videos, opinions, and reviews, all of which are used for marketing purposes (Chan and Guillet 2011; Dixit 2017; Pirnar 2014; Safko 2010; Thevenot 2007). Thus, people or groups joining any of these social networks first create a personal or company profile and start building their network by connecting to friends and by inviting real-world contacts within the circle of the network (Subramaniam and Nair 2013). Social media marketing has been widely used by many industry professionals, and even the wine tourism industry has lagged in adapting it, the usage rate has been increasing rapidly (Forbes, Goodman, and Dolan 2015; Viana 2016). Many wineries, wine routes, wine clubs, and other wine tourism companies have started social media usage as one of their main marketing and communication tools, and it is understood that the Facebook platform is very commonly adapted. Statista data states that Facebook has 2.9 billion monthly active users in the 2022 first quarter, which ranks it as the highest social network used; and '3.59 billion people are using at least one of the company's core products which are Facebook, WhatsApp, Instagram, or Messenger each month' (Statista 2022b). Among these, both Facebook and Instagram are very suitable social media platforms to exchange and update information quickly and flexibly, and to share visual content and stories easily. In addition, video- and photo-sharing characteristics of Facebook, YouTube, and Instagram are one of the reasons that these platforms' usage rates are increasing rapidly. In addition to these, Twitter is also a very popular social network for journalists and the media since it is a quick, simple, and very popular news channel. Twitter updates take place on mobile devices, and people update news anywhere, at anytime; therefore, positive or negative news on topics are distributed very quickly. Thus, Twitter is also another social media platform that is gaining importance among wine tourism marketers. TripAdvisor and Foursquare are also popular platforms, which allow users and fans to post pictures, videos, and reviews as opinions on their experiences.

Social media platforms like Facebook, Instagram, Pinterest, YouTube, and Twitter are modern marketing tools applicable for wine tourism companies (Thach, Lease, and Barton 2016), though the usage rates change among countries. Social media is a common and popular communication tool that helps wine marketers' marketing efforts and is gaining popularity. Though this is the case, when the social media usage of wine tourism establishment is analyzed, one realizes that a variation exists among European countries and overseas countries, and overseas countries use social media more often comparatively (Szolnoki et al. 2014). The social media utilization rate may also vary according to the history of wine production as Galati et al.'s (2017) study indicates that 'in the US, 94.0% of wine producers have a Facebook page, while in the traditional producing countries, such as France, the proportion is as low as 53.0%' (p. 41). The detailed literature on the social media adaptation level of wineries will be covered in the following section.

Advantages of social media usage in wine tourism marketing

Looking into the characteristics of these social media platforms, it is understood that Facebook is commonly used as an event promotional tool, along with a communication tool for wine marketing (Lagos, Kloppenborg, and Alebaki 2013). Among the other advantages of Facebook, the ease in forecasting wine consumers' purchasing intents and the purchase behavior that takes place. Facebook is also used as a source of information on specific wine brands and wine categories. Wine-specialized Facebook users form special groups to discuss and share opinions of various wines and wine brands. The results of various tasting experiences and related stories are also shared among these wine lovers' groups. Thus, the study on Millennials indicates that they prefer to use technology and to use Facebook to obtain special information about wine and the wine-related special groups to engage and collect data and learn about the wine research (Higgins, Wolf, and Wolf 2016). They prefer to be informed by the opinions of reference groups, friends, family members, and shelf-talkers, whereas elders prefer to get professional wine information from wine store staff and wine stewards. Atkin and Thach's (2012) study reveals that the brand of wine is very important to both Millennials and the elderly. Thus, to figure out the wine quality, the elderly give importance to geographical wine cues like the origin of the wine production region, whereas the wine brand's medals, label imagery, and alcohol content are more vital elements of wine-quality determinants for Millennials.

One of the other advantages of social media platforms of wine marketers is their impact on improvement in brand loyalty and provide opportunities for long-term relationships with developed customer relationship management of the wine tourism companies. Further, it improves the feeling of wine consumer's protection since social media users can share their negative or positive reviews as opinions which are open to all the group members (Capitello, Agnoli, and Begalli 2013). In addition, social media platforms and wine blogs may be used as a strategic, innovative wine marketing tool to increase sales volume, increase and improve wine brand awareness, and develop the wine establishment's reputation (Dolan and Goodman 2017). Further, social media platforms provide an easy share of price comparison and wine tasting notes among members. Since the wine market is highly fragmented, consists of a huge variety of brands, and has high product differentiation (Lockshin 2000; Lockshin and Corsi 2012), social media provides optimized search capabilities to reach unique and special wine brands.

To collect, create, process, distribute, and share a large number quantity of knowledge and information related to wine is relatively easy when compared with traditional methods (Sogari et al. 2017). As it goes, it is also relatively easy and costs less to collect data on wine customers' purchases and habitual preferences on social media. In addition, on social media platforms, the social influence and social interactions of wine consumers can easily be observed and traced. The wine customer's involvement pattern and level, including wine brand posts and corresponding responses as likes, shares, negative or positive opinions, reviews, shared photos and videos, and other views, provide a clear insight into their social media engagement (Dolan and Goodman 2017). Further, social media provides two-way communication between the wine marketer and consumers, offering opportunities for interesting storytelling, creative designing, and using interesting content in wine blogs and wine-related social media pages. Wine customers become more active in the process when they share their own individual wine tourism stories and personal experiences on social media platforms. Further on in this process, they may act as co-creators or co-marketers of the wine company or the wine brand (Sigala and Haller 2019).

Wine blogs provide opportunities for wine experts to act as opinion influencers when they share their views and experiences and exchange information with other members through social networks and may lead to wine repurchases and winery revisits (Fiore et al. 2016). The results of related forums and discussions may lead to solid managerial suggestions for wine product improvements and higher customer satisfaction possibilities. In addition, 'the wine industry is characterized traditionally by the culture of socializing and sharing of knowledge, especially among winemakers entering the wine industry because of the lifestyle' (Tomazic 2017, p. 829). Wine-related social media platforms with the increasing number of wine-related customers as members provide them the opportunity to socialize, be part of the discussion, and engage in the wine experience. Thus, personal opinions, ranks, and reviews shared on platforms like TripAdvisor provide a trustworthy wine information source on various price and quality comparisons of products and facilities for wine tourists (Brochado et al. 2020). To summarize, the advantages of social media usage in wine tourism marketing may be briefly stated as follows (Canovi and Pucciarelli 2019; Dolan and Goodman 2017; Scorrano et al. 2019; Tomazic 2017): the ability to sell wines online, higher positive WOM, increased visits to wineries and revisits, building wine brand awareness, increase in brand recognition and brand loyalty, improvement in CRM, establishing wine region loyalty, having a two-way and continuous relationship with customers, ability to use emotional cues like storytelling, opportunities for customer-led content generation and customer-created innovative products, ability to promote intangible wine products, building trust and credibility, ease in determining customer behavior, following feedbacks and reviews, and determining customer satisfaction.

Social media terminology in wine tourism marketing

Some of the most common terminology used in wine tourism marketing include the following (Cristófol, Aramendia, and de-San-Eugenio-Vela 2020; Laverie et al. 2011; Marlowe et al. 2017; Rodríguez-Fernández, Sánchez-Amboage, and Martínez-Fernández 2017; Thach, Lease, and Barton 2016):

Social networking sites are social sites where wine customers create personal or company profiles, share information, visuals like photos and videos, and connect with the members of sites such as Facebook, Twitter, YouTube, and Instagram. As stated before, at present, Facebook is the largest social networking site, and in China, Weibo and WeChat are among the popular social media platforms.

Blogs (abbreviation of 'web-logs') are internet sites where members may share individual news, short articles, and opinions or criticism on wineries and wine brands. Wine blogs may have creative content and may attract many wine consumers. Vlogs are online videos that websites like YouTube allow their subscribers to upload and share their individual videos. Wine blogs help potential customers in their wine country region or winery selection decisions.

Visibility refers to the number of shares, contacts, fans, posts, likes, and comments on the wineries or wine products. It is an important aspect showing the effectiveness of the social media efforts since the greater the visibility, the more valuable the related social media page is. Hence, the usage of visual aids like the photos of wineries, infographics on wine tasting and wine production procedures, using hashtags and icons, and promoting wine products in different social media platforms increase the winery's and wine products' visibility.

Interactivity relates to two-way communication like social media engagement and the wine brands' response to the fans. With the help of higher interactivity, wineries and wine marketers can promptly answer their customers' questions and help to increase their curiosity and satisfaction. Two-way dialogues are always a better way for communication, and interactivity of social media increases the value of the communication between wineries, wine marketers, and customers.

Mobile relates to the accessibility of online wine information from a mobile device. It may include geographic location tracking ability to find wineries and nearby wine stores. Foursquare and Foodspotting are examples of mobile applications in which users can 'check in' to inform their friends about their location, make a comment or recommendation, or rate or review the wine region or the wine company.

The story may be about wine product information, a photo or video of a winery or series, and/or a combination of different photos and videos, which take place on Instagram or on Facebook that is visible for 24 hours to all the friends, to special groups, or to the public according to the choice. Social media stories first appeared on Instagram, but Facebook and many other social media platforms followed in a short time due to their popularity. Though the stories may be saved, they are temporary for 24 hours in nature, which creates a feeling of scarcity and curiosity catching the users' attention.

Impressions are the views that the wine posts or ads are seen by other users on the social media site. It is a metric value showing the digital views or engagements of the wine-related ads or posts, and the higher degree implies higher popularity.

Engagement indicates the social media users' actions, such as sharing, viewing, recommending, reacting, and commenting, on the wine-related posts. The higher the engagement, the higher the popularity of the post, web page, or advertisement.

Shares indicate the number of times that the social media user's message or content is reposted. Share occurs when other users of the same social network 'post' the content of another member in the same network. The higher the number of shares, the higher the potential of the viewers of the post will be. In addition, higher shares indicate that the wine or winery related post is interesting and gets attention.

Twitterati indicates the members and users of Twitter who have a very high number of followers. Twitterati are followed by high numbers of followers, their news and posts get immediate attention, and the engagement levels are rather high.

Ephemeral content is content with a limited lifespan, like stories that disappear in 24 hours on Facebook or Instagram. Snapchat is also another social media platform with ephemeral content.

Relationship relates to the interaction between or among the users of the website. As in real life, the users or groups with similar interests interact and communicate with each other, leading to a relationship. The advantage of social media relationships is that many news on the subject matter like wine may be shared promptly and quickly among huge crowds.

Post relates to the content of the message/shared material of the social media user. There are various types of posts. Most popular posts consist of information, guides, news, articles, photographs or videos, links to external contents, announcements, invitations, reviews, electronic books, infographics, video stories, blogs, or a combination of these.

The *response* indicates the social media user's positive (like or recommend) or negative (dislike) reaction on some post on the same social network. They provide important feedback and valuable information on the customer reaction and satisfaction level to the wine marketer.

A *mention* happens when users refer to another user or social profile in their post or social media content. Social mentions or brands mention act as a PR tool, impacting overall wine brand awareness and popularity.

As an interesting example from the social media platform Facebook, the 'One Drop Wine Association' group was created on February 4, 2014. The One Drop Wine Association is a nongovernmental organization formed in 2002 by a group of wine producers, wine marketers, wine-related academicians, and wine lovers with the aim of improving the wine culture as a whole. This Facebook site is a general and private group, with 887 members by June 2021. It is a private group, meaning that only the members can see who is in the group (members of the group) and what the members post, the posts' content and the shared stories. It is a visible group meaning that if searched, nonmembers can find this group. The logo of the group is shared as an image on the group profile picture. Topics and hashtags shared in the group include: wine, many wine types and grape types like Merlot and kalecik karasi, wine city, wine and city, wine culture, wine tasting, wine house, winery, wine types like semidry, semi-sweet, wine, and food pairing, white wines, and red wines. The group page has six admins and moderators. The shared posts consist mainly of wine products with visual photographs and videos and the text content on ingredients and wine tasting information on these wine products. An events subsection and many posts provide info and links to upcoming wine-related events like wine tasting, winery visits, and wine and food pairing seminars taking place at nearby locations. Wine-related articles are also shared on the site. Mentions take place as links to wineries, wine tasting events, and various wine brands. Many posts are liked and shared, providing positive feedback to users. In addition, comments on the posts provide meaningful information and views to other members. The group page provides an interaction opportunity to create a chat room for interested members, which is a video chat platform enabling members' online interaction and two-way communication. The link to this group's web page on Facebook is at https://www.facebook.com/groups/biryudumsarap/about.

Wineries' adoption of social media

E-Commerce reached US$4.9 trillion in worldwide sales in 2021, and it is estimated to reach US$7.4 trillion by 2025 (Statista 2022a). The main reason for this rapid acceleration is that the internet has become the most viable purchasing option for consumers. In terms of electronic sales, the share of wine in this total sale is negligible compared to other retail products. Reasons such as the need for meticulous and extra precautionary operation design for wine in supply chain and logistics operations and restrictions for alcohol in transportation are the main reasons why wine cannot benefit from the electronic sales benefit (Szolnoki, Thach, and Kolb 2016). However, digital wine marketing tools, basically social media, have been used increasingly by the wine industry to generate inter- and intra-customer communication in addition to building awareness and positioning themselves in consumers' minds.

It is understood that 'the adoption of Web 2.0 and social media components varies by winery region' (Kolb and Thach 2016, p. 226). Thus, the social media adaptation processes of regions that develop their economy with wine tourism, that is, regions that cannot exist without tourists, are slightly different from wine sales. The wine industry is much slower in social media adoption than other industries. Academic studies examining the social media adoption processes of wineries provide very clear information on this subject. In addition, because this research was conducted in different years, it is clearly observed how the use of social media in the wine industry developed on a yearly basis. A study (Thach 2009) investigating the digitalization and social media adoption processes of 208 wineries in the United States stated that these wineries, in 2009, had not yet adopted social media use, and that 61% offered an online shopping infrastructure, 11% used vlogs, and 2% used blogs. According to Reyneke, Pitt, and Berthon (2011), French wineries also had not yet adopted social media usage in 2011. Duarte Alonso et al. (2013) revealed that in 2011, wineries involved in wine tourism located in Italy, Spain, Australia, New Zealand, the United States, Canada, Argentina, Uruguay, Chile, and South Africa still had not fully exploited the benefits of social media. This was because they believed that social media platforms were not very effective and influential in terms of business development, and they had limited time to manage these platforms. Whereas, in the same study, wineries that adopted social media usage found that increasing wine tourism and wine sales are the basic reasons for using social media. In specific, they adopted the reasons of communication, contact, interaction, exposure, information, marketing, promotion, education, positive involvement, and getting satisfied and loyal customers in using social media.

Looking at the recent studies, we see that the social media issue is still not fully spread across the wine industry. Szolnoki et al. (2018) investigated 1,173 wineries located in Germany, the United States, New Zealand, and Australia in terms of social media adoption and activities. According to the results, New Zealand, Australia, and the United States use multiple social media platforms, for example, Facebook, Twitter, and blogs related to vineries and wine clubs. German wineries mostly use a single social media channel (generally Facebook) to engage with customers. In 2014, another study that examined the German wineries also mentioned the same thing about Facebook being the most important social media tool (Szolnoki et al. 2014). Galati et al. (2019) searched the main internal resources of wineries that pushed them towards the social media networks adoption in Sicily. The results indicated that large wineries have a high number of qualified staff who have the ability to use various social media tools and are familiar with social network management; therefore, they are highly motivated towards the adoption of social media and perceive their social media investment as more effective. On the other hand, the wineries that are not using social media stated as their reasons time constraints, lack of effectiveness, and the effort not being worthwhile (Forbes, Goodman, and Dolan 2015).

Trends in social media usage in wine tourism

Wine tourism provides significant gains, especially to wine-producing regions. Considering its contribution to the development of wine production and tourism in those regions, wine tourism provides important contributions to the region, especially in economic terms. Especially recent cases in 2020 provoke the wineries to focus more on social media tools to communicate with customers and build brand awareness. COVID-19 presents a fruitful ground to present the current social media usage trends of wineries. This is because, as in all sectors that do not address basic needs such as health and logistics, social media has provided an incredible opportunity for the wine tourism industry to compensate for unplanned revenue losses, to keep relations with customers, and to attract new customers.

In general, the trends in social media usage in wine tourism can be listed and summarized as follow. The social media literature states that customers prefer and are attracted very much by entertaining online marketing (e.g., Phelps et al. 2004). In Thach's (2009) study to prove this statement, the conclusion reached by using the expressions of the wineries is quite striking. They said that live broadcasting on the internet via a webcam placed on a vineyard increases the social media traffic of that account because customers want to see if the 'noble rot' started in the vineyard (Thach 2009). In addition, wineries' online and short videos (vlogs) on their social media accounts is a highly preferred wine marketing and public relations tool by wineries. It is a beneficial tool for differentiation and for evoking customers to visit and purchase wine. Personal contents, good presentation, and simple, closer, and aesthetic posts are the content trends among wineries because they know that through those methods they can capture greater attention (Viana 2016). Moreover, some strategic posting timings are popular among wineries. For example, being active on social media platforms on Friday surges customers to increase their wine consumption on weekends (Dolan et al. 2016). Tracking the major number of interactions of the followers is also the other hint for wineries for timeliness posting. Mentioning the followers' conversations through hashtags, sharing recipes, harvest notes, 'when to drink' and 'how to' videos are popular in social media marketing of wineries. Online fingerprints of wine consumers also present opportunities for wineries to send similar content and advertisements on their account pages (Barber 2010). This is a subject of behavioral segmenting and targeting of marketing science. In addition to this, geographical targeting is applied by wineries on social media platforms. By focusing on lifestyles of generations in developing, developed, and transition economies, wineries manage the contents of their posts (Fiore et al. 2016). Moreover, the other trend is to create their own winery communities on social media rather than participating in other social communities. In terms of post contents, the recent trend topics are winemaking, viticulture, culinary, food pairings, hospitality, and design (Viana 2016). Apart from the social media management trends of the wineries, it would not be appropriate to ignore the social media sharing of the customers. Especially in terms of wine tourism, social media posts of users and visitors are very important in terms of promoting wine regions to wider segments. The sharing of their own experiences and suggestions to other people contributes greatly to wineries to expand their customer portfolio. This situation, which goes into word-of-mouth, is very vital for wine tourism. In addition, social media posts of opinion leaders, bloggers, and influencers have a strong influence on the promotion of wineries. That is why wine brands introduce campaign deals with celebrities and opinion leaders for being on their own social accounts as the latest trend.

Winery tasting rooms, bars, and restaurants were closed to the public in major wine tourism regions (e.g., Italy, France) due to the COVID-19 measures taken in all countries because of the pandemic that had existed for a year, and therefore. wine tourism came to a screeching halt. Wineries all over the world began to look for innovative ways to continue their sales and keep the existing customers. At this point, social media began to be actively used by wineries to keep their customers connected with their brands and to reach new customers. In addition, wineries have started using online meeting platforms such as FaceTime, Zoom, GoToMeeting, Skype, and Join.me with the desire of remaining less isolated and connected with customers and prospects. For the wine industry, the use of social media and online meeting platforms that were not very active until a year ago laid the groundwork for creating virtual wine tasting. For example, Napa and Sonoma, traditional wine regions in California that rely heavily on the wine and wine tourism industry, started digital wine tasting processes, virtual tasting packages (e.g., personalized tasting, live stream/group tasting, and the happy hour), updated websites, and social media platforms after the pandemic.

Successful social media strategies and communication strategies in wine marketing

As indicated before, social media platforms are important tools for wine marketers since they can impact the purchase behavior and loyalty of wine customers. Managing them effectively and strategically makes an important difference in wine sales volumes, related return of investment, loyalty rates, and customer satisfaction levels (Blazheska and Nickova 2016). In order to optimize social media marketing strategies, wineries should segment their prospects according to demographic and behavioral characteristics. An efficient segmentation includes targeting expert consumers for wine increasing sales, and have a positive word-of-mouth (WOM) wine brand advertisement among less expert consumers. In addition to these benefits, expert customers involving emotional communications such as storytelling through social media helps to increase wine brand awareness and loyalty in the long run (Pucci et al. 2019). Emotional cues and communications related to wine images, mind associations, lifestyles, and their representations through wine products are intangible product attributes of wine and may help to attract customers from developed countries which are more attracted by intangible attributes than tangible ones (Fiore et al. 2016). For positioning, Fernández, Vriesekoop, and Urbano (2017, p. 271) states that 'larger wineries are targeting "fun," "terroir" and "quality" to attract potential consumers to their brands, while the rest of the wineries are conveying sophisticated corporate values associated with their brands, especially to young consumers.'

Branding on social media also provides opportunities to managers to identify consumer purchase intentions and preferences, aiding them through more effective brand-building (Capitello, Agnoli, and Begalli 2013). Virtual wine experiences, video displays of wineries, webcast tasting, and explaining details about viticulture are all effective social media features that enable wine customers to remember the pleasurable memories associated with the experiences and should be used optimally in social media marketing in wine tourism (Viana 2016). In addition, since message relevance and passionate delivery of the message is vital for mass influence, live, authentic messages should be used to have a high brand relevance with consumers (Laverie et al. 2011). Finally, it is understood that sustainability is an important factor in social media strategy building. Encouragement of sustainable wine programs and delivery of sustainable, eco-friendly messages on social media are important marketing activities for wine producers (Sogari et al. 2017). Therefore, wine marketers should communicate and share positive environmental and sustainable practices for their visibility, which impacts positive brand positioning towards potential customers. In addition, the wine company's or winery's website page content is also very important for successful social media application. Table 45.1 shows important social media page contents of wineries, wine companies, and wine regions which may act as a guiding roadmap for wine tourism marketers.

Implications and suggestions

Google Analytics showed that the most searched word related to beverages is wine, and wineries are the third-most popular subject on Pinterest. At the same time, the number of tweets posted on the basis of an hour and a day about wine is too high to ignore. In addition to all these, as we mentioned in this chapter, many empirical studies have revealed that social media posts greatly affect wine vineyard visits and wine purchases. In addition to vineyards, the use of social media has a huge potential as a marketing tool for wine routes which are important attraction factors for national and international tourists.

A better interface, reaching larger customer segments, promoting the business, developing demand and interest, word-of-mouth, informing about activities, organizations, changes in infrastructure, and products are the potential benefits of using social media platforms by wineries and other wine tourism companies. Therefore, the wine industry, which is advancing with baby steps in terms of social media management, should include social media in marketing and customer relationship management in a way that will turn this opportunity into an advantage.

In this chapter, the reasons for wineries and other wine tourism companies' rejection of adaptation of social media tools to manage their brands are also covered, yet, it was stated that the biggest reason was the lack of time to manage social media. To cope with this problem, wineries can hire professionals to put forward a plan that includes the determination of the audience, management of the frequency of online interaction, creating the contents that people want to talk about, and building a social network of fans and followers. The necessity of using social media has now become a must-do rather than a suggestion. Suggestions that can be given on this subject should be more specific rather than generic. For example, segmentation, positioning, innovative content production, brand orientation, and social brand involvement all can be managed in social media specific to wineries. The wineries should optimize value creation through social media and should engage with customers and prospects who are talking about wine on social media platforms. They should go beyond sending information to customers without getting feedback from them actively. Specifically, it can be offered for the wine tourism industry to adapt factors of participation of wine tourists into social media content.

Table 45.1 Important social media page contents of wineries, wine companies, and wine regions

Company web page's contents and features	*Detailed information*
Ease of web navigation/surfing, ease of contact with the winery, wine company, wine destination management office	Availability of 'Home' button from all pages Availability of company links to various other social media; for instance, having icons suggesting opportunities for interaction
Information about the winery/wine company/wine region ('About Us' part)	Availability of history or story about winery/wine company/wine region Availability of social media profile of wine producer Updating 'dead ends' or inactive links
Updated facilities and product information	Availability of up-to-date and interesting website content on the winery or wine company.
Public relations	Cretaive design and page contents to attract website surfers' interest and full informative content on wineries' products and their region Availability of a company's/region's newsletter, media updates Availability of gift shop information, detailed product information, etc.
Opportunity for customer feedback and two-way communication	Availability of 'Contact Us' email Availability of recommendations and rating, available section for sharing customer experiences and expert opinions Availability for consumers to start and participate in online dialogues and discussions.
Operational Information on the winery, wine company, or the wine region	Availability of winery operation hours Availability of winery tour and wine tasting
Branding/positioning/image	Availability of logo, slogan, and other related information of the wine brand and having winery name on every page Desing of the web page according to the positioning Usage of company colors
Partnerships	Availability of the links to local accomodation, hotels and restaurants, tours, festivals, events, and other local attractions
Educational material	Availability of wine production information Availability of educational material such as recipes on food pairing with the winery's wine, educational information on wine tasting, etc.
Online shopping	Availability of product prices, identified paying method, available online shopping cart, etc.
Strategic marketing and professional look	Availability of the mission statement and the vision statement and, if available, past and future strategic plans and shareholder and stakeholder information of the wine company or winery
Other promotional material	Availability of winery-related sustainable applications, eco-friendly stories, etc.

Source: Adapted from Laverie et al. (2011); Duarte Alonso et al. (2013); Lagos, Kloppenborg, and Alebaki (2013)

The contents may focus on socializing, winery atmosphere, virtual wine tasting experiences, online purchasing, introducing different and interesting wines, production information, relaxing contents, restaurants, and eating choices. These contents should focus more on soft issues like feelings, experiences, storytelling, and soft innovation aspects like color and design usage rather than on hard issues like new product ideas.

References

Atkin, T., and Thach, L. (2012). Millennial wine consumers: risk perception and information search. *Wine Economics and Policy*, 1(1): 54–62. https://doi.org/10.1016/j.wep.2012.08.002

Barber, N. (2010). 'Green' wine packaging: targeting environmental consumers. *International Journal of Wine Business Research*, 22(4): 423–444. https://doi.org/10.1108/17511061011092447

Blazheska, D., and Nickova, M. (2016). Creating marketing strategies for development of the wine tourism. *International Journal of Academic Research in Accounting, Finance and Management Sciences*, 6(4): 177–184.

Brochado, A., Troilo, M., Rodrigues, H., and Oliveira-Brochado, F. (2020). Dimensions of wine hotel experiences shared online. *International Journal of Wine Business Research*, 32(1): 59–77. https://doi.org/10.1108/IJWBR-12-2018-0072

Canovi, M., and Pucciarelli, F. (2019). Social media marketing in wine tourism: winery owners' perceptions. *Journal of Travel & Tourism Marketing*, 36(6): 653–664. https://doi.org/10.1080/10548408.2019.1624241

Capitello, R., Agnoli, L., and Begalli, D. (2013). A new approach to the analysis of visitor perceptions towards a tourism destination: the role of food and wine experiences. *Scientific Papers Series Management, Economic Engineering in Agriculture and Rural Development*, 13(1): 57–64.

Chan, N. L., and Guillet, B. D. (2011). Investigation of social media marketing: how does the hotel industry in Hong Kong perform in marketing on social media websites? *Journal of Travel & Tourism Marketing*, 28(4): 345–368. https://doi.org/10.1080/10548408.2011.571571

Cristófol, F. J., Aramendia, G. Z., and de San-Eugenio-Vela, J. (2020). Effects of social media on enotourism. Two cases study: Okanagan Valley (Canada) and Somontano (Spain). *Sustainability*, 12(17): 6705. https://doi.org/10.3390/su12176705

Dixit, S. K. (ed.). (2017). *The Routledge Handbook of Consumer Behaviour in Hospitality and Tourism*, London: Routledge.

Dolan, R., Conduit, J., Fahy, J., and Goodman, S. (2016). Facebook for wine brands: an analysis of strategies for Facebook posts and user engagement actions. In *Proceedings of the 9th AWBR International Conference* (pp. 457–465). Adelaide, Australia: University of South Australia.

Dolan, R., and Goodman, S. (2017). Succeeding on social media: exploring communication strategies for wine marketing. *Journal of Hospitality and Tourism Management*, 33: 23–30. https://doi.org/10.1016/j.jhtm.2017.09.001

Duarte Alonso, A., Bressan, A., O'Shea, M., and Krajsic, V. (2013). Website and social media usage: implications for the further development of wine tourism, hospitality, and the wine sector. *Tourism Planning & Development*, 10(3): 229–248.

Fernández, R. F., Vriesekoop, F., and Urbano, B. (2017). Social media as a means to access millennial wine consumers. *International Journal of Wine Business Research*, 29(3): 269–284. https://doi.org/10.1108/IJWBR-08-2016-0030

Fiore, M., Vrontis, D., Silvestri, R., and Contò, F. (2016). Social media and societal marketing: a path for a better wine? *Journal of Promotion Management*, 22(2): 268–279. https://doi.org/10.1080/10496491.2016.1121755

Forbes, S. L., Goodman, S., and Dolan, R. (2015). Adoption of social media in the Australian and New Zealand wine industries. *Journal of New Business Ideas and Trends*, 13(2): 1–14.

Galati, A., Crescimanno, M., Tinervia, S., and Fagnani, F. (2017). Social media as a strategic marketing tool in the Sicilian wine industry: evidence from Facebook. *Wine Economics and Policy*, 6(1): 40–47. https://doi.org/10.1016/j.wep.2017.03.003

Galati, A., Tinervia, S., Tulone, A., and Crescimanno, M. (2019). Drivers affecting the adoption and effectiveness of social media investments. *International Journal of Wine Business Research*, 31(2): 260–278. https://doi.org/10.1108/IJWBR-04-2018-0016

Higgins, L. M., Wolf, M. M., and Wolf, M. J. (2016). Wine on Facebook: a look at millennials' wine information search. In Szolnoki, G., Thach, L., and Kolb, D. (eds.), *Successful Social Media and Ecommerce Strategies in the Wine Industry* (pp. 13–29). Verlag: Palgrave Macmillan.

Kolb, D., and Thach, L. (2016). Analyzing German winery adoption of Web 2.0 and social media. *Journal of Wine Research*, 27(3): 226–241. https://doi.org/10.1080/09571264.2016.1190324

Lagos, E., Kloppenborg, P., and Alebaki, M. (2013). Tweet me a drink: the use of social media in Victoria's (Australia) wine tourism industry. *Journal of Marketing Vistas*, 3(2): 11–23.

Laverie, D. A., Humphrey, W. F., Velikova, N., Dodd, T. H., and Wilcox, J. B. (2011, June). Building wine brand communities with the use of social media: a conceptual model. In *Proceedings of the 6th AWBR International Conference* (pp. 9–10).

Lockshin, L. (2000). Wine marketing in the new millennium. *Australian & New Zealand Wine Industry Journal*, 15(1): 89–90.

Lockshin, L., and Corsi, A. M. (2012). Consumer behaviour for Wine 2.0: a review since 2003 and future directions. *Wine Economics and Policy*, 1(1): 2–23. https://doi.org/10.1016/j.wep.2012.11.003

Marlowe, B., Brown, E. A., Schrier, T., and Zheng, T. (2017). Beverage bloggers: a developing relationship between wine blogger expertise and Twitter followers. *International Journal of Hospitality Beverage Management*, 1(1), Article 2. https://doi.org/10.34051/j/2019.2

Phelps, J. E., Lewis, R., Mobilio, L., Perry, D., and Raman, N. (2004). Viral marketing or electronic word-of-mouth advertising: examining consumer responses and motivations to pass along email. *Journal of Advertising Research*, 44(4): 333–348. https://doi.org/10.1017/S0021849904040371

Pirnar, I. (2014). *Successful Hospitality Marketing by Social Media Applications in Izmir*. Presented at the *7th World Conference for Graduate Research in Tourism, Hospitality and Leisure*, Istanbul.

Pucci, T., Casprini, E., Nosi, C., and Zanni, L. (2019). Does social media usage affect online purchasing intention for wine? The moderating role of subjective and objective knowledge. *British Food Journal*, 121(2): 275–288. https://doi.org/10.1108/BFJ-06-2018-0400

Reyneke, M., Pitt, L., and Berthon, P. R. (2011). Luxury wine brand visibility in social media: an exploratory study. *International Journal of Wine Business Research*, 23(1): 21–35. https://doi.org/10.1108/17511061111121380

Rodríguez-Fernández, M. M., Sánchez-Amboage, E., and Martínez-Fernández, V. A. (2017). The emergent nature of wine tourism in Ecuador and the role of the social medium Facebook in optimising its positioning. *Revista ESPACIOS*, 38(14): 24–37.

Safko, L. (2010). *The Social Media Bible: Tactics, Tools, and Strategies for Business Success*. Hoboken, NJ: John Wiley & Sons. https://books.google.com.tr/books?id=2c0Yrp1X-o8C&printsec=frontcover&dq=%E2%80%A2%09Safko

Scorrano, P., Fait, M., Maizza, A., and Vrontis, D. (2019). Online branding strategy for wine tourism competitiveness. *International Journal of Wine Business Research*, 31(2): 130–150. https://doi.org/10.1108/IJWBR-06-2017-0043

Sigala, M., and Haller, C. (2019). The impact of social media on the behavior of wine tourists: a typology of power sources. In Sigala, M., and Robinson, R. N. S. (eds.), *Management and Marketing of Wine Tourism Business* (pp. 139–154). Cham: Palgrave Macmillan.

Sogari, G., Pucci, T., Aquilani, B., and Zanni, L. (2017). Millennial generation and environmental sustainability: the role of social media in the consumer purchasing Behavior for Wine. *Sustainability*, 9(10): 1911.

Statista (2022a). *Key Figures of E-Commerce*. https://www.statista.com/statistics/379046/worldwide-retail-e-commerce-sales/

Statista (2022b). *Number of Monthly Active Facebook Users Worldwide as of 1st Quarter 2022*. https://www.statista.com/statistics/264810/number-of-monthly-active-facebook-users-worldwide/

Subramaniam, K., and Nair, T. G. (2013). Transitions in marketing communications from traditional architectures to social network patterns. *International Journal of Internet Marketing and Advertising (IJIMA)*, 8(2): 69–85. https://doi.org/10.1504/IJIMA.2013.058585

Szolnoki, G., Dolan, R., Forbes, S., Thach, L., and Goodman, S. (2018). Using social media for consumer interaction: an international comparison of winery adoption and activity. *Wine Economics and Policy*, 7(2): 109–119. https://doi.org/10.1016/j.wep.2018.07.001

Szolnoki, G., Taits, D., Nagel, M., and Fortunato, A. (2014). Using social media in the wine business. *International Journal of Wine Business Research*, 26(2): 80–96. https://doi.org/10.1108/IJWBR-09-2013-0031

Szolnoki, G., Thach, L., and Kolb, D. (2016). Current status of global wine ecommerce and social media. In Szolnoki, G., Thach, L., and Kolb, D. (eds.), *Successful Social Media and Ecommerce Strategies in the Wine Industry* (pp. 1–12). New York: Palgrave Macmillan.

Thach, L. (2009). Wine 2.0 – the next phase of wine marketing? Exploring US winery adoption of wine 2.0 components. *Journal of Wine Research*, 20(2): 143–157. https://doi.org/10.1080/09571260903169548

Thach, L., Lease, T., and Barton, M. (2016). Exploring the impact of social media practices on wine sales in US wineries. *Journal of Direct, Data and Digital Marketing Practice*, 17(4): 272–283. https://doi.org/10.1057/dddmp.2016.5

Thevenot, G. (2007). Blogging as a social media. *Tourism and Hospitality Research*, 7(3–4): 287–289. https://doi.org/10.1057/palgrave.thr.6050062

Tomazic, T. (2017). The importance of social media from the wine marketing perspective. *Lex localis – Journal of Local Self-Government*, 15(4): 827–844. https://doi.org/10.4335/15.4.827-844(2017)

Viana, N. A. (2016). Digital wine marketing: Social media marketing for the wine industry, In *BIO Web of Conferences* (Vol. 7, p. 03011). EDP Sciences.

46

BRANDING WINE PRODUCTS AND WINE TOURISM DESTINATIONS THROUGH E-STORYTELLING

Examining contemporary tourism interpretation

Amnaj Khaokhrueamuang

Introduction

The wine industry is increasingly competitive, leading many wineries to add value and distinction to their brands. Creating points of diversity through unique, attractive stories about their firms, products, events, activities, and destinations plays a significant role in increasing profit, building a credible brand, and improving a wine region's image and brand identity (Bonarou, Tsartas, and Sarantakou 2019). However, such stories cannot succeed without story providers and consumers co-creating the wine experience. 'Storying' the lived experience is constructing meaning – not just information, but also narrative interpretation (Garcia and Rossiter 2010). Inspiring winery experiences to benefit wine destinations (Lee et al. 2022) and wine products thus involves creative wine stories and tourism interpretations.

Tourism interpretation reveals something more profound than simple facts. It aims to communicate meanings and relationships using original objects, firsthand experiences, and illustrative media (Tilden 1977) on both sides of the tourism system: tourist-generating regions (TGRs) and tourist destination regions (TDRs) (Leiper 1979). Research on tourism interpretation in TDRs has focused on approaches such as tourist guides (El-Menshawy 2016), museum display objects (Dumbraveanu, Craciun, and Tudoricu 2016), and on-site signs (Ababneh 2016), but research on marketing approaches to tourism interpretation through consumer products available in a TGR to promote a destination has received less attention; only Suorsa (2017) and Florek and Connejo (2006) have studied them.

E-storytelling on winery websites, a content marketing strategy in a TGR to create a distinctive brand image, is significant (Bonarou, Tsartas, and Sarantakou 2019), but the few studies that have examined it have not focused on tourism interpretation. Iaia et al.'s (2017) survey tried to understand family-run Italian winery website contents, but it did not focus on their e-storytelling, and although Bonarou, Tsartas, and Sarantakou (2019) examined e-storytelling on the Wine Roads of Northern Greece winery websites, story interpretation was not their

DOI: 10.4324/9781003143628-53

focus; that is, e-storytelling is a tool to establish, reinforce, and even change the wine destination's image.

This chapter explores the contents of wine-related e-storytelling, an original contribution that examines how such contents interpret meaning to brand wine products and wine tourism destinations. A qualitative content analysis of winery websites by the Japan Wineries Association is the study's focus.

Tourism interpretation for branding wine products and wine tourism destinations through E-storytelling

Wine tourism (also known as 'enotourism') refers to tourism whose purpose is visiting vineyards and wineries to taste, consume, and/or purchase wine often at or near its source (Bruwer and Lesscheve 2012; Thanh and Kirova 2018; Lee et al. 2022). By definition, wine tourists are interested not only in learning about a particular product, but also in discovering more about the wine's origins and context (Croce and Perri 2017). This notion raises the importance of linking wine branding with wine-region branding, an innovative marketing tool called 'co-branding' that requires a deeper understanding of how the relationship provides value to both the winery (Velikova, Canziani, and Williams 2019) and its region.

One innovative marketing instrument in branding strategy is e-storytelling. It significantly promotes a TGR's consumer products and tourism destination as a two-way transformation of tourism interpretation. Focusing on wine, effective wine-product branding at the TGR can create a 'tourist image' of the destination that the brand interprets through e-storytelling. The tourist image, or tourism destination image (Bruwer and Lesscheve 2012), deeply impacts potential tourists' behavior patterns (Peng 2013). Tourism promotion-related branding is interpreted by integrated marketing communication (IMC), such as advertising, public relations, and blogs. It requires clients and stakeholders to work together (Mortimer and Laurie 2019), enabling product co-creation.

This co-creation reflects the efficiency of the brand interpretation of the product and the destination. Building a reliable brand for a TGR's product, therefore, is not simply presenting the tourism information but also the meaning of the components used in branding, such as logos, paintings, and words, including storytelling.

Storytelling has deep roots in human history, such as myths, legends, and folktales that have passed on wisdom, knowledge, and culture for thousands of years (Bassano et al. 2019). In wine tourism, storytelling is an important concept in interactions with tourists (Frost et al. 2020). For branding a destination, storytelling concerns more than simply 'telling a story' or presenting facts (Bassano et al. 2019); it also serves to interpret the place's meaning, such as the wine tourism interpretation of staff or guides at the winery about the place's history. Storytelling can thus become a powerful branding tool (Hope 2018) to motivate tourists and local residents to visit (Bassano et al. 2019).

However, storytelling as an effective place-branding strategy requires a platform that can communicate the brand identity and its assets to target markets (Rein and Shields 2007). In the digital age, e-storytelling is common in both branding products in TGRs and creating tourist images of TDRs on platforms such as websites and social network services (SNSs).

When consumers understand the meanings that the brand interprets, the expected tourist images of the destination encourage the desire to visit. The next process of tourism interpretation will occur during their visit, such as during a site interpretation by winemakers. At this stage, the destination marketing organization (DMO) or tourism enterprise plays a significant role in supplying the interpretation of the TDR's tourism products. The experience of consuming those products will enable co-creation by tourists and locals as well, particularly through social media, which can influence the brand experience by comments, content sharing, websites, and social networking (Foley, Fahy, and Ivers 2017).

Not only wine tourism, but also the concept of tourism interpretation in branding a TGR's product and a TDR through e-storytelling is generally applicable to all kinds of commodities and agritourism. Wine tourism, which is a tourist image, is motivated by an interest in wine consumption and the winescape, which is associated with vineyards, winemaking, and the winery (Southall 2012). Tourism interpretation through a branded wine product is yet another factor to inspire wine lovers to travel to wine regions. However, this combination of branding a TGR's product and a TDR's region lacks research.

Therefore, this study's conceptual framework consists of a two-way transformation of tourism interpretation through e-storytelling as the IMC approach between a TGR's branded wine product and a TDR's brand (Figure 46.1) to examine the contents of contemporary e-storytelling and tourism interpretations related to wine products and wine regions.

Methodology

This chapter examines the contents of e-storytelling about wine as an innovative IMC approach to tourism interpretation to brand TGR wine products and wine tourism destinations. We collected data in two ways: studying winery websites (to access wine products provided at the TGR) and observing wine regions in the field (as the TDR). To access wine products, 14 wineries' English websites made by the Japan Wineries Association were studied using a qualitative content analysis of their written works and videos in both English and Japanese. The field observations were then conducted in Katsunuma town, Yamanashi Prefecture, on December 23, 2020. Data were collected through informal interviews with staff and wine tourism experiences at two wineries, Marufuji Winery and Lumiere, whose wine product and wine region content were qualitatively analyzed. The main research questions were as follows:

1 What e-storytelling content did they use to market their wine products and create the wine region's tourism image?
2 How did e-storytelling as an innovative IMC approach interpret its message to brand the wine products and wine region?

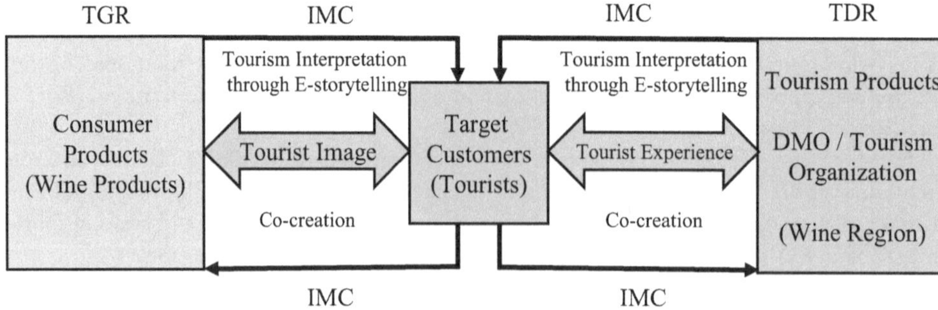

Figure 46.1 The conceptual framework of contemporary tourism interpretation for branding wine products and wine tourism destinations.

Source: Author

Results

Winery website survey

According to the English version of the Japan Wineries Association website (https://www.winery.or.jp/en/), 14 winery websites can be accessed, but most of the information is in Japanese. All these companies were coded as WB1–WB14 (Table 46.1).

The winery websites were mostly in Yamanashi Prefecture (WB4–WB12; 9 of the 14), a capital of the Japanese wine region. All the winery websites' pages and links were checked, such as SNS and tourism information. The messages revealed on each website were then categorized by 13 codes (Table 46.2).

The most common content is wine products (100%), followed by wine tourism activity (86%), wine shop (86%), wine company (79%), and wine history (71%). Interpretations of wine brand and wine news also appeared in more than half of all the websites (57% for both), while information on wine knowledge (50%), wine cuisine (50%), and wine restaurant (50%) was found on exactly half of the websites.

The data in Table 46.2 were provided by winemakers, so not all the information was presented as e-storytelling. However, some written content was associated with facts, emotive language, feelings, and interpretations, which can be classified as storytelling (Bonarou, Tsartas, and Sarantakou 2019). The e-storytelling on each winery website was categorized by codes denoting written work, content, and links to branded wine tourism destinations (Table 46.3).

Table 46.1 Name list of wine brands, wineries, locations, and their websites

Code	*Wine brand*	*Company name*	*Wine region*	*Company website*
WB1	Cave D'occi	European Viticulture Institute Co., Ltd.	Niigata	https://www.docci.com/
WB2	Aizu wine	Honda Winery	Fukushima	https://www.hondawinery.co.jp/index-en/
WB3	Coco-wine	Coco Farm & Winery	Tochigi	https://www.cocowine.com/contents/english
WB4	Alps wine	Alps Wine Co., Ltd.	Yamanashi	https://www.alpswine.co.jp*
WB5	Adega Vinnicola D'Aruga	Katsunuma Winery Co., Ltd.	Yamanashi	https://www.katsunuma-winery.com/english/index.html
WB6	Kurambon	Kurambon Wine Co., Ltd	Yamanashi	https://www.kurambon.com*
WB7	L'orient	Shirayuri Winery Co., Ltd.	Yamanashi	https://shirayuriwine.com*
WB8	Grace wine	Chuo Budoshu Co., Ltd.	Yamanashi	https://www.grace-wine.com/f-english/english.html
WB9	Rubaiyat	Marufuji Winery Co., Ltd.	Yamanashi	https://www.rubaiyat.jp/en/
WB10	Lumiere	Lumiere Co., Ltd.	Yamanashi	https://www.lumiere.jp/en/
WB11	Budoubatake	Budoubatake, Ltd.	Yamanashi	https://budoubatake.net/*
WB12	Fujiclair wine	Fujicco Winery Co., Ltd.	Yamanashi	https://fujiclairwine.jp/
WB13	Katashimo wine	Katashimo Wine Foods Co., Ltd.	Osaka	https://www.kashiwara-wine.com/index.html
WB14	Fujimaru	Papilles Co., Ltd.	Osaka	https://www.papilles.net/

Note:
WB = wine brand.
* = An updated web address different from that of the Japan Wineries Association's English webpage

Table 46.2 Checklist of wine-related written work categories on websites

Code	*WL*	*WC*	*WH*	*WT*	*WSH*	*WR*	*WA*	*WSE*	*WE*	*WK*	*WF*	*WB*	*WN*
WB1	✓	×	×	✓	✓	✓	✓	✓	✓	✓	✓	×	✓
WB2	✓	✓	×	×	✓	×	×	✓	×	×	✓	×	×
WB3	✓	✓	×	✓	✓	✓	×	×	✓	✓	×	×	×
WB4	✓	✓	✓	✓	✓	×	×	✓	×	×	✓	✓	×
WB5	✓	×	✓	✓	×	✓	×	×	×	✓	✓	✓	✓
WB6	✓	✓	✓	✓	✓	×	×	×	×	✓	×	✓	×
WB7	✓	✓	✓	✓	✓	×	×	✓	×	✓	×	✓	✓
WB8	✓	✓	✓	✓	✓	✓	×	×	×	✓	×	✓	✓
WB9	✓	✓	✓	✓	✓	×	×	×	✓	✓	×	✓	×
WB10	✓	✓	✓	✓	✓	✓	×	×	×	×	✓	✓	✓
WB11	✓	×	×	✓	✓	×	×	×	×	×	×	×	✓
WB12	✓	✓	✓	✓	✓	×	×	×	×	×	✓	×	✓
WB13	✓	✓	✓	✓	✓	✓	×	×	✓	×	×	✓	✓
WB14	✓	✓	✓	×	×	✓	×	×	×	×	✓	×	×
Total	14	11	10	12	12	7	1	4	4	7	7	8	8
(%)	(100)	(79)	(71)	(86)	(86)	(50)	(7)	(29)	(29)	(50)	(50)	(57)	(57)

Abbreviations: WL = Wine list (production), price, and campaign; WC = Wine company story related to winery/vineyard/winescape; WH = History of wine product/winery; WT = Wine tourism activity; WSH = Wine shops and stores; WR = Wine restaurants; WA = Winery-related accommodation; WSE = Wine-related service; WE = Wine-related event; WK = Wine knowledge/learning course; WF = Wine and cuisine/recipe; WB = Wine brand; WN = News about wine products and related activities.

E-storytelling can be used to interpret wine products in all categories of written works. However, only some categories include e-storytelling that can interpret the meaning of a place (wine region) and create the tourist image of a wine tourism destination (the italic phrases in Table 46.3): codes WK, WT, WL, WR, WH, WC, WA, and WF.

Regarding the content of e-storytelling by sharing videos, the results found that interpreting the linkage of wine products and wine region was similar to that of written work analysis, but that wine event was included in the e-storytelling patterns, as shown in WB3 in Table 46.4.

Besides the stories told by winemakers, storytelling can conduce value co-creation through communication networks by customers (Table 46.5); all the winery websites offered customers the chance to tell their own stories about the wine products and wine region on various social media platforms, particularly Facebook (71%).

Field observation

Based on the Japanese winery website survey, most wineries were in Yamanashi, particularly Katsunuma town, which is part of the Koshu Valley. The winery field observation, therefore, took place in Katsunuma at two wineries, Marufuji Winery and Lumiere, due to their attractive e-storytelling content and techniques.

Marufuji Winery

The Marufuji Winery's distinctive e-storytelling consists of wine company, wine brand, and wine tourism activity (WB9 in Table 46.3). Regarding the wine company, it is one of the oldest wineries in Katsunuma, as it was established in 1890. The estate has been run by the

Table 46.3 Analysis of distinctive written works classified as storytelling and their combinations to brand the wine region

Code	*Distinctive written work categories (classified as storytelling)*	*Content of e-storytelling*	*Link to the wine region's brand*
WB1	1 WK 2 WR 3 WF 4 *WA*	1 Winery-management school, winemaking and environmentally friendly cultivation methods 2 Historical Western-style building of wine restaurant 3 Japanese Mediterranean cuisine 4 *Winery stay with wild animal series collection rooms and Mt. Kakuda view*	Niigata wine coast, Niigata
WB2	1 *WL* 2 WSE	1 *Small winery at the foot of Mt. Bandai using pure local grown fruits in Aizu and Inawashiro towns for winemaking* 2 Workshop on making a handmade wine label for visitors	Mt. Bandai, Aizu, and Inawashiro towns in Fukushima
WB3	1 *WR* 2 WE	1 *Wine restaurant and café serving seasonal food with homemade wines, presenting the harvesting calendars of local grown vegetables in Tochigi* 2 Wine event: cerebrating St. Vincent, the guardian deity of grapes	Tochigi
WB4	1 *WC* 2 WF	1 *Winery located in the foot of Mt. Fuji, Kofu basin* 2 Wine with Japanese dietary 'Washoku'	Mt. Fuji, Kofu basin, and Fuefuki city in Yamanashi
WB5	1 *WK*	1 *Koshu wine's uniqueness*	Koshu Valley, Katsunuma town, Yamanashi
WB6	1 WC 2 WT 3 *WK* 4 WB	1 Story of winery owner, who studied wine in France 2 Questions draw the readers' attention to the winery tour 3 *Kushu grape and the terroir of Koshu Valley* 4 Brand meaning 'Kurambo' in a fairy story written by the famous Japanese author Kenji Miyazawa	Koshu Valley, Katsunuma town, Yamanashi
WB7	1 *WK* 2 WN	1 *Grape varieties in Koshu Valley* 2 News report related to wine projects, such as wine tourism and firework	Koshu Valley, Katsunuma town, Yamanashi
WB8	1 *WL* 2 WC 3 *WH* 4 WSH 5 *WK* 6 WB	1 *Stories of wine product characteristics linked to the terroir of Yamanashi* 2 History of company from the first to fourth generations 3 *History of Koshu grape and Katsunuma town* 4 Wine shop and wine tasting with new normal practices to prevent the spread of COVID-19 5 *Evolution of Koshu wine to the world* 6 Grace wine, a brand's name from the 'Three Graces' in Greek mythology	Koshu Valley, Katsunuma town, Yamanashi

(*Continued*)

Table 46.3 (Continued)

Code	*Distinctive written work categories (classified as storytelling)*	*Content of e-storytelling*	*Link to the wine region's brand*
WB9	1 WC 2 *WT* 3 WB	1 Company profile presented in a historical timeline with pictures 2 *Winery route tour in Katsunuma described with pictures and captions telling the story* 3 Story of brand 'Rubaiyat,' an ancient poem by the Persian author Omar Khayyam	Koshu Valley, Katsunuma town, Yamanashi
WB10	1 WSH 2 *WR*	1 Wine chart comparing the tastes of wine products 2 *Wine restaurant 'Zelkova' offering selected farm-fresh and locally grown ingredients of Katsunuma*	Katsunuma town, Yamanashi
WB11	1 *WT*	1 *Agricultural tour and volunteer tourism activity in Katstunuma, such as grape picking and working on a vineyard*	Katsunuma town, Yamanashi
WB12	1 *WF*	1 *Wine and culture of Japanese cuisine in Katsunuma*	Katsunuma town, Yamanashi
WB13	1 *WH* 2 *WT* 3 WN	1 *Historical story of a 100-year-old vineyard, Osaka wine* 2 *Wine tourism walking-route map with short stories of wine community–related places* 3 News report on wine stories on printed materials	Osaka
WB14	1 *WC*	1 *Philosophy of the company*	Kashiwara city, Osaka

Note: The italic phrases indicate the e-storytelling that links to the wine destination's branding.

Omura family for over four generations and is rich with history. The storytelling of the company profile is well presented in the historical timeline with easily understood pictures. Some parts of its story connect to the wine brand.

The winery named its wine brand 'Rubaiyat' after the eponymous poem by the famed poet Konosuke Hinatsu, who visited the Marufuji Winery with the Ministry of Interior Sekikawa in 1957. At that time, Tadao Omura, the third-generation owner, asked Hinatsu for brand-name suggestions. Hinatsu gave three names: Rubaiyat, Enchante, and Chariot D'or, described in his letter sent to Omura a few days later. Omura preferred the first suggested name, 'Rubaiyat,' which is the name of a famous ancient Persian poetry series by Omar Khayyam in the 13th century, whom most Japanese intellectuals know as one of the world's greatest poets. Hinatsu's original letter copy is displayed in the cellar gallery with one poem of the Rubaiyat related to wine:

> If thou hast Wine, forswear not the Drinking of it;
> for an hundred Regrets attend such Resolutions.
> Roses burst in Bloom, Bulbuls fill the air with Song –
> would it be wise to forswear Wine at such a Time?
>
> *(Rubaiyat No. 133 by Omar Khayyam*
> *Translated by Hamzeh Carr in 1925)*

Table 46.4 Checklists and analysis of video sharing on websites interpreting the linkage of wine-region branding with e-storytelling content

Code	*VDO sharing*	*E-storytelling content*	*Link to the wine region's brand*
WB1	✓	Wine brand's recipe used local grown ingredients	Niigata wine coast, Niigata
WB2	×	×	×
WB3	✓	Violin and music in the winescape and wine festival 'Harvesting festival in your heart: Itsudemo! Dokodemo! Daredemo!'	Tochigi
WB4	✓	Winescape, winery, and wine processing	Fuefuki city, Yamanashi
WB5	×	×	×
WB6	✓	Winescape of Koshu Valley, winery, and wine processing	Katsunuma town, Yamanashi
WB7	✓	Winescape Koshu Valley, winery, wine processing, and wine products	Katsunuma town, Yamanashi
WB8	×	×	×
WB9	×	×	×
WB10	✓	Wine restaurant, wine product, and wine tourism route	Katsunuma town, Yamanashi
WB11	✓	'Budoubatake Channel,' presenting the terroir of Katsunuma, cultivating Kosyu grape, and winemaking	Katsunuma town, Yamanashi
WB12	✓	Grape cultivation	Katsunuma town, Yamanashi
WB13	✓	Winescape and winery	Osaka
WB14	✓	Wine merchant travel sharing experience	×
Total (%)		10 (71)	9 (64)

Table 46.5 E-storytelling checklist through written communication networks linked to websites

Code	*Communication network availability*	*Social media integration*					
		Facebook	*Instagram*	*Twitter*	*Note*	*Line*	*Other platforms*
WB1	✓	✓	✓	✓	×	✓	×
WB2	✓	✓	×	×	×	×	×
WB3	✓	✓	✓	✓	×	✓	Mail magazine
WB4	✓	✓	×	✓	×	×	Web blog
WB5	✓	×	×	×	×	×	Mail magazine
WB6	✓	✓	✓	×	×	×	×
WB7	✓	✓	×	×	×	×	×
WB8	✓	×	✓	×	✓	×	Mail magazine
WB9	✓	×	✓	×	×	×	TripAdvisor
WB10	✓	✓	✓	×	×	×	×
WB11	✓	×	×	×	×	×	Web blog (Budobatake Monogatari, a grape basket story)
WB12	✓	✓	✓	×	×	×	Web blog (with wine)
WB13	✓	✓	×	×	×	×	×
WB14	✓	✓	×	✓	×	×	×
Total	14	10	7	4	1	2	7
(%)	(100)	(71)	(50)	(29)	(7)	(14)	(50)

This story, posted on the winery website, allures visitors to participate in wine tasting and the winery tour. At the winery, the wine shop is very attractive due to its traditional Japanese architectural style. The wine-tasting room is decorated like a historical museum exhibit of the company history, with displayed objects related to Omar Khayyam and his masterpieces, such as books, pictures, and pieces of poetry. Besides selling wine products, a Rubaiyat poetry book published in Japanese by the winery is available for purchase at the wine shop.

Concerning WB storytelling, Marufuji's winery tour is also attractive. It invites visitors to participate in its tour free of charge for 15 minutes. The staff take groups of visitors to visit the highlights of the winery starting from the tasting room, proceeding to the cellar and vineyard, and ending at the factory. During a tour, visitors can learn about not only the history of the winemaking and wine brand, but also the linkage of the Rubaiyat brand with the wine region brand 'Katsunuma' through the story of the wine's birthplace and its founders.

In August 1877, the first Japanese wine company, Dai-Nihon Yamanashi Budoshu, was established in Katsunuma with governmental support. In the same year, two young men, Masanari Takano and Sukejiro Tsuchiya, were sent to France to study winemaking before returning to Japan to make wine using French methods with the Koshu grape variety for white wine. Since then, winemaking in Katsunuma has expanded, making it a wine town with more than 30 wineries. A photo of Takano and Tsuchiya pruning a Koshu grape (Figure 46.2) represents this story. It can be found in wineries and wine-related places in Katsunuma, such as the Budounokuni Bunkakan, the community's grape culture center. This photo is also used as a brand logo to promote wine tourism in Katsunuma. Thus, many wineries and shops in Katsunuma use this logo with the products and promoting materials. Moreover, those wineries have started co-branding as 'Wine Prefecture, Yamanashi,' a recent brand logo of Yamanashi prefecture as a wine tourism destination. Rubaiyat (BW9) is an example of using both its own brand on wine products and co-branding with Katsunuma and Wine Prefecture, Yamanashi (Figure 46.2).

Figure 46.2 Co-branding logos of Katsunuma and Wine Prefecture Yamanashi.

Source: Author

Lumiere

The boutique winery Lumiere's distinctive e-storytelling focuses on its wine shop and wine restaurant (WB10 in Table 46.3). Established in 1855, Lumiere's history spans over a century.

Focusing on e-storytelling to interpret wine products in the wine shop, the winery uses its own wine chart to compare the characteristics of wine tastes produced from various grape varieties. Among them, Koshu, the local grape developed in Kasunuma, is presented through the staff's interpretation at the wine-tasting machine in the wine shop.

Koshu, a pink-skinned grape, is among the most popular grapes in Japan. Approximately one thousand years ago, it traveled via the Silk Road from the Caucasus, across Central Asia, and to China before being introduced to Japan by Buddhist monks in the 6th or 7th century during the Nara period. This story is also exhibited in the Budounokuni Bunkakan. Koshu was registered as *Vitis vinifera*, an official winemaking variety, by the International Organization of Vine and Wine (OIV) in 2010. The wine's distinctive characteristic is the refreshing aroma of citrus fruits, such as grapefruit and lemon, along with a light acidity and relatively low alcohol.

To promote Koshu wine, the restaurant Zelkova in the winery presents Japanese wine and food culture by offering French cuisine with local, seasonal ingredients. This connects the Lumiere brand with the wine region brand of Katusunuma, which is characterized by Koshu wine and food made with locally produced ingredients.

Discussion and conclusion

Based on the analyzed written works and videos on the wineries' websites, their e-storytelling content to brand their wine products and wine tourism destinations can be divided into nine categories: knowledge-based story, tourism activity–based story, products-based story, history-based story, hospitality-based story, winescape-based story, brand-based story, event-based story, and news-based story.

Knowledge-based story

E-storytelling based on wine includes primarily winery management, wine processing, grape varieties, grape cultivation, and sustainable agriculture and environmental conservation. Because wine tourists are mostly independent, well-educated travelers with excellent cultural knowledge (Croce and Perri 2017), this type of content can target allocentric wine lovers, educators, and students to learn from the topics that e-storytelling interprets, such as this interpretation of sustainable terroir by WB6.

> There is a typical climate and soil in Katsunuma, and there is also each terroir in each vineyard. To preserve the typicity of each soil, we haven't fertilized or plowed our vineyards to leave weeds to grow naturally since 2007.

This excerpt explains how to preserve the soil and environments of vineyards and describes the terroir's relationship to the wine region, adding the value of WB6's brand and the tourist image of Katsunuma.

Tourism activity–based story

This type of story concerns special-interest tour activities with recreational and volunteer experiences, such as wine tasting, winery tours, grape picking, working in vineyards, and walking or cycling along wine routes. These stories on wine tourism activities interpret the relationship between the wine region, wineries, and visitor-consumers (Bruwer and Lesscheve 2012), resulting in a good tourism experience.

Product-based story

This story regards the characteristics and quality of wine products, focusing on aspects such as the uniqueness of taste, color, and design. Visitors can share these stories from their impressions of wine products via social media, or winemakers can post them on winery web pages. Jancis Robinson's comment on WB8's web page is an example of this type of story:

> The best Japanese wines I have tased have been from Koshu, a grape imported centuries ago from the Caucasus along the Silk Route. Grace Vineyards' unoaked, less-matured 1999 Koshu is a wonderfully confident, smoky, rather Alsatine white.

Feedback like this from visitors interprets the quality of the products and influences other consumers' decisions both to select products and to visit the winery because of trust, confidence, and sense of belonging (Bassano et al. 2019).

History-based story

Creating an online story of a brand commodifies wine-related history, including the company background, family heritage, ethnic and geographical heritage (Frost et al. 2020), the winemaker's biography, the evolution or development of grape cultivation and wine products, and the brand creation. Each wine region has its own history and culture (Kerma 2012) that can be used to interpret the 'heritage brand' (Frost et al. 2020) and 'place identity' (Ageeva and Foroudi 2019) to attract potential consumers.

Hospitality-based story

This online content pattern involves the wine restaurant and café, wine and food culture, accommodation, and hospitality-related services, such as a winery wedding studio and handicraft workshop. According to Schmitt (cited in Santos et al. 2019), the tourist's holistic experience should be considered the convergence of cuisine, wine, and tourism. Thus, hospitality-based content regarding food is the focus of most wineries (WB1, WB2, WB4, WB5, WB10, WB12, and WB14 in Table 46.2) as the primary influencer of travel and eating behaviors (Rachão et al. 2020), and most customers come to experience food and wine because they wish to know the story (Velikova, Canziani, and Williams 2019). This kind of content needs social interaction; for example, the content of wine cuisine recipes by WB8's customers, which is the interpretation of wine value co-creation. Furthermore, it is a value of co-branding because most winery properties view restaurant partnerships as branding strategies (Velikova, Canziani, and Williams 2019).

Winescape-based story

The most critical dimension of the winescape is the region's natural beauty or setting (Bruwer and Lesscheve 2012). Territories can be 'powerful markets' and may be expressed as 'geosymbols' that confer meaning (Mitchell, Charters, and Albrecht 2012) on the terroir's uniqueness and the wine region's branding, such as the memory of WB8's winery owner of the Koshu grape:

> I remember when I was a child, my grandfather would let me sniff Koshu grapes. … My father would bring me to the sun-bathed vineyards, showed me a bunch of Koshu grapes, lucid with a light purplish shade under the sun.

Brand-based story

As Bruwer and Lesscheve (2012) stated, once a destination's (brand) image has formed, a brand identity can be linked with a marketing strategy. Constructing the story of a winery's brand is integral to destination branding to attract repeat customers influenced by brand loyalty (Gómez, Lopez, and Molina 2015), such as the brand story presented in WB9:

> Rubaiyat is the name of the poetry of Omar Khayyam, a Persian poet and astronomer in the 11th century. Khayyam was born where grapes originated and was a long lover of wine. He wrote many poems expressing his love of wine and beautiful women. Most Japanese intellectuals know Omar Khayyam as one of the best poets in the world; therefore, if you name your wine Rubaiyat, it would delight them.

Event-based story

The event-based story focuses on wine concerts and musical festivals, such as the violin performance in the harvesting festival of WB3. As noted by Kruger, Rootenberg, and Ellis (2013), hosting events and festivals is the best way to provide wine tourism experiences that reward wineries with repeat business and word-of-mouth recommendations by tourists who attend them. Thus, creating stories about wine events and festivals is crucial to draw attention from wine lovers, who tend to be relatively young, allowing winery and tourism marketers to use social media channels, such as Facebook, Twitter (Kruger, Rootenberg, and Ellis 2013), Instagram, and YouTube (Strickland et al. 2016), including wine tourism mobile apps (Dimitrovski et al. 2019), to interact (Iaia et al. 2017) and co-create wine event–based stories with winemakers or event managers and participants. As noted by Ageeva and Foroudi (2019), event promotion can effectively enhance place image, as e-storytelling about events does for wine region branding.

News-based story

Iaia et al. (2017) grouped wine news with wine events based on content. Wine news concerns broadly relevant content, ranging from knowledge, product, and tourism experiences to wine industry–related business, awards, contests, campaigns, and promotions. The style of e-storytelling about news is relatively simple—a report and announcement—while event-based wine storytelling and other patterns take various forms, such as social interaction, experience sharing (Rachão et al. 2020), advertising, and 'edutainment' (Iaia et al. 2017). Thus, news-based content interprets the updated information and movements of wineries.

These nine categories of wine story content and their interpretative approaches to branding products and destinations support the hypothesis that e-storytelling is created by the enterprise itself and its customers and visitors (Bonarou, Tsartas, and Sarantakou 2019). A wine enterprise can provide all kinds of wine-related content, while customers can share experience-based stories and create content, such as food recipes based on wine products, on social media platforms. Therefore, e-storytelling contributes to interpreting the value of social interaction and the value co-creation of wine products and regions.

The study's main limitation is the small number of examined websites and winery visits, which may not cover all forms of e-storytelling content. However, it contributes to filling the literature gap around e-storytelling and tourism interpretations of wine products and wine tourism.

Case Study 46.1 Wine Prefecture, Yamanashi – branding wine tourism destination with the TGR's restaurants

Japanese wine production developed over 140 years ago as knowledge of French winemaking expanded from Koshu Valley in Yamanashi in the 1870s. Nowadays, Japanese wine is made in a variety of locations, including valleys, mountains, hills, and coastal areas; the grape is cultivated all over the country. The major production areas, however, are Yamanashi, Nagano, Yamagata, and Hokkaido (Japan Wineries Association, n.d.). Of these, Yamanashi is the most well-known destination, winning a Geographical Indication from the Japanese government and being the first regional wine brand of Japan since July 2013.

As a famous wine tourism destination, Yamanashi attracts wine lovers seeking the authenticity and uniqueness of Japanese wine, as 70% of the wines on the Japanese market are imported. Only 6% are proudly Japanese wine, referred to as 'made in Japan.' This small amount of Japanese wine is made chiefly in Yamanashi, the wine country of Japan that contains over 60 wineries around the cities of Fuefuki, Kofu, and Yamanashi (Petkoska 2019).

To promote the uniqueness of high-quality Japanese wines made in Yamanashi, branding the destination plays a significant role in wine tourism. On August 7, 2019, the Yamanashi prefectural government official launched the destination brand 'Wine Prefecture, Yamanashi.' The brand logo is framed by the Japanese character pronounced 'Wine Ken Yamanashi,' meaning 'Wine Prefecture, Yamanashi' in English. On top, it is illustrated with a drawing of Mount Fuji and a sun-shrine in the background, which is the tourist image of Japan and Yamanashi. A bunch of Koshu grapes is on the left, and a bottle of wine with a wine glass is on the right, indicating Yamanashi as the birthplace of Japanese wine (Figure 46.2). This brand logo significantly interprets Yamanashi as the wine tourism destination of Japan. Other wine brands produced in Yamanashi are communicated through e-storytelling at https://wine-ken.jp, the branding project's website.

Using the language of contemporary tourism interpretation in this case, the website introduces wineries and wine restaurants in Yamanashi as the TDR. Meanwhile, on the TGR website, where Yamanashi wines are available, a web-based map marks the locations of restaurants in Tokyo with links to their websites to draw tourist attention. The project's website showcases brand transformation in two ways. First, Yamanashi's branded wine products in the TGR's Tokyo restaurants shape the tourist image of Yamanashi to wine lovers by co-creating e-storytelling. For example, Yamanashi wine consumers can share their experiences and stories with the restaurant's SNS. Second, wineries in the TDR's Yamanashi wine region offer a tourist experience, including wine tasting and winery tours, to patrons of Tokyo's restaurants who notice the wine tourism image in the logo of 'Wine Prefecture, Yamanashi' on wine products and in restaurants' e-storytelling.

The 'Wine Prefecture, Yamanashi' logo not only represents wine products but also communicates stories about wine and wine tourism in Yamanashi through various media, including printed materials, television, and videos.

Implications

Nine categories of e-storytelling content on winery websites can be applied to brand wine products and wine tourism destinations. This study bears significance to the winemakers and marketers in branding and promoting the wine products at the TGR through IMC. On the other hand, it plays an essential role for the policymakers, government bodies, or tourism organizations such as DMO at the TDR in branding the destination as a wine region through any type of e-storytelling.

References

Ababneh, A. (2016). Heritage interpretation: analysis study of the signage system used at the archeological site of Umm Qais in northern Jordan. *Tourism Planning & Development*, 14(3): 297–317.

Ageeva, E., and Foroudi, P. (2019). Tourists's destination image through regional tourism: from supply and demand sides perspectives. *Journal of Business Research*, 101: 334–348. https://doi.org/10.1016/j.jbusres.2019.04.034

Bassano, C., Barile, S., Piciocchi, P., Spohrer, J. C., Iandolo, F., and Fisk, R. (2019). Storytelling about places: tourism marketing in the digital age. *Cities*, 87: 10–20. https://doi.org/10.1016/j.cities.2018.12.025

Bonarou, C., Tsartas, P., and Sarantakou, E. (2019). E-storytelling and wine tourism branding: insights from the 'Wine Roads of Northern Greece.' In Sigala, M., and Robinson, R. N. S. (eds.), *Wine Tourism Destination Management and Marketing* (pp. 77–98). Cham: Palgrave Macmillan.

Bruwer, J., and Lesscheve, I. (2012). Wine tourists' destination region brand image perception and antecedents: conceptualization of a winescape framework. *Journal of Travel & Tourism Marketing* 29: 611–628. https://doi.org/10.1080/10548408.2012.719819

Croce, E., and Perri, G. (2017). *Food and Wine Tourism*. Oxfordshrine: CABI.

Dimitrovski, D., Joukes, V., Rachão, S., and Tibério, M. L. (2019). Wine tourism apps as wine destination branding instruments: content and functionality analysis. *Journal of Hospitality and Tourism Technology*, 10(2): 136–152. https://doi.org/10.1108/JHTT-10-2017-0115

Dumbraveanu, D., Craciun, A., and Tudoricu, A. (2016). Principles of interpretation, tourism and heritage interpretation-the experience of Romanian museums. *Human Geographies – Journal of Studies and Research in Human Geography*, 10(1): 59–75. https://doi.org/10.5719/hgeo.2016.101.4

El-Menshawy, S. (2016). Effective rapport in tourist guiding (interpretation of themes). *Journal of Socialomics*, 5(3): 1–5.

Florek, M., and Connejo, F. (2006). Export flagships in branding small developing countries: the cases of Costa Rica and Moldova. *Place Branding and Public Diplomacy*, 3(1): 53–72.

Foley, A., Fahy, J., and Ivers, A. (2017). Brand experience in tourism in the Internet age. In McCabe, S. (ed.), *The Routledge Handbook of Tourism Marketing* (pp. 140–150). London: Routledge.

Frost, W., Frost, J., Strickland, P., and Maguire, J. S. (2020). Seeking a competitive advantage in wine tourism: heritage and storytelling at the cellar-door. *International Journal of Hospitality Management*, 87: 1–9. https://doi.org/10.1016/j.ijhm.2020.102460

Garcia, P., and Rossiter, M. (2010). Digital storytelling as narrative pedagogy. In Gibson, D., and Dodge, B. (eds.), *Proceedings of SITE 2010 – Society for Information Technology & Teacher Education International Conference* (pp. 1091–1097). San Diego, CA: Association for the Advancement of Computing in Education (AACE). https://www.learntechlib.org/primary/p/33500/

Gómez, M., Lopez, C., and Molina, A. (2015). A model of tourism destination brand equity: The case of wine tourism destination in Spain. *Tourism Management*, 51: 210–222. https://doi.org/10.1016/j.tourman.2015.05.019

Hope, J. (2018). Seven types of stories: which are you telling? https://www.echostores.com/seven-types-of-stories/

Iaia, L., Scorrano, P., Fait, M., and Cavallo, F. (2017). Wine family business and web: marketing strategies to complete effectively. *British Food Journal*, 119(11): 2294–2308. https://doi.org/10.1108/BFJ-02-2017-0110

Japan Wineries Association (n.d.) *What Is Japanese Wine?* https://www.winery.or.jp/en/#:~:text=The%20main%20characteristic%20of%20Japanese,such%20as%20Chardonnay%20and%20Merlot

Kerma, S. (2012). Wine tourism as a factor of regional development: the case of Primorska wine region, Slovenia. In Jordan, P. (ed.), *Regional Development and Regionalisation in the Adriatic Space: Proceedings of the 3rd Conference of the Adriatic Forum Vienna*, September 23–25, 2010, ISR-Forschungsberichte, 38: 127–1396.

Kruger, S., Rootenberg, C., and Ellis, S. (2013). Examining the influence of the wine festival experience on tourists' quality of life. *Social Indicators Research*, 111: 435–452. https://doi.org/10.1007/s11205-012-0013-0

Lee, K., Kladou, S., Usakli, A., and Shi, Y. (2022). Inspiring winery experiences to benefit destination branding? Insight from wine tourists at Yantai, China. *Journal of Hospitality and Tourism Insights*, 5(1): 116–137. https://doi.org/10.1108/JHTI-06-2020-0109

Leiper, N. (1979). The framework of tourism: towards a definition of tourism, tourists, and the tourist industry. *Annals of Tourism Research*, 6(4): 390–407.

Mitchell, R., Charters, S., and Albrecht, J. N. (2012). Cultural systems and the wine tourism product. *Annals of Tourism Research*, 39: 311–335. https://doi.org/10.1016/j.annals.2011.05.002

Mortimer, K., and Laurie, S. (2019). How to achieve true integration: the impact of integrated marketing communication on the client/agency relationship. *Journal of Marketing Management*, 35(3–4): 231–252.

Peng, X. (2013). A study on the tourism destination image of Japan in the Chinese market. *AGI Working Paper Series*, 9, 1–17.

Petkoska, Z. (2019). Yamanashi: the wine country of Japan, from western tradition to local production. https://metropolisjapan.com/yamanashi-wine-country-japan/

Rachão, S. A. S., Breda, Z., Fernandes, C., and Joukes, V. (2020). Food-and-wine experiences towards co-creation in tourism. *Tourism Review*, 76(5): 1050–1066. https://doi.org/10.1108/TR-01-2019-0026

Rein, I., and Shields, B. (2007). Place branding sprots; Strategy for differentiating emerging, transitional, negatively viewed and newly industrialised nations. *Place Branding and Public Diplomacy*, 3(1): 73–85.

Santos, V. R., Ramos, P., Almeida, N., and Santos-Pavón, E. (2019). Wine and wine tourism experience: a theoretical and conceptual review. *Worldwide Hospitality and Tourism Themes*, 11(8): 718–730. https://doi.org/10.1108/WHATT-09-2019-0053

Southall, C. (2012). Wine tourism. In Robinson, P. (ed.), *Tourism: The Key Concepts* (pp. 282–284). London: Routledge.

Strickland, P., Williams, K. M., Laing, J., and Frost, W. (2016). The use of social media in the wine event industry: a case study of the High Country Harvest in Australia. In Szolnoki, G., Thach, L., and Kolb, D. (eds.), *Successful Social Media and Ecommerce Strategies in the Wine Industry* (pp. 74–92). New York: Palgrave Macmillan.

Suorsa, E. (2017). *Country Branding in Export Promotion: Case Food from Finland* [MA thesis]. Jyväskylä University School of Business and Economics. https://jyx.jyu.fi/bitstream/handle/123456789/56646/URN:NBN:fi:jyu201801091118.pdf?sequence=1

Thanh, T. V., and Kirova, V. (2018). Wine tourism experience: a netnography study. *Journal of Business Research*, 83: 30–37. https://doi.org/10.1016/j.burses.2017.10.008

Tilden, F. (1977). *Interpreting Our Heritage* (3rd ed.). Chapel Hill: University of North Carolina Press.

Velikova, N., Canziani, B., and Williams, H. (2019). Small winery-restaurant relationship building: challenges and opportunities. *International Journal of Wine Business Research*, 31(1): 5–11. https://doi.org/10.1108/IJWBR-07-2018-0038

47

HYBRID WINE GRAPES AND EMERGING WINE TOURISM REGIONS

Dan McCole

Introduction

Not every wine-producing region offers or is reliant on wine tourism. Many regions produce large volumes of wine for international distribution. Others simply produce wine for local consumers. Both of these types of regions might attract a small number of visitors; however, wine tourism is a much less important economic driver for the region than wine production. Other regions are more reliant on wine tourism. These include well-known areas that produce large volumes of wine as well as regions where overall wine production is not significant but still attracts many tourists.

One way to categorize wine regions is by the relationship between a region's wine production and its reliance on tourism. Figure 47.1 shows four general categories within this relationship.

The 'Global Recognition Region' quadrant is comprised of regions that are well known for wine production but have come to rely on wine tourism as an important contributor to winery revenues and brand equity, as well as the regions' overall economies. Examples of regions in this quadrant include Tuscany, Bordeaux, Cape Town, and Napa Valley.

'Bulk Production Regions' play important roles in the wine industries of wine-producing countries. These regions have many vineyards that supply large volumes of wine but are not among the top wine tourism destinations. Examples in this category include Languedoc-Roussillon in France, the Riverina in Australia, and the Central Valley in California. Wine tourism is not absent in these regions, but it is not nearly as important to the economy as wine production.

The world has many examples of places that fall into the category of 'Regional Production Regions.' These regions typically have a long tradition of winemaking but tend to account for lower production numbers and export less of their wine outside of the region. The wines in these regions are the everyday wines for local residents. Many of these regions might be pursuing efforts to enhance their global recognition, or attract more wine/culinary tourists, but have had only modest success in doing so (or are still in the early stages).

The final category is 'wine experience regions.' These wine-producing regions account for a small percentage of wine production but have established very successful wine tourism

DOI: 10.4324/9781003143628-54

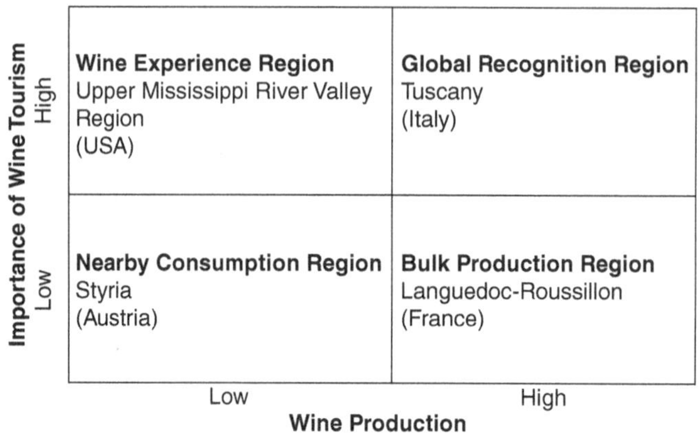

Figure 47.1 Wine regions categorized by production and dependence on tourism.

offerings. The wineries in these regions sell most of the wine they produce directly to their visitors, usually at the tasting room, but also via clubs. Despite the low production volumes, these wineries thrive by offering wine-themed experiences, often in rural areas that can draw people from large population areas for day and weekend trips. These regions differ from the 'Nearby Consumption Region' category because they typically don't have a long history of wine production, as they are less likely to be located in areas typically considered optimal for wine production. Wine regions in this category have benefitted from the two important trends since the 1990s: increased research and development of hybrid wine grape viticulture and winemaking, and increased demand for culinary tourism.

This chapter is largely focused on regions in the 'wine experience' category, and more specifically, those that rely on hybrid wine grapes. In some places, particularly in the eastern, central, and northern regions of the United States and eastern and southern Canada, developments in hybrid wine grape cultivars, and hybrid winemaking techniques, have allowed for wine production in areas where it had not previously been feasible. Although these regions do not produce large volumes of wine, they support thousands of wineries that attract millions of visitors.

This chapter will begin by providing a history of hybrid wine grape development from the days of the Great French Wine Blight of the late 1800s. It will then describe the role of these grapes in the recent development of wine tourism in rural areas, where wine production has not previously been feasible, and the evolving need for new winery business models that are dependent on the provision of wine experiences as much as wine production. Finally, a case study will be presented focused on the Upper Mississippi River Valley American Viticultural Area (AVA), North America's largest designated wine region.

Hybrid wine grapes

Over 70 species of grapes can be found on Earth. Some of these species remain wild, and others have been domesticated. Of the world's cultivated grapes, about 90% are from just one species: *Vitis vinifera* (Mattia et al. 2008). The reason? Wine. *V. vinifera* grapes (e.g., Cabernet Sauvignon, Pinot Noir, Merlot, Chardonnay, Sauvignon Blanc, and Riesling) have historically accounted for almost all of the wine produced throughout the world (McGovern 2003).

Although wine is sometimes made with other species of grape, winemakers and consumers clearly prefer wines made from *vinifera* grapes.

Despite the dominance of *vinifera* grapes for winemaking, there has been a significant increase in the planting of hybrid wine grapes over the past two decades (Sabbatini and Howell 2014). Hybrid grapes are created by cross-breeding two different species of grapes. In the case of wine grapes, *vinifera* grapes have been bred with other grape species that are better suited for the environments in which they are grown. In other words, these hybrids combine the excellent wine quality of *vinifera* with the hardiness and resistance to insects and diseases of the non-cultivated species (Atucha, Hedtcke, and Workmaster 2018). Although the wines made from hybrid grapes represent a small proportion of overall wine production, hybrid wine grapes have become increasingly appreciated for their ability to reduce the amount of environmentally destructive fungicides and pesticides used in grape-growing (Robinson, Harding, and Vouillamoz 2012). Moreover, hybrid grapes have become vital to the success of wine tourism in regions where wine production (and wine tourism) had previously not been feasible.

Origin of hybrid wine grapes

Archeological evidence suggests that *Vitis vinifera* was first cultivated between the Black Sea and Iran about 9,000 years ago (McGovern 2003). Over the centuries, *vinifera* vine plantings spread to the Middle East, Near East, and Europe. During the age of European exploration and colonization, *vinifera* vines were brought to new lands and planted in European colonies. Soon, wines made from *vinifera* grapes were produced on all continents. In the eastern parts of the United States and Canada, however, *vinifera* grapes struggled, so North American winemakers turned to grapes native to the region, such as *Vitis labrusca, Vitis riparia, and Vitis rotundifolia*. Unfortunately, the wines made from these grapes had different taste profiles than ones made with *vinifera* and therefore did not meet the expectations of wine drinkers (Sabbatini and Howell 2014). In time, North American scientists discovered that they could graft *vinifera* vines to native grape rootstocks. This seemed to help the *vinifera* vines to better deal with the North American climate and offered some protection from North American pests and diseases to which *vinifera* vines had never been exposed (Barber, Hutchins, and Dodd 2007).

By the 19th century, wines were produced all over the world. However, after the French Revolution, France began to emerge as the world's top wine producer, as French winemakers embraced new wine technologies and took advantage of the emerging global trade routes (Tattersall and Desalle 2015). By the 1860s, wine was a source of national pride for the French, wine exports were an important part of the French economy, and French wines were considered to be among the best in the world. But the French wine world was about to be exposed to its greatest challenge before or since.

During this same period, the world's scientific community became increasingly interested in studying botanical specimens from around the world. As plants were brought from one country to another, so too were diseases and pests, albeit inadvertently. North America, in particular, was a source of a number of ailments that impacted European vineyards (Tattersall and Desalle 2015). In the 1850s, powdery mildew, a fungus from North America, started appearing in French vineyards, reducing grape yields and impacting wine quality. Within a few years, a fungicide was developed that helped combat powdery mildew, but just as the wine industry began to recover, French vineyards started to suffer from an unknown 'disease.' Thus, began the Great French Wine Blight of the late 1800s, that eventually spread throughout Europe and most of the world's wine-producing regions.

Over time, as scientists studied this blight, they discovered that European vineyards were being destroyed not by disease, but rather by a small insect called Phylloxera, which came from the United States on some of the imported vines (Tattersall and Desalle 2015). Grape varieties native to the United States had evolved to be resistant to Phylloxera, but French grape vines were vulnerable, and the pests devoured the roots of nearly all of Europe's grape vines.

Over the next 30 years, French and European vineyards continued to be devastated until grape breeders developed two strategies to fight Phylloxera. One approach was to replicate what American breeders had done. They imported American root stocks and grafted *vinifera* vines onto them. The other approach was to cross-breed *vinifera* varieties with native grapes to create new varieties of grapes that became known as French-American hybrids. The original hybrids included varieties called Seibel, Couderc, Kuhlmann, and Bertille Seyve. These hybrid cultivars never gained favor in France, where instead, the French, and most of the rest of the winemaking world, committed fully to growing *vinifera* vines on American root stocks (Tattersall and Desalle 2015).

The French-American hybrids did, however, find a home in a relatively small number of vineyards in parts of North America, as they offered resistance not only to Phylloxera and American grape diseases but also to the cold, which was a constant challenge facing *vinifera* vines in eastern vineyards. In the decades since, the original hybrids were bred with other grapes (*V. vinifera* and other species) to make other varieties such as Baco Noir, Maréchal Foch, Seyval Blanc, Vidal Blanc, Chambourcin, Chancellor, and Vignoles. By the 1970s, the original French-American hybrids, as well as their descendant cultivars, could be found scattered in various areas throughout the world, particularly in the United States, but have never accounted for a large portion of wine production (Sabbatini and Howell 2014).

Recent changes in the American wine industry

Throughout the 20th century, the US state of California has been the dominant producer of American wine, consistently producing around 90% of US wine (Barber, Hutchins and Dodd 2007). By 2000, almost 92% of American wine was produced in California (WineAmerica 2020). The remaining wine production was scattered around the United States, with most production coming from *V. vinifera* grapes. In the last 20 years, however, the American wine industry has seen dramatic changes.

First, there has been a large increase in overall wine production and consumption. In 2000, the United States produced 568 million gallons of wine, but just 17 years later, that number had increased to 959 million gallons (WineAmerica 2020). US wine consumption has also risen during this period, growing from 2.01 gallons per resident in 2000 to 2.95 gallons per resident in 2018 (WineAmerica 2020).

During this period of growth, California experienced strong increases in production (+26%), however its share of total US wine production actually dropped from around 92% to just over 87%, showing that there was even greater growth in wine production in the other states.

Even more notable than the increases in production and consumption has been the increase in the number of wineries in the United States. In the year 2000, there were about 2,000 wineries in the United States (Sumner et al. 2001). In 2020, the number of US wineries skyrocketed to over 11,000 (Conway 2020). Again, California saw significant increases in this category, growing from 850 wineries in 2000 (UC Davis 2000) to 4,763 in 2020 (WineAmerica

2020). Although that level of increase in one state is impressive, it also means that there was an increase of over 5,000 wineries in states other than California.

Wine is produced in all 50 US states; however, outside of California, three states rise above the others: The other two West Coast states of Washington and Oregon, and New York on the East Coast. In 2020, these three states accounted for 2,128 wineries and about 9% of wine production (WineAmerica 2020). That means that in the remaining 46 states, there were over 3,000 wineries that together produced only 4% of all US wine. Of these remaining states, a number of them rely on hybrid wine grapes, and a renewed interest in hybrid wine grapes is at least in part responsible for the remarkable growth in the number of wineries throughout the United States – wineries that arguably count not survive without tourism.

Renewed interest in hybrid grapes

As described earlier, the first hybrid wine grapes were developed in response to the Great French Wine Blight. Throughout most of the 20th century, breeders made some additional advancements in hybrid cultivars, but at a much slower pace than in the late 1800s. Over the past three decades, however, environmental concerns and a desire to produce wine in new regions have led to renewed interest in hybrid wine grapes, and a number of breeding programs have produced new cultivars.

In European vineyards, which once rejected and destroyed hybrid vines in favor of *vinifera* vines grafted onto American root stocks (Reynolds 2015), hybrid wine grapes are being reconsidered for their environmental benefits. In Germany, home to some of the world's best Rieslings, some growers are experimenting with a category of hybrid cultivars called Pilzwiderstandsfähige, or 'PIWI' for short. PIWIs are bred to be resistant to the fungal diseases that are so common with *vinifera* grapes, thus reducing the use of fungicides, and they also require less pesticide use. Yet, PIWI grapes are still 85% *V. vinifera*, so the wines produced from them are very similar to the traditionally produced varietals. Even the French have bred disease-resistant hybrids in recent years in an effort to reach national goals for the reduction of fungicide use (Weinmann et al. 2019).

Perhaps the greatest interest in hybrid grapes has been in the United States. As with the original French-American hybrids, many new cultivars are more resistant to disease and pests and therefore require fewer agricultural inputs. This makes them cheaper and more environmentally friendly to grow. Some of these grapes are being bred specifically to be resistant to disease. For example, Pierce's Disease is a bacterial disease the affects grapevines, and some newer cultivars are being developed to be resistant to Pierce's Disease. These vines are generating excitement among wine producers in regions across the American South and Southeast, where Pierce's disease has been very damaging to grape-growing efforts (Kamas, Stein, and Nesbitt 2010).

In regions that have very cold winters, such as the American Midwest and northern states, *vinifera* vines are frequently damaged, and their fruit is too often destroyed by spring frosts. A number of extreme cold-hardy cultivars have been developed for these climates, and they have sparked a sharp increase in the number of wineries across areas of the United States and Canada, where wine production was not previously possible. Moreover, oenology experts have been developing new techniques for improving the quality of the wines made from these grapes. In the United States, hybrid grapes are the most commonly grown wine grape in at least 20 states, mostly midwestern states and northern states that do not produce large volumes of wine, such as Nebraska, Illinois, Iowa, Indiana, Minnesota, Wisconsin, and Vermont. Growers and winemakers in other states are also beginning to show an interest. Many states that grow predominantly *vinifera* grape varieties have begun to plant some hybrid vines because growing *vinifera* grapes in these locations can be expensive, risky, or only possible in some regions (Martinson et al. 2016).

New wine regions require new business models

With the newfound ability to grow wine grapes in nontraditional climates, aspiring winemakers have begun opening wineries in large numbers (Eustice, McCole, and Rutty 2019). However, these new wine businesses were forced to create different business models than wineries in established regions such as California. Several factors make these winery businesses different from producers in well-known regions. First, unlike many other products, wine has always had a connection with specific places. Where a grape is grown makes a difference to consumers and influencers. The great wine regions of Europe gained their reputations over centuries. In the New World, regions such as California, South Africa, Australia, New Zealand, Argentina, and Chile have created good reputations more quickly, but despite wide acclaim for their wines, they still lag behind Europe with regard to global respectability. Even the best wines made from grapes grown in Iowa, Minnesota, Illinois, and Wisconsin would always have to overcome a negative bias of terroir. Add to this challenge the facts that the grape varieties are new and unknown, and that wine has long been subject to an element of snobbery, and there has naturally been little demand at restaurants and retail.

The challenges of demand are not, however, limited to perception. With so many new and unexperienced winemakers and new varieties of grapes that winemakers have yet to master, there have been significant quality issues that need to be addressed with hybrid wines. Several research programs have tried to accelerate the learning curve for growers and winemakers, but mastering the techniques required to make the best possible wines from hybrid grapes, particularly from the newer cultivars, will take time. Additionally, winemakers who use hybrid grapes must decide if they want to make wines that try to mimic *vinifera* wines or make the best wines possible, given the chemistry of the respective hybrid cultivars.

Because of the low demand for wines made from hybrid grapes, production levels for the relatively high number of wineries have been, understandably, low. This challenge explains the small relative proportion of wine production compared to the number of wineries. Consistent with the laws of economics, small levels of wine production lead to low economies of scale, so the production cost per bottle is significantly higher than for the bulk wine producers of the world. This forces the price of cold-hardy wines to be higher. So, if these wines are to compete in retail and restaurant settings, wineries would need to not only convince buyers and consumers that the quality of wine is acceptable but that it is worth paying a higher price than a wine that is made in a recognized wine region, with familiar grape varieties, that may better meet the consumer's taste expectations. So far, most hybrid wineries have not had success with this approach on any large scale.

Tourism-reliant business models

Since high per-bottle production costs cause wines made with hybrid grapes to be sold for a higher price, even if they do not meet or exceed consumers' satisfaction levels compared to cheaper wines, hybrid wineries have been forced to find other methods of distribution for their wines. Most have done this by selling wine directly to consumers at the winery (Lee, McCole, and Holecek 2020). This not only ensures that a winery's products do not need to compete on a shelf with competitors (as with retail) but also avoids the need to share the mark-up portion of the price with distributors. But this approach relies on the ability of wineries to attract enough visitors. Fortunately for new wineries, demand for wine tourism has been very high during this 20-year period of rapid growth.

Millions of people who might never have prioritized a long trip to Napa Valley, Bordeaux, or Tuscany now have the opportunity to have similar experiences closer to home. For wine businesses in nontraditional wine regions, it's much easier to replicate the quality of a wine-themed experience from a famous wine region than it is to replicate the quality of the wine itself. When someone looks at a beautiful vineyard, it doesn't matter which grape is growing. For those interested in touring a winery to view cellars crammed with barrels and learn the basics of how wine is made, the varietal of wine being made is less important. In the tasting room, the sampling of wines is as much about the experience and the comparisons between the offered wines as it is about the wine itself. Some experienced wine drinkers may turn their noses up at hybrid wines, but others might appreciate the wines for their uniqueness or because they are locally produced. Also, research has shown that wine drinkers in nontraditional wine regions have less experience with wines than retail consumers (McCole et al. 2018), so many are less able to compare the wine with ones they have formed a preference for over many years.

Shapiro and Gómez (2014) have shown that the more a visitor is satisfied with their winery experience, the more wine they will purchase. And the factors that led to winery satisfaction had less to do with the quality of the wine itself and more to do with things like customer service, ambiance, and atmosphere. Other studies of tasting room visits have shown that the primary reason for visiting a winery in emerging wine regions has more to do with leisure reasons – such as a chance to socialize, try something new, explore a rural area, or have a relaxing day out – than with reasons related to wine (McCole et al. 2018). This suggests that the experience is a bigger draw than the wine, and it could be argued that the wineries in these regions are tourism providers more than they are wine producers. It is doubtful, however, that many winery owners see themselves in this way. Although winery owners tend to understand that the experience they provide is important, most opened their wineries to become winemakers rather than tourism businesses. Regardless of how hybrid wineries classify themselves, it is clear that they could not exist without tourism.

Case Study 47.1 The Upper Mississippi River Valley American Viticultural Area

The Mississippi River is the second-largest drainage in North America and runs north to south, bisecting the United States. The river's origin is Lake Itasca near the Canadian border in the northern part of the US state of Minnesota. Winter in Minnesota brings extremely low temperatures, making the state the third-coldest of the 50 US states. In January, the average daily low temperature at the origin of the Mississippi River is −4°F (−20°C), and it is not uncommon for temperatures to drop as low as −30°F (−34°C) in many parts of Minnesota. As the Mississippi River travels southward past the Twin Cities of Minneapolis and St. Paul, the river serves as the eastern border of Minnesota with Wisconsin, another very cold state. Further south, the river forms the border between Iowa and Illinois, which are adjacent to Minnesota and Wisconsin, respectively (see Figure 47.2).

Figure 47.2 The Upper Mississippi River Valley American Viticultural Area.

Source: By Drdpw - Own work, CC BY-SA 3.0, https://commons.wikimedia.org/w/index.php?curid=28987184

Because of the extreme cold and late frosts, this region of the United States is not a place most people might think of for wine production. However, the area on both sides of the Mississippi River in these four states is home to the Upper Mississippi River Valley American Viticultural Area (UMRV). An American Viticultural Area (AVA) is a designated wine grape–growing region, essentially the American version of an appellation. There are 252 AVAs in the United States, with more than half (141) located in California. Covering an area of 77,477 square kilometers, the UMRV is the largest AVA in the United States and was certified in 2009.

The relatively recent designation of this region is indicative of the dramatic growth in the number of wineries in this area, a phenomenon seen throughout the United States. In the year 2000, the four states that comprise this AVA had a total of 50 wineries. Just 20 years later, these states have over 500 wineries, a number that represents about 5% of all US wineries, despite the fact that wine production here is only 1/3 of 1% (0.003%) of total U.S. wine production (Alcohol and Tobacco Tax and Trade Bureau 2020). These two facts lead to two questions: Why has this region experienced such dramatic growth in the number of wineries? And, how can a region with such low wine production support so many wineries?

Cold-hardy hybrid wine grapes

One benefit of the French-American hybrids was that they tended to fare better in cooler regions than pure *vinifera* varieties. However, they still had their limits, and many regions with harsh winters and short growing seasons were still not suitable for most of the original hybrids. In Minnesota and Wisconsin, breeders (mostly associated with the University of Minnesota), continued to cultivate wine grapes that could endure the harsh winters and ripen in a shorter growing season. At the same time, oenologists worked with these new grapes to improve the quality of the wine made from cold-hardy hybrid grapes. Beginning in 1996, the University of Minnesota released six cultivars of new cold-hardy hybrid grapes: Frontenac (1996); La Crescent (2002); Frontenac Gris (2003); Marquette (2006); Frontenac Blanc (2012); and Itasca (2017) (Clark, Hemstad, and Luby 2017). These new grapes, as well as a few others that had been cultivated throughout the second half of the 20th century, sparked the rise of wine industries throughout the northern parts of the United States, Canada, and other pockets around the world.

Importance of wine tourism

The development of new cold-hardy hybrid grape cultivars has enabled wine to be produced where it wasn't previously feasible to do so, and is the main reason for the dramatic increase in the number of US wineries. But being able to produce wine in a region doesn't mean there will be a market for it. The region and the grape varieties grown there are not recognized by many wine drinkers, not to mention the distributors and influencers needed to create demand. Instead, wineries in the region have found success by attracting visitors to their wineries and tasting rooms.

The wineries in the UMRV are small producers and are able to sell the relatively small amount of wine they produce to visitors who come to sample wines. Some wineries distribute through local retailers and restaurants as well, but the majority of wine produced in this region is sold directly from the tasting room (Eustice et al. 2019). Like the age-old chicken-and-egg discussion, it's difficult to know if the rise in wineries satisfied an unmet demand for wine tourism experiences, or if their emergence created demand, but thus far, the success of a rapidly growing number of wineries has been able to attract enough visitors to sell their wine. The visitors seem very happy to spend a weekend visiting wineries, driving by beautiful vineyards, learning about how wine is made, and sampling tasty wines. To help enhance the experience, and generate additional revenues, many wineries offer food, live entertainment, and accommodations. And most host events such as weddings, corporate retreats, and parties.

Although there are some complaints from residents about increased traffic, driving under the influence, and strange visitors, most of the communities in the UMRV are happy for the benefits these new wineries bring. From an agricultural perspective, the per-acre production of wine grapes is more lucrative than many other crops. Compared to other agricultural crops that are sent out of the region, the value-added processing of wine grapes occurs locally. Finally, the wineries in the UMRV attract visitors, and those visitors often shop at local retailers, eat at local restaurants, spend the night at local hotels and B&Bs. Moreover, new business opportunities have been created throughout the UMRV for tour operators, driver services, galleries, local food shops, and musicians. In some places, the wineries have sparked brand new opportunities for tourism, and in other places, the wineries complement existing tourism offerings. Although economic figures are not available specifically for the wineries in the UMRV AVA, a recent study commissioned by the National Association of American Wineries (WineAmerica) found that wine tourism in the four states included in the UMRV (Minnesota, Wisconsin, Iowa, and Illinois) generates over 1 million winery visits annually by more than 330,000 unique visitors (Jones-McKee 2017).

The case of the UMRV is a common one throughout the United States as more wine regions have been emerging in the past few decades. Although not all of the emerging regions rely on hybrid grapes, many do. But they all rely on tourism. What the future holds for the UMRV remains to be seen. It's possible that wine tourism here will be a fad that dissipates, forcing wineries out of business. It's possible that demand for local products will increase, and if consumers acquire a taste for wines made from hybrid grapes, wines from the UMRV might find stronger distribution opportunities in the region's retailers and restaurants. It's also possible that climate change could impact the area, either directly through changes in the region's climate or indirectly by changing other wine-producing regions, such as California, which has been experiencing increasingly more severe droughts and wildfires in recent years. Whatever the future holds for the UMRV, it is clear that tourism will have at least played an important role during the region's first couple of decades as a wine region.

Conclusion and implications

Although they were first developed due to the Great French Wine Blight in the late 1800s, hybrid wine grapes gained only minimal acceptance around the world for a century. In recent years, however, new cultivars have been developed, and hybrid wine grapes are finding a receptive audience. Growers and winemakers, particularly in the United States, are turning to these grapes to produce wines in regions where doing so was previously not feasible. Despite the increase in winery numbers, however, wine production from hybrid wine grapes is still very low, but the provision of wine-based experiences attracts significant visitors, making an important impact on communities.

Although many wine regions worldwide have developed wine tourism 'products,' emerging hybrid wine regions are different because tourism is primary to their success rather than complimentary. Selling wine to visitors at the tasting room is the main distribution model for most wineries that sell wines made from hybrid wine grapes, and thus far, demand has matched the relatively small amounts of wine production. However, if production increases, it remains unclear whether there would be enough of a market for hybrid wines through traditional distribution means, such as through retail and restaurants.

Other questions about hybrid wine production remain as well. Will regions not known for wine production in other parts of the world adopt hybrid wine grapes, following the model of places like the Upper Mississippi River Valley? Will demand for wine tourism remain high once the novelty wears off in these new wine regions? Will large-scale wine production shift to regions currently producing hybrid wines as climate change presents challenges to traditional wine regions?

Over the past two decades, wine tourism, and tourism in general, have been enjoying a period of high demand and rapid growth. And like any tourism product, wine tourism suppliers will need to be flexible and able to adapt to evolving consumer trends and external opportunities that present themselves. In the meantime, it seems likely that breeders will continue to develop new hybrid wine grape cultivars, growers will improve their growing methods for new cultivars in new terroirs, and winemakers will continue to learn how to produce quality wines from hybrid wine grapes.

References

Alcohol and Tobacco Tax and Trade Bureau (2020). Statistical report – wine. *Reporting Period: January 2019–December 2019.* https://www.ttb.gov/wine/wine-statistics

Atucha, A., Hedtcke, J., and Workmaster, B. A. (2018). Evaluation of cold-climate interspecific hybrid wine grape cultivars for the upper Midwest. *Journal of American Pomological Society*, 72(2): 80–93.

Barber, N., Hutchins, L., and Dodd, T. (2007). *A History of the American Wine Industry* [Report]. https://www.depts.ttu.edu/hs/texaswine/docs/Wine_History.pdf

Clark, M., Hemstad, P., and Luby, J. (2017). 'Itasca' grapevine, a new cold-hardy hybrid for white wine production. *HortScience* 52: 649–651.

Conway, J. (2020). *Producing Wineries in the U.S. 2009–2021.* https://www.statista.com/statistics/259353/number-of-wineries-in-the-us/

Eustice, C. L., McCole, D. T., and Rutty, M. (2019). The impact of different product messages on wine tourists' willingness to pay: a non-hypothetical experiment. *Tourism Management*, 72: 242–258. https://doi.org/10.1016/j.tourman.2018.11.022

Jones-McKee, L. (2017). National impact of the wine industry: $219.9 billion. *Wines and Vines*, September 27. https://winesvinesanalytics.com/news/article/190064/National-Impact-of-the-Wine-Industry-2199-Billion

Kamas, J., Stein, L., and Nesbitt, M. (2010). Pierce's disease-tolerant grapes. Texas Fruit and Nut Production. Texas AgriLife Extension. Texas A&M University: College Station, Texas. http://aggiehorticulture.tamu.edu/fruit-nut/files/2010/10/pd-grapes.pdf

Lee, J. H., McCole, D., and Holecek, D. (2020). Exploring winery visitors in the emerging wine regions of the north central United States. *Sustainability*, 12(4): 1642. https://doi.org/10.3390/su12041642

Martinson, T. E., Mansfield, A. K., Luby, J. J., Gartner, W. C., Dharmadhikari, M., Domoto, P., and Fennell, A. (2016). The Northern Grapes Project: integrating viticulture, enology, and marketing of new cold-hardy wine grape cultivars in the Midwest and Northeast United States. *Acta Hortic*, 1115: 3–12. https://doi.org/10.17660/ActaHortic.2016.1115.2

Mattia, F., Imazio S., Grassi, F., Doulati, H., Scienza, A., and Labra, M. (2008). Study of genetic relationships between wild and domesticated grapevine distributed from Middle East regions to European countries. *Rendiconti Lincei*, 19: 223–240.

McCole, D. T., Holecek, D., Eustice, C., and Lee, J. (2018). Understanding wine tourists in emerging wine regions: an examination of tasting room visitors in the Great Lakes region of the U.S. *Tourism Review International*, 22(2): 153–168. https://doi.org/10.3727/154427218X15319286372306

McGovern, P. E. (2003). *Ancient Wine: The Search for the Origins of Viniculture*. Princeton, NJ: Princeton University Press.

Reynolds, A. G. (2015). Grapevine breeding in France – a historical perspective. In Reynolds, A. G. (ed.) *Grapevine Breeding Programs for the Wine Industry* (pp. 65–76). Cambridge: Woodhead Publishing, Elsevier.

Robinson, J., Harding, J., and Vouillamoz, J. (2012). *Wine Grapes: A Complete Guide to 1,368 Vine Varieties, Including Their Origins and Flavours*. London: Penguin.

Sabbatini, P., and Howell, S. (2014). Vitis hybrids: history and current status. *Wines and Vines*, January 2014.

Shapiro, M., and Gómez, M. I. (2014). Customer satisfaction and sales performance in wine tasting rooms. *International Journal of Wine Business Research*, 26(1): 45–60.

Sumner, D. A., Bombrun, H., Alston, J., and Heien, D. (2001). *An Economic Survey of the Wine and Winegrape Industry in the United States and Canada*. https://aic.ucdavis.edu/research1/Winegrape.pdf

Tattersall, I., and Desalle, R. A. (2015). *Natural History of Wine* (2nd ed.). New Haven, CT: Yale University Press.

UC Davis (2000). Napa: wine, vineyards, housing. *Rural Migration News*, 6(4).

Weinmann, E., Boos, M., Ehret, P., Flubacher, L., Schneider, C., and Veith, L. (2019). Breeding of new disease-tolerant grape varieties – viticulture in times of climatic change. *BIO Web Conference*, 15.

WineAmerica (2020). Industry statistics. *US Wine Consumption*. https://wineinstitute.org/our-industry/statistics/us-wine-consumption/

48

POST-COVID-19 DEVELOPMENTS IN THE WINE TOURISM SECTOR

Jane Eastham

Introduction

This chapter considers issues of managing risk and, in particular, focuses on COVID-19 and its impact on wine tourism and wine tourism destinations. The curtailment of travel has presented challenges to the wine industry, particularly those who have extended their market through wine tourism. Evidence from international research suggests that losses to wine businesses across continents are extensive; it is also apparent that in certain instances, wine businesses have been able to adapt their business through drawing upon local markets and through online virtual events.

In the context of wine tourism, where tourism has traditionally been required for redistribution of consumers in time and space, lockdown required a redefinition of what constituted tourism and, in this case, wine tourism. This chapter provides an overview and insight into the measures implemented by wineries to maintain their businesses. The chapter outlines the issues of management of demand risk within tourism supply chains, introduces the issues faced by wineries within the context of COVID-19, and looks at initiatives undertaken by a number of wineries, including two case studies of initiatives introduced both in Argentina and in France.

Managing risk in demand chains

The management of risk in supply chains has become and is of increasing concern for businesses, not least for those in the agribusiness sectors (Behzadi et al. 2017). Food supply chains are perhaps more vulnerable to disruption, where food is both perishable and subject to the impact of weather, pests, and diseases. Risks of supply chain disruption can be seen to have been augmented over recent decades with the acceleration of globalization, bringing extended and more complex supply chains, trends such as outsourcing, and leaner supply networks (Behzadi et al. 2017; Fahimnia et al. 2015; Narasimhan and Talluri 2009). Yet, similarly, globalization presents corresponding demand risks as yet to be addressed in commensurate terms within the academic narrative.

In tourism, the resilience of a business has been traditionally dependent upon the facility to relocate consumers to other environs in order they may consume and experience an event or activity. The success of the business activity is contingent on the demand, the availability,

DOI: 10.4324/9781003143628-55

and the capacity of the infrastructure, both to and within the tourism destination. Tourism, whether global, regional, or local, has become an increasingly valuable source of revenue, particularly where there have been global economic pressures on traditional indigenous forms of trade. Developed and developing countries have become more and more reliant on tourism and particularly on international tourists. Yet, with the event of a natural disaster or disease as experienced with COVID-19 and the curtailment of demand, this reliance offers a considerable risk to business income and, indeed, gross domestic product as a whole. Managing demand risk is thus of equal importance for businesses as that of supply.

Supply disruptions adversely affect the operational, market, and financial performance of businesses and supply chains as well as stock prices and shareholder value (Narasimhan and Talluri 2009). The emergent area of study and supply chain risk management (SCRM) has been recognized as a strategic management activity in firms, enabling robustness and resilience (Ivanov and Sokolov 2013; Ivanov, Sokolov, and Dolgui 2014). Such strategies should take account of: the sources and magnitude of risk, impact on business objectives, and potential for supply disruption as well as the degree of exposure to such adverse events. The magnitude of such a demand disruption can be seen to be quite as severe as one of supply, with equal adverse effects on financial performance, requiring similar mechanisms to mitigate risk.

Thus, whilst the essence of SCRM is to make decisions that optimally align organizational processes and decisions to exploit opportunities while simultaneously minimizing risk (Jüttner 2005), so too is the case in the management of demand risk.

As identified in Figure 48.1, the extent to which an event can be anticipated and the severity of its impact determines the nature of the strategic action – proactive or reactive. A supplier bankruptcy may be a surprise but may be mitigated by switching to dual sourcing. Similar strategies of expanding markets can be implemented in the case of risk of demand disruption. Where low risks can be anticipated, such as the cultural preferences of specific tourists (see Figure 48.1), staff should be adequately trained and empowered to manage the customer experience.

Alternatively, there are high-impact events, which are either easy or hard to anticipate. Hard to anticipate could include natural disasters and pandemics where there could be a level of proactive management of such risks and their potential with respect to both the prediction of demand disruption. Whilst the prediction of high-impact events and their impact on businesses can be more problematic, they are not unprecedented, and some degree of scenario planning

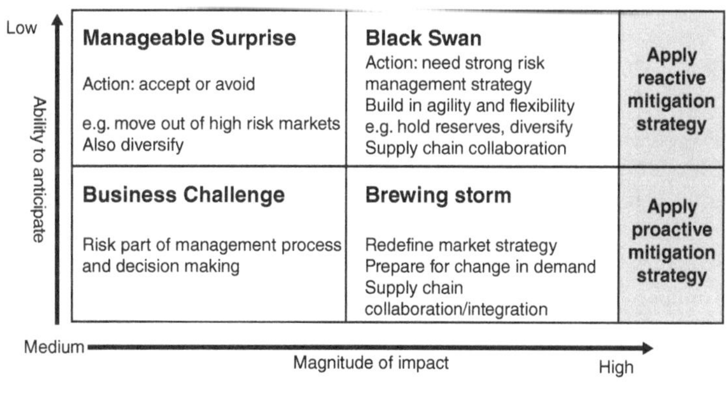

Figure 48.1 Managing demand disruption.

Source: Adapted from Alicke and Strigel (2020)

could have been delivered prior to the event of COVID-19, particularly given the nature of international supply and demand chains.

With reference to Alicke and Strigel (2020), the events of COVID-19 in 2020 can be seen to have the characteristics of a 'Black Swan' due to the failure of businesses to plan and effectively deploy proactive strategies to managing risk within supply chains (Aven 2015; Cao et al. 2010). Where the literature examines proactive strategies to manage supply disruption, they suggest that there is a need for greater supply chain flexibility and agility and potential for easier access to alternative suppliers (Merschmann and Thonemann 2011), although it is apparent that the level of flexibility required depends on each business and its supply chain, and the optimum level of flexibility is debatable (Tang and Tomlin 2008). Where we consider the parallel of demand risk, the adaptation of business practices and models to manage supply, manufacturing, and delivery/sales risk is critical to survival (Hendricks and Singhal 2005; Sreedevi and Saranga 2017; Tang 2006).

Wine tourism – its nature and the COVID-19 risk

The tourism sector has a large impact on the world economy. In 2019, the sector contributed US$8.9 billion, around 10.3% of world global GDP; US$1.7 billion in visitor exports, 6.8% of total exports, and 28.3% of the world's service exports. Tourism remains one of the most important global employers, with 1 in 10 jobs directly related to tourism. It was projected to grow by 21% by 2023 to value over US$429 billion. The sector has seen considerable growth, particularly in the Newly Developed Countries (UNWTO 2020a; WTTC 2020).

As a major GDP contributor for many countries, tourism featured strongly in all international discussions related to COVID-19. In many discussions it was believed that the pandemic offered a transformative opportunity for the sector (WTTC 2020). COVID-19 had a significant impact on all business activities, and the tourism sector has been one of the most severely punished. Health communication strategies and measures (e.g., social distancing, travel and mobility bans, community lockdowns, stay at home campaigns, self or mandatory quarantine, curbs on crowding) have halted global travel, tourism, and leisure.

Tourism of any kind has been often depicted as a 'fair weather sector,' and thus, there is a level of predictability as to the potential impact of a natural disaster on global or, indeed, national tourism (Keller 2020). Although it is argued that COVID-19 was the first worldwide supply shock affecting international tourism, it was by no means the first major event to have an impact on the sector (Keller 2020). The tourism industry in recent years has seen a variety of significant events; the Eyjafjallajökull volcano eruption greatly affected international travel for several weeks in 2010, and in the context of pandemics, examples can be seen in both the outbreaks of SARS in Asia (Au, Ramasamy, and Yeung 2005; Min, Lim, and Kung 2011) and Foot and Mouth in 2001 in the UK (Butler and Airey 2005; Leslie and Black 2006) and the considerable impact on economies. These events highlight the need to put in place contingency plans to mitigate such risks to the business.

With the number of tourism arrivals 'plunging' by 93% in June 2020 alone when compared with 2019 and with a 65% drop in international arrivals in the first six months of 2020, a loss of 440 million international arrivals from export revenues from JOCM international tourism amounted to US$460 billion (UNWTO 2020b). This is around five times the loss in international tourism receipts recorded in 2019 amid the global economic and financial crisis (UNWTO 2020b).

As a sector, tourism, and indeed wine tourism, is heavily dependent upon international travelers, and it is notable that a further fall of 87% in international travel was registered in January

2021 compared with the previous year. Furthermore, the drop in the volume of domestic tourism exacerbated the financial position of wineries during periods of national lockdown.

Sales of wine through direct sales and mainstream channels continued, but there is a symbiotic relationship between wine and tourism; wine tourism may, on average, represent just 19.5% of the income of viticulturists (Keller 2020), but the expansion of demand at an international level has offered stimulation to other business activity with their locality.

Yet, beyond an additional source of revenue for the wine producers in that region, which offered a low-cost distribution channel for the sale of wine direct to the consumers (Getz and Brown 2006), there is considerable potential for the valorization of the local economy through the stimulation of contingent business activity such as food processing and manufacturing, foodservice, and hospitality, as well as a potential stimulant for other leisure and sporting activities.

Furthermore, from its humble beginnings of visits to local caves, the wine tourism sector has emerged as a multifaceted event, which has attracted a wide range of customers, young and older, and gastronomic tourists. This has both been driven by and is driving the expansion of the range of experiences offered by the wineries themselves.

In the context of COVID-19, from May 2020, the predictions were that the impact of the global lockdown globally would amount to 850 million–1.1 billion fewer wine tourists, a loss in export revenues of US$910 billion to US$1.2 trillion, and 100–120 million direct tourism losses, (UNWTO 2020c, 2020d). This impacts not only the sector itself, but also the network of industries which rely on this source of income – including cheese producers, hotels, bed and breakfasts, and ancillary activities (Blake, Sinclair, and Sugiyarto 2003).

It is notable, however, that the economic and social impact of COVID-19 on wine tourism is greater for some regions than for others. Those vineyards and areas that are most reliant upon international tourism are certainly the most heavily hit. The wine industry itself has remained relatively stable. Even in the midst of COVID-19, the global value for wine was estimated at around US$326.6 billion in 2020 (Reportlinker 2020); yet initial reports based on a survey of 34 countries in November 2020, with an 11.9% response rate from 10,080 wineries, showed that wine tourism was negatively impacted by COVID-19. These results reflected an earlier study undertaken by Wine Tourism Think Tank in July 2020.

Some 83% of the businesses surveyed had suffered a loss of income during the tourism season, although 80% of respondents expected that wine tourism would return to similar growth rates over the next ten years. Some 53% of respondents experienced losses of over 53% of income in comparison to the previous year. Only 16% of respondents did not see any significant change in income, with some 6% experiencing an income increase. Those wineries who saw no change or an increase on the previous year were those from wineries in Italy, France, Spain, Germany, and Australia, whose common feature was a business model that drew upon their augmented domestic markets and reimagined their offering in the new 'normal' economic order (Gössling, Scott, and Hall 2020; Hall, Scott, and Gössling 2020).

There are many instances wineries have re-evaluated their business model. Direct delivery formats through merchants and mainstream channels have persisted, however direct sales to end consumers have grown, and a range of different ways of engaging with the wine lover have emerged with the aid of the internet and internet sales. The application of ICT to wine tourism business activities, have helped organizations to develop, manage and publish their tourism-related products and allowed customers to search and purchase tourism products and services with a high degree of personalization (Bethapudi 2013). Initiatives have included wine clubs, virtual events, e.g., the Hongkong Wine and Dine Festival, where 'tourists' were invited to participate virtually and experience workshops on a variety of wines over a number of weeks. Wines and Food were provided by many wine merchants and food by many chefs; sent by courier to enable the participation of customers within the online event.

According to the World Tourism Think Tank, in the nine countries surveyed Inc.: Mexico, Italy, France, Portugal, Chile, Brazil, Argentina, and Spain (see Table 48.1. Garibaldi 2020) Traditional initiatives such as tours around the winery, tastings around the winery declined during COVID-19 (Garibaldi 2020), but special events either virtual or on location grew in popularity in the most renown destinations and enabled networks of businesses the opportunity to reimagine and reform the nature of the vitivinicultural experience (Gössling, Scott, and Hall 2020; Hall, Scott, and Gössling 2020).

Digital experiences also abound, not just in Hong Kong but in wineries across the world. These have emerged in the form of digital tours, virtual wine clubs or *Oeno Gastronomique* clubs, and charity events. Where visitors have made physical visits post lockdown, wineries have very much focused upon the culture, the terroir, and the open space. In Bordeaux, wineries, for instance, have been taking advantage of the outside space and backdrop of the chateau to encourage afterwork and small-group social meetings whilst effecting social distancing – a new role for vineyards, and potentially one that is more sustainable over the longer term. Gastropubs and bistro venues are seen to be a way forward for many, as well as outside sports such as riding trips in many wine regions, including Napa Valley and Bordeaux, as well as outside specially distance eating and bistro venues. In Andalusia, Tio Pepe and Gonzael offer a gastronomic and equestrian experience in conjunction with the Roya Andalusian School of Spanish Equestrian Art.

Across continents, whilst international demand continues at the time of writing to be held in abeyance due to COVID-19, there has been considerable effort by many wineries to cultivate the local market. Efforts have been made particularly on the use of social media and a high degree of personalization of the message to local customers (Bethapudi 2013), allowing

Table 48.1 Experiences offering at wineries – the changing shape of wine tourism activities

Activities and events	*Before COVID-19%*	*With COVID-19%*
Guided tourism inside the company	91	74
Wine tastings inside the company	89	73
Sale of products in store	84	88
Guided tours in the vineyards	64	65
Special events (weddings, etc.)	39	23
Non-guided tours inside the company	30	24
Wine tastings in the vineyards	28	34
Lunch or dinner in the vineyards	26	22
Non-guided tours in the vineyards	23	24
Harvests for tourists	23	13
Tours and specific activities for children	22	10
Picnics in the vineyards	20	29
Wine club	19	23
Guided visits to company museum	19	16
Bicycle tours or other transportation	18	13
Accommodation	17	17
Hiking, trekking, or other sports activities in the vineyards	13	12
Artistic activities in the vineyards	11	4
Yoga, pilates, or similar classes in the vineyards	9	6
Balloon flight in the vineyard	3	1
Spa treatments	3	6

Source: Adapted from Garibaldi (2020)

a small group of wineries to increase their wine tourism revenues. Innovation has been key to such wineries, enabling them to adapt to changes in demand and the curtailment of international travel.

Case Study 48.1 Moving Argentina forward: expanding the experience

Argentina is the eighth-largest country with 1.1 million square miles and the fifth-largest producer of wine in the world. It has six climatic and geological zones across its 23 provinces, from the tropical climate of the northeastern zone to the region of the Andes and the plateau in the northern zone, which borders Brazil, Paraguay, and Bolivia.

Wine and grapes were not indigenous to Argentina. Viticulture was introduced there during the Spanish colonization of the Americas and later again by Christian missionaries. In 1556, Father Juan Cedrón established the first vineyard in Argentina when cuttings from the Chilean Central Valley were brought to what is now the San Juan and Mendoza wine region. From these original cultivars, the Criolla grape varietals emerged, which were high yielding and thus had a lower color and flavor intensity.

The first industrial wineries were founded in the early 19th century following independence, but their expansion was limited to a certain extent by civil wars in Argentina, which took place between 1814 and 1880 (Yeomans 2019).

The major expansion of the sector took place after the establishment of the Transandine Railway in Mendoza in 1885, which allowed wines to be exported from Mendoza to the greater market of Buenos Aires, reducing the time to market from a month to one day. The increased demand led to a much greater supply, and vineyards were planted at a rapid rate. Wineries started to pop up alongside the tracks, and from 1880 to 1900, planted vineyard hectarage grew by more than fivefold.

The heyday of wine production and consumption was in the 1970s, and wine consumption was also at an all-time high at 92 liters per capita. However, at the beginning of the 1980s, crisis struck the wine industry in the guise of a combined social and economic crisis. Grape prices and wine prices dropped by 80% between 1979 and 1982 and resulted in a glut of cheap table wine left in the wineries, and no affluent market left to buy it, and at the time very few bottles made it overseas. This was to change – wine producers started to look to tailor their wines to export markets, and the desire to grow exports resulted in the improvement of wine quality. Currently, Argentina has some 190,000 hectares under cultivation producing around 11.8 million hectoliters (2017 figures).

In 2019, exports amounted to 312 million liters, whereas home consumption accounted for 19.63 liters per capita, with a population size of around 41.5 million. The most important wine regions of the country are located in the provinces of Mendoza, San Juan and La Rioja, Salta, Catamarca, Río Negro, and, more recently, southern Buenos Aires, with Mendoza producing 60% of the Argentinian wine. Innovation in the Argentinian wine industry has been promoted by a Triple Helix Model of Innovation – a partnership between academic institutions, government, and industry. In Mendoza, these consist of the Universidad Nacional de Cuyio and the Universidad Juan Agustin Maza, as well as organizations such as the National Agricultural Technology Institute, an agriculture experiment station that carries out research and forms strategic partnerships with cooperatives and the provincial government. The provincial government has a critical role in the encouragement of university and industry interaction (Liberman 2013). The engagement of the three parties has resulted in the Pro Mendoza Foundation. The public–private nature of these institutions gradually fostered new professional ties between the associations and the regional government.

These initiatives supported the Argentinian wine industry through the 2008 financial crisis and fueled the emergence of Argentinian wine tourism. In 2020, there were 245 wineries offering a tourism experience in Argentina (see Table 48.2). These were located predominantly in Mendoza, with some found in Salta and Rio Negro, and Cordoba. Mendoza's wineries received over 1.1 million visitors in 2018, 36% of whom were foreigners, with their leading international visitors arriving from Brazil, (31%) the United States and Canada (27%), France (7%), Chile (5%), and the United Kingdom (5%).

Wine tourism has not only become important in terms of selling wine at the cellar door, but more importantly, their experience of the Argentine wine country encourages Argentinian wine purchases on the return of the visitor to their home country. Argentina created a very strong brand for itself thanks to its world-renowned Malbec wine, grown particularly in Mendoza, but the opportunity of tasting Argentine wine in situ is also opening people's eyes to the many different terroirs and grape varieties of Argentina today. Argentina has a strong culture of which 46% offer Oeno Gastronomy experiences; held in most of the total number of winery restaurants across the three regions, with the predominant number found within the Centre West. Pre-COVID, the sector employed over 900 permanent and 350 temporary staff (Ceppara 2020).

In addition, the three regions offered a range of experiences to the international tourist throughout the year, including traditional tours of the vineyards, visits to the winery and visitor centers, and gastronomic experiences where wine was tasted, accompanied by cheeses and cooked meats. Educational programs to enhance visitors' knowledge of harvesting, pruning, wine tasting, and cooking were also a standard offer to tourists (Schlüter and Norrild 2015).

As a result of the COVID-19 outbreak, Argentina locked down on March 19, 2020, a period that marks the peak of the grape harvest and the wine tourism season. The measures were relaxed on May 8, but lack of access to both the international market and, to a large extent, the domestic market, where domestic flights were at this point were yet to be resumed, forced viticulturalists to reconsider and reimagine their business model.

Table 48.2 Location of wineries in Argentina

	Region/Zone	*Region/Zone*	*Region/Zone*
	Center West	Northeast	East/South
Wineries open to tourism	165 open	44	36
Specifically, in:	146 in Mendoza	21 in Salta	16 Cordoba
			11 Rio Negro

Whilst many of the Oeno restaurants reopened from May onwards, they did so according to distancing and sanitary measures, which reduced capacity by 50%. They also needed to rethink menus and menu pricing structure in order to appeal to the local market, in place of the international and non-local domestic market. Menus were simplified, and communication to customer bases had traditionally been delivered through travel agents. Wineries needed to develop a stronger online presence and connect – indeed, reconnect – with local populations through social media, developing strong local relationships.

With an increased presence on the internet and social media, wineries have developed online wine tasting, charity events where as many as 770 households have participated in both the preparation of menus with the food ingredient sent and consumption with the wines chosen to complement the meal.

One such company, Susana Balbo Wines in the town of Agrelo, Luján de Cuyo, Mendoza Province, offered virtual events during lockdown and have since reopened their restaurants and winery for open-air consumption between Wednesday and Sunday, their restaurant Osadía de Crear (OSΛDΛ de Crear) offered Argentine cuisine based on seasonal Argentine products and producers, which, they stated, were 'inspired by the landscape that surrounds the winery.' These products were paired with Susana Balbo and Ben Marco Wines. The restaurant offered a four-course menu or lunch à la carte and an executive menu – afternoon tea was also served, all of which took place under the shadow of the Cordon del Plata range. In addition to outside dining, they offer picnics at the winery with a bottle of Crios wine, and on Saturdays, Sundays, and holidays, they offer a traditional Asado Argentino (Argentine barbecue) paired with Crios and Ben Marco Wines. A new venture, in the form of the restaurant Espacio Crios, services finger food which can be offered in liaison with their wine and cocktail menu during the afternoon, again using the open space and view of the mountains.

Case Study 48.2 The French experience

The tradition of tourists visiting caves and vineyards to taste and select vintages has had a long tradition in France. Tourists – or visitors to wine regions would be visiting a vineyard to taste the wines with the view toward immediate purchase. My own first recollection of this was in the 1960s –when I visited elite houses in the Bordeaux and other wine regions of France. The practice has without a doubt always been present wherever visitors traveled within the much-vaunted wine regions of France. The term wine tourism was a much later term to denote the experience, and the 'wine tourism product' has expanded beyond recognition.

The emergent offering became a far more coordinated wine tourism experience and stemmed from the realization that there was a need to make wine more accessible to a wider clientele. In order to achieve this, the French wine region, with the support of the public sector, started to offer a range of visitor attractions. These included: tours of the winery and vineyards; visitor centers; wine routes; wine museums, many of which were located on the wine route; and 'Maison de Vins,' which operated as publicly run information centers along the wine routes. Celebrations and fetes focused on specific events within wine regions that were established and marketed to draw in customers, such as those used in areas such as Beaujolais as means of promoting the Vin Nouveau.

Wine routes became a significant feature of the sector from the 1990s, with museums set up along the route introducing the consumer to insights into the production of wine. Even prior to COVID-19, a new business model had begun to emerge. Visits to wineries and vineyards have remained an integral part of the tourist experience, but Oeno restaurants have also become widespread across the sector. It is apparent that in France, their resilience in the face of the pandemic can be strongly linked to the level of innovation before and during the crisis.

France has a structured approach to facilitating change and developing alternative business models in the wine sector, and it is clearly demonstrated by the emergence of bodies that focus on the promotion of clusters and prioritize the need to develop a coherent integration of varied partnerships in the promotion of regions as a whole. *Atout France*, the French tourism agency, came into existence as a consequence of a merger between the Maison de la France and ODIT France (*Agence Française d'ingenienie Touristique*), and it is a public–private alliance with members from both the public and private sectors. Members include ministries and state-level parties, local authorities, and tourism professionals, and it has had an active role in the promotion of wine tourism regions. However, there is a long tradition in the French wine industry of integrative private collective activity, not least under the wine cooperative body of Vigneron de France. Other collective initiatives include a network of brotherhoods (Frochot 2001), which, despite a chequered history, are now thought to number around 300–350 across France. These confrères or brotherhoods are grouped into regional and national bodies and have been seen to take an active role in the promotion of wine tourism.

Perhaps with such a network of facilitating bodies, it is perhaps understandable that countries such as France are indeed the most optimistic with respect to the recovery of the sector (Garibaldi 2020). Although some €250 million delivered by the French government may have helped many to weather the storm.

During lockdown, direct sales and sales to restaurants, which for some represented up to 80% of product sales, were considerably reduced. Wineries also were at risk of nonpayment by both restaurants and *négociants* for wine already delivered. Sales to households and individuals continued despite the confinement laws and the need to carry an *attestation de déplacement*; wine was considered a primary necessity, and thus wineries and wine merchants were permitted to remain open. Yet the economic and agricultural ministers put in place measures to support wine businesses in the form of a solidarity fund to support their businesses, in effect a job retention scheme.

As with other regions across the world, France experienced the loss of international tourism, and as French vineyards and cavists are hosts to millions of visitors a year, this had a significant impact on the €5 billion in revenue that is normally generated from such sources. During lockdown, the function of the wineries was predominantly in local retail; but as with other countries, the digital experience played a part in accessing remotely the international customer base. However, vineyards in France had to an extent widened their offering to the visitor even prior to COVID-19, encouraged by Atout and other collective networks; thereby they had already developed a huge range of outdoor experiences, including hiking and horseback riding as well as wellness activities such as yoga, all set in a backdrop of the terroir and in foreground of the chateaux.

Despite social distancing, as the curtain of lockdown was lifted, these activities could continue, alongside talks introducing exhibitions and biodiversity, slow tourism, and visits around the vineyards to admire the art curiosities. Food experiences have been featured as part of the product offering for some years, and over the post-COVID-19 period, lockdowns have continued within the requisite safeguarding rules. Furthermore, many vineyards offer longer stays to the tourist and provide holiday accommodation in the splendor of the terroir.

Festive events were canceled, but for many chateaux, these have been replaced by virtual events and by socially distanced bistro-type activities where after work visitors can enjoy their aperitifs in the environ of the chateau. Visits and events outside, with smaller groups of visitors, have generated income with a predominantly French clientele, who, unlike international travelers, use their own mode of transport and are thus have fewer capacity constraints, the average spend per visitor has been higher than in 2019.

The emphasis placed by the French authorities on staycation as expressed under the logo of '*cet été je visite la France*' offered vineries further opportunities to extend and enhance their domestic market. Some 94% of French tourists traveled in France, with 37% holidaying at the coast and 30% in the countryside. Urban tourism declined, and the hotels that were successful were small boutique establishments; yet even this was to the benefit of wineries. The accommodation on those French vineyards who can accommodate resident visitors is chalet-like with open spaces, ideal for safe distancing.

Concluding comments

The impact of the COVID-19 pandemic has undeniably had a negative effect on global wine tourism. Border closures, lockdowns, and social distancing have had a large financial impact on wineries operating in the tourism sector. Yet, whilst many operators have suffered losses, even in France, on the whole, the wine tourism sector is not dead, with those who have proved most resilient honing onto their domestic market. Indeed, it is evident from research noted by Garibaldi 2020 that a small group of wineries have paradoxically increased their wine tourism revenues.

A closer look at these gives an understanding as to what happened and how they managed to grow their business during a pandemic. Research undertaken by WineTourism.com in November 2020 (https://www.winetourism.com/impact-covid-19-wine-tourism/) suggests that, where alternative market channels have been established – local, national, and international – those respondents expected their businesses to recover by 2021. Business models that promote local and domestic trade appear to have facilitated the survival of wineries, as indeed has government or quasi-government intervention. This was the case in France, according to

41% of the wineries surveyed. Due to the designation of wine as a primary necessity during the lockdown, many were able to continue to trade, albeit at a reduced rate, to local customers. Similarly, having steered their tangible offering towards virtual interaction, particularly during the lockdown, businesses kept afloat in the main, as also was the case with many Argentinian firms. In France, as concluded by the wine tourism survey 'Designing innovative wine tourism experiences by implementing digital tools helped them to maintain links to their customers,' and 'online sales, wine tastings and digital tours.'

Following lockdown, they were able to encourage their domestic/local market with an innovative, largely preestablished portfolio of offerings. Already, encouraging the domestic tourism, the nature of their offerings, and the stay at home message of the French authorities provided wineries with a more robust and greater resilience as indeed would a relatively wide spread established online presence by many in the sector.

Yet for some wineries, investment in time, effort and expertise was and would be required, and it may be telling that the March 2020 survey showed that 68% of wineries survey had no clear desire to increase their investment (https://www.winetourism.com/impact-covid-19-wine-tourism/). Indeed, it can be assumed that small enterprises who are already suffering from reduced income investment and development would find access to capital extremely problematic. In some countries, initiatives have been put in place to address such hurdles.

Indeed, research in Italy suggests that the success of a wine tourism territory cannot belong to a single operative with a region (Festa et al. 2020), and that the development of strategies to manage risk in the context of our 'fair weather friend' will depend on the availability of sufficient support, financial and otherwise. Tourism 'is clearly associated with different forms of crisis' (Hall 2010, p. 406), and there is clearly a need to understand and consider contingent plans for the various threats, but in order to attain this goal – through reconnection with the local clientele and through virtual events of conferences, tastings, and indeed, educational experiences – support is required to upskill and facilitate investment. The widening of markets can be seen as desirable even when restrictions and the risk from the pandemic are lifted. Hitherto we have been unable to eliminate the problem of time and space, but depending on the product or offering and the virtual world, now we can.

As is evident in the two case studies identified in this chapter, the emergence, and development of alternative business models can be driven by the private and indeed the public sector. The promotion of clusters, promoted by public and private bodies, has been long since seen to be desirable. Research has examined the potential to operate in synergistic, symbiotic relationships both in real-time (Hall et al. 1997; Lavandoski et al. 2018; Mitchell and Hall 2006; March and Wilkinson 2009; Telfer 2001) and virtual. Whilst there have been questions as to the efficacity of cooperative networks and bodies (Giuliani 2007; Martin and Sunley 2003), a perusal of literature suggests that there are a number of successful examples, or examples which could be successful, in mitigating such a risk of demand disruption. In Italy, the municipalities have a link with the National Association of Wine Cities,' which promotes the emergence of clusters and collaborative and integrative relationships between businesses associated with tourism.

Implications

Tourism has become a significant source of GDP across developed and developing countries and until now has predominantly involved the dislocation of people from their point of abode to the point of consumption. Wine tourism, which although not a new form of tourism, has significantly grown in importance and has offered an important additional market for winegrowers, both for wine sales and range of 'event' and 'activities' offered as well as an additional source of income to rural areas, thereby allowing improved economic and social conditions in these environs. Yet, the impact of COVID-19 drove the change and development of alternative and additional markets. We can anticipate that the proactive virtual experience, particularly in the light of a continued risk of pandemics, has changed the landscape of tourism, and indeed, specifically wine tourism. There likely also will be a long-term shift towards staycations and a larger proportion of local as well as international visitors. In terms of the longer view, businesses have been able to build resilience through innovative solutions, and there have emerged many cases the mechanisms and infrastructure to instigate and support them, yet the journey is not over. Innovation in products and governance need to be perpetuated.

References

Alicke, K., and Strigel, A. (2020). *Supply Chain Risk Management Is Back*. McKinsey & Company.

Au, A. K., Ramasamy, B., and Yeung, M. C. (2005). The effects of SARS on the Hong Kong tourism industry: an empirical evaluation. *Asia Pacific Journal of Tourism Research*, 10(1): 85–95. https://doi.org/10.1080/1094166042000330236

Aven, T. (2015). Implications of black swans to the foundations and practice of risk assessment and management. *Reliability Engineering & System Safety*, 134: 83–91. https://doi.org/10.1016/j.ress.2014.10.004

Behzadi, G., O'Sullivan, M. J., Olsen, T. L., Scrimgeour, F., and Zhang, A. (2017). Robust and resilient strategies for managing supply disruptions in an agribusiness supply chain. *International Journal of Production Economics*, 191: 207–220. https://doi.org/10.1016/j.ijpe.2017.06.018

Bethapudi, A. (2013). The role of ICT in tourism industry. *Journal of Applied Economics and Business*, 1(4): 67–79.

Blake, A., Sinclair, M. T., and Sugiyarto, G. (2003). Quantifying the impact of foot and mouth disease on tourism and the UK economy. *Tourism Economics*, 9(4): 449–465. https://doi.org/10.5367%2F000000003322663221

Butler, R., and Airey, D. (2005). The foot-and-mouth outbreak of 2001: impacts on and implications for tourism in the United Kingdom. In Aramberri, J., and Butler, R. (eds.), *Tourism Development: Issues for a Vulnerable Industry* (pp. 215–238). Clevedon: Channel View Publications.

Cao, M., Vonderembse, M., Zhang, Q., and Ragu-Nathan, T. (2010). Supply chain collaboration: conceptualisation and instrument development. *International Journal of Production Research*, 48(22): 6613–6635. https://doi.org/10.1080/00207540903349039

Ceppara, L. (2020). Innovating in Wine Tourism in the Context of COVID: The Reinvention of Oeno-Gastronomy in Wineries during and post COVID-19. UNWTO-OIV Webinar, September 22. https://webunwto.s3.eu-west-1.amazonaws.com/s3fs-public/2020-09/2.Lorena%20Cepparo_The%20Reinvention%20of%20Oeno-Gastronomy%20in%20wineries%20during%20COVID%20-19%20PDF.pdf

Fahimnia, B., Tang, C. S., Davarz'ani, H., and Sarkis, J. (2015). Quantitative models for managing supply chain risks: a review. *European Journal of Operational Research*, 247(1): 1–15. https://doi.org/10.1016/j.ejor.2015.04.034

Festa, G., Shams, S. R., Metallo, G., and Cuomo, M. T. (2020). Opportunities and challenges in the contribution of wine routes to wine tourism in Italy – a stakeholders' perspective of development. *Tourism Management Perspectives*, 33: 100585. https://doi.org/10.1016/j.tmp.2019.100585

Frochot, I. (2000). Wine tourism in France: a paradox? In Hall, C. M., Sharples, L., Cambourne, B., and Macionis, N. (eds.), *Wine Tourism around the World: Development, Management and Markets* (pp. 67–80). Oxford: Butterworth-Heinemann.

Frochot, I. (2001). French wine brotherhoods and wine tourism: a complex relationship. *Tourism Recreation Research*, 26(2): 53–62. https://doi.org/10.1080/02508281.2001.11081343

Garibaldi, R. (2019). *Rapporto Sul Turismo Enogastronomico Italiano 2019*. https://webunwto.s3.eu-west-1.amazonaws.com/s3fs-public/2019-12/2019_12_06_UNWTO-GARIBALDI_per-UNWTO.pdf last accessed 7 May, 2021.

Garibaldi, R. (2020). *The Impact of COVID-19 on Wine Tourism*.

Getz, D., and Brown, G. (2006). Critical success factors for wine tourism regions: a demand analysis. *Tourism Management*, 27(1): 146–158. https://doi.org/10.1016/j.tourman.2004.08.002

Giuliani, E. (2007). The selective nature of knowledge networks in clusters: evidence from the wine industry. *Journal of Economic Geography*, 7(2): 139–168.
Gössling, S., Scott, D., and Hall, C. M. (2020). Pandemics, tourism and global change: a rapid assessment of COVID-19. *Journal of Sustainable Tourism*, 29(1): 1–20. https://doi.org/10.1080/09669582.2020.1758708
Hall, C. M. (2010). Crisis events in tourism: subjects of crisis in tourism. *Current Issues in Tourism*, 13(5): 401–417. https://doi.org/10.1080/13683500.2010.491900
Hall, C. M., Cambourne, B., Macionis, N., and Johnson, G. (1997). Wine tourism and network development in Australia and New Zealand: review, establishment and prospects. *International Journal of Wine Marketing*, 9(2): 5–31. https://doi.org/10.1108/eb008668
Hall, C. M., Scott, D., and Gössling, S. (2020). Pandemics, transformations and tourism: be careful what you wish for. Tourism *Geographies*, 22(3): 577–598. https://doi.org/10.1080/14616688.2020.1759131
Hendricks, K. B., and Singhal, V. R. (2005). An empirical analysis of the effect of supply chain disruptions on long-run stock price performance and equity risk of the firm. *Production and Operations management*, 14(1): 35–52. https://doi.org/10.1111/j.1937-5956.2005.tb00008.x
Ivanov, D., and Sokolov, B. (2013). Control and system-theoretic identification of the supply chain dynamics domain for planning, analysis and adaptation of performance under uncertainty. *European Journal of Operational Research*, 224(2): 313–323. https://doi.org/10.1016/j.ejor.2012.08.021
Ivanov, D., Sokolov, B., and Dolgui, A. (2014). The Ripple effect in supply chains: trade-off 'efficiency-flexibility-resilience' in disruption management. *International Journal of Production Research*, 52(7): 2154–2172. https://doi.org/10.1080/00207543.2013.858836
Jüttner, U. (2005). Supply chain risk management: understanding the business requirements from a practitioner perspective. *The International Journal of Logistics Management*, 16(1): 120–141. https://doi.org/10.1108/09574090510617385
Keller, P. (2020). Corona pandemic as exogenous shock for international tourism: a context analysis. In Burini, F. (ed.), *Tourism Facing a Pandemic: From Crisis to Recovery* (pp. 15–24). Bergamo: Università degli Studi di Bergama. https://aisber.unibg.it/retreive/handle/10446/164998/3722337
Lavandoski, J., Vargas-Sánchez, A., Pinto, P., and Silva, J.A. (2018). Causes and effects of wine tourism development in organizational context: the case of Alentejo, Portugal. *Tourism and Hospitality Research*, 18(1): 107–122. https://doi.org/10.1177%2F1467358416634159
Leslie, D., and Black, L. (2006). Tourism and the impact of the foot and mouth epidemic in the UK: Reactions, responses and realities with particular reference to Scotland. *Journal of Travel & Tourism Marketing*, 19(2–3): 35–52. https://doi.org/10.1300/J073v19n02_04
March, R., and Wilkinson, I. (2009). Conceptual tools for evaluating tourism partnerships. *Tourism Management*, 30(3): 455–462. https://doi.org/10.1016/j.tourman.2008.09.001
Martin, R., and Sunley, P. (2003). Deconstructing clusters: chaotic concept or policy panacea? *Journal of Economic Geography*, 3(1): 5–35. https://doi.org/10.1093/jeg/3.1.5
Merschmann, U., and Thonemann, U. W. (2011). Supply chain flexibility, uncertainty and firm performance: an empirical analysis of German manufacturing firms. *International Journal of Production Economics*, 130(1): 43–53. https://doi.org/10.1016/j.ijpe.2010.10.013
Min, J.C., Lim, C., and Kung, H. H. (2011). Intervention analysis of SARS on Japanese tourism demand for Taiwan. *Quality & Quantity*, 45(1): 91–102. https://doi.org/10.1007/s11135-010-9338-4
Mitchell, R., and Hall, C. M. (2006). Wine tourism research: the state of play. *Tourism Review International*, 9(4): 307–332. https://doi.org/10.3727/154427206776330535
Narasimhan, R., and Talluri, S. (2009). Perspectives on risk management in supply chains. *Journal of Operations Management*, 27: 114–118. https://doi.org/10.1016/j.jom.2009.02.001
Novelli, M., Burgess, L. G., Jones, A., and Ritchie, B. W. (2018). 'No Ebola … still doomed' – the Ebola-induced tourism crisis. *Annals of Tourism Research*, 70: 76–87. https://doi.org/10.1016/j.annals.2018.03.006
Peregrine, A. (2018). The 10 best destinations in France for wine. *The Telegraph*, 1/2/2018, https://www.telegraph.co.uk/travel/destinations/europe/france/articles/best-vineyards-france/
Reportlinker (2020). Global wine industry, market report, 2020. https://www.reportlinker.com/p05010580/Global-Wine-Industry.html
Schlüter, R. G., and Norrild, J. (2015). Enotourism in Argentina: the power of wine to promote a region. In Netto, A. P., and Luiz Trigo, L. G. (eds.), *Tourism in Latin America* (pp. 71–84). Cham: Springer.
Sreedevi, R., and Saranga, H. (2017). Uncertainty and supply chain risk: the moderating role of supply chain flexibility in risk mitigation. *International Journal of Production Economics*, 193: 332–342. https://doi.org/10.1016/j.ijpe.2017.07.024

Tang, C. S. (2006). Perspectives in supply chain risk management. *International Journal of Production Economics*, 103(2): 451–488. https://doi.org/10.1016/j.ijpe.2005.12.006

Telfer, D. J. (2001). From a wine tourism village to a regional wine route: an investigation of the competitive advantage of embedded clusters in Niagara, Canada. *Tourism Recreation Research*, 26(2): 23–33. https://doi.org/10.1080/02508281.2001.11081340

UNWTO (2020a). *International Tourism Highlights*. https://www.e-unwto.org/doi/pdf/10.18111/9789284422456

UNWTO (2020b). International tourist numbers down 65% in first half of 2020. https://www.unwto.org/news/international-tourist-numbers-down-65-in-first-half-of-2020-unwto-report

UNWTO (2020c). *COVID-19 and Tourism Statistics*. https://www.unwto.org/covid-19-and-tourism-statistics

UNWTO (2020d). World Tourism Barometer and statistical annex May 2020. *UNWTO World Tourism Barometer* 18: (2), May.

WTTC (2020). *Economic Impact Report*. https://wttc.org/Research/Economic-Impact

Yeomans, C. (2019). A journey through Argentine wine history. *South American Wine Guide*. https://southamericawineguide.com/a-journey-through-argentine-wine-history/

49

WINE, TOURISM AND THE GLOBAL PANDEMIC

Realizing opportunities for new wine markets and experiences

Joanna Fountain, Rory Hill, and Nicholas Cradock-Henry

Introduction

The global economy is currently facing unprecedented challenges from the COVID-19 pandemic sweeping the world (Folke et al. 2021). While countries and regions have taken different measures to limit the spread of the virus, the consequences for the tourism and hospitality sectors are far reaching. The pandemic has curtailed international tourist movements and expenditure, while the hospitality sector has been particularly badly affected, due to the role of bars and restaurants in spreading the virus. Primary sectors, including the wine industry, have also been affected by the requirements for lockdowns and social distancing, by disruptions to supply chains, and due to difficulties in accessing a global labor force. Many wineries are also facing significant interruptions to two important distribution channels – the cellar door (or tasting room) and the hospitality trade. This chapter begins by presenting a brief overview of the impacts of the COVID-19 pandemic on the global wine industry and the tourism and hospitality sectors. This overview is followed by a case study of the New Zealand experience and response to the virus, with a particular focus on the response of wine producers and wine tourism operators to the 'new normal'; a largely domestic wine tourism market, keen to explore the country and support local producers. A number of examples of ways in which wine tourism providers are meeting the needs of this market are outlined, including the opportunities to provide more memorable experiences to winery visitors while also being conscious of the need to offer great value for money to a price-conscious market.

The impact of COVID-19 on the global wine value chain

The profound impact caused by the COVID-19 pandemic is one of the biggest challenges facing the global wine industry, with implications across the entire wine value chain. As 2020 dawned, many wine regions of the world were facing a challenging environment due to the impact of climate-related events over the previous few months. In Australia and the western seaboard of North America, wine producers were counting the costs of devastating wildfires

DOI: 10.4324/9781003143628-56

which had burned vineyards and caused smoke taint in wine (Fitchette 2021; Irving 2021; McDermaid and Newton 2020; McIntyre 2021), reinvigorating debates about the impacts of climate change on the global wine industry (e.g., Cabré and Nuñez 2020; Fraga et al. 2020). European wine regions faced additional challenges, including the prospect of a decline in Chinese wine consumption and the imposition of new US tariffs (Masset and Weisskopf 2020).

Restrictions caused by the pandemic from the first quarter of 2020 had spatially distributed impacts corresponding to the stage of the grape-growing cycle. In the Northern Hemisphere, lockdown and social distancing requirements coincided with the winter months, when there is little activity in the winery, and it is relatively simple to maintain social distancing in vineyard activities such as pruning. In the Southern Hemisphere, the global pandemic coincided with the grape harvest. As the impact of the pandemic continued through 2020, food and drink supply chains were significantly disrupted. A significant decrease in airfreight capacity, due to a drop-off in air travel, was particularly problematic for time-sensitive perishable goods (Ministry of Primary Industries [MPI] 2020). Global delays and increasing shipping costs caused by backlogs on docks affected profit margins and cash flow and caused logistical challenges (MPI 2020).

Crises in tourism are not new; however, the global reach and complete shutdown of tourist travel caused by COVID-19 is unprecedented (Benjamin, Dillette, and Alderman 2020; Higgins-Desbiolles 2020). While some countries tentatively opened borders to tourists during the Northern Hemisphere summer of 2020, this activity resulted in second and third waves of infection, causing further severe – and often longer – lockdowns. This has effectively resulted in a cessation of any international tourism, including wine tourism, although in some countries and regions, control of the virus has meant domestic tourism has been able to operate with few restrictions.

Early in the pandemic, it was recognized that the weddings, parties, and other events held in hospitality venues, such as bars and restaurants, were playing a significant role in spreading the virus. For this reason, hospitality establishments have been some of the first to close and the last to reopen when a community outbreak of the virus occurs. In some jurisdictions, takeaway food and drink services have been available during these restrictions (Alonso et al. 2020), and in some places, including Australia and parts of Canada, temporary alcohol policy changes have enabled establishments without off-licenses to sell wine and other alcohol this way (Colbert et al. 2020). In other countries, however, restaurants and bars had to close completely at the strictest lockdown level. This was the case in New Zealand (RNZ 2020), while in South Africa the sale of alcohol was banned entirely, and in France, a curfew effectively limited night-time socializing (Alonso et al. 2020).

The closure – or severe restriction on activity – of hospitality establishments has had a devastating effect on the mostly smaller wine producers who rely on these distribution channels to sell their wines (Gilbertson 2020). These are often the same wine producers who prioritize direct-to-consumer sales at winery cellar doors. By comparison, wine producers operating at a scale suitable for supermarket or other retail distribution, both in a domestic or export market, have generally faced fewer negative impacts from the pandemic, and in some cases have seen a growth in demand and sales (Squire 2020). The news is not all bad for small wine producers, however; globally, small-scale, local, and 'authentic' producers and products have been prioritized by many consumers in recent years (e.g., Ateljevic 2020). The experience of the pandemic has intensified this trend, encouraged by politicians and economists keen to support the economy at a time of crisis (Fountain 2021a; Hall, Scott, and Gössling, 2020; Higgins-Desbiolles 2020), and reflecting a degree of patriotism as consumers rally to get their nation back on its feet (Irving 2021; Kilgallon 2020; Sibley et al. 2020; Squire 2020).

Wine producers around the world have responded to the challenges of the global pandemic to ensure continued wine sales and cash flow. While there have been various responses, depending on legislative frameworks and lockdown restrictions, two key strategies can be identified. First, there has been renewed development of, and surge in, the capacity for online wine sales (Alonso et al. 2020); For some producers, this represents the first foray into this distribution channel, while many others have intensified their efforts in the online environment (Alonso et al. 2020), often supported by regional online initiatives (Irving 2021). In the United States, wineries reported that online sales were up, but 'value brands' were seeing the biggest sales increase (Van Dam and Johnson 2020), due in part to consumers purchasing cheaper wines for home consumption (Irving 2021; McIntyre 2021) and seeking out familiar brands (McIntyre 2021; Squire 2020). While online sales may provide some financial relief, it will not replace the opportunities for relationship building that occur at cellar doors (Squire 2020). Recognizing this, wineries have increased their social media presence to connect with the consumer at a deeper level. One popular strategy has been the development of virtual wine tastings, whereby customers are sent bottles or cases of wine, after which they meet with the winemaker online to complete a structured wine tasting, either in a group or by booking a private session (Coyne 2020; Irving 2021; Van Dam and Johnson 2020).

The rollout of effective COVID-19 vaccines during the first half of 2021, and the elimination of widespread community transmission of the virus in some countries, means wine producers in some regions are beginning to welcome visitors back to the cellar door. This is the case in New Zealand; although with international borders are effectively closed, these experiences are limited to domestic wine tourists. The following section outlines the situation and experiences of New Zealand wine producers as they navigate a 'new normal' of domestic wine tourism in a post-pandemic context.

Case Study 49.1 A 'new normal': wine tourism and new markets in post-pandemic New Zealand

The New Zealand experience of COVID-19

New Zealand's first case of COVID-19 was reported in late February 2020. Within a month, fears of exponential growth in cases of the virus resulted in a strict pre-emptive lockdown of the country and most of the economy for a five-week period, with restrictions gradually easing over the next two months (Baker and Wilson 2020; RNZ 2020). At the strictest level, Alert Level 4, all residents were required to stay home in their 'bubble' (usually a household) to break the chain of transmission of the virus. During this lockdown, the only reasons to leave one's bubble were to attend 'essential' employment, make trips for food and medical supplies, or exercise in the local neighborhood. These restrictions were more severe than in many countries, with restaurants, cafés, and bars closed even for takeaway, contactless service. The only food retail outlets deemed essential were supermarkets and liquor stores, so that other food retailers, such as butcheries, bakeries, and winery cellar doors, remained closed at Alert Level 4. There was some panic buying in the early days of lockdown due to at least the perception of limited availability through the supply chain; however, food and alcohol sales returned to normal levels relatively quickly (Hall et al. 2020; see also Gerritsen et al. 2020).

The goal of the strict lockdown was not only to slow the spread of COVID-19 ('flatten the curve'), but to eliminate the virus from the community entirely (Baker and Wilson 2020). For this reason, New Zealand's borders closed to all but New Zealand citizens and permanent residents who quarantined in managed facilities for two weeks on their arrival. While these strict lockdown measures have been highly effective for the most part in eliminating the virus from the community, the need to maintain strict border controls means that international borders have only fully reopened from mid-2022.

At the time of the lockdown, New Zealand's main exports were the dairy, meat, and horticulture industries, international education, and tourism, with the latter being the largest export earner (Tourism Industry Aotearoa 2020). International tourism and education effectively ceased with the onset of the pandemic; however, the primary sector remained robust, and many of the workers and industries deemed 'essential' during Alert Level 4 were involved in the primary sector – growing, processing, transporting, and selling food and drink. The media highlighted the economic importance of these activities to New Zealand's economy during this time, and positive publicity resulted in greater public recognition and appreciation of the primary sector, including wine producers (MPI 2020; Sivignon 2020).

On a global scale, the New Zealand wine industry is very small and relatively recent in origin; however, it is New Zealand's sixth-largest export earner (MPI 2020). The first commercial wine grapes were planted in Marlborough in the 1970s. Since then, the industry has experienced rapid growth, so that that by 2020, there were 717 wine producers across ten major wine regions (New Zealand Winegrowers 2020, p. 33; see Figure 49.1). Representing only 1% of global wine production, New Zealand's premium wines have gained a strong global reputation, particularly Sauvignon Blanc and Pinot Noir, with the former dominating production and wine exports (New Zealand Winegrowers 2020).

Most New Zealand wine production is relatively small scale. Small producers – those with annual sales of less than 200,000 liters per year – comprise 87% of all wineries. More than half the country's wineries (380 producers) sell an average of just 4,000 liters, or fewer than 500 cases a year, with 75% of this wine sold domestically, primarily via restaurants, direct sales through the cellar door, or online (Cooper 2021, p. 25). Over the past decade, there has been some consolidation of ownership into the hands of a few large global wine and alcohol companies, with the number of large wineries (those with sales over two million liters per year) increasing from 6 wineries to 20 between 2010 and 2020 (New Zealand Winegrowers, 2019, p. 33). The result of this is that 93% of New Zealand wine is made by only 7% of producers (Cooper 2021, p. 25). Evidence of consolidation is difficult for the average consumer to notice, as new owners frequently retain the original winery name and branding (Cooper 2021).

At the time of New Zealand's Alert Level 4 lockdown, the 2020 wine harvest was underway. Despite initial concerns over safety and spread of the virus, the vintage was completed without incident (Arthur 2020). The impact of lockdown varied significantly for New Zealand wine producers, with the effects corresponding to producers' scale. For example, those large enough to benefit from supermarket distribution were largely unaffected, or lockdown had a net positive effect. Supermarkets have been the most important distribution channel for wine for the previous three decades in New Zealand, but this share

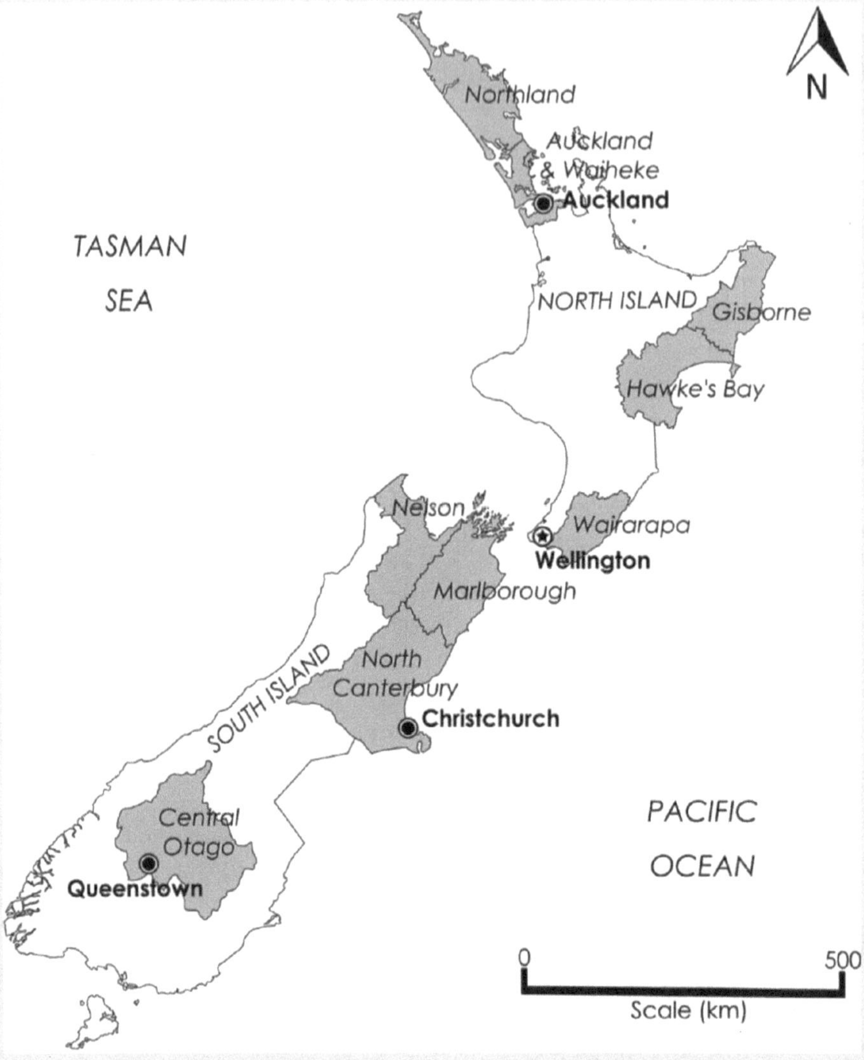

Figure 49.1 New Zealand wine regions.

Source: Map drawn by Rory Hill, based on WikiMedia Commons map by AMitchell125, shared under Creative Commons Attribution-Share Alike 4.0 International license (URL: https://commons.wikimedia.org/wiki/File:Geographical_indications_register_for_New_Zealand_wine.svg).

increased during lockdown as supermarkets and liquor stores were the only brick-and-mortar stores to remain open (Arthur 2020; Cooper 2021).

Other, smaller wineries that relied on direct sales to restaurants and cafés faced significant pressures. As in other countries, the hospitality and restaurant sectors were some of the last to have restrictions on trading removed as lockdown restrictions were lifted (RNZ 2020). Some small wine producers have done what their international counterparts did and increased their online presence, through both online portals for purchases and 'telling the story behind New Zealand wine' (Arthur 2020) and virtual tastings (Preece 2020b). Despite these efforts, challenges remain. A number of community outbreaks have resulted in an escalation of alert levels by region, or across the country, and each time this occurs, restrictions on hospitality venues and cellar doors return. There are concerns that many boutique producers will not survive, a situation that could damage the reputation of the New Zealand wine industry: 'We don't want to be referenced as a one stop shop for large retail brands,' said Alistair King (cited in Preece 2020c, p. 34). This is a challenge acknowledged by New Zealand Winegrowers:

> Positively, for those wineries whose focus is on supermarket and 'big' retail channels, sales have increased … For other wine businesses, the story is more challenging. Physical distancing measures have either halted or significantly slowed on-premise consumption in many markets, but online sales have surged, while the closure of the border has cut-off the flow of overseas visitors to cellar doors
>
> *(cited in Arthur 2020)*

Despite the challenging year, New Zealand wine exports grew by 9%, passing the $NZ2 billion mark for the first time late in 2020 (MPI 2020; Taunton 2020). However, this growth obscures some concerning trends in the export market; while sales are up, price per liter is falling at an average of 6% (Cooper 2021, p. 20). In 2020, shipments of bulk wine increased by 17%, accounting for close to half of all exports in 2020, while exports of premium varietal wines, including Chardonnay, Pinot Noir, and Merlot, are down, Merlot by a staggering 47% and Pinot Noir by 17% (Cooper 2021, p. 19). Part of this trend has been put down to the increased supermarket demand from UK consumers for their favorite Sauvignon Blanc as they faced a wave of extended lockdowns (MPI 2020 and personal communication); however, this trend cannot be blamed on COVID-19 alone, as it has been apparent in the market since 2016 (Cooper 2021, p. 20). These changing patterns of consumption negatively impact small producers; as one well-regarded boutique wine producer put it: 'there is very little positive news, unless you are one of the 20 largest producers (and therefore probably overseas-owned), rather than the 697 small or medium-sized players' (cited in Cooper 2021, p. 19).

The news is not all bad for the New Zealand hospitality and tourism sector and these smaller wine producers. The effectiveness of the country's elimination strategy has resulted in a somewhat surreal situation, where life has returned to a semblance of normality for those living in New Zealand within a 'national bubble,' while the country remains isolated from much of the rest of the world. Restaurants and bars are open, large-scale events and festivals are scheduled, and people are free to travel throughout the country for leisure or business purposes. Amongst many New Zealanders, there seems to be a strong desire to strengthen the national and regional economy by 'supporting local' in hospitality and tourism experiences (Kilgallon 2020; Sibley et al. 2020). There is evidence also that the amount of local wine being purchased in restaurants is increasing (Hill and Fountain 2022), in part due to this conscious desire to support local producers (Hill and Fountain 2020). In a survey of 500 New Zealanders conducted in the summer months of 2020–2021, more than half of respondents reported that as a result of COVID-19 they were purchasing more food and drink made locally (53.2%), buying more New Zealand wine (54.5%), and choosing to support small local businesses rather than large corporations where possible (63.7%; Fountain 2021b). The presence of more local wines on wine lists may also be due to restaurant managers facing difficulties in accessing imported wines, with increased freight costs making domestic wine a more attractive option (Hill and Fountain 2022).

As wine tourism providers look to a largely domestic market for the foreseeable future, many of them have reason to feel some optimism. Wine features in a national campaign exhorting the audience to 'Do Something New, New Zealand' (Tourism Industry Aotearoa 2020), as New Zealanders are encouraged to 'Climb a mountain and drink a Riesling.' Recent research shows a desire amongst domestic holidaymakers to support and explore smaller regional communities (One Picture 2020), in so doing resulting in a recognition that many of the country's regions have gone from quiet and dull places to cultural hubs with a vibrant culinary scene (One Picture 2020, p. 19). Domestic travelers are also keen to find 'non-touristy' – and by extension, inexpensive – experiences as they really get to know the country without paying the earth, with food and drink experiences featuring strongly in sought-out experiences (One Picture 2020; see also Fountain 2021a). But what experiences will they seek? And how will the threat of further lockdowns and restrictions on gatherings impact these experiences? The following section outlines examples of innovation in wine tourism offerings in response to COVID-19, the demand for wine tourism amongst domestic consumers, and the experiences that are resulting from these encounters.

New Zealand wine tourism in the shadow of a global pandemic

Wine tourism activities emerged in New Zealand concurrent with the development of the wine industry in the 1970s. These activities have evolved with the expansion of the wine and tourism industries to become an important sales outlet and vehicle for brand recognition and brand loyalty, particularly for small producers (Baird and Hall 2014; Tourism New Zealand 2017). The winery cellar door is the ubiquitous form of wine tourism in New Zealand. Depending on the size and priorities of the business, cellar doors may be architecturally impressive, or involve little more than an office attached to an existing building or home (Fountain 2021c). Despite the name, cellar doors rarely include actual underground cellars, and unlike other countries (Fountain et al. 2020), winery tours are not an expected part of a visit, which, prior to the pandemic, generally involved standing at a counter tasting wine and chatting informally with staff as they moved between groups of visitors, pouring wine. Winery restaurants and cafes are also relatively common in New Zealand, with wine and food matching opportunities increasing, and some winery premises including areas for picnics. A number of wine producers also collaborate to provide walking or cycling trails for visitors and work together to host many wine-related events and festivals held annually around the country (Fountain 2021c).

In the decade immediately prior to the global pandemic, interest in wine tourism had been increasing at the organizational level of the New Zealand wine industry. In 2014, the national tourism marketing organization, Tourism New Zealand, announced a partnership with the national association of New Zealand Winegrowers to collaborate in promoting the country as a visitor destination for wine tourists and a producer of exceptional wines (Tourism New Zealand 2014a, 2014b). At a regional level, close to half of the regional tourism organizations in the country promote wine tourism opportunities in their region through products and visual imagery of vines and wineries (Fountain 2021c). Wine tourism occurs in all ten wine regions of New Zealand, but the intensity of involvement and level of development differs by region. Research conducted in 2017 shows that there are 279 wineries in New Zealand offering more than 400 wine tourism experiences (Nicholson 2018, p. 14), with these experiences showcased on a recently developed 'visit' wine tourism portal on the New Zealand Winegrowers' website (New Zealand Winegrowers 2020). Research findings and anecdotal evidence suggest that producer interest in engaging with wine tourism was also increasing prior to the pandemic. For example, a survey of wine producers in North Canterbury in 2018–2019 found that 43.8% of respondents were involved in wine tourism, with 29.3% of respondents indicating that their wine tourism involvement had increased significantly in the previous two years. Two-thirds of respondents (68.5%) reported that they planned to increase their involvement in wine tourism in the following two years; a figure including six of the nine respondents not currently involved in wine tourism (Fountain 2019).

The COVID-19 pandemic occurred in this context. As outlined earlier, the effect of lockdown and national border closures was particularly devastating for small wine producers, who lost their two primary sales channels overnight. New Zealand Winegrowers, who had been investing heavily in wine tourism initiatives, cut back their expenditure in this area immediately. The lifting of domestic travel restrictions and reopening of hospitality venues provided a lifeline for many of these small wine producers, but there is recognition also that additional precautions and different wine tourism experiences are required to meet the needs of the domestic market. New Zealand Winegrowers also acknowledge that they will need to 're-imagine tourism activities' (Arthur 2020); a message echoed at the regional level (personal communication).

Not all wine tourism experiences were affected equally. Wine tour operators have been particularly badly impacted by the closure of New Zealand's borders. Prior to COVID-19, tour operators relied heavily on the international market, with inbound tourists making up 80% to 90% of their clientele (Webb 2020, p. 24). This included the cruise ship sector, whose passengers frequently took wine tours. The fact is that domestic tourists, with private vehicles and confidence to drive on the open road, do not require transportation between wineries and do not recognize the added value of visiting wineries with a knowledgeable tour guide (Webb 2020). High-end boutique wine experiences aimed at the international market are not viable in the short term, and some winery restaurants and winery centers who have relied primarily on the international market have ceased operations or changed the nature of their operations (Brookes 2020; Downes 2020; Duhalde 2020). For example, the iconic and award-winning Pegasus Bay restaurant closed and planned to reinvent itself as a function and event space for hire (Burzynska 2020). This is the approach Pernod Ricard planned to introduce when their Brancott Estate Heritage Centre reopens in Marlborough

(Burzynska 2020). At Elephant Hill in Hawkes Bay, the award-winning restaurant ceased to be viable with the absence of international tourists, including cruise ship visitors; however, they planned to reinvent themselves for the domestic market (Burzynska 2020).

While there is no doubt that international wine tourists are sorely missed from the sector, domestic visitors have traditionally made up at least half of all winery visitors in New Zealand (Tourism New Zealand 2014a). The proportion is potentially substantially higher, as national tourism data does not include local residents visiting wineries in their wine tourist statistics (Fountain and Thompson 2019). As stated earlier, New Zealanders have emerged from lockdown with a desire to see the country and support local food and wine producers, in many cases surprising even the most optimistic commentators (Preece 2020a). Regions close to major centers have seen very good engagement in wine tourism, with the Wairarapa Region, 90 minutes from Wellington, experiencing a large influx in visitor numbers (Burzynska 2020). In Central Otago – a region previously highly dependent on international visitors – wine producers have been pleasantly surprised by the patronage of domestic ski tourists visiting Queenstown, making the trip into the region. While visitor numbers are still significantly down, sales were generally higher, as these visitors stayed longer and spent more on wine (Preece 2020a). The trend through lockdown to buy products online may also result in more post-visit wine purchasing directly from the winery (Cooper 2021, p. 23; O'Mahony et al. 2006).

There are a number of explanations for this pattern of domestic behavior. First, many New Zealanders are traveling domestically rather than taking planned overseas trips, so they have more disposable income to spend, whilst still looking for good value. As one wine producer explained: 'People who would usually have spent some time and money in Europe or some other long-distance destinations are now visiting the Wairarapa and are also not afraid to buy a case or two of more expensive wines' (cited in Burzynska 2020, p. 20). This quotation also alludes to a second reason for the increased sales; New Zealanders are generally traveling independently, giving them the freedom to make last-minute travel decisions whilst also being less limited in the amount of wine they can purchase. As has been noted in Central Otago, domestic visitors are taking away boxes of wines in their car, rather than worry about luggage capacity and duty-free limits (Preece 2020a, p. 23).

A third factor relates to a change in the nature of the wine tourism experience being offered. At Alert Level 2, which is when most domestic travel began to occur, restrictions were applied to the serving of food and wine in hospitality settings. In particular, all service had to be seated, and all groups had to be served by one staff member. This has had some interesting, and perhaps unanticipated, consequences for the wine tourism experience. A number of wine producers are reporting positively on the length of time domestic tourists are spending in the cellar door. As one Central Otago producer observed: 'Kiwis are coming in and chatting and have time to hear the stories. They are connected and buy more wine' (cited in Preece 2020a, p. 23). Wine producers in other regions have reported similar positive benefits of this arrangement for both the staff and the wine tourists, with the structured tastings resulting in greater depth of engagement, better interactions, more personal experiences, and ultimately better opportunities for relationship building and brand loyalty (Burzynska 2020).

Wine producers continue to explore innovative ways to extend and enhance the wine tourism experience for domestic visitors, whilst maintaining the affordability and perception of value so critical to this market (One Picture 2020). Many producers are looking at ways to enhance opportunities for combining food and wine experiences in an affordable way (Burzynska 2020). For example, pandemic restrictions and the domestic market's price sensitivities saw Black Estate in North Canterbury introduce a 'cellar door pie cart' in place of its high-end restaurant experience. Gibbston Valley Winery in Central Otago created new packages and options for cheese and wine matching (Burzynska 2020), and Elephant Hill, in Hawkes Bay, planned to relaunch their winery restaurant with food and wine matching options for domestic visitors (Burzynska 2020).

Some wineries were well positioned to take advantage of this trend when the pandemic struck, having already ventured into wine and food matching. Take Terrace Edge, for example, a small family-owned winery in Waipara, North Canterbury. In 2015, the two generations of the family made the conscious decision to focus more deliberately on wine tourism as a distribution strategy and looked at ways they could expand this side of the business (Fountain 2021c). The first evidence of transformation was physical, with the cellar door extended and redesigned. The opening of this new tasting room coincided with the introduction of gourmet toasted sandwiches with an array of seasonal fillings, including braised beef and sticky pork on ciabatta bread. The inclusion of these very reasonably priced food options meant it was possible for Terrace Edge to apply for a liquor license allowing the winery to sell wine by the glass, while not committing to an extensive,

Figure 49.2 Wine and food experience.

Source: Terrace Edge

or expensive, menu where wastage can be a significant problem. Most recently, in August 2019, the winery introduced a 'Food and Wine Experience' (Figure 49.2). In a region where tasting fees are relatively rare, this food and wine pairing sees visitors enjoy a seated tasting of canape-sized food creations matched with five wines for a NZ$20 charge (Terrace Edge 2020). Their ability to be nimble and innovative means that as New Zealand came out of COVID-19 lockdown, they were one of the first wineries to reopen, offering seated tastings for a NZ$10 charge to enable appropriate physical distancing. Now, with many wine tourism facilities still closed as a consequence of the global pandemic and a difficult road ahead for many businesses, Terrace Edge positioned themselves as well as possible to meet the coming challenges head on.

Conclusion

The COVID-19 pandemic has had a massive impact on all aspects of the global wine value chain, with significant implications for the hospitality and tourism sectors. However, as this chapter has highlighted, out of adversity has come some opportunities. In the case of the New Zealand wine industry, successful control of the virus has resulted in the opportunity to open the country to domestic wine tourism. Early evidence suggests that, while not able to replace the international wine tourist market, New Zealanders are getting out to see their country and support local producers, and this includes visiting winery cellar doors. The situation remains difficult for many small producers, however, with export markets and the restaurant sector vulnerable to further lockdowns. Similarly, high-end wine tourism operators face a difficult and uncertain future while borders remain closed. For many smaller wine producers, however, the domestic market has provided important financial relief, as well as the opportunity for new forms of wine tourism experiences, particularly those involving greater depth of experience, length of engagement, and potential long-term brand loyalty.

Implications

The current global pandemic is a time of great uncertainty for wine tourism providers around the world. However, as this case study illustrates, it is possible to pivot to changing markets and consumer (and legislative) demands for safe, socially distanced experiences. In the case of New Zealand, a focus on the domestic market has meant recognition that these visitors want more engaged and less commercial experiences. While these visitors may not be as lucrative in the short term as international markets, by building a relationship through a cellar door experience, wine producers may be able to build brand loyalty and repeat visitation, as well as word-of-mouth recommendations more readily with domestic clientele. The success of small operators such as Terrace Edge shows it is possible to develop memorable wine tourism experiences on a modest scale by connecting with their consumers and creating accessible and affordable food and wine matching experiences.

References

Alonso, A. D., Kok, S. K., Bressan, A., O'Shea, M., Sakellarios, N., Koresisf, A., Solis, M. A. B., and Santoni, L. J. (2020). COVID-19, aftermath, impacts, and hospitality firms: an international perspective. *International Journal of Hospitality Management*, 91. https://doi.org/10.1016/j.ijhm.2020.102654

Arthur, R. (2020, 1 September). New Zealand on coronavirus wine strategies. *Beverage Daily*. https://www.beveragedaily.com/Article/2020/09/01/New-Zealand-adjusts-wine-strategy-around-coronavirus-impact

Ateljevic, I. (2020). Transforming the (tourism) world for good and (re)generating the potential 'new normal.' *Tourism Geographies*, 22(3): 467–475. https://doi.org/10.1080/14616688.2020.1759134

Baird, T., and Hall, C. M. (2014). Between the vines: wine tourism in New Zealand. In Howland, P. J. (ed.), *Social, Cultural and Economic Impacts of Wine in New Zealand* (pp. 191–208). New York: Routledge.

Baker, M., and Wilson, N. (2020, 10 April). Elimination: what New Zealand's coronavirus response can teach the world. *The Guardian*. https://www.theguardian.com/world/2020/apr/10/elimination-what-new-zealands-coronavirus-response-can-teach-theworld

Benjamin, S., Dillette, A., and Alderman, D. H. (2020). 'We can't return to normal': committing to tourism equity in the post-pandemic age. *Tourism Geographies*, 22(3), 476–483. https://doi.org/10.1080/14616688.2020.1759130

Brookes, E. (2020, May 6). Coronavirus: 15 jobs go as top winery restaurant at Pegasus Bay shuts. *Stuff*. https://www.stuff.co.nz/life-style/food-wine/121427112/coronavirus-15-jobs-go-as-top-winery-restaurant-at-pegasus-bay-shuts

Burzynska, J. (2020). Open doors: agility, invention and opportunity at cellar doors. *New Zealand Winegrowers*, 124(October/November): 17–21.

Cabré, F., and Nuñez, M. (2020). Impacts of climate change on viticulture in Argentina. *Regional Environmental Change*, 20: 12. https://doi.org/10.1007/s10113-020-01607-8

Colbert, S., Wilkinson, C., Thornton, L., and Richmond, R. (2020). COVID-19 and alcohol in Australia: industry changes and public health impacts. *Drug and Alcohol Review*, 39(5): 435–440. https://doi.org/10.1111/dar.13092

Cooper, M. (2021, February 27). Tipping point. *New Zealand Listener*, 275(4184): 18–25.

Coyne, M. (2020). Three Sticks Wines: digital marketing, branding and hospitality in a crisis. *Wine Business Journal*, 4(2), 29–53. https://doi.org/10.26813/001c.22071

Downes, S. (2020, April 29). Coronavirus: Top Hawke's Bay winery Elephant Hill restaurant closes for good. *Stuff*. https://www.stuff.co.nz/life-style/food-wine/food-news/121317646/coronavirus-top-hawkes-bay-winery-restaurant-closes-for-good

Duhalde, M. (2020, April 21). How the coronavirus disrupts food supply chains. *South China Morning Post*. https://multimedia.scmp.com/infographics/news/world/article/3080824/covid19-disrupts-food-supply

Fitchette, T. (2021, February 3). Challenges remain: wine industry survives pandemic and smoke. *Western FarmPress*. https://www.farmprogress.com/grapes/challenges-remain-wine-industry-survives-pandemic-and-smoke

Folke, C., Polasky, S., Rockström, J., Galaz, V., Westley, F., Lamont, M., Scheffer, M., Österblom, H., Carpenter, S. R., Chapin, F. S., III, Seto, K. S., Weber, E. U., Crona, B. I., Daily, G. C., Dasgupta, P., Gaffney, O., Gordon, L. J., Hoff, H., Levin, S. A., ...Walker, B. H. (2021). Our future in the Anthropocene biosphere. *Ambio*, 50: 834–869. https://doi.org/10.1007/s13280-021-01544-8

Fountain, J. (2019). Business resilience in the North Canterbury wine region: challenges and opportunities. Presentation for the North Canterbury Winegrowers Association AGM, August 15, Waipara, New Zealand.

Fountain, J. (2021a). The future of food tourism in a post-COVID-19 world: insights from New Zealand. *Journal of Tourism Futures*. ahead-of-print. https://doi.org/10.1108/JTF-04-2021-0100

Fountain, J. (2021b). Wine consumption and wine tourism in a post-COVID New Zealand. *Presentation at Marlborough Research Centre*, April 14, Blenheim, New Zealand.

Fountain, J. (2021c). Wine tourism in New Zealand. In Compes Lopes, R., and Szolnoki, G. (eds.), *Innovative and Sustainable Wine Tourism: National and International Successful Models* (pp. 349–362). Spain: Cajamar. https://publicacionescajamar.es/

Fountain, J., Charters, S., and Cogan-Marie, L. (2020). The real Burgundy: negotiating wine tourism, relational place and the global countryside. *Tourism Geographies*, 23(5–6): 1116–1136. https://doi.org/10.1080/14616688.2020.1713880

Fountain, J., and Thompson, C. (2019). Wine tourists' perception of winescape in Central Otago, New Zealand. In Sigala, M., and Robinson, R. N. S. (eds.), *Wine Tourism Destination Management and Marketing: Theory and Cases* (pp. 29–44). London: Palgrave Macmillan.

Fraga, H., Molitor, D., Leolini, L., and Santos, J. A. (2020). What is the impact of heatwaves on European viticulture? A modelling assessment. *Applied Sciences*, 10: 3030. https://doi.org/10.3390/app10093030

Gerritsen, S., Egli, V., Roy, R., Haszard, J., De Backer, C., Teunissen, L., Cuykx, I., Decorte, P., Pabian S., Van Royen, K., and Te Morenga, L. (2020). Seven weeks of home-cooked meals: changes to New Zealanders' grocery shopping, cooking and eating during the COVID-19 lockdown. *Journal of the Royal Society of New Zealand.* https://doi.org/10.1080/03036758.2020.1841010

Gilbertson, G. M. (2020, April 16). Coronavirus: horticulture and wine sector face challenges once Level 4 lowers. https://www.stuff.co.nz/national/121058203/coronavirus-horticulture-and-wine-sector-face-challenges-once-level-4-lowers

Hall, C. M., Scott, D., and Gössling, S. (2020). Pandemics, transformations and tourism: be careful what you wish for. *Tourism Geographies*, 22(3): 577–598. https://doi.org/10.1080/14616688.2020.1759131

Higgins-Desbiolles, F. (2020). Socialising tourism for social and ecological justice after COVID-19. *Tourism Geographies*, 22(3): 610–623. https://doi.org/10.1080/14616688.2020.1757748

Hill, R., and Fountain, J. (2020). *Challenges and Opportunities for Wine Tourism in the Shadow of a Pandemic: the case of North Canterbury, New Zealand.* Presented at the 2nd International Research Workshop on Wine Tourism [online conference], University of Strasbourg/University of South Australia, September 17–18.

Hill, R., and Fountain, J. (2022). Wines from the hinterland: comparing availability of local wines on wine lists in two New Zealand tourism destinations. *International Journal of Wine Business Research.* https://www.emerald.com/insight/content/doi/10.1108/IJWBR-06-2021-0035/full/html

Irving, K. (2021, February 26). How Victoria's wine industry survived bushfires, lockdowns and China tariffs. *Herald Sun (Online).* https://www.heraldsun.com.au/show-victorias-wine-industry-survived-bushfires-lockdowns-and-china-tariffs

Kilgallon, S. (2020, 3 May). Covid-19 has sparked a 'Buy NZ Made' resurgence – will it last? *Stuff.* https://www.stuff.co.nz/business/industries/300002812/covid19-has-sparked-a-buy-nz-made-resurgence-will-it-last

Masset, P., and Weisskopf, J.-P. (2020). Covid-19 and fine wines: the perfect blend for a severe headache? *EHL Insights.* https://hospitalityinsights.ehl.edu/covid-19-wines

McDermaid, M., and Newton, S. K. (2020). Paradise Ridge Winery: can a tragedy strengthen a brand? *Wine Business Journal*, 4(2), 2–28. https://doi.org/10.26813/001c.22075

McIntyre, D. (2021, January 6). In 2020, the wine industry faced the twin crises of the pandemic and wildfires. *The Washington Post.* https://www.washingtonpost.com/lifestyle/food/2020-year-in-wine/2020/12/31/3

Ministry of Primary Industries (2020). *Economic update for the Primary Industries*, June 2020. MPI, Wellington. https://www.mpi.govt.nz/dmsdocument/40808-Economic-Update-for-the-Primary-Industries-June-2020

New Zealand Winegrowers (2019). *Annual Report.* New Zealand Winegrowers.

New Zealand Winegrowers (2020). *Visit the vines.* https://www.nzwine.com/en/visit/

Nicholson, T. (2018). Tapping into wine tourism. *New Zealand Winegrowers*, 112(Oct/Nov): 12–15.

O'Mahony, B., Hall, J., Lockshin, L., Jago, L., and Brown, G. (2006). Understanding the impact of wine tourism on post-tour purchasing behaviour. In Carlsen, J., and Charters, S. (eds.). *Global Wine Tourism: Research, Management and Marketing* (pp. 123–137). Wallingford, UK: CAB International.

One Picture (2020). *Unlocking Future Tourism.* Prepared for the Tourism Futures Taskforce. https://www.mbie.govt.nz/dmsdocument/13598-appendix-b-1-unlocking-future-tourism-onepicture-and-tnz-research-october-2020

Preece, S. (2020a). Misha's Vineyard: cellar door centre stage. *New Zealand Winegrowers*, 124(October/November), 22–23.

Preece, S. (2020b). Sales soar: wine exports defy expectations. *New Zealand Winegrowers*, 124(October/November), 30.

Preece, S. (2020c). Findex Report: Covid-19 a big hit to small wineries. *New Zealand Winegrowers*, 124(October/November), 34–35.

Radio New Zealand (2020). COVID-19 pandemic timeline. https://shorthand.radionz.co.nz/coronavirus-timeline

Shaw, L. (2021, 9 February). Oregon vineyard reopens with wine pod. *The Drinks Business* https://www.thedrinksbusiness.com/2021/02/oregon-vineyard-reopens-with-wine-pods/

Sibley, C. G., Greaves, L. M., Satherley, N., Wilson, M. S., Overall, N. C., Lee, C. H. J., Milojev, P., Bulbulia, J., Osborne, D., Milfont, T. L., Houkamau, C. A., Duck, I. M., Vickers-Jones, R., and Narpw, F. K. (2020). Effects of the COVID-19 pandemic and nationwide lockdown on trust, attitudes toward government, and well-being. *American Psychologist*, 71(5): 618–630.

Sivignon, C. (2020, July 16). Horticulture aims to 'spearhead' New Zealand's post-Covid recovery. *Stuff*. www.stuff.co.nz/business/122152018/horticulture-aims-to-spearhead-new-zealands-postCovid-recovery
Squire, S. (2020). Adapting wine sales in the time of pandemic. *Grapegrower and Winemaker*, 676: 81–83. https://winetitles.com.au/gwm/articles/may-676/adapting-wine-sales-in-the-time-of-pandemic/
Taunton, E. (2020, November 27). New Zealand wine exports hit record $2 billion. *Stuff*. https://www.stuff.co.nz/business/farming/123528554/new-zealand-wine-exports-hit-record-2-billion
Terrace Edge (2020). *Terrace Edge*. https://terraceedge.co.nz/
Tourism Industry Aotearoa (2020, December 22). New statistics show importance of tourism to NZ recovery. *Scoop Business*. https://www.scoop.co.nz/stories/BU2012/S00446/new-statistics-show-importance-of-tourism-to-nz-recovery.htm
Tourism New Zealand (2014a). Special Interest Tourism: Wine Tourism. http://www.tourismnewzealand.com/media/1132209/wine-tourism-profile.pdf
Tourism New Zealand (2014b, August 29). TNZ and NZ winegrowers sign MOU. https://www.tourismnewzealand.com/news/tnz-and-nz-winegrowers-sign-mou
Tourism New Zealand (2017, December 7). Wine tourism ripe for the picking. https://www.tourismnewzealand.com/news/wine-tourism-ripe-for-the-picking/
Van Dam, B., and Johnson, S. (2020, October 3). Plonk goes the wine industry: Sonoma's wineries embrace onlline sales and budget pricing to woo pandemic drinkers. *Edmonton Journal*.
Webb, B. (2020). Wine tour woes: tough times for wine tourism guides, *New Zealand Winegrower*, 124(October/November): 24.

PART VII

Terroir sustainability and cultural constructs in wine tourism

50

A SUSTAINABLE PERSPECTIVE FOR WINE TOURISM MANAGEMENT

Elena S. Rotarou and Humberto Rivas Ortega

Introduction

The Sustainable Development Goals (SDGs) are 17 goals that are guiding global development efforts during the period 2015–2030. Goal 8, which calls for the promotion of a 'sustained, inclusive and sustainable economic growth, full of productive employment and decent work for all,' includes Target 8.9, whose objective is that by 2030, policies are devised and implemented that would promote sustainable tourism that would create jobs and support local culture and products (United Nations Development Program 2020, n.p.). Resolution A/RES/69/233, adopted by the UN General Assembly on December 19, 2014, regarding the promotion of sustainable tourism, underlines the importance of sustainable tourism strategies and practices for the fight against poverty, the protection of the environment, the promotion of sustainable, rural development, the creation of decent jobs, the promotion of education, the protection of natural and cultural heritage, and the inclusion of women, persons with disabilities, the youth, and older persons in tourism initiatives (United Nations 2015).

Enotourism, oenotourism, or wine tourism has received increasing attention as a sustainable type of tourism that can bring together the primary sector (wine production) and the tertiary sector (tourism services) in a symbiotic relationship. At the Fourth Global Conference on Wine Tourism, hosted by the World Tourism Organization (UNWTO), which was held in December 2019 in the Colchagua Valley – home to some of Chile's most prominent wine producers and open vineyards – the role of enotourism for the promotion of the sustainability agenda was highlighted (UNWTO 2019). The conference underlined the wine tourism sector's ability to support rural communities, create jobs and entrepreneurship opportunities, and promote the regional economy through its linkages to handicrafts, gastronomy, and culture. It also stressed its importance for the socioeconomic development of remote destinations, which, by employing sustainability practices, can achieve economic growth, diversify themselves, and adjust to meet changing consumer demands (UNWTO 2019).

A wine tourism destination can be viewed as successful and sustainable if it manages to gain the loyalty of returning wine tourists and attract new tourists (Yuan et al. 2008) while at the same time promote and integrate local communities. This can be achieved through the adoption of sustainable business practices and fruitful collaboration among stakeholders: the

DOI: 10.4324/9781003143628-58

development of a comprehensive and sustainable tourism plan coupled with sustainable wine production practices can lead to the flourishing of the wine tourism destination.

This chapter explores the interrelationship between enotourism on the one hand, and local community, governance aspects, and sustainability issues on the other, as a key component for the successful development of a wine tourism destination. It employs the case of Chile to demonstrate the importance of these factors for the establishment of a sustainable enotourism sector. The chapter presents the importance of the balance between the economic profits of the tourism sector, the carrying capacity of the wine-producing region, the cooperation among stakeholders, and the existence of effective network development, and the involvement of local communities in the relevant economic activities. It also offers recommendations that can apply to regions and countries that are in the process of developing their wine tourism industry and may face challenges for the sustainable development of this sector.

Literature review

Enotourism has been defined as 'visitation to vineyards, wineries, wine festivals and wine shows for which grape wine tasting and/or experiencing the attributes of a grape wine region are the prime motivating factors for visitors' (Hall et al. 2000, p. 3). In recent years, wine tourism has been on the rise, with an increasing number of tourists traveling to wine-producing countries and visiting wine regions in search of more authentic experiences. It is currently viewed as a type of sustainable tourism development that can provide many benefits for the local, regional, and national economy. According to the World Tourism Organization (2016), enotourism can contribute to the promotion of sustainable tourism by supporting the destination's heritage, by generating substantial economic and social benefits for the actors and local communities involved, and by playing an important role in terms of cultural and natural resource preservation.

The concept of enotourism is based on the following six pillars (Figure 50.1):

1. *Wine culture*: it is the thematic axis of the product/service, which tourists should perceive during all stages of their trip and in any component of the wine tourism value chain.
2. *Tourism*: tourism and tourists are two key factors that clearly determine the demand and supply of the wine tourism product and should be at the center of any measure or development of the wine tourism offer.
3. *Territory*: it plays a very important role since it forms the base or frame of reference for the development of the ecotourism offer at the destination.
4. *Sustainability*: wine tourism should be developed under the principles of sustainable tourism and sustainable development.
5. *Authenticity*: authentic tourism experience is a factor highly valued by the experienced wine tourist, who is looking for authentic, traditional and/or innovative experiences.
6. *Competitiveness*: wine tourism destinations aim at increasing their competitive advantage through the adaptation of products and services to market changes and the maintenance and preservation of local traditions and heritage (Vintur 2005).

Several critical success factors (CSFs) have been suggested for the sustainable development of wine tourism that would promote local economic growth, preserve local traditions and heritage, and offer an authentic tourism experience while at the same time ensure the competitiveness of tourism destinations. Getz et al. (1999) grouped CSFs into quality (of wine, service, and experiences), wine country appeal, winery appeal, and developmental and marketing factors, with quality emerging as the most important success factor. Williams (2001) emphasized the

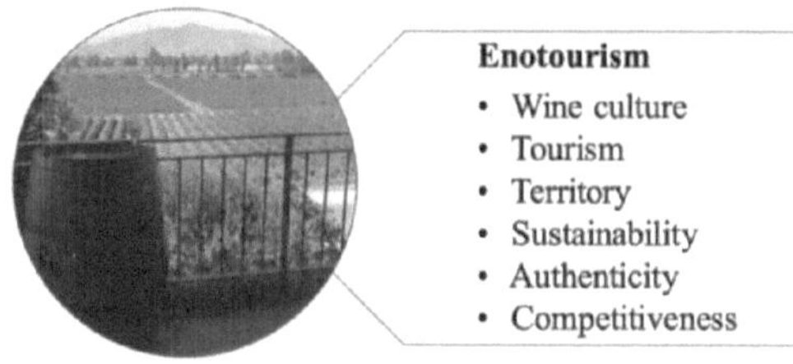

Figure 50.1 Six pillars of wine tourism.

key role of quality wine, coupled with natural landscape, culinary, educational, event hosting, and cultural factors. Other important CSFs include product consolidation and reduction in the number of intermediaries between production and consumption (Mora 2006), the existence of a unique product combined with well-trained staff (Dodd 1995), good customer service (Byrd et al. 2016), and branding and image enhancement (Getz and Brown 2006).

According to Poitras and Getz (2006, p. 426), sustainable wine tourism depends first on a general approach to tourism development that is based on the three pillars of economic, environmental and social sustainability as discussed by Hall (2000) regarding tourism planning, or the four pillars (political, economic, socio-cultural and ecological) proposed by Ritchie and Crouch (2003) in the context of competitive destinations.

On a similar note, Sigala (2008, p. 1590) adds that sustainability in tourism is a 'multi-sectoral (hotels, restaurant, transport companies, etc.) and a multi-disciplinary (economic, social and environmental) concept.'

As tourism is characterized by its dynamism and very rapid transformation in the short term (as can be seen in beach destinations, for example), the impact that tourist activity generates can alter territorial conditions, and thus deteriorate the destination (Poitras and Getz 2006). Table 50.1 presents key issues related to sustainability aspects of the enotourism sector.

Enotourism sustainability is intrinsically linked with environmental sustainability concerns, not least nowadays when factors such as climate change and urbanization have a direct impact on wine production, and therefore on tourism as well (Figueroa and Rotarou 2018). Conflicts regarding land and water usage and concerns over the economic profits of the tourism sector versus the carrying capacity of the wine-producing regions can lead to a decline in the quality of the wine produced, with direct negative effects for the tourism sector (Carlsen and Ali-Knight 2004; Montella 2017). Sustainable practices have been generally introduced in the wine-producing industry, such as energy-efficient technologies, biodiversity protection, organic production processes, and sustainable water usage and management. Such practices have been recently launched in the wine tourism industry as well – for example, energy-saving facilities, recycling, and the purchase of local food and other products. Nevertheless, large disparities in this aspect can be observed among different vineyards, wine-producing regions, and countries. Bearing in mind today's environmentally conscious tourist and consumer, the need for the inclusion of sustainable practices in both wine-producing and enotourism sectors is imperative.

Community involvement in the enotourism sector has been generally acknowledged as a key factor in promoting local economic growth, and avoiding conflicts stemming from the unequal distributions of gains from enotourism activities (Baker and Cameron 2008; Matarrita-Cascante, Brennan, and Luloff 2010). Sustainable enotourism is strongly linked to the culture and characteristics of the local community: a sustainable wine tourism sector needs to combine ethical and resource conservations aspects with the needs and culture of the local community (Lane 1994). The involvement of local community in the enotourism sector is considered an

Table 50.1 Sustainability issues in the enotourism sector

Scopes of sustainability	*Key issues*
Environmental sustainability	• Environmental impacts are associated with viticulture (grape growing), winery development, and wine production, services and infrastructure developed to support tourism, as well as to tourists' activities and travel patterns • Preservation of natural resources upon which sustainable wine tourism relies (soils, land for vineyards and wineries, water for irrigation) is paramount, as in many places there is high competition for these resources • Externalities of viticulture and wine production can be viewed as costs or detriment to the sustainability of other resource users
Economic sustainability	• Long-term sustainability of wine and wine tourism businesses, including the influence of market forces, and the nature and distribution of costs and benefits that are generated, need to be considered • In rural areas, the synergy created between wine production and wine tourism is fundamental for the increase in business profits • A long-term sustainable plan that takes into account business goals and community needs is vital for maintaining and increasing financial gains stemming from enotourism
Sociocultural sustainability	• Wine tourism contributes or should contribute to enhance the habits, traditions, and cultural heritage of rural areas • Employment opportunities are needed for local residents, but there are multiple issues to consider (necessity of training or retraining people for new service-sector and winery-specific positions) • In-migrants might take many of the new jobs, or the wineries might be small, family-operated businesses • Distribution of wealth might be an issue when outside capital dominates the wine industry and locals feel excluded • Displeased residents if low-paying service sector jobs are all that arise, especially if they are largely part-time and seasonal in nature

Source: Author elaboration with data from Poitras and Getz (2006); Tafel and Szolnoki (2020); Torres et al. (2021)

important element for the diversification of local – and, often, rural – economies, as it ensures greater conservation of natural and cultural resources and improves the socioeconomic well-being of local residents.

Another important aspect for a sustainable wine tourism industry is the introduction and maintenance of collaborative networks among the many stakeholders involved (Simpson and Bretherton 2004; Wargenau and Che 2006), inter-industry linkages (Cambourne et al. 2000), and collaboration between wine and tourism industries, and between public and private organizations (Getz et al. 1999). Strategic drivers, such as the need for competitive advantage, positive public image, differentiation, and cost-efficiency, have stressed even further the need for network development and cooperation. Enotourism clustering initiatives can greatly benefit from integrative and collaborative practices through horizontal integration that helps increase market share, vertical integration that facilitates the control of various stages of production, and diagonal integration that connects many related services (Karagiannis and Metaxas 2020).

While network development in the enotourism industry is crucial for the sustainable development of the sector, studies have shown that there are various obstacles in the elaboration and promotion of such networks; lack of communication, financial obligation issues, and long-standing previous arrangements between businesses have been named as some of the impediments to fruitful collaboration in the wine tourism sector (Schreiber 2004). For this reason, government support – at a national, regional, and local level – is crucial in order to achieve coordinated planning of the public and private investment, support sustainable practices, and promote tourism-destination and place-branding initiatives through the introduction and development of wine routes or trails. Overall, collaboration and value co-creation among the various stakeholders – both public and private – coupled with local community involvement and wine tourism clustering initiatives are required for the long-term sustainability of the wine tourism industry (Karagiannis and Metaxas 2020).

Case Study 50.1 Chile

In the last few decades, tourist destinations in Chile have experienced remarkable growth in visitor demand, which has contributed to boosting the economy of local communities. Since 2000, the government has implemented a process of diversification of the tourism sector by developing tourism products with greater territorial identity, which aim at: (a) the development of new and innovative tourism experiences that take into consideration cultural and identity aspects; (b) the encouragement of more Chileans to travel, get to know, and enjoy Chile; (c) the generation of greater added-value per tourist, through key actions that improve tourist offer and through the promotion of such services in priority international markets; and (d) the strengthening of sustainable development of destinations (Subsecretaría de Turismo 2015, p. 9).

In 2016, the Production Development Corporation - CORFO (the Chilean economic development agency) created the *Enoturismo Chile* program, by taking into account the objectives of the National Plan for Sustainable Tourism Development. This program aims at using Chile's comparative advantage as a wine country in order to improve its competitive position at a national and international level. A factor that has encouraged the development of wine tourism offer is the promotion of tourist wine routes, which have been heavily promoted by state agencies such as CORFO and ProChile, the latter being responsible for tourism promotion, and for micro and small business support. Currently, 80% of open vineyards are associated with a wine route, and 75% work with tourist operators (Universidad de Talca and Enoturismo Chile 2020).

As a result of these efforts, there are currently 146 open vineyards (see Table 50.2), an increase of 52% since 2006 (Universidad de Talca and Enoturismo Chile 2020). There are also nearly 8,000 companies that operate wine tourism services in the country's wine valleys, particularly operators and agencies, thematic restaurants, and hotels (Rojas 2019).

Table 50.2 Number of vineyards open to tourist activities (2006–2019)

Year	2006	2010	2012	2013	2016	2019
No. of vineyards	96	64	73	78	119	146

Source: Universidad de Talca and Enoturismo Chile (2020)

Note: The 2010 decrease in the number of open vineyards was due to the effects of the February 2010 earthquake.

The majority (72%) of the open vineyards have received state support, 45% of them have obtained a sustainability certificate, while 90% employ corporate social responsibility practices, including practices related to the environment, commitment to the community, and employment of local personnel and purchase of local products (Universidad de Talca and Enoturismo Chile 2020).

In 2019, the country's open vineyards registered 540,865 paid visits; 26% of international visits arrived trough the international airport of Santiago. The most important markets were Brazil (47.3%), North America (15,1%), and Europe (14.9%) (Subsecretaría de Turismo 2020). It should be mentioned here that one distinctive feature of enotourists in Chile is that they are primarily foreigners: only 25% of total enotourists are Chileans, making Chile one of the few wine-producing countries with many more foreign visitors to its vineyards than local visitors (Enoturismo Chile 2019).

Governance for wine tourism management

Historically, wine tourism management in Chile has been characterized by low cooperation between public actors at ministerial and municipal levels, and weak public–private networks (Enoturismo Chile 2017). Nevertheless, during the last few years, public–private partnerships have been strengthened. More particularly, the design of a public policy to promote tourism, led by CORFO, has been one of the factors that have contributed to the positioning of the wine-growing valleys of the central area as wine tourism destinations. In 2017 various actors at the regional and national level created a partnership in order to promote local enotourism development; these actors include the associations of grape producers from the central regions, the Federation of Tourism Companies (FEDETUR), the Chilean Gastronomy Association (ACHIGA), the Fundación Imagen de Chile, the National Training Institute, the Office of Agrarian Studies and Policies, CORFO, the Undersecretariat of Tourism, and the National Tourism Service. These public–private actors have developed a marketing strategy for wine tourism, incorporating the sustainability certification as a differentiating element in the wineries.

Integration of rural communities in the Chilean wine routes

While the lack of qualified and certified personnel among local residents is still an issue, rural communities have slowly been incorporated as complementary service providers on wine routes. The predominant model on these routes includes a high-end hotel in the surroundings and/or inside the vineyards, which integrates the traditional rural cuisine of the region with cultural activities that value local traditions. The experience of the most consolidated wine routes in the Maipo and Colchagua valleys or in emerging routes, such as the Cachapoal River valley, has shown that the development of a wine tourism destination involves the entire local community, not only rural inhabitants involved in providing direct services for tourists.

From a territorial perspective, this is an example of how a wine tourism destination is managed as a tourism product in its entirety. Responsibility for the growth in sales, income, and its competitive positioning in the international tourism market is the result of collective networks between entrepreneurs and the community.

The challenge of moving towards sustainability

For the Chilean Wine Association, sustainability is one of the four strategic pillars and is addressed through various projects of research, development and innovation, which are led by the Consortium I + D Vinos de Chile together with universities and national entities. One of the most emblematic projects in the Code of Sustainability of the Wine Industry, which guides wine companies towards working in a sustainable way, on the basis of fulfilling requirements in three complementary areas: vineyard, winery-bottling plant, and social area (ProChile 2017). To date, 76 vineyards have received this sustainability certificate; these vineyards account for more than 80% of exports of bottled wine (Sustentavid, n.d.). While sustainable practices have been widely adopted in wine production, they have generally not been introduced in the wine tourism sector as well; most local tourism operators and providers do not have much knowledge of sustainable practices (Subsecretaría de Turismo 2014).

Conclusion

As a result of the tremendous wine tourism increase, rural areas have experienced a cycle of economic development, which provides immediate economic benefits, complementary to traditional production, which has especially favored rural communities surrounding the vineyards. Various bottlenecks still need to be addressed – for example, infrastructure gaps and insufficient and unqualified personnel at the local level. However, the recent, integrated public–private effort between government agencies and wine producers from all valleys across the country has been decisive in positioning Chile as an attractive destination for wine tourism.

As a subsector of the service sector of the economy, enotourism could help the reorientation of the Chilean economy, which is heavily dependent on extractive and environmentally polluting industries, towards a more sustainable type of development (López and Figueroa 2016).

Recommendations

Sustainability: a determining factor for the competitiveness of wine tourism destinations

Wine tourism development in rural areas has contributed significantly to generating income opportunities for resident communities, integrating them into productive chains associated with natural resources, diversifying their economy, and generating new jobs. Although wine tourism offers many benefits for local communities, one of the greatest challenges consists in applying concrete actions that ensure that the demand pressure on the territory, especially in fragile areas with high landscape value, does not become in the long term a negative factor for the visitors themselves and for the communities as well, due to the eventual deterioration of its resources. The challenge of positioning a region as a sustainable wine tourism destination involves different stakeholders facing strategic decisions regarding production and access to markets, as well as introducing technological improvements and innovation aimed at promoting more sustainable forms of design, management, and commercialization of tourism activities (United Nations Environmental Programme and World Tourism Organization 2006).

Sustainable tourism should meet 12 basic objectives, which apply to all forms of tourism, including wine tourism (United Nations Environmental Programme and World Tourism Organization 2006). These objectives are economic viability, local prosperity, quality of employment, social justice, visitor satisfaction, local control, community welfare, cultural wealth, physical integrity, biological diversity, resource efficiency, and environmental purity. According to these objectives:

- The economic viability of tourism depends heavily on maintaining the quality of the local environment.

- Visitor satisfaction is fundamentally about meeting their needs and providing opportunities (a social objective), although it is also very important for economic sustainability.
- Cultural wealth is often considered within the social sphere of sustainability, but it has a strong relationship with environmental aspects in terms of the architectural setting and the cultural dimensions of society's interaction with nature.
- The well-being of the community, which can be considered mainly as a social objective, is strongly related to the management of environmental resources, for example, with regard to access to drinking water.
- The quality of employment and social equity problems, such as poverty reduction, are closely related to economic and social sustainability problems.

In order for the management of wine tourism destinations to move towards more sustainable development, the following types of instruments should be incorporated: measurement, control, economic, aid, and volunteers (United Nations Environmental Programme and World Tourism Organization 2006).

Regarding measurement instruments at the local level, systematic data on enotourism impacts are needed in order to establish acceptable limits and conditions of use to safeguard its competitiveness. The periodic record of sustainability indicators can help decision-makers to evaluate the impacts of the enotourism activity on the environment, employment, or the perception of rural residents. According to Fernández, Herrero, and Vidal's (2017) views on wine tourism in Spain,

> it is important to implement knowledge management actions that promote (i) research and analysis; (ii) the dissemination of knowledge through high-quality education and training programs, and (iii) the application of knowledge through the creation of regulatory and normative frameworks (voluntary or not) and innovation as a dynamic element of transversal applicability. These actions will establish the basis for coordination between the main actors from the public, private and civil sectors involved in the wine tourism value chain, around a shared vision, mission and strategy.
>
> *(p. 802)*

Concerning control instruments, ordinances and other regulations are required to guide the territorial organization of enotourism investments and activities, and to protect the land in natural areas that today are opening up rapidly to enotourism activities. According to Karagiannis and Metaxas (2020, p. 3), 'the pressure for sustainability practices, products and services starts from external drivers such as customer demands, investors, communities, government, and the need for compliance with regulations.'

With regards to economic instruments, in many wine-producing countries, the state has become heavily involved in financing enotourism companies and start-ups in the sector. However, it is necessary to expand this support at a regional level, and particularly in the case of protected areas, in order to promote local development more efficiently. This implies that territories with more disadvantaged conditions from an economic and environmental point of view require a more focused and coordinated investment by the state.

Regarding voluntary instruments, a series of sustainability seals and quality standards have been introduced that acknowledge sustainable business practices; such certifications have been received with increasing enthusiasm by the private sector, as they are seen as a means for the long-term sustainable development of wine tourism businesses, as well as a matter of corporate responsibility that gives a certain guarantee for the environmentally conscious tourist. For example, the Global Code of Ethics for Tourism – which in practice is a voluntary agreement signed by the international community – constitutes a reference framework that could be employed by enotourism destinations.

Nowadays, it is not enough for a tourist destination to offer good food or accommodation services, and that local companies are efficient from a business point of view. It is important that each decision to promote wine tourism is associated with a measure to support the strengthening of the local community's human capital, the provision of eco-efficient equipment and facilities, the protection of cultural heritage, and the protection of ecosystems.

The role of municipalities in wine tourism management

Communal territory constitutes the basic management unit for local tourism development. The challenge of the territory positioning itself as a sustainable wine tourism destination involves different stakeholders at the local level. These stakeholders need to make strategic decisions regarding production and access to markets, as well as to introduce technological improvements and innovation aimed at promoting more sustainable forms of design, management, and commercialization of tourism activities (Crouch 2007; Ejarque 2005; Ros Development and Planning 2008; United Nations Environmental Programme and World Tourism Organization 2006).

In a tourist destination, multiple actors, such as businesses, public organizations, and the local community, interact and establish a link with the visitor, which – even if it is circumstantial and brief – will have a decisive effect on their perception of the trip. For this reason, visitor satisfaction on a wine route is a shared responsibility. To achieve this goal, municipal leadership and commitment is key to disseminating and implementing public policy instruments aimed at strengthening local entrepreneurial capacities and helping to strengthen the competitive environment for more consolidated companies. New road and basic services infrastructure are needed in rural areas, together with the development of a portfolio of project ideas at the local level that would promote sustainable wine tourism. Likewise, local authorities need to implement promotion programs that integrate local communities into initiatives aimed at rescuing local customs and traditions, and to enhance the architectural heritage of rural areas.

Municipal tourism management in enotourism destination requires a comprehensive vision

The complexity and heterogeneity of the tourism sector and its subsectors is the first challenge that local actors face in order to implement a comprehensive management strategy in wine tourism destinations (Altés 1993; Pearce 2016; Dixit 2019; Pearce and Schanzel 2013; Zeithaml, Parasuraman, and Berry 1993). Although facilities, equipment, and supply services might be available, such services represent only a first step towards the provision of a certain experience – in this case, wine tourism – which is the final object of consumption.

Rural municipalities face several difficulties in their effort to address all the challenges of sustainable tourism development, especially wine tourism, which is a relatively new entrant into the tourism scene. Many municipalities do not have planning instruments that guide local initiatives in terms of promotion and development of wine tourism products in the long term, or their tourism promotion initiatives often do not respond to a comprehensive vision of the destination. They may have little basic information on the behavior of tourism demand, or access to periodic and updated information on the tourism market, including up-to-date qualitative and quantitative information. They may also have limited knowledge on the environmental, economic, or sociocultural impacts linked to the tourism development of their territory. Furthermore, they may experience personnel issues, for instance, inadequate qualifications of municipal professional tourism staff, or budgetary constraints, for example, low budgetary allocations for tourism action plans.

An important management aspect for municipal tourism authorities is the coordination with other areas of municipal management that are key to responding to visitor demand, such as the Departments of Cleaning, Transit, Health, Education, and Public Works and Environment (Arriagada 2002). On the other hand, municipal authorities need to implement interventions in various areas related to tourism management, such as: (a) institutional management; (b) tourist market; (c) territory and natural heritage; (d) sociocultural environment; (e) business management, entrepreneurship support, and economic impact of tourism; (f) integration of the local community; (g) infrastructure and equipment; (h) product development; (i) tourism promotion and marketing; and (j) innovation and technological development. Municipal authorities need to adopt a comprehensive approach to sustainable wine tourism development, since the quality of the enotourism experience is conditioned upon the complementarity of all services offered at the destination.

Coordination among stakeholders is a critical management factor

Coordination between municipal authorities and relevant stakeholders – regional and national government, vineyards, service providers, and local community – is also essential for the long-term sustainable development of a wine tourism destination. Communication, cooperation, and a shared vision of the challenges and opportunities that wine tourism in a particular wine destination entail is necessary in order to maximize the benefits for all stakeholders, and to provide a unique and differentiating experience to the enotourist.

The ability to respond to or satisfy consumer expectations is crucial for the competitiveness of wine tourism destinations, as it ensures competitive advantage. In order to achieve this goal, the following strategies, among others, should be considered:

- *Product specialization* allows for better satisfaction of tourist expectations, since it concentrates efforts on the differentiating services that the destination offers.
- *Diversification of the wine destination's service offer* complements the first strategy, as it provides a more ample range of tourism services, with the destination's particular seal.
- *Efficient articulation of related offers* allows for consolidation of the wine tourism destination, through the provision of a strong and attractive image.

Consequently, to ensure that the 'product/destination' supports the wine tourism experience, it is necessary to implement new management strategies at the local level, under an integrating approach and articulating decisions, led by the local authority and with the participation of local actors. From a functional and territorial perspective, the destination should be conceived and fully managed as a tourism product. The responsibility for such management cannot be delegated to the private sector; this is evident from the way in which the relationship between companies, tourism demand, and the territory develop. While for the individual company the maximization of profits is based on management's strategic actions that depend mainly on the company's corporate decisions, for a tourist destination, entrepreneurs' individual decisions do not make sense unless they respond to the vision defined by the different interest groups at the destination. In this context, a wide offer of resources and programs that support micro entrepreneurs can constitute an opportunity to form solid partnerships that allow for an efficient use of resources, better cooperation among actors, and effective capture of the market segments that local actors have identified.

Participation of local communities in wine tourism development

The development of a tourist destination requires a commitment from the entire local community, not just from members providing direct services to tourists. Therefore, the opportunity to work under a collective commitment approach should also be understood as a stimulating factor to strengthen the link between state agencies, businessmen, and the community itself. It is crucial that the benefits from wine tourism development are shared among all those who are directly or indirectly involved or affected. Promoting local participation in wine tourism development strategies, including hiring and training of local personnel, purchase of local goods, and tourism-related micro-funding opportunities, should be at the top of the agenda for wine tourism sustainability.

The competitiveness of wine tourism production chains depends on having personnel trained in matters as diverse as public service, gastronomy, elaboration of handicraft products, and guides who are specialized in wine tourism. New education and training programs in wine tourism-related topics, specifically directed at improving the knowledge and skills of local community members, can have a major positive impact with regards to community socioeconomic development and cohesion. A population enjoying the benefits of a particular economic activity – in this case, wine tourism – is much more likely to be an active supporter of sustainability practices.

Assuming sustainability as a permanent work strategy is the only option in order to address challenges stemming from tourism development, such as increase in visitor arrivals and its consequences on public space, as well as increasing pressures on the environment. The quality of employment (whether it is low-paid or seasonal) and the impact of tourism activities on poverty reduction are decisive factors in the acceptance of any wine tourism initiatives by the local community. Local community support should be fomented through the fruitful cooperation between public and private stakeholders that will demonstrate that the sustainable development of the wine tourism destination is a clear benefit for the local community.

Conclusion

Sustainability is not an attribute of a particular type of tourism: any process of tourism growth, both in emerging and consolidated destinations, should be managed with sustainability criteria. Taking into account sustainability aspects – environmental, economic, and sociocultural – in the development strategies for wine tourism is indispensable for all stakeholders involved: tourists, vineyards, related businesses, and local communities. For this reason, on the one hand, it is important to have precise information regarding available tourism infrastructures, such as accommodation, restaurants, tours, quality of roads, and transport, so as to be able to intervene efficiently in case of bottlenecks.

On the other hand, it is equally essential to look beyond infrastructure needs when trying to introduce a sustainable model of wine tourism. The expansion of the enotourism sector exerts increasing pressure on the environment, especially in fragile ecosystems; hence, the option of applying a work strategy associated with sustainable development objectives is the only alternative to address challenges, such as an increase in visitor arrivals, and the associated economic, social, and environmental consequences. Responding to the challenges of sustainability means that tourism destinations establish limits, for example, regarding the use of natural public spaces.

Tourism is an economic activity, dependent on the identity of the territory and, therefore, the natural and cultural resources (physical environment, rural landscape, historical and architectural heritage, folklore, local customs and values, among others) constitute the basis for its sustainability and competitiveness in the long term. The local community is a direct beneficiary of tourist activity, so it needs to be involved in the management of wine tourism activities. Therefore, their commitment, prior training, and involvement are factors that directly contribute to the sustainability of the tourism development process.

The private sector, on the other hand, represented by the providers of accommodations, food, transport, recreation services, and all those responsible for offering goods and services for the wine tourism industry, requires stable conditions for its operation, which implies safeguarding the basic conditions (economic, environmental, sociocultural, and managerial) that attract visitors. It is not possible to satisfy tourists' expectations without introducing quality and sustainability measures in each of the links of the value chain of the tourism industry. For this reason, the growth cycle of any wine tourism destination is conditioned by its ability to compete in market segments that demand sustainable goods and services, and by its capacity to offer unique and differentiating experiences, while showing commitment to all sustainability aspects.

Wine tourism is a niche type of tourism that addresses the expectations of a demanding market with high spending capacity but also with a higher exigency for a sustainability approach undertaken at the wine tourism destination. It can be argued that sustainability is by now THE critical success factor for wine tourism development.

Implications

All roads to economic development and well-being lead towards sustainability. Sustainability – economic, sociocultural, and environmental – is the key for the successful development of wine tourism. One can only think of the consequences if policies and practices in the wine tourism sector are unsustainable: decrease in visitors' satisfaction, decline in local residents' well-being, environmental degradation, and eventually economic stagnation and/or decline of the wine tourism destination.

In order to avoid this dire scenario, it is imperative that both emerging and consolidated enotourism destinations adopt sustainable business practices in wine-producing and wine tourism activities alike, develop strong networks among public and private stakeholders, as well as involve the local community in the process. Ensuring a more sustainable management of wine tourism destinations will be one of the key challenges to face in the new post-pandemic scenario.

References

Altés, C. (1993). *Marketing y Turismo*. Madrid: Editorial Síntesis.

Arriagada, R. (2002). *Diseño de un sistema de medición de desempeño para evaluar la gestión municipal: una propuesta metodológica*. https://repositorio.cepal.org/bitstream/handle/11362/5582/1/S0260445_es.pdf

Baker, M. J., and Cameron, E. (2008). Critical success factors in destination marketing. *Tourism and Hospitality Research*, 8(2): 79–97. https://doi.org/10.1057/thr.2008.9

Byrd, E. T., Canziani, B., Hsieh, Y. C., Debbage, K., and Sönmez, S. (2016). Wine tourism: motivating visitors through core and supplementary services. *Tourism Management*, 52: 19–29. https://doi.org/10.1016/j.tourman.2015.06.009

Cambourne, B., Macionis, N., Hall, C. M., and Sharples, L. (2000). The future of wine tourism. In Hall, C. M., Sharples, L., Cambourne, B., and Macionis, N. (eds.), *Wine Tourism around the World: Development, Management and Markets* (pp. 297–321). New York: Routledge.

Carlsen, J., and Ali-Knight, J. (2004). Managing wine tourism through demarketing: the case of Napa Valley, California. In Carlsen, J., and Charters, S. (eds.). *International Wine Tourism Research. Proceedings of the International Wine Tourism Conference, Margaret River, Western Australia*, May 2004, Perth, Australia: Vineyard Publications.

Crouch, G. (2007). *Modelling Destination Competitiveness. A Survey and Analysis of the Impact of Competitiveness Attributes*. https://sustain.pata.org/wp-content/uploads/2014/12/Crouch_modelDestnComp-v2.pdf

Dixit, S. K. (ed.). (2019). *The Routledge Handbook of Gastronomic Tourism*. Oxon: Routledge.

Dodd, T. H. (1995). Opportunities and pitfalls of tourism in a developing wine industry. *International Journal of Wine Marketing*, 7(1): 5–16. https://doi.org/10.1108/eb008636

Ejarque, J. (2005). *Destinos Turísticos de Éxito, Diseño, Creación, Gestión y Marketing*. España: Editorial Pirámide.

Enoturismo Chile. (2017). *Programa estratégico mesoregional de enoturismo sostentable*. http://www.chiletrans forma.cl/wpcontent/uploads/2017/09/ENOTURISMO-CHILE-CPIE-07-07-2017.pdf

Enoturismo Chile. (2019). *Perfil del enoturista internacional*. https://www.enoturismochile.cl/wp-content/uploads/2018/06/3.-Cristobal-ForttesPresentacion-Seminario-Enoturismo.pdf

Fernández, A., Herrero, L., and Vidal, J. (2017). Los organismos de gestión de destino en enoturismo: Casos internacionales de éxito. *Pasos*, 15(4): 793–805. https://doi.org/10.25145/j.pasos.2017.15.054

Figueroa, E., and Rotarou, E. S. (2018). Challenges and opportunities for the sustainable development of the wine tourism sector in Chile. *Journal of Wine Research*, 29(4): 243–264. https://doi.org/10.1080/09571264.2018.1532880

Getz, D., and Brown, G. (2006). Critical success factors for wine tourism regions: A demand analysis. *Tourism Management*, 27(1): 146–158. https://doi.org/10.1016/j.tourman.2004.08.002

Getz, D., Dowling, R., Carlsen, J., and Anderson, D. (1999). Critical success factors for wine tourism. *International Journal of Wine Marketing*, 11(3): 20–43. https://doi.org/10.1108/eb008698

Hall, C. M., Cambourne, B., Sharples, L., and Macionis, N. (2000). *Wine tourism around the world – development, management and markets*. Amsterdam: Elsevier.

Karagiannis, D., and Metaxas, T. (2020). Sustainable wine tourism development: case studies from the Greek region of Peloponnese. *Sustainability*, 12: 5223. https://doi.org/10.3390/su12125223

Lane, B. (1994). What is rural tourism? *Journal of Sustainable Tourism*, 2: 7–21. https://doi.org/10.1080/09669589409510680

López, R., and Figueroa, E. (2016). On the nexus between fiscal policy and sustainable development. *Sustainable Development*, 24: 201–219. https://doi.org/10.1002/sd.1622

Matarrita-Cascante, D., Brennan, M. A., and Luloff, A.E. (2010). Community agency and sustainable tourism development: the case of La Fortuna, Costa Rica. *Journal of Sustainable Tourism*, 18(6): 735–756. https://doi.org/10.1080/09669581003653526

Montella, M. M. (2017). Wine tourism and sustainability: a review. *Sustainability*, 9(1): 113. https://doi.org/10.3390/su9010113

Mora, P. (2006). Key factors of success in today's wine sector. *International Journal of Wine Marketing*, 18(2): 139–149. https://doi.org/10.1108/09547540610681112

Pearce, D. (2016). Destination management: plans and practitioners perspectives in New Zealand. *Tourism Planning & Development*, 13(1): 52–71. https://doi.org/10.1080/21568316.2015.1076511

Pearce, D., and Schänzel, H. (2013). Destination management: the tourists perspective. *Journal of Destination Marketing & Management*, 2(3): 137–145. https://doi.org/10.1016/j.jdmm.2013.06.001

Poitras, L., and Getz, D. (2006). Sustainable wine tourism: the host community perspective. *Journal of Sustainable Tourism*, 14: 425–448. https://doi.org/10.2167/jost587.0

ProChile. (2017) Sustentabilidad. *Práctica y compromiso*. https://www.prochile.gob.cl/wp-content/uploads/2016/09/Informe-Sustentabilidad-Vinos-de-Chile.pdf

Ritchie, J. R. B., and Crouch, G. I. (2003). *The Competitive Destination: A Sustainable Tourism Perspective*. Oxon, UK, and Cambridge, MA: CABI Publishing.

Rojas, G. A. (2019). *Estado del arte del turismo del vino a nivel global*. Santiago: Infyde – Vinifera.

Ros Development and Planning (2008). *Modelo de gestión turística local. Principios y prácticas*. http://femp.femp.es/files/566-1483-archivo/Modelo%20de%20Gesti%C3%B3n%20tur%C3%ADstica%20Local%20II.pdf

Schreiber, C. (2004). *The Development of Wine Tourism Networks: An Exploratory Case Study into Established Networks and Issues* [Unpublished dissertation]. Department of Tourism, University of Otago, Dunedin, New Zealand.

Sigala, M. (2008). A supply chain management approach for investigating the role of tour operators on sustainable tourism: the case of TUI. *Journal of Cleaner Production*, 16: 1589–1599. https://doi.org/10.1016/j.jclepro.2008.04.021

Simpson, K., and Bretherton, P. (2004). Co-operative business practices in the competitive leisure destination: lessons from the wine tourism industry in New Zealand. *Managing Leisure*, 9(2): 111–123. https://doi.org/10.1080/13606710410001709635

Subsecretaría de Turismo (2014). *Plan de acción: Formulación estratégica, marco de acción y gobernanza.* http://www.subturismo.gob.cl/wp-content/uploads/2014/03/Plan-de-Accion-Casablanca.pdf

Subsecretaría de Turismo (2015). *Plan nacional de desarrollo turístico sustentable.* http://www.subturismo.gob.cl/wp-content/uploads/2015/10/20150701-PLAN-DE-DESARROLLO-SUSTENTABLE_28_XPAG.pdf

Subsecretaría de Turismo (2020). *Cuadros turismo receptivo. Perfil del turista que ingresa vía aérea. Año 2019.* http://www.subturismo.gob.cl/turismo-receptivo/

Sustentavid (n.d.). *Código Nacional de Sustentabilidad.* https://www.sustentavid.org

Tafel, M., and Szolnoki, G. (2020). Estimating the economic impact of tourism in German wine regions. *International Journal of Tourism Research*, 22: 788–799. https://doi.org/10.1002/jtr.2380

Torres, J., Barrera, J., Kunc, M., and Charters, S. (2021). The dynamics of wine tourism adoption in Chile. *Journal of Business Research*, 127: 474–485. https://doi.org/10.1016/j.jbusres.2020.06.043

United Nations (2015). Resolution adopted by the General Assembly on 19 December 2014. https://undocs.org/A/RES/69/233

United Nations Development Programme (2020). *Sustainable Development Goals.* https://www.undp.org/content/undp/en/home/sustainable-development-goals.html

United Nations Environmental Programme and World Tourism Organization. (2006). *Making Tourism More Sustainable: A Guide for Policy Makers.* http://www.unep.fr/shared/publications/pdf/DTIx0592xPA-TourismPolicyEN.pdf

Universidad de Talca and Enoturismo Chile (2020). *Catastro nacional de enoturismo 2020.* Santiago: Universidad de Talca and Enoturismo Chile.

UNWTO (2019). *World's Biggest Conference on Wine Tourism Celebrates: Rural Transformations and Jobs.* https://www.unwto.org/worlds-biggest-conference-wine-tourism-celebrates-rural-transformation-jobs

VINTUR (2005). *European Enotourism Handbook – Vintur Project.* http://www.culturaydeporte.gob.es/dam/jcr:18c703e4-32f5-42f5-90c0-8b4eea8c5c53/enoturismo-ruta-del-vino-europea.pdf

Wargenau, A., and Che, D. (2006). Wine tourism development and marketing strategies in Southwest Michigan. *International Journal of Wine Marketing*, 18(1): 45–60. https://doi.org/10.1108/09547540610657678

Williams, P. (2001). Positioning wine tourism destinations: an image analysis. *International Journal of Wine Marketing*, 13(3): 42–58. https://doi.org/10.1108/eb008726

World Tourism Organization (2016). Georgia Declaration on wine tourism. Fostering sustainable tourism development through intangible cultural heritage. *UNWTO Declarations*, 26(2). Madrid: UNWTO. https://doi.org/10.18111/unwtodeclarations.2016.25.02

Yuan, J. J., Morrison, A. M., Cai, L. A., and Linton, S. (2008). A model of wine tourist behaviour: a festival approach. *International Journal of Tourism Research*, 10: 207–219. https://doi.org/10.1002/jtr.651

Zeithaml, V., Parasuraman, A., and Berry, L. (1993). *Calidad total en la gestión de servicios.* http://www.editdiazdesantos.com/libros/zeithaml-valarie-a-calidad-total-en-la-gestion-de-servicios-L03000610801.html

51

THE FLUIDITY OF TERROIR

The delusion of sustainability and authenticity in wine tourism

Willy Legrand, Clare Hindley, Roland Vogels, and Andrea Levis

Introduction

The concept of terroir is deeply anchored into the physical and cultural space of a particular region (Barham 2003). Even small parcels of land within that region (in the case of winemaking) potentially each have a unique terroir linked to the soil, climate, topography, and winemaking tradition, to name a few key terroir components (Bonfante et al. 2011). While the meaning of terroir differs across cultures (Legrand et al. 2019), the concept is based on static views such as geographical identity or 'sense of place' (Jones 2013). The idea of 'fluidity' in terroir results from a series of micro and macro influences pertaining in particular to the effect of wine tourism (micro) and the Anthropocene (macro) on terroir. Wine tourism induces micro forces onto the terroir concept due to, most notably, consumer demands and previous product knowledge or perception of wine as a tourism 'merchandise.' The Anthropocene induces macro forces, in terms of climate change and biodiversity loss, affecting a vineyard and its terroir components. At the same time, wine tourism is linked to numerous anthropogenic impacts resulting, for example, from increased traffic on rural roads and the development of facilities in agricultural areas to accommodate tourists (Poitras and Getz 2006). This results in a series of feedback loops with a cyclical reaction chain impacting the very concept of terroir. This chapter, conceptual in its approach, discusses these complex relationships and provides a theoretical model of 'terroir fluidity' based on feedback loops. We use case studies from the wine regions of Tuscany, Italy and the Rhine Valley, Germany, to exemplify the discussion and raise questions concerning the contradictory impulses effecting wine tourism in both production and consumer perception. We examine the call for authenticity and local in consumer purchase, the consumer and producer attitude to sustainability, tourist motivation, and the discrepancies which have emerged through macro and micro changes in the interpretation of these concepts and the phenomena of terroir fluidity.

Terroir: a sense of place

The impact of globalization on the world of wine business is felt every time consumers enter a supermarket. The wine *aficionado* living in Bordeaux has access to the best wines from the Yara Valley (Australia), while the wine buff in Melbourne can sip the best from Ningxia

DOI: 10.4324/9781003143628-59

(China) wines. Wine is in the top 100 (98th position) most traded products globally, with a total trade value of US$37 billion in 2018 (OEC n.d.). There may be a temptation to define and develop a so-called 'international' style of wine, thereby forfeiting local origins to go hand in hand with the ever-growing world of multinational wine producers, flying winemakers, and global wine critics. Another side of the coin exists, however: the often small, single-estate wineries boasting wines with an individual character driven by their terroir. This is supported by an increasing interest toward authenticity, honest, and often local products by consumers around the globe. This movement has been evident for more than a decade, where consumers view 'local food as fresh, healthy, environmentally friendly and economically sustainable for rural communities' (Legrand et al. 2019, p. 30). The movement received significantly increased attention throughout the pandemic year of 2020, where consumers, under a forced travel pause, favored local food. A 2020 survey in the UK found 35% of consumers to be buying more local food since the pandemic (Ewing-Chow 2020). The consumer increasingly links 'local' as meaning higher in quality with rapidly increasing demand in many countries (Shveda 2021; Specialityfood 2020). The concept of terroir is invariably linked to geography (e.g., local) (Barham 2003), but it is a complex term which may include aspects such as 'specific soil, topography, climate, landscape characteristics and biodiversity features' (Bonfante et al. 2011, p. 103) and also includes history such as agricultural traditions and practices. In the minds of wine consumers, terroir, on the same par as food tourism, is a component of authenticity, providing the end product with a unique personality (Sloan, Legrand, and Hindley 2015). Wine critics understand terroir as a key driver of quality, too (Robinson 2018). In the wine world, discussing quality without geography and geology is similar to talking about tourist destinations without landscape and attractions. Terroir, closely related to the word 'territory,' designates a 'sense of place.' Three components form this sense of place: (1) geographical distinctiveness, (2) authenticity, and (3) ancestral skill and practices (Legrand et al. 2019). At first sight, one may assume those components to be more or less static in time. However, we argue the contrary. Terroir is a dynamic or fluid concept mostly due to a series of macro influences such as climate change and biodiversity loss as well as micro influences such as wine tourism and infrastructure development. We, therefore, speak of 'terroir fluidity.'

Case Study 51.1 Wine tourism in Tuscany: authenticity and sense of place

Tuscany is a major wine tourism region for both domestic and international visitors and is Italy's prime wine destination combining the elements that tourists prize most: 'spectacular landscape, high culture, excellent food, and great wines' (Colombini 2015, p. 32). This case study is based on in-depth interviews carried out in March 2020 with professionals in the wine and gastronomy tourism industry in Tuscany, focused on the role of wine and gastronomic tourism in the region. The four interviewees are two Michelin-star restaurant owners, one of whom is also the president of the regional sommelier association, a luxury hotel general manager, and a winery chief executive officer (CEO).

The winery CEO stresses the role of authenticity in the tourism experience: 'We want to produce wines that would give the authentic feel to people looking for Tuscan wines.

People come to us expecting that Tuscan experience, and we try to give them the best experience we could possibly deliver. When people look for authenticity and get it, they'll be happy, and they could tell others about us or come back and purchase from us again.'

The interviewees agree that many tourists travel to Tuscany with a positive image of Italian wine and gastronomy, yet little local knowledge of Tuscan specialties. The 'authentic' is key to their perception, and it is the local wine and food specialties which are often crucial to fulfilling this need. It is these local specialties that will attract the repeat visitor. With reference to consumers' knowledge of a region or location, the sommelier association president stated: 'What people know about a certain location can be important, but what can be shown to them once they get here may be even more important.' He added: 'Authenticity is what all customers are looking for. Without authenticity there is no sense of culture and no sense of feeling at home.' This sense of 'feeling at home' with components such as authenticity and ancestral skills and practices (Legrand et al. 2019) are key components of the 'sense of place' which terroir provides. One of the restaurant owners points out: 'to me authenticity is everything.' The interviewees further argue the importance of authenticity in Tuscan wine and food tourism with economic data showing that food and gastronomy, along with agriculture, are the main employers in the rural communities. As such, tourists' desire for the 'authentic' and 'local' experience is mostly supported by local employees and community with knowledge of the region as summarized by one interviewee: 'wineries in this region contribute a great amount to the local economy. I believe that without the wineries and vineyards, the local economy would be non-existent.' These attitudes clearly represent the arguments and evidence from the literature (Sloan, Legrand, and Hindley 2015; Csurgó, Smith, and Hindley 2019). One of the Michelin-star restaurant owners states:

> Authenticity is probably the most important part of Italian food. Everyone expects Italian food to be super authentic when they come here, and people expect an experience that they've never had before. Authenticity is what all customers are looking for. Without authenticity, there is no sense of culture and no sense of feeling at home. Authenticity should be the most important part of the culinary arts. There is no authenticity in fast food or junk food, but in real home cooked food such as ours, that is the real authentic flavours that people seek in life.

The interviews highlight, as essential in their products' success, the combined impact of a sense of place, authenticity, traditional techniques, history, and geographical location.

Macro influences

In the world of viti- and viniculture, a series of criteria are considered to explain the quality of a wine. As mentioned, the terroir, including geological, geographical, and climate factors as well as winemaking traditions, grape variety, maturity of the fruit, winemaking techniques, and quality of aging all play intricate roles in defining the end product. Climate change and biodiversity loss are two of the greater external macro forces altering the winemaking industry.

Climate change

First and foremost, terroir is linked to the soil. The vineyard location is selected according to the quality of the soil. While a detailed review of soil properties and influence on vines and grape quality is beyond the scope of this chapter, as a generic statement from Lanyon, Cass, and Hansen (2004), who conducted an in-depth literature review on the topic, the 'effects of soil type on wine quality [are] attributed to differences in mineral, thermal and physical properties of the soil' ((2004), p. 1). The intrinsic soil qualities are a cornerstone to the concept of terroir. Equally important are geographical and climate factors such as, for example, the vineyard altitude, daily weather patterns, the seasonal changes, the amount of sunshine, or the water supply (van Leeuwen 2010; Willwerth, Reynolds, and Lesschaeve 2010). The anthropogenic climate change results in three factors impacting viticulture: (1) increase in temperature, (2) increase in the length and intensity of water deficits in the vineyard, and (3) an increase in sun radiation (van Leeuwen and Darriet 2016). The consequence, on a global scale, has been a change in the vegetative cycle of the vine with a faster, hence earlier, ripening season (van Leeuwen and Darriet 2016). While a multitude of factors affect yield performance, in terms of quantity of grape harvested, lower yields have been recorded, due to the combined factors of length and intensity of water stress in the vineyards (van Leeuwen and Darriet 2016). Frequent and unpredictable weather pattern changes may also equal more humidity in certain parts of the world, resulting in fruit rot. On a global scale, vines are cultivated 'between latitudes 4° and 51° in the Northern Hemisphere and between 6° and 45° in the Southern Hemisphere' (Santos et al. 2020). Taking into consideration climate change, traditional winemaking regions of Southern Europe will see a decline in production and require intensive irrigation to continue producing, while Central and Northern Europe will see new areas suitable for viticulture due to an increase in temperature (Santos et al. 2020). In fact, detailed modeling undertaken by researchers suggest 'an extension of the climatic suitability for grapevine cultivation up north to the 55°N parallel, thus creating the necessary conditions for the emergence of new winemaking regions at higher latitudes in Europe' (Santos et al. 2020, p. 9). The 55°N parallel is roughly where Copenhagen in Denmark is situated. The situation is similar across various parts of the world. Therefore, some regions may benefit from changes in climate 'due to higher fruit maturity and opening new areas for cultivation, whereas the impact might be detrimental in other regions by challenging the ability for adequate grape cultivation and wine production' (Santos et al. 2020, p. 8). Uncontrollable climate parameters add to the complex mix of factors affecting the terroir outside the direct control of winemakers in the search to create wines with a unique personality.

Biodiversity loss

Biodiversity is considered a critical regulator in agricultural functions with the many ecosystem services provided (Stefanucci et al. 2018). Biodiversity encompasses all living organisms in terms of variety, species, and diversity existing and interacting in an ecosystem (DeLong 1996). Ecosystem services are the multiple benefits humans derive from natural resources (i.e., food, water) and processes (i.e., water purification, biological controls, crop pollination) (Stefanucci et al. 2018). In viticulture, farming practices can lead to either a deterioration of the vineyard biodiversity (e.g., excess use of fertilizers or herbicides) or an enhancement of biodiversity (e.g., greening the rows between vines). According to Stefanucci et al. (2018) in a detailed review of the interconnections and interaction between vineyard agricultural practices and biodiversity, 'there is compelling evidence that wildlife-friendly farming practices, aimed at reducing the negative impacts of intensive agriculture by implementing conservation actions in farmed landscapes, can be effective in conserving and restoring biodiversity' ((2018), p. 7). The services provided by biodiversity are multifaced and include 'improvement of soil fertility, increasing organic matter, improvement of soil structure, storage of carbon, management of undesirable organisms (conservation biological control) and regulation of hydrological cycle and microclimate' (Stefanucci et al. 2018). On a global scale, however, biodiversity is under attack, and scientists qualify the situation as a biodiversity collapse (de Oliveira et al. 2018). One of the strong drivers of biodiversity loss is farming practices which affect natural habitat. The world of viticulture is characterized by mechanization, use of inorganic fertilizers, pesticides, and large monocultures with limited space for a diverse natural and rich habitat (Porto Protocol 2020). The consequence for winemakers is an increased use of insecticides to combat the various pests which are left without natural enemies. The artificial intervention is a self-defeating cycle as increased input is required over time to ensure a productive vineyard. However, winemakers can actively participate in the restoration of biodiversity, boosting vineyard resilience to changing climate situation.

Case Study 51.2 Wine tourism in Tuscany: sustainability as a component of authenticity

All four interviewees mention sustainability as an essential part of their perception of authenticity. In the winery, sustainability is seen as an important part of the image and of the consumer search for authenticity:

> There cannot be authenticity without sustainability. We need to respect nature by knowing how to pick food, when to do it and how to preserve our land to produce again, in its life circle. Spontaneous, biodiversity, plant chemical free, water sustainably, all matters that count to save and preserve our world today.

As one interviewee states: 'Sustainability is quite important to gastronomy because if someone is sustainable, you could be saving money or energy or even the world!' The winery CEO argues:

> Sustainability for us is important; we must involve sustainable practices in everyday operations. We use less water than most crops, we have energy efficient pumps, alternative energy such as solar power energy, and we use chickens from time to time as natural pesticides. I believe sustainability is also important just for the future of gastronomy as well, with rising temperatures it will affect crops and vineyards.

These quotations show an awareness of the issues facing wine production in Tuscany from climate change to biodiversity loss, as discussed earlier, and correspondingly the issues which consumers may perceive as defining their call for or interest in sustainable and authentic products. In answering how consumers perceive sustainability in food and wine, the hotel owner states, 'Gastronomy is destination. Food and wine are the first ambassador of any geological area and as such they need to be cared for and presented, to send a message out.'

Micro influences

Besides the macro influences, a set of micro influences affect the vineyard environment and its terroir, most notably the human impacts. The winemaker's decision on the types of grapes to be cultivated, the methods used in viticulture, the tools and technology implemented to obtain the best fruit, when to plant, when to prune, how to protect, and when to harvest all have an impact on the way a vineyard looks and feels. One can talk about the art and science of 'growing a terroir' which leads to a unique product with all its own specificities and personality. This is where consumer demands play a role in many of the decisions taken by winemakers, which ultimately influence the terroir.

Consumer demands

Wine is an extremely popular consumer good with an expected growth of up to 9.2% annually in the coming five years (Statista 2020). The average price per liter paid by end-consumers amounts to US$13.48, tendency growing (DWI 2017). This demand is sustained by a strong and affluent base of consumers across many parts of the world. Countries with non-centralized sales systems in place experience a pricing situation which is quite different. German supermarkets reported a 5-cent decrease per liter sold in 2016, averaging a price of €2.92 per liter (DWI 2017). Interestingly, while terroir transcends a sense of place, a significant increase in production has been visible in what are considered international grape varieties, such as Cabernet Sauvignon, Sauvignon Blanc, and Chardonnay. Those can be found scattered across the world, from Hebei in China to Napa Valley in California. This evolution is, in parts, taking place at the cost of more regional varieties such as Airén and Ugni Blanc (IOVW 2017). Some varieties are very flexible in terms of soil and climate requirements, making them attractive to many winemakers. Chardonnay is a primary example: the taste ranges from easy to drink to highly complex; the prices from cheap to expensive; and it is a grape that can adapt to cold and hot climate regions. The Chardonnay variety is an all-rounder and, arguably, fit to every taste. Along the same lines, the market has become more transparent, and the access to wine from all regions around the world is in reach for the end-consumer. With internet assistance, it is possible to access a wide range of information about the once-exotic regions and ultimately purchase those at the local shop – terroir from all over the world at one's reach. Much of the 'Old World,' mainly European winemakers, traditionally focus on the representation of terroir in the final product. The 'New World,' innovating and benefitting with improved cellar techniques, develop end products centralizing or focusing on the grape itself (Robinson 2006). As a ripple effect, this method has in turn affected the modern winemakers in Europe. Conscious that consumers may be neither interested nor willing for a wine to age years to perfection before consuming it, winemakers also adapted and experimented with viti- and viniculture practices.

As discussed, however, terroir equates to a sense of place and, in a world in search of authenticity, a point of value to consumers. Traditionalists, those winemakers betting on the value of terroir rather than the grape only, were one step ahead in satisfying this ever-increasing demand for local products that have a traceable origin or a *sense of place*. For example, saline and mineral notes are the invisible advocates and reflection of terroir, as one would identify drinking a Mosel Riesling from a slate soil, for example (Deneulin, Le Fur, and Bavaud 2016). Understanding these tasting or aroma aspects (e.g., minerality) resonated throughout much of the modern wine world, but is understood by few in terms of the complexity behind terroir and influence on the final product. This situation provided an opportunity to replicate the taste artificially, even though some grape varieties lack the ability to show a mineral note because of natural fruity character, for example. The question remains: To what extent should the end product reflect the wants and needs of the consumer versus what the terroir is capable of producing?

Wine: augmented tourism experience

Since the beginning of this century, wine tourism has increased in popularity as an alternative to mass tourism (Afonso et al. 2018). It is thought that the smaller scale of this form of tourism invites guests to experience a lifestyle vacation in combination with a luxury product that is linked to social status. Several motivational factors rotate around hedonistic desires, such as special food and beverage experiences, but also increase the knowledge about a local culture and nature. The wine offers an augmented experience based on its complex and sophisticated nature. The increasing numbers of tourists and supporting infrastructure requirements have caused a pressure on viticultural areas such as Napa Valley, the Okanagan Valley, Porto, and Bordeaux (Kerr 2020). Wine tourism was perceived to be advantageous for these regions until the scale became such that management of tourist flows and infrastructure became problematic. Wine regions are agricultural areas. Any further commercial development impacts the natural environment and those living and working in the winemaking business. In Germany, vineyards attract 50 million annual tourists, but only 10% are international visitors (DWI 2019). Local travel is popular, and this is exemplified with tourist' desire to be close to the production and development of wine, as a lifestyle product, and to enjoy it in authentic surroundings.

Wine tourism is often a part of an extended vacation with another purpose (i.e., leisure travel along a river) and thus as an add-on beside the main trip motivation (Carlsen and Charters 2007). To also attract visitors who are interested in a wine-related experience, adequate infrastructure, such as roads for transport or restaurants for catering needs, is imperative. So, small, relatively unknown regions boasting a unique terroir can benefit from an increased tourism attractiveness with a second income stream, besides winemaking. This wine tourism development infrastructure, in turn, has a series of impacts on biodiversity of the region and is inevitably linked to anthropogenic climate change, the same changes that affect the uniqueness of the terroir, as discussed previously. In a series of feedback loops, terroirs are being altered at an ever-increasing rate.

Case Study 51.3 Wine tourism in Tuscany: tourist motivation

The interviews revealed contradictory opinions as to how much wine and gastronomy provide principal motivations to visit Tuscany. The owner and general manager of a Michelin-star restaurant stressed that wine and gastronomy are certainly an attraction for tourists, but fundamentally, they do not generally represent the main travel motivation for the majority. This is supported by the winery CEO: 'I believe that it can affect an individual's motivation to travel but it's not the most important thing. … I wouldn't say it has the greatest effect on the motivation to travel somewhere.' He points out that 'beautiful surroundings' of course influence someone's decision of which wine and/or gastronomic region to visit. This clearly corresponds to literature analyzing gastronomic and wine tourism (Baker and Kim 2019; Carlsen and Charters 2007; Dixit 2020). It is naturally important to remember that the highly popular tourist destinations of Florence and Pisa are in Tuscany, and many tours provide a visit to a winery as an attractive add-on.

The second restaurant owner states, 'I know many people that come to Italy just to try the wine,' and the winery CEO adds that food and wine do play a role 'the reason many people come to Tuscany is for the food and wine and the experiences.' Here the 'local' (Csurgó, Hindley, and Smith 2019) plays a role as Tuscany is known to be 'the land of Italy's wine,' and this image brings lots of clients looking for what they perceive as the best wine in the land. It is highly unlikely that tourists will visit Tuscany unaware of its reputation as a wine region.

The hotel owner states:

> Local approach, genuine food, sustainability and local produces are top factors, leading clients' decisions. I would say that after the destination, food would play a major role in anyone decisions to select a specific location. People are looking for memories, when returning home form an authentic journey and food is the second memory generator, after staff friendliness.

The interviews clearly show that although travel motivation to visit Tuscany includes world-famous sites, the wine and gastronomy, despite not being the only or main reason to visit, do play a significant role. As stated by the hotel owner, often small towns where hotels or wineries are located are unknown to visitors, and the wine and gastronomy play a large role in visitor motivation. Correspondingly, wineries and other tourist based enterprises are developing their offers with additional guest houses, tours, and even helicopter trips. However, despite these developments, the destination, Tuscany, is perceived as static, and it is this constant which attracts the gastronomic and wine tourist searching for their authentic and memorable experience.

Feedback loops

As discussed in the previous sections, climate change and biodiversity loss are phenomena affecting vineyards and winemaking across many regions of the globe. There are a series of factors, in turn, which affects climate change and biodiversity loss split into three categories: (1) forcings, (2) tipping points, and (3) feedback loops (Lenton et al. 2008; Marques 2020; NASA 2021). Forcings are the direct influences on climate, such as the sun's radiation or greenhouse gas emissions such as carbon dioxide emissions from burning fossil fuel, for example (NASA 2021). Climate tipping points are considered changes in terms of crossing thresholds, 'implying some degree of irreversibility' (Lenton et al. 2008, p. 1786), such as ice sheet loss impacting ocean circulation and contributing to sea-level rise, for example (NASA 2021). The last of the three factors influencing climate change are feedback loops. Feedback loops, already present in ecosystems, are similar to a cyclical chain reaction. Feedback loops are considered positive when increasing the effects of a phenomenon or negative when reducing the imbalance (NASA, 2021). For example, the ice sheets are white and reflective, while oceans are dark and absorb heat from the atmosphere. As temperatures rise, more ice melts. The oceans capture that heat due to their dark surfaces, causing yet more ice to melt – a chain reaction. A similar situation can be observed with forests and forest fires, for example. In the vineyard environment, macro-feedback loops are in action where the outputs are influencing the inputs, which

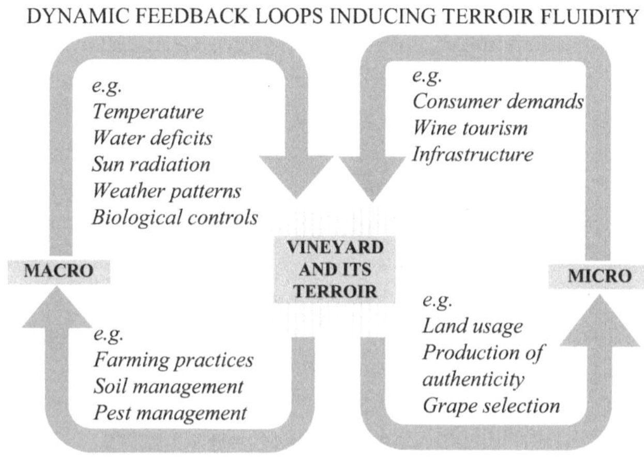

Figure 51.1 Feedback loops and terroir fluidity.

in turn influence the outputs (see Figure 51.1). Additionally, micro-feedback loops are also present, impacting the vineyard and its terroir in a continuous cyclical path (see Figure 51.1). The consequence of feedback loops are changes in the composition of the terroir of a particular vineyard and ultimately the end product, being the wine (and food). Wine tourists traveling to Tuscany in three to four generations from now are likely to experience a very different terroir from what one does today. As Bernetti et al. (2012), in a detailed assessment of climate change adaptation strategies for the Tuscan wine sector, summarized:

> the toughest challenge in the adaptation process will be how farmers respond 'culturally' to changes in the 'identity' of certain regions caused by the mutation of the ampelographic base (identification grapevines) and of quality characteristics (both technical and functional) of the wines
>
> *(2012, p. 74)*

Case Study 51.4 Wine tourism at the Drachenfels: feedback loops in action

The following case study highlights the Drachenfels wine production area in the Rhine Valley near Bonn, Germany. The Drachenfels area is a popular day tourism destination in the Rhine Valley. The wine production area includes several small wineries along the Rhine River, some of which have been producing wine over several centuries. Historically the wines were consumed locally, but for over a hundred years, both domestic and international tourism has become more important and adds to the economy. The region is popular for river trips on the Rhine, hiking, and in combination with visits to Bonn and Cologne.

Initially, the region offered wines made from the grapes suited to the area, which performed well in the relatively cool climate. The grapes reacted well to the local soil, the stony Trachyte, and produced wine of at least medium quality to the taste of a broad group of wine lovers. Many wineries have been producing wines in this manner for generations but started to experience difficulties during the past decades. Changes in weather patterns and more extreme weather events caused extensive damages during the winter season when the river would flood more frequently, and the easily accessible flat vineyards had to be abandoned. The summer heatwaves, more intense and of longer duration, impacted the overall grape yield with a notable decrease in the harvest from the steep hillside′s grape varieties. In order to cover the deficits associated to a lower output, some wineries decided to increase the price of their wine bottles. As a chain reaction, local inhabitants started to increasingly purchase cheaper wines produced elsewhere, either in different German wine regions or from other countries. This drop in local consumption has partly been compensated by an increase in wine tourism, with travelers looking for the authentic local wines. In a sort of ironic outcome, the tourists - and not the local consumers - are the ones keeping the winemaking business alive. Other wineries were not so successful in their attempts to adapt to changes in climate affecting the vineyards' output, and several have closed. This is also in part due to a change in consumer demands. A local businessman, seeing an opportunity, bought one of the wineries and replanted the vineyards with internationally popular and widely grown grape varieties to meet the needs of the more inexperienced palate. The winery is seeing an increased number of tourists who sample and purchase these wines. The feedback loops are in full swing at the Drachenfels, ultimately transforming the terroir.

Conclusion

This chapter has shown how the search for authenticity is a key component of wine tourism, yet the concept itself, which inherently includes a static and tradition-based quality, is bound to become more complicated in the years to come. Similarly, the macro and micro impacts on sustainability, and thus terroir, raise a question as to how long both consumer and producer understandings of the concept can remain. Whether terroir fluidity is desired or unwelcomed is irrelevant to the extent that it is undeniably already happening.

Implications

The implications for winemakers, being at the forefront of the impacts of climate change, are multifaceted. Winemakers are faced with complex daily operational decisions such as when best to prune or harvest, how long the alcoholic fermentation should take place, or how to build the best cuvée possible. The Anthropocene is making long-term, strategic decisions very complex for winemakers. Changes in climate are felt with increased unexcepted weather patterns. Considering the slow development of vines from seedlings to full-grown producing vines, a winemaker may consider investing in drought-resistant plants better suited for vineyards under water stress during summer months, for example. This would be considered an adaptation strategy to climate change. Of course, these new grape varieties alter the composition of the terroir. Authenticity, in this case, becomes a matter of interpretation. Consequently, winemakers are encouraged to undertake educational marketing towards consumers (locals or wine tourists). Beyond winemaking or wine tasting, communication should also include a description of the impacts of climate change and biodiversity collapse on the vineyards and surrounding environment. Mitigation and adaptation strategies set in place by winemakers should be equally highlighted. It is equally a matter of raising awareness about sustainability issues in winemaking amongst consumers as it is keeping the art of winemaking in many regions alive.

References

Afonso, C., Silva, G. M., Gonçalves, H. M., and Duarte, M. (2018). The role of motivations and involvement in wine tourists' intention to return: SEM and fsQCA findings. *Journal of Business Research*, 89: 313–321.

Baker, M. A., and Kim. K. (2019). Heritage and authenticity in gastronomic tourism. In Dixit, S. K. (ed.), *The Routledge Handbook of Gastronomic Tourism* (pp. 252–259). London: Routledge.

Barham, E. (2003). Translating terroir: the global challenge of French AOC labeling. *Journal of Rural Studies*, 19: 127–138. https://doi.org/10.1016/S0743-0167(02)00052-9

Bernetti, I., Menghini, S., Marinelli, N., Sacchelli, S., and Sottini, V. A. (2012). Assessment of climate change impact on viticulture: economic evaluations and adaptation strategies analysis for the Tuscan win sector. *Wine Economics and Policy*, 1: 73–86.

Bonfante, A., Basile, A., Langella, G., Manna, P., and Terribile, F. (2011). A physically oriented approach to analysis and mapping of terroirs. *Geoderma*, 167: 103–117. https://doi.org/10.1016/j.geoderma.2011.08.004

Carlsen, J., and Charters, S. (2007). *Global Wine Tourism: Research, Management and Marketing*. Wallingford: CABI.

Colombini, D. (2015). Wine tourism in Italy. *International Journal of Wine Research*, 7: 29–35. https://doi.org/10.2147/IJWR.S82688

Csurgó, B., Hindley, C., and Smith, M. K. (2019). Imagined idylls and nostalgic idealization: gastronomic tourism in rural Hungary. *Journal of Gastronomy and Tourism*, 4: 13–28. https://doi.org/10.3727/216929719X15657857907761

de Oliveira, F., Menezes, J. F. S., Northfield, T., Ochoa-Quintero, J. M., Campbell, M. J., and Laurence, W. F. (2018). Warning signals of biodiversity collapse across gradients of tropical forest loss. *Scientific Report*, 8(1622): 1–7.

DeLong, D. C., Jr. (1996). Defining biodiversity. *Wildlife Society Bulletin*, 24(4): 738–749.

Deneulin, P., Le Fur, Y., and Bavaud, F. (2016). Study of the polysemic term of minerality in wine: segmentation of consumers based on their textual responses to an open-ended survey. *Food Research International*, 90: 288–297.

Deutsches Wein Institut (2017). *Trend zum Weinkauf in Supermärkten hält an*, Bodenheim: Deutsches Wein Institut.

Deutsches Wein Institut (2019). 5,5 milliarden Euro durch Weintourismus Pressemitteilung. https://www.deutscheweine.de/presse/pressemeldungen/details/news/detail/News/55-milliarden-euro-durch-weintourismus/

Dixit, S. K. (2020). Marketing gastronomic tourism experiences. In Dixit, S. K. (ed.), *The Routledge Handbook of Tourism Experience Management and Marketing* (pp. 322–335). London: Routledge.

Ewing-Chow, D. (2020). Covid 19 has given consumers five new reaasos to eat local. *Forbes*. https://www.forbes.com/sites/daphneewingchow/2020/07/30/covid-19-has-given-consumers-five-new-reasons-to-eat-local/?sh=58445c483ccc

International Organisation of Vine and Wine (2017). *Distribution of the World's Grapevine Varieties*, Paris: OIV. http://www.oiv.int/public/medias/5888/en-distribution-of-the-worlds-grapevine-varieties.pdf

Jones, G. V. (2013). Climate, terroir, and wine: what matters most in producing a great wine? *Earth Magazine*, December 12. https://www.earthmagazine.org/article/climate-terroir-and-wine-what-matters-most-producing-great-wine

Kerr, L. (2020). *Sour Grapes: Mitigating the Risk of Overtourism in British Columbia's Eno-tourism*. Summit Simon Fraser University. https://summit.sfu.ca/item/20130

Lanyon, D. M., Cass, A., and Hansen, D. (2004). *The Effect of Soil Properties on Vine Performance*. CSIRO Land and Water, Technical Report No. 34/04.

Legrand, W., Sloan, P., Fett, M., and Manten, T. (2019). The roles of terroir, food and gastronomy in destination authenticity. In Dixit, S. K. (ed.), *The Routledge Handbook of Gastronomy Tourism* (pp. 260–271). Oxford: Routledge.

Lenton, T. M., Held, H., Kriegler, E., Hall, J. W., Lucht, W., Rahmstorf, S., and Schellnhuber, H. J. (2008). Tipping elements in the Earth's climate system. *Proceedings of the National Academy of Sciences*, 105(6): 1786–1793. https://doi.org/10.1073/pnas.0705414105

Marques, L. (2020). *Capitalism and Environmental Collapse*. Cham: Springer.
NASA (2021). *The Study of Earth as an Integrated System*. https://climate.nasa.gov/nasa_science/science/
OEC (n.d.). *Wine: About*. https://oec.world/en/profile/hs92/wine?redirect=true
Poitras, L., and Getz, D. (2006). Sustainable wine tourism: the host community perspective. *Journal of Sustainable Tourism*, 14: 425–448.
Porto Protocol (2020, November 6). *Monoculture Wars – The Vineyards Killing Ecosystems*. https://www.portoprotocol.com/monoculture-wars-the-vineyards-killing-ecosystems/
Robinson, J. (2006). *The Oxford Companion to Wine* (3rd ed.) Oxford: Oxford University Press.
Robinson, J. (2018). *When Terroir Trumps Quality*. https://www.jancisrobinson.com/articles/when-terroir-trumps-quality
Santos, J. A., Fraga, H., Malheiro, A. C., Moutinho-Pereira, J., Dinis, L.-T., Correia, C., Moriondo, M., Leolini, L., Dibari, C., Costafreda-Aumedes, S., Kartschall, T., Menz, C., Molitor, D., Junk, J., Beyer, M., and Schultz, H. R. (2020). A review of the potential climate change impacts and adaptation options for European viticulture. *Applied Sciences*, 10(9): 1–28. https://doi.org/10.3390/app10093092
Shveda, K. (2021). *How Coronavirus Is Changing Food Shopping*. https://www.bbc.com/future/bespoke/follow-the-food/how-covid-19-is-changing-food-shopping.html
Sloan, P., Legrand, W., and Hindley, C. (2015). *The Routledge Handbook of Sustainable Food, Beverages and Gastronomy*. Oxford: Routledge.
Specialityfood (2020). *Coronvirus* Leads to Rise in Locally Conscious Consumer. https://www.specialityfoodmagazine.com/news/coronavirus-leads-to-rise-in-locally-conscious-consumer
Statista (2020). Market Revenue of Wine Worldwide. https://www.statista.com/outlook/10030000/100/wine/worldwide#market-revenue
Stefanucci, S., Graca, A., Novelle, V., Belda, I., Carlos, C., and Gautier, J. (2018). *Functional Biodiversity in the Vineyard*. Paris: Intenational Organisation of Vine and Wine.
van Leeuwen, C. (2010). Terroir: the effect of the physical environment on vine growth, grape ripening and wine sensory attributes. In A. G. Reynolds (ed.), *Managing Wine Quality* (pp. 273–315). Woodhead Publishing Series in Food Science, Technology and Nutrition. Cambridge: Woodhead Publishing.
Van Leeuwen, C., and Darriet, P. (2016). Le changement climatique en viticulture: les leviers d'adaptation au vignoble. *Acte de Colloque. Les 3èmes Assises Des Vins Du Sud-Ouest. Maîtriser Le Profil Des Vins Dans Un Contexte de Changement Climatique. 2016; 3. Assises Des Vins Du Sud-Ouest, Toulouse, FRA*, 2016-01-14, 15–18.
Willwerth, J. J., Reynolds, A. G., and Lesschaeve, I. (2010). Terroir factors: their impact in the vineyard and on the sensory profiles of Riesling wines. *Le Progrès Agricole et Viticole*, 8: 159–168.

52
SUSTAINABLE WINE TOURISM
The case for natural wines

Camilo Peña-Moreno and Annamma Joy

Introduction

In recent years, a small but growing subset of winemakers has embraced natural winemaking, which employs organic or biodynamic vineyard practices, features minimal human intervention in the cellaring process, and avoids additives. Variations between natural and conventional winemaking fall on a continuum, with the former term often loosely defined. This year, France adopted a charter, label, and trade syndicate for low-intervention wines; wine labels for such wines now define them as 'Vin Méthode Nature.' This formal recognition may spur consumer interest in such wines and the concomitant growth of a wine industry devoted to natural wines. Wine tourism, in response, will be able to offer new and unique experiences tailored to the idiosyncrasies of natural wines, connecting the characteristics of each terroir (the various elements of soil, climate, and place of a given vineyard that give wines their distinct flavors) with the organoleptic and symbolic characteristics of each wine.

In this chapter, we discuss and analyze the potential of natural wines in the development of sustainable wine tourism. Additionally, we address the interrelationship of nature, terroir, and sustainability as showcased through natural wines.

Pairing wine with sustainability

Consumers committed to sustainability demand certain characteristics in the products they purchase and the brands to which they are loyal, whether in the context of clothing, coffee (Thompson and Arsel 2004), chocolate, or other types of food; they want non-GMO, organic, fair-trade products that have been ethically and, if possible, locally sourced (Joy et al. 2012). The list of agricultural products that feature in consumer moralizing of daily choices, as Giesler and Veresiu (2014) suggest, is both extensive and growing. Yet even environmentally aware consumers do not always have strict requirements for some products; among those products is wine (Rahman, Stumpf, and Reynolds 2014). Although, as of this writing, winemakers are under no pressure from consumers to produce sustainable wine, some do so nonetheless, based on personal values and commitments. Such winemakers incorporate organic and/or biodynamic techniques into their winemaking (Legeron 2014), following environmental and health concerns related to agriculture.

DOI: 10.4324/9781003143628-60

Environmental awareness is more relevant than ever, with growing concerns about global climate change, biodiversity loss, and scarcer natural resources (Martin and Schouten 2011). In parallel, consumers have become increasingly sensitive to social concerns in response to tragedies caused by exploitative product manufacture supply chains in third-world and developing countries (Joy and Peña 2017; Rueda and Lambin 2013). Such concerns have had a growing impact on consumer purchase decisions (e.g., purchasing jewelry with diamonds certified as 'conflict-free'; Tamo 2016). The issue of wine sustainability, however, has been largely focused on environmental concerns only (Christ and Burritt 2013; Forbes et al. 2009; Gabzdylova, Raffensperger, and Castka 2009; Point, Tyedmers, and Naugler 2012; Soosay, Fearne, and Dent 2012).

Through a deeper understanding of the environmental and social impact of their purchase decisions, consumers can make their concerns known and potentially significantly affect sales (Dickinson and Carsky 2005). For this reason, a product's backstory is key to the construction and use of sustainability discourses to attract sustainability-conscious consumers. Was a given product manufactured without pesticides? Did its production have a low-carbon footprint? Were flavors manipulated through additives? Consumers expect to learn answers to these questions and will make purchases based on those answers.

The concept of terroir in the world of wine

Schreiner (2006) argues that terroir, a concept first developed in France, is only fully understood by the French. Today, however, terroir is a globally recognized element in the wine industry and is thus a clear example of a global concept applied to local contexts (Kjeldgaard and Askegaard 2006; Robertson 1992; Thompson and Arsel 2004).

With special relevance in French culture, terroir is used to characterize a diverse array of products, including Rousillon peaches, Limousin beef, Noir de Bigorre ham, and Bresse chicken (Erlanger 2013). Because terroir signifies the specific varietals and sensory characteristics in a wine from a particular region, it can be invoked as a guarantee of quality tied to place (e.g., the fine wines of Burgundy, renowned for the memorable depth of their flavors). As a further example, the concept of farm-to-table has become a signifier of freshness and improved flavor. Thus, farmers' markets and community-supported agriculture (CSA) in the United States and elsewhere offer locally grown and seasonally available products (Thompson and Coskuner-Balli 2007a, 2007b). When consumers engage with the idea of terroir, they may seek out a product that, while perhaps not locally produced, was produced under very specific conditions, as indicated by the product's terroir.

Similar to the oppositional nature of locally sourced consumption (Thompson and Coskuner-Balli 2007a, 2007b), terroir-based products offer an alternative to industrialized, mass-manufactured products (Erlanger 2013). Wine marketing strategies have over time employed the concept of terroir as indicative of the rootedness of specific wines and a means for localities to respond to globalization (Demossier 2011). Much like other branches of agriculture, winemaking is a commercial endeavor (some wine connoisseurs' near quasi-religious views of wine notwithstanding; Phillips 2000, p. 324) and is similarly subject to corporate takeovers, which have become common in both the New and Old Worlds of winemaking (Phillips 2000). These changes in ownership, as seen in Bordeaux châteaux wines or Burgundy domaines, have not affected the wines' original integrity, however (Phillips 2000).

Because terroir is embedded within the dynamics of the human/nature relationship, this concept is relevant not only to understanding discourses related to regional winemaking and wine consumption, but also to understanding how consumers relate to nature in diverse ways and contexts. As Varriano (2010, p. 8) explains, '[t]he grapevine itself was the source of wine's mystique. Vegetation cults celebrating the earth's fecundity are the oldest and most deeply embedded in human consciousness.'

In its connection with nature and terroir, wine plays a unique role among alcoholic beverages. Artisanal craft beers, for example, are famed for the brewing techniques and ingredient blends used in their production rather than for the soil in which the barley and hops were grown. Indeed, the natural process of fermentation continues to be perceived in sharp contrast to the artificial process of distillation (Phillips 2000) and serves to further distinguish wine from beverages such as gin, rum, vodka, and whiskey, which all feature very high alcoholic contents. Moreover, the perception of wine as embodying terroir, and thus as a natural product, places it on the continuum of the natural food and drinks category.

It is because of wine's connection to the land and soil where the grapes are grown that wine is perceived as a unique product, intrinsically connected to its region of origin. The identification of wine through its grape varietal is the most widely used method for labeling and distinguishing wines, with such identifications continuing to evolve; some of the terms developed in the New World are quite recent, as with the case of 'Meritage,' a term coined in California in 1988, to refer to red or white wine made from blends of grapes traditionally used in making Bordeaux wines (Phillips 2000). This grape varietal naming method for labeling wine is quite different from the method traditionally used in Europe, where a wine would be labeled as simply coming from a particular geographical location, or terroir; anyone purchasing a wine from a specific region would be expected to know what grape varietals were used (Ibid.).

Nature is key to popularizing the concept of sustainability

Given consumer perceptions of wine as strongly connected to the land and terroir, and thus to the natural environment, we need to take a closer look at why the concept of nature is central to promoting sustainability. Gaining a deeper understanding of why nature is linked to sustainability will, in turn, allow us to discuss the new phenomenon of natural wine tourism.

Contradictions exist within consumers' experiences of nature (Canniford and Shankar 2013) and within their conception of nature itself (Soper 1995). These contradictions unfold in an era with no clear separation between natural-oriented and mass-produced food markets. Some five decades ago, as the organic foods movement emerged and positioned itself as the antithesis of the corporate industrial food production system, the distinction between the two was clear, and each segment served markedly different markets. Today, the concept of 'natural' foods no longer embodies oppositional values and now simply represents a sizable marketing segment, subsumed within the larger agribusiness industry, with product lines oftentimes produced by some of the biggest food corporations in the world (Thompson and Coskuner-Balli 2007a). This reality has resulted in the development of new forms of food production that promote responsible and sustainable practices (e.g., CSA; Thompson and Coskuner-Balli 2007b). These forms of expression encourage a sentiment of enchantment, with an idealized, romantic view of nature, in which local farms and farmers are admired and perceived as universally virtuous (Thompson and Coskuner-Balli 2007b).

In any discussion of sustainability, it is important to understand how nature can be interpreted through different conceptual lenses and which of these are relevant for the sustainability–nature connection. Recent research examining the concepts of sustainability and nature highlights the connection between them (e.g., de Burgh-Woodman and King 2013). This assertion that sustainability and nature are intertwined, although obviously true, is belied by the fact that humans routinely view sustainability and protecting nature only through an anthropocentric/utilitarian perspective (Moisander 2007).

Among multiple possible perspectives, nature may be viewed as objective or subjective. Consumer researchers study the latter from the perspective of a person (e.g., Canniford and Shankar 2013; de Burgh-Woodman and King 2013), in which nature is seen as a self-created symbolic construction that a person uses for identity creation purposes when feeling part of a specific agenda and seen in opposition to another discourse (Luedicke, Thompson, and Giesler 2010). The human–nature connection has also been tied to the symbolically sustainable discourse, which typically does not extend beyond a material/practical statement (de Burgh-Woodman and King 2013). This human–nature connection is a construct in the Western cultural discourse, traceable to Greek mythology and its turbulent relationship with the natural elements, represented by each God (de Burgh-Woodman and King 2013). The relationship with nature, and the human–nature connection, has evolved over time; today's discourses around nature are necessarily tied to the development of the modern environmental movement.

Environmental movements and nature

In 1962, Rachel Carson's groundbreaking and prescient warning of the dangers of chemical pesticides, *Silent Spring*, made her one of the most influential figures in environmental movements since Henry David Thoreau (Griswold 2012); her work arguably gave rise to the modern environmental movement (Michaud 2010). In addition to having a deep understanding of ecology, Carson was a direct and accessible writer: 'Carson combined a scientist's ability to see with a novelist's ability to imagine' (Rothman 2012).

Carson was among the first to warn the public of the toll of environmental pollution, an issue previously recognized only within the scientific community. *Silent Spring*'s message was blunt: if humans poison nature, nature, in turn, will poison humans. Carson opposed the Christian dominionist belief that nature existed to serve man and argued that technological innovation employed to control nature (e.g., pesticides such as DDT) would disrupt the natural system (Rothman 2012).

The environmental movement has evolved to include radical opposition to genetically modified organisms (GMOs). Vandana Shiva, a contemporary leader of the ecofeminist movement and anti-globalist, for example, has campaigned against GMOs; she has been criticized as having embraced pseudoscience from both media (e.g., Entine 2015; Specter 2014) and academic writers (e.g., Gilbert 2013; Herring 2006). Shiva argues that the pre-colonial subcontinent (having unity in feminine and masculine principles) has more meaningful ecological values than those brought to India by Western culture (McGurty 2019). Shiva states that men – without the balance of feminine values – would break away from both women and nature, causing ecological disaster (McGurty 2019). According to McGurty (2019), Shiva proposes that the nonviolent, primarily female forest conservation group known as the Chipko movement (active in 1973 in northern India), embodied the necessary pre-colonial feminine principle of women tree huggers, a principle required for preventing further ecological catastrophe.

From Rachel Carson branded as a hysteric and a fanatic (Lear 1997) to ecofeminists dismissed as hippie tree-huggers, environmental activists have routinely faced derision. Perhaps in a reflection of today's consumer distrust of environmental discourses produced by corporate media (e.g., greenwashing) and of today's mainstream distrust of environmental discourses deployed by the scientific community, the term 'tree hugger' is now an official term meaning 'someone who is regarded as foolish or annoying because of being too concerned about protecting trees, animals, and other parts of the natural world from pollution and other threats'

(Merriam-Webster n.d.). Moreover, what once was considered as a clear environmental model of sustainable agricultural production – organic food – has become a set of standards and restrictions adapted to the demands of agribusiness (Thompson and Coskuner-Balli 2007a). Through co-optation of various environmental discourses (Thompson and Coskuner-Balli 2007a, 2007b), corporations increasingly promote themselves as engaged in sustainable and environmentally conscious practices, aiming to satisfy the needs of specific consumer niches. In many cases, to some degree, such promotions are accurate—but may occur concurrently with non-sustainable and environmentally damaging policies carried out by the same company.

While corporations and governments can play important roles in promoting green and sustainable products and services, many create such products or related services and delegate responsibility for ecological accountability to the citizen/consumer, rather than implementing state- or corporation-based models of such accountability (Giesler and Veresiu 2014). (The concepts of 'green' and 'sustainable' differ in that green products use resources efficiently without damaging the environment in the present, while sustainable products are designed to safeguard the future as well as the present. These dynamics can encourage consumers to make responsible purchasing decisions by accommodating broad ideological views and fostering the view of responsible consumption as a moral imperative from each individual's perspective (Giesler and Veresiu 2014).

To go beyond the adjudication of responsibilities to consumers, corporations must demonstrate how they contribute to sustainable goals in the context of both social and environmental impacts. Notably, once a corporation makes a clear commitment to sustainability, positive consumer perceptions of their products can rise in tandem (Chernev and Blair 2015).

Given these factors, for wineries to showcase their own efforts at achieving sustainability makes sense from both a moral and a business standpoint. To protect the land they benefit from, and to safeguard the grapes that enable their production of wine, is simply a responsible course of action.

In the next section, we address how different consumer concepts of nature can be interpreted and which of these concepts could be most aligned with a sustainable discourse in the wine industry.

Understanding how nature is perceived and experienced through a conceptual lens

Some individuals view nature as an entity subject to human influence, readily available for human interaction and enjoyable activities (Canniford and Shankar 2013). This view of nature is defined in terms of the concept of 'otherness' to humanity, in which nature is seen as opposed to culture, history, or anything defining of the order of humanity (Soper 1995). Newer generations of commentators have begun to set aside this conception that nature is the antithesis of society, and have proposed nature as a social construction (Castree and MacMillan 2001) and as a discursive-material assemblage (Canniford and Shankar 2013).

According to Whatmore (1999), two types of nature are conceptualized as a social construction. The first claims that economic interests have reconstituted nature to every level, using technology and science. The second argues that nature is culturally constructed, in which the concept of nature is created via systems of signification and meaning. In this view, words, concepts, and explanations of nature are all affected by biases and prejudices inherent in each person's culture. This idea has gained strength in considerations of nature as an experience that is socially constructed via cultural discourses and other practices (Arnould and Price 1993; Belk and Costa 1998; Kozinets 2002; Tumbat and Belk 2011).

Nature can also be seen as an external place that is socially/culturally constructed as a beautiful and pristine site where one's troubles can be forgotten and one can 'get away from the daily routine' (Arnould and Price 1993). From this perspective, nature offers extraordinary, absorbing, and joyful experiences that will not be easily forgotten and will open the mind to new perspectives and processes (Csikszentmihalyi 1990). These experiences of nature are embedded into social construction, in many cases, to make the experience intelligible and to help create a life-changing feeling that might otherwise not be achievable without the collaboration of fellow participants in the experience (Belk and Costa 1998). In the particular case of the modern mountain man (Belk and Costa 1998), the experience is considered a consumption enclave that creates an alternative reality for many of its participants. In other cases, such experiences can be prepared in advance, as in widely publicized gatherings such as the Burning Man event in Nevada's Black Rock Desert, where the natural setting is altered to provide one such enclave experience (Kozinets 2002).

How natural wines fit into the sustainability discourse around nature

When nature is perceived as being a culturally and socially constructed idea beyond the materiality of a place, natural wines can address both the symbolic and material aspects of safeguarding nature while making wine. It is a given that winemakers are using nature for production purposes; the importance of the land and climate in making wine notwithstanding, successful and good-quality wine will still need the skills of a knowledgeable winemaker (Lapsley 1996). Thus, natural winemakers accept and embrace their role as artisans even as they grapple with the question of how much human intervention and technology is acceptable in pursuit of that artisanship.

Black (2013) proposes three themes as part of the contestation and construction of the concept of nature in the wine world; these themes, discussed in this section, are essential in understanding why natural wine tourism can be an ideal example of sustainable wine tourism.

First, the romantic marketing theme suggests that wine marketing campaigns should use the stereotypical idyllic and pristine vineyards to promote the romance of drinking wine while avoiding the 'dirty' parts of winemaking (such as machines, tanks, the use of hoses, etc.); natural wine counters this artificially romantic view by showcasing an actual reality: the processes that need to happen to make wine while avoiding excessive use of technology and additives (Legeron 2014).

Second, the sensorial experience theme (Black 2013) refers to the ways in which natural wines have a different visual, olfactory, and gustatory aesthetic from conventional wines. This difference can be experienced by having consumers taste the wines and through consumer education about natural wines; these wines' unique characteristics are derived from both terroir and the winemaker's craft. Although some wine consumers might be more accepting of what could be perceived as sensorial faults in wine (Grohmann, Peña, and Joy 2018), in general winemakers, as well as connoisseurs, tend to be less tolerant of wine faults than the average consumer, despite the fact that some of these faults can be considered part of a natural wine's personality and unique expression of terroir (Goode and Harrop 2011). This uniqueness provides an additional sensorial differentiation from what other wineries are trying to achieve in homogenizing vintages and replicating other famous wine regions (Ibid.).

The third theme Black (2013) proposes is simply the idea of honesty implicit in the discourse around nature and natural wines. While consumers are becoming ever more aware of the ways in which the products they purchase are produced and have more readily available options for researching purchase choices (Berry and McEachern 2005), many remain unaware of the various processes involved in mass-market food production. For their part, corporate winemakers remain as secretive as possible about their various methods of production and flavor manipulation. Given natural wine's association with honesty and transparency, wine tours of such wineries can reveal with a clear conscience the processes in the vineyard and the cellar that result in natural wines.

Case Study 52.1 Natural wine tourism at Bella Wines

Bella Wines Ltd. is a small family-owned Canadian winery in British Columbia's Naramata Bench. The winery produces and sells natural sparkling Gamay Noir and Chardonnay wines and offers impromptu tours of their winery and farm to all interested visitors (bellawines.ca). During 2017–2020, the authors investigated how the winery's tours encompassed both the realms of winemaking and that of nature, with the two clearly interconnected. All data were collected using ethnographic research methods such as depth interviews and participant observations (verbatim excerpts from interviews are italicized).

During 2020, motivated by the need to maintain social distancing in response to the pandemic and public health guidelines followed in the province of British Columbia, Bella began offering outdoor walking tours around their vineyard and farm, rather than traditional indoor tours focused on winery production facilities and tasting rooms.

These tours show all the essential steps required in making a good wine without the use of conventional additives, fertilizers, and pesticides. Visitors first arrive to a designated area (a marked barrel) and learn from their guide what the experience will cover and safety guidelines. Tour participants are invited to grab a wine glass, and the tour starts. Although the cellar and vineyard are always visited, depending on what is happening at the moment on the farm or cellar, the flow of the tour might change from tour to tour (e.g., if it is disgorging time [disgorgement is critical to aroma development in the production of champagne wine; the process eliminates deposits found in the necks of the wine bottles after maturation], visitors will learn about the process and how it affects the final sparkling wine). During their tour of the vineyard, visitors learn about the different organic or conscientious farming makes in the final product and to the health of those who make the wine as well as those who ultimately drink it. Moreover, such farming involves a more hands-on approach compared to conventional methods that rely on pesticide use because of the need to regularly assess vine health. With organic farming, Bella owners see their vineyard and farm as a complete ecosystem unto itself, a holistic circle of life dependent on the diversity of crops and animals.

Visitors see animals – pigs and chickens – as part of the tour and learn why they are essential for maintaining the vineyard ecosystem; they control certain pests, fertilize the soil, and contribute to the overall diversity of the farm. Additionally, visitors see how Bella ensures that their farm animals have good living conditions where they can behave like animals and not prisoners (in stark comparison to how many industrial meat and poultry factory farms keep their animals).

For Bella, there is a clear connection between the health of the vine and grapes and the health of the ferment, as well as, ultimately, the taste of the wine. Having beneficial worms and microbes in the soil has a sizable impact on all of the above; as Bella co-owner, Jay

Drysdale said: *'the healthier the fruit, the better success you have in the winery without having to fix the wine.'* Starting with natural soil is key to running a natural, low-intervention operation. Vine health further flourishes thanks to diverse plantings, including various fruits. By using inter-planting, Bella Wines grows a secondary crop such as a root vegetable; this diversity of cover crops will feed the chicken and pigs, who in turn lower mowing costs by eating the grass. As Drysdale noted, *'[b]eing in different environments and hands-on, we need that level of diversity versus stability.'* By achieving a balance of diverse animals and crops, the vines and soil thrive and produce healthy fruit; with a monoculture bereft of other crops and the presence of various animals, the foundation for healthy vines and grapes is simply missing. For winemakers with a monoculture approach, the answer to insect infestations is typically a high level of human intervention involving synthetic pesticides.

On the other hand, Bella takes a preventative approach to pest control, making sure the soil is healthy by trying to achieve a balance with microbes, microfauna, and flora present in the vineyards. By having a healthy farm and flowering cover crops, Bella's vineyards attract natural predators that will deal with insects such as leaf hoppers and cutworms. Moreover, different types of weeds might give a warning sign about fertility and compaction levels or about the deficiency of certain vitamins or minerals in the soil. After this, Bella can then amend the soils naturally with cover crops, natural manures and fertilizers, natural/biodynamic sprays, and/or teas made from local plants like chamomile, mint, and horsetail. Additionally, Drysdale also purchases fruit from growers that produce organic grapes and thus will only use a small amount of certified organic sprays life sulfur and mineral/vegetable oils.

Bella's natural wine/farm tours offer a hands-on approach throughout the experience. Visitors are encouraged to touch and taste different fruits plucked straight from the trees on the property, including plums, apricots, and cherries. Visitors thus get the opportunity to not only taste wine and compare it to other fruit flavors, but also to experience eating fresh, juicy, unwashed organic fruit, warm from the sun in cooperative weather. Such experiences can be eye-opening for visitors, as they come face-to-face with farm life, breathe fresh fragrant air, and realize that what they consume affects their health in positive or negative ways, depending on how their food was grown. During the tour, Drysdale describes the various bacteria present in some of the farm's organic fruit that are beneficial to humans. By allowing visitors to explore and eat the fruit directly from the trees, Bella highlights the safety of the earth, vines, and trees free of harmful pesticides.

For Drysdale, when an individual consumes nutritious foods and wines, the act carries resonance; the more one experiences a natural approach to food and wine, the harder it is to go back to regular products. As Drysdale noted,

> Everyone's palate is unique and different, and the more you expose yourself to more wines with less things added, you find it harder to go back to the wines with additives. …everything remains clean and it is a pure expression.

The winery's ethos centers on purity and honesty; the company has attracted a consumer base that shares their mindset and values. This connection, they contend, cannot be replicated with synthetic wines and wines made with additives (e.g., enzymes, nutrients, tannins, and fining agents – a substance, typically animal-derived, added during production to prevent cloudiness – among others). After visitors leave a wine tour at Bella, they have the tools to distinguish between a truly sustainable wine-growing process and one that is not but poses as such; for example, they have now seen how a vineyard looks when no herbicides are used (as at Bella), and how they look when the vines have been sprayed with glyphosate-based herbicides.

Finally, by offering visitors the opportunity to taste wines as they walk around the property (with four designated tastings throughout the tour), Drysdale can explain why each wine tastes as it does, stressing the connection between the end product and the natural processes used to create it. The uniqueness of each wine tasted and the differences between vintages are associated with each year's unique climate characteristics and with the low-intervention methods Bella Wines staff employed in tending the vineyard in a given year. These wines, Drysdale argues, reflect that year's specific terroir and, therefore, will differ from wines produced by mainstream wineries, where wines from different years will taste and look the same. As Drysdale stated,

> if we're going to manipulate the wine in the winery, terroir means nothing. So, these guys [mainstream winemakers] want to talk terroir and yet they've manipulated the wine so much that it tastes a certain way for them rather than to express the actual terroir.

For Drysdale, providing this knowledge will leave an impression on visitors and perhaps change how they perceive winemaking.

By offering wine tastings during a farm tour and showing what the process of winemaking really looks like, Bella Wines presents visitors with a sense of honesty and transparency that is not always the case during a more traditional winery tour. This honesty showcases the role of nature as a guiding force in how their wines are developed. The rural setting implicitly presents nature as a central element and also highlights how nature affects the sensorial experience that each visitor will have when tasting the wines.

Conclusion

In a globalized world and wine industry, where most consumers can access wines from a vast range of locations (Mora 2016), the ability of wineries to represent regional/local qualities in their wines can be a key differentiating factor. Moreover, with growing consumer awareness of the eventual costs, both human and ecological, of unsustainable food and drink choices (Bonn, Cronin, and Cho 2015), wineries offering sustainable wines can attract a new cohort of environmentally conscious wine consumers.

Natural wines are, by definition, a more environmentally friendly alternative to mainstream high-intervention wines, albeit riskier for the winemaker given the increased vulnerability of crops and the potential for volatility from one season to the next in vintage profiles (Smith Maguire 2018). These risks could make some wineries decide to produce only a small number of natural wines while making the majority of their wines with a conventional and, therefore, less risky process. Given the growing relevance of sustainability to the wine industry (Santini, Cavicchi, and Casini 2013) and following the lack of consumer trust in what some consider a trend and a co-opted movement (Thompson and Coskuner-Balli 2007a), wineries that sell only a small number of sustainably produced wines, while keeping the majority of their products in their business-as-usual model, might be perceived as less than fully committed to their stated ideals of sustainability (a perception amplified by the lack of a clear definition of natural wine) (Black 2013).

However, wineries whose entire brand mission and vision is based on the concept of sustainability are in a prime position to foster the trust of consumers. Moreover, with more informed consumers, thanks to broader access to information and to educational wine tours such as that described in this chapter's case study, consumer understanding of sustainability in winemaking has inevitably increased, as has wariness. Consumers who detect inconsistencies between what a winery touts in marketing materials and what it actually does in the vineyard and the cellar might lose trust in the brand. As more and bigger wineries worldwide jump onto the sustainability bandwagon, more consumers might question how much of their talk is actually true and demand a transparent view of the companies' winemaking processes.

The way to reassure consumers that a natural wine purchase is indeed in line with their commitment to sustainability is to showcase the typically hidden processes of wine production. In other industries, lifting the curtain on manufacturing processes is challenging, given a global workforce, and thus certification and clear traceability standards are required (Joy and Peña 2017). Although certification and traceability are also used in the wine industry, natural wine tourism allows visitors to see the difference themselves between sustainable and unsustainable practices, in essence opening a window onto otherwise unseen procedures.

Natural wine is thus an oppositional force to the global market of wine, where wineries from diverse locations try to emulate the same organoleptic profiles seen in the world's most renowned terroirs. Even as some winemakers try to imitate much-admired wines, however, they simultaneously aim for homogenous profiles that will be consistent from one vintage to the next (Goode and Harrop 2011). Consumers are simply unaware of the interventions required for such tactics (Legeron 2014). Natural wine tourism can be positioned as revealing the truth behind the widely accepted perceptions of winemaking: not everything that goes into a glass of wine comes from the idyllic landscapes of wine country—unless that glass contains natural wine.

Implications

This chapter proposes an analysis of natural wine tourism that helps in understanding how the concept of natural wines can be used for the promotion of sustainable wine tourism. In theoretical terms, it adds to the analysis and discussion around natural wines by reflecting on the possibilities of using three key concepts to foster natural wine tourism: honest and transparent wine tourism, a showcasing of the labor of nature as opposed to an overtly artificial process, and openness to and education about different and sometimes unusual wine sensorial profiles.

In practical terms, this chapter provides possibilities for winery managers and wine region authorities to consider the use of natural wine as a way to promote a new kind of sustainable wine tourism. By better understanding where this wine movement comes from and its theoretical underpinnings, practitioners can better target specific niches and demographics that might be more interested in such wines and ideals. This case study and analysis can be further extended to other wine regions and wineries to compare and see what similarities and differences are evidenced from different natural wineries and winemakers across the wine world.

References

Arnould, E. J., and Price, L. L. (1993). River magic: extraordinary experience and the extended service encounter. *Journal of Consumer Research*, 20(1): 24–45. https://doi.org/10.1086/209331

Belk, R. W., and Costa, J. A. (1998). The mountain man myth: a contemporary consuming fantasy. *Journal of Consumer Research*, 25(3): 218–240. https://doi.org/10.1086/209536

Berry, H., and McEachern, M. (2005). Informing ethical consumers. In Harrison, R., Newholm, T., and Shaw, D. (eds.), *The Ethical Consumer* (pp. 69–87). Thousand Oaks, CA: Sage. https://doi.org/10.1002/cb.36

Black, R. E. (2013). Vino naturale: tensions between nature and technology in the glass. In Black, R. E., and Ulin, R. C. (eds.), *Wine and Culture: Vineyard to Glass* (pp. 279–294). London: Bloomsbury Academic. https://doi.org/10.5040/9781350042254

Bonn, M. A., Cronin, J. J., and Cho, M. (2015). Do environmental sustainable practices of organic wine suppliers affect consumers' behavioral intentions? The moderating role of trust. *Cornell Hospitality Quarterly*, March: 1–17. https://doi.org/10.1177/1938965515576567

Canniford, R., and Shankar, A. (2013). Purifying practices: how consumers assemble romantic experiences of nature. *Journal of Consumer Research*, 39(5): 1051–1069. https://doi.org/10.1086/667202

Castree, N., and MacMillan, T. (2001). Dissolving dualisms: actor-networks and the reimagination of nature. In Castree, N., and Braun, B. (eds.), *Social Nature: Theory, Practice and Politics* (pp. 208–224). Boston: Blackwell.

Chernev, A., and Blair, S. (2015). Doing well by doing good: the benevolent halo of corporate social responsibility. *Journal of Consumer Research*, 41(6): 1412–1425. https://doi.org/10.1086/680089

Christ, K. L., and Burritt, R. L. (2013). Critical environmental concerns in wine production: an integrative review. *Journal of Cleaner Production*, 53: 232–242. https://doi.org/10.1016/j.jclepro.2013.04.007

Csikszentmihalyi, M. (1990). *Flow: The Psychology of Optimal Experience*. New York: Harper & Row.

de Burgh-Woodman, H., and King, D. (2013). Sustainability and the human/nature connection: a critical discourse analysis of being 'symbolically' sustainable. *Consumption Markets & Culture*, 16(2): 145–168. https://doi.org/10.1080/10253866.2012.662834

Demossier, M. (2011). Beyond terroir: territorial construction, hegemonic discourses, and French wine culture. *Journal of the Royal Anthropological Institute*, 17(4): 685–705. https://doi.org/10.1111/j.1467-9655.2011.01714.x

Dickinson, R. A., and Carsky, M. L. (2005). The consumer as economic voter. In Harrison, R., Newholm, T., and Shaw, D. (eds.), *The Ethical Consumer* (pp. 25–36). Thousand Oaks, CA: Sage. https://doi.org/10.1002/cb.36

Entine, J. (2015). 'Eco-Warrior' Vandana Shiva, at $40,000 a speech, rejoins Hawaii anti-GMO crusade, but truth is the victim. *The Huffington Post*, January. http://www.huffingtonpost.com/jon-entine/ecowarrior-vandana-shiva-_b_6528032.html

Erlanger, S. (2013). Vive le terroir. *The New York Times*. https://www.nytimes.com/2013/09/01sunday-review/vive-le-terroir.html?_r=1

Forbes, S. L., Cohen, D., Cullen, R., Wratten, S., and Fountain, J. (2009). Consumer attitudes regarding environmentally sustainable wine: an exploratory study of the New Zealand marketplace. *Journal of Cleaner Production*, 17(13): 1195–1199. https://doi.org/10.1016/j.jclepro.2009.04.008

Gabzdylova, B., Raffensperger, J. F., and Castka, P. (2009). Sustainability in the New Zealand wine industry: drivers, stakeholders and practices. *Journal of Cleaner Production*, 17(11): 992–998. https://doi.org/10.1016/j.jclepro.2009.02.015

Giesler, M., and Veresiu, E. (2014). Creating the responsible consumer: moralistic governance regimes and consumer subjectivity. *Journal of Consumer Research*, 41(3): 840–857. https://doi.org/10.1086/677842

Gilbert, N. (2013). A hard look at GM crops. *Nature*, 497(7447): 24–26.

Goode, J., and Harrop, S. (2011). *Authentic Wine: Toward Natural and Sustainable Winemaking*. Berkeley and Los Angeles: University of California Press.

Griswold, E. (2012). How 'Silent Spring' ignited the environmental movement. *The New York Times*, September. http://www.nytimes.com/2012/09/23/magazine/how-silent-spring-ignited-the-environmental-movement.html

Grohmann, B., Peña, C., and Joy, A. (2018). Wine quality and sensory assessments: do distinct local groups of wine experts differ? *Journal of Wine Research*, 29(4): 278–289. https://doi.org/10.1080/09571264.2018.1532882

Herring, R. J. (2006). Why did 'Operation Cremate Monsanto' fail? Science and class in India's great terminator-technology hoax. *Critical Asian Studies*, 38(4): 467–493. https://doi.org/10.1080/14672710601073010

Joy, A., and Peña, C. (2017). Sustainability and the fashion industry: conceptualizing nature and traceability. In Henninger, C. E., Alevizou, P. J., Goworek, H., and Ryding, D. (eds.), *Sustainability in Fashion: A Cradle to Upcycle Approach* (pp. 31–54). Cham, Switzerland: Palgrave Macmillan.

Joy, A., Sherry, J. F., Jr., Venkatesh, A., Wang, J., and Chan, R. (2012). Fast fashion, sustainability, and the ethical appeal of luxury brands. *Fashion Theory: The Journal of Dress, Body & Culture*, 16(3): 273–296. https://doi.org/10.2752/175174112X13340749707123

Kjeldgaard, D., and Askegaard, S. (2006). The glocalization of youth culture: the global youth segment as structures of common difference. *Journal of Consumer Research*, 33(2): 231–247. https://doi.org/10.1086/506304

Kozinets, R. V. (2002). Can consumers escape the market? Emancipatory illuminations from Burning Man. *Journal of Consumer Research*, 29(1): 20–38. https://doi.org/10.1086/339919

Lapsley, J. T. (1996). *Bottled Poetry: Napa Winemaking from Prohibition to the Modern Era*. Berkeley: University of California Press.

Lear, L. J. (1997). *Rachel Carson: Witness for Nature*. New York: Henry Holt and Company.

Legeron, I. (2014). *Natural Wine: An Introduction to Organic and Biodynamic Wines Made Naturally*. London and New York: Cico Books.

Luedicke, M. K., Thompson, C. J., and Giesler, M. (2010). Consumer identity work as moral protagonism: how myth and ideology animate a brand-mediated moral conflict. *Journal of Consumer Research*, 36(6): 1016–1032. https://doi.org/10.1086/644761

Martin, D., and Schouten, J. (2011). *Sustainable Marketing*. New Jersey: Pearson Prentice Hall.

McGurty, E. M. (2019). Environmentalism and identity-politics. In Elliott, A. (ed.), *Routledge Handbook of Identity Studies* (2nd ed., pp. 316–331). New York: Routledge.

Merriam-Webster. (n.d.) *tree hugger, Dictionary*. http://www.merriam-webster.com/dictionary/treehugger

Michaud, J. (2010). Eighty-five from the archive: Rachel Carson. *The New Yorker*, April. http://www.newyorker.com/books/double-take/eighty-five-from-the-archive-rachel-carson

Moisander, J. (2007). Motivational complexity of green consumerism. *International Journal of Consumer Studies*, 31(4): 404–409. https://doi.org/10.1111/j.1470-6431.2007.00586.x

Mora, P. (2016). *Wine Positioning: A Handbook with 30 Case Studies of Wine Brands and Wine Regions in the World*. New York: Springer International Publishing.

Phillips, R. (2000). *A Short History of Wine*. New York: Harper Collins.

Point, E., Tyedmers, P., and Naugler, C. (2012). Life cycle environmental impacts of wine production and consumption in Nova Scotia, Canada. *Journal of Cleaner Production*, 27: 11–20. https://doi.org/10.1016/j.jclepro.2011.12.035

Rahman, I., Stumpf, T., and Reynolds, D. (2014). A comparison of the influence of purchaser attitudes and product attributes on organic wine preferences. *Cornell Hospitality Quarterly*, 55(1): 127–134. https://doi.org/10.1177/1938965513496314

Robertson, R. (1992). *Globalization: Social Theory and Global Culture*. London: Sage.

Rothman, J. (2012). Rachel Carson's natural histories. *The New Yorker*, September. http://www.newyorker.com/books/page-turner/rachel-carsons-natural-histories

Rueda, X., and Lambin, E. F. (2013). Linking globalization to local land uses: how eco-consumers and gourmands are changing the colombian coffee landscapes. *World Development*, 41(January): 286–301. https://doi.org/10.1016/j.worlddev.2012.05.018

Santini, C., Cavicchi, A., and Casini, L. (2013). Sustainability in the wine industry: key questions and research trends. *Agricultural and Food Economics*, 1(9). https://doi.org/10.1186/2193-7532-1-9

Schreiner, J. (2006). *John Schreiner's Okanagan Wine Tour Guide*. Vancouver: Whitecap Books.

Smith Maguire, J. (2018). Taste as market practice: the example of 'natural' wine. In Cross, S. N. N., Ruvalcaba, C., Venkatesh, A., and Belk, R. W. (eds.), *Consumer Culture Theory* (Vol. 19, pp. 71–92). West Yorkshire: Emerald Publishing Limited. https://doi.org/10.1108/S0885-211120180000019005

Soosay, C., Fearne, A., and Dent, B. (2012). Sustainable value chain analysis: a case study of Oxford Landing from 'vine to dine.' *Supply Chain Management: An International Journal*, 17(1): 68–77. https://doi.org/10.1108/13598541211212212

Soper, K. (1995). *What Is Nature?* Cambridge, MA: Blackwell Publishers.

Specter, M. (2014). Seeds of doubt. *The New Yorker*, August. http://www.newyorker.com/magazine/2014/08/25/seeds-of-doubt

Tamo, A. (2016). Conflict diamonds are forever in southern Africa: the case for a human rights-based approach to the Kimberley process. *South African Journal on Human Rights*, 32(2): 272–292. https://doi.org/10.1080/02587203.2016.1210886

Thompson, C. J., and Arsel, Z. (2004). The Starbucks brandscape and consumers' (Anticorporate) experiences of glocalization. *Journal of Consumer Research*, 31(December): 631–642. https://doi.org/10.1086/425098

Thompson, C. J., and Coskuner-Balli, G. (2007a). Countervailing market responses to corporate co-optation and the ideological recruitment of consumption communities. *Journal of Consumer Research*, 34(2): 135–152. https://doi.org/10.1086/519143

Thompson, C. J., and Coskuner-Balli, G. (2007b). Enchanting ethical consumerism: the case of community supported agriculture. *Journal of Consumer Culture*, 7(3): 275–303. https://doi.org/10.1177/1469540507081631

Tumbat, G., and Belk, R. W. (2011). Marketplace tensions in extraordinary experiences. *Journal of Consumer Research*, 38(1): 42–61. https://doi.org/10.1086/658220

Varriano, J. (2010). *Wine: A Cultural History*. London: Reaktion Books

Whatmore, S. (1999). Culture-nature. In Cloke, P., Crang, P., and Goodwin, M. (eds.), *Introducing Human Geographies* (pp. 4–11). London: Arnold.

53

RURAL LANDSCAPES, CULTURE, AND WINE TOURISM

Michelle Thompson and Bruce Prideaux

Introduction

The growing interest in wine tourism has led to the emergence of specialized wine destinations ranging from the well-known Old World European wine regions, which include Champagne in France and Mosel in Germany, to New World wine regions in Australia, New Zealand, South Africa, Chile, the United States, and more recently, China. The growing demand for wine tourism has been matched by growing academic interest in issues related to wine tourism. Examples of the diversity of this research include sustainability (Montella 2017), image (Williams 2001), location (Van Leeuwen and Seguin 2006), the landscapes in which wine tourism takes place (Thompson et al. 2016), events (Cavicchi and Santini 2014), profiles of wine tourists (Galloway et al. 2008), and the role of host communities (Poitras and Getz 2006).

This chapter explores how the interactions between regional culture (internal), visitor culture (external culture), and the landscape have influenced the emergence of the Barossa and Margaret River, two of Australia's best-known yet culturally different wine tourism destinations. Although both are significant domestic wine tourism destinations that produce world-class wines, neither region has been able to leverage its international reputation for quality wine to attract large numbers of international visitors. In this chapter, landscape refers to a geographic space where the interaction of soil, climate, geology, drainage, and human factors, including culture and socioeconomic structures, produces a unique pattern of land use. Using this definition, the landscape may be categorized into natural (pre-settlement plant and animal communities, geology, and drainage), agricultural (patterns of post-settlement modification of the natural landscape by agriculture, forestry, and pastoral use of the land), and cultural (human use of the landscape including settlements, transport corridors, and other spaces modified for human use) (Thompson et al. 2016). Culture is used in a collective sense to describe regional culture (internal), visitor culture (external culture), and the patterns of interaction that occur between these aspects of human activity. Given the unique characteristics of both destinations, a place-based approach was adopted on the grounds that this approach recognizes the uniqueness of individual regions and the opportunities that may be derived by viewing a particular region through the lens of its competitive advantages (Barca, McCann, and Rodríguez-Pose 2012; Turnour et al. 2014).

 DOI: 10.4324/9781003143628-61

Literature review

Wine tourism is a type of special interest tourism in which tourists seek wine-related experiences in the region where it is produced. Key elements in the development and operation of wine tourism in a specific wine region include demand for wine-focused experiences, a landscape that supports viticultural production, the local climate that nurtures the landscape, and the communities and their traditions that support the wine industry. The following discussion does not try to capture the entirety of the literature that relates to wine tourism but does briefly consider several issues that directly relate to the success of wine tourism in the two wine destinations considered in this chapter.

Hall et al. (2000) describe wine tourism as an experience of 'visitation to vineyards, wineries, wine festivals and wine shows for which grape wine tasting and/or experiencing the attributes of a grape wine region are the prime motivating factors for visitors' (p. 3). While Hall et al.'s (2000) definition provides a useful starting point for understanding wine tourism, it doesn't capture the complexities that influence the choices made by wine tourists, the centrality of the landscape in the production of wine, and the role that local communities play in creating the destinations where wine tourism occurs.

A growing number of studies have sought to identify the role that various consumer-related components of the wine experience play in delivering a quality wine tourism experience. Carlsen and Boksberger (2015, p. 133) extend the understanding of wine tourism provided by Hall et al. (2000) by suggesting that the nature and value of the wine tourism experience has several universal values irrespective of location and include 'pleasurable, memorable, educational, and sensorial experiences' that collectively create the conditions necessary for wine tourism to thrive. From a different perspective, Roberts and Sparks (2006) identified eight factors which they considered as the key factors that comprise a 'total' wine tourism experience: authenticity of experience, value for money, service interactions, setting and surroundings, product offerings, information dissemination, personal growth (learning experiences), and indulgence (lifestyle).

Charters, Fountain, and Fish (2009) identified gender, generation membership, and cultural affiliation as significant factors that influence how individuals experience wine tourism. They also identified a desire by wine tourists to engage in an authentic experience based on being able to connect with a particular winery, based on its commitment to the wine it produced. Other factors that influence how an individual experiences wine tourism include staff friendliness, understanding and knowledge, and the atmosphere of tasting rooms (Griffin and Loersch 2006). The role of sensation seeking in the wine tourism experience was investigated by Galloway et al. (2008), who found that up to half of the wine tourists they surveyed could be classified as high sensation seekers who had a different range of behaviors and preferences from the remaining survey participants whom they classed as low sensation seekers. These studies demonstrate that there are several contributing factors to wine tourism experiences, highlighting the interaction between the demands of wine tourists, and the cultural and natural attributes of the wine tourism setting in the destination.

The role of the landscape in wine tourism has often been ignored in wine tourism research despite the central role landscapes play in determining the terroir of a wine region. Terroir is a French term that relates to the environmental conditions of the landscape that determine the sensory attributes of wine. Van Leeuwen and Seguin (2006) describe terroir as 'the relationship between the characteristics of an agricultural product (quality, taste, style) and its geographic origin, which might influence these characteristics' (p. 1). Seguin (1986) adds to this understanding by describing terroir as an interactive ecosystem that occurs in a specific place where the quality of the wine grown is a function of the interaction of soil, climate, the vine, and human factors including culture, socioeconomics, and viticultural and oenological techniques. Terroir also links the sensory attributes of wine to the environmental conditions in the areas where the grapes are grown and is used for identifying the position of a particular wine within a hierarchy of high-quality wines. In this chapter, particular attention is paid to the landscape aspect of terroir given the role that landscapes play in determining human settlement patterns, agricultural opportunities, and aesthetic attractiveness from a touristic perspective.

Aside from terroir, other aspects of landscapes that influence the demand for wine tourism include scenic values, the pattern of human settlement shaped by the physical character of the landscape, the soils that occur within the landscape, and the distribution of natural vegetation, crops, and pastures that the soil supports. From both scenic and categorical perspectives, the landscape can be decomposed into natural landscapes (Terkenli 2002), cultural landscapes, and agricultural landscapes (Thompson et al. 2016). As Urry (2002) observed, landscapes provide scenic patterns that can be consumed through the tourist gaze (Urry 2002) to produce experiences that appeal to and attract tourist interest.

Collectively, the landscape, cultural traditions, and opportunities to consume wine create a sense of place that plays an important role in developing wine tourism destinations. The regional identities of wine tourism destinations are in part represented in their tangible outputs, including wine and food produced (Lin, Pearson, and Cai 2011). The production of these outputs is a reflection of how the local community views and values the landscape and the agricultural production it supports. The internal culture of wine-growing communities also has a significant bearing on the capacity and willingness of a community to support wine tourism. In this chapter, internal culture refers to a range of sociocultural factors that contribute to a region's development, identity, and sense of place (Thompson 2015). These factors include the agricultural history of the region, the skill sets of the region's agricultural workforce, willingness to innovate with new grape varieties, farming techniques adopted, as well as community aspects such as local lifestyle, values, and cultural heritage and customs that may reach back to the time when the region was first settled.

External culture also has a significant impact on the demand for wine tourism. When discussed here, external culture refers to aspects of demand for wine that are external to wine destinations and reflect aspects of national culture (Thompson 2015). Examples include changes in relation to the preference for and consumption of different styles of wine, demand for wine experiences in wine regions, and the trend for complementary food and wine tourism experiences (Hall and Sharples 2003). For wine tourism destinations, shifts in the external culture can influence the development and delivery of wine tourism experiences, presenting both challenges and opportunities for wine tourism destinations to respond with desirable wine varieties and desirable wine tourism experiences.

Methodology

Given that the aim of this research was to explore how the interactions between demand for wine-related tourism, the rural landscape, and early settler traditions have influenced the emergence of the Barossa and Margaret River, a multiple case study design using a place-based approach was adopted. A place-based approach facilitates recognition of the uniqueness of individual regions (Barca, McCann, and Rodríguez-Pose 2012; Dixit 2020; Turnour et al. 2014) and provides a platform for comparison between regions that have similarities and unique differences. Two fundamental aspects underpin a place-based approach: the importance of geographical context, including social, cultural, and institutional characteristics, and a focus on knowledge in policy intervention (Barca, McCann, and Rodríguez-Pose 2012). In this research, the employment of a place-based approach recognizes that tourism is a system influenced by a range of components within the system, as well as the geographical context (place) that tourism develops and operates. In this way, a place-based approach acknowledges the resources available to a specific geographic area and can be adopted where tourism is considered an activity that occurs in, and can contribute to the economy of, agricultural regions. Previous research into the development of agriculture has demonstrated that place-based approaches can be used to explore numerous factors affecting competitiveness, so that a region's potential may be realized and maximized (Turnour et al. 2014).

Multiple case studies are used to explore how interactions between the rural landscape, and internal and external culture influenced the development of the Barossa and Margaret River wine destinations. A two-stage approach was used. In the first stage, historical research methods were used to identify the cultural history of each region, particularly those aspects related to the settlement, viticulture, and wine tourism development. Next, semi-structured interviews were undertaken with representatives from the wine, food, and tourism industries, as well as public sector authorities. A total of 54 interviews were conducted, with 27 in each wine tourism destination. Transcripts from interviews were analyzed using NVivo (Version 10) software to identify keywords and phrases (Bazeley and Jackson 2013). Findings were subjected to content, historical, and thematic analyses and triangulated to establish converging lines of inquiry.

Several limitations encountered in this research should be acknowledged. The research design focused on supply-side perspectives rather than demand-side, which limits the generalizability of the findings. Moreover, the findings relate to the two Australian case regions and may not have the same significance in wine tourism destinations internationally.

Results

Results are presented in two case studies to identify the interactions between the landscape, regional culture (internal), and visitor culture (external), demonstrating how this has contributed to the development of two very different wine tourism experiences.

Case Study 53.1 Barossa

The Barossa is one of Australia's oldest wine-growing zones, encompassing the Barossa Valley and Eden Valley wine regions, and is located approximately one hour's drive north of Adelaide, the state capital of South Australia. In 2019, the Barossa attracted approximately 260,000 overnight visitors per year, 95% of which are domestic visitors, with an additional 997,000 day trips (SATC 2020). The area's settlement pattern is best described as a hierarchy of towns that provide the bulk of services to the region (Nuriootpa) and several smaller settlements that have touristic appeal and community services (Tanunda and Angaston). The Barossa's landscape and natural resources, including fertile soil and suitable climate for agriculture, attracted an initial wave of free settlers from Britain and Silesia (now Poland) in 1842. The area's original Indigenous occupants were displaced from their traditional lands as European settlers established their colonial settlements (Munchenberg et al. 2001). British settlers tended to acquire large blocks of land and focused on pastoralism, grain and fruit growing, and viticulture and wine making. Germanic-speaking Silesian settlers, who immigrated as entire communities of Lutheran worshippers escaping persecution, established subsistence farms on smaller land holdings, planting fruit trees, vines, and grain crops, and raising some cattle and pigs. In the decades following European settlement, new forms of agricultural production, including apples, pears, and stone fruits (peaches, plums, and apricots), were

Figure 53.1 A comparison of how the Barossa (a) and Margaret River (b) landscape has been modified to develop the wine tourism experience.

Source: Authors

introduced (Chinner 2010; Heuzenroeder 2002). More recently, growing national demand for wine allied with the quality of wine produced in the Barossa, led to the conversion of large sections of former farmland to vine production (see Figure 53.1).

The wine industry is now the single most important industry in the region and is considered a world leader in both technique and practice (Erlich, Riddell, and Wahlqvist 2005). Wine varieties synonymous with the region include Shiraz, Cabernet Sauvignon, and Riesling. The key factors in the region's success have been specialization in specific varieties, innovation, and the region's role as a hub for grape growing (production), processing and bottling (manufacture), and sales (including domestic and international exports). This innovative flair provides an insight into the local community's values, work ethic, and ability to adapt to changes in demand.

As a tourism destination, its key attractions include wine, regional food culture, settlement history, and cultural heritage set against the natural and built landscape that continues to shape the region's story. Many of the street names, buildings and architecture, festivals, shops, and cellar doors have a distinctive Germanic flavor, reflecting the region's heritage. Many families involved in the wine industry have strong historical connections to the initial settlers and are sixth-generation family members who continue to live by the cultural values of their forebears. Although wine tourism experiences, including visitation to cellar doors, festivals, and events, are an important element of the local economy, its share of overall regional GDP is small compared to the industrial scale of wine production.

Landscape

The Barossa's distinctive landscape is one of its key assets, with diverse soil types and microclimates that support a variety of agriculture, including viticulture (Munchenberg et al. 2001; Webb 2005). As one interviewee commented:

> We have some of the best produce here that grows in this environment. We have a great climate, four clear seasons. So we don't just do one thing all year round.
>
> *(Food representative 3)*

The importance of the region's soils, a key element of the landscape, has been featured in a recent marketing campaign. One interviewee referred to this specifically, saying:

> But now, the real shift is to recognize that – the value of the Barossa – and that's what the ad is trying to get across. It's the land and the fact that this wine is made from this fruit in this vineyard. That's becoming rarer and rarer in the world.
>
> *(Wine representative 2)*

The landscape has also enhanced the retention of elements of the Barossa's settler culture and heritage (Ioannou 2000), described by one interviewee as follows:

> One of the things that is extraordinary about the Barossa is that you cannot discuss the Barossa food culture without talking about what happened 160–170–180 years ago … Nothing here has happened in a vacuum …We have a food culture which is dynamic

> and living and alive and can embrace businesses like this. This doesn't sit in isolation from its environment. It's very much part of the broader food culture that we have. And that's because of the way that this region was settled.
>
> *(Food representative 5)*

The role of the landscape has been central to the development of a specific sense of place, which also resonates with visitors. One interviewee commented that the region's sense of place provided a connection between the natural and cultural landscapes, by suggesting:

> And the sense of place is what people remember … They can look at a view but the easiest way to get it into their being and their memory banks is through their bellies I think.
>
> *(Food representative 9)*

Internal culture

The internal culture of the Barossa encapsulates its regional identity and sense of place, comprised of its settler heritage, the people and their stories, local support, and sense of community. A strong sense of identity is felt among some community members, who referred to themselves as being 'Barossan.' This term tended to be used by fifth- and sixth-generation descendants of the original settlers and is based on traditional cultural customs and values that continue to influence their way of life. Examples of early settler values that continue to influence the region's contemporary internal culture include their faith, a strong work ethic, mutual respect, importance of family and community, and philosophies towards food practices. As one interviewee observed:

> With our heritage, we are talking about a region that was settled in 1842 … so quite unique in the way the Barossa was settled and developed. Many of those traditions carry on to today which gives the Barossa its strength.
>
> *(Wine representative 1)*

Identifying as being Barossan highlights how an individual's sense of identity and belonging to the region continue to contribute to the region's internal culture. However, over time, the regional culture has embraced external cultural influences, resulting in a fusion of modernity with tradition. This fusion of old and new was described by one interviewee who observed:

> There is a wine culture, a food culture. It's very much the artisan type food –what they brought in from Silesia and Europe all those many years ago. But of course, there has been a fusion since then. Freshness and locale is what the Barossa stands for.
>
> *(Wine representative 3)*

The internal culture has also enhanced a sense of community, where residents have a feeling of connectedness to each other, reinforced through their cultural heritage, story, and industry linkages. While businesses actively compete with each other, there is an underlying understanding of the interconnectedness of the region's wine, food, and heritage that form part of its tourist appeal and a shared sense of purpose. One interviewee commented on this perspective by stating:

> I've always long had a view that the wine industry is a little bit unique in that whilst they are competing, they also have strong synergies to each other. There is almost a

> symbiotic relationship going on. I think that, in itself, reflects community and there is a strong connection.
>
> *(LGA representative 1)*

External culture

The Barossa's development as a wine tourism destination has also been influenced by external cultural shifts in Australia's drinking and eating habits. For example, early recognition of a shift in Australia's drinking culture from fortified wines to reds to whites encouraged the region's producers to follow suit and transition to new varieties over time. Commenting on the Barossa's ability to adapt, one interviewee observed:

> One of the reasons they did that, though – prior to the [19]70s the grapes were grown for fortified wines. A lot of people think the Shiraz has always been the Barossa's wine … Once that started to swing around where people were drinking more table wine and less fortified.
>
> *(Wine representative 2)*

A growing interest in food and cooking globally has provided an opportunity for the region to complement its wine experiences with food, as one interviewee highlighted:

> Things like *MasterChef* [a popular Australian TV show] have been pivotal changes in Australian food … But really good for the industry and good for the Barossa. The Barossa has been on all those shows and getting highlights and showing what we actually have because the Barossa is worthwhile looking at.
>
> *(Food representative 4)*

This interest in food has also enabled the Barossa to showcase its regional food culture and heritage, previously practiced only behind the closed doors of the community, by developing authentic products and experiences that are shared with visitors. One interviewee discussed this connection to culture through food by saying:

> I think there is a strong element of regional food and regional culture and local ingredients … People are asking for that more. People want to connect to that regional food story more.
>
> *(Wine representative 1)*

Shifts in the external culture have allowed the Barossa to strengthen its wine industry, and tell the region's story by developing wine and food experiences, which are tangible expressions of its internal culture and are framed by landscape that has shaped its settlement, cultural heritage and development. This sentiment was observed by one interviewee who stated:

> But there is a lot for tourists to enjoy and experience here. That includes local culture. The traditions and the history, the food, the wine and landscapes.
>
> *(Economic development representative 1)*

While aware of the changes in demand, the Barossa is unlike other wine tourism destinations, in that the region draws on its considerable cultural heritage, including its wine, food, and natural and built landscapes, to determine the types of authentic experiences it can offer visitors.

Case Study 53.2 Margaret River

Margaret River is a relatively new wine tourism destination situated a three-hour drive south of Perth, the capital city of Western Australia. The region has a diverse landscape that includes beaches, limestone caves, soils suitable for viticulture and other agricultural outputs, and natural areas dominated by native forests. Prior to the introduction of viticulture in the 1960s, Margaret River was a struggling agricultural area that had remained largely unchanged for 100 years. The rapid expansion of the region's viticulture sector and production of award-winning wines began to attract national and international attention and provided the stimulus for the emergence of the region as a major domestic wine tourism destination. Aside from its attractiveness as a wine tourism destination, Margaret River offers a bundle of tourism opportunities based on its natural, agricultural, and cultural landscapes that include surfing, beaches, coastal walking trails, lighthouses, rainforest walks, guided cave tours, and built heritage.

Until the mid-1960s, Margaret River was a quiet, impoverished, and quaintly rural area that was characterized by beef and dairy production with a small timber sector. There was little tourist interest in the area with the exception of surfing. In 1965, a report by Dr. John Gladstones recognized the suitability of the region's soils to support a profitable viticulture industry. The warm, Mediterranean-style climate provides near-perfect growing conditions for Chardonnay, Cabernet Sauvignon, and Cabernet blends. Viticulture was pioneered by a trio of medical practitioners from Perth, and in 1967 the first commercial plantings of vines were undertaken. From the outset, Margaret River's winemakers focused on making premium-quality wine. Success by the early winemakers in winning a number of notable wine awards during the 1970s and 1980s helped to build the region's reputation and establish Margaret River as one of Australia's leading wine regions (Thompson 2015). By the late 2010s, the number of wine producers had grown to approximately 150, with 5,500 hectares under vine, making viticulture the region's leading, primary industry and underpinning a tourism sector that attracted 1,559,200 overnight visitors in 2019, including 141,200 international visitors (MRBTA 2020).

During the 1960s, Margaret River gained recognition as a surfing destination. However, its geographic isolation, compounded by poor highways linking the region to Perth, hampered efforts to develop a tourism sector beyond surfing. The introduction of viticulture in the late 1960s and expansion of wine production during the 1970s laid the foundation for rapid expansion of the tourism sector during the 1980s. A contributing factor behind this expansion was to build the region's competitive advantage in the wine market. As a comparatively small wine region in terms of production, the winemakers realized that they had an opportunity to develop wine tourism experiences that leveraged the region's natural environment. Taking advantage of visitor interest in regional wine tourism experiences, many vineyard owners invested heavily in cellar doors (which have a winery restaurant), kitchen gardens, unique architecture, art galleries and sculpture, and boutique giftware (see Figure 53.1).

While growth in tourism from the 1980s onwards can be largely attributed to an interest in wine, the diversity of the regional landscape provided additional points of interest that

enhanced the region's competitive advantage as a tourism destination. Key features in the natural environment include native jarrah and karri forests, limestone caves, coastline, beaches, and surf. As tourist interest grew, the region attracted investment in tourism infrastructure, including accommodation, restaurants, retail shops, and second-home ownership, while diversifying into local food experiences in response to visitor demand.

Landscape

The region has a Mediterranean-style climate with maritime influences that create ideal growing conditions for viticulture and a variety of other agricultural produce. As one interviewee stated:

> You can grow anything down here because the climate is so mild ... It is Mediterranean.
> *(Tourism representative 9)*

The landscape is also an attractive natural backdrop that is central to the region's sense of place and identity. The importance of the landscape was highlighted by several interviewees, who referred to the unique, diverse, and pristine environment as key geographical features. These sentiments are reflected in the following statement:

> And I think it's the uniqueness – I mean you just sit out at some of the beautiful beaches or the forest and you just think, 'Well, where else in the world would you see this?' To have all this in one small region.
> *(Tourism representative 5)*

The linkages between the attraction of the environment and the activities that have developed were described by one interviewee who stated:

> It's about exploring the fact that there are so many different wine styles. There are so many different cellar doors. There are so many different ways of doing wine and exploring what's happening in the region, and as an offshoot of that you get the experience of driving through the countryside that's relatively green because of all the vines, and the breweries that are being developed.
> *(Tourism representative 6)*

In addition to viticulture, the natural landscape provides a resource base from which a diversity of experiences has been developed. The combination of landscape, industry (viticulture) and tourism activity provides Margaret River with a competitive advantage based on the ability of the landscape to enhance the total tourist experience.

Internal culture

Lifestyle is an important part of the appeal of Margaret River as a place to live and play, which has been influenced by the natural landscape and environment. The laidback, surfing culture that emerged in the 1960s, combined with the rural lifestyle of the period, produced a unique local culture that has endured into the present. Surfers contributed to building a strong community presence and later established businesses, such as surf clothing and supply shops, which in turn spurred the region's tourism development. Margaret River continues to exhibit a laid-back atmosphere that is overlayed with environmental awareness and conservation values that fit well with the surfing culture (Erlich, Riddell, and Wahlqvist 2005). The connection between surfing and tourism development is highlighted by the following reflection:

> The first vines got put in the 60s, and that was when the tourism first started as well, mainly by surfers exploring the area. So it went through that exploration phase of tourism. And the surfers doubled as workers in the early days, so that worked off itself as well.
>
> *(Tourism representative 9)*

The unique Margaret River lifestyle has also attracted passionate and highly motivated individuals into its (business) community, some of whom share similar environmental values. The shared passion and motivation was expressed by many interviewees from across the food and wine industries, with one stating:

> Businesses across the food, wine and tourism industries share a passion for delivering a quality product or experience.
>
> *(Tourism representative 1)*

This passion for the Margaret River lifestyle and the environmental values it embodies is demonstrated in the community's attitudes towards protecting the environment and key industries (wine and tourism) from other forms of development. Proposals put forward in the early 2010s to approve on-land and off-shore mining prompted a fierce community backlash, leading one interviewee to comment:

> We are sitting on large reserves of coal – gas and coal. And there is oil offshore here. So that's going to be a major issue. Everyone says we fought them off, we beat them off. But they have very good lawyers and they will come back.
>
> *(Tourism representative 7)*

External culture

Changes in national food and drinking culture can exert a powerful influence on local communities that are members of the supply chains for these agricultural products. The growth in Margaret River's wine production and wine tourism paralleled national level changes in wine consumption, travel behavior, and growing interest in food culture. One interviewee described the parallels with changes in national drinking culture by stating:

> There was also a change in the drinking habits of Australians. There was a shift in that from just beer, to beer and wine.
>
> *(Economic development representative 1)*

Another interviewee added to this perspective, commenting on changes in travel, saying:

> The golden years of the 1990s were incredible for the wine industry of Australia and I think that was certainly embraced in WA and Margaret River. Hence, we saw that development into, I think, maybe close to 100 cellar doors in Margaret River ...And

> what's gone along with that side of things has been the development of accommodation and that ranges from backpacking in places to stay up to boutique places to stay like Cape Lodge. It's all developed over that period of time.
>
> *(Wine representative 2)*

The development of Margaret River as a wine tourism destination is a result of a combination of factors, many of which can be traced back to the region's landscape. Aside from the role of the landscape as the key factor in supporting wine production, it also supported the region's surfing culture, which, combined with the region's rural lifestyle, produced a unique local culture that continues to attract lifestyle migration into the region. Unfortunately, there are indications that the region's increased popularity for retirees and investors (second homes) may pose a threat to this unique lifestyle in the future.

Discussion and conclusion

This chapter has examined how the interactions between the landscape, regional culture (internal), and visitor culture (external) have influenced the emergence of two different wine tourism destinations. In the Barossa, aspects of the early settler traditions and cultural heritage continue to be strong influences on the community's way of life and work. This internal culture has also shaped how the landscape was transformed to support agricultural and large-scale viticultural production and manufacturing that continues to dominate the landscape and regional economy into the present. Together, the internal culture and rich landscape has enhanced the region's ability to adapt to changes in demand for wine styles and experiences.

As one of Australia's newest wine regions, Margaret River has experienced rapid changes in a relatively short timeframe. The region's landscape, and recognition of its ability to support viticulture, served as the catalyst behind the development of its wine and tourism industries. In addition to supporting viticulture, the diverse landscape is both attractive and pivotal to the diversity of tourism experiences that have enhanced its competitive advantage, whether based on wine, surfing, beaches, caves, or forests. This natural landscape is recognized and protected by the value-based internal culture that started with early agriculture and was nurtured by the 'surfies' who later settled in the region. The meteoric rise of the region's reputation for wine, wine tourism, and complementary tourism experiences has been enhanced due to an alignment between the development of the region's wine and tourism industries and the growing demand for these experiences within the national culture.

The cases of the Barossa and Margaret River highlight how regional wine tourism experiences are a function of the landscape, and of cultural influences from both within the local community and from changing trends in external visitors attracted to the region. While existing definitions (Hall et al. 2000) of wine tourism have focused on a demand-side perspective, this research demonstrates that supply-side factors also contribute to the wine tourism experience and should be considered when defining wine tourism.

The landscape is the backdrop on which all tourist activities and experiences take place and can be considered inextricably linked to tourism (Terkenli 2005). From a tourism perspective, the landscape, including terroir (Seguin 1986), forms an essential part of the wine tourism experiences in both the Barossa and Margaret River. The landscape contributes to the sense of place, which in turn encourages visitation because of its touristic appeal, where it may be consumed by gazing upon it (Urry 2002) as well as engaging with it (Thompson et al. 2016). In the Barossa and Margaret River, the landscape was an aesthetically pleasing setting and formed part of the 'in region' experience at cellar doors and winery restaurants.

In addition to setting the scene, the landscape provides the resources for the development of wine and complementary tourism experiences. However, the degree to which this comparative advantage is harnessed is dependent on a region's internal culture. Comparatively, the tourism experience in the Barossa is less developed than Margaret River. While both destinations have wine tourism experiences, the diversity of environmental assets in Margaret River has enhanced the development of a bundle of tourism attributes that include wine, food, heritage, and a range of nature-based tourism experiences. However, the popularity of Margaret River's lifestyle elements may threaten its internal culture, in which the sense of community is lost due ongoing investment (through home ownership) from outside the region. Lifestyle is considered a key element of the wine tourism experience, but it cannot detract from other factors that enhance that experience, such as experience authenticity and the setting and surroundings (Roberts and Sparks 2006).

In comparison, the Barossa is fiercely protective of its rich cultural heritage and history that is shared with visitors but on a smaller scale, with a particular focus on its wine and regional food culture. This has been a conscious decision by the industry leaders to maintain the authenticity and essence of the region's cultural heritage and values. This lack of tourism development may threaten future tourism growth and experience development, where the destination continues to cater to day visitors and remains relatively small compared to the industrial scale of wine production in the Barossa. Interestingly, it may also add to its touristic appeal, as visitors are seeking authentic, engaging, and meaningful wine tourism experiences (Charters, Fountain, and Fish 2009; Gunasekar et al. 2021; Roberts and Sparks 2006) based on wine traditions and customs that form part of the Barossa's internal culture.

The internal culture of each wine tourism destination – especially its capacity to innovate – has been central to its ability not only to transform regional assets for tourism purposes, but to respond to changes in demand (external culture) that influence how individuals experience wine tourism (Charters, Fountain, and Fish 2009). It may be assumed that upholding traditions and customs comes at the cost of innovation. However, the Barossa demonstrates how the internal culture provides a unique ability to strive for innovation and efficiencies that are aligned with the underlying values of their cultural heritage and traditions. The uniqueness of the Barossa is its ability to draw people into the past, with a living history of the region and its people, while continuing to innovate and evolve the wine tourism experience to current tastes that are a fusion of old and new, showcasing a dynamic regional and cultural identity that provides a sense of being 'Barossan.'

Meanwhile, Margaret River has demonstrated innovation through the fusion of wine and tourism, developing several wine tourism experiences that incorporate food, music, and visual arts set against a pristine natural setting that all add value to the total wine tourism experience (Roberts and Sparks 2006). The region's innovative culture extends to the development of complementary tourism experiences that draw on the natural landscape and add to the diversity of products and experiences offered, from coastal hikes and surfing to shopping and food experiences. In addition to developing tourism experiences, the innovative qualities of the internal culture have enhanced each region's ability to adapt to trends in the external culture, identifying and catering to changes in demand. This is particularly evident in shifts from fortified to red and white wines in the Barossa and the development of wine tourism experiences in Margaret River.

In addition to the landscape influencing wine tourism destinations from a resource perspective, by supporting agriculture and viticulture production, this influence extends to location. Actual or perceived distance from a source market contributes to a destination's ability to attract the market and the size of that market. While being located close to a key source market may be perceived as an advantage, in the case of the Barossa it may become a hindrance. Located one hour's drive from Adelaide, the Barossa is an attractive destination for day visitors, which has limited its capacity to develop an overnight market and the supporting infrastructure required, such as accommodations and restaurants. On the other hand, located a three-hour drive from its main source market (Perth) has enhanced the ability of Margaret River to develop a range of supporting infrastructure to attract and accommodate an overnight market, catering for long weekends away with a range of accommodation options, winery restaurants, shopping, and touristic activities. Although each destination has a position within the domestic market, relatively few international tourists are drawn to these destinations, highlighting the influence of their regional locations and the importance of accessibility to the destination.

Implications

The place-based approach adopted in this research has shown how a region's landscape, internal culture, and external visitor culture, and the interaction between these factors, influenced the emergence of culturally different wine tourism destinations. Although wine is a dominant function in the Barossa and Margaret River, these wine tourism destinations have developed distinct regional cultures, with a unique sense of place and regional identity that appeals to domestic target markets. This is in part due to a response to market demand, but also based on the degree to which the wine and other regional assets have been transformed into tourism experiences. The unique combination of these factors differentiates wine regions and broadens the appeal of wine tourism through marketing and delivering regionally based experiences.

References

Barca, F., McCann, P., and Rodríguez-Pose, A. (2012). The case for regional development intervention: place-based versus place-neutral approaches. *Journal of Regional Science*, 52(1): 134–152. https://doi.org/10.1111/j.1467-9787.2011.00756.x

Bazeley, P., and Jackson, K. (2013). *Qualitative Data Analysis with NVIVO* (2nd ed.). London: Sage.

Carlsen, J., and Boksberger, P. (2015). Enhancing consumer value in wine tourism. *Journal of Hospitality and Tourism Research*, 39(1): 132–144. https://doi.org/10.1177/1096348012471379

Cavicchi, A., and Santini, C. (2014). *Food and Wine Events in Europe: A Stakeholder Approach*. Oxon: Routledge.

Charters, S., Fountain, J., and Fish, N. (2009). 'You felt like lingering ...' Experiencing 'real' service at the winery tasting room. *Journal of Travel Research*, 48(1): 122–134. https://doi.org/10.1177/0047287508326508

Chinner, B. (2010). *The Fruits of Angaston*. Angaston, Australia: Author.

Dixit, S. K. (ed.). (2020). *Tourism in Asian Cities*. London: Routledge.

Erlich, R., Riddell, R., and Wahlqvist, M. (eds.). (2005). *Regional Foods: Australia's Health and Wealth*. Australian Government – Rural Industries Research and Development Corporation (RIRDC). Australia: Kingston.

Galloway, G., Mitchell, R., Getz, D., Crouch, G., and Ong, B. (2008). Sensation seeking and the prediction of attitudes and behaviours of wine tourists. *Tourism Management*, 29(5): 950–966. https://doi.org/10.1016/j.tourman.2007.11.006

Griffin, T., and Loersch, A. (2006). The determinants of quality experiences in an emerging wine region. In Carlsen, J., and Charters, S. (eds.), *Global Wine Tourism: Research, Management and Marketing* (pp. 80–92). Wallingford: CABI.

Gunasekar, S., Das, P., Dixit, S. K., Mandal, S., and Mehta, S. R. (2021). Wine-experienscape and tourist satisfaction: through the lens of online reviews. *Journal of Foodservice Business Research*. https://doi.org/10.1080/15378020.2021.2006039

Hall, C. M., and Sharples, L. (2003). The consumption of experiences or the experience of consumption? An introduction to the tourism of taste. In Hall, C. M., Sharples, L., Mitchell, R., Macionis, N., and Cambourne, B. (eds.), *Food Tourism around the World: Development, Management and Markets* (pp. 1–24). Oxford: Butterworth-Heinemann.

Hall, C. M., Sharples, L., Cambourne, B., and Macionis, N. (eds.) (2000). *Wine Tourism around the World: Development, Management and Markets*. Oxford: Butterworth-Heinemnn.

Heuzenroeder, A. (2002). *Barossa Food*. Adelaide: Wakefield Press.

Ioannou, N. (2000). *Barossa Journeys: Into a Valley of Tradition*. Australia: New Holland Publishers.

Lin, Y.-C., Pearson, T. E., and Cai, L. P. (2011). Food as a form of destination identity: a tourism destination brand perspective. *Tourism and Hospitality Research*, 11(1): 30–48. https://doi.org/10.1057/thr.2010.22

Margaret River Busselton Tourism Association (MRBTA). (2020). Margaret River region latest visitation. *MRBTA*. https://corporate.margaretriver.com/resources/margaret-river-region-latest-visitation-2/

Montella, M. (2017). Wine tourism and sustainability: a review. *Sustainability*, 9(1): 113. https://doi.org/10.3390/su9010113

Munchenberg, R. S., Proeve, H. F. W., Ross, D. A., Hausler, E. A., Saegenschnitter, G. B., Ioannou, N., and Tuesner, R. E. (2001). *The Barossa: A Vision Realised: The Nineteenth Century Story*. Adelaide: Openbook Publishers.

Poitras, L., and Getz, D. (2006). Sustainable wine tourism: the host community perspective. *Journal of Sustainable Tourism*, 14: 425–448. https://doi.org/10.2167/jost587.0

Roberts, L., and Sparks, B. (2006). Enhancing the wine tourism experience: the customers' viewpoint. In Jackson, J., and Charters, S. (eds.), *Global Wine Tourism: Research, Management and Marketing* (pp. 47–55). Wallingford: CABI.

Seguin, G. (1986). 'Terroirs' and pedology of vine growing. *Experientia*, 42: 861–873.

South Australian Tourism Commission (SATC). (2020). Barossa regional profile. *SATC* https://tourism.sa.gov.au/media/j5rdu13p/barossa_dec19.pdf

Terkenli, T. S. (2002). Landscapes of tourism: towards a global cultural economy of space? *Tourism Geographies*, 4: 227–254. https://doi.org/10.1080/14616680210147409

Terkenli, T. S. (2005). Human activity in landscape seasonality: the case of tourism in Crete. *Landscape Research*, 30: 221–239. https://doi.org/10.1080/01426390500044408

Thompson, M. (2015). Tourism in agricultural regions in Australia: Developing experiences from agricultural resources [PhD thesis]. James Cook University.

Thompson, M., Prideaux, B., McShane, C., Dale, A., Turnour, J., and Atkinson, M. (2016). Tourism development in agricultural landscapes: the case of the Atherton Tablelands, Australia. *Landscape Research*, 41(7): 730–743. https://doi.org/10.1080/01426397.2016.1174839

Turnour, J., McShane, C., Thompson, M., Dale, A., Prideaux, B., and Atkinson, M. (2014). A place-based agriculture development framework. *RIRDC*. https://rirdc.infoservices.com.au/items/15-003

Urry, J. (2002). *The Tourist Gaze* (2nd ed.). London: Sage.

Van Leeuwen, C., and Seguin, G. (2006). The concept of terroir in viticulture. *Journal of Wine Research*, 17(1): 1–10. https://doi.org/10.1080/09571260600633135

Webb, M. (2005). South Australia: the Barossa region. In Erlich, R., Riddell, R., and Wahlqvist, M. (eds.), *Regional Foods: Australia's Health and Wealth* (pp. 105–115). Barton, Australia: RIRDC.

Williams, P. (2001). Positioning wine tourism destinations: an image analysis. *International Journal of Wine Marketing*, 13(3): 42–58. https://doi.org/10.1108/eb008726

54

EXPLORING THE ROLE OF RESTAURANT WINE LISTS IN COMMUNICATING LOCAL SENSE OF PLACE TO TOURISTS

Rory Hill and Joanna Fountain

Placing the wine list

Enjoying a 'taste of place' is an important component of an authentic tourism experience; whether savoring a Burgundy *grand cru* in a medieval cellar surrounded by vineyards, or eating street food at a hawker's market in Malaysia. The close connection between food, wine, and tourism is ubiquitous in popular and academic writing on tourism, with print publications and innumerable blogs and websites dedicated to the topic, and scholars in many disciplines exploring dimensions of the phenomenon (see, for example, Fountain and Thompson 2019; Fusté-Forné and Berno 2016; Gunasekar et al. 2021; Sidali, Kastenholz, and Bianchi 2015; Sims 2009). Furthermore, many consumers feel growing concern over global food homogenization and are seeking out sustainably and ethically produced food and drink products, prioritizing locally grown and made produce, and exploring the people, practices, and places of production (Corsi 2018; Fountain et al. 2021).

The discovery, sale, and enjoyment of local wine within the restaurant setting contributes to the visibility, reputation, and success of a wine region and contributes something to the culture of wine within that region. Within wine regions of the Old World, tourists and local residents expect to see local wines listed on the menus of restaurants, and stocking these wines may be an obvious business decision for the restaurateur. In emerging wine regions of the New World, however, this cannot be taken for granted, and it has been argued that restaurants which make an effort to stock local wines help to promote those regions (Gultek, Dodd, and Guydosh 2005; Perla, Rickard, and Schmit 2014). For many wine producers – particularly those making small volumes of wine – restaurants are a vital sales channel (Wine Australia 2020; Deloitte 2017). Increasing the availability of local wines on wine lists may be of particular importance in regions that have relatively short histories of wine production, and where enjoying a taste of place can open the door to a wider sense of the location, and embed wine production within the region's economy and imaginary. It has been a point of pride and an act of cultural diplomacy for little-known English wines to figure on state dinners in the United Kingdom, for example, just as it has long been for wines from the famous French regions to figure on theirs (Quinn 2015; Picon and Kenedi 2019).

Yet wine, as a cultural and commercial object, exists within its own circuits of globalization, and there is a set of expectations around the kind of wines, and wine regions, that should be

DOI: 10.4324/9781003143628-62

available in a hotel or restaurant context. These expectations are shared to a greater or lesser extent by tourists and by restaurateurs. They are propagated through the media, rehearsed by wine importers, wholesalers, and suppliers, and reinforced by travel. Champagne is perhaps the best example. It would be difficult to find a Michelin-starred restaurant, or five-star hotel, anywhere in the world, that does not stock champagne. This builds on several centuries of global consumption of that wine (Guy 2010; Neiman 2020). There are also wines which, whilst not quite as universally retained or renowned as Champagne, nonetheless figure widely in hospitality businesses throughout at least the OECD countries (which accounted for over half of all tourist arrivals in 2018 (World Bank 2021)). Red wines from Bordeaux, Australia, and California, sparkling wines from Italy, rosé wines from Provence, and white wines from New Zealand: all of these have achieved global notoriety and are marketed and sold through global supply chains. Globally popular wines have even been accused of accelerating a globalization of taste at the expense of lesser-known grape varieties, styles, and regions of origin (Bailly 2000; Pitte 2001; Goodman 2009; Inglis and Almila 2019). Yet rarely do scholars ask: how do hoteliers, restaurateurs, and sommeliers meet these globally-crafted expectations? And, when tourists travel and drink wine, do they want what they know from home, do they want what they can expect wherever they go, or do they want what is local?

This issue can be examined through the wine lists of hospitality providers in tourist destinations, which offer clues to the availability and market positioning of local wines, as well as the prevalence of imported wines. Wine lists exist at the interface of a number of factors. They are created with a market in mind, with a certain kind of food offering and dining experience intended, and they must mediate the wines available for purchase by the business owner with the probable desires of the customers who will enter the premises. The cultural and commercial contexts of wine lists may be visualized in a model (see Figure 54.1). This model is not programmatic, nor does its dimensions indicate any relative size or scale, but it is indicative of the influences that are mediated through the construction of a wine list, and offers an analytical tool to investigate the relationship between these wider issues and the wine list itself.

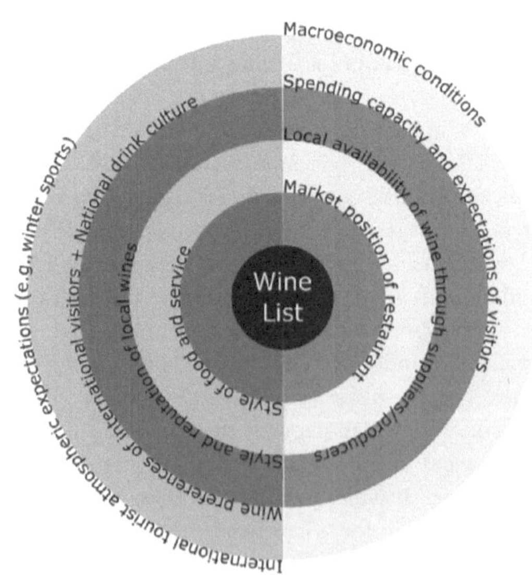

Figure 54.1 Cultural and commercial contexts of the restaurant wine list.

Wine lists have thus far received limited academic attention, but there are some key papers, written from a wine business perspective, that draw upon wine lists to make their arguments. Berenguer, Gil, and Ruiz (2009) studied wine lists as a differentiation strategy amongst fine dining restaurants in Spain, while Corsi, Mueller, and Lockshin (2012) used wine lists to explore how consumers choose wine in the restaurant context. From a management perspective, both Sirieix et al. (2011) and Hwang and Kunc (2015) incorporated wine lists in their analysis of how restaurateurs go about choosing which wines to stock. Thorsen and Hall (2001) undertook an exploratory study of New Zealand restaurateurs' attitudes and policies in regard to stocking local wines, and Livat and Rimaud (2018) have more recently called for more attention to be paid to how the origin of wines is considered in the assembly of wine lists. These papers examine wine in hospitality settings of diverse national contexts, and it is worthwhile studying this in emerging wine regions of the New World, where customs and traditions surrounding both the production and consumption of wine are of more recent origin than the Old World (Black and Ulin 2013; Howland 2014; Hill and Fountain 2022).

Something that is alluded to, but rarely elaborated on, in those articles is how wine lists can be used as springboards for wider conversations. Within the lively, convivial, and performative atmosphere of the restaurant (Crang 1994), referral to wine lists by customers and staff can serve as the starting point for making connections to wider wine knowledge and wine tourism experiences; for 'upselling' and persuasion; for building and maintaining a wine region's reputation; and, we will contend, for conveying a sense of place. The agricultural or viticultural evidence in the surrounding landscape, as well as the interior décor of the restaurant, can whet the appetite and begin to evoke the association of place and product (Fountain and Thompson 2019; Miele and Murdoch 2002); the discussion of a wine identified in the menu may intensify this appreciation. Indeed, as we highlight in this chapter, restaurateurs and restaurant staff members whom we interviewed told us that once people had a taste of a wine, they often wanted to hear much more about where and how it was made. In this way, the experience of tasting wine in the restaurant setting could lead to the appreciation of a wider sense of place, mediated by staff members.

In this chapter, we will analyze the extent to which local wines figure on the menus of restaurants in the popular destination of Queenstown, New Zealand. Building on that analysis, we will explore the degree to which wine lists serve as a starting point for wider discussions about wine, the terroir – or sense of place – of the local wine region, and the connections to other wine tourism experiences within that region.

Case Study 54.1 Queenstown: where the high country wines flow

Case study, materials, and methods

Queenstown is a year-round tourist destination in the Central Otago region of New Zealand's South Island. It is particularly popular amongst adventure sports and winter sports tourists, with the verticality of its nearby mountains, passes, gorges, and cliffs providing ample opportunity and inspiration for harnessing gravity to have fun. Its resident population of around 40,000 (Stats 2020) is swelled by considerable numbers of tourists, with the last pre-COVID-19 figures showing 3.7 million annual guest nights in 2019 (Stats 2019). Queenstown has an international airport, and it is estimated that two-thirds of tourist spending in the town in 2019 was from overseas tourists (Stats 2019). Some of these tourists come to Queenstown as part of a wider New Zealand itinerary; others come specifically for winter sports or other activities available in the Queenstown area, or may repair to the town from other parts of the region for its hospitality offering and renowned nightlife. In common with other parts of New Zealand, the largest proportion of overseas tourists come from Australia, followed by China, the United States, and the United Kingdom (MBIE 2020). Potential tourist experiences in Queenstown and its surrounding region range from hiking trails and staying in hostels or campsites, to heli-skiing and staying in luxurious lodges and hotels. Within New Zealand, Queenstown and the Central Otago region have a reputation for hosting affluent visitors and residents (Newport 2017), and a number of fine dining restaurants operate in the town, some with dedicated sommeliers and substantial wine lists.

Queenstown's town center sits on the shore of Lake Wakatipu, and it is framed by the Remarkables range of mountains. The region's scenic value is considerable, and it has been suggested that 'the seductive physical beauty of Central Otago has played a key role in drawing consumers to the region's wines, the alleged thought sequence being that such God-given beauty must surely provide God-given wine' (Saker 2010, p. 228). The landscapes of Central Otago featured in many scenes of the *Lord of the Rings* films (Carl, Kindon, and Smith 2007), and the region hosts over a hundred wine producers and some 2,000 hectares of vineyard (Central Otago Winegrowers Association 2019; New Zealand Winegrowers 2020). Three quarters of all vineyards in the region grow Pinot Noir grapes; the varietal for which the region has gained a global reputation from the 1990s on (Central Otago Winegrowers Association 2019; Moran 2016). With its international exposure, and its position within a successful wine region, Queenstown is an interesting case study to explore the nexus of globally prevalent wine tastes and locally specific wine offering. As would be common to any study carried out in 2020, however, we must offer the caveat that the COVID-19 pandemic, and government responses to it, dramatically reduced the number of tourists who visited Queenstown. Indeed, it may be that the 2019 tourism figures mentioned here are a high water mark that won't be reached again for some years.

The research methods employed in this study were twofold. First, wine lists from a sample of Queenstown restaurants were analyzed, with 'restaurant' defined as any licensed premises where meals can be purchased and eaten. Some of the premises we visited were in hotels or stand-alone restaurants, while others might be better described as pubs or cafés. Second, in-depth interviews were carried out with owners, managers, or other staff members at an overlapping sample of Queenstown restaurants. The analysis of wine lists was carried out between April and July 2020. In common with Ruiz-Molina, Gil-Saura, and Berenguer-Contri (2010), we used an official directory as a starting point to identify restaurants (Destination Queenstown 2020), followed by a TripAdvisor list (TripAdvisor 2020). The official list displayed 101 restaurants and the TripAdvisor list displayed 195 restaurants, with substantial overlap. An online search for each restaurant was conducted, and where a restaurant website was found with its wine list available, the list was downloaded. In this way, we gathered 43 wine lists. This sample likely represents somewhere between 22% and 43% of the population of restaurants in Queenstown. Menus were analyzed to ascertain the number of different wines offered for sale, their places of origin, their prices, and any further information of note, such as tasting notes, food pairing recommendations, and illustrations. A spreadsheet was used to enumerate the number and origins of wines on each wine list, and this formed the basis for the figures presented in the next section, and acted as a prompt for the subsequent interviews that followed.

Initially, random sampling was used to determine which of the 43 restaurants to approach for interview. Sets of five randomly selected restaurants were contacted by telephone or email, and where contact could not be made, or an interview could not be arranged, further sets were contacted. 'Snowballing,' or recommendations by respondents of other people to interview, led to further interviews at restaurants which did not figure in the original sample, but wherein wine lists were obtained and analyzed during the visit. In total, 16 interviews were carried out at premises in Queenstown between August and October 2020 (including one interview that was carried out over the telephone). The persons interviewed occupied various positions, including sommeliers, restaurant owners, managers, and other staff members involved in the selection, sale, and/or service of wine. Interviews were semi-structured, with questions concerning the choice of wines available on the menu, the most popular wines, the questions customers ask, supply chain challenges, interactions with other tourism operators and wine industry stakeholders, sense of place in the Central Otago wine region, and the effects of the COVID-19 pandemic on the business. Interviews allowed us to gather perspectives from those involved in the provision of wine in the hospitality setting, and they allowed us to ask why the wine lists looked the way they did and contained the wines they did. In line with much work in economic geography, tourism, and business studies (Clark 1998; Fountain, Charters, and Cogan-Marie 2020; Sutherland and Burton 2011), we have left respondents' names out of this work, in the interests of maintaining commercial confidentiality.

Menu analysis

The findings from the Queenstown wine list analysis show a substantial preference for wines from the Central Otago region. Around a third of all wines listed on the sampled wine lists come from Central Otago (838 out of the 2,690 wines, or 31.2%). Furthermore, Central Otago is the most represented single place of origin on three quarters of the wine lists sampled (ahead of Marlborough, Hawke's Bay, South Australia, California, Champagne, etc.). Given the country's proximity to Australia, it is not a surprise that wine lists in New Zealand have long featured Australian wines (Beaven 1977; Thorsen and Hall 2001, p. 95). Australia is the most frequent place of origin for wines imported to New Zealand (New Zealand Winegrowers 2020), and powerful, often inexpensive red wines from South Australia are particularly popular. That region is better represented in our sample than some of the New Zealand wine regions, such as Nelson or Gisborne. The same can also be said for French wines from the famous regions of Bordeaux, Burgundy, and Champagne. Wines from other parts of Europe, and the world, were much less represented, with only one wine each from Chile and South Africa, for example.

Figure 54.2 shows the spread of places of origin for all 2,690 wines listed across the 43 menus. We have distinguished regions within countries in the cases of New Zealand, Australia, France, and the United States, to allow more detailed analysis. Immediately it is evident from the graph that Central Otago is the most common region of origin. That region has over twice as many appearances on Queenstown wine lists as the next region (Marlborough). Central Otago wines were present on all of the 43 restaurants sampled – which ranged from fine dining to fast casual and pub-style establishments – and a Central Otago Pinot Noir red wine was the most expensive wine available on 20 of the 43 wine lists (Champagne was the most expensive wine on 21 of the 43 lists). This speaks to the general popularity of Central Otago wine within Queenstown, as well as to the premium, 'trophy' status of the region's Pinot Noir red wine.

When asked about the rationale for their wine lists, our respondents attested to stocking local wines because they are what they feel visitors to the region want. One succinctly argued that 'most customers want local wines' and that 'everyone knows Central Otago is famous for Pinot Noir.' They also described wanting to showcase the region's wine, and the relative ease with which they could obtain stocks of local wine, building in some cases on longstanding relationships with wine producers. As one restaurateur put it:

> We wouldn't bother with anything else because it's, well it's 15 minutes down the road, and I think even locals know that our Pinot is good. We live in this area, we support local, we go local. But anybody coming to this region knows that Central Otago Pinot is up there with the best in the world, so it's just a no-brainer to promote that for us.

It is also significant that the popularity of Central Otago wines sits within an overall preference for New Zealand wines amongst Queenstown wine lists around three quarters of all wines (1,967 out of 2,690, or 73.1%) were from the country (Figure 54.2). One respondent explained

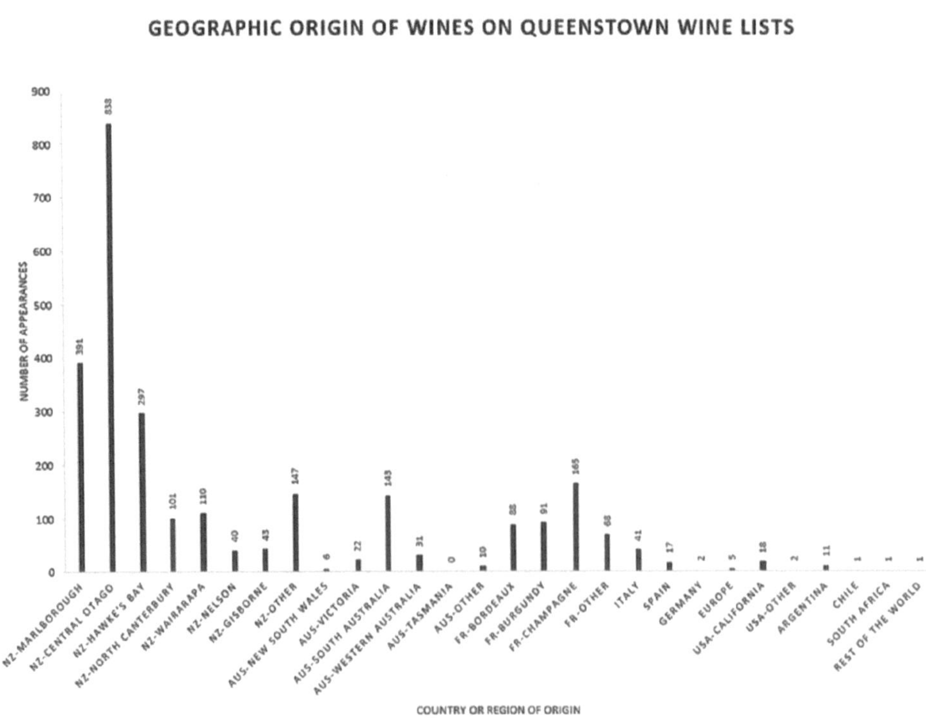

Figure 54.2 Place of origin of wines enumerated in sample of Queenstown wine lists.

Source: Graph is drawn with the assistance of Madeleine Griffith.

that 'Sauvignon Blanc is always very popular, especially with overseas customers, because they know it so well,' and in her restaurant, all of the Sauvignon Blanc was from the Marlborough region of New Zealand. Those white wines have become the backbone of the New Zealand wine industry (New Zealand Winegrowers 2020), enjoying fame throughout the major export markets of Australia, the United Kingdom, and the United States. Meanwhile, in another restaurant, the desire to sell 'local' wine was not confined to the borders of Central Otago:

> I mean 95% of the time, the guest wants something local. And when I say local, I'll kind of include the North Island in that because if they want a Cabernet, or they want a Syrah, or a big Chardonnay, you can't have it from here and you need to go up North.

Whilst wines from overseas do figure on Queenstown wine lists, the large numbers of international tourists described in the previous section do not seem to have driven restaurateurs to favor an international wine offering over a local one; the reverse is true. The findings outlined here indicate the influence of tourism – including wine tourism – in Central Otago on the curation of wine lists in Queenstown hospitality businesses. In some cases, restaurateurs did attest to stocking one or two wines from the places tourists come from:

> We sell a little bit of Australian 'cause obviously people come over quite a lot from Australia and they [might be] set in their ways and they want an Australian Shiraz, and you're like, 'well hey that's fine we've got them.'

But respondents could also cater to the different tastes and preferences of different groups of international tourists *within* their local or national wine offering. They described changing their wine lists not only according to typical seasonal weather (with more white and rosé wines in the summer, for instance), but also according to the seasonal fluxes of incoming tourists:

> A lot of it does depend on, well, pre-COVID, [it] depended on for example, if Australians are here, I get a lot more sort of bigger reds – obviously the Pinot Noir list, everyone [likes], that doesn't change – but I'll get more Rieslings for Australians … especially drier Rieslings. For the American markets, if [it's that] certain time of year when they come, I'll get a few more Chardonnays and your bigger reds again. So [then] it sort of leans more towards Hawke's Bay.

Furthermore, one respondent felt that the local wine had a distinct advantage over its best-known international competitor, by virtue of its approachable taste:

> If you were comparing a Central Otago Pinot Noir to French Burgundy or something like that, people find that the Central Otago is much more approachable, especially if you've not had a lot of Pinot in the past … the French [wine] can be a bit more on the austere side.

Our respondents felt that Queenstown, being within the Central Otago wine region, and near to many winery cellar doors, received a significant number of wine tourists, which maintained the interest in stocking local wines. But they also felt that consuming local wine – and, to some extent, other local produce – fitted within the desires of many tourists who were not heavily wine-involved. One identified 'lodge stays' and 'adventure tourism' as distinctively

Queenstown experiences that attracted certain market segments and paired well with opportunities to also try local wines. The popularity of winter sports tourism in Queenstown, too, of at least 70 years standing (Troup 2007), drives appetites for wine experiences such as a cheese fondue dinner, or the preparation of mulled wine. This speaks to the globalization of tourism and culinary experiences, and the propagation of the expectation of a certain kind of winter sports atmosphere amongst tourists with the disposable income to take part in it (Hill 2022; Howland 2014). Fondue and wine at the fireside have long exceeded the confines of the Swiss Alps and are increasingly an Instagrammable symbol of regions with slopes that support grapevines, pastures, and ski fields, such as the Okanagan Valley in Canada and Central Otago in New Zealand.

Transmitting a sense of place

The wine list, as a site of encounter between visitors and the wine offering of the restaurant, was frequently raised by respondents as a starting point for discussions about the quality of the local wines, the presence of vineyards and wineries in the surrounding region, and any amount of storytelling in their regard. As one respondent put it:

> I would have to say that [customers] want to recognize something – whether it's a variety or a brand – on the [wine] list so that if they don't get the information they want from the staff, they'll default to something that they recognize. But they're, generally speaking, ...wide open to being taken on a journey, so we have to be prepared to take them on that journey.

The wine list can serve as a kind of road map for a journey into taste, varieties, places, traditions, vintages, and more. Our respondents often said the most common question they are asked about wine is 'What do you recommend?' From this starting point, the sommelier or other table server can guide the customer not only towards making a purchase decision, but also towards a greater understanding of some of the (viti)cultural background of the wines under discussion. Another respondent agreed that:

> So ... to me that's probably one of the biggest parts of selling wine....I try and choose wines that do have ... that sense of place. Like most of my favorite wines, or my biggest sellers, are ones that have stories relating to the gold rush, and, you know, glacial deposits from glaciers that came through 200,000 years ago before the last one, etcetera, etcetera. Yeah, all about creating that idea of a sense of space, and a sense of where you are, and, you know, a bit more about the history involved in what that wine is, not just 'here's a glass of wine to go with your [meal].'

That sense of place is made up of the climatic and physiographic influences on wine production, but also the region's landscape features, and its tangible or intangible heritage. In the previous example, the respondent mentions Central Otago's gold mining heritage of the nineteenth century – a set of activities, stories, and inscriptions in the landscape that have been used frequently in the marketing of wines from the region (Carpenter 2016; Fountain and Dawson 2014). The pioneering stories of hard work, colorful characters, and gold rush finds are easily assimilated into narratives of wine production, despite the intervening decades. They add a certain charm and stimulate the imagination, and wines bearing names such as 'Mt. Difficulty,' 'Pick and Shovel,' and 'Mysterious Diggings' speak strongly of this Central Otago heritage. The winegrowing conditions of the region, meanwhile, are also noteworthy, given that it has New Zealand's only quasi-continental climate, with substantial diurnal and seasonal temperature variation, distinctive schist geology, and the status of the world's most southerly wine region. One restaurant we visited included a map of New Zealand on the first page of their wine menu, to facilitate conversations about the climatic and physical sense of place that lie behind the country's wines:

> It means people with no geographical knowledge of the area at all can actually still appreciate the wine and where everything is coming from. And if I'm talking to them about … well hey look the further north you get, the warmer it tends to get, so the type of flavours you tend to get and the thicker-skinned grapes. So, they can then themselves be looking through the wine list and be like, 'oh OK that's from Marlborough so that's from up there,' and 'that's from Waipara so that's down there,' and they can actually help themselves a bit more with the wine list. And actually, learn a little bit whilst, whilst looking through it.

Whilst a staff member would often give guidance as to the taste or style of wine to be expected, or which wine might best match the meal the customer has ordered, time and again the wider context of the wine was raised by respondents as something they would also talk to customers about, often prompted by a question from the customer. As one respondent explained:

> We talk about the style of wine in relation to … the soil and the growing conditions. We might mention things like vine density, we might mention things like …. lower cropping years and things like that to tell the story, and I think that when you, when you are facing a customer it's the story of the wine that gets them interested, not 'this is going to smell and taste like that.'

These wider introductions to the region would also frequently lead to recommendations of other tourist experiences, often related to wine. For one hotel restaurant, upselling guided wine tours was sometimes the result of listing local wines from businesses that were open to visitors: a customer's question about a listed wine might lead, under the staff members' recommendation, to 'a whole day experience of exploring the wineries around the region.' Even where there is no direct commercial interest for the restaurant, sommeliers and staff members will maintain close ties with wineries in the region by virtue of informal advice to customers. One respondent explained that:

> Most nights, I end up writing a list of vineyards for people to go and visit while they're here. Not so much at the moment [with borders closed due to COVID-19] 'cause a lot of Kiwis already know a lot about what's going on, but it was more of that international market, I used to find just about every night three, four, five times a night, I'd be writing out a list of vineyards to go and visit, you know, a few of those ones off the beaten track people wouldn't have heard of that aren't on the tourist bus drop-off.

Such testimony shows us that through the conversations sparked by the wine list and its denotation of place, staff in the hospitality sector transmit a local sense of place, share information about wine cultivation, and act as informal guides to the wider region. This activity is often overlooked in the tourism literature, and it is supplementary to the staff's main duties of selling and serving wine and food. But, taking place the day after, it is important in raising the region's profile, and maintaining links between its producers, hospitality businesses, and tourists. Such links are reinforced in other ways, too, as one of our respondents put it:

> We've worked kind of hand-in-hand with our suppliers for many, many years now, so we've got a real close relationship with them … I'd say the staff are more confident with the stories about the wine than they are about where the food comes from.

Another respondent concurred:

> What makes a unique selling point is that we've been to the vineyards, we can talk about the vineyards themselves. Producers, they do like to get us out, they know that sales will always get boosted if people have been to the vineyard. So a lot of them do a hospo [hospitality sector] lunch or a hospo day … or sometimes we'll organise it ourselves. People can really sell what they know: especially if you're not all that confident selling wine, it gives you a lot of confidence when you have been there.

These comments echo Hwang and Kunc's (2015) finding that a focused wine list, featuring wines that staff have themselves gotten to know, allows staff to effectively describe and sell wine. But our data extends that finding to show how narrating and promoting the place of origin of the wine can not only help to achieve a sale, but also to facilitate further wine tourism experiences, and incrementally raise the profile of the local wine region. In the case of Queenstown, this adds further value to a Central Otago wine region already heavily favored by restaurants in the composition of their wine lists.

Conclusion

In this chapter we have demonstrated how hospitality businesses in the destination of Queenstown support their surrounding wine region by heavily stocking local wines. Through key informant interviews, we have also been able to show how restaurant wine lists offer a starting point for storytelling, the transmission of a sense of place, and the informal promotion of further wine tourism experiences in the wider region by staff. Restaurateurs, sommeliers, and other staff members told us that once people had a taste of a wine, they often wanted to hear much more about where it was made and the way in which it was made. In this way, the experience of tasting wine in the restaurant setting can result in a wider appreciation of sense of place, mediated by the member of staff. We found staff members had significant discretion not only to suggest a wine, but also to add value to the dining experience by narrating details of the wine's origin, its terroir, and regional context. In this way, our chapter has advanced the conceptual understanding of wine lists, further developed the role of place in the dining experience, and identified sommeliers, restaurateurs, and restaurant staff as key informal actors in transmitting a sense of place and generating traffic to wider (wine) tourism businesses.

The Queenstown case study demonstrates the support given to an emerging New World wine region by the local hospitality sector. The Central Otago wine region became commercially active only in the 1980s (Central Otago Winegrowers Association 2019), but our research shows that 40 years later, most Queenstown restaurants have more wines from that region on their wines lists than any other. It is significant that in that same period of time, tourism has grown in Queenstown and the Central Otago region. Throughout history, thriving cities have made nearby wine regions commercially successful (Dion 1959; Hill 2016; Overton 2020), and interest from overseas visitors can stimulate the conversion of other land uses to wine production (Gade 2004; Perkins, Mackay, and Espiner 2015). As we have endeavored to show in this chapter, restaurateurs and the wine lists they curate are key intermediaries in this process. In Queenstown, two-thirds of visitor spending come from overseas tourists, but less than one-third of the wines available on restaurant wine lists are imported; thus, restaurateurs can cater to international tastes whilst still offering a local product. The offering can be calibrated through the choice of grape variety, wine style, or even in telling a story that would be particularly compelling for certain international audiences.

Wine lists have received limited scholarly attention, though what they reveal about the market positioning of wine is of consequence to wine producers and regional wine organizations, such that they are included in proprietary research (see Wine Business Solutions 2021, for example). In this chapter we have sought to show how such lists offer a useful starting point for research into the wider cultural and commercial contexts of wine, including the emergent sense of place, and the relationships that exist amongst wine producers, the hospitality industry, and other tourism operators in the region. In the case of Queenstown and Central Otago, the strength of those relationships, the reputation and distinctiveness of the local wine, and the proclivities of visiting consumers make for a commercial success story. More research would be welcome from those many other parts of the world where international tourism, hospitality, and wine production overlap.

Implications

Beyond the conceptual and academic contributions of this chapter, a couple of practical implications for the hospitality, wine, and tourism industries are worth reiterating. Firstly, we found the most common question our respondents reported being asked in regard to wine was: 'What do you recommend?'; therefore there is significant discretion for restaurant staff to influence consumer choice. Accordingly, training, tastings, and winery visits for restaurant front-line staff can be strategic sales tools for local wineries to increase their restaurant sales. By familiarizing staff with one's wine, place, and story, they will be able to more easily craft a wine and winery narrative to influence consumers. This is particularly relevant for lesser-known wineries, or those with wines made from uncommon or unfashionable grape varieties. Our research also provides evidence of the maturing of New Zealand's wine and food industries as a whole, as measured by the prevalence of domestic wine in the wine lists of hospitality businesses, and the growing tendency to offer local food and wine matches. Developing the findings of Thorsen and Hall (2001) from 20 years earlier, we have found that today, New Zealand wines are offered alongside imported wines without detriment or apology. Our research is further evidence of the remarkable success story of New Zealand wine over the last two decades, a period that has also seen its remarkable growth and success as a tourism destination.

References

Bailly, A. S. (2000). Le temps des cépages: du terroir au système-monde. *Annales de Géographie*, 614: 516–524.

Beaven, D. W. (1977). *Wines for Dining*. Christchurch (NZ): Whitcoulls.

Berenguer, G., Gil, I., and Ruiz, M. E. (2009). Do upscale restaurant owners use wine lists as a differentiation strategy? *International Journal of Hospitality Management*, 28: 86–95.

Black, R. E., and Ulin, R. C. (eds.). (2013). *Wine and Culture: Vineyard to Glass*. London: Bloomsbury.

Carl, D., Kindon, S., and Smith, K. (2007). Tourists' experiences of film locations: New Zealand as 'Middle-Earth.' *Tourism Geographies*, 9: 49–63.

Carpenter, L. (2016). The use of gold rush nostalgia on wine labels: brief history of New Zealand's Central Otago wine region. In Muratovski, G. (ed.), *Consumer Culture: Selected Essays*. (pp. 99–117). Bristol: Intellect.

Central Otago Winegrowers Association. (2019). *Central Otago Wine Map (Edition 8)*. Cromwell (NZ): Central Otago Winegrowers Association.

Clark, G. L. (1998). Stylized facts and close dialogue: methodology in economic geography. *Annals of the Association of American Geographers*, 88: 73–87.

Corsi, A., Barbera, F., Dansero, E., and Peano, C. (eds.). (2018). *Alternative Food Networks: An Interdisciplinary Assessment*. Cham: Palgrave Macmillan.

Corsi, A. M., Mueller, S., and Lockshin, L. (2012). Let's see what they have ...: what consumers look for in a restaurant wine list. *Cornell Hospitality Quarterly*, 53: 110–121.

Crang, P. (1994). It's showtime: on the workplace geographies of display in a restaurant in southeast England. *Environment and Planning D: Society and Space*, 12(6): 675–704.

Deloitte (2017). Ripening opportunities: wine industry benchmarking and insights 2017. *New Zealand*. https://www2.deloitte.com/content/dam/Deloitte/nz/Documents/wine/nz-en-2017-wine-industry-benchmarking-insights.pdf

Destination Queenstown (2020). *Places to Eat and Drink*. https://www.queenstownnz.co.nz/places-to-eat-and-drink/

Dion, R. (1959). *Histoire de la vigne et du vin en France des origines au XIXe siècle*. Paris: Chez l'auteur.

Fountain, J., Charters, S., and Cogan-Marie, L. (2020). The real Burgundy: negotiating wine tourism, relational place and the global countryside. *Tourism Geographies*. https://doi.org/10.1080/14616688.2020.1713880

Fountain, J., Cradock-Henry, N., Buelow, F., and Rennie, H. G. (2021). Agri-food tourism, rural resilience, and recovery in a post-disaster context: insights and evidence from Kaikōura-Hurunui, New Zealand. *Tourism Analysis*, 26(2). https://doi.org/10.3727/108354221X16079839951420

Fountain, J., and Dawson, D. (2014). The new gold: the role of place and heritage in the marketing of the Central Otago wine region. In Harvey, M., White, L., and Frost, W. (eds.), *Wine and Identity: Branding, Heritage, Terroir* (pp. 43–56). Abingdon: Routledge.

Fountain, J., and Thompson, C. (2019). Wine tourist's perception of winescape in Central Otago, New Zealand. In Sigala, M., and Robinson, R. N. S. (eds.), *Wine Tourism Destination Management and Marketing: Theory and Cases* (pp. 45–65). Cham (CH): Palgrave Macmillan.

Fusté-Forné, F., and Berno, T. (2016). Food tourism in New Zealand: Canterbury's foodscapes. *Journal of Gastronomy and Tourism*, 2: 71–86.

Gade, D. W. (2004). Tradition, territory, and terroir in French viniculture: Cassis, France, and appellation contrôlée. *Annals of the Association of American Geographers*, 94: 848–867.

Goodman, S. (2009). An international comparison of retail consumer wine choice. *International Journal of Wine Business Research*, 21: 41–49.

Gultek, M. M., Dodd, T. H., and Guydosh, R. M. (2005). Restaurateurs' attitude toward local wines and its influence on local wine purchases. *International Journal of Wine Marketing*, 17: 5–24.

Gunasekar, S., Das, P., Dixit, S. K., Mandal, S., and Mehta, S. R. (2021). Wine-experienscape and tourist satisfaction: through the lens of online reviews. *Journal of Foodservice Business Research*. https://doi.org/10.1080/15378020.2021.2006039

Guy, K. (2010). *When Champagne Became French: Wine and the Making of a National Identity*. Baltimore, MD: Johns Hopkins University Press.

Hill, R. (2016). Local, loyal and constant'? On the dynamism of terroir in sustainable agriculture [Unpublished DPhil thesis]. University of Oxford.

Hill, R., and Fountain, J. (2022). Expressing sense of place and terroir through wine tourism encounters: antipodal reflections from France and New Zealand. In Charters, S., Demossier, M., Dutton, J., Harding, G., Maguire, J. S., Marks, D., and Unwin, T. (eds.), *Routledge Handbook of Wine and Culture* (pp. 90–98). Abingdon: Routledge. https://doi.org/10.4324/9781003034711-14

Hill, R. A. D. (2022). 'Le terroir, c'est la vie': Re-animating a concept among Burgundy's wine producers. *Environment and Planning E: Nature and Space*, 5(1): 447–472. https://doi.org/10.1177/2514848621989610

Howland, P. (ed.). (2014). *Social, Cultural and Economic Impacts of Wine in New Zealand*. Abingdon: Routledge.

Hwang, J., and Kunc, M. (2015). Business dynamics of on-premise wine trade: cases from South Korea. *International Journal of Wine Business Research*, 27: 239–254.

Inglis, D., and Almila, A.-M. (eds.). (2019). *The Globalization of Wine: the Trans-Nationalization and Localization of Production and Pleasure*. London: Bloomsbury.

Livat, F., and Rimaud, H. (2018). Factors affecting wine price mark-up in restaurants. *Journal of Wine Economics*, 13: 144–159.

MBIE. (2020). Key tourism statistics, February 12 2020. *New Zealand*. https://www.mbie.govt.nz/dmsdocument/3438-key-tourism-statistics-pdf

Miele, M., and Murdoch, J. (2002). The practical aesthetics of traditional cuisines: slow food in Tuscany. *Sociologia Ruralis*, 42: 312–328.

Moran, W. (2016). *New Zealand Wine*. Auckland: Auckland University Press.

Neiman, O. (2020). Les voyages forment le champagne. *Le Monde des Vins* [*Le Monde*], December 15, 2–3.

New Zealand Winegrowers. (2020). Annual Report 2020. *New Zealand*. https://www.nzwine.com/media/17492/annual-report_2020_final_web.pdf

Newport, P. (2017). A very, very important person's guide to Queenstown. *The Spinoff*, April 20. https://thespinoff.co.nz/society/20-04-2017/a-very-very-important-persons-guide-to-queenstown/

Overton, J. (2020). Landscapes of failure: why do some wine regions not succeed? In Myles, C. (ed.), *Fermented Landscapes* (pp. 57–82). Lincoln: University of Nebraska Press.

Perkins, H. C., Mackay, M., and Espiner, S. (2015). Putting pinot alongside merino in Cromwell District, Central Otago, New Zealand: rural amenity and the making of the global countryside. *Journal of Rural Studies*, 39, 85–98.

Perla, J. M., Rickard, B. J., and Schmit, T. M. (2014). Looking for Locapours: using Zagat survey data to examine restaurant demand for local wine. *Agricultural and Resource Economics Review*, 43, 69–86.

Picon, G., and Kenedi, N. (eds.). (2019). *Menus de légende*. Paris: Editions Dilecta.

Pitte, J.-R. (2001). La géographie du goût: entre mondialisation et enracinement local. *Annales de Géographie*, 621, 487–508.

Quinn, B. (2015). On menu at Queen's banquet for Xi Jinping: Balmoral venison and The Spy Who Loved Me. *Guardian*, October 20.

Ruiz-Molina, M. E., Gil-Saura, I., and Berenguer-Contri, G. (2010). Instruments for wine promotion in upscale restaurants. *Journal of Foodservice Business Research*, 13: 98–113.

Saker, J. (2010). *Pinot Noir: The New Zealand Story*. Auckland: Random House.

Sidali, K. L., Kastenholz, E., and Bianchi, R. (2015). Food tourism, niche markets and products in rural tourism: combining the intimacy model and the experience economy as a rural development strategy. *Journal of Sustainable Tourism*, 23(8–9): 1179–1197.

Sims, R. (2009). Food, place and authenticity: local food and the sustainable tourism experience. *Journal of Sustainable Tourism*, 17(3), 321–336.

Sirieix, L., Remaud, H., Lockshin, L., Thach, L., and Lease, T. (2011). Determinants of restaurant's owners/managers selection of wines to be offered on the wine list. *Journal of Retailing and Consumer Services*, 18: 500–508.

Stats, N. Z. (2019). Commercial accommodation monitor, September 2019: Queenstown. *New Zealand*. https://www.mbie.govt.nz/dmsdocument/4020-commercial-accommodation-monitor-september-2019-queenstown-pdf

Stats, N. Z. (2020). Queenstown-Lakes District. *New Zealand*. https://www.stats.govt.nz/tools/2018-census-place-summaries/queenstown-lakes-district

Sutherland, L.-A., and Burton, R. J. F. (2011). Good Farmers, Good Neighbours? The role of cultural capital in social capital development in a Scottish farming community. *Sociologia Ruralis*, 51: 238–255.

Thorsen, E. Ø., and Hall, C. M. (2001). What's on the wine list? Wine policies in the New Zealand restaurant industry. *International Journal of Wine Marketing*, 13: 94–102.

TripAdvisor (2020). *Restaurants in Queenstown*. https://www.tripadvisor.co.nz/Restaurants-g255122-Queenstown_Otago_Region_South_Island.html

Troup, A. (2007). Skiing. *Te Ara – The Encyclopedia of New Zealand*. http://www.TeAra.govt.nz/en/skiing

Wine Australia (2020). How have wine DTC channels fared in 2019–20? Results of the Wine Direct-to-Consumer survey 2020. *Australia*. https://www.wineaustralia.com/getmedia/b2cbbca9-3ded-4bdf-9845-c2e9e8021e26/DTC-report-2020_v2.pdf

Wine Business Solutions (2021). *Wine on Premise UK 2021*. https://winebusinesssolutions.com.au/shop/wine-on-premise-uk-2021/ [Accessed: 12th February, 2021].

World Bank (2021). International tourism, number of arrivals. *World Bank Data*. https://data.worldbank.org/indicator/ST.INT.ARVL

55

VITAL ELEMENTS TO FOSTER TERRITORIAL DEVELOPMENT BASED ON WINE TOURISM

Maria del Pilar Leal Londoño

Introduction

Undoubtedly, in the past decades, wine-related tourism has become increasingly important to a number of rural areas around the world (Hall and Mitchell 2001). Its activities encourage new opportunities for economic growth and drive the development of the sector in which it operates. Di Gregorio and Licari (2006) argue that wine, together with other factors as local typicality, natural resources, cultural heritage, and sustainable tourism promotion, represents a valid system to stimulate the socioeconomic and territorial expansion of vine-growing areas.

Moreover, tourism can facilitate the diversification of the rural community, and for marginal, isolated, and peripheral areas, it can also serve as a vehicle for economic and social regeneration and development (Boniface 2003). In the field of economics, the development of wine tourism might contribute to the horizontal integration of the traditional primary production function with that of a specialized tertiary function, increasing the sources of income and improving employment for local people.

As point out by Ramukumba et al. (2012), the key features of local development or Local Economic Development (LED) are to encourage economic growth and to diversify the local economic base. For some authors (see Bessière 1998), local development is more closely linked with the intrinsic value of a territorial area's cultural and regional identity, in which the heritage can be seen as a social resource structure for local achievement. For instance, in Europe, local development, according to Hjalager (1996), has been promoted through many rural development programs and policies where tourism has been seen as a key driver for it.

Wine production linked to tourism has raised not just the interests of the winemakers and local authorities as a means of local development, but also the interest of tourists. This is because, and according to Getz (2000), wine tourism can offer a full sensory experience. The tourist can experience the pleasures of taste, smell, touch, sight, and sound. For tourists, all the production system becomes an element of motivation; thus, the growing techniques, the grape varieties, the methods, the landscape, and the cellar, among other elements, become part of the touristic experience.

As Nella and Christou (2014) highlighted, wine tourism can be viewed as a local and regional development tool, which allows for the integration of primary (agriculture), secondary (wine industry), and tertiary (tourism) sectors, highlighting the respective wine territory's landscape attributes and the uniqueness of its regional 'touristic terroir.'

DOI: 10.4324/9781003143628-63

However, for wine tourism to become a genuine option for local and regional development, there have to be good links between wine production, the tourism supply chain, and, in general, the whole tourism industry. According to Salom Carrasco and Albertos Puebla (2009), achieving local and regional development does not depend solely on business or economic variables; rather, it is essential to have a development strategy that is common to the social and institutional actors and which is upheld by a given social and territorial unit.

As Pike, Rodriguez-Pose, and Tomaney (2006, 2016) state, the current debate regarding local and regional development incorporates concepts of innovation, knowledge, and learning as essential ideas to explain and understand contemporary local and regional development. Some authors, such as Lüthi (2011), mention that knowledge has become the main source of economic development in advanced regions.

However, these concepts are not only applied to economic growth and development but also can contribute to the broader concept of development – that is, one that links social, economic, and environmental aspects. Thus, the chapter is grounded on concepts related to innovation, knowledge, creativity, and learning as key elements that can help to promote wine tourism as a driving force for local and regional development.

The chapter takes a case study conducted in 15 wine cellars classified as innovative across Spain and seeks to examine the concepts of knowledge, innovation, creativity, and learning production as a key mechanism for fostering territorial development based on wine tourism. The information that supports the analysis is based on quantitative information gathered that had been confronted to the main concepts and characteristics of local and regional development based on literature review. The chapter is organized as follows. The next section elaborates the conceptual framework. The case study is then presented, followed by a discussion of the empirical results and some conclusions.

Finally, the chapter addresses some key theoretical and practical issues that reveal how knowledge and innovation generation depends on the actors' experiences and is adjusted to local cultures. Their experiences allow actors to develop products that reinvent, reinforce, or contribute to the creation of new modes of wine tourism experiences. The results obtained allow us to identify not only knowledge, innovation, and creativity as a toolkit for developing wine tourism, but also how wineries have an impact in regions and contribute to territorial dynamization due to the strategies implemented.

Conceptualizing innovation, knowledge, creativity, and learning linked to local and regional development

It might be argued that the various theoretical approaches taken to explain local and regional development can vary markedly; however, as Pike, Rodriguez-Pose, and Tomaney (2006, 2016) argue, the concepts and theories that can be used in its interpretation are closely related to responses to questions about the kind of local and regional development and for whom it is intended.

In the quest of finding answers to that questions, Lüthi (2011) argues that local and regional development might be seen as forming part of an endogenous approach that incorporates the institutional and socioeconomic focuses that seek to explain the characteristics and underlying forms of development, of a unique set of local assets and the endogenous economic capabilities on which local and regional competitiveness are constituted and established.

In addition, local and regional development also includes concepts related to innovation, knowledge, and learning as ideas that can explain and further our understanding of contemporary local and regional development. According to Esparcia (2008, Esparcia, Escribano, and

Serrano 2016), these concepts allow us to interpret development as the processes of change and enhancement of what is local in a context of globalization and external competition that lead to innovation and knowledge generation through processes of training and learning.

Innovation

According to Dicken (2007), the concept of innovation refers not only to the creation of things but also to the transformation of inventions into products that can be marketed, what Schumpeter (1943) calls 'new combinations.' These new combinations correspond to the introduction of new products, new methods, the development of new markets, or the implementation of new organizational structures. Hjalager (2010) identifies four main innovation categories in tourism: product, process, organizational/managerial, and market innovations. Moreover, Lüthi (2011) adds two more categories: social and technical innovation.

However, innovation linked to local and regional development, as Lundvall (1992) notes, has been built from the transition in understanding innovation from a linear to an interactive model. The linear model is characterized according to Fagerberg (2005) by the linear conception of the innovation process and its different stages, but it focuses on the interests of science, researchers, and institutions. In contrast, and as stated by the author, the interactive model emphasizes the interactive processes that are generated in the different phases of the innovation process between companies and scientific institutions.

In the case of tourism, as Hall (2005) points out, tourism innovation has diverse forms, heterogeneous drivers, and are manifested at different scales, whether it is the individual, the firm, the destination, or the tourism system. Therefore, the author highlights that it is important to see innovation as systemic or as integral to the tourism system as a whole. However, local and regional institutional contexts are essential to explain the potential and role of innovation; some places are clearly more innovative and capable of producing and adapting innovations than others, which, according to Armstrong and Taylor (2000), is reflected in different levels of local and regional economic dynamism.

On the other hand, Salom Carrasco and Albertos Puebla (2009) and Esparcia, Escribano, and Serrano (2016) highlight social innovation as a central concept that emerges associated with local and regional development. This is defined as the satisfaction of human needs alienated through the transformation of social relations, thus expanding the role of innovation in development. Following the authors, they suggest an analysis of innovation applied to forms of cooperation, communication, and good governance. According to the authors, economic innovation often has no chance of bearing fruit unless it is accompanied by a similar innovation in the socio-institutional sphere that allows the construction of consensus and the formation of cooperation networks around joint strategies more or less explicit.

It is important to recognize the relevance of socio-institutional networks in the innovation and development processes of the territories. Processes that have been considered as important elements of local and regional development (Bathelt and Glückler 2003; Esparcia, Escribano, and Serrano 2016; Maillat 1995; Sánchez, Aparicio-Amador, and Alonso Santos 2010; Storper 1997), as well as networks of relationships (networks of business relationships, both institutional and business networks and socio-institutional networks) that appear as a strategic element.

Knowledge

Knowledge never stands still but is constantly being created. It is this continual process that drives economic evolution (Boschma and Martin 2010). As argued by Bathelt and Glückler

(2003), successful innovations are usually associated with the creation of new knowledge or the modification of existing knowledge. According to the authors, the process of generating new technologies and knowledge depends on the firms and actors' past experiences and their capacities to modify existing knowledge or create new knowledge.

On the one hand, Nonaka (1991) argued that knowledge could be explicit (know-that), which is relatively easily codified and transferred, and tacit knowledge, which is more difficult to formalize and, therefore, less easy to interpret and transfer from one organization to another. On the other hand, tacit knowledge, according to Polanyi (1966), is learned through both individual reflection and collaborative experience.

In the case of firms such as wine cellars, knowledge is a decisive asset, and knowledge creation is the key mechanism through which firms produce and sustain competitiveness (Bathelt and Glückler 2011). In tourism firms, knowledge is a critical determinant in innovative capability (Martínez-Román et al. 2015). Moreover, knowledge accumulation leads to business innovation, and sometimes the creation of a knowledge network resulting from business interaction derived from the trade of inputs, services, or memberships (Giuliani 2007).

Previous work by Shaw and Williams (2009) on knowledge transfer and management in tourism confirms knowledge as a competitive advantage for tourism firms. The authors stress the importance of knowledge in innovation. However, and according to Hjalager (2010), in order to influence innovation processes, knowledge has to be captured, made explicit, and properly understood. It also needs to be interpreted, restored, adapted into specific innovations, and recorded.

Therefore, it might be argued based on Tribe and Liburd (2016) that tourism knowledge might be seen as a tourism knowledge system that is complex and dynamic and where the co-creation of tourism and its knowledge needs the collaboration of all the stakeholders involved. Thus, knowledge associated with tourism activities might be recognized as an important element for local and regional development. For instance, Pike, Rodriguez-Pose, and Tomaney (2006, 2016) mention that for localities and regions, the role played by institutions, companies, and public bodies, among others, in the knowledge economy and in the management of their knowledge assets (carried out by individuals and institutions), have implications for its trajectory of development and relative prosperity.

Creativity

Although creativity was historically associated with the creative person, the contemporary debate emphasizes the social context and the broader environment of creativity (Richards 2011). Richards (2011), by citing Scott (2010), argues that socially embedded creativity implies much more than the activities of gifted individuals or members of the 'creative class.' The 'creative class,' according to Florida (2002), is now measured through a range of quasi-scientific rankings using indices for determinants of 'creativity.' It can be argued, as Richards (2011) states, that creativity has become a strategy to be followed by territories on different scales in a search for growth, as well as a strategy from promoting innovation and individual skill development.

Scott (2004) highlights a growing new creative economy and rising levels of optimism surrounding 'cultural-industrial districts' as drivers of local economic development at selected locations, above all in large cosmopolitan cities, but also in many other kinds of geographical contexts. In fact, and according to Richards and Wilson (2007), the economic, social, and cultural impacts of creativity depend to a large extent on the specific places in which creative strategies are applied. Moreover, the authors emphasize that a number of different spatial dimensions can be identified as having an impact on creative development, including national contexts, the urban–rural dimension, and the micro scale of clusters and enclaves within both urban and rural areas.

In fact, and according to Cloke (2007), creativity in rural tourism, for instance, may suggest new ways of understanding creativity, such as in terms of hybrid nature–society performances based around eco-experience or adventure. According to Richards (2011), a way of creating distinctive and creative places relies on creative resources that are employed to generate more distinctive identities, offering regions and cities a symbolic edge in an increasingly crowded marketplace. Therefore, the emphasis on such strategies has also shifted from tangible to intangible cultural resources because more places lacking a rich built heritage are now competing for tourism business (Richards and Wilson 2007). As Richards (2011) argues, this is part of a broader shift from comparative to competitive advantage in destination competitiveness.

Indeed, creativity is attractive as a policy option for stimulating a range of economic, cultural, and social outcomes. It is also attractive because of the argued advantages produced by networking and knowledge spillover, which stimulate further creative activity (Richards 2011). Creative tourism, based on Richards's research (2009), is a new form of tourism that has the potential to change existing models of tourism development and to make a contribution to diversifying and innovating the tourist experience. By doing so, creative tourism can help to stimulate local economic, social, and cultural development.

Learning

The ability to learn, according to Lundvall (1992), is essential to establish an adaptive capacity of localities and regions in development processes. Learning corresponds to a collective, social, and geographical process that affects a change in the capacity or understanding of an individual or an organization (Cooke and Morgan 1998).

However, as noted by MacKinnon, Cumbers, and Chapman (2002), learning must be differentiated between collective learning and simple learning, where the latter involves bilateral cooperation. The establishment of trust between actors is considered critical in collective learning activities. Trust is an important coordination mechanism between actors and, as suggested by the authors, is conceptualized as a key form of agglutination that unites networks among themselves and supports the participation of companies in collective learning processes.

Learning processes require, for the case of companies, according to MacKinnon, Cumbers, and Chapman (2002), a degree of stability and continuity in relationships in the company, which is facilitated by spatial proximity. Pike, Rodriguez-Pose, and Tomaney (2006, 2016) point out that local proximity, for example, reinforces learning. According to the authors, the transfer and fast application of knowledge generated by externalities for companies and institutions are considered crucial for local and regional development.

Spatial proximity may be sufficient as long as there is cognitive proximity for there to be room for interactive learning. Short distances, as Salom Carrasco and Albertos Puebla (2009) point out, facilitate information contacts and knowledge exchange. Localized forms of collective learning can be sustained over time only if companies continue to participate in this open exchange of knowledge and ideas. As a result, and according to Storper (1997), the set of relationships or 'non-commercialized interdependencies' are seen as key learning resources that allow certain regions to respond and adapt effectively to changes in an external market environment.

In the case of tourism, Leal Londoño (2015) identifies the importance of spatial proximity as a catalyst for the establishment of trust among stakeholders participating in food tourism activities. Trust as a coordination mechanism leads to a successful exchange of knowledge and to a reciprocal learning process based on interaction which might have an impact on local development. Therefore, it might be argued that in the case of wine tourism, spatial proximity and trust might play an important role in fostering local and regional development, too. Therefore, institutions and organizations might help to settle face-to-face activities that promote knowledge among stakeholders as a first step for the establishment of trust.

Moreover, when it comes to learning, and as Hall and Mitchell (2000) note, wine tourism also facilitates producer–consumer interaction and involves education about and experience of wine products and wine regions, including local cultures and winescapes. Therefore, wine tourism might drive the learning process across all the stakeholders participating due to face-to-face interactions resulting from the tourism experience.

Case Study 55.1 Knowledge, innovation, creativity, and learning: a toolkit for developing wine tourism

According to *La Asociación Española de Ciudades del Vino* ACEVIN (2021), in Spain, the number of visitors on the wine routes as one of the main wine tourism offers grew by 3.9% during 2019. The upward trend in the sector in recent years is confirmed with a record number of 3,076,334 visitors in 2019, pointing to a clear consolidation of the activity as an important complement to the Spanish tourist offer.

The latest data from ACEVIN (2021) reveal that the top three activities carried out by wine tourists in Spain correspond to visiting local wine cellars (65.41%), visiting towns (62.62%), and enjoying local gastronomy (44.46%). In this context, discussing the activities developed by wine cellars in the framework of tourism based on the concepts of innovation, knowledge, creativity, and learning processes is important for understanding wine tourism and its contribution to local and regional development.

Thus, the current case study takes place in Spain, and the primary data come from an online email survey conducted in 2017 to 15 out of 52 wine cellars identified as innovative and distributed across 14 provinces of the country, obtaining a response rate of 28%. The survey was addressed to wine cellar managers. The database was a compilation based on the list created by the wine tourism expert Manuel Colmenero Larriba (2012) published in his book *Bodegas de España para Visitar y Recomendar.*

In addition to the survey's results, the 52 wine cellars were analyzed based on the information available on their website and publish in Colmenero's book.

Innovation, knowledge, creativity, and learning

Undoubtedly, based on the information obtained, the most common activities offered by wine cellars in Spain are wine tastings (33%) and visits to facilities (40%), two activities that might not be considered a 'new combination' based on Schumpeter's ideas (1943). Despite that 73.3% of the wine cellars surveyed consider that they develop or perform different activities from the other wineries of the region. The tourism activities linked to wine identified are seen in Table 55.1.

Nevertheless, the most innovative activity discovered based on the information gathered corresponds to 'Wine & Sex,' a performative activity developed on the Canary Islands, one

Table 55.1 Main identified tourism activities linked to wine

County region	*Wine activities linked to tourism developed by the wine cellars surveyed*	*Common tourism activities in the region developed by wine cellars*	*Regional tourism features and wine production in 1000 hectoliters (2019)*
Basque Country	Experimental wine tastings Master class Visits with menu Wine therapy	Wine tastings Events	Medium-size wine production 601.98 hl 1.6 million tourists
Catalonia	Flights around the vineyard Gourmet dinner Hotel golf MICE tourism Visit to historical places Wine spa	Wine tastings Training in wine Trekking Events Accommodation	3rd region in wine production 3,148.6 hl 19,3 million visitors Most visited region of Spain
Canary Islands	Cheese and wine tastings Submarine wine tasting Vineyards visits Vineyards visits 'Wine & Sex' Wine house Wine museum	Wine tastings	Second-most visited region 13.1 million visitors Wine is barely recognized in the world wine 32.1 hl
Castile–la Mancha	Quiz game Walking through vineyards Wine shop	Wine tastings	1st wine producer in Spain 16,800 hl 262,044 tourists
Galicia	Wine tastings Wine visits	Wine production but rarely wine tastings or vineyards visits	373.8 hl 1.6 million of tourists
La Rioja	Customized visits Museum visits Photo exhibitions Vineyards visits Wine bar	Wine tastings Vineyard visits	The most famous wine region in Spain and abroad 1,894.3 hl 139,497 tourists
Navarre	Art center Picnics in vineyards Wine games	Wine tastings	Medium-size wine production 573 hl 415,054 tourists

Source: Own elaboration based on INE's data and Statista, 2019

of the regions with lower wine production in Spain. However, this activity is a product and process innovation delivered to the market since 2010. The performance is based on the family knowledge that has been accumulated throughout the fifth generation of winemakers. The show 'Wine & Sex' is performed three times per year, and since the opening in 2010, every season, the tickets have been sold out (see Figure 55.1). Another important element identified as a part of their success is creativity, based on Richards and Wilson (2007), where there is a shift from tangible (wine production) to intangible experiences (the performance), leading to a destination's competitive advantage based on companies' performance.

Tourists and local visitors are gathered in a cabaret ambiance where the main protagonist is the wine. Indeed, it is a way to visit the facilities and to taste wine and good

Figure 55.1 Merchandising of Wine & Sex in Tenerife.

Source: Maria del Pilar Leal Londoño

food. However, it is a creative and unique wine activity linked to tourism that drives in a systemic way the territory where it takes place, since other local wineries and wine cellars are becoming spin-offs of creativity in Tenerife Island. As Armstrong and Taylor (2000) point out, local and regional institutional contexts are essential to explain the potential and role of innovation; Canary Islands, when it comes to wine tourism, has been identified as more innovative and capable of producing innovations than other regions with a higher wine production such as La Rioja, or with a long touristic tradition such as Catalonia. This fact might be based on the limited territorial resources of isolated territories such as the Canary Islands that might drive creativity in winemakers in order to call tourists' and visitors' attention.

Based on Salom Carrasco and Albertos Puebla's (2009) ideas, the innovation identified in the case study has allowed a cooperation network among the wine cellars. This has been led by the Wine & Sex winemaker, which might lead to learning processes based on cooperation, since 34.7% of the respondents in the Canary Islands manifested to interact with other wineries for the implementation of tourism activities.

Knowledge is confirmed as an important asset for wine cellars and as a competitive advantage for them if they want to perform tourism activities linked to wine. There is no doubt that tourism activities are a good mechanism to diversify their productive activity, the fact was confirmed by 40% of the respondents, and 93% of the wine cellars have increased their revenues since they incorporated touristic activities besides the wine production. However, and according to Hjalager (2010), knowledge has to be captured, made explicit, and properly understood in order to influence wider innovation processes.

Discussion and conclusion

This chapter explored the concepts of innovation, knowledge, creativity, and learning as key elements for fostering territorial development based on wine tourism, drawing theoretical and practical insights. In the case of firms such as wine cellars, and according to Denicolai, Zuchella, and Ciocarelli (2010), firms develop specific know-how stored in organizational routines, rituals, and habits, and it is in the storage and use of this knowledge that innovation can be unlocked.

As argued by Bathelt and Glückler (2003), successful innovations are usually associated with the creation of new knowledge or the modification of existing knowledge. The process of generating new knowledge depends on the firms' and actors' past experiences and their capacities to modify existing knowledge or create new knowledge. This fact was observed in the case of the Canary Islands. Indeed, in tourism firms, knowledge is a critical determinant in innovative capability (Martínez-Román et al. 2015).

Moreover, and according to Gammak (2006), interorganizational networking and cooperation across sectors, diffusion of knowledge and learning, and innovation are the basis of competitive advantage. Based on the information gathered, the rise of different activities might be seen as an example of businesses building competitive advantage as a response to different drivers of change, such as new customers and travelers, alternative ways to explore destinations, and the widespread use of technology, among others. Thus, innovations in tourism enterprises respond to and are inspired by a range of external and internal factors (Hjalager 2010).

The territorial resources as a factor do not necessarily mean innovation in wine tourism; this is the case of Castile–la Mancha, which has higher wine production but the lowest differentiation in wine tourism activities. Therefore, another element in order to unlock innovation is creativity, which, based on the case study, has been manifest when there is a shift from tangible to intangible resources.

This chapter has argued that knowledge production and transfer stimulate creativity and lead to learning processes based on spatial proximity and the build of trust among stakeholders. Moreover, the chapter highlights that the ability to learn by individuals or organizations is essential to establish an adaptive capacity of localities and regions in development processes. However, learning processes are born based on trust as an important coordination mechanism; therefore, the establishment of trust facilitated by spatial proximity among stakeholders participating in wine tourism might drive cooperation, resulting in learning ecosystems for wine regions that help to stimulate local economic, social, and cultural development.

In the effort to answer Pike, Rodriguez-Pose, and Tomaney's (2006) question about for whom it is intended the local and regional development, in the case of wine tourism, wine cellars and stakeholders involved are a key driver for local and regional development. Based on Leal Londoño's research for food tourism (2015, 2018) and extrapolated to the case of wine tourism, it might be argued that wine tourism may fulfill the basic elements of local and regional development – it uses endogenous resources such as wine, encourages cooperative learning strategies, and promotes sociocultural heritage, and more recently, wine cellars are walking towards an enhancement of natural resources with their development of sustainable tourism activities linked to wine. However, and in order to explore and unlock innovation based on 'new combinations,' wine cellars in general need to search and explore creativity, share knowledge, and increase cooperation as a way to achieve local and regional development.

The case study suggests that wine tourism is an opportunity for innovation, for destinations as well as for companies. Furthermore, it suggests a creative and innovative wine tourism environment by knowledge production and transfer around wine. On the other hand, it might be claimed that wine tourism may fulfill the basic assumptions for local development identified by Leal Londoño (2015, 2018); therefore, institutions should help to strengthen interactions between wine associations, organizations, and wine cellars. Undoubtedly, the potential of wine tourism as a territorial catalyst is recognized not only in economic terms, but also in social, cultural, and environmental terms. Moreover, wine tourism brings an opportunity to achieve sustainable development because it has the capacity to play a significant role in national and regional sustainable tourism-development plans through its contribution to sustaining the economic and social bases of regions, as well as environmental dimensions (Hall and Mitchell 2000, p. 449).

However, and as Gammak (2006) mentions in the case of Australia's wine tourism but with application to the current chapter, the significant potential of wine tourism for regions and localities will be conditioned by the effectiveness of the partnerships formed by operators for branding, knowledge sharing, and sustainable practices. Moreover, and according to Karagiannis and Metaxas (2020), cooperation, trust, and commitment to environmental preservation, customer satisfaction, and quality is required by all participating members in order to make a wine region sustainable.

Implications and future research directions

Undoubtedly, wine tourism might be considered as a touristic phenomenon that might contribute to foster local and regional development. However, in order to achieve it, the chapter has highlighted how innovation, knowledge, learning production, and creativity are key elements that lead to territorial development based on wine tourism activities. Commonly, local and regional development has been predominantly analyzed and conceptualized to the economic dimension (Esparcia, Escribano, and Serrano 2016). However, the social dimension discussed in the chapter and represented on Wine & Sex as a specific example of innovation, knowledge, creativity, and learning production reveals that pivot from tangible to intangible experiences is a successful pathway for bringing memorable experiences to tourists. It might be argued that the chapter's case study illustrates a very specific situation; however, it reveals that in wine tourism, a higher wine production of wine cellars and territories does not mean innovative or creative wine tourism experiences. In fact, it has illustrated how wine cellars, independently of the territories, produce and reproduce the same wine tourism experiences that rely on wine production, such as wine tastings, and visit their facilities. Moreover, the case study exposes the importance of performative activities linked to wine in order to produce rich experiences, because it encompasses aspects of entertainment, aesthetic, and escapist realms and balance elements of active and passive consumer participation in which customers are both absorbed and immersed (Thanh and Kirova 2018).

Thus, the analysis of performative experiences in wine tourism, its aesthetic, its creativity level, and its knowledge sharing of wine cellars within specific wine destinations would place on the research agenda. Moreover, the chapter emphasized the need for further research on how wine tourism activities and experiences drive spillovers that might lead to a wider territorial dynamization where trust and spatial proximity were identified as an important coordination mechanism. However, a wider investigation on coordination mechanisms for knowledge transfer and sharing within the wine tourism sector is needed, specifically how wine cellars interact and cooperate among them, leading to wine destination competitiveness.

Based on the previous information, wine territories and wine cellars interested in exploring and implement wine tourism experiences should think and design their experiences based on wine knowledge accumulation that should unleash creativity and innovation in order to produce rich and memorable experiences. However, that creativity and knowledge should be shared among different stakeholders of a specific destination in order to achieve local and regional development. Last but not least, researchers interested in wine tourism might contribute to providing new knowledge regarding territorial development based on wine tourism innovation, knowledge, creativity, and learning processes among wine cellars and among wine tourism destinations.

References

ACEVIN (Asociación Española de Ciudades del Vino). (2021). *Análisis de la demanda turística Rutas del vino de España 2020–2021*. https://wineroutesofspain.com/wp-content/uploads/2021/10/informe-analisis-de-la-demanda-rve-2020-2021-.pdf

Armstrong, H., and Taylor, J. (2000). *Regional Economy and Policy* (3rd ed.). Oxford: Blackwell.

Bathelt, H., and Glückler, J. (2003). Toward a relational economic geography. *Journal of Economic Geography*, 3(2): 117–144. https://doi.org/10.1093/jeg/3.2.117

Bathelt, H., and Glückler, J. (2011). *The Relational Economy: Geographies of Knowing and Learning*. Oxford: Oxford University Press.

Bessière, J. (1998). Local development and heritage: traditional food and cuisine as tourist attractions in rural areas. *Sociologia Ruralis*, 38(1): 21–34. https://doi.org/10.1111/1467-9523.00061

Boniface, P. (2003). *Tasting Tourism: Travelling for Food and Drink*. Burlington, VT: Ashgate.

Boschma, R., and Martin, R. (2010). The aims and scope of evolutionary economic geography. In Boschma, R., and Martin, R. (eds.), *The Handbook of Evolutionary Economic Geography* (pp. 3–43). Cheltenham, UK, and Northampton, MA: Edward Elgar.

Cloke, P. (2007). Creativity and tourism in rural environments. In Richards, G., and Wilson, J. (eds.), *Tourism, Creativity and Development* (pp. 37–47). London: Routledge.

Colmenero Larriba, M. (2012). *Bodegas de España para visitar y recomendar*. Barcelona: Lunwerg.

Cooke, P., and Morgan, K. (1998). *The Associational Economy: Firms, Regions and Innovation*. Oxford: Oxford University Press.

Denicolai, S., Zuchella, A., and Ciocarelli, G. (2010). Reputation, trust and relational centrality in local networks: an evolutionary geography perspective. In Boschma, R., and Martin, R. (eds.), *The Handbook of Evolutionary Economic Geography* (pp. 280–297). Cheltenham, UK, and Northampton, MA: Edward Elgar.

Di Gregorio, D., and Licari, E. (2006, August). Rural development and wine tourism in southern Italy. In ERSA (eds.), *Enlargement, Southern Europe and the Mediterranean 46th Congress of the European Regional Science Association Conference Proceedings* (pp. 1–26). Volos, Greece: European Regional Science Association (ERSA), Louvain-la-Neuve.

Dicken, P. (2007). *Global Shift. Mapping the Changing Contours of the World Economy* (5th ed.). London: Sage.

Esparcia, J. (2008, June). La Formación para el desarrollo local. In García Rodríguez, L. (ed.), *La organización territorial del desarrollo local en España VII Coloquio de Desarrollo local Conference Proceedings* (pp. 271–310). La Laguna: Asociación de Geógrafos Españoles.

Esparcia, J., Escribano, J., and Serrano, J. (2016). Una aproximación al enfoque del capital social y a su contribución a los procesos de desarrollo local. *Investigaciones Regionales – Journal of Regional Research*, 34: 49–71. https://ideas.repec.org/

Fagerberg, J. (2005). Innovation: a guide to the literature. In Fagerberg, J., Mowery, D., and Nelson, R. (eds.), *The Oxford Handbook of Innovation* (pp. 1–26). Oxford: Oxford University Press.

Florida, R. (2002). *The Rise of the Creative Class: And How It Is Transforming Work, Leisure, Community, and Everyday Life*. New York: Basic Books.

Gammak, J. (2006). Wine tourism and sustainable development in regional Australia. In Carlsen, J., and Charters, S. (eds), *Global Wine Tourism: Research, Management and Marketing* (pp. 59–66). Oxon: CABI.

Getz, D. (2000). *Explore Wine Tourism, Management, Development and Destinations*. London: Cognizant Communication Corporation.

Giuliani, E. (2007). The selective nature of knowledge networks in clusters: evidence from the wine industry. *Journal of Economic Geography*, 7(2): 139–168. https://doi.org/10.1093/jeg/lbl014

Hall, C. M., and Mitchell, R. (2000). Wine tourism in the Mediterranean: a tool for restructuring and development. *Thunderbird International Business Review*, 42(4): 445–465.

Hall, C. M., and Mitchell, R. (2001). Wine and food tourism. In Douglas, N., and Derrett, R. (eds.), *Special Interest Tourism: Context and Cases* (pp. 307–329). Brisbane: John Wiley.

Hall, M. (2005). Rural wine and food tourism: cluster and network development. In Hall, D., Patrik, I., and Mitchell, M. (eds.), *Rural Tourism and Sustainable Business* (pp. 149–164). Frankfurt, New York, Ontario: Channel View Publications.

Hjalager, A. (1996). Agricultural diversification into tourism: evidence of a European community development program. *Tourism Management*, 17(2): 103–111. https://doi.org/10.1016/0261-5177(95)00113-1

Hjalager, A. (2010). Regional innovations systems: the case of angling tourism. *Tourism Geographies*, 12(2): 192–216. https://doi.org/10.1080/14616681003725201

Karagiannis, D., and Metaxas, T. (2020). Sustainable wine tourism development: case studies from the Greek region of Peloponnese. *Sustainability*, 12(12): 5223. https://doi.org/10.3390/su12125223

Leal Londoño, M. P. (2015). *Turismo Gastronómico, impulsor del comercio de proximidad*. Barcelona: Editorial UOC.

Leal Londoño, M. P. (2018). Culinary associations as a driving force for local development: the case of the Baix Llobregat gastronomy association (AGT). In Medina, X., and Tresserras, J. (eds.), *Estudios del hombre series antropología de la alimentación* (pp. 85–99). Guadalajara: Editorial Universidad de Guadalajara.

Lundvall, B. (1992). *National Systems of Innovation: Towards a Theory of Innovation and Interactive Learning*. London: Printer.

Lüthi, S. (2011). *Interlocking Firms Networks and Emerging Mega-City Regions, the Relational Geography of the Knowledge Economy in Germany* [PhD thesis]. Technische Universität München. https://mediatum.ub.tum.de/doc/1072569/1072569.pdf

MacKinnon, D., Cumbers, A., and Chapman, K. (2002). Learning, innovation and regional development: a critical appraisal of recent debates. *Progress in Human Geography*, 26(3): 293–311. https://doi.org/10.1191/0309132502ph371ra

Maillat, D. (1995). Territorial dynamic, innovative milieu and regional policy. *Entrepreneurship and Regional Development*, 7(2): 157–165. https://doi.org/10.1080/08985629500000010

Martínez-Román, J., Tamayo, J. A., Gamero, J., and Romero, J. E. (2015). Innovativeness and business performances in tourism SMEs. *Annals of Tourism Research*, 54: 118–135. https://doi.org/10.1016/j.annals.2015.07.004

Nella, A., and Christou, E. (2014). Segmenting wine tourists on the basis of involvement with wine. *Journal of Traveling & Tourism Marketing*, 31(7): 783–798. https://doi.org/10.1080/10548408.2014.889639

Nonaka, I. (1991). The knowledge creating company. *Harvard Business Review*, 69(6): 96–104. http://www.hbr.org

Pike, A., Rodriguez-Pose, J. T., and Tomaney, J. (2006). *Local and Regional Development*. London: Routledge.

Pike, A., Rodriguez-Pose, J. T., and Tomaney, J. (2016). *Local and Regional Development* (2nd ed.). London: Routledge.

Polanyi, M. (1966). *The Tacit Dimension*. London: Routledge & Kegan Paul.

Ramukumba, T., Mmbengwa V. M., Mwamayi, K. A., and Groenewald, J. A. (2012). Analysis of local economic development (LED) initiated partnership and support services for emerging tourism entrepreneurs in George municipality, Western Cape Province, RSA. *Tourism Management Perspectives*, 2–3: 7–12. https://doi.org/10.1016/j.tmp.2011.12.006

Richards, G. (2009). Creative tourism and local development. In Wurzburger, R., Pattakos, A., and Pratt, S. (eds.), *Creative Tourism: A Global Conversation* (pp. 78–90). Santa Fe, NM: Sunstone Press.

Richards, G. (2011). Creativity and tourism: the State of the Art. *Annals of Tourism Research*, 38(4): 1225–1253. https://doi.org/10.1016/j.annals.2011.07.008

Richards, G., and Wilson, J. (2007). Creativities in tourism development. In Richards, G., and Wilson, J. (eds.). *Tourism, Creativity and Development* (pp. 255–288). London: Routledge.

Salom Carrasco, J., and Albertos Puebla, J. M. (2009). El papel de las redes en el desarrollo territorial. In Salom Carrasco, J., and Albertos Puebla, J. M. (eds.), *Redes socioinstitucionales, estrategias de innovación y desarrollo territorial en España* (pp. 13–42). Valencia: Publicaciones de la Universidad de Valencia-Instituto Interuniversitario de Desarrollo Local.

Sánchez, J. L., Aparicio-Amador, J., and Alonso Santos, J. L. (2010). The shift between worlds of production as an innovative process in the wine industry in Castile and Leon (Spain). *Geoforum*, 41(3): 469–478. https://doi.org/10.1016/j.geoforum.2009.12.004

Schumpeter, J. A. (1943). *Capitalism, Socialism and Democracy*. London: Allen and Unwin (originally published in the USA in 1942; reprinted by Routledge, London in 1994).

Scott, A. J. (2004). Cultural-products industries and urban economic development: prospects for growth and market contestation in a global context. *Urban Affairs Review*, 39(4): 461–490. https://doi.org/10.1177/1078087403261256

Scott, A. J. (2010). Cultural economy and the creative field of the city. *Geografiska Annaler: Series B, Human Geography*, 92: 115–130.

Shaw, G., and Williams, A. (2009). Knowledge transfer and management in tourism organisations: an emerging research agenda. *Tourism Management*, 30(3): 325–335. https://doi.org/10.1016/j.tourman.2008.02.023

Storper, M. (1997). *The Regional World: Territorial Development in a Global Economy*. New York, London: Guilford.

Thanh, T. V., and Kirova, V. (2018). Wine tourism experience: a netnography study. *Journal of Business Research*, 83(C): 30–37. https://doi.org/10.1016/j.jbusres.2017.10.008

Tribe, J., and Liburd, J. J. (2016). The tourism knowledge system. *Annals of Tourism Research*, 57(C): 44–61. https://doi.org/10.1016/j.annals.2015.11.011

56
INTEGRATING WINE TOURISM WITH RURAL TOURISM EXPERIENCE

Keith H. Mandabach

Introduction

The passion that drives travelers to picturesque rural landscapes with unique cultures/heritage, as well as their impact and interaction promoting wine tourism, is examined in this chapter. A brief discussion describes rural agricultural tourism, as well as the current and potential impact on a rural area economy. This type of travel can be a major driver of rural economic development with a substantial impact worldwide. Definitions and explanations of the phenomenon of wine, heritage, cultural, rural, scenic, landscape, and agricultural tourism, their interrelationships, and marketing approaches to use this phenomenon to promote tourism are provided.

Theories explaining the variety of motivations that lead to this type of travel focusing on the experiential nature of rural tourism have been especially spotlighted during the COVID-19 pandemic. Our increasingly urbanized world during lockdown makes rural tourism especially attractive to a wider number of market segments. While destination wine tourism was a driver for tourist visits to rural areas, now it might be just an afterthought in a pandemic-evolved industry. Consumers' goals have now shifted to escape from the city or restrictions while seeking open-air experiences that allow them to create positive emotional responses such as surprise, joy, and love missing during lockdowns. The 'living in the moment' mentality presents wine tourism with the opportunity to showcase their operations to visitors enjoying rural experiences. Examples of how wine tourism operations can benefit from this type of visitor are provided in a unique case study. The chapter's purpose is to help wine tourism operators understand ways to integrate tourists experiencing outstanding natural landscapes, unique historical perspectives, interesting cultural backgrounds, agritourism, art, and interaction with residents into their businesses. Their goal should be to use these experiences to build synergy resulting in emotions of joy, love, and adoration for the rural tourism destination experience as well as wineries that happen to be there.

Wine tourism is traveling to vineyards or tasting rooms, touring the wineries, purchasing wine often at or near the source, walking and learning about the vineyards, which might also include dining, staying at the vineyard, or taking an active part in the planting, pruning, or harvest. The vineyards are normally located in rural areas that have a variety of other types of

DOI: 10.4324/9781003143628-64

attractions for visitors (WTO 2021). One type is *heritage tourism*, which is a segment of cultural tourism and is focused on the heritage of a tourist destination. Another is *cultural tourism*, which attracts visitors whose motivation is to learn, discover, experience, and consume tangible and intangible products and attractions in a destination. These products and attractions allow the tourist to experience and connect intellectually and emotionally at multiple levels with the features of society through arts, architecture, historical heritage, culinary traditions, food specialties, wines/beverages literature, music, creative industries, and the living cultures with their lifestyles, value systems, beliefs, and traditions (Tourism Notes 2021).

Wine, heritage, and cultural tourism markets are usually components of the larger *Rural Tourism* market. This market includes any form of tourism that showcases the rural life, art, culture, and heritage at rural locations, thereby benefiting the local community economically and socially as well as enabling interaction between the tourists and the locals for a more enriching tourism experience. This can be termed as rural tourism. (Tourism Notes 2021). *Rural landscape/scenic tourism* includes travel whose purpose is to view rural landscapes which are dynamic, living systems encompassing places produced and managed through traditional methods, techniques, accumulated knowledge, and cultural practices, as well as those places where traditional approaches to production have been recently changed (Lennon 2021). The natural beauty of special sunsets, sunrises, beautiful vistas, spacious skies, and majestic mountains are a few of these, but there are many more. The dramatic scenic landscape photo of the Chiricahua Mountains serves as an example of this type of tourism (see Figure 56.1).

Figure 56.1 Spectacular view of the Chiricahua Mountains.

Source: Keith Mandabach

Definitions

Agritourism is a more specific type of rural tourism and often is used to describe all tourism activities in rural areas. More frequently, either term relates to tourism products that are directly connected with the agrarian environment, agrarian products, or agrarian stays or staying at the farm, whether in rooms or camping, educational visits, 'farm to table' meals, recreational activities, and the sale of farm product or handicrafts (Tourism Notes 2021). *Agricultural tourism* is a term for a commercial enterprise at a working farm or ranch conducted for the enjoyment and education of visitors that generates supplemental income for the owner or operator (California Agritourism 2021).

Understanding the development in rural agricultural tourism

Adding rural agricultural tourism to agriculture or rural business is an afterthought to build a new revenue stream. Often the unique cultural heritage and history of the area is often overlooked, as is the landscape and architecture. Those living in the destination may realize it is a beautiful place to live but need a fresh perspective to see aspects that could draw visitors. To integrate and build the infrastructure needed to make a rural area a sustainable tourism destination is usually a slow process if it is to ensure long-term success (Whitt, Low, and Van Sandt 2019). This growth of rural tourism is a worldwide phenomenon, and in many areas integrating wine tourism into rural agriculture tourism is an added bonus that is not always originally planned.

Profitably operating an independent agribusiness (farm or ranch) today is extremely difficult. Massive international agribusinesses with reduced costs and vertically integrated operations with the crops presold with seamless transportation, storage, and delivery make it hard for independent operators to compete. Often this is made worse by the intrusion of world trade wars affecting prices and demand. Farms and ranches also utilize less labor today and are more mechanized than in the past (Kneafsey 2001). The first person on the food supply chain, even in wealthy nations, receives the lowest return because of the many links in the supply chain, such as transportation and processing food as well as multiple brokerage levels, which all add to the overhead cost. Farmers and ranchers in the United States have for many years looked to create new income by offering corn mazes, hayrides, petting zoos, harvest-your-own, or selling crops in stands on the road. Added now are educational tours, recreational activities, restaurants, and lodging that focus on building a relationship with the visitor and marketing agricultural specialty crops. Sometimes they may not contribute substantially to the profit of the farm or ranch. However, the increase in visitors occurring and a better knowledge of how to integrate their business into the rural tourism marketplace, provides greater opportunity for evolvement (California 2020). Statistics on the impact of the pandemic are currently incomplete, but it appears that rural tourism is still able to draw local visitors and is a positive force in a bleak economic segment. Worldwide tourism is down an estimated \$1.2 trillion (UNCTAD 2021). Since 2002, US agritourism revenue has grown from \$704 million to over \$1 billion in 2019 (Whitt, Low, and Van Sandt 2019).

Since 1970, repurposing the land to grow grapes has been driven by a variety of factors. The perception that the increased demand for wine has driven some to either repurpose their land or sell their properties to those wanting to enter the wine business. The most important factor in winemaking is the right sort of terroir or soil, sun, and moisture, and wine growers and makers search the world for it. It is fortunate that the right combination of these factors is often found in scenic, rural landscapes. The process of developing a vineyard and then building infrastructure takes a minimum of four years before the first bottle of wine is bottled and gives the vineyard owner time to plan to make wine tourism part of the total rural agriculture, cultural, culinary and historical tourism package. The tasting room and direct sales in the US wine business have been estimated to be $17.9 billion, with 43 million visits to 11,000 wineries on almost 700,000 acres with a total economic impact for local communities with retail sales to be over $220 billion (WineAmerica 2019). But COVID has greatly impacted the business. The wine business is difficult, and both agritourism businesses and wineries need to work together to help each other.

Theories of motivation: factors driving rural landscapes, cultural/historical tourism

Different individuals have multiple expectations, and destination image is crucial to the tourist development of their mental images. In some cases, the intangible experiential aspects may even outweigh tangible quality in satisfying tourist's experiential needs. This the core of the tourist's evaluation of the experience, and it may be behavioral, cognitive, or emotional (Oh, Fiore, and Jeoung 2007). People compare the reality to their images to decide if they are at the appropriate satisfaction level. Post-consumption experiences could be photos, video, or souvenir products as well as the memories that present the special features of the natural environment and the diversity featured in everyday life of the rural population. This occurs through tourist engagement with operators and employees of rural tourism businesses. This is one way businesses might develop brand and destination loyalty, triggering tourism returns and recommendations to their friends and relatives and the public.

Tourism opens a window for locals to introduce their history, culture, heritage, and cuisines to outsiders and draw people from other regions. Visitors digest multiple attributes, including agriculture, communication, economy, geography, history, law, literature, medicine, morals, religion, technology, trade, philosophy, and policy. Sometimes culinary or gastronomic tourism, which is broadly used to describe visitors who visit a place partially or mainly motivated by learning about, appreciating, or consuming local, regional, or national cuisine, heritage, culture, tradition, or culinary techniques of that region, is the driving force. Other times the rural tourist is motivated to visit food producers and to purchase locally produced food products and specialty crops. The uniqueness of the local environmental factors that affect local products, often referred to as terroir, plays a driving role in marketing to tourists as they physically travel to the region and spend. This results not only in developing the local economy but also in job creation (Zahari et al. 2009). Wineries in rural areas should search out the rural agritourism businesses to develop cooperative marketing. A collective shift in cultural conversations allows the tourism window to open wider and allows previously underrepresented groups opportunities to share their heritage with others to create authentic connections that could have felt limited before. This also highlights a crucial aspect in reimagining business or target markets, not limiting your business goals to confines of past perception.

Tourists also dine at restaurants, tasting novel dishes in special places, and attend food, wine, spirit, cultural, and nature festivals (UNWTO 2017). They visit a wide variety of businesses, including culinary retail outlets, culinary tours, cooking classes and schools, wine tours, vineyard/distillery tours as well as related agritourism activities on farms and processors. Many explore travel through print, online visitor content, and social media sites. Rural tourism consists of multiple segments, and the manner in which they are connected is what drives rural tourism. Understanding the factors that drive memorable experiences can assist in managing and marketing the tourism product (Kim, Ritchie, and McCormick 2011).

Building tourist loyalty through authentic customer engagement is an important factor in increasing tourism visitors (Mandabach 2017). The COVID pandemic has also presented new insight into the virtual aspect of building loyalty through virtual engagement. The experiential nature of rural tourism draws multiple segments of visitors and livens the tourist travel experience. With this in mind, literature shows that respecting the importance of residents' quality of life, values, and social systems will improve the visitor experience (Rivera, Croes, and Lee 2016). Tourism promotions should maintain or improve the quality of life, integrate local social and cultural values, and avoid intrusion into resident life when possible (Bachleitner and Zins 1999). Lessening the impact on the resident of the rural community is essential to improve the quality of rural, agricultural, and culinary tourism. Wineries being aware of the impact visitors have on the rural areas can lead to proactive approaches to maintain good relations with their neighbors.

Case Study 56.1 Escape from the city for a pandemic adventure

Riley and Kelly were tired of being stuck in their house. They were especially concerned with the pandemic because they were both over 70 years old. The cruise they had booked for Spain, France, and Italy in April was canceled, as was another trip to Argentina for the fall. They did not visit family, and they took precautions ordering food for pickup/delivery or Amazon to limit exposure. They walked in their neighborhood every day wearing a mask and sometimes took long drives in the country hiking in remote areas. When people approached, they were socially distanced. Most of their interactions with humans were phone, FaceTime, or Zoom.

Riley took an online exercise class three days a week and was intrigued when one of the participants related that they traveled the past weekend to a ranch three hours away in the western part of the state and stayed in a cabin next to a small river with great views of the mountains and wonderful birds. Riley was surprised to hear that mobile phones did not have service on this ranch. There was a traditional phone at the office, and the owners provided spotty Wi-Fi outside the office. The friend said to bring plenty of food and drink, a book to read, and binoculars to watch the birds because there were no services for over 20 miles. Riley called the ranch to book a cabin and was disappointed it was booked for the next three months but asked to be waitlisted anyway. Five minutes later, the ranch called back and said they just had a cancellation for the coming Saturday, Sunday, and Monday. Riley and Kelly said yes and were excited they were finally going to get out of the house.

It was a beautiful day to drive in the country, and they were pleased to watch the scenery change. On a whim, they stopped at the Chatterbox Café in a little town for a delightful brunch on the patio. The food was excellent; their server Marie was very pleasant. They told her about their mini-vacation. She was very excited and told them they will have a beautiful time in the Chiricahua Mountains, which was nice because they did not even realize that was the proper name of the area where they were going. She told them about a winery she knew, encouraging a visit because they had a unique variety of Austrian grapes and their lunches featured locally sourced products. She worked on the ranch as a child, and the owners were friends. Riley and Kelly felt great about their choice and their trip to the country.

The drive to the ranch and their cabin went through a beautiful valley and then snaked through foothills following a dry riverbed lined with cottonwood trees. They passed occasional picturesque cabins and ranches until they came to the end of a paved road. The rough, graded gravel road was fun, but their map on the mobile phone stopped working because they lost mobile service. They saw a sign for the ranch and turned in. Riley slowed as a deer ran across the road, and the deer seemed to be directing them to an impressive ranch house and office built from indigenous stone. There were a number of bird feeders in a rustically fenced area with chairs scattered in properly social distance spaces in front, which had a view of the surrounding mountains, which climbed precipitously, forcing one to raise one's eyes to the sky. Scattered from the office were eight cabins, each with unique characteristics. Their hosts greeted them with a packet of information for their stay, and they found their three-room wrap-around porch cabin with a view of the river and mountains. The cabin was decorated with homemade crafts/art by local artists. A visitor book provided tips from previous guests, birding books, binoculars, and brochures on attractions. Riley and Kelly unpacked, had a cup of coffee, and planned their afternoon. They packed a lunch and left for a short hike in what was supposed to be a great birding area that had the smallest hawk in the United States, the rare American Kestrel.

The sunlit hike followed the rushing narrow river and multitudes of chattering beautiful birds and mountain vistas. They came upon some serious birders with huge lenses taking pictures of something. Kelly politely started to speak when they turned, and the serious birders shushed and asked them to please leave. Feeling bad, they returned to the cabin, went to watch the birds by the office, and were pleased the people they saw were more friendly. The staff gave input on ideas about attractions in the area. They suggested a drive to the historic Apache Stronghold, a nearby agritourism farm site with a tour, crafts/art, specialty honeys, preserves, pies, and fresh corn, and then a stop at a winery for lunch. But more importantly, Riley and Kelly heard a narrative of the area's history and geology, the architect of stone constructed office/ranch house/outbuilding, as well the details of family ownership with its transition from the ranch, agritourism business, and bird/animal sanctuary. They watched the sunset on the porch, made a wonderful dinner, and relaxed with their books.

The rest of the trip, they learned that most birders were not rude and even pointed out where to see the elusive American Kestrels and other birds. The Kestrel was magnificent and had a special call. Riley and Kelly hiked the Apache Stronghold, toured the farm, and the hospitality was so wonderful they spent more than planned on the products for sale. Moving onto the winery for lunch, they sat on the patio with a pleasant view of the valley and ordered a bottle of the Gruner Veltliner wine, local cheese, and fresh bread with a local ranch-raised salad with fresh chicken. They heard tales of the area and how wonderful it was to live in this beautiful place. A fun story was told about the fact that just up the road had been a hideout for 19th-century outlaws. They bought a few more bottles of wine to take home and give to friends. Figure 56.2 illustrates how the sky darkened on the drive back to the cabin. They sat on their porch watching the storm move with spectacular flashes of lightning and accompanying thunder, and the air smelled like rain. The creek in front of the cabin began to fill with water and suddenly, the skies opened; fierce rain fell, soaking them instantly and

Figure 56.2 Chiricahua Mountains storm moving in.
Source: Keith Mandabach

chasing them inside. Riley and Kelly had a great romantic dinner, talked about their day, and, exhausted, fell asleep listening to the rain on the tin roof.

In the morning, the sun was warm, and the creek was all the way up to the side of its banks. They walked around the property, saw numerous deer, rabbits, a fox, and javelinas as well as a very large bluebird. They reflected on the wonderful time on their getaway and headed home to the city. They felt refreshed, and when their mobile phones finally received service, they called their daughter to share the joy of their memorable experiences in the country.

The experiential integration of wine tourism with rural tourism

The most important factor to remember for wine tourism is the relationship between tourist emotions and their perspective of the tourist experience. Most if not all wineries are part of the larger rural tourism experience. This chapter's main message to the operator is to emphasize the importance of including rural tourism components such as accentuating cultural/heritage, landscape/scenic, and agritourism that makes up the total experience for the visitor when planning marketing and the business plan. The winery and/or tasting room business operator, like any business, should have a narrative or summary of their operational goals. Making award-winning wine, producing wine varietals that are marketable, and doing this in as profitable a manner as possible is all-important in this narrative. Any, if not all, winery operators will incorporate wine tourism and sales to visitors as part of this plan. Often for smaller wineries, varietals are a major component of the business narrative, and how to retain the customers by meeting visitor satisfaction should be as important in business plan goals. The best measure of customer satisfaction is whether the customer will recommend your product to others (Reichheld 2006).

Emotion

Emotions are affective states, and the intensity of feelings connected with a specific event cause specific response behavior. These are measured by psychologists cognitively focusing on dimensional (positive or negative) and/or categorical (joy, fear, surprise, and love are the most common) approaches. Prior perceptions of the destination's image are an important part of the components that build visitor satisfaction, but the actual experience and the remembered experience can change those perceptions. This is another way in which virtual opportunity is growing to be utilized in the digital age but accelerated with the COVID pandemic. Creative ways to utilize virtual marketing or experiences add new opportunities in building perception or emotional connection. The role of emotions has been studied in detail, utilizing cognitive appraisal theory and visitor appraisals which are often complex in multi-componential response patterns of emotions. What one can or may want to explain when studying visitor emotional responses determines the nature and number of variables to study. Up to 15 different emotions have been identified and most often categorized using pleasant or unpleasant as the variable (Moors et al. 2013). Obviously, the rural tourist visiting the winery is seeking positive experiences and does not want unpleasantness.

Later these experiences are categorized and analyzed by the visitor. Rural tourists usually have goals when they travel to their destinations. During the pandemic, the goal of escaping from familiar areas to new areas and seeking new memorable experiences or reconnecting with past emotions was prevalent. Evaluating the congruence of the tourist emotional experience, impacts on appraisals of pleasantness as well as self-congruency with their goals have been found to be the antecedent of tourists experiencing love, joy, and positive surprise (Hosany 2011). As tourists travel to rural destinations to escape and possibly replace a more distant unattainable 'bucket list' destination during the pandemic, wineries finding ways to capitalize on the emotions by fulfilling the tourist goals could build satisfaction and lead to longer-term, more sustainable business.

One way to do this is to understand the subgroups of tourists. Some classifications by emotional terms include delighted, unemotional, mixed and 'passionates' (Dixit 2017; Hosany and Prayag 2013). Every winery has the opportunity to make themselves unique in some way. Hospitality has always been about meeting needs while using intangible elements to provide experiences to make people happy. Tastes for wine varietals vary, and the challenge for a winery is to transcend a visitors' preconceptions and build place attachment into the experience. This is possible if the operator realizes that the rural tourist arriving at the venue is traveling not just for the wine but for the rural experience. This includes experiencing and understanding the rural setting of which the winery is a part. All facilities do not have incredible ambiance or architecture, but the people working and serving can provide joy by communicating their love of the rural landscape and their own unique lifestyles. In the narrative written for the goals of the business, communicating this might help their guests achieve their goals of understanding and loving the rural experience.

Conclusion and summary

This chapter provided an understanding of how wine tourism is integrated into the greater rural tourism experience. Rural agricultural tourism has a major impact on the rural area economy and is becoming a major driver of rural economic development with a substantial impact worldwide. The emotional components of the rural tourism goals are travel to picturesque rural landscapes with unique cultures/heritage and passionately experiencing this beauty with the culture and history of the areas. Understanding the experience of all aspects are crucial for the wine tourism operator. Communicating the joy of living the lifestyle of the area they are visiting, the tourist might improve tourist post-experience evaluation of their travel and extend marketing opportunities. To take time to evaluate and understand how emotions such as joy, surprise, and love truly affect tourist evaluations of their travel experience. Attempting to better match tourist travel goals congruently is a method to improve satisfaction. This method can also increase the tourist tendency to recommend a travel experience, which is the best measure of tourist satisfaction and continued marketing. Integrating the outstanding natural landscapes, unique historical perspectives, sharing cultural backgrounds, agritourism, art, and interaction with residents/employees into the wine tourism experience can all work together to produce the emotional connections for a rural tourism destination and symbiotically benefit all tourism segments.

Implications

These pandemic rural tourists will continue to impact the rural economies and will share their positive experiences through social media and word-of-mouth. It is important that the rural tourism businesses deliver a quality, joyful product. While day and short-term rural tourism growth will be positive, the potential for the extended-stay market may be larger. Many of the rural tourists may have been working remotely as a result of the pandemic and may continue to do so after it is over. Many are looking to relocate either permanently or remotely to rural areas, and rural tourism operators might consider marketing their businesses to these new part or full-time residents. This will be successful only if they have the appropriate connections to the internet that are so important to remote work. Finding ways to accomplish this will be important.

References

Bachleitner, R. A., and Zins, A. H. (1999). Cultural tourism in rural communities: the resident's perspective. *Journal of Business Research*, 44(3): 199–209. https://doi.org/10.1016/S0148-2963(97)00201-4

California Agritourism (2021). Agritourism. https://ucanr.edu/sites/agritourism/

Dixit, S. K. (ed.). (2017). *The Routledge Handbook of Consumer Behaviour in Hospitality and Tourism*. London: Routledge.

Hosany, S. (2011). Appraisal determinants of tourists emotional responses. *Journal of Travel Research*, 51(3): 303–314. https://doi.org/10.1177/0047287511410320

Hosany, S., and Prayag, G. (2013). Patterns of tourists' emotional responses, satisfaction, and intention to recommend. *Journal of Business Research*, 66(6): 730–737. https://doi.org/10.1016/j.jbusres.2011.09.011

Kim, J.-H., Ritchie, B. R., and McCormick, B. P. (2011). Development of a scale to measure memorable tourism experiences. *Journal of Travel Research*, 51(1): 12–25. https://doi.org/10.1177/0047287510385467

Kneafsey, M. (2001). Rural cultural economy: tourism and social relations. *Annals of Tourism Research*, 28(3): 762–783. https://doi.org/10.1016/S0160-7383(00)00077-3

Lennon, J. (2021). *World Rural Landscape Principles: Principle One, Definition and Values of Rural Landscapes*. Living Landscape Observer. https://livinglandscapeobserver.net/world-rural-landscape-principles-principle-1-definition-and-values-of-rural-landscapes/

Mandabach, K. H. (2017). Customer engagement in building customer loyalty. In Dixit, S. K. (ed.), *The Routledge Handbook of Consumer Behaviour in Hospitality and Tourism* (pp. 286–296). London: Routledge

Moors, A., Ellsworth, P. C., Scheer, K. R., and Frijda, N. (2013). Appraisal theories of emotion: state of the art and future development. *Emotion Review* 5(2): 119–124. https://doi.org/10.1177/1754073912468165

Oh, H., Fiore, A. M., and Jeoung, M. (2007). Measuring experience economy concepts: tourism applications. *Journal of Travel Research*, 46(2): 119–132. https://doi.org/10.1177/0047287507304039

Reichheld, F. (2006). *Ultimate Question: Driving Good Profits and True Growth*, Boston: Harvard Business Review Press.

Rivera, M., Croes, R., and Lee, H. L. (2016). Tourism development and happiness: a residents' perspective. *Journal of Destination Marketing and Management*, 5(1): 5–15.

Tourism Notes (2021). *Rural Tourism*. https://tourismnotes.com/rural-tourism/ accessed 10/03/2021.

UNCTAD (2021). *Coronavirus Will Cost Global Tourism at Least $1.2 Trillion*. https://unctad.org/news/coronavirus-will-cost-global-tourism-least-12-trillion accessed 10/01/2021.

UNWTO (2017). Definition adopted during the 22nd Session of the General Assembly, Chengdu, China, 09/15/2017. https://www.unwto.org/archive/global/press-release/2017-09-18/22nd-unwto-general-assembly-china-week-important-achievements

Whitt, S. A., Low, S. A., and Van Sandt, A. (2019). *Agritourism Allows Farms to Diversity and has Potential Benefits for Rural Communities*. USDA Economic Research Service. https://www.ers.usda.gov/amber-waves/2019/november/agritourism-allows-farms-to-diversify-and-has-potential-benefits-for-rural-communities/

Wine America (2019). https://wineamerica.org/impact/ accessed 10/4/2021

World Tourism Organization (2021). https://www.unwto.org/gastronomy-wine-tourism accessed 10/05/2021

Zahari, M. S. M., Jalis, M. H., Zulfifly, M. I., Radzi, S. M., and Othman, Z. (2009). Gastronomy: an opportunity for Malaysian culinary educators. *International Education Studies Journal*, 2(2): 67–77.

57

CONCLUSION

Uncorking the emerging realms for wine tourism research

Saurabh Kumar Dixit

The emergence of wine tourism as a research arena has evolved considerably during the last few years due to the growth and spread of wine tourism in new, old, and emerging wine regions. The research implications of wine tourism go beyond mere consumption and experiencing winery products and offerings and focus further on establishing linkage amongst cuisine, place, customer, and enduring experiences (Dixit 2020). The wine production sector, including the associated tourism experiences, cannot be separated from the social, economic, environmental, and cultural issues of wine-producing regions and territories (González 2017). According to Gómez and Molina (2012), contemporary wine tourism research has been considered to be developing in four key directions, i.e., studies that analyze wine tourism from the supply side, from the demand side, from both sides together (supply and demand), and other studies. Hence, modern scholarship has widened the scope of wine tourism as a transformed and diversified wine tourism demand reveals that wine tourism is no longer an activity strictly reserved for wine lovers wishing to enjoy wine tasting. Nor is wine tourism a pursuit of the elite and wine connoisseur alone. Instead, wine tourists of all manner go to wineries to also visit and immerse themselves into the winescape. (Sigala and Robinson 2019, p. 1).

Therefore, the present *Handbook* will be an indispensable resource to integrate the diverse facets of wine tourism management and marketing from both demand and supply perspectives. This *Handbook* has underscored an up-to-date collection of research that has focused on global wine tourism. In this concluding chapter, the editor aims to reflect a convergence of the *Handbook*'s overall aims and the varied research presented by the authors in this volume, highlighting key issues and relating them to wine tourism. The following sections offer a comprehensive outlook on the 57 *Handbook* chapters' core issues and important foci (see Figure 57.1). This *Handbook* is expected to contribute immensely in developing a better approach to this interdisciplinary field by reviewing both the theory and the practice. I hope this *Handbook* will advance our comprehension of wine tourism's existing body of knowledge to enhance further empirical research.

Historical evolution of wine tourism

Wine is considered one of the oldest consumed alcoholic staple beverages of our society. Wine is also considered as a constituent of many religions, cultures, and traditions besides serving as a consumption product for the tourists. The oldest record (dating back to 7000 BCE) of

DOI: 10.4324/9781003143628-65

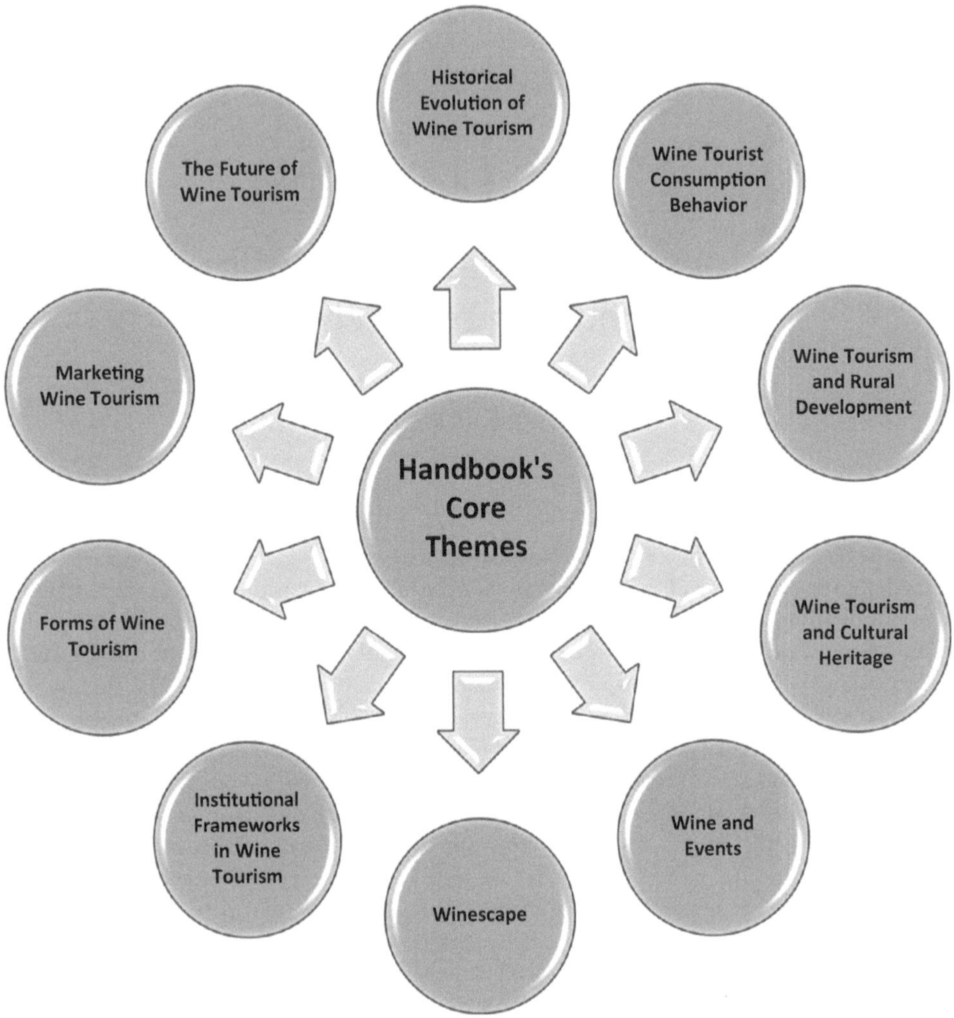

Figure 57.1 Core themes pondered upon in the *Handbook*.

winemaking is found in Jiahu in the Yellow River Basin of the Henan province of northern China, where the traces of a fermented drink made from rice, honey, and either grapes or Hawthorne fruit are discovered (Terziyska 2020). Chapter 1 presented the historical upbringing of wine tourism besides discussing the prevailing trends globally. Chapter 5 narrated the Indian wine story of India, with a special focus on the state of Maharashtra. Chapter 7 highlighted the emergence and evolution of the wine industry in the principal wine regions of Canada. Chapter 42 showed the historical evolution of wine tourism by highlighting the wine routes and trails of Slovenia. Chapter 47 discussed the emergence and spread of the hybrid grapes being used to manufacture hybrid wine in the hybrid wine regions.

Wine tourist consumption behavior

The wine consumption behavior is a pretty complex construct and difficult to anticipate (Dixit 2021). The motivation to consume wine is multifaceted; it is more than just the quest of 'a beverage that tastes good.' Most of the wine tourism literature has focused on comprehending the behavioral patterns of winery visitors and has consequently endeavored to categorize or classify these visitors in terms of motivation. Identifying the needs, preferences, and characteristics of wine tourists and the wine market is significant in developing wine tourism in a region. Chapter 2 threw light on developing an in-depth understanding of wine tourism from a senior management and business owner's perspective. Chapter 9 enlightened readers about the key characteristics of 'Old World,' 'New World,' and 'Emerging World' wine tourism markets. The understanding of key wine tourism markets enables destination marketers to plan and strategize to grab opportunities. Chapter 10 detailed the wine travelers' behavior that can influence their pre-trip decision making, onsite behavior and sharing behavior on social media platforms. Chapter 11 advanced the themes of consumer behavior, consumer needs, and preferences, and profiles of wine consumers from involvement, personality, and lifestyle. Chapter 13 elaborated on the motivations of wine tourism, which embrace the perception of wine tourism destination attributes. Chapter 14 looked into the consumers' knowledge of wine and its linkage with their self-confidence to choose wine. Chapter 15 studied the expectation and behavioral intentions of culturally diverse wine students' first-time visit to a winery. Chapter 16 posited the behavior and experiences of senior travelers in the arena of wine tourism. As wine is considered a lifestyle product, it is a consumer choice rather than a necessity, with its consumption regarded as experiential (Bruwer and Alant, 2009). Understanding the wine tourists' experiences and demands is one of the most discussed topics in global wine tourism studies. Chapters 32, 33, 34, 35, 36, 37, 38, 39 and 41 dealt with vivid facets of wine tourism experiences.

Wine tourism and rural development

Rural or agricultural tourism has a major stimulus on the rural area economy and is becoming a major driver of rural economic development with a substantial global impact. Wine tourism is closely associated with terroir, territory, landscape, and culture, as most of the wineries are located in rural areas. Viticulture and winemaking are, therefore, significantly endorsed by its territorial heritage, local history, and place identity. The Fifth UNWTO Global Conference on Wine Tourism, held on September 8–10, 2021, in Reguengos de Monsaraz (Alentejo), Portugal, endorsed the view that innovation, partnerships and enhanced coordination are essential for making wine tourism a driver of rural development. Chapter 21 investigated how the entrepreneurs in rural Japan practice wine tourism using the local community to stimulate the economy. Therefore, wine tourism is envisioned as an integrated instrument to address pastoral challenges with local entrepreneurs and government collaboration. Chapter 22 explained the transformation of rural wine tourism destination areas in the United States through the lens of gentrification. Chapter 53 used the place-based approach to comprehend interlinkages of regional culture (internal), visitor culture (external culture), and the landscape contributing to the rise of the Barossa and Margaret River as wine tourism destinations in Australia. Chapter 55 probed the concepts of knowledge, innovation, creativity, and learning production for fostering territorial development based on wine tourism. Chapter 56 detailed the phenomenon of rural agricultural tourism and showcases the experiential integration of wine tourism with rural tourism.

Wine tourism and cultural heritage

Wine tourism is growing as an emerging form of tourism, benefitting local, regional, and national economies. Therefore, wine tourism comprises of tourism and leisure activities devoted to finding out the cultural and recreational avenues arising from the vine, wine, and its territory. Historical, geographical, and cultural factors shape the core of wine production and consumption. Wine is considered as a socializing drink, and its consumption overwhelmingly makes people mingle together. Chapter 19 explored residents' attitudes toward wine tourism development and wine tourism-related social capital in the Piedmont region of North Carolina. Chapter 28 elaborated on the main features of a creative star wine district by highlighting the contribution of the wine industry to other creative industries and local institutions. Chapter 40 shed light on the constructs of authenticity in wine tourism, which is closely linked with sustainable development, primarily in preserving local cultural and social identity. Chapter 42 offered insights on the liaison between gastronomic tourism and wine tourism. Chapter 46 delved into wine-related e-storytelling in wineries' websites of the Japan Wineries Association and explored how such contents interpret meaning to brand wine products and wine tourism destinations.

Wine and events

Wine festivals and events are the vital constructs of the wine tourism experience, complementing other customary tourism activities. Such events feature hedonism (i.e., mainly from the consumption of authentic wine), facilitate socializing, and appeal to escapism and novelty-seeking for the visitors (Sigala and Robinson 2019). Chapter 24 discussed the 'Reus viu el vi' (Reus lives wine) event celebrated in Reus, southern Catalonia, northeastern Spain. The event involves the participation of both wine and vermouth producers. Drawing on a conversation with the event's organizer and supported with other secondary data, this chapter revealed the role of such events in protecting and promoting local heritage and cultural identity. Chapter 26 revisited the contribution of wine festivals in Swedish wine tourism marketing. It proposed that a wine region's marketing and promotion efforts and characteristics are essential facets to be targeted in this context. Chapter 29 offered insights into the Italian folk food festival, often called *sagra*, by depicting the characteristics and history of these events besides their impacts on the local community. Chapter 37 explored the impact of wine involvement on motivations in a wine festival context with the help of the case of the Christchurch/South Island Wine and Food Festival.

Winescape

Winescape is one of the most researched concepts of wine tourism, dealing primarily with the supply side of wine tourism. Winescape centers on the cultural/viticultural landscape of vineyards, wineries, and supporting activities encompassing landscape, cultural landscape, and servicescape. The winescape is seen as providing an aesthetic experience in which 'the beauty of the winescape, vineyards, picturesque main streets and typical architecture' (Vo Thanh and Kirova 2018, pp. 31–32) supplement the sensory experience for the tourists. Wine is intrinsically linked to the environment, or terroir (of which people are integral), and tourism and travel are also replete with sustainability challenges to both the environment and the people (Sigala and Robinson 2019). Chapter 3 suggested constructing a winescape framework using the publications listed in Web of Science and Scopus databases. Chapter 9 highlighted the key characteristics and winescape features of the Old World, New World, and Emerging World wine markets. Chapter 40 offered a conceptual model of authenticity markers in wine tourism, based on the winescape concept. Chapter 51 offered reflections on the theoretical model of 'terroir fluidity' based on feedback loops.

Institutional frameworks in wine tourism

There are enormous studies on the supply side of wine tourism; there is still much to be revealed about the nature of the development to closely understand the wine tourism business's organizational framework (Carlsen 2004). Wineries adapt the contemporary frameworks to match the needs and demands of new market opportunities and to cope with future challenges. Chapter 5 appreciates the relationship and interlinkages between ethical and regulatory influences on the growth of wine tourism. Chapter 6 explored wine distributors'/producers' personal and organizational social capabilities to appreciate relationship quality. Chapter 12 informed readers about the wine tourism institutional framework and governance system of Portugal by explicitly integrating the same with the Douro Wine Region (DWR). The chapter also observed the online institutional structure and governance model of Italy, France, the United States, Australia, South Africa, and New Zealand as reference markets in wine tourism. Chapter 17 presented the liaison between wine and place and probed how these relate to building strong wine tourism destinations with supportive narratives. Chapter 18 offered insights on strategy formulation for internationalization marketing strategy and action planning for wine tourism in the Douro Valley region. Chapter 23 focused on positioning wine tourism as a competitive factor of territory with the support of the Great Wine Capitals (GWC) network.

Forms of wine tourism

Wine tourism is categorized as special interest tourism comprising visits to vineyards, wineries, wine festivals, and wine shows, experiencing local attractions and the local community's lifestyle. Because of the perception linked with wine consumption as both sacred and disrespectful is the paradox that wine can therefore be considered to be both healthy and unhealthy. The *Handbook* chapters ponder upon varied global facets of wine tourism, including its emerging forms and features. Chapter 4 offered a framework that enables understanding the role of wine trails/routes in creating the enduring wine tourist experience. Chapter 8 debated the special interest form of wine tourism. Chapter 27 presented the entrepreneurial skirmish in the new field of modern fruit wine production in Denmark to emerge as a new form of fruit wine. Chapter 47 detailed the emergence of hybrid wines and their relevance and importance to the wine tourism industry. Chapter 52 shed light on the sustainable form of natural wines characterized by organic or biodynamic vineyard practices and minimal human intervention in the cellaring process. Therefore, natural wines are considered a more environmentally friendly alternative to mainstream high-intervention varieties of wines.

Marketing wine tourism

Wine tourism marketing focuses primarily on matching the available wine resources and winery's competencies with tourists' needs and requirements. According to Torres and Kunc (2016), winery managers may have two approaches for marketing strategies applicable to wine: traditional and relational. The traditional way of wine tourism marketing is something the manager does for the consumer by coordinating all marketing mix elements. Marketing from a relational perspective has been defined as the process of identifying, establishing, maintaining, and enhancing relationships with customers by interacting with them (Grönroos, 1997). Hojman and Hunter-Jones (2012) observed that vineyards employ relational marketing tactics (such as wine festivals and events, traditional cultural activities, food tasting, and cellar door experience) by integrating wine tourism into their business models. Chapter 25 highlighted the strategic marketing approaches in wine tourism destinations by identifying wine tourism destination attributes and experiences. Chapter 28 threw light on the features of a creative star wine district and its marketing efforts. Chapter 30 offered linkages between wine tourism and business/MICE tourism. Chapter 31 profiled wine visitors based on their perceived level of memorability towards the visitor experience and concluded that tourism managers should plan and implement to deliver memorable, rather than 'merely' satisfactory, tourism experiences. Chapter 45 discussed the importance of social media marketing for wine tourism organizations and wineries' adoption of social media platforms. Chapter 54 offered insights on the contribution of wine lists in marketing wine tourism destinations effectively.

The future of wine tourism

In the post-pandemic era, the global wine industry accelerates to adopt digitalization, technology use, and new markets. The role of e-commerce and social media keeps growing, and the wine tourism industry needs to understand more about how it can cope with the changes. Wine tourism may be considered an essential sector of the whole tourism value chain, which complements the vineyards and wineries' activities to generate more jobs and businesses, promote local products, and advance the visitor experience. This section pinpointed some emergent themes reflected in the present title, which will enlighten scholars to uncover the future avenues of research in wine tourism. Chapter 1 highlighted the emerging trends in the global wine tourism marketplace. Chapter 43 underlined the issue of the adoption of technologies in enhancing the wine tourism experience. Chapter 44 offered an overview of the VR and AR technologies in wine tourism to offer future avenues of research to both practitioners and academicians. Chapters 48 and 49 offered insights into the impacts of the COVID-19 pandemic on wine tourism. Chapter 48 shared the concerns faced by wineries due to COVID-19 and looked at initiatives undertaken by wineries to handle them. Chapter 49 briefed about how the New Zealand wine industry has perceived the COVID-19 situation as an opportunity and promoted domestic wine tourism in the country. Chapter 50 addressed the issues of sustainability of wine tourism destinations. The study also advised that it would be possible to assess general rules to develop a standard process to implement sustainable practices in wine tourism.

I hope that this *Handbook* will offer significant research contributions for its readers and serve as an essential resource for present and future wine tourism research.

References

Bruwer, J., and Alant, K. (2009). The hedonic nature of wine tourism consumption: an experiential view. *International Journal of Wine Business Research*, 21(3): 235–257. https://doi.org/10.1108/17511060910985962

Carlsen, J. (2004). A review of global wine tourism research. *Journal of Wine Research*, 15(1): 5–13. https://doi.org/10.1080/0957126042000300281

Dixit, S. K. (ed.). (2020). *The Routledge Handbook of Tourism Experience Management and Marketing*. London: Routledge.

Dixit, S. K. (2021). Tourist consumption behavior: an unsolved puzzle. *International Journal of Hospitality & Tourism Administration*, 22(5): 475–480. https://doi.org/10.1080/15256480.2021.1982104

Gómez, M., and Molina, A. (2012). Wine tourism in Spain: denomination of origin effects on brand equity. *International Journal of Tourism Research*, 14: 353–368. https://doi.org/10.1002/jtr.868

González, M. L. (2017). Enoturismo y entornos sostenibles. *Arbor*, 193(785): a399. https://doi.org/10.3989/arbor.2017.785n3005

Grönroos, C. (1997). Value-driven relational marketing: from products to resources and competencies. *Journal of Marketing Management*, 13(5): 407–419. https://doi.org/10.1080/0267257X.1997.9964482

Hojman, D., and Hunter-Jones, P. (2012). Wine tourism: Chilean wine regions and routes. *Journal of Business Research*, 65(1): 13–21. https://doi.org/10.1016/j.jbusres.2011.07.009
Sigala, M., and Robinson, R. N. S. (2019). Introduction: wine destination management and marketing-critical success factors. In Sigala, M., and Robinson, R. N. S. (eds.), *Management and Marketing of Wine Tourism Business* (pp. 1–8), Cham, Switzerland: Palgrave Macmillan.
Terziyska, I. (2020). *Wine Tourism: Critical Success Factors for an Emerging Destination*. Sofia: Gea Libris.
Torres, J. P., and Kunc, M. H. (2016). Market opportunity recognition in the Chilean wine industry: traditional versus relational marketing approaches. *Journal of Wine Research*, 27(1): 19–33. https://doi.org/10.1080/09571264.2016.1144584
Vo Thanh, T., and Kirova, V. (2018). Wine tourism experience: a netnography study. *Journal of Business Research*, 83: 30–37. https://doi.org/10.1016/j.jbusres.2017.10.008

INDEX

Note: Page numbers in **bold** refer to tables; and page numbers in *italics* refer to figures